METHODS OF MATHEMATICAL PHYSICS
VOLUME II

METHODS OF MATHEMATICAL PHYSICS

By R. COURANT and D. HILBERT

VOLUME II
PARTIAL DIFFERENTIAL EQUATIONS
By R. Courant

Wiley Classics Edition Published in 1989

WILEY

INTERSCIENCE PUBLISHERS

JOHN WILEY & SONS,
New York • Chichester • Brisbane • Toronto • Singapore

TO

KURT OTTO FRIEDRICHS

BOOKS BY R. COURANT
Differential and Integral Calculus. Second
Edition. Volumes 1 and 2.
Dirichlet's Principle, Conformal Mapping
and Minimal Surfaces (Pure and Applied
Mathematics, Volume 3)

BY R. COURANT AND F. JOHN
Calculus. Volume 1

BY R. COURANT AND K. O. FRIEDRICHS
Supersonic Flow and Shock Waves
(Pure and Applied Mathematics,
Volume 1)

BY R. COURANT AND D. HILBERT
Volume 1
Volume 2. Partial Differential Equations

PREFACE

The present volume is concerned with the theory of partial differential equations, in particular with parts of this wide field that are related to concepts of physics and mechanics. Even with this restriction, completeness seems unattainable; to a certain extent the material selected corresponds to my personal experience and taste. The intention is to make an important branch of mathematical analysis more accessible by emphasizing concepts and methods rather than presenting a collection of theorems and facts, and by leading from an elementary level to key points on the frontiers of our knowledge

Almost forty years ago I discussed with David Hilbert the plan of a. work on mathematical physics. Although Hilbert could not participate in carrying out the plan, I hope the work, and in particular the present volume, reflects his scientific ethos, which was always firmly directed towards the relevant nucleus of a mathematical problem and averse to merely formal generality. We shall introduce our topics by first concentrating on typical specific cases which are suggestive by their concrete freshness and yet exhibit the core of the underlying abstract situation. Individual phenomena are not relegated to the role of special examples; rather, general theories emerge by steps as we reach higher vantage points from which the details on a "lower level" can be better viewed, unified, and mastered. Thus, corresponding to the organic process of learning and teaching, an inductive approach is favored, sometimes at the expense of the conciseness which can be gained by a deductive, authoritarian mode of presentation.

This book is essentially self-contained; it corresponds to Volume II of the German edition of the "Methoden der Mathematischen Physik" which appeared in 1937. The original work was subsequently suppressed by the Ministry of Culture in Nazi Germany; later my loyal friend Ferdinand Springer was forced out as the head of his famous publishing house. The reprinting by Interscience Publishers under license of the United States government (1943) secured the survival of the book. Ever since, a completely new version in the English language has been in preparation. During this long period, knowledge in the field has advanced considerably, and I too have been struggling towards more comprehensive understanding. Naturally the book reflects these developments to the extent to which I have shared in them as an active and as a learning participant.

The table of contents indicates the scope of the present book. It differs in almost every important detail from the German original. For example, the theory of characteristics and their role for the theory of wave propagation is now treated much more adequately than was possible twenty-five years ago. Also the concept of weak solutions of differential equations, clarified by Sobolev and Friedrichs and already contained in the German edition, now appears in the context of the theory of ideal functions which, introduced and called "distributions" by Laurent Schwartz, have become an indispensable tool of advanced calculus. An appendix to Chapter VI contains an elementary presentation of this theory. On the other hand, the material of the last chapter of the German edition, in particular the discussion of existence of solutions of elliptic differential equations, did not find room in this volume. A short third volume on the construction of solutions will treat these topics, including an account of recent mathematics.

The book as now submitted to the public is certainly uneven in style, completeness and level of difficulty. Still, I hope that it will be useful to my fellow students, whether they are beginners, scholars, mathematicians, other scientists or engineers. Possibly the presence of various levels in the book might make the terrain all the more accessible by way of the lower regions.

I am apologetically conscious of the fact that some of the progress achieved outside of my own sphere may have been inadequately reported or even overlooked in this book. Some of these shortcomings will be remedied by other publications in the foreseeable future such as a forthcoming book by Gårding and Leray about their fascinating work.

The present publication would have been impossible without the sustained unselfish cooperation given to me by friends. Throughout all my career I have had the rare fortune to work with younger people who were successively my students, scientific companions and instructors. Many of them have long since attained high prominence and yet have continued their helpful attitude. Kurt O. Friedrichs and Fritz John, whose scientific association with me began more than thirty years ago, are still actively interested in this work on mathematical physics.—That this volume is dedicated to K. O. Friedrichs is a natural acknowledgment of a lasting scientific and personal friendship.

To the cooperation of Peter D. Lax and Louis Nirenberg I owe much more than can be expressed by quoting specific details. Peter Ungar has greatly helped me with productive suggestions and criticisms. Also, Lipman Bers has rendered most valuable help and, moreover, has contributed an important appendix to Chapter IV.

Among younger assistants I must particularly mention Donald Ludwig whose active and spontaneous participation has led to a number of significant contributions.

Critical revision of parts of the manuscript in different stages was undertaken by Konrad Jörgens, Herbert Kranzer, Anneli Lax, Hanan Rubin. Proofs were read by Natascha Brunswick, Susan Hahn, Reuben Hersh, Alan Jeffrey, Peter Rejto, Brigitte Rellich, Leonard Sarason, Alan Solomon and others. Jane Richtmyer assisted in preparing the list of references and in many other aspects of the production. A great deal of the editing was done by Lori Berkowitz.

Most of the technical preparation was in the hands of Ruth Murray, who typed and retyped thousands of pages of manuscript, drew the figures and altogether was most instrumental in the exasperating process of transforming hardly legible drafts into the present book.

To all these helpers and to others, whose names may have been omitted, I wish to extend my profound thanks.

Thanks are also due to my patient friend Eric S. Proskauer of Interscience.

Finally I wish to thank the Office of Naval Research and the National Science Foundation, in particular F. Joachim Weyl and Arthur Grad, for the effective and understanding support given in the preparation of this book.

New Rochelle, New York　　　　　　　　　　　　　　　R. Courant
November 1961

CONTENTS

The present volume, essentially independent of the first, treats the theory of partial differential equations from the point of view of mathematical physics. A shorter third volume will be concerned with existence proofs and with the construction of solutions by finite difference methods and other procedures.

CHAPTER I

Introductory Remarks

We begin with an introductory chapter describing basic concepts, problems, and lines of approach to their solution.

A partial differential equation is given as a relation of the form

$$(1) \qquad F(x, y, \cdots, u, u_x, u_y, \cdots, u_{xx}, u_{xy}, \cdots) = 0,$$

where F is a function of the variables x, y, $\cdots$, u, u_x, u_y, $\cdots$, u_{xx}, u_{xy}, $\cdots$; a function $u(x, y, \cdots)$ of the independent variables x, y, $\cdots$ is sought such that equation (1) is identically satisfied in these independent variables if $u(x, y, \cdots)$ and its partial derivatives

$$u_x = \frac{\partial u}{\partial x}, \qquad u_y = \frac{\partial u}{\partial y}, \qquad \cdots,$$

$$u_{xx} = \frac{\partial^2 u}{\partial x^2}, \qquad u_{xy} = \frac{\partial^2 u}{\partial x \partial y}, \qquad \cdots,$$

$$\cdots\cdots\cdots\cdots\cdots\cdots\cdots\cdots\cdots$$

are substituted in F.

Such a function $u(x, y, \cdots)$ is called a *solution of the partial differential equation* (1). We shall not only look for a *single "particular" solution* but investigate the *totality of solutions* and, in particular, characterize *individual solutions* by further conditions which may be imposed in addition to (1).

The partial differential equation (1) becomes an ordinary differential equation if the number of independent variables is one.

The order of the highest derivative occurring in a differential equation is called the *order* of the differential equation.

Frequently we shall restrict the independent variables x, y, $\cdots$ to a specific region of the x, y, $\cdots$-space; similarly, we shall consider F

1

only in a restricted part of the x, y, $\cdots$, u, u_x, u_y, $\cdots$-space. This restriction means that we admit only those functions $u(x, y, \cdots)$ of the basic region in the x, y, $\cdots$-space which satisfy the conditions imposed on the corresponding arguments of F. *Once and for all we stipulate that all our considerations refer to regions chosen sufficiently small. Similarly, we shall assume that, unless the contrary is specifically stated, all occurring functions F, u, $\cdots$ are continuous and have continuous derivatives of all occurring orders.*[1]

The differential equation is called *linear* if F is linear in the variables u, u_x, u_y, $\cdots$, u_{xx}, u_{xy}, $\cdots$ with coefficients depending only on the independent variables x, y, $\cdots$. If F is linear in the highest order derivatives (say the n-th), with coefficients depending upon x, y, $\cdots$ and possibly upon u and its derivatives up to order $n - 1$, then the differential equation is called *quasi-linear*.

We shall deal mainly with either linear or quasi-linear differential equations; more general differential equations will usually be reduced to equations of this type.

In the case of merely two independent variables x, y, the solution $u(x, y)$ of the differential equation (1) is visualized geometrically as a surface, an *"integral surface"* in the x, y, u-space.

§1. *General Information about the Variety of Solutions*

1. Examples. For an ordinary differential equation of n-th order, the totality of solutions (except possible "singular" solutions) is a function of the independent variable x which also depends on n arbitrary integration constants c_1, c_2, $\cdots$, c_n. Conversely, for every n-parameter family of functions

$$u = \phi(x; c_1, c_2, \cdots, c_n),$$

there is an n-th order differential equation with the solution $u = \phi$ obtained by eliminating the parameters c_1, c_2, $\cdots$, c_n from the equation $u = \phi(x; c_1, c_2, \cdots, c_n)$ and from the n equations

$$u' = \phi'(x; c_1, c_2, \cdots, c_n),$$
$$\cdots\cdots\cdots\cdots\cdots\cdots\cdots\cdots$$
$$u^{(n)} = \phi^{(n)}(x; c_1, c_2, \cdots, c_n).$$

[1] Also, when systems of equations are inverted, we will always consider a neighborhood of a point in which the corresponding Jacobian does not vanish.

For partial differential equations the situation is more complicated. Here too, one may seek the totality of solutions or the *"general solution"*; i.e., one may seek a solution which, after certain "arbitrary" elements are fixed, represents every individual solution (again with the possible exception of certain "singular" solutions). In the case of partial differential equations such arbitrary elements can no longer occur in the form of constants of integration, but must involve arbitrary functions; in general, the number of these arbitrary functions is equal to the order of the differential equation. These arbitrary functions depend on one independent variable less than the solution u. A more precise statement of the situation is implied in the existence theorem of §7. In the present section however, we merely collect information by studying a few examples.

1) The differential equation

$$u_y = 0$$

for a function $u(x, y)$ states that u does not depend on y; hence,

$$u = w(x),$$

where $w(x)$ is an arbitrary function of x.

2) For the equation

$$u_{xy} = 0,$$

one immediately obtains the general solution

$$u = w(x) + v(y).$$

3) Similarly, the solution of the nonhomogeneous differential equation

is $$u_{xy} = f(x, y)$$

$$u(x, y) = \int_{x_0}^{x} \int_{y_0}^{y} f(\xi, \eta)\, d\xi\, d\eta + w(x) + v(y)$$

with arbitrary functions w and v and fixed values x_0, y_0.

More generally, one may replace the integral by an area integral if one takes, as the region of integration ⊓ , a "triangle" such as that in Figure 1, whose curved boundary consists of a curve $C: y = g(x)$ or $x = h(y)$ which is not intersected more than once by any of the

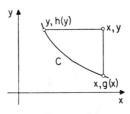

Figure 1

curves x = const. or y = const. Then,

$$u(x, y) = \iint_{\triangleleft} f(\xi, \eta)\, d\xi\, d\eta + w(x) + v(y),$$

(2)

$$u_x = \int_{g(x)}^{y} f(x, \eta)\, d\eta + w'(x), \qquad u_y = \int_{h(y)}^{x} f(\xi, y)\, d\xi + v'(y).$$

The special solution of the differential equation for $w(x) = v(y) = 0$ satisfies the condition $u = u_x = u_y = 0$ for all points (x, y) on the curve C.

4) The partial differential equation

$$u_x = u_y$$

is transformed into the equation

$$2\omega_\eta = 0$$

by the transformation of variables

$$x + y = \xi, \qquad x - y = \eta, \qquad u(x, y) = \omega(\xi, \eta).$$

The "general solution" of the transformed equation is $\omega = w(\xi)$; therefore

$$u = w(x + y).$$

Similarly, if α and β are constants, the general solution of the differential equation

$$\alpha u_x + \beta u_y = 0$$

is

$$u = w(\beta x - \alpha y).$$

5) According to elementary theorems of the differential calculus, the partial differential equation

$$u_x g_y - u_y g_x = 0,$$

where $g(x, y)$ is any given function of x, y states that the Jacobian $\partial(u, g)/\partial(x, y)$ of u, g with respect to x, y vanishes. This means that u depends on g, i.e., that

$$(3) \qquad u = w[g(x, y)],$$

where w is an arbitrary function of the quantity g. Since, conversely, every function u of the form (3) satisfies the differential equation $u_x g_y - u_y g_x = 0$, we obtain the totality of solutions by means of the arbitrary function w.

It is noteworthy that the same result holds for the more general—quasi-linear—differential equation

$$u_x g_y(x, y, u) - u_y g_x(x, y, u) = 0,$$

where g now depends explicitly not only on x, y but on the unknown function $u(x, y)$ as well. For, as one sees, the Jacobian of any solution $u(x, y)$ and $\gamma(x, y) = g[x, y, u(x, y)]$ vanishes since

$$u_x \gamma_y - u_y \gamma_x = u_x g_y - u_y g_x + u_x g_u u_y - u_y g_u u_x = 0.$$

Thus, even in this case, the solution is given by the relation

$$(4) \qquad u(x, y) = W[g(x, y, u)],$$

which is an implicit definition of u by means of the arbitrary function W.

For instance, the solution $u(x, y)$ of the differential equation

$$\alpha(u)u_x - \beta(u)u_y = 0$$

is implicitly defined by

$$(5) \qquad u = W[\alpha(u)y + \beta(u)x],$$

(or by $\alpha(u)y + \beta(u)x = w(u)$), so that u depends on the arbitrary function W in a rather involved way. (An application will be given in §7, 1.)

A special case of the differential equation $\alpha(u)u_x - \beta(u)u_y = 0$ is

$$u_y + uu_x = 0;$$

the solution is given implicitly by

$$u = W(-x + uy),$$

where W is arbitrary. If $u = u(x(y), y)$ is interpreted as the velocity of a particle at a point $x = x(y)$ moving with the time y, then the differential equation states that the acceleration of all the particles is zero.

6) The partial differential equation of second order

$$u_{xx} - u_{yy} = 0$$

is transformed into

$$4\omega_{\xi\eta} = 0$$

by the transformation

$$x + y = \xi, \qquad x - y = \eta, \qquad u(x, y) = \omega(\xi, \eta).$$

Hence, according to example *2*), its solutions are

$$u(x, y) = w(x + y) + v(x - y).$$

7) In a similar way the general solution of the differential equation

$$u_{xx} - \frac{1}{t^2} u_{yy} = 0$$

for any value of the parameter t is

$$u = w(x + ty) + v(x - ty).$$

In particular, the functions

$$u = (x + ty)^n$$

and

$$u = (x - ty)^n$$

are solutions; i.e.,

$$t^2 u_{xx} - u_{yy}$$

vanishes for all x, y and for all real t.

8) According to elementary algebra, if a polynomial in t vanishes for all real values of t, then it vanishes for all complex values of t as well. Thus, if we substitute $t = i = \sqrt{-1}$, the differential equation of example 7) is transformed into the *potential equation*

$$\Delta u \equiv u_{xx} + u_{yy} = 0;$$

for this equation we obtain solutions of the form

$$(x + iy)^n = P_n(x, y) + iQ_n(x, y),$$

$$(x - iy)^n = P_n(x, y) - iQ_n(x, y),$$

where P_n and Q_n are polynomials with real coefficients which must themselves satisfy the potential equation.[1] Letting n range over the numbers 0, 1, 2, $\cdots$, then, we have found infinitely many solutions of the potential equation but, in contrast to the previous examples, so far only denumerably many solutions.

In polar coordinates r, θ defined by $x = r \cos \theta$, $y = r \sin \theta$, we have

(6) $\qquad P_n(x, y) = r^n \cos n\theta, \qquad Q_n(x, y) = r^n \sin n\theta.$

For any arbitrary real α, the functions

$$P_\alpha(x, y) = r^\alpha \cos \alpha\theta, \qquad Q_\alpha(x, y) = r^\alpha \sin \alpha\theta$$

also satisfy the potential equation in any region of the x, y-plane excluding the origin $x = y = 0$. This is immediately verified after transforming Δu into polar coordinates (cf. Vol. I, p. 226):

$$\Delta u = u_{rr} + \frac{u_r}{r} + \frac{1}{r^2} u_{\theta\theta}.$$

If we choose two functions $w(\alpha)$ and $v(\alpha)$ in such a way that the first and second order derivatives of the integrals

$$\int_a^b w(\alpha)r^\alpha \cos \alpha\theta \, d\alpha \qquad \text{and} \qquad \int_a^b v(\alpha)r^\alpha \sin \alpha\theta \, d\alpha$$

can be obtained by differentiation under the integral sign, then we can construct a family of solutions, depending on two arbitrary func-

---

[1] These solutions illustrate the general fact that the real and the imaginary parts of an analytic function of a complex variable $x + iy$ satisfy the potential equation, i.e., are "harmonic" functions.

tions w and v, of the form

$$\int_a^b r^\alpha(w(\alpha) \cos \alpha\theta + v(\alpha) \sin \alpha\theta)\, d\alpha.$$

9) As an example of a differential equation of higher order, we consider

$$u_{xxyy} = 0$$

and find that

$$u(x, y) = w(y) + xw_1(y) + v(x) + yv_1(x)$$

is its general solution.

10) If the number of independent variables is greater than two, then arbitrary functions depending on two or more variables occur in the general solution. For example, the differential equation

$$u_z = 0$$

for the function $u(x, y, z)$ has the general solution

$$u = w(x, y).$$

2. Differential Equations for Given Families of Functions. In article 1 we mentioned that ordinary differential equations can be constructed which are satisfied by a given family of functions depending on a number of arbitrary parameters. We now pose the question: Can one construct a partial differential equation in n independent variables which is satisfied by a family of functions depending on an arbitrary function of $n - 1$ independent variables?

Consider, for example, a set of functions of the form

$$(7) \qquad u = f[x, y, w(g(x, y))],$$

where f is a given function of the arguments x, y, w and where $g(x, y)$ is a given function of x, y, e.g., $g = xy$. In order to obtain a partial differential equation for this set of functions, we differentiate equation (7) with respect to x and y:

$$u_x = f_x + f_w w' g_x,$$

$$u_y = f_y + f_w w' g_y.$$

The elimination of w' then yields the desired differential equation

(8) $$(u_x - f_x)g_y - (u_y - f_y)g_x = 0,$$

where the arbitrary function w, which still appears in f_x and f_y, is to be expressed by means of equation (7) in terms of x, y, u.

The partial differential equation thus obtained is of a special type, namely quasi-linear, since it involves the derivatives linearly. Hence, such a set of functions (7) is not general enough to lead to every differential equation of first order.

If, however, we start with a two-parameter family of functions

$$u = f(x, y; \alpha, \beta)$$

rather than with a set depending on an arbitrary function, and if we form the derivatives

$$u_x = f_x(x, y; \alpha, \beta),$$
$$u_y = f_y(x, y; \alpha, \beta),$$

then we have three equations from which we usually can eliminate α and β (certainly if $f_{x\alpha}f_{y\beta} - f_{x\beta}f_{y\alpha} \neq 0$). We obtain a partial differential equation $F(x, y, u, u_x, u_y) = 0$ which, in general, is no longer linear in u_x and u_y.

The paradox that a more restricted class of prescribed solutions leads to a more general type of differential equations will be resolved in §4.

Examples: 1) For the set of functions

$$u = w(xy)$$

we have, by elimination of w' from the equations,

$$u_x = yw', \qquad u_y = xw',$$

the differential equation

$$xu_x - yu_y = 0.$$

If we interpret x, y, u as rectangular coordinates, every function in this set represents, geometrically, a surface whose intersections with horizontal planes are equilateral hyperbolas.

2) The totality of *surfaces of revolution* generated by the rotation of a plane curve about the u-axis is given by

$$u = w(x^2 + y^2).$$

The corresponding differential equation is

$$yu_x - xu_y = 0.$$

3) Similarly

$$xu_x + yu_y = 0$$

is the differential equation for ruled surfaces generated by horizontal straight lines through the u-axis, i.e., represented by

$$u = w(x/y).$$

4) The differential equation for all *developable surfaces* is deduced from the definition of such a surface as *the envelope of a one-parameter family of planes*. With the exception of cylinders perpendicular to the x, y-plane, all such surfaces are given by the functions

(9) $$u = \alpha x + w(\alpha)y + v(\alpha),$$

where α is implicitly defined as a function of x and y by equation

(10) $$0 = x + w'(\alpha)y + v'(\alpha).$$

Thus, the function u depends on two arbitrary functions in a rather involved way. The first derivatives u_x, u_y are immediately obtained from (9), (10):

$$u_x = \alpha,$$
$$u_y = w(\alpha),$$

hence

(11) $$u_y = w(u_x).$$

In order to eliminate the arbitrary function w, we differentiate once more:

$$u_{yy} = w'u_{xy}, \qquad u_{xy} = w'u_{xx};$$

we obtain

(12) $$u_{xx}u_{yy} - u_{xy}^2 = 0$$

as the desired differential equation for all developable surfaces except cylinders perpendicular to the x, y-plane.

In all these examples it can easily be shown that the converse is also true, i.e., that all solutions of the corresponding differential equations belong to the given sets of functions.

5) All *homogeneous functions* $u(x_1, x_2, \cdots, x_n)$ of degree α in the variables $x_1, x_2, \cdots, x_n$ are characterized by the condition

$$(13) \qquad u(tx_1, tx_2, \cdots, tx_n) = t^\alpha u(x_1, x_2, \cdots, x_n),$$

identically satisfied in t. If we set $t = 1/x_n$, we have

$$u(x_1, x_2, \cdots, x_n) = x_n^\alpha u\left(\frac{x_1}{x_n}, \frac{x_2}{x_n}, \cdots, \frac{x_{n-1}}{x_n}, 1\right);$$

hence u is represented by

$$(14) \qquad u = x_n^\alpha w\left(\frac{x_1}{x_n}, \frac{x_2}{x_n}, \cdots, \frac{x_{n-1}}{x_n}\right)$$

with some function w. Since, conversely, every such function u formed by means of an arbitrary function w of $n - 1$ arguments satisfies the above condition of homogeneity, the expression (14) represents the totality of homogeneous functions of degree α.

In order to obtain a partial differential equation for this set of functions, we take the derivatives of equation (14) with respect to the variables $x_1, x_2, \cdots, x_n$ and eliminate the function w. This yields *Euler's homogeneity relation*

$$(15) \qquad x_1 u_{x_1} + x_2 u_{x_2} + \cdots + x_n u_{x_n} = \alpha u.$$

Incidentally, we could have obtained relation (15) directly by differentiating equation (13) with respect to t and setting $t = 1$.

Conversely, the homogeneity relation (15) for the function $u(x_1, x_2, \cdots, x_n)$ implies

$$\frac{\partial}{\partial t}\left(\frac{1}{t^\alpha} u(tx_1, tx_2, \cdots, tx_n)\right)$$

$$= \frac{1}{t^{\alpha+1}}\left\{\sum_{i=1}^{n} tx_i u_{x_i}(tx_1, tx_2, \cdot\ , tx_n) - \alpha u(tx_1, tx_2, \cdots, tx_n)\right\} = 0,$$

so that the expression $u(tx_1, tx_2, \cdots, tx_n)/t^\alpha$ is a function not depending on t; hence it is equal to its value for $t = 1$, which is

$u(x_1, x_2, \cdots, x_n)$. But this means, according to (13), that u is homogeneous.

§2. Systems of Differential Equations

1. The Question of Equivalence of a System of Differential Equations and a Single Differential Equation. For ordinary differential equations the theory of single differential equations is equivalent to the theory of systems; for partial differential equations the situation is different.

An ordinary differential equation of second order

$$(1) \qquad F(x, y, y', y'') = 0$$

can be reduced, by the substitution $y' = z$, to a system of two first order differential equations for two functions $y(x)$, $z(x)$:

$$(2) \qquad \begin{aligned} F(x, y, z, z') &= 0, \\ y' - z &= 0. \end{aligned}$$

Every solution of the differential equation (1) leads to a solution of the system (2) and conversely. More generally, a system of two first order ordinary differential equations

$$(3) \qquad f(x, y, z, y', z') = 0, \qquad g(x, y, z, y', z') = 0$$

for two functions $y(x)$ and $z(x)$ can be reduced to one second order differential equation for the one function $y(x)$ provided that, in the domain considered, $f_z g_{z'} - f_{z'} g_z \neq 0$. It is then possible to solve equations (3) for z' and z in the form

$$(3a) \qquad z' = \phi(x, y, y'), \qquad z = \psi(x, y, y').$$

By differentiating the second equation and eliminating z' we obtain immediately

$$(3b) \qquad \phi(x, y, y') - \psi_x - \psi_y y' - \psi_{y'} y'' = 0,$$

a differential equation of second order for $y(x)$ alone. If we substitute a solution of this differential equation (3b) in the relation $z = \psi(x, y, y')$, then we have the corresponding function z which, together with y, solves the original system (3) or (3a).

Hence, if we assume $f_z g_{z'} - f_{z'} g_z \neq 0$, the system (3) is indeed equivalent to one single differential equation.

We now consider a partial differential equation of the second order:

(4) $$F(x, y, u, u_x, u_y, u_{xx}, u_{xy}, u_{yy}) = 0$$

for a function $u(x, y)$. The substitution $u_x = p$, $u_y = q$ leads us to a system of three first order partial differential equations for three functions u, p, q:

$$F(x, y, u, p, q, p_x, p_y, q_y) = 0,$$

(5) $$u_x - p = 0,$$

$$u_y - q = 0.$$

Every solution u, p, q of this system yields u as a solution of the differential equation (4) and, conversely, every solution u of (4) leads to a set of solutions u, u_x, u_y of (5).

Thus, a partial differential equation of the second order is equivalent to a system of three differential equations of the first order (but to one of a very special form).

The converse, however, is by no means true. Not every system of two first order partial differential equations—let alone one of three first order differential equations—is equivalent to a second order differential equation.[1] In general, it is not possible by differentiation and elimination to obtain from the system of two partial differential equations

(6) $$\begin{aligned} f(x, y, u, v, u_x, v_x, u_y, v_y) &= 0, \\ g(x, y, u, v, u_x, v_x, u_y, v_y) &= 0 \end{aligned}$$

for two unknown functions $u(x, y)$ and $v(x, y)$ an equivalent second order partial differential equation for u alone. Differentiation with respect to x and y leads to four more equations. In order to replace the system (6) by one equivalent second order partial differential equation for u, one would have to eliminate the six quantities v, v_x, v_y, v_{xx}, v_{xy}, v_{yy} from the six equations. However, the elimination of six

[1] However, we shall see in §7 that such an equivalence often is obtained by adding to the system of differential equations certain "initial conditions" which restrict the set of solutions.
Concerning the problem of equivalence see further Appendix 2.

quantities from six equations is in general not possible as may be shown by constructing counterexamples.[1]

Differentiating further and comparing the number of equations with the number of the quantities to be eliminated we see that we cannot expect to find a single equation which replaces the system (6) even if we do not restrict its order. For example, if we differentiate each of the six equations we obtain twelve relations; in order to obtain a third order differential equation for u alone we would now have to eliminate ten quantities v, v_x, v_y, v_{xx}, v_{xy}, v_{yy}, v_{xxx}, v_{xxy}, v_{xyy}, v_{yyy}. Since the elimination of ten quantities from twelve equations usually results in two independent relations, we must expect two distinct third order equations for u alone as a consequence of the elimination,[2] except in special cases.

2. Elimination from a Linear System with Constant Coefficients. It is noteworthy that, in contrast to the general situation, the following theorem holds for an important special case:

From any system[3] of linear differential equations in n unknowns with constant coefficients a single linear differential equation with constant coefficients can be obtained for any of the unknown functions.

Let u, v, w, $\cdots$ be the unknown functions of the independent variables x, y, z, $\cdots$, and let P_i, Q_i, $\cdots$ be formal polynomials in the differentiation symbols $\partial/\partial x$, $\partial/\partial y$, $\partial/\partial z$, $\cdots$, e.g.,

$$P_i\left(\frac{\partial}{\partial x}, \frac{\partial}{\partial y}, \frac{\partial}{\partial z}, \cdots\right) = \sum \alpha^i_{\nu_1\nu_2\cdots} \frac{\partial^{\nu_1+\nu_2+\nu_3+\cdots}}{\partial x^{\nu_1}\partial y^{\nu_2}\partial z^{\nu_3}\cdots}$$

with constant coefficients $\alpha^i_{\nu_1\nu_2\cdots}$; then we write the system formally as

---

[1] An example of a system of equations for which elimination and differentiation do not lead to an equation of second order, is the system $u_x + v_y = -yu$, $u_y + v_x = yv$. One obtains here two equations of third order for the one (overdetermined) function u:

$$\frac{\partial}{\partial y}(y^2u + u_{yy} - u_{xx}) + u_x + yu = 0,$$

$$\frac{\partial}{\partial x}(y^2u + u_{yy} - u_{xx}) + u_y - y(y^2u + u_{yy} - u_{xx}) = 0.$$

[2] See previous footnote.
[3] See article 3 for definitions.

$$P_1\left(\frac{\partial}{\partial x}, \frac{\partial}{\partial y}, \cdots\right) u + Q_1\left(\frac{\partial}{\partial x}, \frac{\partial}{\partial y}, \cdots\right) v + \cdots = g_1(x, y, \cdots),$$

$$P_2\left(\frac{\partial}{\partial x}, \frac{\partial}{\partial y}, \cdots\right) u + Q_2\left(\frac{\partial}{\partial x}, \frac{\partial}{\partial y}, \cdots\right) v + \cdots = g_2(x, y, \cdots),$$

$$\cdots\cdots\cdots\cdots\cdots\cdots\cdots\cdots\cdots\cdots\cdots\cdots\cdots\cdots$$

with given right-hand sides g_1, g_2, $\cdots$. Now formal algebraic elimination (Cramer's rule) yields differential equations for single functions:

$$Du = G^1, \qquad Dv = G^2, \qquad \cdots,$$

where D is the determinant of the symbols P_i, Q_i, $\cdots$, and G^j is a suitable symbolic linear combination of the functions g_ν. Clearly D is a linear differential operator; its order is the degree of the symbolic polynomial D (and the degree of D depends on the degrees of the P_i, Q_i, $\cdots$). The symbols G^j likewise denote differential operators corresponding to the minors of the determinant D. If, in particular, the original system consists of n first order equations, i.e., if the polynomials P_i, Q_i, $\cdots$ are linear, then the resulting equations are, in general, of order n.

Suppose u is a solution of one of the resulting higher order equations; then after substituting u in the given system, we may omit one of the original equations, since the system now is dependent. Thus a system of fewer equations for v, w, $\cdots$ is obtained. This can be treated like the original one and by elimination leads to a differential equation $D^*v = G^{2*}$, where D^* is a minor of the determinant D. Continuing in this way, the original system can be replaced by a sequence of independent equations $Du = G^1, D^*v = G^{2*}, \cdots$ of decreasing order which corresponds to the original system.

3. Determined, Overdetermined, Underdetermined Systems. The general form of a system of partial differential equations in two independent variables is

(7)
$$F_i(x, y, u^{(1)}, u^{(2)}, \cdots, u^{(m)}, u_x^{(1)}, u_y^{(1)}, \cdots,$$
$$u_x^{(m)}, u_y^{(m)}, u_{xx}^{(1)}, \cdots) = 0 \quad (i = 1, 2, \cdots, h),$$

i.e., a system of h equations for m functions $u^{(1)}$, $u^{(2)}$, $\cdots$, $u^{(m)}$ of the independent variables x and y. We assume that these h equations are independent, i.e., that none of them can be deduced from the others by differentiation and elimination.

If $h = m$, we speak of a *determined system*; if $h > m$ the system is called *overdetermined*, and if $h < m$, *underdetermined*.

The Cauchy-Riemann differential equations for two functions $u(x, y)$, $v(x, y)$,

$$u_x - v_y = 0, \qquad u_y + v_x = 0$$

are an example of a determined system. For this special system it follows easily by differentiation and elimination that u and v separately satisfy the partial differential equations $\Delta u = 0$ and $\Delta v = 0$ (cf. article 1), that is, that u and v are "harmonic".

The simplest example of an overdetermined system for a function $u(x, y)$ is

$$u_x = f(x, y), \qquad u_y = g(x, y)$$

which, as is well-known, can be solved if and only if

$$f_y = g_x .$$

A more interesting example is furnished by the theory of *analytic functions* $f(z_1, z_2)$ of *two complex variables*

$$z_1 = x_1 + iy_1, \qquad z_2 = x_2 + iy_2 .$$

The Cauchy-Riemann differential equations which express the analytic character of the function $f(z_1, z_2) = u + iv$ are

$$(8) \qquad \begin{aligned} u_{x_1} &= v_{y_1}, & u_{x_2} &= v_{y_2}, \\ u_{y_1} &= -v_{x_1}, & u_{y_2} &= -v_{x_2}. \end{aligned}$$

These lead, by differentiation, to the following overdetermined system for u alone:

$$(8') \qquad \begin{aligned} u_{x_1x_1} + u_{y_1y_1} &= 0, & u_{x_1x_2} + u_{y_1y_2} &= 0, \\ u_{x_2x_2} + u_{y_2y_2} &= 0, & u_{x_1y_2} - u_{x_2y_1} &= 0. \end{aligned}$$

The fact that this system is highly overdetermined indicates that the theory of functions of several complex variables is inherently more complicated than the classical theory of functions of one complex variable.

We obtain a third example of an overdetermined system if we introduce $n + 1$ "homogeneous variables" x_1, x_2, $\cdots$, x_{n+1} instead of the n variables x, y, $\cdots$ by means of the relations

$$x = \frac{x_2}{x_1}, \qquad y = \frac{x_3}{x_1}, \qquad \cdots .$$

A function $u(x, y, \cdots)$ then becomes a function $\omega(x_1, x_2, \cdots)$, which is homogeneous of degree zero in the new variables and which therefore satisfies the Euler homogeneity relation

$$x_1\omega_{x_1} + x_2\omega_{x_2} + \cdots = 0.$$

The first partial derivatives of the function $u(x, y, \cdots)$ with respect to x, y, $\cdots$ can be expressed in terms of the derivatives of the function $\omega(x_1, x_2, \cdots)$:

$$u_x = x_1\omega_{x_2},$$
$$u_y = x_1\omega_{x_3},$$
$$\cdots\cdots\cdots\cdots$$

Thus a given first order partial differential equation for u,

$$f(x, y, \cdots, u, u_x, u_y, \cdots) = 0,$$

is transformed into a first order partial differential equation of the form

$$\phi(x_1, x_2, \cdots, \omega, \omega_{x_2}, \omega_{x_3}, \cdots) = 0$$

for the function $\omega(x_1, x_2, \cdots)$; furthermore, the homogeneity relation

$$x_1\omega_{x_1} + x_2\omega_{x_2} + \cdots = 0$$

appears as an additional equation. Instead of one differential equation, we obtain an overdetermined system of two equations. If we transform a system of differential equations by introducing homogeneous variables we have, of course, a similar situation.

The equation

$$u_x v_y - u_y v_x = 0,$$

which, as one can see, expresses the identical vanishing of the Jacobian of the two functions $u(x, y)$ and $v(x, y)$, is an example of an *underdetermined* system. This equation implies[1] that a relation

$$w(u, v) = 0$$

[1] Cf. §1, 1, example 5).

exists between u and v, which does not explicitly contain the independent variables x and y; it represents the "general solution" of the underdetermined system of differential equations.[1] For a system of n functions $u^{(1)}$, $u^{(2)}$, $\cdots$, $u^{(n)}$ of the variables x_1, x_2, $\cdots$, x_n the vanishing of the Jacobian

$$(9) \qquad \frac{\partial(u^{(1)}, \cdots, u^{(n)})}{\partial(x_1, \cdots, x_n)} = \begin{vmatrix} u_{x_1}^{(1)} & \cdots & u_{x_1}^{(n)} \\ \cdots & \cdots & \cdots \\ u_{x_n}^{(1)} & \cdots & u_{x_n}^{(n)} \end{vmatrix} = 0$$

generally characterizes adependence of the n functions $u^{(1)}$, $u^{(2)}$, $\cdots$, $u^{(n)}$:

$$(10) \qquad w(u^{(1)}, u^{(2)}, \cdots, u^{(n)}) = 0.$$

Hence the relation (10) may be regarded as the general solution of the underdetermined system of differential equations (9). Later, in Chapters II and III, we shall return to the problem of solving various types of underdetermined systems of differential equations.

§3. *Methods of Integration for Special Differential Equations*

1. Separation of Variables. For many differential equation problems in mathematical physics families of solutions depending on arbitrary parameters can be obtained by special methods, although these methods do not give the totality of solutions directly.

The most important of these methods is *separation of variables*; it will be illustrated by several examples.

1) Consider the equation

$$u_x^2 + u_y^2 = 1;$$

---

[1] In a similar way, the underdetermined equation $u_x v_y - u_y v_x = 1$ which characterizes area-preserving transformations from the x,y-plane to the u,v-plane is solved by

$$x = \alpha + \omega_\beta, \qquad u = \alpha - \omega_\beta,$$
$$y = \beta - \omega_\alpha, \qquad v = \beta + \omega_\alpha,$$

where ω is an arbitrary function for which

$$\frac{\partial(x, y)}{\partial(\alpha, \beta)} = \frac{\partial(u, v)}{\partial(\alpha, \beta)} = 1 + \omega_\alpha \omega_{\beta\beta} - \omega_{\alpha\beta}^2 \neq 0.$$

assuming

$$u(x, y) = \phi(x) + \psi(y),$$

we obtain

$$(\phi'(x))^2 + (\psi'(y))^2 = 1$$

or

$$(\phi'(x))^2 = 1 - (\psi'(y))^2.$$

Since the right side is independent of x and the left side independent of y, both sides are independent of x as well as y, hence equal to the same constant α^2; thus we obtain immediately the family of solutions

(1) $$u(x, y) = \alpha x + \sqrt{1 - \alpha^2}\, y + \beta$$

with the two arbitrary parameters α and β.

2) Similarly, the differential equation

$$u_x^2 + u_y^2 + u_z^2 = 1$$

for a function u of the three variables x, y, z, leads, if we assume that $u = \phi(x) + \psi(y) + \chi(z)$, to the family of solutions

(2) $$u = \alpha x + \beta y + \sqrt{1 - \alpha^2 - \beta^2}\, z + \gamma$$

depending on three arbitrary parameters α, β, γ.

3) The tentative hypothesis

$$u = \phi(x) + \psi(y)$$

applied to the differential equation

$$f(x)u_x^2 + g(y)u_y^2 = a(x) + b(y)$$

leads, as in previous examples, to

(3) $$u(x, y) = \int_{x_0}^{x} \sqrt{\frac{a(\xi) + \alpha}{f(\xi)}}\, d\xi + \int_{y_0}^{y} \sqrt{\frac{b(\eta) - \alpha}{g(\eta)}}\, d\eta + \beta,$$

where α and β are arbitrary constants.

4) A transformation of the independent variables frequently makes separation of variables successful afterwards. For example, the equation

$$u_x^2 + u_y^2 = \frac{k}{r} - h \quad (r^2 = x^2 + y^2;\ k, h \text{ constants}),$$

for $u(x, y)$ occurring in the *two body problem* of celestial mechanics is transformed into the equation

$$u_r^2 + \frac{1}{r^2} u_\theta^2 = \frac{k}{r} - h \qquad \text{or} \qquad r^2 u_r^2 + u_\theta^2 = kr - hr^2$$

for $u(r, \theta)$ in polar coordinates r, θ. Hence, formula (3) yields the family of solutions

(4) $$u = \int_0^r \sqrt{\frac{k}{\rho} - h - \frac{\alpha^2}{\rho^2}} \, d\rho + \alpha\theta + \beta$$

which depends on two arbitrary parameters α, β.

5) In the case of linear differential equations, particularly those of the second order, it is often profitable to set

$$u(x, y) = \phi(x)\psi(y)$$

(examples are given in Vol. I, Ch. V, §§3–9). For the heat equation

(5) $$u_{xx} - u_y = 0$$

we have

$$\phi''(x):\phi(x) = \psi'(y):\psi(y)$$

and therefore both the right and the left sides must be constant. One may assume this constant to be either positive or negative and denote it, accordingly, by ν^2 or $-\nu^2$; thus one obtains the two families of solutions

$$u = a \sinh \nu(x - \alpha)e^{\nu^2 y},$$

$$u = a \sin \nu(x - \alpha)e^{-\nu^2 y}.$$

The latter plays a particular role in mathematical physics; if u is the temperature, y the time, and x a space coordinate, it describes a temperature distribution which tends to zero as time progresses.

2. Construction of Further Solutions by Superposition. Fundamental Solution of the Heat Equation. Poisson's Integral. From the solutions of linear differential equations containing parameters, further solutions may be obtained by summation, integration, and differentiation processes. Since many such examples are given in Vol. I, Ch. V, only a few more will be discussed here.

In order to obtain another solution of the heat equation, we integrate the solution $e^{-\nu^2 y} \cos \nu x$ with respect to the parameter ν between the limits $-\infty$ and ∞ and find the new solution

$$u = \int_{-\infty}^{\infty} e^{-\nu^2 y} \cos \nu x \, d\nu \qquad (y > 0).$$

The integral on the right can easily be computed[1] and yields

(6)
$$u = \sqrt{\frac{\pi}{y}} \, e^{-x^2/4y},$$

the "fundamental solution" of the heat equation.

As a second example for the principle of superposition, we give the solution of the *boundary value problem for the potential equation* $\Delta u = 0$ for the circular disk $r^2 = x^2 + y^2 < 1$; for $r = 1$, the boundary values of u are given as a (continuously differentiable) function $g(\theta)$ of the polar angle θ. Let

$$a_n = \frac{1}{\pi} \int_{-\pi}^{\pi} g(\phi) \cos n\phi \, d\phi, \qquad b_n = \frac{1}{\pi} \int_{-\pi}^{\pi} g(\phi) \sin n\phi \, d\phi$$

be the coefficients of the Fourier series for $g(\theta)$; then

$$u(x, y) = \frac{a_0}{2} + \sum_{\nu=1}^{\infty} (a_\nu \cos \nu\theta + b_\nu \sin \nu\theta) r^\nu$$

$$= \frac{a_0}{2} + \sum_{\nu=1}^{\infty} (a_\nu P_\nu(x, y) + b_\nu Q_\nu(x, y))$$

---

[1] To evaluate this integral we substitute $\nu^2 y = \lambda^2$ and obtain the integral

$$\frac{1}{\sqrt{y}} J(a), \quad \text{where} \quad a = \frac{x}{\sqrt{y}} \quad \text{and} \quad J(a) = \int_{-\infty}^{\infty} e^{-\lambda^2} \cos (a\lambda) \, d\lambda.$$

To determine $J(a)$, we find $J'(a)$ by differentiating under the integral sign:

$$J'(a) = -\int_{-\infty}^{\infty} e^{-\lambda^2} \lambda \sin (a\lambda) \, d\lambda;$$

integrating by parts, we find almost immediately $J'(a) = -aJ(a)/2$, while direct calculation shows that

$$J(0) = \int_{-\infty}^{\infty} e^{-\lambda^2} \, d\lambda = \sqrt{\pi}.$$

From this it follows that $J(a) = \sqrt{\pi} e^{-a^2/4}$, and hence (6) is established.

converges uniformly for $r \leq q < 1$. This series, which may be differentiated twice term by term for $r \leq q$, represents a superposition of the potential functions P_n and Q_n considered in example 8) of §1, 1. Hence it is a harmonic function, and moreover solves the boundary value problem. In the interior of the circle, we may interchange summation and integration and obtain

$$u(x, y) = \frac{1}{\pi} \int_{-\pi}^{\pi} g(\phi) \left[\frac{1}{2} + \sum_{\nu=1}^{\infty} r^{\nu} \cos \nu(\theta - \phi) \right] d\phi.$$

Writing $2 \cos \alpha = e^{i\alpha} + e^{-i\alpha}$ and summing the geometric series thus obtained under the integral sign we arrive, after a trivial manipulation, at the expression

$$(7) \qquad u(x, y) = \frac{1}{2\pi} \int_{-\pi}^{\pi} \frac{1 - r^2}{1 - 2r \cos (\theta - \phi) + r^2} g(\phi) \, d\phi,$$

which represents the solution of the boundary value problem by means of the Poisson integral (cf. Ch. IV, and also Vol. I, p. 514).

§4. *Geometric Interpretation of a First Order Partial Differential Equation in Two Independent Variables. The Complete Integral*

1. Geometric Interpretation of a First Order Partial Differential Equation. Geometric intuition is of great help in the theory of integration of first order partial differential equations for a function $u(x, y)$ of two independent variables. Given the differential equation

$$(1) \qquad\qquad F(x, y, u, p, q) = 0,$$

$F_p^2 + F_q^2 \neq 0$, with the abbreviation $p = u_x$, $q = u_y$. Then for every integral surface through the point P with coordinates x, y, u the quantities p and q, which determine the position of the tangent plane at this point, must satisfy condition (1). The tangent plane of an integral surface at the point P^1 is restricted to positions which belong to the manifold characterized by equation (1).[2] For a given point $P\colon (x, y, u)$, this manifold is in general a one-parameter family (for

---

[1] To emphasize the fact that, in the tangent planes considered, only the immediate neighborhood of the point P of contact plays a role, it is convenient to consider P together with an arbitrarily small neighborhood of P in the tangent plane as a "surface element" and to operate with such surface elements (in the case of ordinary differential equations, line elements are used similarly).

[2] For a more detailed discussion, see Ch. II, §3,1.

example, for $p^2 + q^2 = 1$ the family is $p = \cos t$, $q = \sin t$ with the parameter t). If F is linear in p and q then this family of possible tangent planes forms an axial pencil of planes through a straight line called the *"Monge axis"*. For the present, we ignore this special case of the first order "quasi-linear" equation which will be discussed in §5; instead we assume that, at every point P considered, our family of planes envelops a genuine cone, the *"Monge cone"*.[1] Thus, in a domain of the x, y, u-space the differential equation is represented geometrically by a "cone field" just as an ordinary differential equation of the first order is represented by a direction field. To find a solution means to find a surface which at each of its points touches the corresponding Monge cone (or "fits" into the cone field).

As in the case of ordinary differential equations, geometric visualization makes the following theorem evident: *If a family of solutions*

$$(2) \qquad u = f(x, y, a)$$

of the differential equation $F(x, y, u, p, q) = 0$ depending on a parameter a possesses an envelope, then this envelope is also a solution.

Indeed, the envelope of a family of integral surfaces has at each point P a tangent plane which touches the Monge cone there; this tangent plane is the same as that of the integral surface of the family which touches the envelope at P.

Analytically, one is led to this statement in the following way: the envelope is obtained by expressing a as a function of x and y from the equation

$$(3) \qquad f_a(x, y, a) = 0,$$

and then inserting this function $a(x, y)$ into f; thus the envelope appears in the form

$$u = f(x, y, a(x, y)) = \psi(x, y).$$

Then, using (3), we have

$$\psi_x = u_x = f_x + f_a a_x = f_x, \qquad \psi_y = u_y = f_y + f_a a_y = f_y.$$

Accordingly, the values of $\psi(x_0, y_0)$, $\psi_x(x_0, y_0)$, $\psi_y(x_0, y_0)$ coincide, at a fixed point (x_0, y_0), with the values of $f(x, y, a_0)$, $f_x(x, y, a_0)$, $f_y(x, y, a_0)$, respectively, where $a_0 = a(x_0, y_0)$. Therefore, since the

[1] After Gaspard Monge, 1746–1818.

differential equation is satisfied by the function $u = f(x, y, a_0)$ at the point (x_0, y_0), it is also satisfied by $u = f(x, y, a(x, y)) = \psi(x, y)$.

2. The Complete Integral. The examples in §3 showed that, in the case of first order differential equations, we can frequently obtain solutions which depend on arbitrary parameters; in particular, let the differential equation (1)

$$F(x, y, u, p, q) = 0$$

for a function $u(x, y)$ of two independent variables, have a solution

$$(4) \qquad\qquad u = \phi(x, y, a, b)$$

which depends on two parameters a, b. (Incidentally, if u does not appear explicitly in F, then the one-parameter family of solutions $u = \phi(x, y, a)$ leads immediately to a family $u = \phi(x, y, a) + b$ which depends on two parameters.)

A two-parameter family of solutions is called a complete integral of (1), *if in the region considered the rank of the matrix*

$$M = \begin{pmatrix} \phi_a & \phi_{xa} & \phi_{ya} \\ \phi_b & \phi_{xb} & \phi_{yb} \end{pmatrix}$$

is 2[1]; in particular, this is true if the determinant

$$(5) \qquad\qquad D = \phi_{xa}\phi_{yb} - \phi_{xb}\phi_{ya}$$

does not vanish.

The significance of the concept "complete integral" comes from the following basic idea: *By the formation of envelopes, i.e., merely by differentiation and elimination processes, one may obtain from a complete integral (4) a set of solutions of the differential equation (3) which*

---

[1] This condition assures that the function ϕ essentially depends on two independent parameters. For if by introduction of an appropriate combination $\gamma = g(a, b)$ it could be brought into the form $\phi(x, y, a, b) = \psi(x, y, \gamma)$ depending only on the one parameter γ one could immediately deduce from the relations

$$\phi_{xa} = \psi_{x\gamma}\gamma_a, \qquad \phi_{xb} = \psi_{x\gamma}\gamma_b, \qquad \phi_{ya} = \psi_{y\gamma}\gamma_a, \qquad \phi_{yb} = \psi_{y\gamma}\gamma_b,$$

$$\phi_a = \psi_\gamma\gamma_a, \qquad \phi_b = \psi_\gamma\gamma_b$$

that the rank of the above matrix cannot be 2.

depends on an arbitrary function.[1] To construct such solutions we select a one-parameter family from the two-parameter family by connecting the as yet independent parameters a and b by an arbitrary function, for example $b = w(a)$, and then form the envelope of this one-parameter family. We consider a as a function of x and y obtained from the equation

$$(6) \qquad \phi_a + \phi_b w'(a) = 0 \qquad (b = w(a))$$

and substitute it into

$$u = \phi(x, y, a, w(a)) = \psi(x, y).$$

Here we make the additional assumption that (6) can be solved for a. Thus we get a manifold of solutions $\psi(x, y)$ depending on an arbitrary function w. Incidentally, the apparent paradox mentioned in §1, 2 is clarified by the situation just described. By giving a two-parameter family of solutions for a partial differential equation, we have furnished at the same time a set of solutions depending on an arbitrary function; but the arbitrary function enters in such a complicated way that the set of functions cannot, in general, be represented as in §1, 2.

The systematic treatment of the theory of first order differential equations presented in the next chapter shows that the theory of the complete integral can be generalized to differential equations for functions of n independent variables and that it is intimately related to the general theory of integration of first order differential equations.

3. Singular Integrals. In addition to the "general" solution obtained in article 2 by the formation of envelopes of one-parameter subfamilies of the two-parameter family $u = \phi(x, y, a, b)$, we can by forming envelopes sometimes find still another, the *singular* solution. For, the two-parameter family u may possess an envelope[2] which is not included among the envelopes formed by the one-parameter subfamilies. This envelope, obtained by eliminating a and

---

[1] Whether this method yields all the solutions will not be discussed here. The difficulty in making general statements is illuminated by the following example: Let $F(x, y, u, p, q) = G(x, y, u, p, q)H(x, y, u, p, q)$, and let ϕ be a complete integral of the equation $G = 0$ which is not simultaneously a solution of $H = 0$. Then, by our definition, ϕ is also a complete integral of $F = 0$; yet there exist families of functions which solve $F = 0$—namely solutions of $H = 0$ —that cannot be constructed as envelopes from ϕ.

[2] This, however, is excluded if u does not occur explicitly in F.

b from the three equations

$$u = \phi(x, y, a, b),$$

(7) $$0 = \phi_a ,$$

$$0 = \phi_b ,$$

must also be a solution; it is called a "singular" solution of (1). Note that here, as in ordinary differential equations, no knowledge of a complete integral is necessary in order to find singular solutions; one finds them directly from the differential equation through differentiation and elimination: *The singular solution is obtained by eliminating p and q from the equations*

(8) $$F(x, y, u, p, q) = 0, \qquad F_p = 0, \quad F_q = 0.$$

The equation

$$F(x, y, \phi, \phi_x , \phi_y) = 0,$$

which holds identically in a and b, can be differentiated with respect to a and b and leads to

$$F_u\phi_a + F_p\phi_{xa} + F_q\phi_{ya} = 0,$$

$$F_u\phi_b + F_p\phi_{xb} + F_q\phi_{yb} = 0.$$

On the singular integral surface, $\phi_a = \phi_b = 0$ and, therefore, we have for all of its points

$$F_p\phi_{xa} + F_q\phi_{ya} = 0,$$

$$F_p\phi_{xb} + F_q\phi_{yb} = 0.$$

If on this surface we assume that the determinant

$$D = \phi_{xa}\phi_{yb} - \phi_{xb}\phi_{ya}$$

does not vanish then the equations

$$F_p = 0, \qquad F_q = 0$$

hold. The equation for the singular integral can therefore be deduced from the equations (8) by elimination of p and q.

Accordingly, a singular solution can be defined, without reference to a specific complete integral, as a solution for which

$$F = F_p = F_q = 0$$

(cf. Ch. II. §4).

4. Examples. We consider the two-parameter family of functions

$$(9) \qquad (x - a)^2 + (y - b)^2 + u^2 = 1,$$

i.e. the totality of spheres of radius 1 in x, y, u-space whose centers lie in the x, y-plane. These functions form a complete integral of the differential equation

$$(10) \qquad u^2(1 + p^2 + q^2) = 1.$$

If we set $b = w(a)$, thus singling out from all the spheres that one-parameter family whose centers lie on the curve $y = w(x)$ in the x, y-plane, then the envelope of this family, i.e., the surface obtained by eliminating a from

$$(11) \qquad \begin{aligned} (x - a)^2 + (y - w(a))^2 + u^2 &= 1, \\ x - a + w'(a)(y - w(a)) &= 0, \end{aligned}$$

yields another solution. Every such envelope is a *tubular surface* whose axis is $y = w(x)$.

The total two-parameter family (9) also has another envelope consisting of the planes $u = 1$ and $u = -1$; this is intuitively clear and can be confirmed analytically by eliminating a and b from

$$(12) \qquad \begin{aligned} (x - a)^2 + (y - b)^2 + u^2 &= 1, \\ x - a &= 0, \\ x - b &= 0. \end{aligned}$$

Since these surfaces satisfy the differential equations (10), they constitute the singular solutions of (10). We are also led to these surfaces if we eliminate the quantities p and q from the equations

$$(13) \qquad \begin{aligned} F &= u^2(1 + p^2 + q^2) = 1, \\ F_p &= 2u^2 p = 0, \\ F_q &= 2u^2 q = 0. \end{aligned}$$

Another example is furnished by *Clairaut's differential equation*

$$(14) \qquad u = xu_x + yu_y + f(u_x, u_y),$$

which occurs frequently in applications. We start with the two-parameter family of planes

$$(15) \qquad u = ax + by + f(a, b),$$

where $f(a, b)$ is a prescribed function of the parameters a and b. Since $u_x = a$, $u_y = b$, this family satisfies the partial differential equation (14). There $D = 1$ (see formula (5)), hence u given by (15) is a complete integral of Clairaut's differential equation.

Again we form envelopes to obtain the general solution of this equation; choosing an arbitrary function $b = w(a)$, we eliminate a from the equations

(16)
$$u = ax + yw(a) + f(a, w(a)),$$
$$0 = x + yw'(a) + f_a + f_b w'(a).$$

The singular solution of Clairaut's equation is of importance. We obtain it as the envelope of the two-parameter family (15), i.e., by eliminating a and b from the equations

$$u = ax + by + f(a, b),$$

(17) $$x = -f_a,$$

$$y = -f_b.$$

If we differentiate the differential equation (14) with respect to $u_x = p$, $u_y = q$, the rule in article 3 leads to the same formulas. (Compare §6, 3 where a different point of view is presented.)

§5. Theory of Linear and Quasi-Linear Differential Equations of First Order

1. Linear Differential Equations. We consider a partial differential equation for $u(x_1, x_2, \cdots, x_n)$ of the form

(1) $$\sum_{i=1}^{n} a_i u_{x_i} = a.$$

If the a_i and a are given continuously differentiable functions of the independent variables $x_1, x_2, \cdots, x_n$ alone, then (1) is called a linear differential equation; more generally, if the a_i and a may also depend on the unknown function u itself, we call the equation quasi-linear. In this section we show that the theory of such quasi-linear partial differential equations is equivalent to the theory of a system of ordinary differential equations (cf. Ch. II, §2).

First we treat the special case of a homogeneous linear differential equation

$$(1') \qquad\qquad \sum_{i=1}^{n} a_i u_{x_i} = 0.$$

In the n-dimensional space of the variables x_1, x_2, $\cdots$, x_n we determine the curves $x_i = x_i(s)$ in terms of a parameter s by means of the system of ordinary differential equations

$$(2) \qquad\qquad \frac{dx_i}{ds} = a_i(x_1, x_2, \cdots, x_n) \qquad (i = 1, 2, \cdots, n).$$

These curves are called the *characteristic curves*. (We shall discuss their general significance in connection with the treatment of quasi-linear differential equations given in Ch. II, §2.) In the case $n = 2$ they are those curves which are tangent to the Monge axes mentioned in §4, 1 as degenerations of Monge cones.

We recall some facts concerning ordinary differential equations. By introducing in (2) one of the quantities x_i instead of s as the independent variable, we may represent the general solution of the resulting system which depends on $n - 1$ parameters, c_i :

$$c_i = \phi_i(x_1, x_2, \cdots, x_n) \qquad\qquad (i = 1, 2, \cdots, n - 1).$$

Here the c_i are the arbitrary constants of integration and the ϕ_i are mutually independent "integrals" of the system. The "integral" $\phi(x_1, x_2, \cdots, x_n)$ means here a function of the independent variables x_i which has a constant value along each curve $x_i(s)$ which solves the system (2).

Equation (1') states: For the values $u(s) = u[x_1(s), x_2(s), \cdots, x_n(s)]$ of a solution u of the partial differential equation along an integral curve of the system of ordinary differential equations, the relation

$$(3) \qquad\qquad \frac{du}{ds} = 0$$

holds. Thus, *along each integral curve of the system (2) of ordinary differential equations every solution of the partial differential equation (1') has a constant value, i.e., a value independent of s.* Every solution of the partial differential equation is an *integral* of the system of ordinary differential equations.

On the other hand, every integral

$$\phi(x_1, x_2. \cdots. x_n)$$

of the system of ordinary differential equations (2) is a solution of the partial differential equation (1'); substituting in this integral for x_i any solution $x_i(s)$ of the system (2), and differentiating ϕ with respect to s, one verifies that (1') holds along each integral curve $x_i(s)$. One of the integral curves passes through each point of a suitably limited region in the x-space; hence ϕ satisfies the differential equation (1') identically in x_1, x_2, $\cdots$, x_n in this region.

Among each set of n integrals

$$\phi_i(x_1, x_2, \cdots, x_n) \qquad (i = 1, 2, \cdots, n)$$

of our system (2) of differential equations, a relation of the form

$$(4) \qquad \omega(\phi_1, \phi_2, \cdots, \phi_n) = 0$$

holds; for, the equations

$$\sum_{k=1}^{n} a_k \frac{\partial \phi_i}{\partial x_k} = 0 \qquad (i = 1, 2, \cdots, n)$$

with some coefficients $a_\nu \neq 0$, can be valid only if the determinant

$$(5) \qquad \frac{\partial(\phi_1, \phi_2, \cdots, \phi_n)}{\partial(x_1, x_2, \cdots, x_n)}$$

vanishes. But this condition is a sufficient condition for a relation of the form (4) to hold. On the other hand, by the elementary existence theorem in the theory of ordinary differential equations, there do exist $n - 1$ mutually independent integrals ϕ_1, ϕ_2, $\cdots$, ϕ_{n-1} of the system (2), so that each integral ϕ must be of the form

$$(6) \qquad \phi(x_1, x_2, \cdots, x_n) = w(\phi_1, \phi_2, \cdots, \phi_{n-1}).$$

Since, conversely, each function $w(\phi_1, \phi_2, \cdots, \phi_{n-1})$ is constant along each integral curve of (2) and hence is itself an integral of (2), all solutions of the partial differential equation (1') are obtained in the form (6) where w is an arbitrary function of the $n - 1$ arguments.

Conversely, the ordinary differential equations (2) can be solved by means of $n - 1$ mutually independent solutions

$$\phi_1, \phi_2, \cdots, \phi_{n-1}$$

of the partial differential equation; they can, for example, be solved by calculating from the equations $\phi_\nu = c_\nu$ the $n - 1$ quantities x_1, x_2, $\cdots$, x_{n-1} as functions of the independent variable x_n and the parameters c_1, c_2, $\cdots$, c_{n-1}.

2. Quasi-Linear Differential Equations. The general case, in which the differential equation (1) is quasi-linear and may have a nonvanishing right-hand side

$$a(x_1, x_2, \cdots, x_n, u),$$

is essentially not more difficult; it can be reduced to the case of homogeneous linear differential equations with one additional independent variable x_{n+1}, and can thus be disposed of completely. (The device by which this reduction is effected will also be useful later in this book.) We introduce $u = x_{n+1}$ as a new independent variable; if we allow the desired solution of (1) to be defined in the implicit form $\phi(x_1, x_2, \cdots, x_{n+1}) = 0$ or, more generally, in terms of a constant c in the form

$$(7) \qquad \phi(x_1, x_2, \cdots, x_{n+1}) = c,$$

the problem is reduced to finding ϕ. Since $\phi_{x_i} + \phi_{x_{n+1}} u_{x_i} = 0$, ϕ must satisfy the partial differential equation

$$(8) \qquad \sum_{\nu=1}^{n+1} a_\nu \phi_{x_\nu} = 0,$$

where we set $a(x_1, x_2, \cdots, x_n, u) = a_{n+1}$. The form of this relation is exactly that of a linear homogeneous differential equation for the function $\phi(x_1, x_2, \cdots, x_{n+1})$ of $n + 1$ variables. However, there is a slight conceptual difficulty: The equation (8) need not hold identically in $x_1, x_2, \cdots, x_{n+1}$ because it is derived from only those sets of values x_ν for which the relation $\phi = 0$ or $\phi = c$ holds. Thus, from this point of view, (8) is not yet a linear homogeneous partial differential equation. But if instead of considering a single solution of the original differential equation we consider a one-parameter family depending on the parameter c and given by $\phi = c$, then equation (8) must hold for all values of $x_1, x_2, \cdots, x_{n+1}$; i.e., it is really a linear differential equation of the type treated. If we select

$$x_1, x_2, \cdots, x_{n+1}$$

arbitrarily and take the value c given by $\phi(x_1, x_2, \cdots, x_{n+1}) = c$, then since (8) must hold for this value of c, it holds identically in $x_1, x_2, \cdots, x_{n+1}$.

Conversely, by finding a solution ϕ of (8) and setting $\phi = c$, we obtain a one-parameter family of solutions of (1).

Thus we have shown that there is a one-to-one correspondence between solutions of (8) and one-parameter families of solutions of the original equation (1). This shows that integration of the general quasi-linear differential equation (1) is equivalent to integration of the system of ordinary differential equations

$$(9) \qquad \frac{dx_i}{ds} = a_i, \qquad \frac{du}{ds} = a.$$

§6. *The Legendre Transformation*

1. The Legendre Transformation for Functions of Two Variables. The integration of certain classes of differential equations can be considerably simplified by applying the "*Legendre* transformation". This transformation is suggested by the geometric interpretation of the differential equation if we represent the integral surface by its tangent plane coordinates instead of by point coordinates.[1]

For the description of a surface in the x, y, u-space, there are two dual possibilities. Either one may give the surface as a point set determined by a function $u(x, y)$, or one may regard it as the envelope of its tangent planes, i.e., set up the equation which a plane must satisfy in order to be tangent to the surface. If $\bar{x}, \bar{y}, \bar{u}$ are the running coordinates of a plane whose equation is

$$\bar{u} - \xi\bar{x} - \eta\bar{y} + \omega = 0,$$

then we call ξ, η, ω the coordinates of this plane. Since the plane tangent to the surface $u(x, y)$ at the point (x, y, u) has the equation

$$\bar{u} - u - (\bar{x} - x)u_x - (\bar{y} - y)u_y = 0,$$

its plane coordinates are

$$\xi = u_x, \qquad \eta = u_y, \qquad \omega = xu_x + yu_y - u.$$

Now, the surface considered is determined also if ω is given as a function of ξ and η, by which the two-parameter family of tangent planes is characterized. We can find the dependence $\omega(\xi, \eta)$ from $u(x, y)$ by determining the values x and y as functions of ξ and η from the equations

$$\xi = u_x, \qquad \eta = u_y$$

and by substituting them into the equation

$$\omega = xu_x + yu_y - u = x\xi + y\eta - u.$$

[1] See Vol. I, pp. 234–235.

Conversely, in order to determine the point coordinates from the tangent plane coordinates, we form the partial derivatives of the function $\omega(\xi, \eta)$. Since $\xi = u_x$ and $\eta = u_y$, we immediately have

$$\omega_\xi = x + \xi \frac{\partial x}{\partial \xi} + \eta \frac{\partial y}{\partial \xi} - u_x \frac{\partial x}{\partial \xi} - u_y \frac{\partial y}{\partial \xi} = x$$

and, similarly,

$$\omega_\eta = y.$$

Thus we obtain the system of formulas

$$\omega(\xi, \eta) + u(x, y) = x\xi + y\eta,$$

(1)
$$\xi = u_x, \quad \eta = u_y,$$

$$x = \omega_\xi, \quad y = \omega_\eta,$$

which demonstrates the dual character of the relation between point and tangent plane coordinates.

This transformation of a surface from point coordinates to plane coordinates is called the *Legendre transformation* for functions of two variables. It is essentially different in character from a mere co-ordinate transformation. For, rather than assigning to a single point another point, the system (1) assigns to every surface element (x, y, u, u_x, u_y) a surface element $(\xi, \eta, \omega, \omega_\xi, \omega_\eta)$.

The Legendre transformation is always feasible if the two equations $u_x = \xi$, $u_y = \eta$ can be solved for x and y; this is possible whenever the Jacobian

(2)
$$u_{xx}u_{yy} - u_{xy}^2 = \rho$$

does not vanish for the points of the surface considered. The Legendre transformation evidently fails for surfaces which satisfy the differential equation

$$u_{xx}u_{yy} - u_{xy}^2 = 0,$$

i.e., for *developable surfaces*. This result can be visualized geometrically. A developable surface possesses by definition a one-parameter family of tangent planes which are tangent along straight lines, not merely at points; thus it is not possible to establish a one-to-one correspondence between the points and the tangent planes of the surface.

Finally, to apply the Legendre transformation to second order differential equations, we calculate the transformation for the second derivatives of the functions $u(x, y)$ and $\omega(\xi, \eta)$. To this end we think of the variables x and y in the equations $\xi = u_x$, $\eta = u_y$ as expressed in terms of ξ and η by means of the relations $x = \omega_\xi$, $y = \omega_\eta$. By differentiating $\xi = u_x$, $\eta = u_y$ with respect to ξ and η, we find

$$1 = u_{xx}\omega_{\xi\xi} + u_{xy}\omega_{\xi\eta},$$

$$0 = u_{xy}\omega_{\xi\xi} + u_{yy}\omega_{\xi\eta},$$

$$0 = u_{xx}\omega_{\xi\eta} + u_{xy}\omega_{\eta\eta},$$

$$1 = u_{xy}\omega_{\xi\eta} + u_{yy}\omega_{\eta\eta},$$

or, in matrix notation,

$$\begin{pmatrix} u_{xx} & u_{xy} \\ u_{xy} & u_{yy} \end{pmatrix} \begin{pmatrix} \omega_{\xi\xi} & \omega_{\xi\eta} \\ \omega_{\xi\eta} & \omega_{\eta\eta} \end{pmatrix} = \begin{pmatrix} 1 & 0 \\ 0 & 1 \end{pmatrix}.$$

If, for brevity, we set

(3)
$$\omega_{\xi\xi}\omega_{\eta\eta} - \omega_{\xi\eta}^2 = \frac{1}{\rho},$$

$$u_{xx}u_{yy} - u_{xy}^2 = \rho,$$

we obtain

(4)
$$u_{xx} = \rho\omega_{\eta\eta},$$

$$u_{xy} = -\rho\omega_{\xi\eta},$$

$$u_{yy} = \rho\omega_{\xi\xi}.$$

2. The Legendre Transformation for Functions of n Variables. For the sake of completeness we mention Legendre's transformation for functions of n independent variables. It is given by the following system of formulas:

$$u(x_1, x_2, \cdots, x_n) + \omega(\xi_1, \xi_2, \cdots, \xi_n)$$

$$= x_1\xi_1 + x_2\xi_2 + \cdots + x_n\xi_n,$$

(5) $u_{x_1} = \xi_1,$ $u_{x_2} = \xi_2,$ $\cdots,$ $u_{x_n} = \xi_n,$

$\omega_{\xi_1} = x_1,$ $\omega_{\xi_2} = x_2,$ $\cdots,$ $\omega_{\xi_n} = x_n.$

In order to give the transformation formulas for the second derivatives, we denote the cofactors of the elements $u_{x_i x_k}$, $\omega_{\xi_i \xi_k}$ of the matrices

$$\begin{pmatrix} u_{x_1 x_1} & \cdots & u_{x_1 x_n} \\ \cdots\cdots\cdots\cdots \\ u_{x_n x_1} & \cdots & u_{x_n x_n} \end{pmatrix} \quad \text{and} \quad \begin{pmatrix} \omega_{\xi_1 \xi_1} & \cdots & \omega_{\xi_1 \xi_n} \\ \cdots\cdots\cdots\cdots \\ \omega_{\xi_n \xi_1} & \cdots & \omega_{\xi_n \xi_n} \end{pmatrix}$$

by U_{ik} and Ω_{ik} and the determinants of the matrices by U and Ω. The transformation formulas then are

(6) $$u_{x_i x_k} = \frac{\Omega_{ik}}{\Omega}, \qquad \omega_{\xi_i \xi_k} = \frac{U_{ik}}{U}.$$

and $\Omega U = 1$.

The applicability of the Legendre transformation depends on the condition $U \neq 0$ (or $\Omega \neq 0$), as one may easily verify.

3. Application of the Legendre Transformation to Partial Differential Equations. We consider a partial differential equation of at most second order

(7) $$F(x, y, u, u_x, u_y, u_{xx}, u_{xy}, u_{yy}) = 0.$$

By means of the Legendre transformation we assign the function $\omega(\xi, \eta)$ to an integral surface $u(x, y)$ of this equation. Then the equation $F = 0$ goes over into a differential equation for the function ω, also of at most second order, namely into

(8) $$G = F(\omega_\xi, \omega_\eta, \xi\omega_\xi + \eta\omega_\eta - \omega, \xi, \eta, \rho\omega_{\eta\eta}, -\rho\omega_{\xi\eta}, \rho\omega_{\xi\xi}) = 0,$$

where

$$\rho = \frac{1}{\omega_{\xi\xi}\omega_{\eta\eta} - \omega_{\xi\eta}^2}.$$

This differential equation, however, in general yields only the nondevelopable integral surfaces of the original differential equation, since the Legendre transformation is not applicable to developable surfaces.

Particularly in the case of first order partial differential equations, the Legendre transformation may be profitably applied if the variables x, y and u occur in a simple way while the derivatives u_x, u_y occur in a more complicated manner.

As an example, we consider the equation

(9) $$u_x u_y = x$$

which by the Legendre transformation goes over into

$$(10) \qquad \xi\eta = \omega_\xi \, ;$$

its solution can be given immediately by

$$\omega = \tfrac{1}{2}\xi^2\eta + w(\eta).$$

On the basis of the transformation formulas, it follows that

$$x = \xi\eta,$$

$$(11) \qquad y = \tfrac{1}{2}\xi^2 + w'(\eta),$$

$$u = \tfrac{3}{2}\xi^2\eta + \eta w'(\eta) - w(\eta).$$

If we eliminate ξ and η from these three equations, we obtain the desired solutions of the given differential equation.[1]

On the other hand, the differential equation

$$(12) \qquad u_x u_y = 1$$

is transformed by the Legendre transformation into

$$\xi\eta = 1.$$

This equation is no longer a differential equation and the transformation fails here; all solutions of $u_x u_y = 1$ are developable surfaces. This is immediately confirmed by differentiating the equation with respect to x and y:

$$u_{xx}u_y + u_{xy}u_x = 0,$$

$$(13)$$

$$u_{xy}u_y + u_{yy}u_x = 0.$$

---

[1] However, the solutions for which the expression $u_{xx}u_{yy} - u_{xy}^2$ vanishes are missing. By differentiating the equation $u_x u_y = x$ with respect to x and y, we obtain

$$u_{xx}u_y + u_{xy}u_x = 1,$$

$$u_{xy}u_y + u_{yy}u_x = 0,$$

i.e., a nonhomogeneous system of equations whose determinant $u_{xx}u_{yy} - u_{xy}^2$ can vanish only if $u_{xy} = u_{yy} = 0$. From this it follows that the missing solutions must have the form

$$u = ay + \frac{1}{2a} x^2 + b,$$

where a and b are arbitrary constants. In fact, this expression constitutes a complete integral of (9) which, by the method of §4, 2 and the introduction of appropriate parameters, gives the system (11).

Since the possibility $u_x = u_y = 0$ is excluded because $u_x u_y = 1$, the condition

$$u_{xx} u_{yy} - u_{xy}^2 = 0$$

must be satisfied by every integral surface $u(x, y)$.[1]

The Legendre transformation fails for every differential equation of the form

(14) $$F(u_x, u_y) = 0$$

in the same way.

A third example is furnished by the *Clairaut equation*

(15) $$u = x u_x + y u_y + f(u_x, u_y)$$

already considered in §4, 4. By the Legendre transformation, (15) goes over into the simple equation

(16) $$\omega = -f(\xi, \eta).$$

From this we deduce that the only nondevelopable integral surface of Clairaut's differential equation is represented by equation (16) or, in point coordinates, by

(17)
$$x = -f_\xi(\xi, \eta),$$
$$y = -f_\eta(\xi, \eta),$$
$$u = f - \xi f_\xi - \eta f_\eta.$$

The following calculation confirms this conclusion: We differentiate the differential equation (15) and obtain the formulas

$$(x + f_p) u_{xx} + (y + f_q) u_{xy} = 0,$$
$$(x + f_p) u_{xy} + (y + f_q) u_{yy} = 0$$

(where $p = u_x$, $q = u_y$); it follows that, for an integral surface, either

$$D = u_{xx} u_{yy} - u_{xy}^2 = 0$$

---

[1] Incidentally, the differential equation (12) can be reduced by the substitution $x = \xi^2/2$ to the form (9) and thus solved; this may also be done by means of the complete integral

$$u = ax + \frac{1}{a} y + b.$$

or

$$x = -f_p, \qquad y = -f_q.$$

But the latter possibility yields precisely the exceptional surface obtained by the Legendre transformation.

As another example, we consider the second order differential equation of *minimal surfaces* (see also Vol. I, p. 193)

(18) $$(1 + u_y^2)u_{xx} - 2u_x u_y u_{xy} + (1 + u_x^2)u_{yy} = 0$$

which is nonlinear in the derivatives of $u(x, y)$. This apparent difficulty can be overcome by transforming (18) by the Legendre transformation into

(19) $$(1 + \eta^2)\omega_{\eta\eta} + 2\xi\eta\omega_{\xi\eta} + (1 + \xi^2)\omega_{\xi\xi} = 0,$$

i.e. into a linear differential equation. Later (cf. Appendix 1 of this chapter, Ch. III, §1, 4 and Vol. III) we shall consider other ways of linearizing the differential equation (18), which will yield a simple approach to the theory of minimal surfaces.

A similar important application of the Legendre transformation occurs in fluid dynamics[1]: Steady flow of a two-dimensional compressible fluid is described by two velocity components u, v as functions of the rectangular coordinates x, y. Suppose the sound speed c is a given function of $u^2 + v^2$. The motion is governed by the first order system of equations

$$u_y - v_x = 0,$$

$$(c^2 - u^2)u_x - uv(u_y + v_x) + (c^2 - v^2)v_y = 0.$$

Accordingly, there exists a velocity potential $\phi(x, y)$ such that

$$u = \phi_x, \qquad v = \phi_y$$

and

$$(c^2 - \phi_x^2)\phi_{xx} - 2\phi_x\phi_y\phi_{xy} + (c^2 - \phi_y^2)\phi_{yy} = 0.$$

A crucial step in dealing with this nonlinear differential equation of second order is the Legendre transformation

$$\Phi + \phi = ux + vy,$$

$$\phi_x = u, \qquad \phi_y = v,$$

$$\Phi_u = x, \qquad \Phi_v = y.$$

[1] See also Chapter V, and R. Courant and K. O. Friedrichs, [1], pp. 247–249.

It yields for $\Phi(u, v)$ a linear differential equation of second order

$$(c^2 - u^2)\Phi_{vv} + 2uv\Phi_{uv} + (c^2 - v^2)\Phi_{uu} = 0$$

which is useful for solving many flow problems.[1]

§7. The Existence Theorem of Cauchy and Kowalewsky

1. Introduction and Examples. We conclude this chapter with the discussion of a fundamental theorem which assures the existence of solutions of partial differential equations and at the same time clarifies the manner in which arbitrary functions enter into the "general" solution. The theorem is due to Cauchy, who initiated the modern theory of partial differential equations. Sophie Kowalewski in her Thesis, inspired by Weierstrass, has carried out the proof in a rather general manner.[2]

The theorem refers to the "initial value problem" as or we often shall call it "Cauchy's problem". It is restricted by the assumption that the differential equations and the initial data as well as the solutions are analytic and it refers to a system of m partial differential equations each of order k for m unknown functions u^1, u^2, $\cdots$, u^m (or sometimes also written u_1, u_2, $\cdots$, u_m) of $n + 1$ independent variables x, and y_1, $\cdots$, y_n. This system is assumed to be in the "normal" form, singling out the variable x:

$$(1) \qquad \frac{\partial^k}{\partial x^k} u^i = f_i(x, y_1, \cdots, y_n, \frac{\partial u^1}{\partial x^1}, \cdots, \frac{\partial^k u^m}{\partial y_n^k}),$$

where the functions f_i depend *analytically* on the quantities x, y_1, y_2, $\cdots$, $\partial^k u^m/\partial y_n^k$ within a certain domain in the multidimensional space of these quantities as variables. (That means, they can be expanded into power series in all these variables converging in a suitably small domain, which may be assumed to contain the origin $x = 0$ $y_i = 0$, $u_i = 0$.) In the "initial plane" $x = 0$ we prescribe km arbitrary analytic functions $\phi_{i,\kappa}(y_1, \cdots, y_n)$ $(i = 1, \cdots, m; \kappa = 0, \cdots, k - 1)$ of the variables y_1, $\cdots$, y_n in a suitably small neighborhood of the

---

[1] This transformation can be obtained directly by inverting the system of functions $u(x,y), v(x,y)$, i.e., by introducing x, y as functions of the independent variables u, v. This procedure is often called the "hodograph" method, since the plane of the velocity vector u, v, called "hodograph plane", is made the frame of reference (see Ch. V, § 2).

[2] See J. Hadamard [2], footnote on p. 11. Hadamard refers to A. Cauchy, S. Kowalewski, G. Darboux, and E. Goursat.

origin $y_i = 0$. Then Cauchy's problem is to construct a solution of the system (1) for which the initial values

$$u^i(0, y_1, \cdots, y_n) = \phi_{i,0}(y_1, \cdots, y_n), \cdots, \frac{\partial^{n-1} u^i}{\partial x^{n-1}} (0, y_1, \cdots, y_n)$$

$$= \phi_{i,n-1}(y_1, \cdots, y_n)$$

are attained at $x = 0$. One should always realize that this initial value problem is posed merely in the small, i.e., for suitably narrow neighborhoods of $x = 0$, $y_i = 0$.

Now the main theorem states: *Cauchy's problem possesses one and only one analytic solution*, $u^1, \cdots, u^m$.

It will be sufficient to assume $k = 1$, that is, to consider systems of equations of first order. Furthermore, we shall carry out the discussion merely for the case $n = 1$, that is, for differential equations with two independent variables x, y, since there is no modification necessary for more independent variables.

In general, the normal form (1) of the system is attainable. However, as we shall see in article 5 there are important exceptional situations, in which this is not the case.[1]

To prove the theorem one first formally constructs power series for the solution and then shows the uniform convergence of these series.

Before carrying out the general discussion we consider some examples.

For the differential equation

$$\alpha u_x + \beta u_y = 0$$

with constant α and β, the totality of solutions is given by

$$u = w(\alpha y - \beta x)$$

with arbitrary w. Let us now suppose that the initial values $u(0, y) = \phi(y)$ are arbitrarily prescribed. Then the function w is determined by setting $x = 0$; we obtain the solution $u = \phi(y - \beta x/\alpha)$.

---

[1] That Cauchy's initial value problem is a most natural question (just as for ordinary differential equations where it immediately clarifies the occurrence of arbitrary constants of integration) is evident from the very meaning of the differential equations (1); they express by the initial data and their implicitly known y-derivatives the one derivative which is not given by the data; as we shall see, the way is thus opened for finding all the derivatives of the unknown functions at the origin.

More generally, let us consider (cf. §1, 1, example 5)) the nonlinear differential equation

$$\alpha(u)u_x + \beta(u)u_y = 0$$

assuming that the coefficients α and β depend on the unknown function u. The initial value problem is to find a solution for which $u(0, y) = \phi(y)$ is prescribed.

As seen before, the totality of solutions of our differential equation is given by $\alpha(u)y - \beta(u)x = w(u)$ where w is an arbitrary function; again w is found from the initial condition. Substituting $x = 0$ and $u = \phi(y)$, we have $w(\phi) = \alpha(\phi)y$; if the function $u = \phi(y)$ is inverted by $y = \chi(u)$, then we have determined the arbitrary function w by the relation $w(\phi) = \alpha(\phi)\chi(\phi)$. Thus, the desired solution satisfies $\alpha(u)y - \beta(u)x = \alpha(u)\chi(u)$ or the equivalent relation

$$u = \phi\left(y - \frac{\beta(u)}{\alpha(u)}\,x\right).$$

If a function $u(x, y)$ is determined from this implicit equation, the initial value problem is solved. (At the end of this section, page 53, we shall discuss a special case which will be important for a later application.)

For the second order differential equation $u_{xy} = f(x, y)$ the triangle integral in formula (2) of §1, 1 solves the initial value problem.

The simplest differential equation of vibration

$$u_{xx} - u_{yy} = 0$$

leads to the initial value problem: find u if, for $x = 0$, the initial state $u(0, y) = \phi(y)$ and $u_x(0, y) = \psi(y)$ is arbitrarily prescribed (cf. Vol. I, Ch. V, §3). From the general solution $u = f(y + x) + g(y - x)$ of the differential equation one obtains the particular form of the functions f and g by adapting them to the initial conditions according to the relations $f(y) + g(y) = \phi(y)$, $f'(y) - g'(y) = \psi(y)$ which yield f and g immediately:

$$2u(x, y) = \phi(y + x) + \phi(y - x) + \int_{y-x}^{y+x} \psi(\lambda)\,d\lambda.$$

In formulating general initial value problems we suppose that the differential equations can be solved for the highest derivatives with respect to x of the unknown function or functions.

Accordingly we consider e.g. the first order differential equation

$$(2') \qquad\qquad F(x, y, u, p, q) = 0,$$

where

$$p = u_x, \qquad q = u_y,$$

and suppose that equation $(2')$ can be solved for p, assuming the form

$$(2) \qquad\qquad p = f(x, y, u, q).$$

The initial value problem now is to find a solution $u(x, y)$ of (2) which becomes a prescribed function $u(0, y) = \phi(y)$ for $x = 0$; geometrically expressed, an integral surface is to be found which meets the plane $x = 0$ in a given initial curve $u = \phi(y)$.

One could pose the more general problem: Find an integral surface of $F(x, y, u, p, q) = 0$ which passes through a given space curve $u = \phi(y), x = \psi(y)$. If we introduce $\xi = x - \psi(y)$ and $\eta = y$ as new independent variables instead of x and y, setting

$$u(x, y) = u(\xi + \psi(\eta), \eta) = \omega(\xi, \eta),$$

the differential equation goes over into

$$F(\xi + \psi(\eta), \eta, \omega, \omega_\xi, \omega_\eta - \psi'\omega_\xi) = G(\xi, \eta, \omega, \omega_\xi, \omega_\eta) = 0$$

with the initial condition $\omega(0, \eta) = \phi(\eta)$. The more general problem is thus reduced to one of the special form originally considered, to which we shall now restrict ourselves.

Let us also consider a second order differential equation

$$(3') \qquad\qquad F(x, y, u, p, q, r, s, t) = 0$$

with the abbreviation

$$r = u_{xx} = p_x, \qquad s = u_{xy} = p_y = q_x, \qquad t = u_{yy} = q_y$$

which will be used frequently. We suppose that this equation can be solved for r in the argument domain considered, i.e., that it can be brought into the form

$$(3) \qquad\qquad r = f(x, y, u, p, q, s, t).$$

The initial value problem for this differential equation is to find a solution $u(x, y)$ for which, at $x = 0$, the initial values of u and u_x

$$(4) \qquad u(0, y) = \phi(y), \qquad u_x(0, y) = \psi(y)$$

are prescribed. Instead of the one arbitrary initial function $\phi(y)$ which appears in a first order differential equation, we have two arbitrarily prescribed functions $\phi(y)$ and $\psi(y)$.

Analogous problems can be posed for differential equations of higher order or for systems of differential equations. In particular we consider for the unknown functions $u_i(x, y)$ (also sometimes denoted by $u^i(x, y)$) a first order system

$$(5) \qquad \frac{\partial u_i}{\partial x} = f_i\left(x, y, u_1, \cdots, u_m, \frac{\partial u_1}{\partial y}, \cdots, \frac{\partial u_m}{\partial y}\right) \qquad (i = 1, 2, \cdots, m)$$

and having the arbitrarily prescribed initial values

$$u_i(0, y) = \phi_i(y).$$

By showing that these initial value problems have uniquely determined solutions, we will clarify the occurrence of arbitrary functions in the general solution.

2. Reduction to a System of Quasi-Linear Differential Equations. All the initial value problems formulated above can be reduced to equivalent problems for systems of quasi-linear differential equations of first order. It was emphasized that the totality of solutions of a system of differential equations is generally not equivalent to the set of solutions of a single equation. However, as we shall see, they are equivalent if we consider the differential equations together with suitable additional initial conditions rather than the differential equations alone. A system of quasi-linear differential equations would yield a wider variety of solutions than the original equations, yet we shall restrict the initial conditions in such a way that the sets of solutions of both initial value problems coincide.

First, we carry out the reduction for the first order differential equation (2). We note that by prescribing $u(0, y) = \phi(y)$, the initial values $q(0, y) = \phi'(y)$ are automatically prescribed also. Moreover, the differential equation (2) yields the initial value for p, namely

$$p(0, y) = f(0, y, \phi(y), \phi'(y)).$$

By differentiating equation (2) with respect to x, we obtain for the three quantities u, p, q the system of quasi-linear first order partial differential equations

$$u_x = p,$$

(6) $$q_x = p_y,$$

$$p_x = f_x + f_u p + f_q p_y$$

and the initial conditions

$$u(0, y) = \phi(y),$$

(7) $$q(0, y) = \phi'(y),$$

$$p(0, y) = f(0, y, \phi(y), \phi'(y)).$$

We assert that this initial value problem is equivalent to the original one.

To justify this statement, it is sufficient to show that, for a solution u, p, q of the system of equations (6), (7), the equations

$$p = f(x, y, u, q), \qquad u_x = p, \qquad u_y = q$$

are satisfied. Since by (6) $p_y = q_x$, we have

$$u_{xy} = q_x ;$$

hence, integrating with respect to x,

$$u_y(x, y) = q + v(y).$$

Substituting $x = 0$ and observing the initial conditions (7) we have

$$u_y = q$$

for all x and y, for certainly $\phi'(y) = u_y(0, y)$ implies $v(y) = 0$. Furthermore, according to (6)

$$u_{xx} = p_x = \frac{\partial}{\partial x} f(x, y, u, q);$$

hence by integration,

$$u_x = f(x, y, u, q) + a(y).$$

But since $u_x = f$ holds for $x = 0$, it follows that $a(y) = 0$ and hence $u_x = f(x, y, u, u_y)$; i.e., $u(x, y)$ solves the original problem.

Similarly the initial value problem for the second order differential equation (3) with the two initial conditions (4) can be replaced by the equivalent initial value problem for the following system of differential equations for six functions u, p, q, r, s, t of the independent variables x, y:

$$u_x = p, \qquad q_x = p_y, \qquad p_x = r,$$

$$s_x = r_y, \qquad t_x = s_y,$$

$$r_x = f_x + f_u p + f_p r + f_q p_y + f_s r_y + f_t s_y,$$

where

$$u(0, y) = \phi(y), \qquad p(0, y) = \psi(y), \qquad q(0, y) = \phi'(y),$$

$$t(0, y) = \phi''(y), \qquad s(0, y) = \psi'(y),$$

$$r(0, y) = f(0, y, \phi(y), \psi(y), \phi'(y), \psi'(y), \phi''(y))$$

are prescribed as initial conditions. From the given initial data ϕ, ψ of the original problem and from the differential equation, we immediately have the additional correctly adapted initial data for q, t, s, r. As above, we can show that p, q, r, s, t and the derivatives u_x, u_y, u_{xx}, u_{xy}, u_{yy} coincide, u and u_x assume the prescribed initial values (4) and the differential equation (3), $r = f(x, y, u, p, q, s, t)$ is satisfied.

Similarly we can replace a higher order differential equation or system of equations by a first order quasi-linear system.

The quasi-linear systems of differential equations obtained above contain the independent variables x and y in the coefficients of the right side. It is often convenient to pass, by means of a slight artifice, to another equivalent quasi-linear system of differential equations in which the independent variables x, y no longer appear explicitly and which, moreover, is homogeneous in the derivatives. To this end we formally introduce two functions $\xi(x, y)$ and $\eta(x, y)$ in place of x and y by means of the equations

(8) $$\xi_x = \eta_y, \qquad \eta_x = 0$$

and initial conditions

(9) $$\xi(0, y) = 0, \qquad \eta(0, y) = y;$$

the solutions of (8) and (9) are $\xi = x$, $\eta = y$. Since $\eta_y = 1$, we may now replace our initial value problem (6), (7) by the obviously equivalent system for the five functions u, p, q, ξ, η:

$$u_x = p\eta_y, \qquad q_x = p_y,$$

$$(10) \qquad \xi_x = \eta_y, \qquad \eta_x = 0,$$

$$p_x = f_q p_y + (f_x + pf_u)\eta_y.$$

We must, however, substitute ξ, η for x, y in f_q, f_x, f_u and require the initial conditions

$$u(0, y) = \phi(y), \qquad q(0, y) = \phi'(y),$$

$$(11) \qquad \xi(0, y) = 0, \qquad \eta(0, y) = y,$$

$$p(0, y) = f(0, y, \phi(y), \phi'(y)).$$

Thus a problem equivalent to the initial value problem for (2) is formulated which has the form described above.

A similar result obtains for the initial value problem of second order. As in the case of the first order problem, we replace x and y artificially by the auxiliary functions ξ and η, which satisfy the differential equations (8) and the initial conditions (9); again, instead of (3), (4), we can formulate an equivalent initial value problem for a system of quasi-linear, homogeneous, first order differential equations for the functions u, p, q, r, s, t, ξ, η.

All the initial value problems arising in this manner have the form of a quasi-linear system of first order

$$(12) \qquad \frac{\partial u_i}{\partial x} = \sum_{j=1}^{m} G_{ij}(u_1, u_2, \cdots, u_m) \frac{\partial u_j}{\partial y} \quad (i = 1, 2, \cdots, m)$$

with prescribed initial conditions of the form

$$(13) \qquad u_i(0, y) = \phi_i(y).$$

In this quasi-linear system the coefficients $G_{ik}(u_1, u_2, \cdots, u_m)$ depend explicitly only on the unknown functions u_i themselves, not on the independent variables x and y.

The relevant general result is: Initial value problems of all orders for systems of differential equations can be reduced without difficulty to initial value problems of this type. Of course, in the case of n

variables $y_1, \cdots, y_n$ (12) has to be replaced by a system of the form

(12a) $$\frac{\partial u_i}{\partial x} = \sum_{\nu=1}^{n} \sum_{j=1}^{m} G_{i,j,\nu}(u_1, \cdots, u_m) \frac{\partial u_j}{\partial y_\nu}.$$

3. Determination of Derivatives Along the Initial Manifold. Decisive is the fact that the differential equations together with the initial data furnish a method for calculating all the derivatives of the desired solution along the initial curve, say $x = 0$, provided that such solutions exist and that the solutions, as well as the differential equations and initial functions, are assumed analytic. First we note that, along the initial curve $x = 0$, all quantities already known (say u and certain derivatives of u) yield, by differentiation with respect to y, more known quantities, i.e., additional derivatives. The derivatives formed by differentiation with respect to x, which are still missing, must then be determined with the help of the differential equations.

Thus, in the case of the differential equation (2), $p = f(x, y, u, q)$, we can determine $q = \phi'(y)$ and $t = u_{yy} = q_y = \phi''(y)$, etc. along $x = 0$ from the prescribed data. The differential equation itself yields the value $p(0, y) = f(0, y, \phi(y), \phi'(y))$. Similarly,

$$q_x = p_y = f_y + f_u q + f_q q_y$$

is known for $x = 0$. In order to determine the still missing second derivative $r = p_x = u_{xx}$ along the initial curve, we differentiate equation (2) with respect to x and obtain $r = p_x = f_x + f_u p + f_q q_x$. On the right-hand side we have quantities already known for $x = 0$ from the above considerations; hence the left side is also determined for $x = 0$.

Further differentiation with respect to x of the quantities so determined and of the differential equation yields all higher derivatives along $x = 0$ as long as the assumption of continuous differentiability of the function f and the solution u is valid.

In a similar manner we can determine the derivatives of u along the initial curve in the case of an initial value problem for the second order differential equation (3). But it is just as simple to discuss the general initial value problem (12), (13), which includes all the special problems considered. For a system of this form, one sees clearly how the derivatives of the functions u_i are determined successively along the initial manifold, i.e., for $x = 0$.

First the derivatives $\partial u_i/\partial y$, $\partial^2 u_i/\partial y^2$, $\cdots$ along the line $x = 0$ are obtained from the functions $\phi_i(y)$ by differentiation, and then the first derivatives with respect to x from the differential equations; differentiating the quantities so found with respect to y, we get the mixed derivatives $\partial^2 u_i/\partial x \partial y$ for $x = 0$. Then, by differentiating the system of differential equations (12) with respect to x, one obtains expressions on the right side which contain only first and mixed second derivatives of the functions u_i with respect to x and y; these expressions therefore are known and determine the values of the left-hand side, that is, the second derivatives $\partial^2 u_i/\partial x^2$ and so on. We emphasize that only differentiations and substitutions are used in this process of successive determination.

According to our assumptions of analyticity an unlimited number of differentiations may be performed; then, from the initial data, all derivatives of the functions u_i are determined along $x = 0$, in particular, therefore, at $x = 0$, $y = 0$.

It is now natural to reverse this procedure. If the process of successive determination of initial derivatives described above can be applied an unlimited number of times—and this is the case if the differential equations themselves and the initial values are *analytic*—one may construct a formal power series using the derivatives so obtained as coefficients. Then one has to prove the power series so constructed converges and solves the original initial value problem.

4. Existence Proof for Solutions of Analytic Differential Equations. In the proof of the fundamental existence theorem for the system (12) we may aim at an expansion about the origin $x = 0$ and $y = 0$ and assume for the initial values $\phi_i(0) = 0$. Otherwise we could introduce the differences $u_i - \phi_i(0)$ as new unknown functions.[1] The analyticity of our data means that the functions G_{ik}, ϕ_i are defined in terms of the power series

$$(14) \qquad\qquad \phi_i(y) = \sum_{\nu=1}^{\infty} a_\nu^i y^\nu,$$

$$(15) \qquad G_{ik}(u_1, \cdots, u_m) = \sum_{\nu_1,\cdots,\nu_m=0}^{\infty} b_{\nu_1,\cdots,\nu_m}^{ik} u_1^{\nu_1} \cdots u_m^{\nu_m}$$

which converge in regions $|y| \leq \rho$ and $|u_i| \leq r$, respectively.

---

[1] We could even attain a further simplification by introducing $u_i - \phi_i(y) = v_i$ as new unknowns for which then all initial values $v_i(0, y)$ are identically zero while the general form of the differential equations is not changed.

We assert: The initial value problem posed for the system of differential equations has a solution which can be expressed by power series

$$(16) \qquad u_i(x, y) = \sum_{\substack{l=0 \\ k=1}}^{\infty} c_{lk}^i x^l y^k.$$

According to article 3, the coefficients of the power series (16) are determined uniquely by means of the differential equations and the initial data. For, the values of the derivatives of the—still hypothetical—solutions u_i at the point $x = 0$, $y = 0$ may be obtained by simply substituting the special value $y = 0$ into the derivatives along the initial curve $x = 0$. Thus the coefficients c_{lk}^i of the series expansion (16) are uniquely determined.

If we assume that the power series which are constructed in this way converge in a certain region about the point $x = 0$, $y = 0$, then, according to well-known theorems, we may differentiate term by term in the interior of the region of convergence; the derivatives thus obtained may be substituted into the differential equations. The resulting expressions may then be rearranged as power series in x and y. By recalling the definition of the successive derivatives of the u_i at the origin, we see that the left- and right-hand sides of each equation together with all their derivatives agree at the point $x = 0$, $y = 0$. Because of the analyticity the differential equations are satisfied identically, i.e., the functions u_i represent a system of solutions. That this system of solutions possesses the prescribed initial values and thus solves our initial value problem follows directly from the construction of the power series (16), and the assumption that these converge. *The proof of the existence theorem will, therefore, be complete as soon as the power series (16) are shown to converge in the interior of some region.*

To prove this convergence we examine the dependence of the coefficients c_{lk}^i on the coefficients a_ν^j, $b_{\nu_1 \ldots \nu_m}^{js}$. We note first that term by term differentiation of any power series produces a new power series whose coefficients are formed from the original coefficients as linear combinations with non-negative integers. In substituting power series into the differential equations (12) only the processes of addition and multiplication occur. The expressions so formed on the right side thus become power series in x and y, whose coefficients

c_{lk}^i are polynomials in the quantities a_ν^j, $b_{\nu_1,\cdots\nu_m}^{js}$:

(17) $$c_{lk}^i = P_{lk}^i(a_\nu^j, b_{\nu_1,\cdots,\nu_m}^{js}).$$

The coefficients of these polynomials are non-negative integers which do not depend on the special form of the functions G_{ik} and ϕ_i.

After this preparation we shall establish convergence using the classical *method of majorants*. In addition to our original initial value problem referring to the expressions G_{ik} and ϕ_i, we consider a new initial value problem in which the functions G_{ik} and ϕ_i are replaced by other, "majorant", functions K_{ik} and ψ_i. In some neighborhood of the origin, we set

(18) $$\psi_i(y) = \sum_{\nu=1}^{\infty} A_\nu^i y^\nu$$

and

(19) $$K_{ik}(u_1, \cdots, u_m) = \sum_{\nu_1,\cdots,\nu_m=0}^{\infty} B_{\nu_1,\cdots,\nu_m}^{ik} u_1^{\nu_1} \cdots u_m^{\nu_m},$$

where

$$A_\nu^i \geq |a_\nu^i| \quad \text{and} \quad B_{\nu_1,\cdots,\nu_m}^{ik} \geq |b_{\nu_1,\cdots,\nu_m}^{ik}|.$$

In other words: The coefficients in the expansion of the new functions K_{ik} and ψ_i are non-negative, and not smaller than the absolute values of the corresponding coefficients of the original functions G_{ik} and ϕ_i. We pose the initial value problem

(20) $$\frac{\partial v_i}{\partial x} = \sum_{k=1}^{m} K_{ik}(v_1, v_2, \cdots, v_m) \frac{\partial v_k}{\partial y},$$

$$(i = 1, 2, \cdots, m),$$

(21) $$v_i(0, y) = \psi_i(y),$$

a "majorant problem" of the original one. If, according to the preceding method, we form the coefficients C_{lk}^i of the hypothetic power series solutions

(22) $$v_i(x, y) = \sum_{\substack{l=0 \\ k=1}}^{\infty} C_{lk}^i x^l y^k$$

of the majorant problem, then we get the new quantities C_{lk}^i from the A_ν^j and $B_{\nu_1,\cdots,\nu_m}^{js}$ in the way the original coefficients c_{lk}^i were obtained from the a_ν^j and $b_{\nu_1, \cdots, \nu_m}^{js}$ (i.e., $C_{lk}^i = P_{lk}^i(A_\nu^j, B_{\nu_1, \cdots, \nu_m}^{js})$). But since these polynomials P_{lk}^i have non-negative coefficients, we immediately obtain

$$C_{lk}^i \geq |c_{lk}^i|.$$

Thus the formal power series (22) is a majorant of the power series (16). If, therefore, we can prove the convergence of such a majorant series (22), the convergence of our original series (16) is assured.

We exploit this remark by forming a majorant problem of a particularly simple nature whose solution we can give explicitly, so that the convergence of the majorant series is shown. To this end we choose, as above, two positive numbers r and ρ such that the power series for $G_{ik}(u_1, u_2, \cdots, u_m)$ and $\phi_i(y)$ converge for $|u_i| \leq r$ and for $|y| \leq \rho$, respectively. Then, by a well-known theorem of the theory of power series, there exists a constant M such that

$$|a_\nu^i| \leq \frac{M}{\rho^\nu} = A_\nu^i$$

and

$$|b_{\nu_1, \cdots, \nu_m}^{js}| \leq \frac{M}{r^{\nu_1 + \cdots + \nu_m}};$$

thus, *a fortiori*, we have

$$|b_{\nu_1, \cdots, \nu_m}^{js}| \leq \frac{M}{r^{\nu_1 + \cdots + \nu_m}} \frac{(\nu_1 + \cdots + \nu_m)!}{\nu_1! \cdots \nu_m!} = B_{\nu_1, \cdots, \nu_m}^{js}.$$

Now we set (cf. (18), (19))

$$(23) \qquad \psi_i(y) = \sum_{\nu=1}^{\infty} A_\nu^i y^\nu = M \sum_{\nu=1}^{\infty} \left(\frac{y}{\rho}\right)^\nu$$

and

$$(24) \quad \begin{aligned} K_{ik}(u_1, \cdots, u_m) &= \sum_{\nu_1, \cdots, \nu_m=0}^{\infty} B_{\nu_1, \cdots, \nu_m}^{ik} u_1^{\nu_1} \cdots u_m^{\nu_m} \\ &= M \sum_{\nu_1, \cdots, \nu_m=0}^{\infty} \left(\frac{u_1}{r}\right)^{\nu_1} \cdots \left(\frac{u_m}{r}\right)^{\nu_m} \frac{(\nu_1 + \cdots + \nu_m)!}{\nu_1! \cdots \nu_m!}. \end{aligned}$$

The series (24) converges if we restrict the arguments

$$u_1, u_2, \cdots, u_m$$

to a region where

$$|u_1| + |u_2| + \cdots + |u_m| < r;$$

its sum can then be expressed as

$$(25) \qquad K_{ik}(u_1, u_2, \cdots, u_m) = \cfrac{M}{1 - \cfrac{u_1 + u_2 + \cdots + u_m}{r}}.$$

The series (23) yields, for $|y| < \rho$,

$$(26) \qquad \psi_i(y) = \frac{My}{\rho - y}.$$

Thus the problem

$$(27) \qquad \frac{\partial v_i}{\partial x} = \cfrac{M}{1 - \cfrac{v_1 + v_2 + \cdots + v_m}{r}} \sum_{k=1}^{m} \frac{\partial v_k}{\partial y},$$

$$(28) \qquad v_i(0, y) = \frac{My}{\rho - y}$$

constitutes a majorant initial value problem for problem (12).

The only remaining task is to explicitly construct solutions $v_i(x, y)$ of this system and show that, at the point $x = 0$, $y = 0$, these solutions can be expanded into power series.

Since all the functions K_{ik} as well as all the ψ_i are identical, it seems plausible to assume tentatively

$$v_i(x, y) = v(x, y),$$

irrespective of i. This leads to the single partial differential equation

$$\frac{\partial v}{\partial x} = \cfrac{mM}{1 - \cfrac{m}{r}v} \frac{\partial v}{\partial y}$$

or

$$(29) \qquad \left(1 - \frac{m}{r}v\right)v_x - mMv_y = 0$$

and the initial condition

$$(30) \qquad v(0, y) = v_0 = \frac{My}{\rho - y}.$$

Therefore, we need only show that this initial value problem possesses a solution $v(x, y)$ which can be expanded into a power series in a sufficiently small neighborhood of the origin.

Our initial value problem coincides with that already considered as the example in article 1, page 41. The argument given there yields the quadratic equation

$$(31) \qquad (v + M)\left[\left(1 - \frac{m}{r} v\right) y + mMx\right] = \rho v \left(1 - \frac{m}{r} v\right)$$

for the solution v. Among the two roots of this equation, we must choose the one which assumes the value 0 at $x = 0$, $y = 0$. The existence of such a solution immediately follows from the equation (31) which, for $x = y = 0$, goes over into $v(1 - mv/r) = 0$. Because of the assumption $r > 0$, the two roots are distinct at $x = y = 0$. Hence, the discriminant of the quadratic equation (31) is different from zero at the origin and thus also in a neighborhood of the origin where the root can now certainly be expanded into a convergent power series in x and y.

Actually, such a solution is given explicitly in the form

$$(32) \qquad v = \frac{1}{2} \frac{\dfrac{M}{\rho} (y - rx) + \dfrac{r}{m}\left(1 - \dfrac{y}{\rho}\right)}{1 - \dfrac{y}{\rho}} - \frac{1}{2} \frac{\sqrt{\left[\dfrac{M}{\rho}(y + rx) - \dfrac{r}{m}\left(1 - \dfrac{y}{\rho}\right)\right]^2 - 4M^2 \dfrac{rx}{\rho}}}{1 - \dfrac{y}{\rho}},$$

as can be easily verified.

Thus, the convergence of the majorant series (22) and, hence, the convergence of the original series (16) is proved in a certain neighborhood of the origin, and the existence of an analytic solution of our initial value problem is completely established.

At the same time it is immediately clear: *For all differential equations and all initial functions which allow the same majorants the power series converge uniformly and the solution exists in a common domain.*

Obviously, the preceding proof shows that there is only one *analytic* solution of the initial value problem for the analytic system (12), provided that the initial data are analytic. In many cases the assumption of analyticity of the data is too strong; often it is even contrary to the nature of underlying physical problems. Therefore, it is of interest that Holmgren[1] established the uniqueness of the solution of the Cauchy problem for an analytic equation or system, regardless of whether the data on solution are analytic. In particular, the analytic solution corresponding to analytic Cauchy data is the only one among all solutions having those Cauchy data; every solution is automatically analytic. (See Ch. III, Appendix 2.)

For the precise statement and proof of Holmgren's theorem, the reader is referred to Appendix 2 of Chapter III.

4a. Observation About Linear Differential Equations. If the coefficients G are fixed, then the preceding argument shows: The power series for the solution u converges within a circle of radius r, which is a function $r(M, \rho)$ which depends on M and ρ only.

Now, if the original differential equation is linear and homogeneous, then αu with arbitrary constant α is a solution with the same radius r of convergence for the initial values $\alpha\phi$ which retain the radius ρ of convergence according to the preceding article. Hence $r = r(\alpha M, \rho)$ is independent of M and depends only on ρ. Similarly we may reason for nonhomogeneous differential equations. This implies the important fact: *If for fixed linear differential equations the initial values converge for all values of their argument, e.g., if they are polynomials, then the expression of the solutions converges within a fixed radius, independently of the specific initial values.*

4b. Remark About Nonanalytic Differential Equations. For the needs of mathematical physics the assumption of analytic differential equations is too narrow. In subsequent chapters we shall construct solutions for wide classes of not necessarily analytic differential equations, for which the existence proof of the present section does not apply without supplementary or entirely different arguments. All the more it is important that Hans Lewy[2] has recently discovered an example, and Hörmander[3] further investigated classes, of linear partial differential equations with entirely smooth, though nonanalytic, coefficients, for which no solutions at all exist. Such unexpected behavior of some differential equations which seem not to

[1] See E. Holmgren [1].
[2] See Hans Lewy [1].
[3] See L. Hörmander [1].

differ appreciably from analytic equations poses a challenge: To seek a simple distinction between solvable and not solvable "abnormal" differential equations. Hörmander's work throws a great deal of light on this question.

5. Remarks on Critical Initial Data. Characteristics. The results of articles 2 and 3 are based on the assumption that one can write a system of differential equations in the form (12). If we examine the conditions under which it is possible to write down such a "normal" form, we are led to the concept of exceptional or "critical" initial manifolds or "characteristics". We define the concept of characteristics, which plays a fundamental role in the theory of partial differential equations and will occur in various contexts throughout this book.

Suppose first that the given system of equations is linear and that the number of independent variables is two:

$$\sum_{j=1}^{n} \left(a_{ij} \frac{\partial u_j}{\partial x} + b_{ij} \frac{\partial u_j}{\partial y} + c_{ij} u_j \right) = d_i \qquad (i = 1, 2, \cdots, n);$$

here the a_{ij}, b_{ij}, c_{ij}, d_i are given functions of x and y. This system can be solved for the x-derivatives at any point (x_0, y_0) if and only if the matrix of coefficients a_{ij} is *nonsingular* at (x_0, y_0). If the matrix a_{ij} is singular at (x_0, y_0) the line $x = x_0$ is called *characteristic* at the point (x_0, y_0); otherwise, it is called *free*.

More generally we may consider an analytic curve C given in the form $\xi(x, y) = 0$ and embedded in a family $\xi(x, y) = $ const. Then we often want to consider the initial value problem for initial values of $u^1, u^2, \cdots, u^m$ given on C. This problem can simply be reduced to the case where the initial manifold is a coordinate line by introducing new independent variables ξ and η instead of x and y where η is any suitable second variable. Then, write the system in terms of ξ and η:

$$(33) \quad \sum_{j=1}^{n} \left[(a_{ij}\xi_x + b_{ij}\xi_y) \frac{\partial u_j}{\partial \xi} + (a_{ij}\eta_x + b_{ij}\eta_y) \frac{\partial u_j}{\partial \eta} + c_{ij} u_j \right] = d_i$$

$$(i = 1, 2, \cdots, n).$$

The curve C is said to be characteristic at the point (x_0, y_0) if the line $\xi = \xi_0$ is characteristic for the transformed system at the corresponding point (ξ_0, η_0).

Now (33) is a system of linear equations for the derivatives $\partial u_j/\partial \xi$ on C; it has one and only one solution if the matrix $a_{ik}\xi_x + b_{ik}\xi_y$ is nonsingular at the point P under consideration, that is, if the determinant $Q = \| a_{ik}\xi_x + b_{ik}\xi_y \|$ is different from zero at P. Provided this is the case, the curve C is called free at P; otherwise it is called *critical or characteristic* at P.

In Chapter III we shall examine the situation thoroughly. Here we merely state that for initial curves which are free at every point C Cauchy-Kowalewski's existence theorem and its proof remain unchanged.

The definitions and statements extend easily to any number of variables, to quasi-linear or other nonlinear systems, and to systems of higher order.

A brief remark should be added concerning *characteristic initial value problems*.

The Cauchy-Kowalewski theorem for linear equations has been extended to the case where the initial manifold is characteristic at each point.[1] In this case, the initial data cannot be prescribed arbitrarily, but must satisfy certain conditions imposed by the differential equation (see Ch. VI, §3). Correspondingly, the solution is not uniquely determined unless certain additional conditions are imposed along a manifold transversal to the initial manifold; this situation is analogous to that for a system of linear algebraic equations with vanishing determinant.

Leray[2] has treated the case where the initial manifold is characteristic along certain curves. In general, the solution is multivalued in the neighborhood of the initial surface; the degree of ramification of the solution is determined from the geometry of the relevant characteristic surfaces.

Appendix 1 to Chapter I

Laplace's Differential Equation for the Support Function of a Minimal Surface

The nonlinear differential equation of a minimal surface $u(x, y)$ was transformed into a linear equation by Legendre's transformation in

---

[1] See J. Hadamard [2] p. 77, G. F. D. Duff [1], and D. Ludwig [1].
[2] See J. Leray [1].

§6, 3. By a slightly different, homogeneous, form of Legendre's transformation, the equation of minimal surfaces can be changed into Laplace's equation for a function (the so-called support function) of three independent variables, as follows:

We first write the equation

$$\frac{\partial}{\partial x}\frac{u_x}{\sqrt{1 + u_x^2 + u_y^2}} + \frac{\partial}{\partial y}\frac{u_y}{\sqrt{1 + u_x^2 + u_y^2}} = 0$$

of a minimal surface M (cf. Vol. I, p. 193) in the form

$$\alpha_x + \beta_y = 0,$$

where α, β and $\gamma = \sqrt{1 - \alpha^2 - \beta^2}$ denote the direction cosines of the normal to M. We restrict our attention to a portion of M such that no two normals at distinct points have the same direction cosines α, β. With α, β as independent variables on the surface and x, y, u as functions of these two independent variables, this equation is equivalent to

(1) $$x_\alpha + y_\beta = 0.$$

Now, instead of the direction cosines, let us consider any set of three "homogeneous" variables α, β, γ on the surface which are proportional to the three direction cosines of the normal. Then the tangent plane to which the vector with components α, β, γ is orthogonal is given by the equation

(2) $$x\alpha + y\beta + u\gamma = \phi(\alpha, \beta, \gamma),$$

where the "support function" $\phi(\alpha, \beta, \gamma)$ is homogeneous of degree 1 in α, β, γ and represents, for $\alpha^2 + \beta^2 + \gamma^2 = 1$, the distance of the plane from the origin. The surface is to be regarded as the envelope of its tangent planes given by (2). The point of tangency corresponding to the normal direction with components proportional to α, β, γ is given by

$$x = \phi_\alpha, \qquad y = \phi_\beta, \qquad u = \phi_\gamma,$$

which is the "inverse Legendre transformation" to

$$\alpha = f_x, \qquad \beta = f_y, \qquad \gamma = f_u$$

if $f(x,y,u) = 0$ is an equation of the surface.

Using the preceding relations and the homogeneity condition, it is now easy to verify that the equation $x_\alpha + y_\beta = 0$ of a minimal surface is simply transformed into Laplace's equation

$$\phi_{\alpha\alpha} + \phi_{\beta\beta} + \phi_{\gamma\gamma} = 0$$

for the support function of the minimal surface.
(For a different version see Ch. V, §2.)

Appendix 2 to Chapter I

Systems of Differential Equations of First Order and Differential Equations of Higher Order

1. Plausibility Considerations. In §§2 and 7 we showed that the solution of single differential equations of higher order can be reduced to that of systems of differential equations of first order, if additional restrictive initial conditions are imposed on the solutions. Thus it is proper to emphasize the theory of systems of first order.

Nevertheless, in general we cannot expect complete equivalence of single differential equations and systems. As was made plausible in §2, a system of partial differential equations in two independent variables cannot in general be reduced by differentiation and elimination processes to a differential equation of higher order for a single function. The considerations of §2 apply similarly to equations with n independent variables.

Of course, these considerations do not conclusively prove the general impossibility of elimination. In fact, the system formed from a given system by differentiations is of a very special nature so that it may be possible to accomplish the elimination in some cases. Therefore, we shall determine in the following article (at least for a special case) the necessary and sufficient conditions under which a system can be reduced to a single differential equation of higher order (see also §2).

2. Conditions of Equivalence for Systems of Two First Order Partial Differential Equations and a Differential Equation of Second Order. The example of the Cauchy-Riemann differential equations

(1)
$$u_x = v_y \,,$$
$$u_y = -v_x$$

shows that in special cases systems of differential equations can be equivalent to a second order differential equation for a single function: Every solution u of (1) satisfies the potential equation $\Delta u = 0$, and to every such potential function a conjugate potential function v can be found such that u and v satisfy the system (1).

More generally, let us ask under what conditions a system

(2)
$$\Phi(x, y, u, v, u_x, u_y, v_x, v_y) = 0,$$
$$\Psi(x, y, u, v, u_x, u_y, v_x, v_y) = 0$$

is equivalent to a second order differential equation $L[u] = 0$ for u alone, in the sense that *every solution u of (2) satisfies the equation $L[u] = 0$ and that, conversely, to every solution u of $L[u] = 0$ a "conjugate" function v can be found such that u and v satisfy the system (2).*

First we consider linear differential equations written in the form

(3)
$$v_x = a(x, y)v + A(x, y, u, u_x, u_y),$$
$$v_y = b(x, y)v + B(x, y, u, u_x, u_y).$$

Here A and B are linear functions of u, u_x, u_y, whose coefficients, as well as the functions $a(x, y)$, $b(x, y)$, depend analytically on their arguments in a neighborhood of the origin; in addition, we assume the coefficient of u_x in B to be different from zero.

It now follows from §7 that, if analytic initial values $u(0, y)$ and $v(0, y)$ are prescribed, (3) possesses *unique analytic solutions $u(x, y)$ and $v(x, y)$* in a neighborhood of the origin. On the other hand, we could arbitrarily prescribe $u(0, y) = \phi(y)$ and $u_x(0, y) = \psi(y)$ instead of $u(0, y)$ and $v(0, y)$; *these initial conditions do not, however, uniquely determine the solutions u and v of the system.* Actually, the second equation in (3) yields an ordinary first order differential equation for $v (0, y)$:

$$v_y(0, y) = b(0, y)v(0, y) + B(0, y, \phi(y), \psi(y), \phi'(y)),$$

i.e., a one-parameter family of initial values $v(0, y)$ and, with it, a one-parameter family of solutions $v(x, y)$ of the system (3).

Keeping this in mind, we prove the following theorem:

The system (3) is equivalent to a second order differential equation for u alone if and only if the condition

$$a_y = b_x$$

is satisfied.[1]

[1] If the system (3) is elliptic and equivalence is understood in a somewhat wider sense, then it can be shown that there always exists an "equivalent" second order equation for a single function and vice-versa.

To prove this we differentiate the equations (3) with respect to y and x and obtain

$$(4) \qquad (a_y - b_x)v = L[u],$$

where $L[u] = -[aB - bA + A_y - B_x]$ is a second order linear differential expression in u alone. (Here the symbols A_y, B_x denote the total y and x derivatives of A and B.)

First, suppose that $a_y - b_x = 0$, then $L[u] = 0$, i.e., u satisfies the second order differential equation

$$L[u] \equiv aB - bA + A_y - B_x = 0.$$

In order to find the function $v(x, y)$ conjugate to u, we substitute the solution u of $L[u] = 0$ into A and B in the system (3). The system will have a solution v if the compatibility condition

$$\partial v_x / \partial y = \partial v_y / \partial x$$

is satisfied. Carrying out the necessary differentiations and making use of $a_y = b_x$ we obtain a nonhomogeneous linear first order equation which can be solved for $v(x, y)$ provided $a(x, y)$ and $b(x, y)$ do not vanish simultaneously (cf. §5, 2). Thus the first part of our theorem is proved.

Suppose now that $a_y - b_x \neq 0$ at $x = 0$ (and hence in a neighborhood of the y-axis). Then, from (4), we obtain the expression

$$(5) \qquad v = \frac{L[u]}{a_y - b_x}$$

which determines v uniquely for every u.

But if all u satisfying (3) would satisfy one and the same second order partial differential equation, then u and, according to (5) also v, could be characterized uniquely by the initial values $u(0, y)$ and $u_x(0, y)$; this, however, contradicts the result obtained earlier in this article.[1]

Finally, we treat the case in which the system (2) can be

---

[1] Note that the theorem is valid also if A and B do not depend linearly on u, u_x, u_y. If, in addition, the coefficients a and b depend also on u, on $u_x = p$ and on $u_y = q$, then, instead of $a_y = b_x$, we have the conditions

$$a_q = b_p = 0, \qquad a_p = b_q, \qquad a_y + a_u q = b_x + b_u p$$

for the possibility of the reduction.

written in the form

$$(6) \qquad \begin{aligned} v_x &= F(x, y, u, v, p, q), \\ v_y &= G(x, y, u, v, p, q), \end{aligned}$$

where F and G depend analytically on their arguments and where, in addition, $\partial G/\partial p \neq 0$. Calculating v_{xy} from both equations, we obtain

$$G_p u_{xx} + (G_q - F_p)u_{xy} - F_q u_{yy} + G_u p - F_u q$$
$$+ G_x - F_y + G_v F - F_v G = 0.$$

This is a second order differential equation in u alone if the expressions

$$(7) \qquad \frac{F_q}{G_p}, \qquad \frac{G_q - F_p}{G_p}, \qquad \frac{G_u p - F_u q + G_x - F_y + G_v F - F_v G}{G_p}$$

do not depend on v. Conversely, it can be shown that a second order differential equation for u which is equivalent to (6) exists if and only if the expressions (7) are independent of v.

CHAPTER II

General Theory of Partial Differential Equations of First Order

To construct solutions of partial differential equations by means of power series a very restrictive assumption is required—the data must be analytic (see Ch. I, §7, 4). This excludes many relevant problems.

However, for partial differential equations of first order a more direct and complete theory of integration can be developed under rather weak assumptions of continuity and differentiability. The main result of this chapter will be the *equivalence of a first order partial differential equation with a certain system of ordinary differential equations.*[1] The key to the theory is the *concept of characteristics,* which will play a decisive part also in higher order problems.

Note that here again all derivatives which occur are assumed to be continuous unless the opposite is specifically stated.

Also, it should be emphasized again that all statements and derivations are "in the small", i.e., they concern merely neighborhoods of points, etc., without necessarily specifying the extension of these neighborhoods.[2]

§1. *Geometric Theory of Quasi-Linear Differential Equations in Two Independent Variables*

1. Characteristic Curves. We shall briefly review quasi-linear differential equations which were treated in Ch. I, §5. First we consider an equation in two independent variables x, y:

$$(1) \qquad au_x + bu_y = c,$$

where a, b, c are given functions of x, y, u which, in the region considered, are assumed to be continuous together with their first derivatives and to satisfy $a^2 + b^2 \neq 0$.

This partial differential equation may be interpreted geometrically

---

[1] See also C. Carathéodory [1].

[2] An exception will be found in Appendix 2, Theory of Conservation Laws.

as follows: The integral surface $u(x, y)$ of the differential equation is required to possess at the point $P:(x, y, u)$, a tangent plane whose normal has direction numbers $u_x = p$, $u_y = q$ and -1 connected by the linear equation $ap + bq = c$. According to this equation the tangent planes of all integral surfaces through the point (x, y, u) belong to a single pencil of planes whose axis is given by the relations

(2) $dx:dy:du = a:b:c$

at the point P; these pencils and their axes are called *Monge pencils* and *Monge axes*.[1]

The point P together with the direction of the Monge axis through P constitutes a *characteristic line element*.

The directions of the Monge axes form a direction field in the x, y, u-space; the integral curves of this direction field are defined by the system of ordinary differential equations (2) and are called the *characteristic curves* of our partial differential equation. If we introduce a parameter s along the characteristic curves the differential equations become

(2′) $$\frac{dx}{ds} = a, \qquad \frac{dy}{ds} = b, \qquad \frac{du}{ds} = c.$$

The projections of the characteristic curves on the x, y-plane are called "characteristic base curves".

To integrate the partial differential equation (1) is the same as to find surfaces which "fit" the Monge field at every point, i.e., surfaces whose tangent plane at every point belongs to the Monge pencil or, in other words, surfaces which are at every point tangent to the Monge axis. Thus we see: *Every surface $u(x, y)$ generated by a one-parameter family of characteristic curves is an integral surface of the partial differential equation.* Conversely, *every integral surface $u(x, y)$ is generated by a one-parameter family of characteristic curves.*

The last statement is easily verified: On every integral surface $u(x, y)$ of the differential equation (1), a one-parameter family of curves $x = x(s)$, $y = y(s)$, $u = u(x(s), y(s))$ can be defined by the differential equations

$$\frac{dx}{ds} = a, \qquad \frac{dy}{ds} = b,$$

where the function $u(x, y)$ is to be substituted for u in a and b.

[1] Cf. p. 23.

Along such a curve, the partial differential equation (1) goes over into the statement $du/ds = c$. Thus our one-parameter family satisfies the relations (2′) and, hence, consists of characteristic curves. Note that the parameter s does not appear explicitly in the differential equations, so that the same integral curves are obtained if s + const. is substituted for s. In this sense, an additive constant in the parameter s is to be considered inessential.

Since the solutions of the system of differential equations (2′) are uniquely determined by the initial values of x, y, u for $s = 0$, we obtain the following theorem:

Every characteristic curve which has one point in common with an integral surface lies entirely on the integral surface. Moreover, *every integral surface is generated by a one-parameter family of characteristic curves.*

2. Initial Value Problem. We obtain the general manifold of solutions of the partial differential equation from the *initial value problem*. We define a space curve C by prescribing x, y, u as functions of a parameter t, where $x_t^2 + y_t^2 \neq 0$ and C has a simple[1] projection C_0 on the x, y-plane. In a neighborhood of the projection C_0, we now seek an integral surface $u(x, y)$ which passes through C, that is, a solution of (1) for which $u(t) = u(x(t), y(t))$ holds identically in t. The data, i.e., the coefficients of the differential equation and the initial quantities $x(t)$, $y(t)$, $u(t)$ are required to be merely continuously differentiable in the region considered; they need not be analytic.

To solve the initial value problem, we draw through each point of C a characteristic curve, i.e., a solution curve of the system of differential equations (2′); this is possible in a unique way within a certain neighborhood. We obtain a family of characteristic curves

$$x = x(s, t), \qquad y = y(s, t), \qquad u = u(s, t)$$

which depend on t as a parameter. These curves generate a surface $u(x, y)$ if, using the first two functions, we can express s and t in terms of x and y. A sufficient condition for this is the nonvanishing of the Jacobian

$$(3) \qquad\qquad \Delta = x_s y_t - y_s x_t = a y_t - b x_t$$

---

[1] Incidentally, if either the projection C_0 of C on the x, y-plane or C itself were to have double points, then one would be led to integral surfaces with self-intersections.

along the curve C. (Here we make use of a well-known theorem on ordinary differential equations, according to which the functions x, y, u depend in a continuously differentiable way on s and t.[1])

If the condition $\Delta \neq 0$ holds on C, u becomes a function of x and y; the third differential equation $du/ds = c$ is then equivalent to the given partial differential equation, since $du/ds = au_x + bu_y$. *Thus the initial value problem for the initial curve C is solved.* The uniqueness of the solution follows immediately from the theorem in article 1 which states that a characteristic curve having one point in common with an integral surface lies entirely on the surface. That is, each solution through C contains the entire one-parameter family of characteristics through C and hence is identical with u.

The condition $\Delta \neq 0$ along C can be interpreted geometrically in the following way: At every point of C, the tangent direction and the characteristic direction should have distinct projections on the x, y-plane.

If the initial value problem has a solution and the exceptional situation $\Delta = 0$ prevails everywhere along the curve C, then C is itself a characteristic curve,[2] for in that case we can choose the parameter t on the curve so that along the curve $a = dx/dt$, $b = dy/dt$. The partial differential equation, moreover, states that $c = du/dt$; hence C must in fact be a characteristic curve. But if C is a characteristic curve, then not only one but infinitely many integral surfaces pass through C as the initial curve. Consider another curve C' which passes through an arbitrary point of C and which has a vanishing Jacobian Δ, then the integral surface through C' surely contains the characteristic curve C and thus the set of solutions of the initial value problem for C is determined by the set of curves C'. All integral surfaces passing through curves of this set contain the curve C. Hence the characteristic curves are the curves in which two integral surfaces meet—*branch curves*—while at most one integral surface can pass through any noncharacteristic curve.

In the following theorem the preceding results are summarized:

If $\Delta \neq 0$ everywhere on the initial curve C, then the initial value prob-

---

[1] See similar theorems with proof in Ch. V, §6.

[2] The case in which Δ vanishes only at isolated points or on some other proper subset of C will not be considered here. See J. Leray [1].

lem has one and only one solution.[1] *If, however,* $\Delta = 0$ *everywhere along C, the initial value problem cannot be solved unless C is a characteristic curve, and then the problem has infinitely many solutions.*

Note that solutions of the initial value problem consist of only those solutions of the differential equation (1) which pass through C and which, in a neighborhood of C—thus also on C itself—are continuously differentiable. Without the assumption that u is continuously differentiable along C, we cannot deduce from the condition $\Delta = 0$ that C is a characteristic curve. There may, in fact, be solutions of the differential equation which pass through a noncharacteristic curve C and for which $\Delta = 0$. However, the derivatives of u on C can then no longer be continuous. In such cases, C may appear as an edge of *regression* (envelope of characteristics) *of the integral surface* $u(x, y)$—in any event, the projection C_0 of C is an envelope of the projections of the characteristic curves. In the neighborhood of C, u can no longer be defined as a single-valued function of x and y.

3. Examples

1) To illustrate our results, we consider the differential equation

$$(4) \qquad uu_x + u_y = 1$$

(a special case of example 5 in Ch. I, §1). The corresponding characteristic differential equations are

$$
(5) \qquad
\begin{aligned}
\frac{dx}{ds} &= u, \\[1mm]
\frac{dy}{ds} &= 1, \\[1mm]
\frac{du}{ds} &= 1;
\end{aligned}
$$

---

[1] The somewhat cumbersome formulation of the existence theorem enumerating all assumptions is as follows: Let G_0 be a domain in the x, y-plane, G the domain in x, y, u-space formed from G_0 by adding the values of u with $|u| < U$. Let a, b, c be continuously differentiable functions of x, y, u in G and $x(t)$, $y(t)$, $u(t)$ continuously differentiable functions of t for $|t| < T$ with $x_t^2 + y_t^2 \neq 0$ which define a curve C in G whose projection C_0 in G_0 is simple. Assume that, along C, $\Delta \equiv ay_t - bx_t \neq 0$. Then there exists a subdomain G_0' of G_0 which contains C_0 and in which a continuously differentiable function $u(x, y)$ is defined satisfying the differential equation $au_x + bu_y = c$ in G_0 and the initial condition $u(x(t), y(t)) = u(t)$ on C_0. This function u is uniquely determined.

they are solved by

$$x = x_0 + u_0 s + \frac{s^2}{2},$$

$$y = y_0 + s,$$

$$u = u_0 + s,$$

where x_0, y_0, u_0 are arbitrary constants of integration. In particular, the family of characteristics intersecting a given initial curve C:

$$x_0 = \phi(t), \qquad y_0 = \psi(t), \qquad u_0 = \chi(t)$$

is given by

(6)
$$x(s, t) = \phi(t) + s\chi(t) + \frac{s^2}{2},$$

$$y(s, t) = \psi(t) + s,$$

$$u(s, t) = \chi(t) + s.$$

The determinant

$$\Delta(s, t) = x_s y_t - x_t y_s = s(\psi_t - \chi_t) + \chi\psi_t - \phi_t$$

has on C the value

(7)
$$\Delta = \Delta(0, t) = \chi\psi_t - \phi_t .$$

If this determinant Δ is not zero along C, then the parameters s and t can be eliminated from (6); u may then be expressed as a function of x and y. In fact, it now follows that

$$u = y + \chi(t) - \psi(t),$$

$$x = \phi(t) + \chi(t)(y - \psi(t)) + \frac{(y - \psi(t))^2}{2};$$

from the second equation, t is obtained as a function of x and y, provided that

$$D = \phi_t - \chi\psi_t + (y - \psi(t)) (\chi_t - \psi_t)$$

is different from zero. As we approach the curve C, $y - \psi(t) \to 0$; thus, since $\phi_t - \chi\psi_t \neq 0$, there exists a neighborhood of C in which $D \neq 0$, hence $t = t(x, y)$ and therefore $u = u(x, y)$.

If for C we choose the *characteristic curve*

$$(8) \qquad x_0 = \tfrac{1}{2}t^2, \qquad y_0 = t, \qquad u_0 = t,$$

then (6) goes over into

$$x = \tfrac{1}{2}(s + t)^2, \qquad y = s + t, \qquad u = s + t$$

(i.e., again into an expression for the same curve C) and does not represent a solution of (4). To solve the initial value problem for such a C, we observe that a solution of (4) is given implicitly by the equation

$$(9) \qquad x = \tfrac{1}{2}u^2 + w(u - y),$$

with an arbitrary function w. If we choose w in such a way that $w(0) = 0$ and that u can be determined uniquely from (9), then all the corresponding integral surfaces $u = u(x, y)$ pass through the characteristic curve C.

Finally, let C be the noncharacteristic curve

$$(10) \qquad x_0 = t^2, \qquad y_0 = 2t, \qquad u_0 = t.$$

The system (6) goes over into

$$(11) \qquad x = \frac{s^2}{2} + st + t^2, \qquad y = s + 2t, \qquad u = s + t.$$

The determinant $\Delta(s, t) = s$ vanishes for $s = 0$ although C is not characteristic. By eliminating s and t we have

$$(12) \qquad u(x, y) = \frac{y}{2} \pm \sqrt{x - \frac{y^2}{4}},$$

i.e., two surfaces through C which, for $x > y^2/4$, satisfy equation (4). The initial value problem is not solved by (12) since the derivatives u_x and u_y do not remain bounded as we approach C.

The curve C is not an edge of regression of the surface $u = u(x, y)$; it is, however, singular on the surface in the sense that in the neighborhood of the projection of C on the x, y-plane u is no longer single-valued.

2) In the case of a *linear differential equation* (1), where the functions a, b, c do not depend explicitly on u, the vanishing of Δ along a noncharacteristic initial curve means, as we shall see, that the manifold determined by the system

$$x = x(s, t), \qquad y = y(s, t), \qquad u = u(s, t)$$

is a cylindrical surface perpendicular to the x, y-plane.   If we assume that

$$y_s u_t - u_s y_t \neq 0$$

on C, then x can be expressed as a function of y and u.   We show that $x = f(y, u)$ is independent of u.

First we note that, for linear differential equations, the relation

$$\Delta_s = (a_x + b_y)\Delta$$

follows from (2′) for the determinant Δ, so that Δ vanishes everywhere provided it vanishes on C.   Now, from

$$x_s = f_y y_s + f_u u_s ,$$

$$x_t = f_y y_t + f_u u_t ,$$

it follows immediately that

$$\Delta = f_u(u_s y_t - u_t y_s);$$

hence

$$f_u = 0 \quad \text{or} \quad x = f(y).$$

§2. *Quasi-Linear Differential Equations in n Independent Variables*

For n independent variables $x_1 , x_2 , \cdots , x_n$ with $n > 2$, there is no essential change in the theory; here, however, we must consider not only the one-dimensional characteristic curves and the n-dimensional integral surfaces but also $(n - 1)$-dimensional characteristic manifolds C.   An integral surface can be constructed either from an $(n - 1)$-parameter family of characteristic curves, or from a one-parameter family of $(n - 1)$-dimensional characteristic manifolds each of which is in turn generated by an $(n - 2)$-parameter family of characteristic curves.[1]

---

[1] Intermediate $(n - 2)$-dimensional characteristic manifolds, etc., can also be defined; they will, however, play no role in the following discussions.

We consider the quasi-linear differential equation

(1)
$$\sum_{i=1}^{n} a_i u_{x_i} = a,$$

where the coefficients a_i and a are given functions of the variables x_1, x_2, $\cdots$, x_n, u with continuous derivatives, and $\sum_{i=1}^{n} a_i^2 \neq 0$. Clearly, (1) states geometrically that at every point of the x, u-space on the surface $u = u(x_1, x_2, \cdots, x_n)$ the "characteristic directions" $dx_1:dx_2:\cdots:dx_n:du = a_1:a_2:\cdots:a_n:a$ are tangent to the surface. Again, as in the case of two independent variables, we use the following definition of characteristic curves: The n-parameter family of curves given in the x, u-space by the system of ordinary differential equations

(2)
$$\frac{dx_i}{ds} = a_i, \qquad \frac{du}{ds} = a \qquad (i = 1, 2, \cdots, n)$$

is called the *family of characteristic curves* belonging to the differential equation; the projection of a characteristic curve on the x-space is called a *characteristic base curve*.

These characteristic curves in the x, u-space are defined by (2) without reference to specific solutions of the partial differential equation (1).[1]

The connection between characteristic curves and integral surfaces is given by the following theorem:

On every integral surface $u = u(x_1, x_2, \cdots, x_n)$ of the partial differential equation there exists an $(n - 1)$-parameter family of characteristic curves which generate the integral surface. Conversely, every surface $u = u(x_1, x_2, \cdots, x_n)$ generated by such a family is an integral surface. Moreover, if a characteristic curve has a point in common with an integral surface, then it lies entirely on the surface.

[1] That the $n + 1$ characteristic differential equations (2) define only an n-parameter family of curves is due to the presence of an inessential additive constant of integration in the parameter s (s does not occur explicitly in (2)).

Note particularly that in the special case of a linear differential equation, i.e., one in which the a_i do not depend explicitly on u, the differential equations $\frac{dx_i}{ds} = a_i$ ($i = 1, 2, \cdots, n$) already form a determined system which represents the "characteristic base curves" as an $(n - 1)$-parameter family in the x_1, x_2, $\cdots$, x_n-space, while in the general case they form an n-parameter family.

To prove the first statement, we consider the system of ordinary differential equations $dx_i/ds = a_i (i = 1, 2, \cdots, n)$, substituting the solution function $u(x_1, x_2, \cdots, x_n)$ for u in the right member of the equation. This system then defines on the integral surface an $(n - 1)$-parameter family of curves which generate the surface. Along a curve of this family u becomes a function of the curve parameter s, and we obtain

$$\frac{du}{ds} = \sum_{i=1}^{n} u_{x_i} \frac{dx_i}{ds} = \sum_{i=1}^{n} a_i u_{x_i} ;$$

thus, since u satisfies the differential equation, we see that $du/ds = a$. The curve is, therefore, a characteristic curve. As in §1, the second statement immediately follows from the definition of characteristic curves and the third from the uniqueness of the characteristic curve through a given point.

From the characteristic curves defined by the characteristic differential equations we can now construct the integral surface by solving the following *initial value problem*: In $(n + 1)$-dimensional x, u-space let an $(n - 1)$-dimensional manifold C,

$$x_i = x_i(t_1, t_2, \cdots, t_{n-1}), \qquad u = u(t_1, t_2, \cdots, t_{n-1})$$

$$(i = 1, 2, \cdots, n)$$

be given by means of $n - 1$ independent parameters $t_1, t_2, \cdots, t_{n-1}$; we assume that the rank of the matrix $(\partial x_i/\partial t_\nu)$ is $n - 1$. We suppose that the projection C_0 of this manifold on the x-space is free of double points, i.e., that different points on C_0 correspond to different sets of values $t_1, t_2, \cdots, t_{n-1}$. In a neighborhood of C_0 we seek a solution $u(x_1, x_2, \cdots, x_n)$ of the differential equation which passes through C, i.e., which goes over into $u(t_1, t_2, \cdots, t_{n-1})$ when the quantities x_i are replaced by $x_i(t_1, t_2, \cdots, t_{n-1})$.

We solve this initial value problem in the following way: For a given set of values $t_1, t_2, \cdots, t_{n-1}$, we find solutions

$$x_i(s, t_1, \cdots, t_{n-1}), \qquad u(s, t_1, \cdots, t_{n-1})$$

of the system of ordinary characteristic differential equations (2), which at $s = 0$ coincide with the prescribed functions of $t_1, t_2, \cdots, t_{n-1}$. These functions depend in a continuously differentiable way not only on s but also on $t_1, t_2, \cdots, t_{n-1}$. We now consider the quantities $s, t_1, \cdots, t_{n-1}$ to be expressed in terms of $x_1, x_2, \cdots, x_n$ (by means of the equations $x_i = x_i(s, t_1, \cdots, t_{n-1}))$, and substitute them into $u(s, t_1, \cdots, t_{n-1})$, so that u appears as a function of $x_1, x_2, \cdots, x_n$. This introduction of the quantities $x_1, x_2, \cdots, x_n$ as new independent variables is certainly possible if the Jacobian

$$(3) \qquad \Delta = \frac{\partial(x_1, x_2, \cdots, x_n)}{\partial(s, t_1, \cdots, t_{n-1})} = \begin{vmatrix} \dfrac{\partial x_1}{\partial s} & \cdots & \dfrac{\partial x_n}{\partial s} \\ \dfrac{\partial x_1}{\partial t_1} & \cdots & \dfrac{\partial x_n}{\partial t_1} \\ \cdots\cdots\cdots\cdots \\ \dfrac{\partial x_1}{\partial t_{n-1}} & \cdots & \dfrac{\partial x_n}{\partial t_{n-1}} \end{vmatrix}$$

does not vanish along C, i.e., for $s = 0$. Because of (2), the elements of the first row can be expressed along C by the relations $\partial x_i / \partial s = a_i(x_1, x_2, \cdots, x_n, u)$; here the prescribed initial values are to be substituted for the x_i and u as functions of $t_1, t_2, \cdots, t_{n-1}$.

Thus the Jacobian (3) is identical with

$$\Delta = \begin{vmatrix} a_1 & \cdots & a_n \\ \dfrac{\partial x_1}{\partial t_1} & \cdots & \dfrac{\partial x_n}{\partial t_1} \\ \cdots\cdots\cdots\cdots \\ \dfrac{\partial x_1}{\partial t_{n-1}} & \cdots & \dfrac{\partial x_n}{\partial t_{n-1}} \end{vmatrix}.$$

Under the assumption $\Delta \neq 0$ we obtain from $u(s, t_1, \cdots, t_{n-1})$ a function $u(x_1, x_2, \cdots, x_n)$. The equation $du/ds = a$ for

$$u(s, t_1, \cdots, t_{n-1})$$

goes over into

$$\sum_{i=1}^{n} u_{x_i} \frac{dx_i}{ds} = \sum_{i=1}^{n} a_i u_{x_i} = a;$$

$u(x_1, x_2, \cdots x_n)$ is, therefore, a solution of the differential equation (1). *Thus, assuming $\Delta \neq 0$, our initial value problem possesses a uniquely determined solution.*

This solution of the initial value problem fails in the exceptional case where $\Delta = 0$ everywhere along C. The question arises: What further conditions are needed to guarantee the existence of a solution of the initial value problem in this case?

We note first that the assumption $\Delta = 0$ is equivalent to the existence of $n - 1$ uniquely determined functions $\lambda_i(t_1, t_2, \cdots, t_{n-1})$ which depend continuously on the parameters, so that along C the linear relations

$$(4) \qquad a_i = \sum_{\nu=1}^{n-1} \lambda_\nu \frac{\partial x_i}{\partial t_\nu}$$

hold. Indeed, this follows immediately from the vanishing of the determinant Δ and the nonvanishing of the cofactor of at least one element of the first row.

To formulate the desired necessary conditions, we introduce the concept of a *characteristic $(n - 1)$-dimensional manifold*. In geometric language, we assign to each point $(x_1, x_2, \cdots, x_n, u)$ of x, u-space a *characteristic vector* $a_1, a_2, \cdots, a_n, a$. *The $(n - 1)$-dimensional manifold C is called characteristic if at each point the corresponding characteristic vector is tangent to the manifold.*

Analytically we formulate this definition by representing the manifold C in terms of $n - 1$ parameters $t_1, t_2, \cdots, t_{n-1}$: C is called a characteristic manifold of the partial differential equation $\sum_{i=1}^{n} a_i u_{x_i} = a$ if there exist $n - 1$ functions $\lambda_i(t_1, t_2, \cdots, t_{n-1})$ $(i = 1, 2, \cdots, n-1)$ such that on C the relations

$$(5) \qquad a_i = \sum_{\nu=1}^{n-1} \lambda_\nu \frac{\partial x_i}{\partial t_\nu} \qquad (i = 1, 2, \cdots, n),$$

$$(5') \qquad a = \sum_{\nu=1}^{n-1} \lambda_\nu \frac{\partial u}{\partial t_\nu}$$

hold, i.e., such that the characteristic vector depends linearly on the $n - 1$ linearly independent tangent vectors with components $\partial x_i/\partial t_\nu$, $\partial u/\partial t_\nu$.

The following theorems hold for a characteristic manifold of a quasi-linear differential equation: *Every characteristic manifold is generated by an $(n - 2)$-parameter family of characteristic curves; conversely, every such family of curves generates a characteristic manifold.*

If a characteristic curve Γ has one point in common with a characteristic manifold C, then it lies entirely on it.

In order to prove the first theorem, we consider in the parameter space of the t_i the $(n - 2)$-parameter family of curves $t_i = t_i(s)$ defined by the system of $n - 1$ ordinary differential equations

$$(6) \qquad \frac{dt_\nu}{ds} = \lambda_\nu(t_1, t_2, \cdots, t_{n-1}).$$

The functions $x_i(t_1, t_2, \cdots, t_{n-1})$ and $u(t_1, t_2, \cdots, t_{n-1})$ then go over into functions of s representing curves on C for which the relations

$$\frac{dx_i}{ds} = \sum_{\nu=1}^{n-1} \frac{\partial x_i}{\partial t_\nu} \lambda_\nu \quad \text{and} \quad \frac{du}{ds} = \sum_{\nu=1}^{n-1} \frac{\partial u}{\partial t_\nu} \lambda_\nu$$

hold. Because of the equations defining the characteristic manifold, we obtain $dx_i/ds = a_i$, $du/ds = a$, which shows that our curves on C are characteristic curves; they form an $(n-2)$-parameter family, and hence generate C. The converse is self-evident since a manifold generated by families of characteristic curves is certainly tangent to the corresponding characteristic vector at each of its points.

The second theorem immediately follows from the fact that the solutions of the characteristic differential equations are uniquely determined by the initial condition requiring each characteristic curve to contain a point of C. As seen above, there exists a characteristic curve lying entirely in C which goes through this point of C.

We now summarize the main results in the following theorem: *If $\Delta \neq 0$ on the initial manifold C, then there exists one and only one solution of the initial value problem. If, however, $\Delta = 0$ everywhere on C, then a necessary and sufficient condition for the solvability of the initial value problem is that C be a characteristic manifold. In this case there exist infinitely many solutions of the initial value problem and C again is a branch manifold across which different solutions can be smoothly continued into each other.*

Only that part of the theorem which concerns the case $\Delta = 0$ remains to be proved. From the condition $\Delta = 0$ we deduced the existence of $n - 1$ functions λ which depend—in a continuous and differentiable manner—on the parameters $t_1, t_2, \cdots, t_{n-1}$, so that the relations (5) are valid. We must also establish the missing relation (5') as a consequence of the assumption that an integral surface $u = u(x_1, x_2, \cdots, x_n)$ passes through C. For a solution u on C we have

$$\sum_{\nu=1}^{n-1} \lambda_\nu \frac{\partial u}{\partial t_\nu} = \sum_{\nu=1}^{n-1} \lambda_\nu \sum_{i=1}^{n} \frac{\partial u}{\partial x_i} \frac{\partial x_i}{\partial t_\nu} = \sum_{i=1}^{n} a_i u_{x_i}$$

and, therefore, since u satisfies the differential equation (1),

$$a = \sum_{\nu=1}^{n-1} \lambda_\nu \frac{\partial u}{\partial t_\nu}$$

which is the relation we wanted to prove.[1]

[1] Geometrically, the result is visualized if we recall that the characteristic

Conversely, if C is a characteristic manifold, we proceed to construct an infinite number of solutions u of the partial differential equation which contain C. We choose an arbitrary $(n - 1)$-dimensional manifold C' which intersects C in an $(n - 2)$-dimensional manifold S and for which $\Delta \neq 0$ everywhere. Then a uniquely determined integral surface J' goes through C'. But, according to the above construction, J' contains all characteristic curves through S, hence also the manifold C generated by them. This completes the proof of our main theorem.

§3. General Differential Equations in Two Independent Variables

1. Characteristic Curves and Focal Curves. The Monge Cone. We consider a general differential equation

$$(1) \qquad F(x, y, u, p, q) = 0,$$

with the abbreviation $u_x = p$, $u_y = q$. According to our usual convention, we assume that F is a continuous function and possesses continuous first derivatives with respect to all its five arguments in the region considered. In addition we require that

$$F_p^2 + F_q^2 \neq 0.$$

The partial differential equation (1) can be interpreted geometrically in the following way (cf. Ch. I, §4): The direction coefficients p, q of the tangent plane to an integral surface in x, y, u-space satisfy the equation $F = 0$ at every point $P: (x, y, u)$. This equation is no longer linear in p and q; hence, in general, the possible tangent planes do not form a pencil of planes through a line but a one-parameter family enveloping a conical surface with P as vertex, called the "Monge cone". (We should not forget that our considerations refer to the conical surface in the small, that is, to a suitably small portion of a conical sheet. In the large, the Monge cone may well consist of separate sheets.[1]) Such a Monge cone is thus assigned by the differential equation to each point (x, y, u) of the region in space considered.

---

vector is tangent to the integral surface; since, by (5), its projection on the x_1, x_2, $\cdots$, x_n-space is tangent to the projection of C on the x_1, x_2, $\cdots$, x_n-space, it is itself tangent to C.

[1] In Chapters V and VI we shall give a more detailed analysis of the Monge cone in the "large". The considerations here refer merely to a suitably small range of tangent planes, e.g., a portion of a sheet of the cone where q can be expressed as a single-valued differentiable function of p.

The problem of integrating the equation consists in finding surfaces which fit this field of cones, i.e., surfaces which are at each point tangent to the corresponding cone.

We may also represent the Monge cones by means of a relation for their generating lines instead of the relation $F = 0$ for their tangent planes. To do this analytically we first represent the Monge cone $F = 0$ parametrically by considering p and q as functions of a parameter λ. A generating line of the cone is then the limit of the line of intersection of the tangent planes belonging to the parameters λ and $\lambda + h$, respectively, as $h \to 0$. If we consider x, y, u along a fixed generator as functions of the distance σ from the vertex of the cone, then we obtain easily the equations

$$\frac{du}{d\sigma} = p(\lambda) \frac{dx}{d\sigma} + q(\lambda) \frac{dy}{d\sigma}$$

and

$$0 = p'(\lambda) \frac{dx}{d\sigma} + q'(\lambda) \frac{dy}{d\sigma}.$$

By differentiating $F = 0$ with respect to λ we also find

$$F_p p'(\lambda) + F_q q'(\lambda) = 0;$$

hence the relation

(2) $$dx:dy:du = F_p:F_q:(pF_p + qF_q)$$

holds for the generators of the cone. We may regard this relation as the representation of the Monge cone dual to that given by the differential equation (1).

We call the directions of the generators of the Monge cone *characteristic directions*. While in the case of quasi-linear equations only one characteristic direction belongs to each point in space, we have here a one-parameter family of characteristic directions at each point. Space curves having a characteristic direction at each point shall be called *focal curves* or *Monge curves*. We can write the conditions (2) for the focal curves in the form

(3) $$\frac{dx}{ds} = F_p, \qquad \frac{dy}{ds} = F_q, \qquad \frac{du}{ds} = pF_p + qF_q$$

by introducing along them a suitable parameter s. The last of these three differential equations is called the *strip condition*. It states that the functions $x(s)$, $y(s)$, $u(s)$, $p(s)$, $q(s)$ not only define a space curve, but simultaneously a plane tangent to it at every point. A configuration consisting of a curve and a family of tangent planes to this curve is called a *strip*. This system of three ordinary differential equations (3) and the relation $F(x, y, u, p, q) = 0$ for the five functions x, y, u, p, q of the argument s represent an *underdetermined* system. Each solution of this system yields a so-called *focal strip*.[1]

The basis of our theory is the following consideration which determines such focal strips as are embedded in integral surfaces, and will be called *characteristic strips*: Every integral surface $u(x, y)$ must have focal curves since at every point the integral surface is tangent to a Monge cone, and hence contains a characteristic direction. The field formed by these characteristic directions yields the corresponding focal curves as its integral curves on the integral surface. The requirement that a focal curve be embedded[2] in an integral surface $u(x, y)$ leads to two additional ordinary differential equations for the quantities p and q as functions of s; this will now be shown.

To find these differential equations on a specific integral surface $u = u(x, y)$, on which the quantities p and q may also be considered specific given functions of x, y, we note that the differential equations

$$\frac{dx}{ds} = F_p, \qquad \frac{dy}{ds} = F_q$$

define a one-parameter family of curves on the surface, along which

$$\frac{du}{ds} = u_x \frac{dx}{ds} + u_y \frac{dy}{ds}$$

and hence

$$\frac{du}{ds} = pF_p + qF_q$$

---

[1] The four conditions for a focal strip become a determined system if we prescribe an additional arbitrary relation between x, y, u, i.e., if we require the focal curve to lie on a prescribed surface. (However, the strip need not be tangent to the surface.) Thus it becomes clear that there exists, in general, a one-parameter family of focal curves on a prescribed surface.

[2] Here embedding means that in the neighborhood of the projection of the focal curve on the x, y-plane u is a single-valued, twice continuously differentiable function of x and y.

holds. Thus our curves form a family of Monge curves and generate the integral surface. By differentiating the partial differential equation first with respect to x and then with respect to y, we obtain ·the relations

$$F_p p_x + F_q q_x + F_u p + F_x = 0,$$

$$F_p p_y + F_q q_y + F_u q + F_y = 0,$$

which hold identically on our surface. Since $F_p = dx/ds$, $F_q = dy/ds$, $p_y = q_x$, we see that for Monge curves given in terms of the parameter s the above two equations immediately go over into the relations

$$\frac{dp}{ds} + F_u p + F_x = 0, \qquad \frac{dq}{ds} + F_u q + F_y = 0.$$

Thus, if a Monge curve is imbedded in an integral surface, the co-ordinates x, y, u of its points and the quantities p and q satisfy, along that curve, a system of five ordinary differential equations

(4)
$$\frac{dx}{ds} = F_p, \qquad \frac{dy}{ds} = F_q, \qquad \frac{du}{ds} = pF_p + qF_q,$$

$$\frac{dp}{ds} = -(pF_u + F_x), \qquad \frac{dq}{ds} = -(qF_u + F_y).$$

This system is called the *characteristic system of differential equations* belonging to equation (1).

We now reverse the process by disregarding the fact that this system of ordinary differential equations was obtained by considering a presumably given integral surface; instead we use the system as a point of departure without reference to solutions of (1). Since there is an irrelevant additive constant in the parameter s, the system defines a *four-parameter family of curves*: $x(s)$, $y(s)$, $u(s)$ with corresponding tangent planes $p(s)$, $q(s)$, i.e., a family of strips.

We note that *the function F is an integral*[1] *of our characteristic system of differential equations.* In other words, along every solution of this system, F has a constant value. Indeed, for such a solution the re-

----

[1] "Integral" as used here is not to be confused with "integral" in the sense of *solution.* By an integral $\emptyset(x_1, x_2, \cdots, x_n)$ we mean a function of the independent variables x_i which has a constant value along each curve $x_i(s)$ which solves the system

$$\frac{dx_i}{ds} = a_i(x_1, x_2, \cdots, x_n) \qquad (i = 1, 2, \cdots, n).$$

lation

$$\frac{dF}{ds} = F_p \frac{dp}{ds} + F_q \frac{dq}{ds} + F_u \frac{du}{ds} + F_x \frac{dx}{ds} + F_y \frac{dy}{ds}$$

holds and, because of the characteristic differential equations, the expression on the right vanishes identically in s.

We now single out from the four-parameter family of solutions of the characteristic differential equations a three-parameter family by using the condition that, along these solutions, F should have the constant value zero as stipulated by the original differential equation.[1] *Every solution of the characteristic differential equations which also satisfies the equation $F = 0$ will be called a "characteristic strip"; a space curve $x(s)$, $y(s)$, $u(s)$ bearing such a strip is called a characteristic curve.*

As for quasi-linear equations, the following theorems follow from the derivation of the characteristic differential equations: *In every integral surface, there exists a one-parameter family of characteristic curves and corresponding characteristic strips.* Moreover, *if a characteristic strip has an element (i.e., values x, y, u, p, q) in common with an integral surface, then this strip belongs entirely to the integral surface.*[2]

2. Solution of the Initial Value Problem. The most important feature of the theory of first order partial differential equations is the equivalence of the problems of integrating the partial differential equation (1) and the characteristic system of ordinary differential equations (4). In other words, the integration of a first order partial differential equation can be reduced to that of the corresponding characteristic system of ordinary differential equations.

To prove this equivalence, we construct the integral surfaces by means of the characteristic strips (a similar method was used in §§1 and 2). We pose again the *initial value problem* for our partial differential equation. For our purpose it is convenient to present it in the following form: Let an *initial strip* C_1 be given by the functions $x(t)$, $y(t)$, $u(t)$, $p(t)$, $q(t)$ in terms of a parameter t such that the projection C_0 of the curve $C:x(t), y(t), u(t)$ has no double points. Assume moreover that

1) the strip relation

---

[1] Without this restriction we would have to consider the whole family of differential equations F = const. simultaneously.

[2] It may be stated again that not every focal curve is characteristic, but that the manifold of focal curves is essentially larger than that of characteristics. See §3, 3.

$$\frac{du}{dt} = p\frac{dx}{dt} + q\frac{dy}{dt} ,$$

and

2) $F = 0$

hold identically in t. Such a strip C_1 is called an *integral strip*.

Now the initial value problem consists in finding a function $u(x, y)$ in a neighborhood of C_0 which satisfies the differential equation there and which, together with $p = u_x$, $q = u_y$, assumes the prescribed initial values on C_0, i.e., which contains the initial strip C_1.[1]

To solve the initial value problem we draw through every element of the prescribed initial strip the characteristic strip (with s as a running parameter), given by the solution of the characteristic differential equation (4), which for $s = 0$ reduces to the given initial element of the strip $x(t)$, $y(t)$, $u(t)$, $p(t)$, $q(t)$. We denote this system of solutions by

$$x(s, t), \qquad y(s, t), \qquad u(s, t), \qquad p(s, t), \qquad q(s, t).$$

Again the unique determination of these solutions and their continuously differentiable dependence on s and on the parameter t is assured by well-known theorems about ordinary differential equations. If the expression

(5) $$\Delta = F_p y_t - F_q x_t = x_s y_t - x_t y_s$$

is different from zero along our initial strip and hence also in a certain s, t-neighborhood of it, then we may introduce the quantities x and y as independent variables in that neighborhood instead of the parameters s and t. This means we can express the quantities u, p, q as functions of x and y; in particular, we obtain a surface $u(x, y)$. We claim that on this surface $p = u_x$, $q = u_y$ and that it is an integral surface and thus solves our initial value problem.

The last fact is self-evident as soon as the relations $p = u_x$, $q = u_y$ are established. For, since F is an integral of the system of equations (4), the quantity $F(x, y, u, p, q)$ certainly vanishes along our surface

---

[1] In posing this problem, one could first consider only the initial curve C as given and the quantities p and q determined by the strip relation as well as by the equation $F = 0$; the form of the initial value problem chosen here is, however, preferable, because it avoids inessential reference to possible multiple-valuedness of solutions of equations for p and q along C.

identically in s and t (because of the second initial condition); it therefore vanishes identically in x and y also.

To verify the relations $p = u_x$ and $q = u_y$ we need only show that the two quantities

(6)
$$U = u_t - px_t - qy_t,$$
$$V = u_s - px_s - qy_s$$

vanish identically on our surface; the relations

(7)
$$0 = u_t - u_x x_t - u_y y_t,$$
$$0 = u_s - u_x x_s - u_y y_s$$

then imply that $p = u_x$, $q = u_y$ since the determinant $x_t y_s - x_s y_t = \Delta$ of this system of linear equations for u_x, u_y is, by hypothesis, not equal to zero.

That the quantity V vanishes is an immediate consequence of the characteristic differential equations. In order to prove the vanishing of U, we regard U and V as functions of s and t. Considering the identity

$$\frac{\partial U}{\partial s} - \frac{\partial V}{\partial t} = -(p_s x_t - p_t x_s + q_s y_t - q_t y_s),$$

using the characteristic differential equations (4) and the fact that the identical vanishing of V also implies $\partial V/\partial t = 0$, we deduce

$$\frac{\partial U}{\partial s} = p_t F_p + q_t F_q + (px_t + qy_t)F_u + F_x x_t + F_y y_t.$$

On the other hand, by differentiating with respect to t the relation $F = 0$, which holds identically in s and t, we have

$$p_t F_p + q_t F_q + u_t F_u + x_t F_x + y_t F_y = 0;$$

hence

(8)
$$\frac{\partial U}{\partial s} = -F_u U.$$

For fixed t this equation represents a linear ordinary differential equation for U as a function of s. Since by hypothesis U vanishes for $s = 0$, we see from the unique determination of a solution of an ordinary differential equation by its initial value $U(0)$[1] that the quan-

---

[1] That is, from the relation

$$U(s) = U(0)e^{-\int_0^s F_u \, ds}.$$

tity U is zero for all values of s. This is just what we wanted to show. The integral surface constructed is unique, since solutions of ordinary differential equations (in this case the characteristic equations) are uniquely determined by their initial conditions.

We summarize our result in the following way: *Given a space curve* $C: x = x(t)$, $y = y(t)$, $u = u(t)$ *which can be completed by* $p(t)$, $q(t)$ *to an initial strip* $C_1: x, y, u, p, q$ *where* C_1 *satisfies the strip relation and the relation* $F = 0$; *if along this strip* $\Delta = F_p y_t - F_q x_t \neq 0$, *then in a neighborhood of* C_1 *there exists one and only one integral surface* $u(x, y)$ *through this strip.*

3. Characteristics as Branch Elements. Supplementary Remarks. Integral Conoid. Caustics. We have yet to clarify the meaning of the exceptional case $\Delta = 0$. If $F_p y_t - F_q x_t = 0$ holds everywhere along a strip C_1 on an integral surface then, according to the discussion on page 78, C_1 must be a characteristic strip on this surface. Hence, in the exceptional case $\Delta = 0$, an integral surface can pass through the curve C only if C is characteristic, i.e., if the functions p and q determined by conditions *1*) and *2*) complete this curve to a characteristic strip. But if this condition is satisfied, there exist not only one but infinitely many integral surfaces which touch each other along this initial strip. Consider a curve C' which intersects C and which is completed to an initial strip in such a way that this initial strip touches the characteristic strip belonging to C at the common point. Then the initial value problem with respect to C' yields an integral surface containing the whole characteristic strip belonging to C, since this integral surface has an element in common with the characteristic strip.

Thus the characteristics on an integral surface are curves along which different integral surfaces meet and are tangent to each other. Hence one may consider these curves (or strips) as *branch elements of integral surfaces.* By crossing one of these curves one can, without destroying the continuity of the first derivatives of u, proceed to another member of the family of integral surfaces instead of continuing on the original integral surface.

Summarizing, we have the following two cases for the initial value problem: *If* $\Delta \neq 0$ *on the initial strip* C_1, *then the initial value problem has a unique solution. However, if* $\Delta = 0$ *along* C_1, *then the initial value problem has a solution only if the initial strip* C_1 *is characteristic; in this case, there are infinitely many solutions.*

One final remark concerning the case in which $\Delta = 0$: If the initial strip C_1 is not characteristic, then it is merely a focal strip, and there exists no solution of the initial value problem through C_1, i.e., *there exists no integral surface which contains this initial strip and has continuous derivatives up to the second order in its neighborhood.* It is, however, possible that there exists an integral surface for which the focal curve C is a singular curve. In fact, if we construct the characteristic strips so that they pass through every element (x, y, u, p, q) of C_1 as initial element and if they do not all coincide (which would have been the case had C_1 been characteristic) then these strips may form an integral surface.

On these integral surfaces, the curve C is necessarily singular and in general appears as the envelope of the characteristic curves which generate the surface. We may expect that it is an *edge of regression of the integral surface,* or at least that in the neighborhood of the projection of C on the x, y-plane, u can no longer be defined as a single-valued function of x and y. We shall illustrate these possibilities by means of examples (cf. §6 and the example discussed for the case of quasi-linear differential equations in §1). In the theory of the propagation of light, the *characteristics* correspond to *light rays*; therefore the *caustic curves* of these rays appear as *focal curves* (which motivates the terminology).

A particular limiting case of the initial value problem, the case where the initial curve degenerates into a point, is of special interest. Again the arguments used above lead to the following conclusion: *All characteristic curves through a fixed point P of x, y, u-space form an integral surface.* This integral surface (which may consist of several sheets) has a conical singularity at P (with the Monge cone as tangent cone) and is called the *integral conoid of the partial differential equation at P.* As we shall see later, it plays the role of *light cone* in the theory of propagation of light.

The following final remark—a remark equally applicable to the case of n variables—shows the essential difference between quasi-linear equations and general nonlinear equations. To construct the solution in the linear and quasi-linear case, it suffices to deal with the characteristic curves which form a two-parameter (or, as the case may be, an n-parameter) family. In the general case, however, one is forced to consider the complete characteristic strip in order to include the tangent directions p and q. The x, y, u-curve of this strip is a characteristic curve. However, these strips now form a three-parameter (or a $(2n - 1)$-parameter) family; the same is in general also true for the corresponding characteristic curves.

§4. The Complete Integral

In Ch. I, §4 a complete integral u of the differential equation $F = 0$ depending on two parameters a and b,

$$u = \phi(x, y, a, b)$$

was used to construct a solution which involved an arbitrary function $w(a)$. The construction consisted in the formation of envelopes by setting $b = w(a)$ and eliminating a from the two equations

$$u = \phi(x, y, a, w(a)),$$

$$0 = \phi_a + \phi_b w'(a).$$

For a fixed value of a these equations represent the contact curves of the integral surface $u = \phi(x, y, a, w(a))$ with the envelope. Since the function $w(a)$ can be so chosen that, for a certain a, it has an arbitrary value b and its derivative $w'(a)$ an arbitrary value c, the two equations

(1) $$u = \phi(x, y, a, b),$$

(1') $$0 = \phi_a + c\phi_b$$

represent a family of curves (depending on three parameters a, b, c) which appear as contact curves in the formation of these envelopes.[1]

Now we show that *the curves represented by equations* (1), (1') *are characteristic curves of our differential equation.* The corresponding strips obtained from $p = \phi_x(x, y, a, b)$, $q = \phi_y(x, y, a, b)$ are then automatically characteristic.

The proof follows intuitively from the fact that along our curves two distinct integral surfaces are tangent. This, as we have seen in §3, is possible only along a characteristic strip.

We easily verify this statement by direct calculation. Considering x instead of s as the independent variable along our curve, we find from equation (1') by differentiating with respect to x

(2) $$\phi_{ax} + c\phi_{bx} = -y_x(\phi_{ay} + c\phi_{by}).$$

---

[1] Of course, all these statements refer to suitably small ranges of the parameters.

This quantity cannot vanish since by definition $\phi_{ax}\phi_{by} - \phi_{ay}\phi_{bx} \neq 0$. Now, differentiating the differential equation $F = 0$ for $u = \phi(x, y, a, b)$ first with respect to a and then with respect to b we obtain the equations

(3)
$$F_u\phi_a + F_p\phi_{ax} + F_q\phi_{ay} = 0,$$
$$F_u\phi_b + F_p\phi_{bx} + F_q\phi_{by} = 0.$$

Multiplying the second equation by c, adding it to the first, using (1') and (2), we obtain the equation $F_p y_x - F_q = 0$. According to §3, 1, (3), if $F_p \neq 0$, this expresses the fact that our curves are characteristic. The essential assumption for our conclusions is that

$$F_p^2 + F_q^2 \neq 0$$

in the region considered.

Hence any complete integral of the partial differential equation yields a three-parameter family of characteristic curves and strips. (This choice of the independent variable x instead of the symmetric representation by a parameter s leads to no essential difficulty.)

Thus we have reversed the arguments of §3; i.e., *we have obtained the solutions of the characteristic differential equations from a complete integral of the partial differential equation*. The same approach will be used again in §7.

In this way we obtain in general all the characteristics, and accordingly the solutions of the partial differential equation. This becomes evident if we assume that at each point of any given integral surface with $F_p^2 + F_q^2 \neq 0$ we can assign a member of the family $u = \phi(x, y, a, b)$ which is tangent to the surface at the point.

A final remark concerning the role of the singular solution is appropriate: According to Ch. I, §4, 3, this solution is obtained by forming the envelope of the two-parameter family $u = \phi(x, y, a, b)$, or, without reference to a special complete integral, by eliminating p and q from the equations

$$F = 0, \qquad F_p = 0, \qquad F_q = 0.$$

None of the considerations of this section remains valid for a singular solution, since we always assumed that the condition $F_p^2 + F_q^2 \neq 0$ holds on our integral surfaces. The exceptional nature of the singular solution is also clear from the fact that the characteristic initial condition

$$\Delta \equiv F_p \, y_t - F_q \, x_t = 0$$

is identically satisfied no matter how the initial curve is chosen. Every strip on a singular solution is characteristic in this sense.

§5. Focal Curves and the Monge Equation

In §3, 1 the focal curves were represented by the system of differential equations (3), where the quantities p and q were subject to the subsidiary condition $F(x, y, u, p, q) = 0$. If, assuming $F_p \neq 0$, we introduce x instead of s as a parameter along the curves, then the three equations

$$(1) \qquad F = 0, \qquad \frac{dy}{dx} = \frac{F_q}{F_p}, \qquad \frac{du}{dx} = \frac{pF_p + qF_q}{F_p}$$

hold. By eliminating p and q from these equations one arrives at the single ordinary differential equation

$$(2) \qquad M\left(x, y, u, \frac{dy}{dx}, \frac{du}{dx}\right) = 0$$

in the two unknowns y and u. This equation is called the *Monge differential equation*. It is a simple example of an underdetermined system of ordinary differential equations and states a condition for the direction of the generators of the Monge cone at the point (x, y, u), while the original partial differential equation $F = 0$ expresses a relation satisfied by the tangent planes of the Monge cone. In terms of the parameter s instead of x, the equation of the cone would be of the form

$$(2') \qquad M\left(x, y, u, \frac{dx}{ds}, \frac{dy}{ds}, \frac{du}{ds}\right) = 0,$$

where M is homogeneous in the last three arguments.

Conversely, to a given Monge equation $M(x, y, u, y', u') = 0$ we can construct a corresponding partial differential equation by eliminating the quantities y', u' from the equation $M = 0$ and from the two equations defining the tangent plane containing the line element dx, dy, du:

$$q = -\frac{M_{y'}}{M_{u'}}, \qquad p = \frac{y'M_{y'} + u'M_{u'}}{M_{u'}}.$$

As a result, we obtain the equation $F(x, y, u, p, q) = 0$ (transition to the tangent plane representation of the Monge cone). Thus, a Monge equation (i.e., an ordinary differential equation of first order in two unknown functions) and a partial differential equation of first order for a function of two independent variables represent the same geometric configuration, namely a cone with (x, y, u) as vertex. The equations $F = 0$ and $M = 0$ are *dual* to each other in the sense of projective geometry.

The solutions of the Monge equation, the focal curves, are those curves which, at each point, are tangent to a characteristic curve. We deduce from the considerations of §3, 3 that these focal curves (except the characteristics themselves) are obtained by forming the envelope of the characteristic curves on an integral surface of the differential equation $F = 0$ (if such an envelope exists).

This leads to a remarkable theory for solving arbitrarily given Monge equations. At first sight the determination of u and y from the Monge equation seems to require that one imposes some arbitrary relation, $W(x, y, u) = 0$, and subsequently integrates an ordinary first order differential equation obtained by eliminating u or y; this would imply an infinite number of such processes of integration. No single representation of all solutions of the Monge equation (2) by means of an arbitrary function is obtained in this way. However, by means of the complete integral, it is possible to give an explicit solution of the Monge equation depending on an arbitrary function and involving no further integrations. We obtain this "explicit" solution by assuming that $u = \phi(x, y, a, b)$ is a known complete integral of the partial differential equation equivalent to the Monge equation. To the two equations

$$
\begin{aligned}
(3) \qquad & u = \phi(x, y, a, w(a)), \\
& 0 = \phi_a(x, y, a, w(a)) + \phi_b(x, y, a, w(a))w'(a)
\end{aligned}
$$

which represent the family of characteristic curves depending on the parameter a on the integral surface (cf. page 84), we must add the equation

$$
(4) \qquad \phi_{aa} + 2\phi_{ab}w'(a) + \phi_{bb}w'^2(a) + \phi_b w''(a) = 0,
$$

obtained by differentiating with respect to a, which expresses the formation of the envelope. These three equations represent a space curve in the parameter a, namely an envelope of characteristics. They constitute the desired solution of the Monge equation once u and y are expressed as functions of x by the process of elimination. To represent all solutions in the form (3), (4) of a given ordinary underdetermined "diophantine" differential equation (2) one first replaces it by the equivalent partial differential equation $F = 0$, and then has to find a complete integral.

§6. *Examples*

The theory just developed will now be illustrated by a few examples, some of which are significant in themselves.

1. The Differential Equation of Straight Light Rays, (grad $u)^2 = 1$. We consider the differential equations

$$(1) \qquad u_x^2 + u_y^2 = 1$$

for a function $u(x, y)$ and

$$(2) \qquad u_x^2 + u_y^2 + u_z^2 = 1$$

for a function $u(x, y, z)$. These differential equations occur, for example, in geometrical optics. The surfaces $u = $ const. represent the *wave fronts* and the characteristics the *light rays*; more generally, the differential equation

$$(3) \qquad u_x^2 + u_y^2 + u_z^2 = n(x, y, z)$$

describes the wave fronts of light in an inhomogeneous medium with a variable index of refraction $n(x, y, z)$.

We shall first discuss the case of two independent variables for which we obtained (see Ch. I, §3) the complete integral

$$(4) \qquad u = ax + \sqrt{1 - a^2}\, y + b$$

and for which the equations

$$(4') \qquad \begin{aligned} u &= ax + \sqrt{1 - a^2}\, y + w(a), \\ 0 &= x - \frac{a}{\sqrt{1 - a^2}}\, y + w'(a) \end{aligned}$$

represent a solution containing an arbitrary function. The two equations

$$u = ax + \sqrt{1 - a^2}\, y + b,$$

$$0 = x - \frac{a}{\sqrt{1 - a^2}}\, y + c$$

and the corresponding relations

$$p = a, \qquad q = \sqrt{1 - a^2}$$

define a family of characteristic strips which depends on the three parameters a, b, c; along these strips we may consider x as the independent variable. The characteristic curves, the "light rays", are straight lines, and along any one of these lines the corresponding tangent plane remains fixed. Both the characteristic lines and the corresponding planes form an angle of 45° with the x, y-plane and are defined by this fact. The Monge cone at the point (x_0, y_0, u_0) is obviously

$$(x - x_0)^2 + (y - y_0)^2 = (u - u_0)^2.$$

The characteristic differential equations, which are solved by the above relations, may be written as

(5) $$dx:dy:du:dp:dq = p:q:1:0:0$$

and can be integrated; we immediately obtain

$$p = p_0, \qquad q = q_0, \qquad u = s + u_0,$$

$$x = p_0 s + x_0, \qquad y = q_0 s + y_0,$$

where the appropriate initial values for the parameter value $s = 0$ are denoted by x_0, y_0, u_0, p_0, q_0.

By eliminating p and q from $p^2 + q^2 = 1$, $y' = q/p$, $u' = 1/p$, we obtain for the functions $u(x)$ and $y(x)$ the Monge equation which belongs to our partial differential equation:

(6) $$\left(\frac{du}{dx}\right)^2 - \left(\frac{dy}{dx}\right)^2 = 1.$$

Its solutions are those curves whose tangents form an angle of 45° with the x, y-plane at all points. These *focal* or "*caustic*" curves (which are the involutes of the curves $u = $ const.) can be represented without integrals by means of an arbitrary function $w(a)$ in the following way:

$$u = ax + \sqrt{1 - a^2}\, y + w(a),$$

(7)
$$0 = x - \frac{a}{\sqrt{1 - a^2}}\, y + w'(a),$$

$$0 = -\frac{1}{\sqrt{(1 - a^2)^3}}\, y + w''(a).^{[1]}$$

These focal curves can be used to characterize the solutions[2] of the differential equation (1) as developable tangent surfaces with an edge of regression which is a focal curve, i.e., a curve whose tangents form an angle of 45° with the x, y-plane.

The solutions of (1) have another important geometric meaning. Consider the family of curves $u(x, y) = c = $ const. in the x, y-plane. We assert: *The value of the function* $u(x, y)$ *at any point in the plane is equal to the distance of this point from the curve* $u(x, y) = 0$. The curves $u(x, y) = c$ are equidistant and parallel to $u(x, y) = 0$ at a distance c; the orthogonal trajectories to the family of curves are straight lines (namely, the projections of our characteristic curves) and the envelope of these straight lines, the common evolute of the curves $u = $ const., is the projection of the edge of regression, i.e., of the focal curve.

This fact may be proved, for example, by solving the initial value problem for a given initial curve $G(x_0, y_0) = 0$ along which the initial condition $u = 0$ is stipulated. To construct a solution we consider the characteristic passing through each point (x_0, y_0) on the initial curve, namely $x = p_0 s + x_0$, $y = q_0 s + y_0$, $u = s$. Since

$$p_0^2 + q_0^2 = 1,$$

s represents the distance from the point (x, y) to the point (x_0, y_0) on the projection of this straight line. To determine p_0 and q_0, we note that if we consider x_0 as the independent parameter along the initial curve, then $du_0/dx_0 = p_0 + q_0\, dy_0/dx_0$. This equation and the initial condition $G_{x_0} + G_{y_0}\, dy_0/dx_0 = 0$ therefore imply $p_0\, G_{y_0} - q_0\, G_{x_0} = 0$. Thus the projection of the above characteristic line is orthogonal

---

[1] Direct verification of this representation by computations is left as an exercise.

[2] Except for the planes (4) and the integral conoids, i.e., the right circular cones whose generators are inclined to the x, y-plane at an angle of 45°.

to the initial curve. Actually, u is then—at least in a sufficiently small neighborhood of the initial curve—the distance of the point (x, y) from the curve $G(x_0, y_0)$. From these remarks it follows immediately that every curve $u = $ const. is again orthogonal to our straight line.

The proof appears in a slightly different form if we start with the following problem: Find the orthogonal trajectories to a given curve $u = $ const. These trajectories are characterized by the system of ordinary differential equations

$$(8) \qquad \frac{dx}{ds} = u_x, \qquad \frac{dy}{ds} = u_y.$$

Squaring and adding the two equations we obtain

$$(9) \qquad \left(\frac{dx}{ds}\right)^2 + \left(\frac{dy}{ds}\right)^2 = 1;$$

thus we see that s represents arc lengths on the trajectories. Differentiating the first of the differential equations (8) with respect to s and considering again the same differential equations, we obtain $d^2x/ds^2 = u_{xx}u_x + u_{xy}u_y$. The right side is identically zero, as can be seen by differentiating the partial differential equation (1) with respect to x. Similarly we have $d^2y/ds^2 = 0$, which shows that the trajectories are straight lines.

In the case of three independent variables the same method shows that the solutions of the partial differential equation (2) are given by a family of equidistant surfaces $u(x, y, z) = $ const. which are parallel to an arbitrary initial surface $G(x, y, z) = 0$. These surfaces possess rectilinear orthogonal trajectories, and the part of these straight lines which lies between the surfaces $u = c_1$ and $u = c_2$ has the constant length $c_1 - c_2$; u itself is the distance of the point (x, y, z) from the initial surface.

2. The Equation $F(u_x, u_y) = 0$. We consider now the differential equation

$$(10) \qquad p \cdot q = \tfrac{1}{2} \qquad \qquad (p = u_x, q = u_y).[1]$$

The equivalent Monge equation for $y(x)$ and $u(x)$ has the form

[1] By the transformation (rotation) $\omega = u, \eta = (x - y)/\sqrt 2, \xi = (x + y)/\sqrt 2$, our differential equation goes over into $\omega_\xi^2 - \omega_\eta^2 = 1$ for $\omega(\xi, \eta)$, which can be treated in a way similar to that used above.

$$(11) \qquad\qquad u'^2 = 2y'.$$

A complete integral containing all surface elements of the differential equation is given by

$$(12) \qquad\qquad u = ax + \frac{1}{2a}\, y + b,$$

which yields the family of solutions

$$u = ax + \frac{1}{2a}\, y + w(a),$$

$$(13)$$

$$0 = x - \frac{1}{2a^2}\, y + w'(a)$$

depending on an arbitrary function $w(a)$. Finally, if we apply

$$(13') \qquad\qquad 0 = \frac{1}{a^3}\, y + w''(a),$$

we obtain a representation free of integrals for the focal curves in terms of an arbitrary function w.

The totality of the characteristics is represented by the equations

$$u = ax + \frac{1}{2a}\, y + b,$$

$$0 = x - \frac{1}{2a^2}\, y + c$$

with the three parameters a, b, c and with x as independent variable. The characteristic equations are given by

$$(14) \qquad\qquad dx:dy:du:dp:dq = q:p:1:0:0.$$

Thus the characteristics are again straight lines and the corresponding tangent planes remain fixed along each entire line. Hence the equations of the characteristic lines are

$$(15) \qquad\qquad y = \frac{p_0}{q_0}\, x + y_0, \qquad u = \frac{1}{q_0}\, x + u_0.$$

Finally, we solve the initial value problem for the initial values $u(0, y_0) = u_0 = v(y_0)$, which we assume to be arbitrarily given. We immediately have

$$q(0, y_0) = v'(y_0), \qquad p(0, y_0) = \frac{1}{2v'(y_0)}.$$

Thus, the equations

(16)

$$u = \frac{1}{v'(y_0)} x + v(y_0),$$

$$y = \frac{1}{2v'(y_0)^2} x + y_0$$

represent the solution of the initial value problem if y_0 is expressed in terms of x and y by means of the second equation and substituted into the first. Comparison with the solution

$$u = 2ax + aw'(a) + w(a),$$

$$y = 2a^2x + 2a^2w'(a)$$

given by (13) shows that the solutions can actually be transformed into each other in the following way: We introduce a new parameter y_0 in place of a by the equation $y_0 = 2a^2w'(a)$, and then obtain a new function $v(y_0)$ from $w(a)$ by the equation

$$v(y_0) = (aw(a))' = aw'(a) + w(a).$$

Because of the relations

$$\frac{dy_0}{da} = 2a[2w'(a) + aw''(a)] = 2a(aw(a))'',$$

$$v'(y_0) = \frac{dv(y_0)}{dy_0} = (aw(a))'' \frac{da}{dy_0} = \frac{1}{2a},$$

the two representations now go over into each other.

The two preceding examples are special cases of the general differential equation

(17) $$F(u_x, u_y) = 0,$$

for which similar relations hold. From the characteristic equations

(18) $$dx:dy:du:dp:dq = F_p:F_q:(pF_p + qF_q):0:0$$

we recognize, as above, that the characteristic strips consist of straight lines with only one corresponding tangent plane and, as a result, that the solutions are developable surfaces. This becomes even clearer if we note that it is possible to construct a complete integral which consists entirely of planes. For this purpose we suppose that the equation $F(p, q) = 0$ is satisfied by two functions $p(a)$ and $q(a)$, where a is a parameter. We obtain the complete solution

$$u = p(a)x + q(a)y + b,$$

which consists entirely of planes.

3. Clairaut's Differential Equation.[1] Again we consider Clairaut's differential equation

$$(19) \qquad u = xu_x + yu_y + f(u_x, u_y).$$

In Ch. I, §4, we found the family of planes

$$(20) \qquad u = ax + by + f(a, b)$$

as a complete integral. The solutions obtained by forming envelopes from the complete integral are given by the formulas

$$(21) \qquad \begin{aligned} u &= ax + w(a)y + f(a, w(a)), \\ 0 &= x + w'(a)y + f_a + f_b w'; \end{aligned}$$

these solutions represent developable surfaces. In the same way we find that all integral surfaces generated by families of characteristics are developable; from the characteristic equations

$$(22) \qquad \begin{aligned} &dx:dy:du:dp:dq \\ &= (x + f_p):(y + f_q):(px + qy + pf_p + qf_q):0:0 \end{aligned}$$

we deduce, as above (cf. articles 1, 2), that the characteristic strips consist of straight lines, each with only one corresponding tangent plane.

The singular solution of the differential equation (19) has already been considered (cf. Ch. I, §§4, 6). Assuming $f_{aa}f_{bb} - f_{ab}^2 \neq 0$, we obtain it by solving the equations

$$x = -f_a, \qquad y = -f_b$$

for a and b and substituting them into

$$u = ax + by + f(a, b);$$

---

[1] See Ch. I, §§4 and 6.

in tangent plane coordinates ξ, η, ω, this solution is represented simply by the support function

$$(23) \qquad \omega = -f(\xi, \eta).$$

All solutions can now be simply related to the singular solution. We note that the planes of the complete integral are the same as the tangent planes of the singular solution and that the characteristics are tangents. Hence the totality of solutions of the Clairaut differential equation consists of those developable surfaces which are tangent to the singular solution. The initial value problem may, therefore, be easily solved, by choosing the planes which touch both the initial curve and the singular solution, and forming their envelopes.

From the differential equation it could have been shown directly that the Monge cone through a point P is the cone which has P as vertex and is tangent to the singular solution. Moreover, *the Monge cone is also an integral conoid*.

4. Differential Equation of Tubular Surfaces. An instructive example is afforded by the differential equation of the tubular surfaces mentioned in Ch. I, §4:

$$(24) \qquad u^2(p^2 + q^2 + 1) = 1.$$

The family of spheres

$$(25) \qquad (x - a)^2 + (y - b)^2 + u^2 = 1$$

is a complete integral for equation (24). It is geometrically obvious that the characteristics are the great circles of the spheres parallel to the u-axis.

Analytically this fact follows from the characteristic differential equations

$$(26) \qquad dx:dy:du:dp:dq = u^2 p:u^2 q:(1 - u^2): -\frac{p}{u}: -\frac{q}{u},$$

from which we find

$$d(x + up) = d(y + uq) = d\left(\frac{p}{q}\right) = 0.$$

We obtain at once the equations

$$x - a = -up, \qquad y - b = -uq, \qquad p = cq,$$

where a, b, c are constants of integration. From these equations and from relation (24) we have the equations

$$(x - a)^2 + (y - b)^2 + u^2 = 1$$

and $(x - a)/(y - b) = c$; thus the characteristic curves are the circles specified above. Moreover, from the relations

$$(x - a):(y - b):u = p:q: - 1,$$

it follows that the normal to a tangent plane at a point of a circle points toward the center of the circle.

The remaining integral surfaces are envelopes of one-parameter families of spheres of radius 1 whose center moves along a curve in the x, y-plane. If the curvature of this curve, the axis of the tubular surface, is less than that of the unit circle, then the tubular surface will actually have the form of a tube. Its characteristic circles then possess no envelope. They do, however, if the radius of curvature of the axis is less than unity; then they generate an edge of regression on the integral surface. These edges of regression are the focal curves of our differential equation. The projection of these focal curves on the x, y-plane is the evolute of the core of our "tube". These relationships can easily be visualized by means of concrete examples or models.

5. Homogeneity Relation. As the last example we consider the homogeneity relation (cf. Ch. I, §1)

$$(27) \qquad\qquad px + qy = hu,$$

where h is a constant. Integrating the characteristic differential equations

$$(28) \qquad\qquad dx:dy:du = x:y:hu$$

we obtain the equations

$$(29) \qquad\qquad \frac{u}{x^h} = a \quad \text{and} \quad \frac{x}{y} = b.$$

Hence the general solution of the differential equation is given by $u = x^h V(y/x)$, where V is an arbitrary function, or by $u = y^h v(y/x)$, where v is an arbitrary function, that is, by a homogeneous function of the h-th degree in x and y.

A different representation of the general solution is obtained from the complete integral

$$u = ax^h + by^h,$$

by means of the equations

$$u = ax^h + w(a)y^h,$$

$$0 = x^h + w'(a)y^h.$$

Since the second equation yields a as a function of the quotient x/y, we again obtain a general homogeneous function of the h-th degree for u.

§7. General Differential Equation in n Independent Variables

The theory for the general first order partial differential equation

$$(1) \qquad F(x_1, x_2, \cdots, x_n, u, p_1, p_2, \cdots, p_n) = 0 \qquad \left(p_i = \frac{\partial u}{\partial x_i} \right)$$

in n independent variables is analogous to that for $n = 2$. Therefore we shall not repeat its geometric motivation but mainly discuss the role of the characteristic strips.

With the differential equation $F = 0$ we associate, in analogy with §3, the system of ordinary differential equations

$$(2) \qquad \frac{dx_i}{ds} = F_{p_i}, \qquad \frac{du}{ds} = \sum_{i=1}^{n} p_i F_{p_i}, \qquad \frac{dp_i}{ds} = -(F_u p_i + F_{x_i})$$

for the $2n + 1$ functions x_i, u, p_i of a parameter s. The system (2) is called the system of *characteristic differential equations* belonging to the partial differential equation (1).

$F(x_i, u, p_i)$ is an integral of this system since

$$\frac{dF}{ds} = \sum_{i=1}^{n} F_{x_i} \frac{dx_i}{ds} + \sum_{i=1}^{n} F_{p_i} \frac{dp_i}{ds} + F_u \frac{du}{ds} = 0$$

for every solution of (2).

All solutions of (2) which also satisfy the condition $F = 0$ are called *characteristic strips*. These strips form a $(2n - 1)$-parameter family. Moreover, precisely as in the case of two variables, *infinitely many characteristic strips lie on every integral surface*

$$u(x_1, x_2, \cdots, x_n)$$

of the differential equation $F = 0$. *Every characteristic strip which has an element, i.e., a system of values* x_i, u, p_i, *in common with an integral surface lies wholly on the integral surface.*

As in §3 we have the following initial value problem: Let an $(n - 1)$ dimensional *initial manifold* C be given by continuously differentiable functions x_1, x_2, $\cdots$, x_n, u of the parameters t_1, t_2, $\cdots$, t_{n-1}, so that the matrix of derivatives $\partial x_i/\partial t_k$ has rank $n - 1$. Let this manifold C be extended to a *strip manifold* C_1 by the further prescription of n functions p_1, p_2, $\cdots$, p_n of the parameters t_i satisfying the *strip conditions*

$$(3) \qquad\qquad u_{t_\nu} = \sum_{i=1}^{n} p_i \frac{\partial x_i}{\partial t_\nu} \qquad (\nu = 1, 2, \cdots, n - 1)$$

identically in the t_ν. Moreover, let the strip quantities satisfy the equation $F(x_1, x_2, \cdots, x_n, u, p_1, p_2, \cdots, p_n) = 0$ identically in the t_i. *Our aim is to find an integral manifold*

$$u = u(x_1, x_2, \cdots, x_n),$$

i.e., a solution of the differential equation $F = 0$, which contains the given initial manifold C_1.

To solve this problem we consider the family of characteristic strips with parameter s whose initial element, for $s = 0$, lies in the given initial strip manifold C_1; i.e., we consider those solutions

$$x_i(s, t_1, t_2, \cdots, t_{n-1}),$$

$$u(s, t_1, t_2, \cdots, t_{n-1}),$$

$$p_i(s, t_1, t_2, \cdots, t_{n-1})$$

of the characteristic differential equations which, for $s = 0$, go over into the given functions of the t_ν. If the Jacobian

$$(4) \qquad\qquad \frac{\partial(x_1, x_2, \cdots, x_n)}{\partial(s, t_1, t_2, \cdots, t_{n-1})},$$

which, because of (2), coincides with

$$\Delta = \begin{vmatrix} F_{p_1} & \cdots & F_{p_n} \\ \dfrac{\partial x_1}{\partial t_1} & \cdots & \dfrac{\partial x_n}{\partial t_1} \\ \cdots\cdots\cdots\cdots\cdots \\ \dfrac{\partial x_1}{\partial t_{n-1}} & \cdots & \dfrac{\partial x_n}{\partial t_{n-1}} \end{vmatrix},$$

does not vanish along the initial manifold C_1 (i.e., for $s = 0$), and therefore does not vanish in a neighborhood of C_1, then the quantities s, t_1, $\cdots$, t_{n-1} can be expressed in that neighborhood in terms of x_1, x_2, $\cdots$, x_n; by substituting these expressions in

$$u(s, t_1, t_2, \cdots, t_{n-1}),$$

we obtain a uniquely determined surface $u = u(x_1, x_2, \cdots, x_n)$ which contains the initial manifold C_1. We have now to show that this function u is a solution of our initial value problem. If the solutions of the characteristic differential equations are substituted for x_i, u, p_i, the quantity $F(x_i, u, p_i)$ vanishes identically on the surface $u = u(x_1, x_2, \cdots, x_n)$. Hence we merely have to show that

$$p_i = \partial u / \partial x_i$$

everywhere on this surface. The verification is similar to that for two independent variables (cf. §3, 2) and may be omitted here.

It remains to discuss the exceptional case: $\Delta = 0$ identically on C_1. As in §2, we may again conclude from the relation $\Delta = 0$ that $n - 1$ functions λ_1, λ_2, $\cdots$, λ_{n-1} exist for which the linear relations

$$(5) \qquad F_{p_i} = \sum_{\nu=1}^{n-1} \lambda_\nu \frac{\partial x_i}{\partial t_\nu}$$

are valid along C_1.

Our object is to find under which additional conditions the initial value problem can be solved in this case. Here again the situation is illuminated by the concept of the *characteristic manifold*, which we shall now define and analyze. In contrast to the quasi-linear case (see §2) where a characteristic manifold was an $(n - 1)$-dimensional manifold C in $(n + 1)$-dimensional x, u-space, we must now consider $(n - 1)$-dimensional strip-manifolds C_1 characterized by $2n + 1$ quantities x_i, u, p_i which may therefore be interpreted in a $(2n + 1)$-dimensional x, u, p-space.

With every point of the $(2n + 1)$-dimensional x, u, p-space (or with every surface element of the $(n + 1)$-dimensional x, u-space) we now associate the system of $2n + 1$ quantities

$$
\begin{aligned}
a_i &= F_{p_i} \\
b_i &= -p_i F_u - F_{x_i} \\
a &= \sum_{\nu=1}^{n} p_\nu F_{p_\nu} = \sum_{\nu=1}^{n} p_\nu a_\nu
\end{aligned}
\qquad (i = 1, 2, \cdots, n)
$$

(6)

as components of a *characteristic strip vector*. Moreover, we adopt the following definition: *An $(n-1)$-dimensional strip-manifold C_1 for which the relation $F(x_i, u, p_i) = 0$ holds is called a characteristic strip-manifold if the characteristic strip vector is tangent to it at every point.* This geometric definition, based on an interpretation in $(2n+1)$-dimensional space, may be formulated analytically by requiring that the vector (a_i, a, b_i) depend linearly on the $n-1$ independent vectors $\partial x_i/\partial t_\nu$, $\partial u/\partial t_\nu$, $\partial p_i/\partial t_\nu$ $(\nu = 1, 2, \cdots, n-1)$, which are by definition tangent to C_1. For the $(n-1)$-dimensional manifold C_1 given by the functions x_i, u, p_i of the parameters $t_1, t_2, \cdots, t_{n-1}$ let both the relation

$$(7) \qquad\qquad F(x_i, u, p_i) = 0$$

and the strip relations

$$(8) \qquad\qquad \frac{\partial u}{\partial t_\nu} = \sum_{i=1}^{n} p_i \frac{\partial x_i}{\partial t_\nu} \qquad (\nu = 1, 2, \cdots, n-1)$$

hold identically in the t_ν. The strip-manifold is then called characteristic if there exist $n-1$ functions $\lambda_\nu(t_1, t_2, \cdots, t_{n-1})$ such that the linear relations

$$(9) \qquad\qquad X_i \equiv a_i - \sum_{\nu=1}^{n-1} \lambda_\nu \frac{\partial x_i}{\partial t_\nu} = 0,$$

$$(10) \qquad\qquad U \equiv a - \sum_{\nu=1}^{n-1} \lambda_\nu \frac{\partial u}{\partial t_\nu} = 0,$$

$$(11) \qquad\qquad P_i \equiv b_i - \sum_{\nu=1}^{n-1} \lambda_\nu \frac{\partial p_i}{\partial t_\nu} = 0$$

are satisfied.[1]

[1] These $2n+1$ relations between the quantities x_i, u, p_i are not independent of each other, since, as can easily be verified,

$$U = \sum_{i=1}^{n} p_i X_i$$

and, in addition,

$$\frac{\partial F}{\partial t_\rho} = -\sum_{\nu=1}^{n-1} \lambda_\nu \left(\frac{\partial U_\nu}{\partial t_\rho} - \frac{\partial U_\rho}{\partial t_\nu} \right) + F_u U_\rho + \sum_{i=1}^{n} \left(X_i \frac{\partial p_i}{\partial t_\rho} - P_i \frac{\partial x_i}{\partial t_\rho} \right),$$

$$(\rho = 1, 2, \cdots, n-1),$$

The following two theorems again hold:

Every characteristic strip-manifold C_1 is generated by an $(n - 2)$-parameter family of characteristic strips lying wholly within C_1 .

Every characteristic strip which has an initial element in common with a characteristic strip-manifold lies wholly in that manifold.

To prove the above theorems we again, as in §2, define the curves $t_i(s)$ on the $(n - 1)$-dimensional t_i manifold by the system of ordinary differential equations

$$(12) \qquad \frac{dt_\nu}{ds} = \lambda_\nu(t_1, t_2, \cdots, t_{n-1}) \qquad (\nu = 1, 2, \cdots, n - 1).$$

These curves form an $(n - 2)$-parameter family which generates the t_i manifold. Along these curves a one-dimensional strip lying in C_1 is defined by $x_i(t_\nu)$, $u(t_\nu)$, $p_i(t_\nu)$ which after the substitution $t_\nu = t_\nu(s)$ is of the form $x_i(s)$, $u(s)$, $p_i(s)$; we have now to prove that this strip is a characteristic strip of our original partial differential equation. Recalling relations (9), (10), (11) we have, in fact,

$$\frac{dx_i}{ds} = \sum_{\nu=1}^{n-1} \frac{\partial x_i}{\partial t_\nu} \lambda_\nu = a_i = F_{p_i}, \qquad \frac{du}{ds} = \sum_{\nu=1}^{n-1} \frac{\partial u}{\partial t_\nu} \lambda_\nu = a = \sum_{i=1}^{n} p_i F_{p_i},$$

$$\frac{dp_i}{ds} = \sum_{\nu=1}^{n-1} \frac{\partial p_i}{\partial t_\nu} \lambda_\nu = b_i = -F_{x_i} - p_i F_u.$$

Thus our functions are solutions of the system of characteristic equations (2) and therefore, since $F = 0$, define a characteristic strip. The $(n - 2)$-parameter family of such strips covers C_1 .

By arguments completely analogous to those used for quasi-linear equations (see §2), the second theorem now follows from the fact that the solutions of the system of characteristic differential equations are uniquely determined by the initial values.

---

where

$$U_\nu = \frac{\partial u}{\partial t_\nu} - \sum_{i=1}^{n} p_i \frac{\partial x_i}{\partial t_\nu} .$$

Thus it follows that, in addition to $F = 0$ and the relation $\partial F/\partial t_\rho = 0$ obtained from it, the strip conditions (8) and the conditions (9), only one of the n conditions (11) need be imposed in order to assure that (10) and the other $n - 1$ conditions of (11) hold. However, as is often the case in geometry and analysis, it is advantageous for reasons of symmetry to retain the above system of dependent relations.

After this analysis of characteristic manifolds we may state and prove the complete results as we did in the quasi-linear case:

The initial value problem for a given initial manifold C_1 possesses one and only one solution if $\Delta \neq 0$ everywhere on C_1. If, however, the relation $\Delta = 0$ holds along C_1, then a necessary and sufficient condition for the solvability of the initial value problem is that C_1 be a characteristic manifold. In this case there are infinitely many solutions.

We have only to prove the assertions for the case $\Delta = 0$. In this case we may immediately infer the existence of $n - 1$ functions

$$\lambda_\nu(t_1, t_2, \cdots, t_{n-1})$$

such that the relations (9) are satisfied. If we now assume that $u = u(x_1, x_2, \cdots, x_n)$ represents an integral surface J through C_1 with $p_i = \partial u/\partial x_i$, then the required relations (10), (11) which assure that C_1 is a characteristic manifold, follow immediately. Since $p_i = u_{x_i}$ and since (9) holds, we have

$$a = \sum_{i=1}^{n} p_i F_{p_i} = \sum_{i=1}^{n} u_{x_i} \sum_{\nu=1}^{n-1} \lambda_\nu \frac{\partial x_i}{\partial t_\nu} = \sum_{\nu=1}^{n-1} \lambda_\nu \frac{\partial u}{\partial t_\nu};$$

hence relation (10) is established. We now use the fact that u must satisfy the relation

$$(13) \qquad \sum_{i=1}^{n} F_{p_i} \frac{\partial p_k}{\partial x_i} + F_u p_k + F_{x_k} = 0$$

identically in the x_i which, when we recall $\partial p_k/\partial x_i = \partial p_i/\partial x_k$, is obviously the partial differential equation (1) differentiated with respect to x_k. Applying (12) we obtain

$$\sum_{\nu=1}^{n-1} \lambda_\nu \frac{\partial p_k}{\partial t_\nu} = -F_u p_k - F_{x_k} = b_k,$$

i.e., the required equation (11).

Thus we have proved that if the initial value problem for C_1 can be solved, then C_1 is a characteristic strip-manifold.

That this property is also sufficient for solvability follows exactly as it did in the quasi-linear case. Let any manifold C_1' not tangent to C_1 be constructed which has an $(n - 2)$-dimensional manifold S in common with C_1 and for which the condition $\Delta \neq 0$ is everywhere satisfied; the initial value problem for C_1' is then uniquely solved by an integral surface J'. All characteristic strips which pass through

S and the manifold C_1 generated by these strips lie on J'. Since C_1' is chosen arbitrarily, infinitely many solutions of the initial value problem for C_1 exist.

In concluding this section we once more emphasize that it deals with situations in the small only. Later in Chapter VI we shall have to consider solutions in their full extension, including singularities and multiple-valuedness; this discussion in the large will entail a much greater effort than our local analysis.

§8. Complete Integral and Hamilton-Jacobi Theory

1. Construction of Envelopes and Characteristic Curves. Consider the partial differential equation

$$(1) \qquad F(x_1, x_2, \cdots, x_n, u, p_1, p_2, \cdots, p_n) = 0$$

with $p_i = \partial u / \partial x_i$ and $\sum_i F_{p_i}^2 \neq 0$. With a specific solution

$$(2) \qquad u = \phi(x_1, x_2, \cdots, x_n, a_1, a_2, \cdots, a_n)$$

depending on n parameters a_i (a complete integral), assume that the determinant condition

$$(3) \qquad D = |\phi_{x_i a_k}| \neq 0$$

is satisfied in the part of the x, u-space considered.[1] Then the envelope of an arbitrary $(n - 1)$-parameter family of these solutions is also a solution. To prove this we set

$$a_i = \omega_i(t_1, t_2, \cdots, t_{n-1}) \qquad (i = 1, 2, \cdots, n)$$

where the ω_i are arbitrary functions of the $n - 1$ parameters t_k. The envelope is obtained by calculating $t_1, t_2, \cdots, t_{n-1}$ from the equations

$$(4) \qquad 0 = \sum_{i=1}^{n} \phi_{a_i} \frac{\partial \omega_i}{\partial t_\nu} \qquad (\nu = 1, 2, \cdots, n - 1)$$

---

[1] We might, as before (cf. Ch. I, §4, 2), impose the more general condition that the n-rowed matrix

$$\begin{pmatrix} \phi_{a_1} & \phi_{x_1 a_1} & \cdots & \phi_{x_n a_1} \\ \cdots\cdots\cdots\cdots\cdots\cdots\cdots\cdots\cdots \\ \phi_{a_n} & \phi_{x_1 a_n} & \cdots & \phi_{x_n a_n} \end{pmatrix}$$

has rank n.

as functions of x_1, x_2, $\cdots$, x_n and substituting these t_ν into

$$u = \phi(x_1, \cdots, x_n, \omega_1(t_1, \cdots, t_{n-1}), \cdots, \omega_n(t_1, \cdots, t_{n-1})).$$

The *curves of contact* of a surface given by the complete integral with the envelope will prove to be *characteristic curves*. Such a contact curve corresponds to a fixed system of quantities t_ν, $\partial \omega_i / \partial t_\nu$ and a_i; furthermore relations (4) hold along the curve, which imply, to within a common proportionality factor λ, certain constant values b_i for the ϕ_{a_i}:

$$(5) \qquad\qquad \phi_{a_i} = \lambda b_i.$$

Using this equation we may assign values λb_i corresponding to given values of the quantities x_i and a_i; then, because of condition (3), we can solve (5) uniquely for the x_i in a neighborhood of the system of values considered and obtain functions

$$x_i(a_1, a_2, \cdots, a_n, b_1, b_2, \cdots, b_n, \lambda).$$

If these functions are substituted into

$$\phi(x_1, x_2, \cdots, x_n, a_1, a_2, \cdots, a_n)$$

we have a curve represented in terms of the parameter λ. Since in the neighborhood considered, any desired values can be given to the quantities a_i and b_i by an appropriate choice of the functions ω_i, we thus obtain a $2n$-parameter family of contact curves of our envelope to our complete integral. These curves are characteristic curves of our partial differential equation (1) and, together with

$$p_i = \phi_{x_i}(x_\kappa(a_\nu, b_\nu, \lambda), a_\kappa),$$

produce characteristic strips. This follows from the geometric meaning of our strips as strips of contact.

To prove the assertion analytically we differentiate equation (5) with respect to the curve parameter λ:

$$(6) \qquad\qquad \sum_{\kappa=1}^{n} \phi_{a_i x_\kappa} x_\kappa' = b_i,$$

where differentiation with respect to λ is denoted by a prime. On the other hand, the differential equation (1) is identically satisfied in the x_i and a_i by $\phi(x_1, x_2, \cdots, x_n, a_1, a_2, \cdots, a_n)$; if we differentiate it with respect to a_i and use (5) we obtain

$$(7) \qquad \sum_{\kappa=1}^{n} \phi_{x_\kappa a_i} F_{p_\kappa} + F_u \lambda b_i = 0.$$

Thus the quantities $-(F_{p_\kappa}/\lambda F_u)$ satisfy the same system of inhomogeneous equations as the x_κ' ; since the determinant of this system does not vanish, we may conclude that

$$x_\kappa' = -\frac{F_{p_\kappa}}{\lambda F_u}.$$

If we set the expression $-1/\lambda F_u$ (which is different from zero) equal to ρ, then

$$(8) \qquad x_\kappa' = \rho F_{p_\kappa}.$$

Furthermore, if we differentiate (1) with respect to x_κ , we have

$$F_u p_\kappa + \sum_{i=1}^{n} F_{p_i} \frac{\partial p_i}{\partial x_\kappa} + F_{x_\kappa} = 0$$

or, since, in view of (8),

$$\sum_{i=1}^{n} F_{p_i} \frac{\partial p_i}{\partial x_\kappa} = \frac{1}{\rho} \sum_{i=1}^{n} \frac{\partial p_i}{\partial x_\kappa} x_i' = \frac{1}{\rho} \sum_{i=1}^{n} \frac{\partial^2 u}{\partial x_i \, \partial x_\kappa} x_i'$$

$$= \frac{1}{\rho} \sum_{i=1}^{n} \frac{\partial p_\kappa}{\partial x_i} x_i' = \frac{1}{\rho} p_\kappa',$$

we obtain

$$p_\kappa' = -\rho(F_u p_\kappa + F_{x_\kappa}).$$

Finally, it follows from (8) that

$$u' = \sum_{i=1}^{n} u_{x_i} x_i' = \rho \sum_{i=1}^{n} p_i F_{p_i}.$$

Since the curve parameter λ can be so chosen that $\rho = 1$, the characteristic equations (2) of §7 are satisfied by the curves considered.[1]

[1] Note, incidentally, that one can obtain solutions also in other ways by the construction of envelopes from a solution $\phi(x_1 , x_2 , \cdots , x_n , a_1 , a_2 , \cdots , a_n)$ depending on arbitrary parameters. For example, one can construct the envelope of the n-parameter family (2) and thus again arrive at a singular solution which, as in the case $n = 2$, can also be obtained by differentiation and elimination processes from the relations $F = F_{p_i} = 0$. Or, one can select any m-parameter family with $m < n$ from the n-parameter family by means of arbitrary functions and construct its envelope. The manifolds of contact will in this case be characteristic manifolds of dimension $n - m$.

2. Canonical Form of the Characteristic Differential Equations.   A more transparent form can be given to the theory of partial differential equations of the first order and the calculations of article 1 can be simplified if the dependent variable u does not appear explicitly in the differential equation.   An arbitrary differential equation can always be brought into this special form by increasing artificially the number of independent variables by one.

For this purpose we need only introduce, e.g. (cf. Ch. I, §5), $u = x_{n+1}$ as an independent variable and express a family of solutions

$$u = \psi(x_1, x_2, \cdots, x_n, c)$$

implicitly in the form

$$\phi(x_1, x_2, \cdots, x_{n+1}) = c.$$

If we replace u_{x_i} by $-\phi_{x_i}/\phi_{x_{n+1}}$ $(i = 1, 2, \cdots, n)$, then we obtain for the new unknown function ϕ a differential equation which does not depend on ϕ explicitly.

For such a differential equation we single out a variable, e.g. $x_{n+1} = x$, and assume the differential equation solved for the derivative of ϕ with respect to this variable.   Thus, if instead of ϕ we again write u, we may without loss of generality consider differential equations of the form

$$
(9) \quad
\begin{aligned}
& p + H(x_1, x_2, \cdots, x_n, x, p_1, p_2, \cdots, p_n) = 0 \\
& p = u_x, \qquad p_i = u_{x_i} \qquad (i = 1, 2, \cdots, n)
\end{aligned}
$$

for a function u of the $n + 1$ variables $x, x_1, x_2, \cdots, x_n$.

Then the system of characteristic differential equations, one of which is $dx/ds = 1$ (or $x = s$), goes over into the system

$$
(10) \qquad \frac{dx_i}{dx} = H_{p_i}, \qquad \frac{dp_i}{dx} = -H_{x_i} \qquad (i = 1, 2, \cdots, n);
$$

moreover,

$$
(11) \qquad \frac{du}{dx} = \sum_{i=1}^{n} p_i H_{p_i} - H, \qquad \frac{dp}{dx} = -H_x
$$

holds.   The equations (10) alone form a system of $2n$ differential equations for $2n$ quantities x_i, p_i.   If the functions $x_i(x)$ and $p_i(x)$ are solutions of (10) then $p(x)$ and $u(x)$ are obtained from equations (11) by simple integrations.

In mechanics and in the calculus of variations (cf. Vol. I, Ch. IV, §9 and the present chapter, §9) one is often led to differential equations of the form (10). A system of ordinary differential equations (10)

$$\frac{dx_i}{dx} = H_{p_i}, \qquad \frac{dp_i}{dx} = -H_{x_i}$$

associated with a function $H(x_1, x_2, \cdots, x_n, x, p_1, p_2, \cdots, p_n)$ of $2n + 1$ variables is called a *canonical system of differential equations*. The results of this article imply that the integration of the partial differential equation (9) may be reduced to the integration of a canonical system with the same function H.

3. Hamilton-Jacobi Theory. Hamilton and Jacobi achieved a major success by recognizing that this relationship may be reversed. To be sure, the integration of a partial differential equation is usually considered as a problem more difficult than that of a system of ordinary differential equations. In mathematical physics one is often led, however, to a system of ordinary differential equations in canonical form. These equations may be difficult to integrate by elementary methods, while the corresponding partial differential equation is manageable; in particular, it may happen that a complete integral is easily obtained, e.g., with the help of the separation of variables (cf. Ch. I, §3). Knowing the complete integral, one can then solve the corresponding system of characteristic ordinary differential equations by processes of differentiation and elimination. This fact, which is contained in the earlier results of §4 and §8, 1, can be formulated in a particularly simple way for the case of canonical differential equations and can be verified analytically, independently of the motivation, by envelope construction.

We first formulate anew the concept "complete integral" for differential equation (9): We remark that, for every solution u of the differential equation, $u + a$ (with an arbitrary constant a) is also a solution. If $u = \phi(x_1, x_2, \cdots, x_n, x, a_1, a_2, \cdots, a_n)$ is a solution depending on n parameters a_i such that the determinant

$$(12) \qquad\qquad |\phi_{x_i a_k}|$$

is different from zero, then the expression

$$u = \phi + a,$$

which depends on $n + 1$ parameters, is called a *complete integral*. The principal content of the theory now to be treated is stated in the following theorem analogous to the facts proved in article 1:

If a complete integral $u = \phi(x_1, x_2, \cdots, x_n, x, a_1, a_2, \cdots, a_n) + a$ is known for the partial differential equation (9)

$$u_x + H(x_1, x_2, \cdots, x_n, x, u_{x_1}, u_{x_2}, \cdots, u_{x_n}) = 0,$$

then, from the equations

$$(13) \qquad\qquad \phi_{a_i} = b_i, \qquad \phi_{x_i} = p_i \qquad (i = 1, 2, \cdots, n)$$

with the $2n$ arbitrary parameters a_i and b_i, one obtains (*implicitly*) *the $2n$-parameter family of solutions of the canonical system of differential equations* (10)

$$\frac{dx_i}{dx} = H_{p_i}, \qquad \frac{dp_i}{dx} = -H_{x_i}.$$

Let us assume that from the first n equations (13) the quantities x_i are expressed as functions of x and the $2n$ parameters a_i, b_i—this is possible because, by assumption, $|\phi_{x_i a_k}| \neq 0$—and let us moreover imagine these values of x_i introduced into the second set of equations (13); we thus obtain functions $x_i(x)$ and $p_i(x)$ which still depend on $2n$ parameters and will be seen to represent the general solution of the system of canonical differential equations. Thereby the solution of the system is reduced to the problem of finding a complete integral of the corresponding partial differential equation.

The shortest proof of this statement is a simple verification similar to that used in article 1.[1] To show that the functions $x_i(x)$ and $p_i(x)$, so determined, satisfy equations (10) we differentiate the equations $\phi_{a_i} = b_i$ with respect to x and the equation

$$\phi_x + H(x_i, x, \phi_{x_i}) = 0$$

with respect to a_i ; we obtain the $2n$ equations

$$\frac{\partial^2 \phi}{\partial x \partial a_i} + \sum_{k=1}^{n} \frac{\partial^2 \phi}{\partial x_k \partial a_i} \frac{\partial x_k}{\partial x} = 0,$$

$$\frac{\partial^2 \phi}{\partial x \partial a_i} + \sum_{k=1}^{n} H_{p_k} \frac{\partial^2 \phi}{\partial x_k \partial a_i} = 0,$$

[1] The difference between this proof and that given in article 1 is that here the unsymmetric notation is retained.

from which the first relation of (10) follows, since the determinant $|\phi_{a_k x_i}|$ does not vanish. To verify the second relation, we differentiate the equations $\phi_{x_i} = p_i$ with respect to x and the equation $\phi_x + H(x_i, x, \phi_{x_i}) = 0$ with respect to x_i and obtain the equations

(14)
$$\frac{dp_i}{dx} = \frac{\partial^2 \phi}{\partial x \partial x_i} + \sum_{k=1}^{n} \frac{\partial^2 \phi}{\partial x_i \partial x_k} \frac{\partial x_k}{\partial x},$$

$$0 = \frac{\partial^2 \phi}{\partial x \partial x_i} + \sum_{k=1}^{n} H_{p_k} \frac{\partial^2 \phi}{\partial x_k \partial x_i} + H_{x_i}.$$

Since we have already proved that $dx_i/dx = H_{p_i}$, the second relation of (10) follows immediately.

4. Example. The Two-Body Problem. The motion of two particles P_1 and P_2 which attract each other is described according to Newton's law of gravitation by the differential equations

(15)
$$m_1 \ddot{x}_1 = U_{x_1}, \qquad m_1 \ddot{y}_1 = U_{y_1}, \qquad m_1 \ddot{z}_1 = U_{z_1},$$

$$m_2 \ddot{x}_2 = U_{x_2}, \qquad m_2 \ddot{y}_2 = U_{y_2}, \qquad m_2 \ddot{z}_2 = U_{z_2},$$

where we set

$$U = \frac{\kappa^2 m_1 m_2}{\sqrt{(x_1 - x_2)^2 + (y_1 - y_2)^2 + (z_1 - z_2)^2}}.$$

As is easily seen, the motion always remains in a plane; we can therefore choose the plane of motion as the x, y-plane of our coordinate system with P_2 placed at the origin. For the position (x, y) of the particle P_1 we then obtain the equations of motion

(16) $\qquad m_1 \ddot{x} = U_x, \qquad m_1 \ddot{y} = U_y, \qquad U = \dfrac{k^2}{\sqrt{x^2 + y^2}},$

where $k^2 = \kappa^2 m_1 m_2$.

If we introduce the Hamiltonian function

(17) $\qquad H = \frac{1}{2}(p^2 + q^2) - \dfrac{k^2}{\sqrt{x^2 + y^2}},$

the system (16) finally goes over into the *canonical* system of *differential equations*

(18)
$$\dot{x} = H_p, \qquad \dot{p} = -H_x,$$

$$\dot{y} = H_q, \qquad \dot{q} = -H_y.$$

for the quantities $x, y, p = \dot{x}, q = \dot{y}$; the integration of these equations is equivalent to the problem of finding a complete integral of the partial differential equation[1]

(19) $$\phi_t + \tfrac{1}{2}(\phi_x^2 + \phi_y^2) = \frac{k^2}{\sqrt{x^2 + y^2}} \cdot$$

If we introduce polar coordinates r, θ, we obtain from (19)

(20) $$\phi_t + \frac{1}{2}\left(\phi_r^2 + \frac{1}{r^2}\phi_\theta^2\right) = \frac{k^2}{r} \, ;$$

this equation clearly possesses the family of solutions

(21) $$\phi = -\alpha t - \beta\theta - \int_{r_0}^r \sqrt{2\alpha + \frac{2k^2}{\rho} - \frac{\beta^2}{\rho^2}} \, d\rho$$

which depend on the parameters α, β. By the main theorem of article 3 we then obtain the general solution of (18) in the form

$$\frac{\partial\phi}{\partial\alpha} = -t_0 , \qquad \frac{\partial\phi}{\partial\beta} = -\theta_0$$

or, explicitly,

(22)

$$t - t_0 = -\int_{r_0}^r \frac{d\rho}{\sqrt{2\alpha + \frac{2k^2}{\rho} - \frac{\beta^2}{\rho^2}}} \, ,$$

$$\theta - \theta_0 = \beta \int_{r_0}^r \frac{d\rho}{\rho^2 \sqrt{2\alpha + \frac{2k^2}{\rho} - \frac{\beta^2}{\rho^2}}} \cdot$$

The second equation gives us the *trajectory* (or particle path); the first determines the motion of the particle on this path as a function of the time t.

If we introduce the variable of integration $\rho' = 1/\rho$, the trajectory may be calculated explicitly and is given by

$$\theta - \theta_0 = -\arcsin \frac{\dfrac{\beta^2}{k^2}\dfrac{1}{r} - 1}{\sqrt{1 + \dfrac{2\alpha\beta^2}{k^4}}} \, ,$$

[1] See also Ch. I, §3, 1, example *4* .

or, if we set

$$p = \frac{\beta^2}{k^2}, \qquad \epsilon^2 = \sqrt{1 + \frac{2\alpha\beta^2}{k^4}},$$

by

$$\theta - \theta_0 = -\text{arc sin} \frac{p/r - 1}{\epsilon^2}$$

i.e.,

$$r = \frac{p}{1 - \epsilon^2 \sin(\theta - \theta_0)}.$$

The path is an ellipse, a parabola, or a hyperbola, depending on whether $\epsilon < 1$, $\epsilon = 1$, or $\epsilon > 1$.[1]

5. Example. Geodesics on an Ellipsoid. The differential equations of the geodesics $u = u(s)$, $v = v(s)$ of a surface

$$x = x(u, v), \qquad y = y(u, v), \qquad z = z(u, v)$$

may, according to Vol. I, Ch. IV, §9, be written as follows in canonical form:

$$(23) \qquad \begin{aligned} u_s &= H_p, & p_s &= -H_u, \\ v_s &= H_q, & q_s &= -H_v, \end{aligned}$$

where we set

$$p = Eu_s + Fv_s,$$
$$q = Fu_s + Gv_s,$$

and

$$H = \frac{1}{2} \frac{1}{EG - F^2} (Gp^2 - 2Fpq + Eq^2)$$

with

$$E = x_u^2 + y_u^2 + z_u^2, \qquad F = x_u x_v + y_u y_v + z_u z_v,$$
$$G = x_v^2 + y_v^2 + z_v^2.$$

[1] For a general discussion of equations (22), see R. Courant [1] pp. 422–428.

Following article 3, we consider the partial differential equation

(24) $$\phi_s + \frac{1}{2}\frac{1}{EG - F^2}(G\phi_u^2 - 2F\phi_u\phi_v + E\phi_v^2) = 0$$

corresponding to (23); our object is to obtain a complete integral of this equation. Now if we set

$$\phi = -\tfrac{1}{2}s + \psi(u, v),$$

then ψ satisfies the equation[1]

(25) $$G\psi_u^2 - 2F\psi_u\psi_v + E\psi_v^2 = EG - F^2.$$

We are interested in the solution curves, not in the special parametric representation of these curves; it suffices, therefore, to find a one-parameter family of solutions $\psi(u, v, \alpha)$ of (25), from which one obtains the two-parameter family of geodesics in the form

(26) $$\frac{\partial\psi}{\partial\alpha} = C$$

according to the main theorem of article 3.

In the special case of the *ellipsoid*

$$\frac{x^2}{a} + \frac{y^2}{b} + \frac{z^2}{c} = 1 \qquad\qquad (a, b, c, > 0),$$

the following parametric representation (cf. Vol. I, p. 226) holds, as can easily be verified:

(27)
$$x = \sqrt{\frac{a(u - a)(v - a)}{(b - a)(c - a)}}\,,$$
$$y = \sqrt{\frac{b(u - b)(v - b)}{(c - b)(a - b)}}\,,$$
$$z = \sqrt{\frac{c(u - c)(v - c)}{(a - c)(b - c)}}\,.$$

From this it follows that

(28)
$$E = (u - v)A(u),$$
$$F = 0,$$
$$G = (v - u)A(v),$$

[1] See also §9, 3.

where, for brevity, we have set

$$A(u) = \frac{1}{4} \frac{u}{(a - u)(b - u)(c - u)}.$$

For $\psi(u, v)$ we thus obtain the partial differential equation

(29) $$A(v)\psi_u^2 - A(u)\psi_v^2 = (u - v)A(u)A(v),$$

and, if we write $\psi(u, v) = f(u) + g(v)$, we immediately obtain the family of solutions

(30)
$$\psi(u, v, \alpha)$$
$$= \int_{u_0}^{u} \sqrt{A(u')(u' + \alpha)} \, du' + \int_{v_0}^{v} \sqrt{A(v')(v' + \alpha)} \, dv',$$

which depends on the parameter α.

From (26) we find that the *equation of the geodesics on the ellipsoid* is

(31) $$\int_{u_0}^{u} \sqrt{\frac{A(u')}{u' + \alpha}} \, du' + \int_{v_0}^{v} \sqrt{\frac{A(v')}{v' + \alpha}} \, dv' = 2C.$$

§9. Hamilton-Jacobi Theory and the Calculus of Variations

The Hamilton-Jacobi theory of partial differential equations of the first order is closely connected with the classical calculus of variations.[1] The theory of partial differential equations of the first order in which the unknown function does not appear explicitly is equivalent to the problem of choosing the functions $u_i(s)$ such that the variation of an integral

(1) $$J \equiv \int_{\tau}^{t} F(\dot{u}_1, \dot{u}_2, \cdots, \dot{u}_n, u_1, u_2, \cdots, u_n, s) \, ds$$

vanishes. Here $u_1(s), u_2(s), \cdots, u_n(s)$ are n functions of a parameter s, the dot denotes differentiation with respect to s, and $F(\dot{u}_i, u_i, s)$ is a twice continuously differentiable function of its $2n + 1$ arguments in the domain considered.[2] We shall now briefly explain this relationship and thereby obtain again the results of §8 and gain a deeper understanding of them.

---

[1] Compare the comprehensive work by C. Carathéodory [1].
[2] See for concepts and notations Vol. I, Ch. IV, §3.

1. Euler's Differential Equations in Canonical Form.   The extremals of the variational problem (1) (cf. Vol. I, Ch. IV) are given by the system of n *Euler differential equations* of the second order for the functions $u_\nu(s)$:

$$(2) \qquad \frac{d}{ds} F_{\dot{u}_\nu} - F_{u_\nu} = 0 \qquad (\nu = 1, 2, \cdots, n).$$

We may now (cf. Vol. I, Ch. IV, §9) replace the variational problem by an equivalent canonical variational problem which leads to a system of $2n$ canonical differential equations of the first order for the extremals. For this purpose we introduce the "moments"

$$(3) \qquad F_{\dot{u}_\nu} = v_\nu \qquad (\nu = 1, 2, \cdots, n)$$

by a *Legendre transformation*. We assume that the quantities $\dot{u}_i$ may be calculated from equations (3), in the appropriate domain of the variables $\dot{u}_i$, u_i, s, as functions of the variables v_i, u_i, s. We make the assumption

$$(4) \qquad | F_{\dot{u}_\nu \dot{u}_\mu} | \neq 0,$$

where $| F_{\dot{u}_\nu \dot{u}_\mu} |$ denotes the n-row determinant with the elements $\partial^2 F / \partial \dot{u}_\nu \partial \dot{u}_\mu$. The system of equations

$$(5) \qquad \begin{aligned} & F_{\dot{u}_\nu} = v_\nu, \qquad L_{v_\nu} = \dot{u}_\nu, \\ & F(\dot{u}_i, u_i, s) + L(v_i, u_i, s) = \sum_{\nu=1}^{n} \dot{u}_\nu v_\nu \end{aligned}$$

then represents a *Legendre transformation and its inverse* in which u_i and s remain untransformed parameters (cf. Ch. I, §6); we immediately obtain the further relation

$$(6) \qquad L_{u_\nu} + F_{u_\nu} = 0.$$

The Euler differential equations go over into the *canonical system*

$$(7) \qquad \begin{aligned} \dot{v}_\nu &= -L_{u_\nu}, \\ \dot{u}_\nu &= L_{v_\nu} \end{aligned}$$

with the *Legendre function* $L(u_1, \cdots, u_n, v_1, \cdots, v_n, s)$ belonging to the variational problem. These canonical differential equations are Euler equations of a variational problem—*canonical form of the variational problem* (cf. Vol. I, Ch. IV, §9)—equivalent to the original one which is

$$\delta \int_\tau^t \left(\sum_{\nu=1}^n \dot{u}_\nu v_\nu - L(v_i , u_i , s) \right) ds = 0$$

or

$$\delta \int_\tau^t \left(\sum_{\nu=1}^n u_\nu \dot{v}_\nu + L(v_i , u_i , s) \right) ds = 0 ,$$

where the $2n$ arguments u_i , v_i are functions of the parameter s. The variables u_i and v_i are called *canonically conjugate*.

Note that no canonical transformation exists if the function F is homogeneous of degree one in the $\dot{u}_i$,[1] for example, if $F = \sqrt{\sum_{\nu=1}^n \dot{u}_\nu^2}$; see, however, article 3. It should be observed that, if condition (4) holds, formula (5) enables us to reverse the procedure which led from the Euler to the canonical representation of the extremals; i.e., *to every variational integrand* $F(\dot{u}_i , u_i , s)$ *there corresponds a Legendre function* $L(v_i , u_i , s)$ *and vice versa*.

The canonical system of Euler differential equations (7) is identical with the system of characteristic differential equations of the partial differential equation of the first order

(8) $$J_s + L(J_{u_i} , u_i , s) = 0$$

for an unknown function $J(u_1 , u_2 , \cdots , u_n , s)$. In articles 2 and 4, we shall see that equation (8) has direct significance for the variational problem.

2. Geodetic Distance or Eiconal and Its Derivatives. Hamilton-Jacobi Partial Differential Equation. We now make the additional assumption that, in an appropriate domain of the $(n + 1)$-dimensional space of the variables u_i , s, each pair of points $A(\kappa_1 , \kappa_2 , \cdots , \kappa_n , \tau)$ and $B(q_1 , q_2 , \cdots , q_n , t)$ may be connected in a uniquely determined way by an extremal. See Figure 2. These extremals and the corresponding moments then can be represented by means of κ_i , τ, q_i, t as parameters in the form

(9) $$u_\nu = f_\nu(s, \kappa_i , \tau, q_i , t),$$

---

[1] The determinant in (4) is then identically zero.

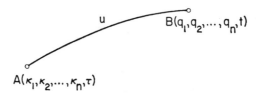

(9')
$$v_\nu = g_\nu(s, \kappa_i, \tau, q_i, t).$$

In particular, we have,

(10)
$$\kappa_\nu = f_\nu(\tau, \kappa_i, \tau, q_i, t)$$
$$q_\nu = f_\nu(t, \kappa_i, \tau, q_i, t),$$

for the points A and B. The direction of the extremals at these points is given by

(11)
$$\dot{\kappa}_\nu \equiv \dot{u}_\nu(A) = \dot{f}_\nu(\tau, \kappa_i, \tau, q_i, t),$$
$$\dot{q}_\nu \equiv \dot{u}_\nu(B) = \dot{f}_\nu(t, \kappa_i, \tau, q_i, t);$$

here the dot denotes differentiation with respect to the first argument s (differentiation along the extremal). The quantities (11) as well as the so-called *field functions* (i.e., the moments v_ν at the end points)

(12)
$$\pi_\nu = g_\nu(\tau, \kappa_i, \tau, q_i, t) = F_{\dot{\kappa}_\nu}(\dot{\kappa}_i, \kappa_i, \tau),$$
$$p_\nu = g_\nu(t, \kappa_i, \tau, q_i, t) = F_{\dot{q}_\nu}(\dot{q}_i, q_i, t),$$

are functions of the $2n + 2$ quantities κ_i, τ, q_i, t.

If we introduce the functions (9) and (9') into the variational integral

$$J = \int_\tau^t F(\dot{u}_i, u_i, s) \, ds = \int_\tau^t \left(\sum_{\nu=1}^n v_\nu \dot{u}_\nu - L(v_i, u_i, s) \right) ds$$

then this integral becomes a function

$$J(\kappa_i, \tau, q_i, t)$$

of the $2n + 2$ variables κ_i, τ, q_i, t.

This function is called the *geodetic distance between the points A and B*, referring to the fact that the variational problem can be understood as a generalization of the problem of finding the shortest curve between two points in space.   The function $J(\kappa_i, \tau, q_i, t)$ also has an optical interpretation.   We consider s as the time variable and set

$$F = \frac{\sqrt{\sum_{\nu=1}^{n} \dot{u}_\nu^2}}{V(\dot{u}_i, u_i, s)}$$

where V is interpreted as the speed of light in u_i-space, depending on the position, direction, and time.   If, according to *Fermat's principle of least time* (cf. Vol. I, Ch. IV, §1), we assume that the light rays are extremals of our variational problem, then the function J measures the time needed by light to traverse the distance from A to B on its path.   In this connection J is referred to as an *eiconal*.

The main point of the theory is to express the derivatives of the eiconal J with respect to its $2n + 2$ independent variables in terms of the function F.

The partial derivatives of the eiconal are given by the formulas

(13)
$$J_t = -L(p_i, q_i, t) = F(\dot{q}_i, q_i, t) - \sum_{\nu=1}^{n} \dot{q}_\nu F_{\dot{q}_\nu},$$

$$J_{q_\nu} = p_\nu = F_{\dot{q}_\nu} \qquad\qquad (\nu = 1, 2, \cdots, n)$$

and

(14)
$$J_\tau = L(\pi_i, \kappa_i, \tau) = -F(\dot{\kappa}_i, \kappa_i, \tau) + \sum_{\nu=1}^{n} \dot{\kappa}_\nu F_{\dot{\kappa}_\nu},$$

$$J_{\kappa_\nu} = -\pi_\nu = -F_{\dot{\kappa}_\nu} \qquad\qquad (\nu = 1, 2, \cdots, n)$$

which can be combined into

(15)
$$\delta J = -L(p_i, q_i, t)\delta t + \sum_{\nu=1}^{n} p_\nu \delta q_\nu$$
$$+ L(\pi_i, \kappa_i, \tau)\delta\tau - \sum_{\nu=1}^{n} \pi_\nu \delta\kappa_\nu;$$

here $\dot{q}_i, \dot{\kappa}_i, p_i, \pi_i$ are given by (11) and (12).

These formulas are obtained most directly from the canonical representation of the variational problem.   Consider the $2n + 2$ coordinates of the initial point A and the end point B as depending, in a continuously differentiable but otherwise arbitrary way, on a parameter ϵ, and let differentiation with respect to this parameter be denoted by the symbol δ.   Remembering the canonical differential equations (7) which hold for the extremals, we then have

$$\delta J = \left(\sum_{\nu=1}^{n} \dot{q}_\nu \, p_\nu - L(p_i \,, q_i \,, t) \right) \delta t - \left(\sum_{\nu=1}^{n} \dot{\kappa}_\nu \pi_\nu - L(\pi_i \,, \kappa_i \,, \tau) \right) \delta\tau$$

$$+ \sum_{\nu=1}^{n} \int_\tau^t [(v_\nu \delta \dot{u}_\nu + \dot{u}_\nu \delta v_\nu) - (L_{u_\nu} \delta u_\nu + L_{v_\nu} \delta v_\nu)] \, ds$$

$$= \left(\sum_{\nu=1}^{n} \dot{q}_\nu \, p_\nu - L(p_i \,, q_i \,, t) \right) \delta t - \left(\sum_{\nu=1}^{n} \dot{\kappa}_\nu \pi_\nu - L(\pi_i \,, \kappa_i \,, \tau) \right) \delta\tau$$

$$+ \sum_{\nu=1}^{n} \int_\tau^t (v_\nu \delta u_\nu)^{\cdot} \, ds.$$

From (10) it now follows immediately that

$$\delta\kappa_\nu = \dot{\kappa}_\nu \delta\tau + \delta u_\nu \mid_{s=\tau} .$$

$$\delta q_\nu = \dot{q}_\nu \delta t + \delta u_\nu \mid_{s=t} .$$

Because of

$$\delta J = \left(\sum_{\nu=1}^{n} \dot{q}_\nu \, p_\nu - L(p_i \,, q_i \,, t) \right) \delta t - \left(\sum_{\nu=1}^{n} \dot{\kappa}_\nu \pi_\nu - L(\pi_i \,, \kappa_i \,, \tau) \right) \delta\tau$$

$$+ \left[\sum_{\nu=1}^{n} v_\nu \delta u_\nu \right]_\tau^t ,$$

we now have

$$\delta J = -L(p_i \,, q_i \,, t)\delta t + L(\pi_i \,, \kappa_i \,, \tau)\delta\tau + \sum_{\nu=1}^{n} p_\nu \delta q_\nu - \sum_{\nu=1}^{n} \pi_\nu \delta\kappa_\nu ,$$

which is the asserted relation (15).

We can immediately eliminate the moments p_i from equations (13).   We thereby obtain the *"Hamilton-Jacobi" differential equation*

(16) $$J_t + L(J_{q_i} \,, q_i \,, t) = 0$$

for the geodetic distance J as a function of the end point B; this equation is also called the *eiconal equation*.   It is the same as (8) of article 1. As noted there, the characteristic equations of (16) are the same as our canonical differential equations; *the characteristics of the Hamilton-Jacobi equation* (16) *are, then, the extremals of the canonical variational problem.*

3. Homogeneous Integrands. In the exceptional case where F is homogeneous of degree one in the quantities $\dot{u}_\nu$, a corresponding argument may also be carried out. Here we have

$$| F_{\dot{u}_\nu \dot{u}_\mu} | = 0 \quad \text{and also} \quad L = -F + \sum_{\nu=1}^{n} \dot{u}_\nu F_{\dot{u}_\nu} = 0,$$

and the Legendre transformation to the canonical form cannot be used. In this case however as in article 2, the equations $J_t = -L = 0$, $J_{q_\nu} = F_{\dot{q}_\nu}$ remain valid, and moreover, the expressions $F_{\dot{q}_\nu}$ are homogeneous of degree zero in the $\dot{q}_\nu$. Hence, one can express the ratios of the numbers $\dot{q}_\nu$ by the derivatives J_{q_ν} , and the homogeneity relation

$$\sum_{\nu=1}^{n} \dot{q}_\nu F_{\cdot_\nu} = F$$

gives us the substitute for the Hamilton-Jacobi partial differential equation.

As an example we consider the case of the *geodetic lines* given by

$$F = \sqrt{Q}, \qquad Q = \sum_{\nu,\mu=1}^{n} a_{\nu\mu} \dot{u}_\nu \dot{u}_\mu ,$$

where the coefficients $a_{\nu\mu}$ of the quadratic form Q are functions of $u_1 , u_2 , \cdots , u_n$. We obtain

$$J_t = 0, \qquad J_{q_\nu} = F_{\dot{q}_\nu} = \sum_{\mu=1}^{n} \frac{a_{\nu\mu} \dot{q}_\mu}{F}$$

or

$$\frac{\dot{q}_\nu}{F} = \sum_{\mu=1}^{n} A_{\nu\mu} J_{q_\mu} ,$$

in which the numbers $A_{\nu\mu}$ form the inverse matrix to the matrix $a_{\nu\mu}$. Because of the homogeneity relation, multiplication by $F_{\dot{q}_\nu} = J_{q_\nu}$ and summation gives us the equation

$$(17) \qquad \sum_{\nu,\mu=1}^{n} A_{\nu\mu} J_{q_\nu} J_{q_\mu} = 1$$

as the Hamilton-Jacobi partial differential equation for the geodetic distance J, and the partial differential equation

$$(17') \qquad \sum_{\nu,\mu=1}^{n} A_{\nu\mu} \Gamma_{q_\nu} \Gamma_{q_\mu} = 4\Gamma$$

for the quantity $\Gamma = J^2$ follows. For example, in the Euclidean case
$F = \sqrt{\sum_{\nu=1}^{n} \dot{u}_\nu^2}$, we obtain the differential equation

$$1 = \sum_{\nu=1}^{n} J_{q_\nu}^2 .$$

Since s does not appear explicitly in F, the same general result (17) may be obtained for the problem of the geodetic curves by choosing the parameter s in such a way that $Q = F^2 = 1$. From the Euler differential equations

$$\frac{d}{ds} F_{\dot{u}_\nu} - F_{u_\nu} = 0, \qquad F = \sqrt{Q}$$

or

(18) $$\frac{d}{ds} \frac{Q_{\dot{u}_\nu}}{\sqrt{Q}} - \frac{1}{\sqrt{Q}} Q_{u_\nu} = 0,$$

we then have

(18′) $$\frac{d}{ds} Q_{\dot{u}_\nu} - Q_{u_\nu} = 0.$$

This system of linear differential equations always has Q as an integral;[1] thus we may impose without contradiction the supplementary restriction $Q = 1$. We can now transform the new differential equations (18′) to canonical form since they belong to the quadratic integrand Q rather than to the integrand $\sqrt{Q}$ which is homogeneous of degree one. In view of the homogeneity of Q, the canonical transformation with H instead of L yields

$$-Q + \sum_{\nu=1}^{n} Q_{\dot{u}_\nu} \dot{u}_\nu = Q = H(p_i , u_i),$$

(19) $$\dot{u}_\nu = H_{p_\nu}, \qquad\qquad (\nu = 1, 2, \cdots , n)$$

$$\dot{p}_\nu = -H_{u_\nu}.$$

[1] Proof: Along an extremal Q becomes a function of s with the derivative

$$\frac{dQ}{ds} = \sum_{\nu=1}^{n} Q_{\dot{u}_\nu} \ddot{u}_\nu + \sum_{\nu=1}^{n} Q_{u_\nu} \dot{u}_\nu .$$

Because of the homogeneity relation, the right side is equal to

$$2 \frac{dQ}{ds} + \sum_{\nu=1}^{n} \dot{u}_\nu \left[Q_{u_\nu} - \frac{d}{ds} Q_{\dot{u}_\nu} \right].$$

From (18′) it then follows that $dQ/ds = 2dQ/ds$, and thus that $dQ/ds = 0$

From the supplementary condition $Q = 1$ we immediately obtain

$$H(J_{u_i}, u_i) = 1,$$

which is equivalent to our equation (17).

4. Fields of Extremals.   Hamilton-Jacobi Differential Equation.   We return to the distance function J considered in article 2.   If we keep the initial point A fixed, J becomes a function of the $n + 1$ coordinates q_ν, t of the end point B alone and satisfies the Hamilton-Jacobi differential equation (16); we assume, as emphasized before, that the end point B ranges over a domain for which the extremals AB and, therefore, also the field functions introduced in formulas (11) and (12) are uniquely determined.   A domain of this sort with a corresponding family of extremals is called a *field*.

The concepts of a field and of the associated *distance function* of $n + 1$ variables which satisfies the Hamilton-Jacobi equation may now be broadened.   We define the *geodetic distance* not only from a fixed point but also *from a fixed initial surface*

$$T(\kappa_1, \kappa_2, \cdots, \kappa_n, \tau) = 0.$$

This concept of geodetic distance arises in the following way: We temporarily consider the end point B of an extremal as fixed and seek an initial point A on the given surface

$$(20) \qquad T(\kappa_1, \kappa_2, \cdots, \kappa_n, \tau) = 0$$

so that the geodetic distance $J(A, B)$ remains stationary under variations of the point A.   Thus we have to introduce in formula (15) the value zero for the variations δq_ν, δt of the end point B and since $\delta J = 0$ we obtain the condition

$$(21) \qquad L(\pi_i, \kappa_i, \tau)\delta\tau - \sum_{\nu=1}^{n} \pi_\nu \, \delta\kappa_\nu = 0$$

for the initial point A.   This condition must always be satisfied regardless of the way in which the initial point is varied on the given surface $T = 0$; i.e., (21) must be a consequence of equation (20) or of its differentiated form

$$\delta T = \sum_{\nu=1}^{n} T_{\kappa_\nu} \, \delta\kappa_\nu + T_\tau \, \delta\tau = 0.$$

This requirement is equivalent to the following condition, the so-called *transversality condition* (cf. Vol. I, Ch. IV, §5),

$$(22) \qquad\qquad -L:\pi_\nu = T_\tau:T_{\kappa_\nu} \qquad\qquad (\nu = 1, 2, \cdots, n)$$

with $L = L(\pi_i, \kappa_i, \tau)$, or to

$$(22') \qquad\qquad \left[F - \sum_{\mu=1}^{n} \dot{\kappa}_\mu F_{\dot{\kappa}_\mu} \right]:F_{\dot{\kappa}_\nu} = T_\tau:T_{\kappa_\nu},$$

with $F = F(\dot{\kappa}_i, \kappa_i, \tau)$. The transversality condition is a relation between the coordinates of a point of the surface $T = 0$ and the derivatives $\dot{\kappa}_\nu$, or the canonically conjugate numbers π_ν, of the extremal. An extremal which satisfies condition (22) at A is called an *extremal transverse* or *transversal* to the surface $T = 0$. If we assign a transversal to every point of such a surface, then these curves form an n-parameter family.

Let us, moreover, assume that for every point of a region of the surface T we can construct such a transverse extremal, and that the family of these extremals covers a certain domain around the surface region, or, in other words, forms an *extremal field* so that through each point of the domain there is exactly one extremal. Then to every point B of this field there corresponds a unique point A on the surface. In the field the quantities $\dot{q}_i$ and, in particular, the quantities $\dot{q}_i$ for the extremals are uniquely determined as functions of position. The value of the eiconal between A and B can thus be regarded as a function of the coordinates q_i, t of the end point B. This eiconal measures the stationary geodetic distance from the point B to points on the surface $T = 0$, in short, the geodetic distance of the point B from the surface.

The case of a fixed initial point A, which we first considered, is the limiting case which arises if the initial surface (e.g., a sphere) shrinks to a point.

As is easily seen, in the special case of the integrand $F = \sqrt{1 + \sum_{\nu=1}^{n} \dot{q}_\nu^2}$ the geodetic distance is the same as the Euclidean rectilinear distance. Here the extremal field consists of the family of lines orthogonal to the surface $T = 0$, a special type of n-parameter family of straight lines. Thus the general concept of an extremal field is merely the generalization of this elementary concept if the Euclidean distance is replaced by the geodetic distance defined by our variational problem and the straight line between two points by the corresponding extremal.

The set of surfaces $J = $ const. is called a *family of parallel surfaces of our variational problem*.

According to the transversality conditions (22), (22'), relation (21) holds for our geodetic distance; therefore, from (15) we immediately obtain for the geodetic distance from a surface the same relation as that obtained for a fixed initial point A:

$$(23) \qquad \delta J = -L(p_i, q_i, t)\delta t + \sum_{\nu=1}^{n} p_\nu \, \delta q_\nu.$$

Thus we arrive at the following general result: *If*

$$\dot{q}_\nu = \dot{q}_\nu(q_1, q_2, \cdots, q_n, t) \ (or \ p_\nu = p_\nu(q_1, q_2, \cdots, q_n, t))$$

belong to an extremal field transverse to a continuously differentiable surface $T = 0$, then in this field the partial derivatives of the geodetic distance $J = J(q_1, q_2, \cdots, q_n, t)$ from the surface $T = 0$ are given by the formulas (13)

$$J_t = F(\dot{q}_i, q_i, t) - \sum_{\nu=1}^{n} \dot{q}_\nu F_{\dot{q}_\nu} = -L(p_i, q_i, t),$$

$$J_{q_\nu} = F_{\dot{q}_\nu} = p_\nu \qquad\qquad (\nu = 1, 2, \cdots, n).$$

The geodetic distance itself satisfies the Hamilton-Jacobi partial differential equation (eiconal equation) (16)

$$J_t + L(J_{q_i}, q_i, t) = 0.$$

This assertion rests on the assumption that the underlying field construction is possible. An exceptional case arises if the initial surface itself is generated by the extremals transverse to it (if it is "characteristic"), i.e., if the extremals lie wholly within the initial surface. It is entirely possible, however, for the transversal curves to touch the initial surface without lying on it; they may then serve in the construction of a field contiguous to the initial surface on one side. The initial surface is then called a *caustic surface* and the above result also holds in this case.

We now formulate the *converse* of the preceding theorem:

If $J(q_1, q_2, \cdots, q_n, t)$ is a solution of equation (16) *then there exists an extremal field whose extremals are transverse to all surfaces of a family $J = $ const., e.g., to the initial surface $J = 0$. In this case J denotes the geodetic distance from the initial surface in this extremal field.*

To prove this converse we start with a given solution J of the differential equation (16) and define the n field quantities p_ν in the appropriate domain by the equations

$$p_\nu = J_{q_\nu}(q_1 , q_2 , \cdots , q_n , t).$$

By (16) we have

$$J_t = -L(p_i , q_i , t).$$

We now define an n-parameter family of curves by the system of ordinary differential equations

$$\dot{q}_\nu = L_{p_\nu}, \qquad\qquad (\nu = 1, 2, \cdots , n),$$

where the quantities

$$p_i = J_{q_i}(q_1 , q_2 , \cdots , q_n , t)$$

are substituted for the variables p_ν on the right side. The quantities p_ν become functions of the parameter t along an integral curve of this system of differential equations; differentiation with respect to this parameter gives

$$\dot{p}_\nu = \sum_{\mu=1}^{n} J_{q_\nu q_\mu} \dot{q}_\mu + J_{q_\nu t}.$$

On the other hand, by differentiating the differential equation (16) with respect to q_ν we obtain the identity

$$J_{q_\nu t} + L_{q_\nu} + \sum_{\mu=1}^{n} L_{p_\mu} J_{q_\nu q_\mu} = 0,$$

and thus

$$\dot{p}_\nu = -L_{q_\nu} .$$

These equations, together with $\dot{q}_\nu = L_{p_\nu}$, characterize our family of curves as an n-parameter family of extremals. If we replace π_ν in the transversality condition (22) by J_{q_ν}, and κ_i , τ, by q_i , t, and finally T by $J(q_1 , q_2 , \cdots , q_n , t)$, we immediately find that the family of these extremals is transversal to all surfaces of the family $J = \text{const.}$

5. Cone of Rays. Huyghens' Construction. In the preceding article we have solved the variational problem and thus constructed solutions of the Hamilton-Jacobi partial differential equation which still depend on an arbitrary function, i.e., solutions which vanish on a given surface $T = 0$. It was, moreover, shown that we thereby exhaust all possible solutions of the partial differential equation. The special case in which the initial surface contracts to a point, i.e., J becomes the geodetic distance from a fixed point, leads to the solutions of the partial differential equation previously called *integral conoids* or *cones of rays*. The corresponding surfaces $J = $ const. $= c$ are naturally described as *geodetic spheres*.

It should be noted that the construction of surfaces $J = c$ parallel to an arbitrarily given surface $T = 0$ is essentially equivalent to the envelope construction of the complete integral of (16) from a solution containing n parameters. For the surfaces $J = c$ are the envelopes of the family of parallel surfaces of radius c around the points A on the initial surface. This construction goes back to Huyghens' notion of considering at a time $t = c$ the "wave front" of light issued from $T = 0$ at the time $t = 0$ as the envelope of spherical wave fronts emanating from the individual points of $T = 0$.

It should be emphasized again that the preceding constructions are meant "in the small", that is, refer merely to sufficiently small neighborhoods of, say, a ray. Considerations "in the large" require further attention as we shall see later.

6. Hilbert's Invariant Integral for the Representation of the Eiconal. The preceding expressions for the derivatives of J permit us to consider the eiconal itself as a line integral of a total differential, independent of the path of integration. In the field of the u_i, s-space we connect a point A with the variable end point B by an arbitrary piecewise smooth curve C given by functions $u_i(s)$, with the parameter s and the derivatives $u_i'(s)$. The derivatives and moments belonging to the points of the field and corresponding to the extremals of the field are functions of u_i, s, which we denote by $\dot{u}_i$, v_i, as in article 1.

For any function J of the position B, the equation

$$(24) \qquad J(B) - J(A) = \int_A^B \left(\sum_{\nu=1}^n J_{u_\nu} \, du_\nu + J_s \, ds \right)$$

holds, where the path of integration C between the points A and B is arbitrary. We consider the geodetic distance J of the point B from the initial surface $T = 0$ in our field (if A lies on the surface $T = 0$, then $J(A) = 0$). Substituting the partial derivatives of J from (13), we obtain the following integral representation for the eiconal between $A:(\kappa_i, \tau)$ and $B:(q_i, t)$,

$$(25) \quad J(q_i, t) - J(\kappa_i, \tau) = \int_A^B \left(F(\dot{u}_i, u_i, s) + \sum_{\nu=1}^n (u'_\nu - \dot{u}_\nu) F_{\dot{u}_\nu} \right) ds$$

or

$$J(q_i, t) - J(\kappa_i, \tau) = \int_A^B \left(\sum_{\nu=1}^n v_\nu u'_\nu - L(v_i, u_i, s) \right) ds.$$

Again the symbols u'_ν denote the derivatives along the curve C, while the symbols $\dot{u}_\nu$ and v_ν denote the field quantities defined before, i.e., the derivatives and moments which at the points in the field belong to the field extremals passing through them. These field quantities are here regarded as given functions of the coordinates in the field.

Conversely, the "Hilbert invariant integral" (25) has the following property: If in a given domain of $(n + 1)$-dimensional u, s-space, the $v_\nu(u_1, u_2, \cdots, u_n, s)$, $\nu = 1, 2, \cdots, n$, are given functions such that the integral

$$\int_A^B \left(\sum_{\nu=1}^n v_\nu u'_\nu - L(v_i, u_i, s) \right) ds$$

extending between two points A and B in this space over a curve is independent of the path, then the functions $v_\nu(u_1, u_2, \cdots, u_n, s)$ are field quantities of an extremal field; the value of the integral as a function of the end point B is the distance function $J(q_1, q_2, \cdots, q_n, t)$ belonging to this extremal field.

This can be proved immediately if we recall that, for an integral of this kind which is independent of the path, the relations

$$J_{q_\nu} = p_\nu, \qquad J_t = -L$$

hold at the end point B. As a function of its upper limit, such an integral therefore satisfies the Hamilton-Jacobi differential equation (16) and our assertion follows by the theorem of article 4, which states that every solution of the Hamilton-Jacobi differential equation is a distance function in an extremal field.

Thus it is clear that the Hamilton-Jacobi partial differential equation, the construction of extremal fields and corresponding distance functions, and the independence of the path for an integral of the type (25) are all equivalent expressions of the same situation.

7. Theorem of Hamilton and Jacobi. From the Hilbert integral, we obtain a new understanding of the theorem of Jacobi (cf. §8). If $J(q_1, q_2, \cdots, q_n, t, a_1, a_2, \cdots, a_n)$ is a solution of the Hamilton-Jacobi partial differential equation for which the determinant $|J_{a_\nu q_\mu}|$ does not vanish, then the equations $J_{a_\nu} = b_\nu$ and $J_{q_\nu} = p_\nu (\nu = 1, 2, \cdots, n)$ yield a $2n$-parameter family of solutions of the canonical equations.

Our earlier results show: The function J defines an extremal field depending on the parameters $a_1, a_2, \cdots, a_n$, and J is represented in this extremal field by the Hilbert integral (25). Moreover, differentiating under the integral sign we obtain the integral representation

$$(26) \qquad J_{a_\mu} = \int_A^B \sum_{\nu=1}^n (u_\nu' - \dot{u}_\nu) F_{\dot{u}_\nu a_\mu} \, ds,$$

which, of course, is also independent of the path C. Now, if the point B moves from an initial position B_0 along an extremal of the field belonging to the system of values a_i, i.e., if the arc of C concerned is an extremal and, therefore, $u_\nu' = \dot{u}_\nu$, then the integrand in (26) vanishes and we have

$$(27) \qquad J_{a_\nu} = b_\nu.$$

Here b_ν is a constant—the value of the integral between A and B_0. Conversely, if a family of curves $q_\nu(t, a_\nu, b_\nu)$ is defined by the equations $J_{a_\nu} = b_\nu$, which is possible in only one way because of the restriction $|J_{a_\nu q_\mu}| \neq 0$ for a certain neighborhood of the system of values a_i, b_i considered, the curves must be extremals. For the integrand in (26) must vanish on an arc C of this family, and we have a linear homogeneous system of equations for the differences $u_\nu' - \dot{u}_\nu$ with the determinant $|F_{\dot{u}_\nu a_\mu}|$. On the other hand, according to article 4, the field quantities are given by $v_\nu = F_{\dot{u}_\nu} = J_{u_\nu}$. Thus our determinant is identical with $|J_{u_\nu a_\mu}|$ and therefore, by assumption, does not vanish. We then have $u_\nu' - \dot{u}_\nu = 0$ which shows that the curves C are extremals.

In the next section, we shall give another proof of the Hamilton-Jacobi theorem.

§10. *Canonical Transformations and Applications*

1. The Canonical Transformation. The canonical representation of the characteristic differential equations of a variational problem or of a partial differential equation of the first order is the starting point of the *theory of canonical transformations*, which has important applications.

Let a function $L(v_\nu, u_\nu, s)$ and the corresponding system of canonical differential equations

$$(1) \qquad \dot{u}_\nu = L_{v_\nu}, \qquad \dot{v}_\nu = -L_{u_\nu}$$

be given. We pose the question whether and in what manner we may transform the canonically conjugate variables v_ν, u_ν into new variables

$$(2) \qquad \begin{aligned} \eta_\nu &= \eta_\nu(u_1, u_2, \cdots, u_n, v_1, v_2, \cdots, v_n), \\ \omega_\nu &= \omega_\nu(u_1, u_2, \cdots, u_n, v_1, v_2, \cdots, v_n) \end{aligned}$$

and thus obtain from the function $L(v_\nu, u_\nu, s)$ a new function $\Lambda(\eta_\nu, \omega_\nu, t)$ in such a way that the solutions of the new canonical differential equations

$$(3) \qquad \dot{\omega}_\nu = \Lambda_{\eta_\nu}, \qquad \dot{\eta}_\nu = -\Lambda_{\omega_\nu}$$

correspond to the solutions of the original canonical differential equations (1). Such a transformation of the variables, or of the canonical system of differential equations, is called a *canonical transformation*. Canonical transformations are easily obtained from the variational problem. Our requirements are satisfied, in fact, if under the transformation (2) the integrand of one canonical variational problem goes over into that of another except for an additive divergence expression (cf. Vol. I, Ch. IV, §3, 5) which has no influence on the Euler differential equations. This is achieved if, for example, the transformation (2) is so chosen that the relation

$$(4) \qquad \sum_{\nu=1}^{n} \dot{u}_\nu v_\nu - L(v_i, u_i, s) \equiv \sum_{\nu=1}^{n} \dot{\omega}_\nu \eta_\nu - \Lambda(\eta_i, \omega_i, s) + \frac{dW}{ds}$$

holds identically in the quantities u_i, ω_i, $\dot{u}_i$, $\dot{\omega}_i$; here

$$W = W(\omega_i, u_i, s)$$

is an arbitrarily chosen differentiable function with

$$\frac{dW}{ds} = \sum_{\nu=1}^{n} W_{\omega_\nu} \dot{\omega}_\nu + \sum_{\nu=1}^{n} W_{u_\nu} \dot{u}_\nu + W_s.$$

Our equation (4) goes over into

$$\sum_{\nu=1}^{n} \dot{u}_\nu (v_\nu - W_{u_\nu}) - \sum_{\nu=1}^{n} \dot{\omega}_\nu (\eta_\nu + W_{\omega_\nu}) - L + \Lambda - W_s = 0$$

and, since it should hold identically in $\dot{u}_\nu$, $\dot{\omega}_\nu$, u_ν, ω_ν, we immediately obtain the following *theorem*: Equations (1) go over into equations (3) by means of a canonical transformation which depends on the arbitrary function $W(\omega_\nu, u_\nu, s)$ and is obtained from the relations

(5)
$$v_\nu = W_{u_\nu}, \qquad \eta_\nu = -W_{\omega_\nu},$$
$$\Lambda = L + W_s.$$

The function Λ must afterwards be expressed in terms of the variables η_ν and ω_ν instead of v_ν and u_ν.

In an altogether similar way, we can obtain other expressions which generate canonical transformations by selecting other variables and, correspondingly, starting with the second form of the canonical variational problem given in §9, 1, p. 115. As an example, let W be an arbitrary function of v_ν, ω_ν, s. Then the equations

(6)
$$u_\nu = W_{v_\nu}, \qquad \eta_\nu = -W_{\omega_\nu},$$
$$\Lambda = L - W_s$$

give us a canonical transformation if we later introduce the quantities ω_ν, η_ν into Λ as variables. In the same way two more corresponding forms for the canonical transformations are obtained with the help of the arbitrary functions

$$W(u_\nu, \eta_\nu, s) \quad \text{and} \quad W(v_\nu, \omega_\nu, s).$$

These arbitrary functions are always characterized by the fact that they depend on one set of the old and one set of the new canonical ·variables.

2. New Proof of the Hamilton-Jacobi Theorem. Our results lead to a simple new proof of the theorem of Hamilton and Jacobi. We attempt to solve the given canonical differential equations (1) by determining a canonical transformation with a function Λ in such a way that this function vanishes identically, so that the two new canonically conjugate variables are constant along each trajectory.

We find this function Λ by assuming that we have a solution $J(u_1, u_2, \cdots, u_n, t, a_1, a_2, \cdots, a_n)$ of the Hamilton-Jacobi

differential equation $J_t + L(J_{u_\nu}, u_\nu, s) = 0$ which depends not
only on the independent variables but also on n parameters
$a_1, a_2, \cdots, a_n$ and for which $|J_{u_\nu a_\mu}|$ is not zero in the domain considered. In constructing a canonical transformation we choose the
function $J(u_\nu, \omega_\nu, s)$ as $W(u_\nu, \omega_\nu, s)$ and immediately obtain from
(5) the canonical transformation

$$v_\nu = \frac{\partial J}{\partial u_\nu}, \qquad \eta_\nu = -\frac{\partial J}{\partial \omega_\nu}, \qquad \Lambda = L(v_\nu, u_\nu, t) + \frac{\partial J}{\partial s}.$$

Since our differential equation holds identically in $v_\nu = \partial J/\partial u_\nu, u_\nu$,
and s, we now have in fact $\Lambda \equiv 0$. The new canonical differential
equations are

$$\dot{\omega}_\nu = 0, \qquad \dot{\eta}_\nu = 0,$$

and their solutions are

$$\omega_\nu = a_\nu = \text{const.},$$

$$\eta_\nu = J_{a_\nu} = b_\nu = \text{const.};$$

this is the assertion of the Hamilton-Jacobi theorem.

3. Variation of Constants (Canonical Perturbation Theory). Another
application of canonical transformations is *canonical perturbation
theory*, which is significant for astronomy and physics. We assume
that the function L appears as a sum

(7) $$L = L_1(v_\nu, u_\nu, s) + L_2(v_\nu, u_\nu, s)$$

and that the integration of the canonical differential equations with
the function L_1 has already been performed, i.e., that we already
possess a complete integral $J(u_\nu, a_\nu, s)$ of the partial differential
equation

$$J_t + L_1\left(\frac{\partial J}{\partial u_\nu}, u_\nu, s\right) = 0.$$

We then transform the canonical differential equations of the problem
corresponding to the function L by choosing $J(u_\nu, \omega_\nu, s)$ instead of
W as the generating function for this canonical transformation. In
other words: We introduce the canonically conjugate variables by
means of

$$v_\nu = J_{u_\nu}, \qquad \eta_\nu = -J_{\omega_\nu}$$

and the new Legendre function

$$\Lambda = L + J_s = L - L_1 = L_2 .$$

If the "perturbation term" L_2 were not present—i.e., if it were equal to zero—then, by article 2, the new canonically conjugate variables for each trajectory of the system of differential equations would be constants. Because of the perturbation term L_2, they go over into new variables which satisfy the canonical "perturbation equations"

$$(8) \qquad\qquad \dot{\omega}_\nu = \frac{\partial L_2}{\partial \eta_\nu}, \qquad \dot{\eta}_\nu = -\frac{\partial L_2}{\partial \omega_\nu}.$$

In some cases it is possible to simplify the problem essentially by such a decomposition of the integration problem.

Appendix 1 to Chapter II

§1. *Further Discussion of Characteristic Manifolds*

In the following section we shall use a slightly different approach and introduce characteristics by a method which can be generalized to differential equations of higher order.

1. Remarks on Differentiation in n Dimensions. In a domain of the independent variables x_1, x_2, $\cdots$, x_n we consider a function $u(x_1$, x_2, $\cdots$, $x_n)$ with continuous derivatives. At the point P, with the coordinates x_1, x_2, $\cdots$, x_n, numbers a_1, a_2, $\cdots$, a_n, representing a vector a, may be given such that

$$a_1^2 + a_2^2 + \cdots + a_n^2 \neq 0.$$

Through the point P, we construct a straight line whose points are given in terms of a parameter s by the expressions

$$x_1 + a_1 s, x_2 + a_2 s, \cdots, x_n + a_n s.$$

Then

$$\frac{\partial u}{\partial s} = \sum_{i=1}^{n} a_i u_{x_i}$$

is defined as the derivative of the function u with respect to s or as the derivative of u in the "direction" given by the vector a. At every point, therefore, the symbol

$$\frac{\partial}{\partial s} = \sum_{i=1}^{n} a_i \frac{\partial}{\partial x_i}$$

denotes differentiation in the direction of the vector a.[1]

---

[1] If the quantities a_i are continuously differentiable functions

$$a_i(x_1, x_2, \cdots, x_n)$$

of position, then the directions defined by the a_i at each point of the space form a direction field whose trajectories are uniquely given by the system of

132

In n-dimensional space consider an $(n-1)$-dimensional surface B: $\phi(x_1, x_2, \cdots, x_n) = 0$ and a function $u(x_1, x_2, \cdots, x_n)$ whose derivatives are continuous in a neighborhood of B. Furthermore, let P be a point of B at which

$$\sum_{i=1}^{n} \phi_{x_i}^2 \neq 0,$$

and let $a \neq 0$ be an arbitrary vector. We consider the derivative of u on B in the direction given by a:

$$(1) \qquad \frac{\partial u}{\partial s} = \sum_{i=1}^{n} a_i u_{x_i}.$$

If the equations

$$a_i = \lambda \phi_{x_i} \qquad (i = 1, 2, \cdots, n)$$

hold, then (1) is called the *derivative "in the direction of the normal"*; if in particular $\sum_{i=1}^{n} a_i^2 = 1$, so that

$$\frac{\partial u}{\partial s} = \sum_{i=1}^{n} \frac{\phi_{x_i}}{\sqrt{\sum_{j=1}^{n} \phi_{x_j}^2}} u_{x_i},$$

then we speak of the *"normal" derivative* of u at P.

If the vector a is tangent to B at P, and therefore perpendicular to the normal at P—i.e., if

$$\sum_{i=1}^{n} a_i \phi_{x_i} = 0$$

---

ordinary differential equations

$$\frac{dx_i}{ds} = a_i \qquad (i = 1, 2, \cdots, n).$$

Thus $\partial/\partial s$ denotes the derivative with respect to this parameter s. Here s is not necessarily the arc length on the trajectory; however, if the arc length is denoted by σ, then σ is connected with s by the equation $(d\sigma/ds)^2 = \sum_{i=1}^{n} a_i^2$. The derivative of the function u on the curve with respect to the arc length σ is given accordingly by

$$\frac{\partial u}{\partial \sigma} = \frac{1}{\sqrt{\sum_{i=1}^{n} a_i^2}} \sum_{i=1}^{n} a_i \frac{\partial u}{\partial x_i}.$$

—then $\partial u/\partial s = \sum_{i=1}^{n} a_i u_{x_i}$ is called a *"tangential"* derivative, or an *"inner"* derivative in B, and is said to "lie on the surface B"; on the other hand, if $\sum_{i=1}^{n} a_i \phi_{x_i} \neq 0$, $\partial u/\partial s$ is called an *"outward"* derivative, and is said to "lead out of B".

For example, the expressions

$$(2) \qquad \phi_{x_i} \frac{\partial}{\partial x_k} - \phi_{x_k} \frac{\partial}{\partial x_i}$$

for each pair of indices $i \neq k$ represent derivatives within the surface; we may interpret (2) to mean differentiation in that direction which is obtained by cutting the surface $\phi = 0$ with the two-dimensional plane extending through P in the x_i- and x_k-directions.

The inner derivatives of u on the surface depend only on the distribution of the values of u on the surface itself; they are, therefore, known if the values of u on the surface B are known. For if, in the neighborhood of B, we introduce instead of x_1, x_2, $\cdots$, x_n the new independent variables ξ_1, ξ_2, $\cdots$, ξ_n such that ξ_2, ξ_3, $\cdots$, ξ_n are $n-1$ independent parameters in B and $\xi_1 = \phi$, then $u_{x_i} = u_\phi \phi_{x_i} + \cdots$, where the dots stand for expressions which contain only derivatives of u with respect to the inner parameters ξ_2, ξ_3, $\cdots$, ξ_n. The expression

$$\sum_{i=1}^{n} a_i u_{x_i} = u_\phi \sum_{i=1}^{n} a_i \phi_{x_i} + \sum_{k=2}^{n} u_{\xi_k} \sum_{i=1}^{n} a_i \xi_{k_{x_i}}$$

is therefore known, under the condition $\sum_{i=1}^{n} a_i \phi_{x_i} = 0$, if the values $u(0, \xi_2, \cdots, \xi_n)$ of u on B are given.

Obviously, from $n-1$ mutually independent inner derivatives of u lying on B (e.g., $\phi_{x_i}\partial u/\partial x_n - \phi_{x_n}\partial u/\partial x_i$, for $\phi_{x_n} \neq 0$, $i = 1, 2, \cdots$, $n-1$) and a single outward derivative (e.g., u_ϕ) we may obtain all the derivatives of u by forming linear combinations. Thus all the derivatives u_{x_i} are known if the function u and one outward derivative of u are given on B.

In particular, if $n = 2$ and $x_1 = x$, $x_2 = y$, then B is a curve in the x,y-plane which may be represented by two functions $x(\tau)$, $y(\tau)$ of a parameter τ. In this case the condition for inner differentiation on B is simply $a_1 dy/d\tau - a_2 dx/d\tau = 0$, or, by the choice of a suitable parameter t instead of τ,

$$a_1 = \frac{dx}{dt}, \qquad a_2 = \frac{dy}{dt}.$$

2. Initial Value Problem. Characteristic Manifolds. We now modify the formulation of the initial value problem already given (cf. §7) by relating all our statements to n-dimensional x-space. Let an $(n-1)$-dimensional basic manifold B be given in this space by a relation

$$(3) \qquad \phi(x_1, x_2, \cdots, x_n) = 0;$$

previously (cf. §7) the manifold was represented by the n coordinates x_i as functions of $n-1$ independent parameters $t_1, t_2, \cdots, t_{n-1}$. Assigning functional values $u = u(t_1, t_2, \cdots, t_{n-1})$ to the points of this manifold B we enlarge B to the x, u-manifold C. In the same way, by the addition of n more functions $p_1, p_2, \cdots, p_n$ of the variables $t_1, t_2, \cdots, t_{n-1}$ which satisfy on B the strip condition

$$(3') \qquad du = \sum_{i=1}^{n} p_i \, dx_i,$$

or, in parameter representation,

$$\frac{\partial u}{\partial t_\nu} = \sum_{i=1}^{n} p_i \frac{\partial x_i}{\partial t_\nu},$$

we can enlarge B to a *strip manifold* C_1.

Without again discussing the actual solution of the initial value problem, we pose the following question: Consider an initial manifold B with given values u, or u and p_i, respectively. Assume that the differential equation $F(x_i, u, u_{x_i}) = 0$ is satisfied in some arbitrarily small neighborhood of B by a function $u(x_1, x_2, \cdots, x_n)$ with the given initial values. We ask: What does the differential equation assert along the initial manifold B for the function u and its derivatives?

First we consider a quasi-linear differential equation

$$(4) \qquad \sum_{i=1}^{n} a_i u_{x_i} = a.$$

At a point of the manifold B on which the initial function u is given, a particular direction of differentiation $\partial/\partial s = \sum_{i=1}^{n} a_i \partial/\partial x_i$ in n-dimensional x-space is defined by the relations $dx_i/ds = a_i$. The differentiation and its direction at this point are called *characteristic differentiation* and *characteristic direction*. Now the differential equation simply states

$$(5) \qquad \frac{du}{ds} = a;$$

i.e., it establishes the value of the *characteristic derivative* of u along B, since the right side is known on B.

We have the following *alternative:* At the point of B considered, either (a) the equation

$$(6) \qquad \gamma = \sum_{i=1}^{n} a_i \phi_{x_i} \neq 0$$

holds, or (b) the equation

$$(7) \qquad \gamma = \sum_{i=1}^{n} a_i \phi_{x_i} = 0$$

holds.

If equation (6) holds, then the characteristic direction leads out of the manifold B at this point. Equation (5), and therefore differential equation (4), yields an outward derivative of u; all the first derivatives of u on B, or at the point of B considered, are thus determined by the value of u on B alone and by the differential equation. If we apply this result to the differential equation after differentiating with respect to the independent variables, e.g., with respect to x_k, we find that the higher derivatives of u along B are also uniquely determined.

If equation (7), called the *characteristic condition*, holds, then $\partial u/\partial s$ is an inner derivative in B, and therefore already known from the assignment of u on B. Relation (5) thus represents a restriction on the assignment of u in B; this restrictive condition must be satisfied if a solution u of the partial differential equation is to exist in the neighborhood of B with the given initial values on B. If the two relations (5) and (7) are satisfied at every point P of B, they characterize B together with the covering u as a *characteristic (basic) manifold.*[1]

In other words: *At the point P of the given basic manifold $\phi = 0$ on which the values of u are arbitrarily prescribed, either the differential*

---

[1] One sees easily that our expression γ agrees with the determinant considered in §7, except for a nonvanishing factor; hence the above characterization of the characteristic manifold is equivalent to the former definition.

equation determines the corresponding derivatives of u in a unique way or it puts a restriction on the given initial values u.[1]

Analogous considerations hold for the general differential equation (1) of §7:

$$F(x_1, x_2, \cdots, x_n, u, p_1, p_2, \cdots, p_n) = 0.$$

We replace (1) of §7 by a system of (quasi-linear) differential equations which are linear in the derivatives $\partial p_\nu / \partial x_i$ by differentiating with respect to the independent variables[2] and inserting $\partial p_\nu / \partial x_i = \partial p_i / \partial x_\nu$:

$$(8) \qquad \sum_{\nu=1}^{n} F_{p_\nu} \frac{\partial p_i}{\partial x_\nu} + F_u p_i + F_{x_i} = 0 \quad (i = 1, 2, \cdots, n).$$

As initial manifold we again take the basic manifold $B: \phi = 0$. We assume the functions $u(x_1, x_2, \cdots, x_n)$, $p_1(x_1, x_2, \cdots, x_n)$, $\cdots$, $p_n(x_1, x_2, \cdots, x_n)$ given on B subject to the condition $F = 0$ and the strip condition (3′)

$$du = \sum_{\nu=1}^{n} p_\nu \, dx_\nu.$$

At the points of B we again define a characteristic differentiation by

$$(9) \qquad \frac{\partial}{\partial s} = \sum_{\nu=1}^{n} F_{p_\nu} \frac{\partial}{\partial x_\nu}.$$

Relations (8) on B then go over into

$$(10) \qquad \frac{\partial p_i}{\partial s} = -F_{x_i} - p_i F_u.$$

At a point of B they lead to the following alternative: Either (a) the equation

$$(11) \qquad \gamma = \sum_{i=1}^{n} \phi_{x_i} F_{p_i} \neq 0$$

holds, or (b) the equation

$$(12) \qquad \sum_{i=1}^{n} \phi_{x_i} F_{p_i} = 0$$

holds.

---

[1] The present alternative recalls a similar one for systems of linear equations (cf. Vol. I, Ch. I).

[2] Such linearization processes will often play an important role.

If equation (11) holds, then the differentiation $\partial/\partial s$ leads out of B and relations (10) yield the outward derivatives of p_i along B, since the right side is known from the initial data. Thus all the second derivatives of u are uniquely determined along B by the differential equation and the initial data.

If equation (12), the *characteristic condition*, is satisfied at points of B; then $\partial/\partial s$ is an inner differentiation. Since now the left side is also known from the data, relation (10) states that the data on B satisfy, in addition to (3'), the further conditions

$$\frac{\partial p_i}{\partial s} = -F_{x_i} - p_i F_u \,.$$

If $\gamma = 0$ everywhere on B and if the additional characteristic conditions (10) and the strip condition (3') are satisfied, then B is called a characteristic basic manifold belonging to the strip manifold which arises by covering B with values u and p_1, p_2, $\cdots$, p_n. This new definition is easily seen to be equivalent to that given in §7.

We note that the characteristic condition might be obtained formally in other ways. For example, we may start with the fact that the expressions

(13) $p_i \phi_{x_n} - p_n \phi_{x_i} = A_i \qquad (i = 1, 2, \cdots, n - 1)$

denote inner derivatives of u if $\phi_{x_n} \neq 0$ everywhere on B, and that they are, therefore, known as soon as u is prescribed. Thus one can attempt to calculate the values p_i along B from the $n - 1$ expressions (13) and the equation

$$F(x_i, u, p_i) = 0.$$

The condition that this is not uniquely possible is the vanishing of the Jacobian of these n equations with respect to p_1, p_2, $\cdots$, p_n:

(14)
$$\begin{vmatrix} F_{p_1} & F_{p_2} & \cdots & F_{p_{n-1}} & F_{p_n} \\ \phi_{x_n} & 0 & \cdots 0 & -\phi_{x_1} \\ 0 & \phi_{x_n} & \cdots 0 & -\phi_{x_2} \\ \multicolumn{5}{c}{\cdots\cdots\cdots\cdots\cdots\cdots} \\ 0 & 0 & \cdots \phi_{x_n} & -\phi_{x_{n-1}} \end{vmatrix} = (-1)^{n+1} \left(\sum_{i=1}^{n} F_{p_i} \phi_{x_i} \right) \phi_{x_n}^{n-2}.$$

Thus, the requirement that the quantities p_i should not be uniquely determinable is equivalent to the characteristic condition (12).

A final remark about the characteristic condition: In the nonlinear case this equation becomes meaningful only after we substitute appropriate functions for u and p_i ; e.g., if we consider characteristic manifolds on a given integral surface J: $u = u(x_1, x_2, \cdots, x_n)$. If we express u and the quantities $p_i = \partial u/\partial x_i$ in F_{p_i} as functions of the independent variables x_i, then the relation (12)

$$\sum_{i=1}^{n} F_{p_i} \phi_{x_i} = 0$$

—if it is satisfied (not necessarily identically in $x_1, x_2, \cdots, x_n$, but merely under the assumption $\phi = 0$)—qualifies the manifold defined on J as characteristic. If the relation is satisfied not only for $\phi = 0$, but identically in $x_1, x_2, \cdots, x_n$, then it becomes a linear homogeneous differential equation for the function $\phi(x_1, x_2, \cdots, x_n)$. It then defines a one-parameter family of characteristic manifolds $\phi = c \stackrel{.}{=}$ const. which generates J (cf. Ch. I, §5, p. 31). If we wish to write relation (12) as a partial differential equation, assuming it to be stipulated only for one manifold $\phi = 0$, then we consider this manifold as expressed in the form

$$\psi(x_1, x_2, \cdots, x_{n-1}) - x_n = 0,$$

where $x_1, x_2, \cdots, x_{n-1}$ are independent variables. If we substitute, everywhere in (12), the function ψ for x_n and write

$$\frac{\partial \phi}{\partial x_1} = \psi_{x_1}, \qquad \frac{\partial \phi}{\partial x_2} = \psi_{x_2}, \quad \cdots, \qquad \frac{\partial \phi}{\partial x_{n-1}} = \psi_{x_{n-1}}, \qquad \frac{\partial \phi}{\partial x_n} = -1.$$

then we obtain the genuine partial differential equation

$$(15) \qquad \sum_{i=1}^{n-1} F_{p_i} \psi_{x_i} - F_{p_n} = 0$$

for ψ in the $n - 1$ independent variables $x_1, x_2, \cdots, x_{n-1}$ alone. Let us finally note that the characteristic curves of the partial differential equations (12), (15), or (7) are the same as those of the original ones.

§2. Systems of Quasi-Linear Differential Equations with the Same Principal Part. New Derivation of the Theory

An approach to the theory of characteristics slightly different from that of Ch. II, §7, is suggested by an investigation of the system of quasi-linear differential equations

(1) $$\sum_{\kappa=1}^{n} a_\kappa \frac{\partial u_\mu}{\partial x_\kappa} = b_\mu \qquad (\mu = 1, 2, \cdots, m).$$

The coefficients a_1, a_2, $\cdots$, a_n which may depend, like the b_μ, on the variables x_1, x_2, $\cdots$, x_n, u_1, u_2, $\cdots$, u_m are identical in all the equations (1). We say that the differential equations of such a system *have the same principal part*. We first prove the following theorem (cf. Ch. I, §5, 2): *A system of m quasi-linear differential equations in n independent variables having the same principal part (of the form (1)) is equivalent to a homogeneous linear differential equation for a function of m + n variables.*

Let a system of solutions u_1, u_2, $\cdots$, u_m of (1) depending on the parameters c_1, c_2, $\cdots$, c_m be given implicitly in the form

(2)
$$\phi_1(x_1, x_2, \cdots, x_n, u_1, u_2, \cdots, u_m) = c_1,$$
$$\cdots\cdots\cdots\cdots\cdots\cdots\cdots\cdots\cdots\cdots\cdots\cdots\cdots\cdots\cdots$$
$$\phi_m(x_1, x_2, \cdots, x_n, u_1, u_2, \cdots, u_m) = c_m.$$

To insure the possibility of calculating the functions u_1, u_2, $\cdots$, u_m, assume that the Jacobian

$$\frac{\partial(\phi_1, \phi_2, \cdots, \phi_m)}{\partial(u_1, u_2, \cdots, u_m)}$$

is everywhere different from zero. By differentiating equations (2) we obtain

$$\frac{\partial \phi_\mu}{\partial x_\kappa} + \sum_{\lambda=1}^{m} \frac{\partial \phi_\mu}{\partial u_\lambda} \frac{\partial u_\lambda}{\partial x_\kappa} = 0 \qquad \left(\begin{matrix} \mu = 1, 2, \cdots, m \\ \kappa = 1, 2, \cdots, n \end{matrix} \right).$$

Multiplying by a_κ and summing over κ, we have

$$\sum_{\kappa=1}^{n} a_\kappa \frac{\partial \phi_\mu}{\partial x_\kappa} + \sum_{\lambda=1}^{m} \frac{\partial \phi_\mu}{\partial u_\lambda} \left(\sum_{\kappa=1}^{n} a_\kappa \frac{\partial u_\lambda}{\partial x_\kappa} \right) = 0;$$

therefore, because of (1),

(3) $$\sum_{\kappa=1}^{n} a_\kappa \frac{\partial \phi_\mu}{\partial x_\kappa} + \sum_{\lambda=1}^{m} b_\lambda \frac{\partial \phi_\mu}{\partial u_\lambda} = 0.$$

We recognize that the functions $\phi = \phi_\mu$ of the system (2) satisfy equation (3) identically in x_1, x_2, $\cdots$, x_n, c_1, c_2, $\cdots$, c_m; i.e., they also satisfy the same linear differential equation

(3')
$$\sum_{\kappa=1}^{n} a_\kappa \frac{\partial \phi}{\partial x_\kappa} + \sum_{\lambda=1}^{m} b_\lambda \frac{\partial \phi}{\partial u_\lambda} = 0$$

identically in x_1, x_2, $\cdots$, x_n, u_1, u_2, $\cdots$, u_m. If we introduce the notation

$$b_\lambda = a_{n+\lambda}, \qquad u_\lambda = x_{n+\lambda}, \qquad r = m + n,$$

then (3') goes over finally into the differential equation

(3")
$$\sum_{\kappa=1}^{r} a_\kappa \frac{\partial \phi}{\partial x_\kappa} = 0$$

for a function $\phi(x_1, x_2, \cdots, x_r)$; thus the first part of our theorem has been proved.

Conversely, let m solutions ϕ_1, ϕ_2, $\cdots$, ϕ_m of the differential equation (3") be given, whose Jacobian

$$\frac{\partial(\phi_1, \phi_2, \cdots, \phi_m)}{\partial(x_{n+1}, x_{n+2}, \cdots, x_r)}$$

vanishes nowhere. We shall show that the functions u_1, u_2, $\cdots$, u_m calculated from the equations

$$\phi_\mu(x_1, x_2, \cdots, x_n, u_1, u_2, \cdots, u_m) = c_\mu$$

satisfy the system (1). First we obtain by differentiation the equations

$$\frac{\partial \phi_\mu}{\partial x_\kappa} + \sum_{\lambda=1}^{m} \frac{\partial \phi_\mu}{\partial u_\lambda} \frac{\partial u_\lambda}{\partial x_\kappa} = 0.$$

Again we multiply by a_κ and sum over κ; using (3), we have

$$\sum_{\lambda=1}^{m} b_\lambda \frac{\partial \phi_\mu}{\partial u_\lambda} = \sum_{\kappa=1}^{n} \sum_{\lambda=1}^{m} a_\kappa \frac{\partial \phi_\mu}{\partial u_\lambda} \frac{\partial u_\lambda}{\partial x_\kappa}$$

or

$$\sum_{\lambda=1}^{m} \frac{\partial \phi_\mu}{\partial u_\lambda} \left(b_\lambda - \sum_{\kappa=1}^{n} a_\kappa \frac{\partial u_\lambda}{\partial x_\kappa} \right) = 0.$$

Since the determinant of the quantities $\partial \phi_\mu / \partial u_\lambda$ does not vanish, the equations

$$b_\lambda - \sum_{\kappa=1}^{n} a_\kappa \frac{\partial u_\lambda}{\partial x_\kappa} = 0$$

hold, i.e., the system (1) is satisfied.

According to Ch. II, §2, the integration of the linear differential equation (3″) is equivalent to the integration of the system of characteristic differential equations

$$\frac{dx_\kappa}{ds} = a_\kappa \qquad (\kappa = 1, 2, \cdots, r).$$

Thus we recognize that *the system* (1) *of partial differential equations with the same principal part is equivalent to a system of m + n ordinary differential equations, namely, to the system*

(4)
$$\frac{dx_\kappa}{ds} = a_\kappa \qquad (\kappa = 1, 2, \cdots, n)$$

$$\frac{du_\lambda}{ds} = b_\lambda \qquad (\lambda = 1, 2, \cdots, m).$$

We use these results to *develop again the theory of characteristics for general differential equations of the first order.* We consider the differential equation

(5) $\qquad F(x_1, x_2, \cdots, x_n, u, u_{x_1}, u_{x_2}, \cdots, u_{x_n}) = 0$

and replace it by the following system of $n + 1$ quasi-linear differential equations for $u, p_1, \cdots, p_n$ with the same principal part, formed from the function $F(x_1, x_2, \cdots, x_n, u, p_1, p_2, \cdots, p_n)$:

(6)
$$\sum_{\nu=1}^{n} F_{p_\nu} \frac{\partial p_i}{\partial x_\nu} + F_u p_i + F_{x_i} = 0 \quad (i = 1, 2, \cdots, n),$$

$$\sum_{\nu=1}^{n} F_{p_\nu} \frac{\partial u}{\partial x_\nu} - \sum_{\nu=1}^{n} F_{p_\nu} p_\nu = 0.$$

The first n of these equations follow formally from (5) by differentiating with respect to x_i and then replacing u_{x_i} by p_i and $\partial^2 u / \partial x_i \partial x_\nu$ by $\partial p_i / \partial x_\nu$. By this substitution the last equation becomes trivial.

Starting from the system of quasi-linear differential equations (6) with the same principal part, we can now develop the theory of the differential equation (5) for the $n + 1$ unknown functions u, p_i. First we see from the foregoing remarks that the integration of (6) is equivalent to that of the system

$$(7) \qquad \frac{dx_i}{ds} = F_{p_i}, \qquad \frac{dp_i}{ds} = -F_{x_i} - p_i F_u, \qquad \frac{du}{ds} = \sum_{\nu=1}^{n} p_\nu F_{p_\nu}$$

of ordinary differential equations, i.e., to the integration of the characteristic differential equations derived for F in a different way in Ch. II, §7. We show further that a suitable specialized initial value problem for the system (6) is equivalent to one for the differential equation (5); this provides a new basis for the solution of an initial value problem carried out in Ch. II, §7, by means of the characteristic differential equations (7).

It is clear, to begin with, that for every solution of the differential equation (5) the functions u and $p_i = \partial u / \partial x_i$ are a solution of (6). Conversely we now consider a system of solutions u, p_i of the system of differential equations (6) which satisfies the following initial conditions: Let C be an $(n - 1)$-dimensional initial manifold in x, u-space which is nowhere characteristic. Let initial values of p_i be given on C such that $F = 0$ everywhere on C and, moreover, such that on C

$$(7\text{a}) \qquad\qquad du - \sum_{\nu=1}^{n} p_\nu \, dx_\nu = 0.$$

Furthermore, let those solutions of the system of differential equations (7) which go through every point of C with the corresponding initial values of p_i form an n-dimensional surface S given by

$$u = u(x_1, x_2, \cdots, x_n)$$

and containing C. This function u—together with the corresponding functions p_i—then is precisely the solution of the corresponding initial value problem for (6).

We have now to show that it also solves the initial value problem for $F = 0$. For this purpose we need only prove that the relations

$$F(x_1, x_2, \cdots, x_n, u, p_1, p_2, \cdots, p_n) = R(x_1, x_2, \cdots, x_n) = 0,$$

$$p_i(x_1, x_2, \cdots, x_n) - u_{x_i}(x_1, x_2, \cdots, x_n) = P_i(x_1, x_2, \cdots, x_n) = 0$$

are satisfied everywhere on the surface S. We bear in mind that the relations

(8)
$$\frac{\partial P_i}{\partial x_\kappa} - \frac{\partial P_\kappa}{\partial x_i} = \frac{\partial p_i}{\partial x_\kappa} - \frac{\partial p_\kappa}{\partial x_i}$$

hold for the functions $P_i(x_1, x_2, \cdots, x_n)$. Moreover, we have

$$R_{x_i} = \sum_{\nu=1}^{n} F_{p_\nu} \frac{\partial p_\nu}{\partial x_i} + F_{x_i} + F_u u_{x_i},$$

and, therefore, on the basis of the first n differential equations in (6) and equation (8),

(9)
$$R_{x_i} = \sum_{\nu=1}^{n} F_{p_\nu} \left(\frac{\partial P_\nu}{\partial x_i} - \frac{\partial P_i}{\partial x_\nu} \right) - F_u P_i.$$

On the other hand, the last differential equation in (6) may be written in the form

(10)
$$0 = \sum_{\nu=1}^{n} F_{p_\nu} P_\nu.$$

Thus we obtain

$$\sum_{i=1}^{n} R_{x_i} F_{p_i} = 0,$$

i.e., using the abbreviation

$$F_{p_i} = a_i,$$

where the functions $a_i(x_1, x_2, \cdots, x_n)$ are to be considered as known coefficients,

(10′)
$$\sum_{i=1}^{n} a_i R_{x_i} = 0.$$

On the integral surface S, consider now the curves, defined by (7), which generate this surface. Equation (10′) asserts that, on each of these curves,

$$\frac{dR}{ds} = 0;$$

i.e., since R vanishes at the initial point on C, we have

(11)
$$R \equiv 0$$

on S. Moreover, from equation (9) we obtain

(12) $$\sum_{\nu=1}^{n} a_{\nu} \frac{\partial P_{i}}{\partial x_{\nu}} - \sum_{\nu=1}^{n} a_{\nu} \frac{\partial P_{\nu}}{\partial x_{i}} + P_{i} F_{u} = 0,$$

while, after differentiation with respect to x_i, equation (10),

$$\sum_{\nu=1}^{n} a_{\nu} P_{\nu} = 0,$$

yields the relation

(13) $$\sum_{\nu=1}^{n} a_{\nu} \frac{\partial P_{\nu}}{\partial x_{i}} + \sum_{\nu=1}^{n} b_{i\nu} P_{\nu} = 0;$$

here $b_{i\nu} = \partial a_{\nu}/\partial x_{i}$ is again a known function of the variables

$$x_1, x_2, \cdots, x_n.$$

Adding (12) and (13) yields equations of the form

$$\sum_{\nu=1}^{n} a_{\nu} \frac{\partial P_{i}}{\partial x_{\nu}} + \sum_{\nu=1}^{n} c_{i\nu} P_{\nu} = 0,$$

where the quantities $c_{i\nu}$ are also known functions of $x_1, x_2, \cdots, x_n$. On each of the characteristic curves $dx_i/ds = a_i$, these equations go over into

$$\frac{dP_i}{ds} + \sum_{\nu=1}^{n} c_{i\nu} P_{\nu} = 0,$$

a system of ordinary linear homogeneous differential equations for the functions P_i. However, from (10) together with the initial conditions (7), we have the following result: Since C is not characteristic, the determinant Δ defined on page 98 does not vanish. The initial values of P_i are zero on C, and hence these functions vanish identically. In this way the required proof of the equivalence of our initial value problems for (6) and (5) is completed.

§3. *Haar's Uniqueness Proof*

The solution of a single nonlinear first order equation as developed in §7 of this chapter is based on the concept of characteristic strips; it had to be assumed that the first derivatives p and q of the solution are differentiable. Yet, the concept of solutions of the differential equation presupposes only continuity of the first derivatives. The previous proof of the existence of the solution is not valid under this weaker but natural condition. It is therefore remarkable that nevertheless for two independent variables at least the uniqueness was established by A. Haar[1]: *The initial value problem has at most one solution if continuity of merely the first derivatives is assumed.*

Specifically, consider the equation $u_y = G(x, y, u, u_x)$, in which G is assumed to satisfy a Lipschitz condition with respect to u and p. Let u and v be two solutions which agree on the interval $y = 0$, $x_1 \leq x \leq x_2$; then they coincide in the whole triangle $T:y \geq 0$, $y \leq (x - x_1)/k$, $y \leq (x_2 - x)/k$, where k is the Lipschitz constant of G with respect to p.

Proof: Denote the difference of u and v by w; u and v are solutions of the given differential equation. Subtracting these and remembering that G satisfies a Lipschitz condition we obtain a differential inequality for w:

$$| w_y | \leq \alpha | w | + k | w_x | ,$$

where α and k are the Lipschitz constants of G. Note that at points where w is *positive* this inequality can be written as

$$| w_y | \leq \alpha w + k | w_x | .$$

If we replace the constant α by a slightly larger one, β, the *strict* inequality

$$(1) \qquad\qquad | w_y | < \beta w + k | w_x |$$

can be achieved.

Define $W = e^{-\beta y} w$; we claim that W—and therefore w—is identically zero in T. For if W were not identically zero, it would assume positive or negative values. Suppose it assumes positive values. Let its maximum occur at P in T. P is not on the base of T since W is zero there. Therefore the directions $(-k, -1)$ and $(k, -1)$ lead *into* T at P, and these directional derivatives are nonpositive:

$$-kW_x - W_y \leq 0, \qquad kW_x - W_y \leq 0;$$

hence

$$W_y \geq k | W_x | \quad \text{at} \quad P.$$

[1] See A. Haar [1].

Rewriting this for the original variable w, we have

$$w_y \geq \beta w + k \mid w_x \mid \; ;$$

this contradicts (1).

If we assume that W takes on negative values in T, we are similarly led to a contradiction. Hence $u = v$ in T as stated.

Plis has shown, furthermore, that for two independent variables every solution merely with continuous first derivatives is generated by characteristic strips.[1]

Appendix 2 to Chapter II

Theory of Conservation Laws[2]

In Chapter II we have solved the noncharacteristic initial value problem for a single first order quasi-linear equation. The existence theorem obtained is local, i.e., it was only shown that solutions exist in some neighborhood of the initial curve. Furthermore, in §1 we constructed integral surfaces with an edge of regression; this shows that smooth solutions need not exist in the large.

In this appendix we shall investigate further the occurrence of discontinuities which terminate the regions of existence. Then we shall show how solutions may nevertheless be continued beyond their singularities, provided that the differential equation is interpreted as a "conservation law".

Consider quasi-linear equations of the form

$$(1) \qquad u_t = a u_x ,$$

$a = a(u)$ being a function of u. Singularities of solutions of such equations may arise from the initial values on the line $t = 0$ in the following manner:

According to the theory of characteristics, every solution u remains constant along each characteristic. The slope of a characteristic

---

[1] See A. Plis [2].

[2] For a more general discussion of this important subject see Ch. V, §9 and Ch. VI, §4, 9.

is $-1/a(u)$, and since u is a constant along the characteristic curve, it follows that the slope is a constant and all characteristic curves are straight lines.

From each point x_1 of the initial line $t = 0$ there issues a characteristic line whose slope is determined by the value of u at x_1. Suppose that there is a pair of points x_1 and x_2 on the initial line, say $x_1 < x_2$, where the prescribed values u_1 and u_2 of u are such that

$$a(u_1) < a(u_2).$$

Then the characteristic lines issuing from x_1 and x_2 intersect at time $t = (x_1 - x_2)/(a(u_1) - a(u_2))$. Since u has different values along the two characteristics, this shows that u cannot be defined as a continuous solution beyond this time.

The presence or absence of singularities can also be seen from the following implicit formula for the solution of equation (1) with initial value $u(x, 0) = \phi(x)$:

$$u - \phi(x + ta(u)) = 0.$$

According to the theorem on implicit functions, u is a regular function of x and t as long as the derivative of

$$u - \phi(x + ta(u))$$

with respect to u is not zero, i.e.,

$$ta'\phi' \neq 1.$$

This condition is satisfied for t small enough, but will be violated if t becomes larger (if a' and ϕ' have the same sign, such a value t is positive, otherwise t is negative). At the point where the condition is violated we expect u to become singular.

The foregoing examples show that solutions of quasi-linear equations in general do not exist in the large. But there does exist a theory in the large for solutions of conservation laws.

A conservation law for a single function u is an equation of the form

$$(2) \qquad \frac{d}{dt} \int_{x_1}^{x_2} u \, dx = f(u(x_2), x_2, t) - f(u(x_1), x_1, t),$$

where f is a given function of u, x, and t. This equation expresses the fact that the total quantity represented by the function u and contained in the interval (x_1, x_2) changes at a rate equal to the "flux" f of u through the end points of the interval. This is the form of those laws of physics which ignore dissipative mechanisms and thus express a "phenomenon of conservation".

If u is a differentiable solution of the conservation law (2) then the conservation law (2) is expressed by the quasi-linear differential equation

$$(2') \qquad u_t = f_u u_x + f_x = \frac{\partial}{\partial x} f(u, x, t),$$

obtained from (2) by differentiating with respect to x_1 and then setting $x_1 = x_2 = x$. However, as we shall confirm, (2) has discontinuous solutions as well. By admitting discontinuous solutions we shall show that within the class of discontinuous solutions the conservation law (2) has a solution in the large, whereas we have seen before that the differential equation $(2')$ does not.

Later in Chapters V and VI we shall study "shock" discontinuities for systems of conservation laws in any number of dimensions. We shall see there that the qualitative properties of discontinuous solutions of single conservation laws are the same as those of the physically more interesting systems.

Assume that f does not depend explicitly on x and t and abbreviate f_u as $a = a(u)$. Let $u(x, t)$ be a piecewise differentiable function which is a solution of the integral equation (2). Then u must be a solution of the differential equation

$$u_t = a u_x$$

whenever u is differentiable. Letting x_1 and x_2 approach a point of discontinuity from opposite sides, we deduce (see details in Ch. V, §9) the "jump relation"

$$(3) \qquad [f] = -U[u],$$

where U is the speed of propagation of the discontinuity and the symbol $[g]$ denotes the jump across the discontinuity of the quantity g. In case of small discontinuities we have

$$U = -\frac{[f]}{[u]} \approx -\frac{df}{du} = -a.$$

Since $-a$ is the speed corresponding to the characteristic line direction, we conclude that the small discontinuities propagate with nearly characteristic speed.

Consider the example

$$(4) \qquad \frac{d}{dt} \int u = \tfrac{1}{2}u^2(x_2) - \tfrac{1}{2}u^2(x_1).$$

When the differentiation is carried out, we obtain

$$u_t = uu_x .$$

Dividing by u we have

$$(5) \qquad \frac{u_t}{u} = (\log u)_t = u_x .$$

Denoting $\log u$ by v, we can rewrite (5) as the conservation law

$$(5') \qquad \frac{d}{dt} \int v = \exp v(x_2) - \exp v(x_1).$$

The jump relation (3) for the conservation law (4) is

$$(6) \qquad \frac{u_1 + u_2}{2} = -U.$$

where u_1 and u_2 denote values of u on either side of the line of discontinuity. For (5') the jump relation is

$$\frac{e^{v_1} - e^{v_2}}{v_1 - v_2} = -U.$$

From these jump conditions we conclude: If u is a discontinuous solution of (4) then $v = \log u$ is *not* a solution of (5'). One could say that jump relations are not invariant under change of dependent variables; two conservation laws, such as (4) and (5), may correspond to the same differential equations for smooth solutions, but as conservation laws for discontinuous solutions they need not be equivalent.

Next we show by an example that *solutions of conservation laws are not determined uniquely by their initial data.* We take again the conservation law (4). The function

$$u(x, t) = \begin{cases} 1 & \text{for} \quad 2x < -t \\ 0 & \text{for} \quad -t < 2x \end{cases}$$

is a discontinuous solution of (4), since on either side of the line $2x = -t$, u is a constant and thus a smooth solution of equation (4), and across the line of discontinuity $2x = t$ the jump condition (6) is satisfied. On the other hand, the function

$$u'(x, t) = \begin{cases} 1 & \text{for} \quad x < -t \\ \dfrac{-x}{t} & \text{for} \quad -t < x < 0 \\ 0 & \text{for} \quad 0 < x \end{cases}$$

is continuous for t positive and satisfies the differential equation everywhere except on the lines $x = 0$ and $x = -t$. From this one can easily show by integration that u' is a continuous solution of (4). The two solutions u and u' have the same value at $t = 0$. More generally it is possible to show that for arbitrarily prescribed initial data there exist uncountably many discontinuous solutions with the same prescribed initial data.

Among all these discontinuous solutions with the same initial value there is only one which has physical significance. This solution we shall call the *permissible* one.

What is needed is a mathematical principle characterizing permissible solutions. Such a principle is suggested by the argument used earlier to deduce the occurrence of discontinuity from the crossing of characteristics. A discontinuity is permissible if it prevents the crossing of characteristics. Thus we have the following criterion: A discontinuous solution is *permissible* if every line of discontinuity is crossed by the *forward* drawn characteristic issuing from either side. Analytically this condition means that for a permissible discontinuity

$$-a(u_L) \geq U \geq -a(u_R),$$

where u_L and u_R denote the value of u to the left and right of the line of discontinuity, and U denotes the velocity of propagation of the discontinuity.

Germain and Bader[1] have shown that two permissible discontinuous solutions which agree at $t = 0$ are identical. A more general definition of permissibility and a more general uniqueness theorem were given by O. A. Oleinik.[2]

[1] See P. Germain and R. Bader [1].
[2] See O. A. Oleinik [2] and [3].

It is possible to state an analogous condition of permissibility for discontinuities of solutions of systems of conservation laws. When applied to the equations of compressible fluid flow this condition turns out to be equivalent analytically to the assertion that upon crossing a discontinuity the entropy of the flow increases.

We shall give now an explicit formula for permissible solutions of a conservation law with arbitrarily prescribed initial values. This formula is due to E. Hopf,[1] O. A. Oleinik,[2] and P. D. Lax.[3] We shall assume that $a(u)$ is a monotonic function of u; this implies that $f(u)$ is convex or concave.

Now $g(s)$ is defined as the conjugate of the convex (or concave) function $f(u)$, given by the formula

$$g(s) = \underset{u}{\text{Max}}\ (\text{Min})\{us + f(u)\};$$

we define $b(s)$ as the derivative of g with respect to s. Let $\phi(x)$ be the prescribed initial value

$$u(x, 0) = \phi(x).$$

Define $\Phi(y)$ as the integral of ϕ, i.e.,

$$\frac{d\Phi}{dy} = \phi(y).$$

Consider the function

$$\Phi(y) + tg\left(\frac{x - y}{t}\right);$$

for fixed x and t this is a continuous function of y. It is easy to show that for t fixed and with the exception of a countable set of values of x, the function has a unique maximum (or minimum) in y, whose position we denote by $y_0(x, t)$. We define

$$(7) \qquad u(x, t) = b\left(\frac{x - y_0}{t}\right)$$

and assert:

---

[1] See E. Hopf [3].
[2] See O. A. Oleinik [1].
[3] See P. D. Lax [3].

The function u defined by formula (7) *is a permissible solution of* (1) *with initial value φ.*

For a verification of this assertion and further properties of solutions furnished by this formula, see, e.g., P. D. Lax.[1]   An existence theorem for conservation laws where f is less restricted has been given by Kalashnikov.[2]

[1] See P. D. Lax [2].
[2] See A. S. Kalashnikov [1].

CHAPTER III

Differential Equations of Higher Order

Partial differential equations of higher than first order present so many diverse aspects that a unified general theory (as in Chapter II) is not possible. There is a decisive distinction between several types of differential equations, called "elliptic", "hyperbolic", and "parabolic", each of which shows an entirely different behavior regarding properties and construction of solutions.[1] In this chapter we shall introduce this classification, guided by examples of physical interest. Moreover we shall discuss in a preliminary way methods of approach toward the solution of relevant problems. The subsequent chapters will be primarily concerned with a systematic theory of elliptic and hyperbolic problems.[2]

Some classical differential equations of second order for a function $u(x, y, z)$ are representative examples:

The Laplace equation (elliptic type): $u_{xx} + u_{yy} + u_{zz} = 0$

The wave equation (hyperbolic type): $u_{xx} + u_{yy} - u_{zz} = 0$

The heat equation (parabolic type): $u_z = u_{xx} + u_{yy}$

§1. *Normal Forms for Linear and Quasi-Linear Differential Operators of Second Order in Two Independent Variables*

For linear and also for quasi-linear differential equations of second order (or corresponding systems of two first order equations) in two independent variables, the classification can be carried out by explicit elementary steps without reference to general theory. It originates from the attempt to find simple normal forms.

---

[1] For a more general and abstract approach see L. Hörmander [4], B. Malgrange [1], and L. Ehrenpreis [1]. For a survey of recent problems see I. G Petrovskii [2].

[2] See also the recent exposition by G. Hellwig [1].

1. Elliptic, Hyperbolic, and Parabolic Normal Forms.   Mixed Types.

A linear differential operator of second order for the function $u(x, y)$ is given by

$$(1) \qquad L[u] = au_{xx} + 2bu_{xy} + cu_{yy} \, ;$$

the coefficients a, b, c are assumed to be continuously differentiable and not simultaneously vanishing functions of x and y in a domain G. We consider

$$(2) \qquad L[u] + g(x, y, u, u_x, u_y) = L[u] + \cdots ,$$

where the differential expression $g(x, y, u, u_x, u_y)$ is not necessarily linear and contains no second derivatives. Our object is to transform the differential operator (2) or the corresponding differential equation

$$(3) \qquad L[u] + \cdots = 0$$

into a simple normal form by introducing new independent variables

$$(4) \qquad \xi = \phi(x, y), \qquad \eta = \psi(x, y).$$

Denoting by $u(\xi, \eta)$ the function into which $u(x, y)$ is transformed, we have the relations

$$u_x = u_\xi \phi_x + u_\eta \psi_x , \qquad u_y = u_\xi \phi_y + u_\eta \psi_y ,$$

$$u_{xx} = u_{\xi\xi} \phi_x^2 + 2u_{\xi\eta} \phi_x \psi_x + u_{\eta\eta} \psi_x^2 + \cdots ,$$

$$u_{xy} = u_{\xi\xi} \phi_x \phi_y + u_{\xi\eta}(\phi_x \psi_y + \phi_y \psi_x) + u_{\eta\eta} \psi_x \psi_y + \cdots ,$$

$$u_{yy} = u_{\xi\xi} \phi_y^2 + 2u_{\xi\eta} \phi_y \psi_y + u_{\eta\eta} \psi_y^2 + \cdots .$$

(Here again the dots mean terms in which no second order derivatives of u appear.) Thus the differential operator (1) assumes the form

$$(5) \qquad \Lambda[u] = \alpha u_{\xi\xi} + 2\beta u_{\xi\eta} + \gamma u_{\eta\eta}$$

with

$$(6) \qquad \begin{aligned} \alpha &= a\phi_x^2 + 2b\phi_x \phi_y + c\phi_y^2 , \\ \beta &= a\phi_x \psi_x + b(\phi_x \psi_y + \phi_y \psi_x) + c\phi_y \psi_y , \\ \gamma &= a\psi_x^2 + 2b\psi_x \psi_y + c\psi_y^2 . \end{aligned}$$

Moreover, a, b, c and α, β, γ are related by

(7) $$\alpha\gamma - \beta^2 = (ac - b^2)(\phi_x\psi_y - \phi_y\psi_x)^2,$$

and by the identity for the "characteristic quadratic form"

$$Q(l, m) = al^2 + 2blm + cm^2 = \alpha\lambda^2 + 2\beta\lambda\mu + \gamma\mu^2,$$

where the variables l, m and λ, μ are connected at a fixed point x, y by the linear transformation

$$l = \lambda\phi_x + \mu\phi_y, \qquad m = \lambda\psi_x + \mu\psi_y.$$

In the transformation (4) two functions ϕ, ψ are at our disposal so that we may impose two conditions on the transformed coefficients α, β, γ aiming at simple normal forms of the transformed differential equation (5).

We consider the following sets of conditions:

I. $\alpha = \gamma$, $\qquad \beta = 0$,

II. $\alpha = -\gamma$, $\qquad \beta = 0$ $\qquad$ (or $\alpha = \gamma = 0$),

III. $\beta = \gamma = 0$.

Which of these conditions can be satisfied by the transformations—of course always real transformations are assumed—depends on the algebraic character of the form $Q(l, m)$, or geometrically speaking on the character of the quadratic curve in the l, m-plane, for fixed x, y: $Q(l, m) = 1$; this curve may be an ellipse, a hyperbola, or a parabola. Accordingly at a point x, y we call the operator $L[u]$

I. elliptic if $ac - b^2 > 0$,

II. hyperbolic if $ac - b^2 < 0$,

III. parabolic if $ac - b^2 = 0$.

The corresponding normal forms of the differential operator are

I. $\Lambda[u] = \alpha(u_{\xi\xi} + u_{\eta\eta}) + \cdots$,

II. $\begin{cases} \Lambda[u] = \alpha(u_{\xi\xi} - u_{\eta\eta}) + \cdots \\ \quad\text{or} \\ \Lambda[u] = 2\beta u_{\xi\eta} + \cdots, \end{cases}$

$$\text{III.} \quad \Lambda[u] = \alpha u_{\xi\xi} + \cdots,$$

and the normal forms of the differential equation are

$$\text{I.} \quad u_{\xi\xi} + u_{\eta\eta} + \cdots = 0,$$

$$\text{II.} \begin{cases} u_{\xi\xi} - u_{\eta\eta} + \cdots = 0, \\ \text{or} \\ u_{\xi\eta} + \cdots = 0, \end{cases}$$

$$\text{III.} \quad u_{\xi\xi} + \cdots = 0.$$

For fixed x, y, such a normal form can always be obtained simply by the linear transformation which takes Q into the corresponding normal form. However, assuming that the operator L is of the same type in every point of a domain G we want to find functions ϕ and ψ which will transform $L[u]$ into *a normal form at every point of* G. Success depends on whether certain first order systems of linear partial differential equations can be solved.

Without loss of generality we may assume a $\neq 0$ everywhere in the domain G; otherwise, either the equivalent assumption $c \neq 0$ would hold or our expression would already be in the normal form II.

To determine transformation functions ϕ and ψ for the whole domain G, we first assume that $L[u]$ is hyperbolic in G, and that the new coefficients must satisfy the condition $\alpha = \gamma = 0$. Equations (6) then lead to the quadratic equation

$$(8) \qquad Q = a\lambda^2 + 2b\lambda\mu + c\mu^2 = 0$$

for the ratio λ/μ of the derivatives ϕ_x/ϕ_y and ψ_x/ψ_y.

If $L[u]$ is hyperbolic in G, then $ac - b^2 < 0$, and thus equation (8) has two distinct real solutions λ_1/μ_1 and λ_2/μ_2. Since $a \neq 0$, we may assume that

$$\mu_1 = \mu_2 = 1;$$

then (8) defines the quantities λ_1 and λ_2 in G as continuously differentiable functions of x and y. Thus in the hyperbolic case we obtain the normal form

$$(9) \qquad \beta u_{\xi\eta} + \cdots = 0$$

by determining the transformation functions ϕ and ψ from the differential equations

(10) $$\phi_x - \lambda_1 \phi_y = 0, \qquad \psi_x - \lambda_2 \psi_y = 0.$$

These two first order linear homogeneous partial differential equations yield, in fact, two families of curves $\phi = $ const. and $\psi = $ const. which also may be defined as the families of solutions of the ordinary differential equations

$$y' + \lambda_1 = 0, \qquad y' + \lambda_2 = 0$$

or

$$a y'^2 - 2b y' + c = 0,$$

where y is considered as a function of x along the curves of the family. The relation

$$\lambda_1 - \lambda_2 = \frac{2}{a} \sqrt{b^2 - ac}$$

shows that curves of the two families cannot be tangent at any point of G, and that $\phi_x \psi_y - \phi_y \psi_x \neq 0$. If $\alpha = \gamma = 0$, equation (7) implies $\beta \neq 0$.

The curves $\xi = \phi(x, y) = $ const. and $\eta = \psi(x, y) = $ const. are called the *characteristic curves of the linear hyperbolic differential operator* $L[u]$.

Since we may divide (9) by β, we can state: *If $L[u]$ is hyperbolic, i.e. $ac - b^2 < 0$, the second order differential equation (3) may be transformed into the normal form*

(11) $$u_{\xi\eta} + \cdots = 0$$

by introducing the two families of characteristic curves $\xi = $ const. and $\eta = $ const. as coordinate curves.

If $ac - b^2 > 0$ holds, then the operator (2) is *elliptic* in G. In this case the quadratic equation (8) has no real solutions, but it has two conjugate complex solutions λ_1 and λ_2 which are continuous complex-valued functions of the real variables x, y. The equations $\alpha = \gamma = 0$ are satisfied by no family of real curves $\phi = $ const.; i.e., there are no characteristic curves. If, however, a, b, c are analytic functions of x, y and if we assume that $\phi(x, y)$ and $\psi(x, y)$ are analytic, then we may consider the differential equations (10) for complex x and y, and, as before, transform them into the new variables ξ

and η which become complex conjugate. Introducing real independent variables ρ and σ by the equations

(12) $$\frac{\xi + \eta}{2} = \rho, \qquad \frac{\xi - \eta}{2i} = \sigma$$

we obtain $4u_{\xi\eta} = u_{\sigma\sigma} + u_{\rho\rho}$. Thus we arrive at *the normal form*

(13) $$\Delta u + \cdots = u_{\sigma\sigma} + u_{\rho\rho} + \cdots = 0$$

in the elliptic case.

For the preceding transformation involving complex quantities we had to require a condition of analyticity for these coefficients, a condition incisive but essentially alien to the problem. To avoid this restriction, we may use the following procedure (not involving complex quantities) of transforming an elliptic expression to normal form: Writing ρ and σ instead of ξ and η in equations (3), (4), we stipulate the conditions

$$\alpha = \gamma, \qquad \beta = 0$$

or, explicitly,

$$a\rho_x^2 + 2b\rho_x\rho_y + c\rho_y^2 = a\sigma_x^2 + 2b\sigma_x\sigma_y + c\sigma_y^2 \,,$$

$$a\rho_x\sigma_x + b(\rho_x\sigma_y + \rho_y\sigma_x) + c\rho_y\sigma_y = 0.$$

These differential equations can be reduced by an elementary algebraic manipulation to the following first order system of linear partial differential equations:

(14) $$\sigma_x = \frac{b\rho_x + c\rho_y}{W}, \qquad \sigma_y = -\frac{a\rho_x + b\rho_y}{W},$$

where

$$W^2 = ac - b^2$$

and either sign is permissible for W. From these so-called *Beltrami differential equations*, we obtain immediately, by eliminating one of the unknowns (e.g., σ), the following second order differential equation for the other quantity:

(15) $$\frac{\partial}{\partial x}\frac{a\rho_x + b\rho_y}{W} + \frac{\partial}{\partial y}\frac{b\rho_x + c\rho_y}{W} = 0.$$

The transformation of the differential equation to normal form (13) in a neighborhood of a point is given by any pair of functions ρ, σ satisfying (14) and having a nonvanishing Jacobian

$$\sigma_x\rho_y - \sigma_y\rho_x = \frac{1}{W}(a\rho_x^2 + 2b\rho_x\rho_y + c\rho_y^2).$$

Such functions are determined once we have a solution of (15) with nonvanishing gradient. We shall see in Ch. IV, §7 that under certain smoothness assumptions on the coefficients, (e.g., existence of continuous derivatives up to second order of a, b, c) such a solution always exists—at least *locally*—and hence a normalizing parameter system ρ, σ may be introduced in a neighborhood of any point.[1]

The third case is the *parabolic case*: $ac - b^2 = 0$. The quadratic equation (8) then has one real root, and we can accordingly introduce one family of curves $\xi = \phi(x, y)$ in such a way that $\alpha = 0$ holds; then, on account of relation (7), we must also have $\beta = 0$, while, for instance, for $\psi = x$ in G, $\gamma = a \neq 0$. In the *parabolic case* we obtain *the normal form*

$$u_{\eta\eta} + \cdots = 0.$$

The theorem stated at the beginning has thus been proved.

Note that the transformation to the normal form is by no means uniquely determined. For example, in the elliptic case the normal form remains unchanged if we subject the ρ, σ-domain to any conformal mapping.

2. Examples. Several examples of the different types of differential

---

[1] As a consequence of the properties of ρ, σ we can now deduce the *existence of a single normalizing parameter system $\rho(x, y)$, $\sigma(x, y)$ in the large*, i.e., in the whole domain G. We use the fact: If two solutions ρ, σ and σ_1, ρ_1 of (14) are defined in some sufficiently small neighborhood and if $W < 0$, then $\rho_1 + i\sigma_1$ is a complex analytic function of $\rho + i\sigma$; indeed, a simple computation shows that the Cauchy-Riemann equations are satisfied. Thus the transformation given by the local normalizing parameters defined in any two neighborhoods is conformal in the intersection of these neighborhoods. Therefore, G, together with the normalizing parameters in a system of neighborhoods covering G forms a Riemann surface (the notion of Riemann surface is presented in H. Weyl [2]; see also R. Courant [2]) on which analytic functions are defined as analytic in the normalizing parameters. The problem of finding a single uniform normalizing parameter system ρ, σ in the whole domain G is thus equivalent to that of finding a complex function $\rho + i\sigma$ which maps G one-to-one into a region of the ρ, σ-plane and which is analytic in the sense just described. This is exactly the problem solved by the general uniformization theorem for plane domains (see references given above), and the existence of such a mapping is therefore insured.

equations have already been discussed in Ch. I, §1. The simplest hyperbolic equation (that of the vibrating string) $u_{xx} - u_{tt} = 0$ was completely solved. The prototype of the elliptic differential equation is the potential equation $\Delta u = u_{xx} + u_{yy} = 0$ (see, e.g., Ch. I, §1). The parabolic equation of heat conduction $u_t - u_{xx} = 0$ was discussed in Ch. I, §3.

From the "type" of the equation we shall deduce important properties which suggest not only methods of solution but also criteria to determine whether or not problems are reasonably posed.

Incidentally, a given differential equation may be of different type in different domains (mixed types); for example, the equation

$$(16) \qquad u_{xx} + y u_{yy} = 0$$

is elliptic for $y > 0$ and hyperbolic for $y < 0$, since $ac - b^2 = y$

In the domain $y < 0$ equation (8), i.e., the equation

$$\lambda^2 + y\mu^2 = 0,$$

possesses the two real roots $\lambda/\mu = \pm \sqrt{-y}$; thus the two differential equations

$$(17) \qquad \phi_x + \sqrt{-y}\,\phi_y = 0, \qquad \psi_x - \sqrt{-y}\,\psi_y = 0$$

hold for ϕ and ψ. They have the solutions

$$\phi = x + 2\sqrt{-y},$$
$$\psi = x - 2\sqrt{-y}.$$

By the transformation

$$(18) \qquad \begin{aligned} \xi &= x + 2\sqrt{-y}, \\ \eta &= x - 2\sqrt{-y}, \end{aligned}$$

(16) assumes the hyperbolic normal form

$$(19) \qquad u_{xx} + y u_{yy} = 4u_{\xi\eta} + \frac{2}{\xi - \eta}\,(u_\xi - u_\eta) = 0$$

for $y < 0$. The characteristic curves are given by the parabolas

$$y = -\tfrac{1}{4}(x - c)^2;$$

in particular, the curves $\phi = $ const. are the branches of the parabolas having positive slope, the curves $\psi = $ const. those having negative slope (cf. Figure 3).

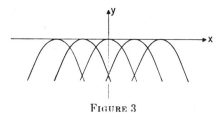

FIGURE 3

For $y > 0$ we write

(20)
$$\xi = x,$$
$$\eta = 2\sqrt{y};$$

by this transformation, equation (16) assumes the elliptic normal form

(21) $$u_{xx} + yu_{yy} = u_{\xi\xi} + u_{\eta\eta} - \frac{1}{\eta} u_\eta = 0.$$

Similarly, the differential equation

(22) $$u_{xx} + xu_{yy} = 0$$

known as "Tricomi's equation"[1] is elliptic for $x > 0$ and hyperbolic for $x < 0$ because $ac - b^2 = x$.

In the half-plane $x < 0$, the equations

(23)
$$\xi = \phi(x, y) = \tfrac{3}{2}y + \left(\sqrt{-x}\right)^3,$$
$$\eta = \psi(x, y) = \tfrac{3}{2}y - \left(\sqrt{-x}\right)^3$$

transform (22) into the normal form

(24) $$u_{xx} + xu_{yy} = 9x\left[u_{\xi\eta} - \frac{1}{6(\xi - \eta)} (u_\xi - u_\eta) \right] \qquad (\xi > \eta)$$

The characteristic curves are the cubic parabolas

$$y - c = \pm\tfrac{2}{3}\left(\sqrt{-x}\right)^3;$$

the branches with a downward direction yield the curves $\phi = $ const., those directed upward yield the curves $\psi = $ const. (cf. Figure 4).

---

[1] This equation is of special interest in gas dynamics. The important work of Tricomi [1] has recently led to an extensive literature. See, for example, L. Bers [5], and P. Germain [1].

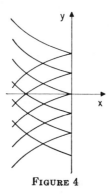

FIGURE 4

For $x > 0$ we write

$$\xi = \tfrac{3}{2}y - i\sqrt{x^3},$$

$$\eta = \tfrac{3}{2}y + i\sqrt{x^3},$$

and set

(25)
$$\rho = \frac{\xi + \eta}{2} = \tfrac{3}{2}y,$$

$$\sigma = \frac{\xi - \eta}{2i} = -\sqrt{x^3};$$

we obtain by this transformation the normal form

(26)
$$u_{xx} + xu_{yy} = \tfrac{9}{4}x\left[(u_{\rho\rho} + u_{\sigma\sigma}) + \frac{1}{3\sigma}u_\sigma\right].$$

The functions (25) satisfy the Beltrami differential equations

(27)
$$\sigma_x = -\sqrt{x}\rho_y,$$

$$\sigma_y = \frac{1}{\sqrt{x}}\rho_x.$$

3. Normal Forms for Quasi-Linear Second Order Differential Equations in Two Variables. The transformation into normal forms will now be generalized to include nonlinear, specifically quasi-linear, differential equations. With the convenient abbreviations

$$p = u_x, \qquad q = u_y, \qquad r = u_{xx}, \qquad s = u_{xy}, \qquad t = u_{yy},$$

quasi-linear differential operators have the form

(28)
$$L[u] = ar + 2bs + ct + d,$$

where a, b, c, d are given functions of the quantities x, y, u, p, q. Such an operator is again called *elliptic* if $ac - b^2 > 0$, *hyperbolic* if $ac - b^2 < 0$, and *parabolic* if $ac - b^2 = 0$. However, since a, b, c depend on the specific function $u(x, y)$, the character of L at a point (x, y) also depends on u and on the derivatives p, q as functions of x, y. For example, the differential operator $uu_{xx} + u_{yy}$ is elliptic in the region where $u(x, y) > 0$, and hyperbolic where $u(x, y) < 0$. Similarly the differential equations for the two families of characteristics depend on u, and hence it is not possible to introduce *a priori* two fixed families of characteristic curves as coordinate curves for all functions u simultaneously. After substituting a specific function $u(x, y)$ and its derivatives p and q in L, we can treat it in the same way as *linear* second order operators were treated. If L is hyperbolic for this function u, one can introduce characteristic parameters $\xi = \phi(x, y)$, $\eta = \psi(x, y)$ which satisfy equations (10):

$$\phi_x - \lambda_1 \phi_y = 0, \qquad \psi_x - \lambda_2 \psi_y = 0.$$

Similarly, in the elliptic case the condition that ρ and σ be characteristic parameters is expressed by

(29)
$$a\rho_x^2 + 2b\rho_x \rho_y + c\rho_y^2 = a\sigma_x^2 + 2b\sigma_x \sigma_y + c\sigma_y^2,$$

$$a\rho_x \sigma_x + b(\rho_x \sigma_y + \rho_y \sigma_x) + c\rho_y \sigma_y = 0.$$

The crucial step in eliminating the reference to a specific function u in equations (10) or (29) consists in considering u, x and y simultaneously as functions of ξ and η instead of u as a function of x and y. If we introduce ξ and η as new independent variables, then $L[u] = 0$ and (10) or (29) go over into differential equations for u, x and y as functions of ξ and η. To obtain these equations we use the differentiation formulas for inverse functions, expressing x, y-derivatives in terms of ξ, η-derivatives:

(30)
$$Dx_\xi = \eta_y, \qquad Dx_\eta = -\xi_y, \qquad p = u_x = D(u_\xi y_\eta - u_\eta y_\xi),$$

$$Dy_\xi = -\eta_x, \qquad Dy_\eta = \xi_x, \qquad q = u_y = D(u_\eta x_\xi - u_\xi x_\eta),$$

where

$$D = \phi_x \psi_y - \psi_x \phi_y = (x_\xi y_\eta - x_\eta y_\xi)^{-1}$$

is the Jacobian of the transformation. Equations (10) and (29) then are transformed into

(10') $$y_\eta + \lambda_1 x_\eta = 0, \qquad y_\xi + \lambda_2 x_\xi = 0$$

and

(29')
$$ay_\rho^2 - 2by_\rho x_\rho + cx_\rho^2 = ay_\sigma^2 - 2by_\sigma x_\sigma + cx_\sigma^2,$$
$$ay_\rho y_\sigma - b(x_\rho y_\sigma + y_\rho x_\sigma) + cx_\rho x_\sigma = 0.$$

In the hyperbolic case (10'), the functions λ_1 and λ_2 depend on u, p, q and on x and y. If, by means of (30), p and q are replaced by their expressions in terms of ξ, η-derivatives, then (10') represents two relations between the quantities x, y, u and their partial derivatives of the first order with respect to ξ and η. As said before, in contrast to the linear case, these two equations do not suffice to determine, independently of $u(x, y)$, the curves $\xi = $ const. and $\eta = $ const. They now represent a system of two first order partial differential equations for three functions $u(\xi, \eta)$, $x(\xi, \eta)$, $y(\xi, \eta)$, i.e., an underdetermined system.

This consideration suggests that we add the original second order differential equation $L[u] = 0$ to our two "characteristic" equations, thus obtaining a system of three differential equations for the three functions u, x, y of ξ, η. Geometrically, this formulation means that we seek the integral surface not in the asymmetric form $u(x, y)$ but in a parametric representation by means of characteristic parameters ξ, η.

From the transformation of the differential expression (28) we obtain without difficulty in the hyperbolic case the following forms in which only the mixed second derivatives with respect to ξ and η appear:

(31)
$$x_{\xi\eta}(y_\xi u_\eta - u_\xi y_\eta) + y_{\xi\eta}(u_\xi x_\eta - u_\eta x_\xi) + u_{\xi\eta}(x_\xi y_\eta - x_\eta y_\xi)$$
$$= (x_\xi y_\eta - x_\eta y_\xi)^2 \frac{d}{2\sqrt{b^2 - ac}},$$

or

(32)
$$\begin{vmatrix} x_{\xi\eta} & y_{\xi\eta} & u_{\xi\eta} \\ x_\xi & y_\xi & u_\xi \\ x_\eta & y_\eta & u_\eta \end{vmatrix} = (x_\xi y_\eta - x_\eta y_\xi)^2 \frac{d}{2\sqrt{b^2 - ac}},$$

or, if we consider the numbers x, y, u as components of a position vector $\mathbf{x}$,

$$(33) \qquad \mathbf{x}_{\xi\eta}(\mathbf{x}_\xi \times \mathbf{x}_\eta) = (x_\xi y_\eta - x_\eta y_\xi)^2 \frac{d}{2\sqrt{b^2 - ac}}.$$

In particular, if $d = 0$, we arrive at the remarkable result: *The second order normal form* (31) *of a hyperbolic equation is independent of a, b, c.*

Again we note that the expressions (30) are to be substituted everywhere in our differential equations for p and q. Our system (10′), (31) of three partial differential equations for the position vector $\mathbf{x}$ is the desired general normal form for the hyperbolic case.

If on a surface u (and hence in a certain neighborhood) the equation is elliptic, i.e., $b^2 - ac < 0$, another normal form arises. We obtain the corresponding transformation either directly from equations (29′) or formally using the above result by the substitution

$$\frac{\xi + \eta}{2} = \rho, \qquad \frac{\xi - \eta}{2i} = \sigma.$$

We conclude: *In the elliptic case the differential equation* (28) *is equivalent to the following system of three differential equations for the quantities* x, y, u (*or for the position vector* $\mathbf{x}$ *as a function of the parameters* ρ *and* σ):

$$(34) \qquad \begin{aligned} &ay_\rho^2 - 2by_\rho x_\rho + cx_\rho^2 = ay_\sigma^2 - 2by_\sigma x_\sigma + cx_\sigma^2, \\ &ay_\rho y_\sigma - b(y_\rho x_\sigma + y_\sigma x_\rho) + cx_\rho x_\sigma = 0, \end{aligned}$$

which can be written in vector notation:

$$(34a) \quad \Delta\mathbf{x}(\mathbf{x}_\rho \times \mathbf{x}_\sigma) = \begin{vmatrix} \Delta x & \Delta y & \Delta u \\ x_\rho & y_\rho & u_\rho \\ x_\sigma & y_\sigma & u_\sigma \end{vmatrix} = (x_\rho y_\sigma - x_\sigma y_\rho)^2 \frac{d}{\sqrt{ac - b^2}},$$

where the vector Δx denotes the Laplace operator on the vector x.

In particular, if $d = 0$, *the second order differential equation is independent of a, b, c; it has the form*

$$\Delta\mathbf{x}(\mathbf{x}_\rho \times \mathbf{x}_\sigma) = 0,$$

where $\times$ means vectorial multiplication.

4. Example. Minimal Surfaces.[1] Let us consider the differential equation of minimal surfaces

$$(35) \qquad (1 + q^2)r - 2pqs + (1 + p^2)t = 0;$$

since $ac - b^2 = 1 + p^2 + q^2 > 0$ this differential equation is everywhere elliptic and can be transformed into a normal system of the form (34). In fact, a short calculation yields the following equations:

$$(36) \qquad x_\rho^2 + y_\rho^2 + u_\rho^2 = x_\sigma^2 + y_\sigma^2 + u_\sigma^2 \quad \text{or} \quad \mathbf{x}_\rho^2 = \mathbf{x}_\sigma^2 ,$$

$$x_\rho x_\sigma + y_\rho y_\sigma + u_\rho u_\sigma = 0 \quad \text{or} \quad \mathbf{x}_\rho \mathbf{x}_\sigma = 0,$$

$$(37) \qquad \Delta\mathbf{x}(\mathbf{x}_\rho \times \mathbf{x}_\sigma) = 0 \quad \text{with} \quad \Delta\mathbf{x} = \mathbf{x}_{\rho\rho} + \mathbf{x}_{\sigma\sigma} .$$

This system can be put into a simpler form: Differentiating equations (36) we have

$$\mathbf{x}_{\rho\rho}\mathbf{x}_\rho = \mathbf{x}_{\rho\sigma}\mathbf{x}_\sigma \quad \text{and} \quad \mathbf{x}_{\sigma\sigma}\mathbf{x}_\rho = -\mathbf{x}_{\rho\sigma}\mathbf{x}_\sigma ;$$

hence,

$$\mathbf{x}_\rho \Delta\mathbf{x} = 0 \quad \text{and} \quad \mathbf{x}_\sigma \Delta\mathbf{x} = 0.$$

On the other hand, equation (37) implies that $\Delta\mathbf{x}$ is a linear combination $\Delta\mathbf{x} = \alpha\mathbf{x}_\rho + \beta\mathbf{x}_\sigma$ of the vectors $\mathbf{x}_\rho$ and $\mathbf{x}_\sigma$. Hence $\alpha = \beta = 0$ and thus $\Delta\mathbf{x} = 0$. Thus, *a minimal surface in parametric representation with suitable parameters ρ and σ may be characterized by the following conditions: Each of the three coordinates u, x, y satisfies the potential equation*, i.e.,

$$(38) \qquad \Delta x = 0, \qquad \Delta y = 0, \qquad \Delta u = 0.$$

Moreover, they satisfy the conditions

$$(39) \qquad \begin{aligned} A &= \mathbf{x}_\sigma^2 - \mathbf{x}_\rho^2 = 0, \\ B &= 2\mathbf{x}_\rho\mathbf{x}_\sigma = 0. \end{aligned}$$

With the usual notation of differential geometry

$$E = \mathbf{x}_\rho^2 , \qquad F = \mathbf{x}_\rho\mathbf{x}_\sigma , \qquad G = \mathbf{x}_\sigma^2 ,$$

(39) implies the conditions

$$E - G = 0, \qquad F = 0$$

for the first fundamental form of the surface.

[1] Compare R. Courant [2].

These additional conditions apparently add two new differential equations to the three equations (38); however, they merely represent a boundary condition. We need not impose the additional restrictions (39) in a whole two-dimensional ρ, σ-domain, but only along some closed curve in the ρ, σ-plane. From equations (38), the two relations

$$A_\rho = B_\sigma, \qquad A_\sigma = -B_\rho$$

follow immediately. They characterize $A + iB$ as an analytic function of the complex variables $\rho + i\sigma$; therefore, $A + iB$ vanishes identically if the real part A vanishes on any closed curve (e.g., the boundary) and if B is zero at some point.

Two immediate conclusions are significant for the theory of minimal surfaces:

(1) The mapping of the ρ, σ-plane on the minimal surface is conformal.

(2) The representation of the minimal surface by harmonic functions is equivalent to the classical Weierstrass representation by means of analytic functions of the complex variable

$$\rho + i\sigma = \omega.$$

To obtain the *formulas of Weierstrass*, we consider the potential functions x, y, u of ρ, σ as the real parts of analytic functions $f_1(\omega)$, $f_2(\omega)$, $f_3(\omega)$. If $\tilde{x}$, $\tilde{y}$, $\tilde{u}$ are the conjugate potential functions, we have

$$x + i\tilde{x} = f_1(\omega), \qquad y + i\tilde{y} = f_2(\omega), \qquad u + i\tilde{u} = f_3(\omega).$$

Since, by the Cauchy-Riemann differential equations,

$$x_\sigma = -\tilde{x}_\rho, \qquad y_\sigma = -\tilde{y}_\rho, \qquad u_\sigma = -\tilde{u}_\rho,$$

we have

$$x_\rho - ix_\sigma = f_1'(\omega), \qquad y_\rho - iy_\sigma = f_2'(\omega), \qquad u_\rho - iu_\sigma = f_3'(\omega),$$

so that conditions (39) become

$$\phi(\omega) = E - G - 2iF = \sum_{\nu=1}^{3} f_\nu'(\omega)^2 = 0.$$

Thus, all minimal surfaces may be represented by

$$x = \mathrm{Re}(f_1(\omega)), \qquad y = \mathrm{Re}(f_2(\omega)), \qquad u = \mathrm{Re}(f_3(\omega)),$$

where the otherwise arbitrary analytic functions $f_\nu(\omega)$ are subject to the condition

$$\sum_{\nu=1}^{3} f_\nu'(\omega)^2 = 0.$$

Instead of ω we may introduce one of the functions f_ν, e.g., $f_3(\omega)$, as the independent variable. Therefore, the totality of minimal surfaces depends essentially on only one arbitrary analytic function of a complex variable.

5. Systems of Two Differential Equations of First Order. A few remarks may be added here about systems of two first order differential equations for two functions u, v, as they occur in important applications, particularly in fluid dynamics. (In Chapter V, a comprehensive theory will be developed.)

In the linear case, in analogy with §1, 2, a hyperbolic system can be brought into the canonical normal form with characteristic coordinates ξ, η as independent variables:

$$au_\xi + bv_\xi + \cdots = 0,$$
$$a'u_\eta + b'v_\eta + \cdots = 0,$$

a, b, a', b' being given functions of ξ, η. Introducing new dependent variables

$$U = au + bv,$$
$$V = a'u + b'v,$$

we obtain finally the normal form

$$U_\xi + \cdots = 0,$$
$$V_\eta + \cdots = 0,$$

where the dots denote given functions of U, V, ξ, η.

In case the two families of characteristics coincide, i.e., $\xi = \eta$, a transformation to the form

$$U_\xi + \cdots = 0,$$
$$V_\xi + \cdots = 0$$

may be possible with a second independent variable ζ, different fro

ξ.   The system is then equivalent to a system of two ordinary differential equations with ζ as a parameter.

By a procedure similar to that in article 1 one obtains in the elliptic case a normal form

$$P_\rho + Q_\sigma + \cdots = 0,$$

$$Q_\rho - P_\sigma + \cdots = 0,$$

where P, Q are the unknown functions and ρ, σ the independent variables.

§2. Classification in General and Characteristics

We shall now turn to a much more general and penetrating approach.

1. Notations. The concept of characteristics is most transparent for systems of first order equations.   For brevity we restrict ourselves largely to linear equations, although the inclusion of quasilinear or general systems does not entail essential complications. (See Chapter V, also Ch. VI, §3.)

Occasionally, as in equation (1) below, we shall use the familiar convenient abbreviation $p_\nu q_\nu$ for $\sum_\nu p_\nu q_\nu$ to avoid writing summation signs when there is no danger of ambiguity.   Moreover, it will be convenient to use vector and matrix notations.

We also recall the concept of interior differentiation of a function $f(x, y)$ in a curve $\phi(x, y) = 0$ with $\phi_x^2 + \phi_y^2 \neq 0$, say $\phi_x \neq 0$.   A differentiation $\alpha f_x + \beta f_y$ is interior if $\alpha \phi_x + \beta \phi_y = 0$; in particular,

$$\phi_y f_x - \phi_x f_y$$

is such an interior differentiation of f.   Similarly, interior differentiation, in a manifold $\phi(x_1, \cdots, x_n) = 0$ with grad $\phi \neq 0$, of a function $f(x_1, \cdots, x_n)$ of n variables $x_1, \cdots, x_n$ (or the vector x) is defined as a linear combination

$$\alpha_\nu f_{x_\nu} = \alpha_\nu f_\nu$$

if the condition

$$\alpha_\nu \phi_{x_\nu} = \alpha_\nu \phi_\nu = 0$$

is satisfied.   Here and later we abbreviate f_{x_ν}, ϕ_{x_ν}, u_{x_ν} by f_ν, ϕ_ν, u_ν, respectively (sometimes also by $D_\nu f$, $D_\nu \phi$, etc.).   Such interior derivatives are known on $\phi = 0$ if the values of f themselves are known. (See Ch. II, App. 1.)

2. Systems of First Order with Two Independent Variables.   Characteristics.

In the case of two independent variables x and y we write a system of k equations for a function vector u with the components $u^1, \cdots, u^k$ in the form

$$(1) \qquad L_j[u] = a^{ij}u_x^i + b^{ij}u_y^i + d^j = 0$$

$$(j = 1, 2, \cdots, k),$$

where the elements a^{ij}, b^{ij} constitute k by k matrices A and B, respectively.

We assume that at least one of these, say B, is nonsingular, i.e., that $\| b^{ij} \| \neq 0$. The coefficients are supposed to possess continuous derivatives. The terms d^j may depend on the unknown functions in a linear or nonlinear way; in the latter case we call our system *semilinear*.

In matrix notation we may write

$$(1a) \qquad L[u] = A u_x + B u_y + d,$$

where L, d and u are vectors.

We now consider the equation $L[u] = 0$ and pose the question underlying Cauchy's initial value problem: From given initial values of the vector u on a curve $C: \phi(x, y) = 0$ with $\phi_x^2 + \phi_y^2 \neq 0$ to determine the first derivatives u_{x_i} on C so that $L[u] = 0$ is satisfied on the strip.

First we realize that on C the interior derivative $u_y \phi_x - u_x \phi_y$ is known. As a consequence we have on C a relation between u_x and u_y, with $\tau = -\phi_y/\phi_x$, of the form[1]

$$u_y = -\tau u_x + \cdots,$$

where the dots denote here and afterwards quantities known on C. Substituting in the differential equations we obtain on C

$$L_j[u] = (a^{ij} - \tau b^{ij})u_x^i + \cdots = 0$$

$$(j = 1, \cdots, k),$$

a system of linear equations for the k derivatives u_x^i. Hence a necessary and sufficient condition for the unique determinacy of all the

[1] Without restricting generality we assume $\phi_x \neq 0$ on the part of C under consideration.

first derivatives along C is

$$(2) \qquad Q = \| \, a^{ij} - \tau b^{ij} \, \| = | \, A - \tau B \, | \neq 0.$$

Q is called the *characteristic determinant* of the system (1).

If $Q \neq 0$ along the curves $\phi = $ const., then these curves are called *free*. Each of these curves can be continued into a "strip" satisfying (1). Initial values are chosen arbitrarily.

If $\tau(x, y)$ is a real solution of the algebraic equation $Q = 0$ of order k for τ, then the curves C, defined by the ordinary differential equation

$$(3) \qquad dx{:}dy = \tau, \qquad \text{or} \qquad Q\left(x, y, \frac{dx}{dy}\right) = 0,$$

are called *characteristic curves*. As we shall presently see, for characteristic curves a continuation of initial values into an integral strip is in general not possible.

If the equation $Q = 0$ does not possess real roots τ, then *all* curves are free; continuation into a strip of initial values is always possible and unique. The system then is called *elliptic*. In the opposite case, that is, when $Q = 0$ possesses k real roots which are all different from each other, the system is called *totally hyperbolic*. Such systems will be systematically studied in Chapter V.

If τ is a real root (maybe the only one) of (2) we can solve along C the system of linear homogeneous equations for the vector l with components $l^1, \cdots, l^k$:

$$l^j(a^{ij} - \tau b^{ij}) = 0 \qquad \text{or} \qquad l(A - \tau B) = 0.$$

Then the linear, "characteristic", combination $l^j L_j[u] = lL[u]$ of our differential equations (1) can be written in the *characteristic normal form*

$$l^j L_j[u] = l^j b^{ij}(u_y^i + \tau u_x^i) + \cdots = 0,$$

or

$$lL[u] = lB(u_y + \tau u_x) + \cdots = 0,$$

where all the unknowns are differentiated in the same direction, i.e., along the characteristic curve corresponding to τ.

Thus in the hyperbolic case, that is, when k such families of characteristic curves exist, we can replace the system by an equivalent one in which each equation contains differentiation only in one, characteristic, direction. We may use this property of a hyperbolic system as a slightly *more general definition of hyperbolicity* (which does not exclude multiple roots τ).

In Chapter V we shall use these definitions as a basis for a complete solution of hyperbolic problems with two independent variables.

A "characteristic combination" of the differential operators L_j is an interior differentiation in C. It implies for the components of u a relation, i.e., a differential equation, along the characteristic C. Therefore it is obviously not possible to prescribe arbitrary initial values of u on a characteristic C. This justifies the distinction between characteristic and "free" curves.

3. Systems of First Order with n Independent Variables.[1] For systems of first order with any number n of independent variables x one may proceed similarly, as will be indicated briefly, again with reference to a more systematic discussion in Chapter VI. The system may be written in the form

$$(4) \qquad L_j[u] = a^{ij,\nu}u_{x_\nu}^i + b^j = 0 \qquad (j = 1, \cdots, k)$$

with $a^{ij,\nu}$ depending on x, and b^j on x and possibly also on u. The index ν is assumed from 1 to n. Using matrix notation and the abbreviation $u_{x_\nu} = u_\nu$, we may also write (4) in the form

$$(4a) \qquad L[u] = A^\nu u_\nu + b = 0,$$

where A^ν are k by k matrices $(a^{ij,\nu})$ and the operator as well as b are vectors.

We consider again a surface C: $\phi(x) = 0$ with grad $\phi \neq 0$, say $\phi_n \neq 0$. On C we consider the *characteristic matrix*

$$(5) \qquad A = A^\nu \phi_\nu$$

and the characteristic determinant or characteristic form

$$(5a) \qquad Q(\phi_1, \cdots, \phi_n) = \| A \| .$$

Initial values of a vector u may be given on C. Then we state:

If $Q \neq 0$ on C then the differential equation (1) uniquely determines along C all derivatives u_ν from arbitrarily given initial values; in this case the *surface C* is called *free*.

[1] See Ch. VI, §3 for more details.

If $Q = 0$ along C we call C a *characteristic surface*. Then there exists a characteristic linear combination

(6) $$lL[u] = l^j L_j[u] = \Lambda[u]$$

of the differential operators L_j such that in Λ the differentiation of the vector u on C is interior; $\Lambda[u] = 0$ establishes a relationship between the initial data, and hence these data cannot be chosen arbitrarily.

To prove these statements, we first realize that $u_\nu \phi_n - u_n \phi_\nu$ is an inner or interior derivative of u in C. Hence u_ν is known in C from the data if only the one (outgoing) derivative u_n is known ($\phi_n \neq 0$ was assumed). Multiplying (4) by ϕ_n we now find

(4a) $$\phi_n L[u] = A^r \phi_\nu u_n + \mathcal{S} = A u_n + \mathcal{S} = 0,$$

where $\mathcal{S}$ is an interior differential operator on u in C. Hence, under the assumption $\| A \| = Q \neq 0$, the system (4a) of linear equations for the vector u_n determines u_n uniquely.

If on the other hand $Q = \| A \| = 0$, then there exists a null vector l such that $lA = 0$. Multiplying (4a) by l yields an equation

(4b) $$l\phi_n L[u] = l\mathcal{S} = 0,$$

expressed by an interior differential operator on the data along C; this operator $l\mathcal{S}$ does not contain u_n. Thus $l\mathcal{S} = 0$ is then a differential relation which restricts the initial values of u on C.

The characteristic equation $Q = 0$ has the form of a partial differential equation of first order for $\phi(x)$. If it is satisfied identically in x, not merely under the condition $\phi = 0$, then the whole family of surfaces $\phi = $ const. consists of characteristic surfaces. The characteristics for $n > 2$ obviously show a much greater variety than the k families of curves in the case $n = 2$. Naturally, therefore, the theory of these systems for $n > 2$ is more involved than that for $n = 2$.

As to the *classification*: If the homogeneous algebraic equation $Q = 0$ in the quantities $\phi_1, \cdots, \phi_n$ cannot be satisfied by any real set of values (except $\phi_\nu = 0$), then characteristics cannot exist, and the system is called *elliptic*.

If, in extreme contrast to the elliptic case, the equation $Q = 0$ possesses k different real solutions ϕ_n for arbitrarily prescribed values of $\phi_1, \cdots, \phi_{n-1}$ (or if a corresponding statement is true after a suitable coordinate transformation), then the system is called *totally hyperbolic*. We shall discuss the concept of hyperbolicity and its significance in §6 and more fully in Chapter VI. The most important goal in these discussions will be the theorem: For hyperbolic equations Cauchy's problem is always solvable.

4. Differential Equations of Higher Order. Hyperbolicity. For single differential equations of higher order and for systems of such equations a similar situation prevails.

Referring for more details to Ch. VI, §3; we restrict ourselves here to a brief remark concerning single differential equations of order m. With the notation D_ν for the differentiation $\partial/\partial x_\nu$, we may write the differential equation in the symbolic form

$$(7) \quad L[u] = H(D_1, \cdots, D_n)u + K(D_1, \cdots, D_n)u + f(x) = 0,$$

where H is a homogeneous polynomial in D of degree m and K is a polynomial of degree lower than m, all the coefficients being continuous functions of x.

The Cauchy data, i.e., the given initial values, consist of the values of the function u and its first $m - 1$ derivatives on the surface $C:\phi(x_1, \cdots, x_n) = 0$, on which we again assume $\phi_n \neq 0$. As before, the basic question is: Under what conditions do such arbitrary data on C determine uniquely the m-th derivatives of u on C? The answer is: It is necessary and sufficient that the *characteristic form*

$$Q(\phi_1, \cdots, \phi_n) = H(\phi_1, \cdots, \phi_n)$$

does not vanish on C. If the surface C is characteristic, i.e., satisfies the equation $Q = 0$, then $Hu + Ku$ is an internal differential operator of order m on C. This means it contains m-th derivatives only in such a way that they combine into internal first derivatives of operators of order $m - 1$ and thus are known on C from the data.

For the proof, one may introduce new coordinates as independent variables. ϕ is chosen as one of these coordinates and $\lambda_1, \cdots, \lambda_{n-1}$ are chosen as interior coordinates in the surfaces $\phi = $ const. Then all the m-th derivatives of a function u are easily expressed as combinations of the m-th "outgoing" derivative $(\partial^m/\partial\phi^m)u$ with terms which contain at most $(m - 1)$-fold differentiation with respect to ϕ and are therefore known from the data. One sees easily that the equation takes on the form

$$Q(\phi_1, \cdots, \phi_n)u_{\phi^m} + \cdots = 0,$$

where the dots denote terms which are known on C from the data. This equation for u has a unique solution if and only if Q does not vanish. If $Q = 0$ on C, the equation represents an internal condition for the data. As to the definition of hyperbolicity, it refers to the characteristic form Q and remains the same as that in article 3.

5. Supplementary Remarks. To properly generalize for more variables we cannot merely repeat the definition of hyperbolicity of article 2. It is sufficient, however, to stipulate the existence of k linearly independent combinations of the equations such that each of these combinations contains only internal differentiations of the unknowns u in an $(n - 1)$-dimensional surface C. A detailed discussion of this important form of the definition will be given later in Ch. VI, §3.

A second remark concerns *quasi-linear systems of equations*. All the essential statements of this section remain valid for quasi-linear equations. The characteristic condition is then dependent on the values of the vector u on C, and therefore characteristics cannot be defined independently of the specific vector u which is under consideration. The complication thus introduced is not relevant for the definition of characteristics, but will be essential later in Chapters V and VI where the solution of Cauchy's problem is constructed.

Finally it should be emphasized that between the elliptic type and the hyperbolic type, indicated above, intermediate types are possible. For example, for two independent variables we may have q real characteristics and p pairs of conjugate complex characteristics, such that $q + 2p = k$. Not much has been done so far to investigate these intermediary types which do not seem to occur in problems of mathematical physics. For more independent variables the typical example of such intermediate types is the "ultrahyperbolic" equation

$$u_{x_1 x_1} + \cdots + u_{x_n x_n} = u_{y_1 y_1} + \cdots + u_{y_n y_n}$$

for a function of $2n$ variables x and y. (See Ch. VI, §16.)

6. Examples. Maxwell's and Dirac's Equations. It will be easy for the reader to identify the wave equation as hyperbolic, Laplace's equation as elliptic, Cauchy-Riemann's equations $u_x - v_y = 0$, $u_y + v_x = 0$ as an elliptic system, $u_x - v_y = 0$, $u_y - v_x = 0$ as a hyperbolic system, and $u_x = v$, $u_y = v_x$ as a parabolic system.

For the elliptic case we give the following additional examples: First

$$\Delta\Delta u = 0 \quad \text{or} \quad \sum_{i,k=1}^{n} \frac{\partial^4 u}{\partial x_i^2 \partial x_k^2} = 0,$$

with the characteristic form

$$Q = \left(\sum_{i=1}^{n} \phi_i^2\right)^2,$$

and second, the differential equation

$$\sum_{i=1}^{n} \frac{\partial^4 u}{\partial x_i^4} = 0$$

with the characteristic form

$$Q = \sum_{i=1}^{n} \phi_i^4.$$

An example of a "parabolic" differential equation is

$$u_t = \Delta\Delta u$$

for $n + 1$ independent variables with the time variable $x_0 = t$ singled out. Here the characteristic form $(\sum_{i=1}^{n} \phi_i^2)^2$ is degenerate since it does not contain the variable ϕ_0.

The operator

$$\left(\Delta - \frac{\partial^2}{\partial t^2}\right)\left(\Delta - 2\frac{\partial^2}{\partial t^2}\right) u = \Delta\Delta u - 3\Delta u_{tt} + 2u_{tttt}$$

is hyperbolic because its characteristic form in the variables $\phi_1, \cdots, \phi_n, \phi_0 = \tau$,

$$Q = \left(\sum_{i=1}^{n} \phi_i^2 - \tau^2\right)\left(\sum_{i=1}^{n} \phi_i^2 - 2\tau^2\right),$$

clearly has the required property.

On the other hand, the operator

$$\left(\Delta - \frac{\partial^2}{\partial t^2}\right)\left(\Delta + \frac{\partial^2}{\partial t^2}\right) u = \Delta\Delta u - \frac{\partial^4 u}{\partial t^4}$$

represents an intermediate type; it is neither elliptic nor parabolic nor hyperbolic since the form

$$Q = \left(\sum_{i=1}^{n} \phi_i^2\right)^2 - \tau^4$$

has only two, not four, real roots τ if the values of the variables $\phi_1, \cdots, \phi_n$ are fixed.

Further examples of systems of first order are the Beltrami differential equations.

$$Wu_x - bv_x - cv_y = 0,$$

$$Wu_y + av_x + bv_y = 0,$$

where the matrix

$$\begin{pmatrix} a & b \\ b & c \end{pmatrix}$$

is assumed to be positive definite. Here the corresponding characteristic form is

$$Q(\phi) = \begin{vmatrix} -W\phi_1 & b\phi_1 + c\phi_2 \\ W\phi_2 & a\phi_1 + b\phi_2 \end{vmatrix} = -W(a\phi_1^2 + 2b\phi_1\phi_2 + c\phi_2^2).$$

In the special case where $W = 1, a = c = 1, b = 0$ (Cauchy-Riemann) it is $Q(\phi) = -(\phi_1^2 + \phi_2^2)$.

Maxwell's system of differential equations is essentially hyperbolic. In the simplest case (for vacuum) these equations are

$$\mathfrak{E}_t - \operatorname{curl} \mathfrak{H} = 0, \qquad \mathfrak{H}_t + \operatorname{curl} \mathfrak{E} = 0$$

if the speed of light is taken as unity; here $\mathfrak{E} = (u_1, u_2, u_3)$ is the electric field vector and $\mathfrak{H} = (u_4, u_5, u_6)$ the magnetic field vector; instead of the fourth independent variable x_0 (the time variable) we write t. Written out in coordinates x, y, z, the equations read:

$$(8) \quad \begin{aligned} -\frac{\partial u_1}{\partial t} &= \frac{\partial u_5}{\partial z} - \frac{\partial u_6}{\partial y}, & \frac{\partial u_4}{\partial t} &= \frac{\partial u_2}{\partial z} - \frac{\partial u_3}{\partial y}, \\[2mm] -\frac{\partial u_2}{\partial t} &= \frac{\partial u_6}{\partial x} - \frac{\partial u_4}{\partial z}, & \frac{\partial u_5}{\partial t} &= \frac{\partial u_3}{\partial x} - \frac{\partial u_1}{\partial z}, \\[2mm] -\frac{\partial u_3}{\partial t} &= \frac{\partial u_4}{\partial y} - \frac{\partial u_5}{\partial x}, & \frac{\partial u_6}{\partial t} &= \frac{\partial u_1}{\partial y} - \frac{\partial u_2}{\partial x}. \end{aligned}$$

The reader can easily show that the characteristic form is

$$Q = \tau^2(\tau^2 - \phi_1^2 - \phi_2^2 - \phi_3^2)^2.$$

Now $Q = 0$ represents essentially the characteristic relation for the wave equation. This reflects the fact that by elimination each com-

ponent of u is seen to satisfy the wave equation, and that the following theorem holds: *Suppose that one single equation is obtained by
elimination from a given system of differential equations with characteristic form Q; then the characteristic form of this single equation is a
factor of Q*[1]. The proof is left to the reader. Strictly speaking,
Maxwell's equations, having multiple characteristics, do not satisfy
the preceding narrow definition of hyperbolicity. A later generalization of the concept will remedy this imperfection.

The characteristic equation belonging to *Dirac's differential equations* is similar to that of the Maxwell equations. The Dirac equations involve a system of four complex-valued functions

$$u = (u_1 , u_2 , u_3 , u_4)$$

of four variables x_1, x_2, x_3, and x_4 (where $x_4 = t$). To formulate
them simply, we introduce the following matrices:

$$\alpha_1 = \begin{pmatrix} 0 & 0 & 0 & 1 \\ 0 & 0 & 1 & 0 \\ 0 & 1 & 0 & 0 \\ 1 & 0 & 0 & 0 \end{pmatrix}, \quad \alpha_2 = \begin{pmatrix} 0 & 0 & 0 & -i \\ 0 & 0 & i & 0 \\ 0 & -i & 0 & 0 \\ i & 0 & 0 & 0 \end{pmatrix},$$

$$\alpha_3 = \begin{pmatrix} 0 & 0 & 1 & 0 \\ 0 & 0 & 0 & -1 \\ 1 & 0 & 0 & 0 \\ 0 & -1 & 0 & 0 \end{pmatrix}, \quad \alpha_4 = \begin{pmatrix} -1 & 0 & 0 & 0 \\ 0 & -1 & 0 & 0 \\ 0 & 0 & -1 & 0 \\ 0 & 0 & 0 & -1 \end{pmatrix},$$

$$\beta = \begin{pmatrix} 1 & 0 & 0 & 0 \\ 0 & 0 & 0 & 0 \\ 0 & 0 & -1 & 0 \\ 0 & 0 & 0 & -1 \end{pmatrix}.$$

The equations then are

$$\sum_{\kappa=1}^{4} \alpha_\kappa \left(\frac{\partial}{\partial x_\kappa} - a_\kappa \right) u - \beta b u = 0.$$

Here the vector (a_1 , a_2 , a_3) is proportional to the magnetic potential,
$-a_4$ to the electric potential, and b to the rest-mass. According to
our rules, the characteristic determinant becomes

$$Q(\phi) = \left| \sum_{\kappa=1}^{4} \alpha_\kappa \phi_\kappa \right| = (\phi_1^2 + \phi_2^2 + \phi_3^2 - \phi_4^2)^2,$$

[1] This theorem has an analogue for characteristic forms of systems of higher
order.

a form of the fourth degree in the variables ϕ_1, ϕ_2, ϕ_3, ϕ_4. Thus the characteristic manifolds are again the same as those for the wave equation.

Finally, by a short calculation we illustrate the equivalence of the definitions of characteristics for a single higher order equation and for a first order system obtained from it. If we replace the second order differential equation

$$\sum_{i,k=1}^{n} a_{ik}\, u_{x_i x_k} + \cdots = 0$$

by the system of first order differential equations

$$\frac{\partial p_l}{\partial x_n} = \frac{\partial p_n}{\partial x_l} \qquad (l = 1, 2, \cdots, n - 1),$$

$$\sum_{i,k=1}^{n} a_{ik}\, \frac{\partial p_i}{\partial x_k} + \cdots = 0,$$

then we obtain for this system the characteristic condition

$$\begin{vmatrix} \sum a_{1k}\,\phi_k & \sum a_{2k}\,\phi_k & \cdots & \sum a_{n-1,k}\,\phi_k & \sum a_{nk}\,\phi_k \\ \phi_n & 0 & \cdots & 0 & -\phi_1 \\ 0 & \phi_n & \cdots & 0 & -\phi_2 \\ \cdots\cdots\cdots\cdots\cdots\cdots\cdots\cdots\cdots\cdots\cdots\cdots \\ 0 & 0 & \cdots & \phi_n & -\phi_{n-1} \end{vmatrix} = 0,$$

i.e.,

$$(-1)^{n-1}\phi_n^{n-2} \sum_{i,k=1}^{n} a_{ik}\,\phi_i\,\phi_k = 0,$$

which agrees with the characteristic condition for a single equation.

For further examples see the following sections and Chapters V and VI.

§3. *Linear Differential Equations with Constant Coefficients*

Linear differential equations with constant coefficients (and others reducible to this category) allow a more complete treatment than is possible in the general case. Moreover, since the classification of the equations at a point P is determined merely by the local values of the coefficients, it is sufficient to consider the case of constant coefficients in order to distinguish the various types. Indeed, near a point

P a linear or quasi-linear system of differential equations may be locally approximated by a linear one with constant coefficients if we replace the values of the coefficients in the neighborhood of P by those at P.

1. Normal Form and Classification for Equations of Second Order. We consider an operator of the second order

$$(1) \qquad L[u] = \sum_{i,k=1}^{n} a_{ik}\, u_{x_i x_k} + \cdots$$

or the differential equation $L[u] = 0$, where the coefficients $a_{ik} = a_{ki}$ are continuously differentiable functions of the independent variables $x_1, x_2, \cdots, x_n$ in a domain G, and where the dots denote expressions of lower than the second order in u. The expression of second order is again called the *principal part of the differential operator*.

The classification of the differential operator (1) depends on the effect which a transformation of variables,

$$(2) \qquad \xi_i = t_i(x_1, x_2, \cdots, x_n) \quad (i = 1, 2, \cdots, n),$$

has on the form of the differential operator at a certain point P^0 $(x^0) = (x_1^0, x_2^0, \cdots, x_n^0)$. Denoting $\partial t_i / \partial x_k$ by t_{ik}, we have

$$u_{x_l} = \sum_{k=1}^{n} t_{kl}\, u_{\xi_k} \quad \text{and} \quad u_{x_l x_s} = \sum_{i,k=1}^{n} t_{kl}\, t_{is}\, u_{\xi_i \xi_k} + \cdots,$$

where the dots again denote expressions containing at most first order derivatives of the function u. The transformation (2) takes the operator (1) into the form

$$(3) \qquad \Lambda[u] = \sum_{i,k=1}^{n} \alpha_{ik}\, u_{\xi_i \xi_k} + \cdots$$

with the coefficients α_{ik} defined by the transformation

$$(4) \qquad \alpha_{ik} = \sum_{l,s=1}^{n} t_{kl}\, t_{is}\, a_{ls} .$$

At the point (x^0), the coefficients of the principal part of $L[u]$ are thus transformed like the coefficients of the quadratic form $Q = \sum_{i,k=1}^{n} a_{ik} y_i y_k$ (the *characteristic form*) if the parameters y_i are subjected to the affine linear transformation

$$(5) \qquad y_i = \sum_{l=1}^{n} t_{il}\, \eta_l .$$

A quadratic form of this type can always be transformed by an affine transformation into *canonical form*

$$Q = \sum_{i=1}^{n} \kappa_i \eta_i^2$$

in which the coefficients κ_i can have only the values $+1$, -1, or 0. The number of negative coefficients, called *index of inertia*, and the number of vanishing coefficients are affine invariants. Accordingly, these numbers have a meaning for the differential operator at the point P^0.

The *differential operator* is called *elliptic* at the point P^0 if all values κ_i are equal either to 1 or to -1. We call it "properly hyperbolic" or simply *hyperbolic* if all values κ_i have the same sign, say positive, except one, say κ_n, which is negative. If several of the signs are positive and several negative, then the operator is called *ultrahyperbolic*. If Q is singular, then one or more of the coefficients κ_i vanish, and the differential equation is called *parabolic*.

If the differential operator is elliptic at a point P^0, then, by a suitable linear transformation, the differential equation can be written at P^0 as $u_{x_1 x_1} + u_{x_2 x_2} + \cdots + u_{x_n x_n} + \cdots = 0$. Similarly, if it is hyperbolic we can transform it into

$$u_{x_1 x_1} + u_{x_2 x_2} + \cdots + u_{x_{n-1} x_{n-1}} - u_{x_n x_n} + \cdots = 0.$$

In general, however, it is impossible to obtain a transformation into one of these normal forms valid for a *whole domain*.[1]

Yet if the coefficients a_{ik} in equation (1) are constant, then a nor-

---

[1] If, as in §1, we want to eliminate the mixed components of the matrix (α_{ik}) in the transformed operator

$$\Lambda[u] = \sum_{i,k=1}^{n} \alpha_{ik} u_{\xi_i \xi_k} + \cdots,$$

then we have to subject the n functions t_i to the $\frac{1}{2}n(n-1)$ conditions (cf. formula (4))

$$\sum_{l,s=1}^{n} a_{ls} t_{kl} t_{is} = 0 \qquad (i \neq k).$$

For $\frac{1}{2}n(n-1) > n$, i.e., for $n > 3$, this system of equations is overdetermined; accordingly it cannot be solved in general. For $n = 3$ it is still possible to eliminate the mixed terms, but the terms on the principal diagonal cannot be subjected to the further restriction of being equal.

mal form valid for the whole domain can be obtained by a single affine transformation of the variables x_i into variables ξ_i:

$$\xi_i = \sum_{k=1}^{n} t_{ik} x_k .$$

This transforms the characteristic form into its canonical form in accordance with (5). If we again write x_1, x_2, $\cdots$, x_n for the new independent variables and equation (1) is homogeneous, then the differential equation takes the form

$$(6) \qquad \sum_{i=1}^{n} \kappa_i u_{x_i x_i} + \sum_{i=1}^{n} b_i u_{x_i} + cu = 0,$$

where κ_i equals 1, -1, or 0.

In the case of constant coefficients, b_i and c are constants and the differential equation may be further simplified as follows by a transformation of the dependent variable u, which eliminates the first derivatives with respect to those variables x_i for which $\kappa_i \neq 0$. We exclude the parabolic case and split off an exponential factor from u by setting

$$(7) \qquad u = v \exp\left\{ -\frac{1}{2} \sum_{l=1}^{n} \frac{b_l}{\kappa_l} x_l \right\}.$$

Then the differential operator becomes

$$(8) \quad L[u] = \exp\left\{ -\frac{1}{2} \sum_{l=1}^{n} \frac{b_l}{\kappa_l} x_l \right\} \left[\sum_{l=1}^{n} \kappa_i v_{x_i x_i} + \left(c - \frac{1}{4} \sum_{l=1}^{n} \frac{b_l^2}{\kappa_l} \right) v \right].$$

Consequently, nonparabolic linear differential equations with constant coefficients can be reduced to differential equations of the form

$$(9) \qquad \sum_{i=1}^{n} \kappa_i v_{x_i x_i} + pv = f(x_1, x_2, \cdots, x_n),$$

where f is a given function of the independent variables and p denotes a constant. Thus *all linear elliptic differential equations of second order with constant coefficients can be reduced to the form*

$$\Delta v + pv = f(x_1, x_2, \cdots, x_n),$$

and all linear hyperbolic differential equations of second order with constant coefficients can be reduced to

$$\Delta v - v_{tt} + pv = f(x_1, x_2, \cdots, x_n, t).$$

(Here we consider $n + 1$ independent variables x_0, x_1, $\cdots$, x_n, set $x_0 = t$, and form the Laplacian operator Δ with respect to $x:(x_1, \cdots, x_n)$.)

2. Fundamental Solutions for Equations of Second Order. Irrespective of order or variability of coefficients, for all linear differential equations, elliptic or hyperbolic, "fundamental solutions" defined by certain singularities play an important role, as will become apparent in the next chapters.[1] Here we shall merely insert a brief preliminary discussion for elliptic second order equations with constant coefficients. We consider the equation

$$L[u] = \Delta u + \rho u = 0$$

and start by asking for "fundamental" solutions which depend only on the distance $r = \sqrt{\Sigma(x_i - \xi_i)^2}$ from the point x to a parameter point ξ. By transforming the Laplacian operator to polar coordinates, we obtain (see Vol. I, p. 225)

$$(10) \qquad u_{rr} + \frac{n-1}{r} u_r + \rho u = 0.$$

As is easily verified, the function

$$w(r) = \frac{u_r}{r}$$

satisfies the same equation as u with $n - 1$ replaced by $n + 1$:

$$(11) \qquad w_{rr} + \frac{n+1}{r} w_r + \rho w = 0.$$

Thus, again writing u instead of w, we obtain the "fundamental solutions" u for any n by recursion as soon as we know u for $n = 2$ and $n = 3$ from the ordinary differential equations

$$u'' + \frac{1}{r} u' + \rho u = 0$$

and

$$u'' + \frac{2}{r} u' + \rho u = 0$$

respectively. For $\rho = 0$, i.e., for the Laplace equation, the solutions,

[1] See in particular Ch. VI, §15.

except for a constant factor at our disposal, are $u = \log 1/r$ and

$$u = \frac{1}{r}.$$

Thus one obtains for any $n \geq 3$ the fundamental solutions

$$u = \text{const. } r^{2-n}.$$

With $\rho \neq 0$, say $\rho = \omega^2$, we find in complex notation for $n = 1$

$$u = e^{i\omega r}.$$

Hence for $n = 3$

$$u = i\omega \frac{e^{i\omega r}}{r}$$

and for $n = 5$

$$u = -\omega^2 \left(\frac{1}{r^2} - \frac{1}{i\omega} \frac{1}{r^3} \right) e^{i\omega r},$$

etc. Thus all solutions for odd n can be expressed in terms of trigonometric functions (or hyperbolic functions if $\omega^2 < 0$).

With even n we have for $n = 2$

$$u = \alpha J_0(\omega r) + \beta N_0(\omega r) + \text{regular function,}$$

where J_0 and $N_0 = (2/\pi) J_0(\omega r) \log r + \cdots$ are the Bessel function and the Neumann function of order zero, respectively, and α, β are constants. If α is chosen as zero we find for $n = 4$ the singular solution

$$u = \frac{J_0(\omega r)}{r^2} + \frac{w}{r} J_0'(\omega r) \log r + \cdots.$$

($J_0'(\omega r)/r$ is regular for $r = 0$.) This solution we call the fundamental solution. One easily ascertains in general: For odd $n > 1$ we have the singular ("fundamental") solutions

$$u = \frac{U}{r^{n-2}} + \cdots,$$

and for even n

$$u = \frac{U}{r^{n-2}} + W \log r + \cdots,$$

where the dots denote regular terms and where U and W are regular solutions of $L[U] = L[W] = 0$.

For $\rho < 0$, i.e., for imaginary ω, corresponding relations also prevail.

In the case of the hyperbolic differential equation

$$(12) \qquad L[u] = u_{tt} - \Delta u - \rho u = 0 \quad (x = x_1, \cdots, x_n)$$

a quite parallel reasoning leads to the following result:

We seek singular "fundamental solutions" of (12) which depend only on the "hyperbolic distance"

$$r = \sqrt{(t - \tau)^2 - \sum_1^n (x_\nu - \xi_\nu)^2}$$

from the point t, x to the parameter point τ, ξ in the space of $n = m + 1$ dimensions. For $u(r)$ we obtain the ordinary differential equation

$$u'' + \frac{n - 1}{r} u' - \rho u = 0.$$

As before, fundamental solutions which are singular on the cone $r = 0$ are found to be of the form described above for the elliptic case. The main difference is that now the singularity is spread over a whole cone and that outside the cone the function u is not defined or may be defined as identically zero, while in the elliptic case only the point $x = \xi$ is singular. The significance of such fundamental solutions (which may be modified by multiplication with a constant and addition of any regular solution of $L[u] = 0$) will become clear in Chapter VI. In Volume I, we have already met such solutions in the form of Green's function (see Vol. I, Ch. V, §14).

It may be stated here that these fundamental solutions $u(x; \xi)$ as functions of the point x and the parameter point ξ have the following basic property:

In the elliptic case the integral

$$v(x) = \int\int_G f(\xi_1, \cdots, \xi_n) u(x, \xi) \, d\xi_1, \cdots, d\xi_n$$

extended over a domain G including the point x satisfies with a suitable constant c the Poisson equation

$$L[v] = cf(x).$$

In particular, for $n = 3$ the integral

$$cv = \int\int\int_G f(\xi) \frac{e^{i\omega r}}{i\omega r} \, dx_1, \cdots, dx_n$$

satisfies the "reduced wave equation"

$$\Delta v + \omega^2 v = 0.$$

In the hyperbolic case it can be proved that $v(x)$ also satisfies the differential equation if the domain of integration G fills the characteristic cone issuing from the point x into the ξ-space. (See Ch. VI, §15.)

3. Plane Waves. Turning to equations of arbitrary order k we write the differential equation in n independent variables $x_1, \cdots, x_n$ again in the symbolic form

$$(13) \qquad (P_k D_i + P_{k-1} D_i + \cdots + P_0) u + f = 0,$$

where P_κ is a homogeneous polynomial with constant coefficients of degree κ in the symbols $D_i = \partial/\partial x_i$ $(i = 1, \cdots, n)$ and f denotes a given function of the independent variables. We need consider only the homogeneous equation;[1] i.e., we assume $f = 0$. The nonhomogeneous equation can then be easily treated (see, e.g., §4).

The basic fact is: For any number of independent variables the homogeneous equation (13) possesses solutions in the form of exponential functions $e^{(ax)}$, where

$$(ax) = a_1 x_1 + a_2 x_2 + \cdots$$

with constants a_ν. (On occasion we shall also write (a, x) or $a \cdot x$.) The necessary and sufficient condition for u to be a solution is that $a = (a_1, \cdots)$ satisfy the algebraic equation of degree k

$$(14) \qquad Q^*(a) = P_k(a) + P_{k-1}(a) + \cdots + P_0 = 0,$$

which defines an algebraic surface $Q^*(a) = 0$ of degree k in the space of the coordinates $a_1, a_2, \cdots$. The classification into types, however, refers more simply to the homogeneous equation

$$Q(a) = P_k(a) = 0;$$

this "characteristic equation" depends only on the principal part of the differential equation; it determines the normals of characteristic surface elements,[2] in agreement with the definitions of §2.

---

[1] Incidentally, constant coefficients imply: If $u(x_1, x_2, \cdots, x_n)$ is a solution of a homogeneous equation, then $u(x_1 - \xi_1, x_2 - \xi_2, \cdots, x_{n+1} - \xi_n)$—with arbitrary parameters ξ_i—as well as the partial derivatives $\partial u/\partial x_i$, are also solutions.

[2] As to the significance of the full equation see L. Gårding [2].

For example, in three dimensions for the Laplace equation $u_{xx} + u_{yy} + u_{zz} = \Delta u = 0$ we obtain the relation $a_1^2 + a_2^2 + a_3^2 = 0$. Hence at least one of the exponents a_ν is imaginary; the corresponding solutions might be written, e.g., in the form

$$e^{xa_1 + ya_2} e^{iz\sqrt{a_1^2 + a_2^2}}.$$

For the wave equation we have solutions $e^{i(a_1x_1 + a_2x_2 + a_3x_3 - a_4t)}$ with $a_1^2 + a_2^2 + a_3^2 - a_4^2 = 0$, and for the "reduced" wave equation $\Delta u + \omega^2 u = 0$ we have the relation $a_1^2 + a_2^2 + a_3^2 = \omega^2$. For the heat equation $u_t = \Delta u$, we have the relation $a_1^2 + a_2^2 + a_3^2 - a_4 = 0$.

If the equation $Q(a) = 0$ cannot be satisfied by real values of $a_1, \cdots, a_n$, then the differential equation is called *elliptic*.

4. Plane Waves Continued. Progressing Waves. Dispersion. In the following sections we shall be primarily concerned with solutions which represent *propagation* phenomena, in particular with *plane waves* arising in the *hyperbolic* case. In addition to the n independent space-variables x we shall consider a further variable $x_0 = t$; we form the inner product $(ax) = a_1x_1 + \cdots + a_nx_n = A$ with the n-dimensional vector $a:(a_1, \cdots, a_n)$ and define the *phase*

$$B = (ax) - bt = A - bt$$

with a constant $a_0 = -b$.

Let us assume first that the differential equation contains only the principal part, i.e., the terms of order k, or in other words that $P_\kappa = 0$ for $\kappa < k$. Then the important fact holds: Not only the exponential functions as above are solutions but quite generally all functions of the form

$$(15) \qquad\qquad u = f(B)$$

are solutions, where the *wave form* $f(B)$ is an *arbitrary function* of the phase $B = A - bt$ and the coefficients a_ν, b are subject to the characteristic equation $Q(-b, a) = 0$. (Compare §2, 4.)

Provided that this equation can be satisfied by real values of $a_1, \cdots, a_n$ and b, the functions $f(B)$ represent *undistorted progressing waves*.

By the term *progressing plane wave* for a homogeneous linear differential equation $L[u] = 0$ we mean a solution of the form (15).

Plane waves of this type have constant values on every phase plane

DIFFERENTIAL EQUATIONS WITH CONSTANT COEFFICIENTS 189

of the family

$$B = (ax) - bt = \text{const.}$$

in the $(n + 1)$-dimensional x, t-space.

. To motivate the term "progressing wave" we consider the n-dimensional space $\mathfrak{R}_n$ of the space variables x_1, x_2, $\cdots$, x_n in which the "field" u varies with the time t. A solution u of the form (15) is constant on a whole plane of *constant phase* B of a family of parallel planes. A plane with constant phase moves with constant speed parallel to itself through the space $\mathfrak{R}_n$.

If we set

$$a_l = \rho \alpha_l, \qquad \sum_{l=1}^n \alpha_l^2 = 1, \qquad \rho^2 = \sum_{l=1}^n a_l^2, \qquad b = \rho\gamma,$$

$$B = A - bt = \rho \left(\sum_{i=1}^n \alpha_i x_i - \gamma t \right) = \rho((\alpha x) - \gamma t) = \rho E,$$

and write

$$u = f(B) = \phi(E),$$

we obtain a representation in which the numbers α_l are the "direction cosines" of the normals to the plane waves and γ denotes the speed of propagation of the waves. E is again called the *phase of the wave*, and the function ϕ, or f, is called the *wave form*.

For example, the ordinary wave equation in n space variables, $\Delta u - u_{tt} = 0$, admits plane waves of the form

$$u = \phi((\alpha x) - t);$$

here the coefficients α_l may form an arbitrary unit vector α with $\alpha^2 = 1$ and the wave form ϕ may be an arbitrary function.

In other words, *the wave equation $\Delta u - u_{tt} = 0$ is solved by plane waves of arbitrary direction and arbitrarily given form; all these waves progress with the speed $\gamma = 1$.*

The waves $f(B)$ are called undistorted or *free of dispersion* because they represent for an *arbitrary* form of the wave or signal $f(B)$ an undistorted translation with speed γ (in the direction α of the normal of the phase planes).

If, for an arbitrary direction α, the characteristic equation $Q = 0$ possesses k real and different roots γ, i.e., if there are k different speeds possible for undistorted waves in every direction, the speeds depending in general on the direction α, then the differential equation (13) is called *hyperbolic*. (We shall later generalize this definition by admitting multiple roots in certain cases.) This definition of hyperbolicity referring to the characteristic equation $Q = 0$ is also retained if the differential equation contains terms of lower order.

For differential equations (13) which contain lower order terms, i.e., for which not all the polynomials P_κ for $\kappa < k$ vanish, the situation is different from that in the case of undistorted waves. Provided that there are progressing waves, they no longer have arbitrary form, nor is their speed determined by the direction of the normal. Instead, their shape is restricted to exponential functions; it depends on the given direction and a given speed.

Let us first consider the example of the differential equation

$$(16) \qquad \Delta u - u_{tt} + cu = 0$$

when $c \neq 0$. If $u = f(B)$ is a plane wave for (16) with

$$B = (ax) - bt,$$

then we immediately obtain the equation

$$(17) \qquad f''(B)(a^2 - b^2) + f(B)c = 0$$

for given a, b. That is, $f(B)$ must satisfy a linear ordinary differential equation with constant coefficients, and this restricts the wave form $f(B)$ to exponential functions. Clearly, for the speed $\gamma = 1$, i.e., for $b^2 = a^2$, there is no longer a progressing wave. However, for any other speed and for an arbitrary direction the possible wave forms are determined from (17) as exponential functions. Hence, *the direction and the speed* (except for the value 1 of the speed) *of the wave belonging to the differential equation* (16) *can be assigned arbitrarily beforehand; but only special wave forms are possible for progressing waves.*

Of course, the shape of the exponential function $f(B)$ depends on the signs of the coefficients in the ordinary differential equation (17).

With a view to physical relevance we exclude solutions which are not uniformly bounded in space; in other words, we consider only waves of the form

$$f(B) = e^{i\rho((ax)-\gamma t)},$$

where ρ is the "frequency". Then in our case, if $c > 0$, such waves exist in arbitrary directions for $a^2 > b^2$, that is, for any speed γ below the limit speed $\gamma = 1$. For speeds exceeding the limit speed the solutions $f(B)$ will no longer be classed as admissible waves since they are not bounded in space.

At any rate, the differential equation (17) represents a phenomenon of *dispersion* in the following sense: If a solution u is a superposition of progressing waves with the same direction, all of a form satisfying (17), then the different components are propagated at different speeds; thus the form of a composite wave will change in time.

In the case of the general differential equation (13), the form $f(B)$ of a progressing wave

$$u = f(B) = f\left(\sum_1^n (ax) - bt\right)$$

is again subjected to an ordinary differential equation

(18) $f^k(B)P_k(-b, a) + f^{k-1}(B)P_{k-1}(-b, a) + \cdots + f(B)P_0 = 0$

whose coefficients are constants for each arbitrary set of parameters $a_0 = -b, a_1, \cdots, a_n$. As before, we restrict admissible waves by stipulating

$$B = i\rho(\alpha_1 x_1, \cdots + \alpha_n x_n - \gamma t)$$

with $a_\nu = \rho\alpha_\nu$ and $a_{n+1} = b = -\rho\gamma$, so that ρ is the frequency, α the direction normal, and γ the speed of the wave. ρ and α are real, while $\gamma = p + iq$ may be allowed to be complex.

For arbitrary ρ and α the equation (18) defines the speeds γ as continuous functions of α and of the frequency ρ except in singular cases, e.g., when the coefficients of (18) all vanish, with the possible exception of P_0.[1]

[1] Equation (16), for example, admits no progressing wave if the prescribed speed is 1 and the prescribed direction arbitrary.

In another example of dispersion, given by the equation

$$\Delta u - u_{tt} + \sum_{i=1}^n u_{x_i} - u_t = 0,$$

the exceptional values for direction and speed are given by the conditions

$$\sum_{i=1}^n a_i^2 - b^2 = 0, \qquad \sum_{i=1}^n a_i - b = 0,$$

and if they are satisfied, progressing waves of arbitrary form exist. They progress with speed 1 and their directions belong to the cone $\sum_{i \neq k} a_i a_k = 0$.

If for given ρ and α the speed γ possesses an imaginary part q, then the wave may be written in the form

$$e^{i\rho((\alpha x)-pt)}e^{-qt}.$$

We speak of *damped waves* exponentially attenuated in time at a fixed point in space. (The solution with the factor e^{qt} for $q > 0$ is usually discarded, being not bounded for increasing t.)

Again we have the phenomenon of distortion or *dispersion*: An initial harmonic component $e^{i\rho(\alpha x)}$ is propagated at a speed depending on the frequency; thus, an initial shape of u, given by superposition of terms $e^{i\rho(\alpha x)}$, is distorted in time (apart from the attenuation by damping), since the different components are propagated at different speeds or "dispersed" according to their different frequencies.

We summarize: The alternative between the case of dispersion and undistorted progressing families of plane waves is inherent in the presence or absence of terms of lower order in the differential equation. In the former case the wave form of progressing plane waves in a given direction is exponential, but the speed can vary continuously with the frequency. In the second case the wave form is arbitrary and the speed is restricted to the discrete roots of the characteristic equation.[1]

5. Examples. Telegraph Equation. Undistorted Waves in Cables.

For the wave equation $\dfrac{1}{c^2}u_{tt} = \Delta u$, progressing undistorted plane waves with the speed c and the arbitrary form

$$\phi\left(\sum_{l=1}^{n}\alpha_l x_l - ct\right), \qquad \sum_{l=1}^{n}\alpha_l^2 = 1$$

are possible in every direction. A more general example is given by the telegraph equation

$$(19) \qquad u_{tt} - c^2 u_{xx} + (\alpha + \beta)u_t + \alpha\beta u = 0,$$

satisfied by the voltage or the current u as a function of the time t and the position x along a cable; here x measures the length of the cable from an initial point.[2]

---

[1] It is an instructive exercise to study the transition from the first to the second alternative if the coefficients P_κ for $\kappa < k$ tend to zero in dependence on a parameter.

[2] This differential equation is derived by elimination of one of the unknown functions from the following system of two differential equations of first order

Unless $\alpha = \beta = 0$, this equation represents dispersion. If we introduce $v = e^{\frac{1}{2}(\alpha+\beta)t}u$, we obtain the simpler equation

$$v_{tt} - c^2 v_{xx} - \left(\frac{\alpha - \beta}{2}\right)^2 v = 0$$

for the function v. This new equation represents the dispersionless case if and only if

$$(20) \qquad\qquad \alpha = \beta.$$

In this case the original telegraph equation, of course, possesses no absolutely undistorted wave solutions of arbitrarily prescribed form. However, our result may be stated in the following way: *If condition (20) holds, the telegraph equation possesses damped, yet "relatively" undistorted, progressing wave solutions of the form*

$$(21) \qquad\qquad u = e^{-\frac{1}{2}(\alpha+\beta)t}f(x \pm ct),$$

with arbitrary f, progressing in both directions of the cable.

This result is important for telegraphy; it shows that, given appropriate values for the capacity and inductance of a cable, signals can be transmitted—damped in time—in a relatively undistorted form (cf. Ch. V, App. 2).

6. Cylindrical and Spherical Waves. The principle of superposition leads to other important forms of solutions for our differential equations, in particular to *cylindrical and spherical waves*.

(*a*) *Cylindrical Waves.* The wave equation in two dimensions

$$(22) \qquad\qquad u_{xx} + u_{yy} - u_{tt} = 0$$

for the current i and the voltage u as functions of x and t:

$$Cu_t + Gu + i_x = 0,$$

$$Li_t + Ri + u_x = 0.$$

Here L is the inductance of the cable, R its resistance, C its shunt capacity, and, finally, G its shunt conductance (loss of current divided by voltage). The constants in equation (19), which arise in the elimination process, have the meaning

$$\frac{1}{c^2} = LC, \qquad \alpha = \frac{G}{C}, \qquad \beta = \frac{R}{L},$$

where c is the speed of light and α the capacitive and β the inductive damping factor.

is solved for arbitrary θ by exp $\{i\rho(x \cos \theta + y \sin \theta)\}$ exp $\{i\rho t\}$, where ρ is a number which can be arbitrarily chosen. Integration of this "plane wave" with respect to the direction angle θ yields the new solution

$$u(x, y, t) = e^{i\rho t} \int_0^{2\pi} \exp \{i\rho r \cos (\theta - \phi)\} \, d\theta = 2\pi e^{i\rho t} J_0(\rho r),$$

where the polar coordinate r is introduced by $x = r \cos \phi$, $y = r \sin \phi$. This solution represents a *standing wave*.

Thus, a *rotationally symmetric solution of the wave equation* (22), a so-called *cylindrical wave, is given by the Bessel function J_0*. This solution is regular at the origin $r = 0$.

By the superposition of plane waves we can also construct a solution which is singular at the origin, corresponding to a *radiation process* (cf. §4) *with a source at the origin*. For this construction we use improper waves. We consider the complex path L of integration in the θ-plane illustrated in Figure 5 (cf. Vol. I, Ch. VII), and form the complex integral

$$u = e^{i\rho t} \int_L e^{i\rho r \cos \theta} \, d\theta = \pi e^{i\rho t} H_0^1(\rho r),$$

where H_0^1 denotes the Hankel function. Then u is a solution of the wave equation.

Both cylindrical waves are periodic in t, of course, and oscillating but not periodic in the space variable r.

(*b*) *Spherical Waves*. In three-dimensional space the situation is slightly different. From the solution

$$\exp \{i\rho t\} \exp \{i\rho(\alpha x + \beta y + \gamma z)\} = \exp \{i\rho t\}w,$$

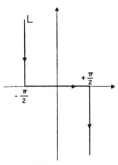

FIGURE 5

we obtain by integrating w over the unit sphere in α, β, γ-space the new function

$$v = \iint_\Omega e^{i\rho(\alpha x + \beta y + \gamma z)} \, d\Omega,$$

where $d\Omega$ is the surface element of the unit sphere. Since this function is evidently invariant under rotation of the coordinate system we may, for purposes of calculation, set $x = y = 0$, $z = r$. Introducing polar coordinates θ, ϕ in α, β, γ-space, we obtain

$$v = \int_0^{2\pi} d\phi \int_0^\pi e^{i\rho r \cos \theta} \sin \theta \, d\theta$$

or

$$v = \frac{4\pi}{\rho} \frac{\sin \rho r}{r} \, .$$

Thus $\exp \{i\rho t\}$ $(\sin \rho r)/r$ is a standing spherical wave, rotationally symmetric, regular at the origin, and derived by the superposition of regular progressing plane waves.

Waves with a singularity at the origin which correspond to radiation phenomena must again be formed by means of improper plane waves. The path of integration L of Figure 6 leads to

$$(23) \qquad v = 2\pi \int_L e^{i\rho r \cos \theta} \sin \theta \, d\theta = 2\pi \frac{e^{i\rho r}}{i\rho r} \, .$$

In terms of real quantities, we have simultaneously constructed the two spherical wave forms $(\cos \rho r)/r$ and $(\sin \rho r)/r$, the second of which is the regular one just computed.

We observe: The spherical wave form (23) may be derived by superposition of the plane waves $\exp \{i\rho(\alpha x + \beta y + \gamma z)\}$ for an arbi-

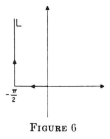

FIGURE 6

trary position of the point (x, y, z), with $z > 0$. Independently of the position of this point,

$$(24) \qquad 2\pi \frac{e^{i\rho r}}{i\rho r} = \int_0^{2\pi} d\phi \int_L e^{i\rho (\alpha x + \beta y + \gamma z)} \sin \theta \, d\theta$$

holds with $x^2 + y^2 + z^2 = r^2$. The easy verification of (24) may be omitted.[1]

Since the wave equation does not contain dispersion terms, we can construct the rotationally symmetric wave

$$u = \int_0^{2\pi} \int_0^{\pi} f(t - \alpha x - \beta y - \gamma z) \sin \theta \, d\theta \, d\phi$$

with an arbitrary function $f(\lambda)$. This expression is invariant under rotation; thus we may evaluate the integral assuming $x = y = 0$. In polar coordinates we obtain

$$u = 2\pi \int_0^{\pi} f(t - r \cos \theta) \sin \theta \, d\theta$$

$$= \frac{2\pi}{r} [F(t + r) - F(t - r)],$$

where F, the indefinite integral of f, is arbitrary. Thus, for arbitrary (twice differentiable) F, the function

$$\frac{F(t + r) - F(t - r)}{r}$$

is a solution.[2] Likewise, each of the functions

$$\frac{F(t + r)}{r} \quad \text{and} \quad \frac{F(t - r)}{r}$$

[1] The identification of the two integrals (23) and (24) is related to Cauchy's integral theorem for two complex variables, since the change from $\rho \neq 0$ to $\rho = 0$ means only a displacement of the path of integration in the complex θ-plane (see H. Weyl [1], where an important application of formula (24) to the problem of propagation of radio waves is made).

[2] For two-dimensional space the analogous simplification of the integral

$$u = \int_0^{2\pi} f(t - r \cos \theta) \, d\theta$$

is not possible. This illustrates the basic difference between problems in even- and odd-dimensional space, which has already been indicated and will become clearer in §4 and in Chapter VI.

in itself is also a solution. This can be easily seen by making appropriate changes in the function f or F as well as by direct verification. These solutions, which obviously are singular at the origin, represent *"progressing spherical waves* attenuated in space".

Moreover, these are the only solutions of the wave equation in three-dimensional space which depend spatially on r alone, because for a function $u(r, t)$ the expression $\Delta u = u_{xx} + u_{yy} + u_{zz}$ becomes

$$\Delta u = u_{rr} + \frac{2}{r} u_r = \frac{1}{r} (ru)_{rr}$$

(cf. Vol. I, p. 225). Hence the wave equation $\Delta u - u_{tt} = 0$ goes over into the equation

$$\frac{1}{r} [(ru)_{rr} - (ru)_{tt}] = 0,$$

whose general solution, according to Ch. I, p. 6, is

$$ru = F(t + r) + G(t - r)$$

with arbitrary F and G.

§4. *Initial Value Problems. Radiation Problems for the Wave Equation*

Linear problems of propagation can often be solved by superposition of explicitly known special solutions of the differential equation. The task always is to find solutions u of the space variables x and the time, for $t \geq 0$, in a space domain G, subject to given initial conditions at $t = 0$ and to boundary conditions at the boundary of G (mixed problems). If no boundary conditions are posed and G is the whole x-space, we have the simpler case of mere initial value problems, or "Cauchy problems". If u does not depend on t, and, correspondingly, no initial conditions are given, but G is bounded, then we have a boundary value problem.

In the present section we shall discuss a number of individual examples, while more general theories will/be taken up more systematically later. (See also §6.)

1. Initial Value Problems for Heat Conduction. Transformation of the Theta Function. For the heat equation

$$(1) \qquad u_{xx} - u_t = 0$$

we consider the initial value problem: Find a solution $u(x, t)$ with continuous derivatives up to the second order for all values of the variable x and for $t > 0$, having a prescribed set of values

$$u(x, 0) = \psi(x) \quad \text{at} \quad t = 0.$$

$\psi(x)$ is assumed to be everywhere continuous and bounded,

$$|\psi(x)| < M.$$

The solution of this initial value problem is given by the function

$$(2) \qquad u(x, t) = \frac{1}{2\sqrt{\pi t}} \int_{-\infty}^{\infty} \psi(\xi) e^{-(x-\xi)^2/4t} \, d\xi,$$

which is derived from the "fundamental solution" previously obtained (Ch. I, §3) by superposition. It describes the heat flow as the superposition of single events for each of which the initial temperature is zero everywhere except at the point $x = \xi$, where we initially have a local concentration of heat proportional to $\psi(\xi)$.

We prove the result by verification: Differentiation under the integral sign shows immediately that (2) satisfies the heat equation for $t > 0$. In order to verify the initial condition (for $t = 0$) we introduce the new variable of integration $\sigma = (\xi - x)/2\sqrt{t}$ instead of ξ and obtain

$$(3) \qquad u = \frac{1}{\sqrt{\pi}} \int_{-\infty}^{\infty} \psi(x + 2\sigma\sqrt{t}) e^{-\sigma^2} \, d\sigma.$$

We split this integral into three parts

$$J_1 + J_2 + J_3 = \frac{1}{\sqrt{\pi}} \int_{-\infty}^{-T} + \frac{1}{\sqrt{\pi}} \int_{-T}^{T} + \frac{1}{\sqrt{\pi}} \int_{T}^{\infty}$$

and choose $T = |t|^{-\frac{1}{4}}$. If t is sufficiently small, then for arbitrarily small given ϵ, $|\psi(x + 2\sigma\sqrt{t}) - \psi(x)| < \epsilon$ holds in the interval $-T \leq \sigma \leq T$, since obviously $|\sigma|\sqrt{t} \leq |t|^{\frac{1}{4}}$ and since the function ψ is continuous by assumption. From the convergence of the integral $\int_{-\infty}^{\infty} e^{-\sigma^2} \, d\sigma = \sqrt{\pi}$ we immediately conclude that for sufficiently small t, the difference between the integral J_2 and the function $\psi(x)$ is arbitrarily small. The integrals J_1 and J_3 may be estimated by

$$J_1 \leq \frac{M}{\sqrt{\pi}} \int_{-\infty}^{-T} e^{-\sigma^2} \, d\sigma$$

and

$$J_3 \leq \frac{M}{\sqrt{\pi}} \int_T^\infty e^{-\sigma^2} \, d\sigma.$$

Since the infinite integral $\displaystyle\int_{-\infty}^\infty e^{-\sigma^2} d\sigma$ converges, these integrals can be made arbitrarily small if t is chosen small enough. Thus the given function is indeed the solution for our boundary value problem.

A similar explicit formula gives the solution of the initial value problem for the heat equation in two and more dimensions. Consider for example the problem: For $t > 0$ find a solution $u(x, y, z; t)$ of

$$u_{xx} + u_{yy} + u_{zz} - u_t = 0$$

which for $t = 0$ coincides with a given continuous function $\psi(x, y, z)$. It is solved by

(4) $u(x, y, z; t)$

$$= \frac{1}{8(\sqrt{\pi t})^3} \iiint_{-\infty}^\infty \psi(\xi, \eta, \zeta) e^{-(1/4t) [(x-\xi)^2 + (y-\eta)^2 + (z-\zeta)^2]} \, d\xi \, d\eta \, d\zeta,$$

as can be easily seen.

Another initial value problem for the heat equation concerns a closed linearly extended heat conductor (a wire, for instance) of length 1. Here the initial value problem for the equation $u_{xx} - u_t = 0$ is stated as before; but in addition it is required that both the function $\psi(x)$ and the solution $u(x, t)$ be periodic in x with period 1. By superimposing the solutions

$$\exp\{-4\pi^2\nu^2 t\} \; (a_\nu \cos 2\pi\nu x + b_\nu \sin 2\pi\nu x)$$

under the assumption that the initial function $\psi(x)$ may be developed in a uniformly convergent Fourier series

$$\psi(x) = \frac{a_0}{2} + \sum_{\nu=1}^\infty (a_\nu \cos 2\pi\nu x + b_\nu \sin 2\pi\nu x),$$

we find a solution in the form

$$u(x, t) = \frac{a_0}{2} + \sum_{\nu=1}^\infty (a_\nu \cos 2\pi\nu x + b_\nu \sin 2\pi\nu x) e^{-4\pi^2\nu^2 t}.$$

Expressing the Fourier coefficients by integrals and interchanging the order of summation and integration (which is certainly permissible for $t > 0$), we obtain

$$(5)\quad u(x, t) = \int_0^1 \psi(\xi) \left\{ 1 + 2 \sum_{\nu=1}^{\infty} e^{-4\pi^2\nu^2 t} \cos 2\pi\nu(x - \xi) \right\} d\xi.$$

On the other hand, we can solve our initial value problem explicitly in a quite different way if we remember that

$$(6)\qquad W(x - \xi, t) = \frac{1}{2\sqrt{\pi t}} \sum_{\nu=-\infty}^{\infty} \exp\left\{ -\frac{(x - \xi - \nu)^2}{4t} \right\}$$

is a periodic solution of the heat equation with the period 1. The previous argument then shows that

$$(7)\qquad u(x, t) = \int_0^1 \psi(\xi) W(x - \xi, t) \, d\xi$$

solves the initial value problem.

Comparing the two solutions and applying the "fundamental lemma" of the calculus of variations,[1] we obtain the identity

$$(8)\qquad \sum_{\nu=-\infty}^{\infty} e^{-\pi\nu^2 t} \cos 2\pi\nu x = \frac{1}{\sqrt{t}} \sum_{\nu=-\infty}^{\infty} e^{-\pi(x-\nu)^2/t}$$

since the function $\psi(\xi)$ is arbitrary. This identity was derived earlier (Vol. I, p. 76) for the special case $x = 0$ as the *transformation formula of the elliptic theta function* and is thus illustrated here by the heat equation.

The conclusion (8) relies on the fact that the two solutions (5) and (7) are identical. To prove that the solution of the initial value problem is unique, we show that a solution belonging to the initial function zero—e.g., the difference between two solutions belonging to the same initial values—is identically zero. Indeed, by multiplying the equation $u_{xx} - u_t = 0$ by u, integrating over the interval and making use of the periodicity, we obtain

$$\frac{1}{2} \frac{d}{dt} \int_0^1 u^2 \, dx + \int_0^1 u_x^2 \, dx = 0,$$

and consequently

$$\frac{d}{dt} \int_0^1 u^2 \, dx \leq 0.$$

---

[1] See Vol. I, p. 185.

Since $u = 0$ holds identically for $t = 0$, it follows that u also vanishes identically for $t > 0$.[1]

2. Initial Value Problems for the Wave Equation. We have already solved the initial value problem for the wave equation in one-dimensional space (cf. Ch. I, §7, 1). Now we develop the important solution of the initial value problem for the wave equation

$$(9) \qquad u_{xx} + u_{yy} + u_{zz} = u_{tt}$$

in three dimensions starting with the previously found solutions $F(r - t)/r$ (with arbitrary F). Here r is defined by

$$r^2 = (x - \xi)^2 + (y - \eta)^2 + (z - \zeta)^2$$

for the parameter point (ξ, η, ζ).

Let $F_\epsilon(\lambda)$ be a non-negative function of a parameter λ, which vanishes outside the interval $-\epsilon < \lambda < \epsilon$ and for which

$$\int_{-\epsilon}^{\epsilon} F_\epsilon(\lambda) \, d\lambda = 1.$$

Clearly the function obtained by superimposing spherical waves,

$$\frac{1}{4\pi} \iiint \phi(\xi, \eta, \zeta) \frac{F_\epsilon(r - t)}{r} \, d\xi \, d\eta \, d\zeta,$$

with arbitrary $\phi(\xi, \eta, \zeta)$, is a solution of the wave equation. Now, letting ϵ tend to zero and passing to the limit under the integral sign, we obtain the expression

$$(10) \quad u(x, y, z; t) = \frac{t}{4\pi} \iint_\Omega \phi(x + t\alpha, y + t\beta, z + t\gamma) \, d\omega = t M_t\{\phi\},$$

where $d\omega$ is the surface element on the sphere $\Omega : \alpha^2 + \beta^2 + \gamma^2 = 1$ and $M_t\{\phi\}$ denotes the mean value of the function ϕ over the surface of the sphere of radius t with center at (x, y, z).

Instead of justifying this limit process, it is simpler to verify directly that the function u expressed by (10) is a solution of the wave equation. We omit this verification at present since it will be obtained systematically in a more general case in Ch. VI, §12.

Obviously the function u satisfies the initial conditions

$$u(x, y, z; 0) = 0, \qquad u_t(x, y, z; 0) = \phi(x, y, z).$$

[1] This method of proving uniqueness, in a widely generalized form, will play an important role later on (Ch. V, §4 and Ch. VI, §8).

Remembering that u_t is a solution of the wave equation as well as u, we easily find that *the function*

$$(11) \qquad u = tM_t\{\phi\} + \frac{\partial}{\partial t} tM_t\{\psi\}$$

solves the initial value problem for the given initial values

$$u(x, y, z; 0) = \psi(x, y, z), \qquad u_t(x, y, z; 0) = \phi(x, y, z).$$

3. Duhamel's Principle. Nonhomogeneous Equations. Retarded Potentials. Once the initial value problem for a homogeneous linear differential equation such as the wave equation has been solved, all the solutions of the corresponding nonhomogeneous differential equation can be found by the simple and general "principle of Duhamel", an analogue of the well-known variation of parameters or method of impulses for ordinary differential equations. We first formulate the principle (which will also occur later in this volume) and then apply it to the wave equation.

We consider the differential equation for a function

$$u(x_1, x_2, \cdots, x_n; t), \quad \text{or briefly,} \quad u(x, t):$$

$$(12) \qquad u_{tt} - L[u] = g(x, t).$$

Here L is an arbitrary linear differential operator which may contain the derivative u_t but no higher derivatives with respect to t. In applications the right side $g(x, t)$ represents external forces acting on a system. The initial value problem to be solved is: Find a solution u of the differential equation (12) which at $t = 0$ satisfies the initial conditions

$$(12') \qquad u(x, 0) = 0, \qquad u_t(x, 0) = 0.$$

The following motivation leads to the solution: We suppose that for fixed τ the right side of (12) is a function g_ϵ vanishing everywhere except in a small interval $\tau - \epsilon \le t \le \tau$ for which

$$\int_{\tau-\epsilon}^{\tau} g_\epsilon(x, t) \, dt = g(x, \tau).$$

We shall formally pass to the limit by letting ϵ go to zero, after first integrating the differential equation with respect to t between the limits $\tau - \epsilon$ and τ. In this way we are led to the following initial value problem for the homogeneous differential equation: With a given parameter value τ, find a solution $u(x; t)$ for $t \geq \tau$ of

$$(13) \qquad u_{tt} - L[u] = 0$$

such that, for $t = \tau$,

$$(13') \qquad u(x, \tau) = 0, \qquad u_t(x, \tau) = g(x, \tau).$$

This solution, which we continue as identically zero for $t \leq \tau$, corresponds to an instantaneous impulse of strength $g(x, \tau)$ affecting a system at rest at time $t \leq \tau$. We denote this solution of (13), (13'), which depends on the parameter τ, by $\phi(x, t; \tau)$; it is defined independently of its heuristic motivation. Now we assert: *The function*

$$(14) \qquad u(x, t) = \int_0^t \phi(x, t; \tau) \, d\tau$$

obtained by a superposition of the impulses ϕ is a solution of the initial value problem of the nonhomogeneous differential equation (12) with the initial conditions (12').

One can easily verify this statement. Since

$$u_t = \int_0^t \phi(x, t; \tau) \, d\tau,$$

$$u_{tt} = \phi_t(x, t; t) + \int_0^t \phi_{tt}(x, t; \tau) \, d\tau,$$

$$L[u] = \int_0^t L[\phi] \, d\tau,$$

and since $\phi_t(x, t; t) = g(x, t)$, u satisfies the differential equation (12) as well as the initial conditions (12').

We now apply this general result to the wave equation in three dimensions. According to article 2, we have

$$\phi(x, y, z, t; \tau) = (t - \tau)M_{t-\tau}\{g(x, y, z; \tau)\}.$$

Clearly, then, the initial value problem for the wave equation

$$u_{tt} - \Delta u = g(x, y, z; t)$$

with the initial conditions

$$u(x, y, z; 0) = 0, \qquad u_t(x, y, z; 0) = 0$$

is solved by the function

$$u(x, y, z; t) = \int_0^t (t - \tau)M_{t-\tau}\{g(x, y, z; \tau)\} \, d\tau$$

$$= \int_0^t \tau M_\tau\{g(x, y, z; t - \tau)\} \, d\tau$$

$$= \frac{1}{4\pi} \int_0^t \tau \, d\tau \int_\Omega g(x + \tau\alpha, y + \tau\beta, z + \tau\gamma; t - \tau) \, d\omega,$$

where α, β, γ are the components of a unit vector. Again introducing the rectangular coordinates $\xi = x + \tau\alpha$, $\eta = y + \tau\beta$, $\zeta = z + \tau\gamma$ instead of the polar coordinates, we have

$$(15) \qquad u(x, y, z; t) = \frac{1}{4\pi} \iiint_{r \le t} \frac{g(\xi, \eta, \zeta; t - r)}{r} \, d\xi \, d\eta \, d\zeta,$$

where $r = \sqrt{(x - \xi)^2 + (y - \eta)^2 + (z - \zeta)^2}$. This expression u is called a *retarded potential*; it is formed, in fact, like the potential of a mass distribution in space with the density g varying in time (cf. Ch. IV, §1). This density is not to be taken, however, at the time t but at an earlier time $t - r$, the difference r being the time interval which a signal moving with speed 1 would need to traverse the path between the center of the sphere and the point (ξ, η, ζ).

3a. Duhamel's Principle for Systems of First Order. The transition from the solution of the Cauchy problem for a homogeneous equation to a solution of the inhomogeneous equation is particularly simple and useful if formulated for a system of equations of first order in the matrix form (see §2, 3, equation (4a)).

We assume the system in the vectorial form

$$(13a) \qquad L[u] = u_t + \sum_{\nu=1}^n A^\nu u_\nu + Bu = g(x, t),$$

where u is a vector with k components, A^ν, B given $k \times k$ matrices, and g a given vector.

Suppose $u = \phi(x, t; \tau)$, depending on the parameter τ, is a solution of the homogeneous equation $L[u] = 0$ for $t > \tau$ under the initial condition $u(x, t) = g(x, \tau)$ for $t = \tau$, then

$$u(x, t) = \int_0^t \phi(x, t; \tau) d\tau$$

is the solution of $L[u] = g(x, t)$ with the initial condition $u(x, 0) = 0$. The proof is immediate and may be omitted.

4. Initial Value Problem for the Wave Equation in Two-Dimensional Space. Method of Descent.

The solution of the initial value problem for the wave equation in two dimensions

$$(16) \qquad u_{xx} + u_{yy} = u_{tt}$$

may be obtained directly from that for three dimensions by the following far-reaching method, called by Hadamard the *method of descent*. (See also Ch. VI, §12.) We consider equation (16) as a special case of the wave equation for three dimensions in which both the initial data and the solution itself are assumed to be independent of the third space variable z. Thus one "descends" from three to two dimensions. This argument immediately furnishes the desired solution if we assume in formula (10) of article 2 that

$$\phi(x, \, y, \, z) \; = \; \phi(x, \, y)$$

does not depend on z. Writing $\xi = t\alpha$, $\eta = t\beta$, $\zeta = t\gamma$ in (10) we find

$$u(x, \, y; \, t) = \frac{t}{4\pi} \iint_{\alpha^2+\beta^2+\gamma^2=1} \phi(x + \xi, y + \eta) \, d\omega,$$

where ξ, η are independent variables. This integral can be written as an integral over the sphere $\xi^2 + \eta^2 \leq t^2$ of radius t:

$$(17) \qquad u(x, \, y; \, t) = \frac{1}{2\pi} \iint_{\sqrt{\xi^2+\eta^2} \leq t} \frac{\phi(x + \xi, y + \eta)}{\sqrt{t^2 - \xi^2 - \eta^2}} \, d\xi \, d\eta.$$

Here the surface element of the sphere $\xi^2 + \eta^2 + \zeta^2 = t^2$ is expressed as

$$t^2 \, d\omega = \frac{t}{\zeta} \, d\xi \, d\eta = \frac{t}{\sqrt{t^2 - \xi^2 - \eta^2}} \, d\xi \, d\eta.$$

Therefore, (17) represents the solution of the initial value problem of the wave equation in two dimensions, if the initial conditions $u(x, y; 0) = 0$, $u_t(x, y; 0) = \phi(x, y)$ are stipulated.

A remarkable difference between two- and three-dimensional space becomes evident when we compare formulas (17) and (10). For the three-dimensional space the solution at one point depends only on the initial values on the surface of the three-dimensional sphere with radius t around this point, while the domain of dependency for two-dimensional space consists of both the boundary and the interior of the corresponding two-dimensional sphere or the circle with radius t. Later we shall analyze the deeper significance of this fact (cf. §4, 6 and Ch. VI, §18).

Moreover, the method of article 3 yields as the solution of the nonhomogeneous equation

$$(18) \qquad u_{tt} - u_{xx} - u_{yy} = f(x, y; t)$$

with the initial conditions

$$(18') \qquad u(x, y; 0) = 0, \qquad u_t(x, y; 0) = 0,$$

the expression

$$u(x, y; t) = \frac{1}{2\pi} \int_0^t d\tau \iint \frac{f(\xi, \eta; t - \tau)}{\sqrt{\tau^2 - (x - \xi)^2 - (y - \eta)^2}} \, d\xi \, d\eta.$$

This can also be written in the form

$$(19) \quad u(x, y; t) = \frac{1}{2\pi} \iiint_K \frac{f(\xi, \eta; \tau)}{\sqrt{(t - \tau)^2 - (x - \xi)^2 - (y - \eta)^2}} \, d\xi \, d\eta \, d\tau,$$

where K is the domain of (ξ, η, τ)-space defined by

$$0 \le \tau \le t; \qquad (x - \xi)^2 + (y - \eta)^2 \le (t - \tau)^2.$$

5. The Radiation Problem. Radiation problems are as important in physics as initial value problems; in fact, they can be treated as limiting cases of initial value problems. A formulation independent of such limiting operations will be given later, in Ch. VI, §18. In a radiation problem, the required function u and its derivatives with respect to t are zero initially (i.e., for $t = 0$, we have a state of rest). At a certain point in space, however, e.g., at the origin $r = 0$, a singularity for the solution u is prescribed as a function of time.

In three-dimensional space we already know solutions of the wave equation with a singularity at a fixed point in space: The functions

$$\frac{F(t - r)}{r}, \qquad \frac{G(t + r)}{r}$$

yield outgoing and incoming waves of this kind (we disregard the initial conditions yet to be satisfied). The solution of a radiation problem corresponds to the following limiting process: We consider the nonhomogeneous differential equation

$$(20) \qquad u_{tt} - \Delta u = f(x, y, z; t),$$

where f is the "density of the external force". The solution of the corresponding initial value problem for $t > 0$, with the state of rest as the initial condition, (cf. equation (15)) is given by

$$u = \frac{1}{4\pi} \iiint_{r \leq t} \frac{f(\xi, \eta, \zeta; t - r)}{r} \, d\xi \, d\eta \, d\zeta \, .$$

Given a small parameter ϵ, we now assume that $f = 0$ for

$$\rho^2 = \xi^2 + \eta^2 + \zeta^2 \geq \epsilon^2,$$

and write

$$\iiint_{\rho \leq \epsilon} f(\xi, \eta, \zeta; t) \, d\xi \, d\eta \, d\zeta = 4\pi g(t).$$

If we set $g(t) = 0$ for $t < 0$ and then pass to the limit as $\epsilon \to 0$, our solution goes over into

$$(21) \qquad u = \frac{g(t - r)}{r} \qquad \text{with} \quad r^2 = x^2 + y^2 + z^2 \, .$$

In this solution of the radiation problem the function $4\pi g(t)$ represents, therefore, the exciting force concentrated at the origin at time $t = 0$. Observe that at a certain point (x, y, z) in space at the time t, the solution u depends only on a single impulse which began at the origin at the time $t - r$ and was propagated to the point (x, y, z) with the speed 1.

The situation is quite different for the radiation problem in two-dimensional space. Here we have the differential equation

$$(22) \qquad u_{tt} - u_{xx} - u_{yy} = f(x, y; t).$$

We assume $f = 0$ for $r^2 = x^2 + y^2 \geq \epsilon^2$, and write

$$\iint_{\sqrt{\xi^2 + \eta^2} \leq \epsilon} f(\xi, \eta; t) \, d\xi = 2\pi g(t) \, .$$

Using the result of article 4 we obtain

$$(23) \qquad u(x, y; t) = \begin{cases} \displaystyle\int_0^{t-r} \frac{g(\tau)}{\sqrt{(t - \tau)^2 - r^2}} \, d\tau & \text{for } r \leq t, \\ 0 & \text{for } r > t, \end{cases}$$

by passing to the limit as $\epsilon \to 0$. In contrast to the three-dimensional case, the solution u at a point (x, y) and at a time t depends not only on a single preceding impulse but on the entire history of the radiation process up to the time $t - r$.

It is also interesting to study the character of the singularity of our solution at $r = 0$ in the two-dimensional case. For this purpose we first integrate by parts, observing that

$$\frac{1}{\sqrt{(t - \tau)^2 - r^2}} = -\frac{d}{d\tau} \log | t - \tau + \sqrt{(t - \tau)^2 - r^2} |,$$

and then expand in terms of r, which leads to the following representation of the function in the neighborhood of the singularity:

$$u(x, y; t) = -g(t - r) \log r + g(0) \log 2t$$

$$+ \int_0^t g'(\tau) \log 2(t - \tau) \, d\tau + \epsilon(t, r);$$

here

$$\epsilon(t, r) \to 0 \quad \text{as} \quad r \to 0.$$

Thus in two-dimensional space the singularity of the solution of the radiation problem is more complicated than that for three dimensions.

6. Propagation Phenomena and Huyghens' Principle. We now examine the character of propagation phenomena somewhat more closely (a discussion of basic principles will, however, be taken up only in Chapter VI). First we consider the homogeneous wave equation in three space dimensions. Suppose that at the time $t = 0$ the initial state is different from zero only in the neighborhood G of one point, for example, the origin. To calculate u at a point (x, y, z) and at time t, we surround (x, y, z) with a sphere of radius t and form certain integrals of the initial values over this sphere. Thus $u(x, y, z; t)$ is different from zero only when the surface of this sphere intersects the initial domain G, i.e., only within an interval of time $t_1 < t < t_2$ whose length is equal to the difference between the greatest and the least distance of the point (x, y, z) from the initial domain G. This fact characterizes our differential equation as an equation for a propagation phenomenon with the velocity 1. The initial state in the domain G is not perceptible at another point (x, y, z) until the time $t = t_1$ which is equal to the shortest distance between the domain G and the point (x, y, z). After the time t_2 which corresponds to the greatest such distance, the initial state ceases to have any effect. This phenomenon is called *Huyghens' principle* for the wave equation. It asserts that a sharply localized initial state is observed later at a different place as an effect which is equally sharply delimited. In the limiting case where the neighborhood G in which the initial state differs from zero shrinks to a point, say where the initial disturbance at $t = 0$ is concentrated at the origin, the effect at another point (x, y, z) will be felt only at a definite instant t, and t depends on the distance from the origin to the point (x, y, z).

In the case of two-dimensional space, however, the situation is altogether different. We again consider a domain G containing the origin and assume that the initial values of u and u_t are different from zero only in this domain. At a point P with the coordinates x, y and with shortest distance t_1 from the domain G, $u = 0$ certainly holds for $t < t_1$. According to formula (17) of article 4, the quantity u no longer is identically zero for $t > t_2$. In fact, if for example the initial function ϕ is not negative, u always remains different from zero at P for $t > t_1$. In other words, we still have a propagation phenomenon for the wave equation in two-dimensional space, in so far as a localized initial state requires a certain time to reach a point in space. However, Huyghens' principle no longer holds, because the effect of the initial state is not sharply limited in time: Once the signal has reached a point in space, it persists there indefinitely as a "reverberation".

If, in a propagation phenomenon, we study the state at a point (x, y, z) and at time t, we note that it depends on the initial values of u in a certain domain of space, the *domain of dependence* belonging to $(x, y, z; t)$. Hence, for the wave equation in three-dimensional space, this domain of dependence is the surface of the sphere of radius t around the point (x, y, z). The disturbance at this point at the time t does not depend on the initial data outside or inside the surface of the sphere.

On the other hand, in the case of two-dimensional space, the domain of dependence consists of the whole interior and the periphery of the circle of radius t around the point (x, y).

The physical difference becomes even clearer for the solutions of the radiation problem in article 5. Suppose a disturbance is propagated from the origin of three-dimensional space. Then at the time t we observe at a point $P(x, y, z)$ only the effect of what was sent out of the origin at the instant $t - r$. In two-dimensional space, however, the effect observed at the point $P(x, y)$ and at the time t depends on the entire radiation which has occurred up to the time $t - r$.

Thus, in a three-dimensional world in which waves are propagated according to the wave equation, sharp signals are transmitted and can be recorded as sharp signals. In a two-dimensional world the recording would be blurred.

In Chapter VI we shall see that arguments of this kind are restricted neither to the wave equation nor to two- and three-dimensional space. We shall note that Huyghens' principle holds for the wave equation in every odd number n of space dimensions except $n = 1$ and that it does not hold in spaces with an even number of dimensions.

§5. Solution of Initial Value Problems by Fourier Integrals

1. Cauchy's Method of the Fourier Integral. We shall now describe a general method for solving initial value problems by the superposition of plane waves. To avoid discussions of the legitimacy of interchanging limiting operations, we shall derive solutions heuristically; afterwards it must be verified directly as valid that the expressions thus obtained do indeed solve the problem.[1]

Again let

$$(1) \qquad\qquad L[u] = 0$$

be a linear homogeneous differential equation of order k with constant coefficients for the function $u(x_1, x_2, \cdots, x_n; t)$, or $u(x; t)$, and let

$$(2) \qquad\qquad u = e^{i(a_1 x_1 + a_2 x_2 + \cdots + a_n x_n - bt)},$$

or briefly $u = e^{i(ax)} e^{-ibt}$, be a solution of (1). We assume that (1) is *hyperbolic*: For every system of real numbers $a_1, a_2, \cdots, a_n$ (or for every vector a) there exist exactly k real distinct values (cf. §3, 4)

$$b = b_\kappa(a_1, a_2, \cdots, a_n) \qquad (\kappa = 1, 2, \cdots, k)$$

which depend algebraically on the parameters a_i and for which (2) is a solution of (1). If $W_1, W_2, \cdots, W_k$ denote k arbitrary functions

---

[1] In article 3, however, we shall not separate the formal construction and the verification.

of a_1, a_2, $\cdots$, a_n then we may formally construct the expression

$$(3) \qquad u = \sum_{\kappa=1}^{k} \int \cdots \int_{-\infty}^{\infty} W_\kappa(a)\, e^{i(ax)} e^{-itb_\kappa(a_1,a_2,\cdots,a_n)}\, da$$

by superposition of plane waves writing da for da_1, da_2, $\cdots$, da_n. Clearly, this formal expression is again a solution of (1) if all the integration processes converge and if it is permissible to apply the differential operator $L[u]$ under the integral sign.

We now use this remark to construct a solution u of (1) which, for $t = 0$, satisfies the initial conditions

$$u(x, 0) = \phi_0(x),$$

$$u_t(x, 0) = \phi_1(x),$$

$$(4) \qquad \cdots \cdots \cdots,$$

$$\frac{\partial^{k-1}}{\partial t^{k-1}}(u(x, 0)) = \phi_{k-1}(x),$$

with arbitrarily prescribed functions ϕ_ν, for $\nu < k$.

Differentiating with respect to t under the integral sign in (3) we obtain from these initial conditions, for $t = 0$, the system of equations

$$\phi_0 = \int \cdots \int_{-\infty}^{\infty} \sum_{\kappa=1}^{k} W_\kappa(a) e^{i(ax)}\, da,$$

$$(5) \qquad \phi_1 = \int \cdots \int_{-\infty}^{\infty} \sum_{\kappa=1}^{k} (-ib_\kappa) W_\kappa(a) e^{i(ax)}\, da,$$

$$\cdots \cdots \cdots \cdots \cdots \cdots \cdots \cdots,$$

$$\phi_{k-1} = \int \cdots \int_{-\infty}^{\infty} \sum_{\kappa=1}^{k} (-ib_\kappa)^{k-1} W_\kappa(a) e^{i(ax)}\, da$$

for the functions W_1, W_2, $\cdots$, W_k. According to Fourier's inversion theorem, these equations are solved by the formulas

$$(6) \qquad \sum_{\kappa=1}^{k} (-ib_\kappa)^l W_\kappa(a) = \frac{1}{(2\pi)^n} \int \cdots \int_{-\infty}^{\infty} \phi_l(\xi) e^{-i(a\xi)}\, d\xi$$

$$(l = 0, 1, \cdots, k - 1);$$

here ξ is the vector ξ_1, ξ_2, $\cdots$, ξ_n, $d\xi = d\xi_1\, d\xi_2 \cdots d\xi_n$ and the expressions on the right are known. Thus we obtain for the k unknown functions W_1, W_2, $\cdots$, W_k a system of k linear equations whose determinant $|\,(-ib_\kappa)^l\,|$ does not vanish since, by assumption, all the

b_i are distinct. The functions W_κ are uniquely determined and our initial value problem is thereby solved formally. The justification is contained in article 3.

2. Example. As an illustration we again consider the wave equation in three-dimensional space

$$u_{tt} - \Delta u = 0$$

with the initial conditions

$$u(x, y, z; 0) = 0, \qquad u_t(x, y, z; 0) = \phi(x, y, z).$$

Here we obtain for b the two values

(7) $$b = \pm\sqrt{a_1^2 + a_2^2 + a_3^2} = \pm\rho.$$

Applying the Fourier integral and using the initial condition $u(x, y, z; 0) = 0$, we obtain the representation

(8) $$u(x, y, z; t) = \iiint_{-\infty}^{\infty} W(a_1, a_2, a_3)e^{i(a_1x+a_2y+a_3z)} \sin \rho t \, da.$$

After differentiating under the integral sign we have, for $t = 0$,

$$u_t(x, y, z; 0) = \phi(x, y, z)$$

$$= \iiint_{-\infty}^{\infty} \rho W(a_1, a_2, a_3)e^{i(a_1x+a_2y+a_3z)} \, da;$$

according to the inversion theorem, W is therefore given by

(9) $$W(a_1, a_2, a_3) = \frac{1}{(2\pi)^3\rho} \iiint_{-\infty}^{\infty} \phi(\xi, \eta, \zeta)e^{-i(a_1\xi+a_2\eta+a_3\zeta)} \, d\xi \, d\eta \, d\zeta.$$

By introducing this value for W into (8) and then interchanging the integrations with respect to a_1, a_2, a_3 and ξ, η, ζ in the six-fold integral

$$u = \frac{1}{(2\pi)^3} \iiint_{-\infty}^{\infty} \frac{da}{\rho} \iiint_{-\infty}^{\infty} \phi(\xi, \eta, \zeta)e^{i[a_1(x-\xi)+a_2(y-\eta)+a_3(z-\zeta)]} \, d\xi \, d\eta \, d\zeta,$$

we might try to obtain a simpler form of the solution. However, this interchange cannot be accomplished directly since then the inner integral would not converge. But a simple, frequently used artifice (e.g., Ch. VI, §12) circumvents this obstacle. We consider not the integral (8) itself but the integral

(10) $\qquad v(x, y, z, t) = -\iiint_{-\infty}^{\infty} W(a_1, a_2, a_3) e^{i(ax)} \frac{\sin \rho t}{\rho^2} da,$

which differentiated twice with respect to t, yields the function

$$u = v_{tt}.$$

If, in (10), we substitute the expression (9) for W, interchange the order of integrations[1] and use suitable notation,[2] we have

$$v = -\iiint_{-\infty}^{\infty} \phi(\xi, \eta, \zeta) \, d\xi \, d\eta \, d\zeta \frac{1}{(2\pi)^3} \iiint_{-\infty}^{\infty} e^{i[a(x-\xi)]} \frac{\sin \rho t}{\rho^3} da,$$

where now the inner integral J converges. A simple calculation yields

$$J = \frac{1}{(2\pi)^3} \iiint_{-\infty}^{\infty} e^{i[a(x-\xi)]} \frac{\sin \rho t}{\rho^3} da$$

$$= \frac{1}{2\pi^2 r} \int_0^{\infty} \frac{\sin \rho r \sin \rho t}{\rho^2} d\rho.$$

Since

$$\sin \rho r \sin \rho t = \sin^2 \frac{t + r}{2} \rho - \sin^2 \frac{t - r}{2} \rho,$$

we immediately obtain

(11)
$$\int_0^{\infty} \frac{\sin \rho r \sin \rho t}{\rho^2} d\rho = \left(\frac{t + r}{2} - \left| \frac{t - r}{2} \right| \right) \int_0^{\infty} \frac{\sin^2 \rho}{\rho^2} d\rho$$

$$= \frac{\pi}{2} \left(\frac{t + r}{2} - \left| \frac{t - r}{2} \right| \right)$$

and

(12)
$$J = \begin{cases} \dfrac{1}{4\pi} & \text{for } r \le t, \\[2ex] \dfrac{1}{4\pi} \dfrac{t}{r} & \text{for } r \ge t \end{cases}$$

[1] The interchange is carried out without proof, since we are concerned with a heuristic method to obtain a solution which will be verified in Ch. VI, §13.

[2] That is, $\rho^2 = a_1^2 + a_2^2 + a_3^2$, $\rho^2 \, d\rho \, d\omega = da$.

and

(13) $$v = -\frac{1}{4\pi} \iiint_{r \leq t} \phi \, d\xi \, d\eta \, d\zeta - \frac{t}{4\pi} \iiint_{r \geq t} \frac{\phi}{r} \, d\xi \, d\eta \, d\zeta.$$

Differentiating with respect to t an integral of the form

$$J_1 = \iiint_{r \leq t} f(\xi, \eta, \zeta) \, d\xi \, d\eta \, d\zeta,$$

over the interior of a sphere of radius t with center at the point (x, y, z), yields the integral over the surface Ω of this sphere; thus

$$\frac{\partial J_1}{\partial t} = \iint_{\Omega} f(\xi, \eta, \zeta) \, d\Omega.$$

Correspondingly, the integral

$$J_2 = \iiint_{r \geq t} f(\xi, \eta, \zeta) \, d\xi \, d\eta \, d\zeta$$

taken over the exterior of the sphere has the derivative

$$\frac{\partial J_2}{\partial t} = -\iint_{\Omega} f(\xi, \eta, \zeta) \, d\Omega.$$

Therefore, we have from (13)

$$v_t = -\frac{1}{4\pi} \iint_{\Omega} \phi \, d\Omega + \frac{1}{4\pi} \iint_{\Omega} \phi \, d\Omega - \frac{1}{4\pi} \iiint_{r \geq t} \frac{\phi}{r} \, d\xi \, d\eta \, d\zeta,$$

and hence

(14) $$v_t = -\frac{1}{4\pi} \iiint_{r \geq t} \frac{\phi}{r} \, d\xi \, d\eta \, d\zeta.$$

Further differentiation finally leads to

(15) $$v_{tt} = \frac{1}{4\pi t} \iint_{\Omega} \phi \, d\Omega$$

or, with the symbol $M_t\{\phi\}$ introduced earlier

(16) $$u = v_{tt} = t M_t\{\phi\}$$

in agreement with the result in §4, 2.[1]

[1] For the generalization of this formula to n space-dimensions see Ch. VI, §12.

3. Justification of Cauchy's Method. Instead of using Cauchy's method merely as a formal construction in need of subsequent verification, one can obtain the solution together with a complete proof and an analysis of its validity. A brief presentation of this analysis will now be given with slight modifications of some of the preceding arguments.

The most general linear homogeneous equation of order k with constant coefficients can be written in the form (see §3)

$$(17) \qquad P\left(\frac{\partial}{\partial x_1}, \frac{\partial}{\partial x_2}, \cdots, \frac{\partial}{\partial x_0}\right) u = 0,$$

where

$$(18) \qquad P(y_1, y_2, \cdots, y_0) = \sum_{\kappa=0}^{k} P_\kappa(y_1, y_2, \cdots, y_0)$$

is a polynomial of order k with constant coefficients decomposed into a sum of homogeneous polynomials P_κ of degrees $\kappa \leq k$.[1]

The *Cauchy problem* for equation (17) with the plane $x_0 = 0$ as initial surface consists in finding a solution of (17) which satisfies the initial conditions

$$(19) \qquad \frac{\partial^\kappa u}{\partial x_0^\kappa} = \phi_\kappa(x_1, x_2, \cdots, x_n)$$

$$\text{for} \qquad x_0 = 0 \qquad \text{and} \qquad \kappa = 0, 1, \cdots, k - 1,$$

where the functions ϕ_κ are given. This problem is *proper*[2] if there exists a number N such that for functions ϕ_κ of class[3] C_N the conditions (17), (19) are satisfied for exactly one function u (which depends continuously on the ϕ_κ and their derivatives of order $\leq N$).

A necessary condition is obviously

$$(20) \qquad p = P_k(0, \cdots, 0, 1) \neq 0$$

since otherwise equation (17) for $x_0 = 0$ would furnish a relation between the data ϕ_κ and their derivatives which is not satisfied identi-

---

[1] We write κ as subscript here, slightly deviating from the notation in §3.

[2] Criteria for when a problem is properly posed are discussed in §6, 2. It is in this sense that we use the word "proper" here.

[3] C_N is the class of all functions for which all partial derivatives of order $\leq N$ exist and are continuous.

cally. (Condition (20) states that the initial manifold $x_0 = 0$ is not characteristic; see §2, 4.)

A less obvious necessary condition is that the equation

$$(21) \qquad P_k(y_1, y_2, \cdots, y_n, \eta) = 0$$

should have only real roots η for any real values of $y_1, y_2, \cdots, y_n$. We shall demonstrate the necessity of this condition here only for the case of a differential equation

$$(17') \qquad P_k\left(\frac{\partial}{\partial x_1}, \frac{\partial}{\partial x_2}, \cdots, \frac{\partial}{\partial x_0}\right) u = 0$$

in which all derivatives occurring are of the same order k. Assume that for certain real values of $y_1, y_2, \cdots, y_n$ equation (21) has an imaginary root η. Since the coefficients of the algebraic equation (21) for η are real, the complex conjugate of η must also be a root, and we can assume without loss of generality that

$$\text{Im}(\eta) = -z < 0.$$

We have then, for any real positive λ, a solution

$$(22) \qquad u = \lambda^{-N-k} e^{i\lambda(x_1 y_1 + \cdots + x_n y_n + x_0 \eta)}$$

of the differential equation (17') with initial values

$$\phi_\kappa = i^\kappa \eta^\kappa \lambda^{\kappa-N-k} e^{i\lambda(x_1 y_1 + \cdots + x_n y_n)} \qquad \text{for } \kappa = 0, 1, \cdots, k-1.$$

For $\lambda \to +\infty$ the functions ϕ_κ and their derivatives of order $\leq N$ tend to 0 uniformly in $x_1, x_2, \cdots, x_n$, whereas

$$|u(0, \cdots, 0, x_0)| = \lambda^{-N-k} e^{\lambda z x_0}$$

tends to infinity. This is incompatible with the continuous dependence of u on the ϕ_κ and their derivatives of order $\leq N$. A similar argument applies to the more general equation (17) and shows that the initial value problem can be proper only if all roots of (21) are real for real $y_1, y_2, \cdots, y_n$, i.e., if equation (17) is *hyperbolic* in the sense defined in §2, 3.[1]

Reality of the roots of (21) is in fact not quite sufficient.[2] As an

[1] See also F. John [4], Chapter II.

[2] For equations with constant coefficients L. Gårding [2] has given conditions which are both necessary and sufficient for the Cauchy problem to be properly posed.

example we consider the parabolic equation

(23) $$u_{x_0 x_0} - u_{x_1} = 0$$

corresponding to $k = 2$, $n = 1$. Equation (21) here reduces to $\eta^2 = 0$, and hence has only real roots. One solution of (23) is given by

$$u = y_1^{-N-1} e^{i(x_1 y_1 + x_0 \eta)},$$

if

$$\eta^2 + i y_1 = 0.$$

The solution u corresponding to the root $\eta = (1 - i)(y_1/2)^{\frac{1}{2}}$ tends to infinity with y_1, although its initial values and their derivatives of order $\leq N$ tend to zero. Consequently the Cauchy problem for (23) is not proper in spite of the reality of the roots of (21).

We shall now give a *sufficient* condition: Cauchy's problem is proper if all roots of (21) are *real and distinct* for all real not simultaneously vanishing values y_1, y_2, $\cdots$, y_n. For the proof we construct the solution in essentially the same way as Cauchy did.[1] First we observe that Cauchy's problem for equation (17) with initial values (19) can be reduced to the problem for the same equation with the special initial values $\phi_0 = \phi_1 = \cdots = \phi_{k-2} = 0$ and with ϕ_{k-1} a given arbitrary function. The desired solution u with the general initial values (19) can be written in the form

$$u = \sum_{r=0}^{k-1} \frac{\partial^r u_r}{\partial x_0^r},$$

where the u_r are solutions of (17) with initial values of the special type

$$\frac{\partial^\kappa u_r}{\partial x_0^\kappa} = \begin{cases} 0 & \text{for} \quad \kappa = 0, 1, \cdots, k - 2 \\ g_r(x_1, x_2, \cdots, x_n) & \text{for} \quad \kappa = k - 1. \end{cases}$$

The functions g_r must be chosen in such a way that u satisfies (19); hence we must stipulate

$$g_{k-1} = \phi_0,$$

$$g_{k-1-\kappa} + \sum_{r=k-\kappa}^{k-1} \left(\frac{\partial^{r+\kappa} u_r}{\partial x_0^{r+\kappa}} \right)_{x_0=0} = \phi_\kappa \quad \text{for} \quad \kappa = 1, 2, \cdots, k - 1.$$

[1] See A. Cauchy [1].

For $\kappa = 0, 1, \cdots, k - 1$ we obtain successively relations determining first g_{k-1}, and hence u_{k-1}, then g_{k-2}, and hence u_{k-2}, etc. Thus it is sufficient to solve (17) for the initial values

$$(24) \qquad \frac{\partial^\kappa u}{\partial x_0^\kappa} = \begin{cases} 0 & \text{for} \quad \kappa = 0, 1, \cdots, k - 2 \\ \phi(x_1, x_2, \cdots, x_n) & \text{for} \quad \kappa = k - 1 \end{cases}$$

on the surface $x_0 = 0$. To complete the construction of the solution we first consider the solution corresponding to a function ϕ of the form $\exp\{i(x_1 y_1 + \cdots + x_n y_n)\}$; according to the theory of ordinary differential equations the solution is given by

$$(25) \qquad \exp\{i(x_1 y_1 + \cdots + x_n y_n)\} Z(y_1, \cdots, y_n, x_0),$$

where

$$(26) \qquad Z(y_1, \cdots, y_n, x_0) = \frac{p}{2\pi} \oint \frac{e^{i\eta x_0}}{P(iy_1, \cdots, iy_n, i\eta)} \, d\eta.$$

Here p is defined by (20), and the path of integration in the complex η-plane should enclose all zeros of the denominator. For y restricted to a bounded part of the y-space, the roots of the denominator are also bounded, so that a path of integration independent of y may be chosen. It follows that Z is an entire analytic function of its arguments. Set $y_\nu = r z_\nu$, where $z_1^2 + z_2^2 + \cdots + z_n^2 = 1$. We show that the quantities

$$(27) \qquad r^{k-1} Z, \; r^{k-2} \frac{\partial Z}{\partial x_0}, \; \cdots, \; r^{-1} \frac{\partial^k Z}{\partial x_0^k}$$

are uniformly bounded for real arguments $y_1, y_2, \cdots, y_n, x_0$ with $r > 1, |x_0| < M$.

To this end we first write the equation

$$(28) \qquad P(iy_1, \cdots, iy_n, i\eta) = 0$$

in the form

$$(29) \qquad P_k\left(z_1, \cdots, z_n, \frac{\eta}{r}\right) + \frac{1}{ir} P_{k-1}\left(z_1, \cdots, z_n, \frac{\eta}{r}\right) + \cdots = 0.$$

Since the roots ζ of

$$(30) \qquad P_k(z_1, \cdots, z_n, \zeta) = 0$$

are by assumption simple, and since the roots of a polynomial with fixed highest coefficient and simple roots depend continuously and differentiably on the other coefficients, it follows that for every root η of (28) there exists a root ζ of (30) such that, for large r, $|\eta/r - \zeta|$ is of the order $1/r$. The roots ζ are distinct and real; hence the roots η are distinct and of bounded imaginary part for sufficiently large r. Using the theory of residues, we can then write Z for large r in the form

$$Z = p \sum_{\kappa=1}^{k} \frac{e^{i\eta_\kappa x_0}}{P'(iy_1, \cdots, iy_n, i\eta_\kappa)}$$

where $\eta_1, \eta_2, \cdots, \eta_k$ are the roots of (28), and P' is the derivative of P with respect to its last argument. Since

$$P'(iy_1, \cdots, iy_n, i\eta_\kappa)$$

$$= (ir)^{k-1} \left[P'_k\left(z_1, \cdots, z_n, \frac{\eta_\kappa}{r}\right) + \frac{1}{ir} P'_{k-1}\left(z_1, \cdots, z_n, \frac{\eta_\kappa}{r}\right) + \cdots \right]$$

$$= (ir)^{k-1} P'_k(z_1, \cdots, z_n, \zeta) + O(r^{k-2}),$$

and since $\exp\{i\eta_\kappa x_0\}$ and η_κ/r are bounded, it follows that the expressions (27) are bounded for $r > 1$ and $|x_0| < M$.

We are now in a position to solve the Cauchy problem with the initial conditions (24). As before we use the abbreviations x, y, ξ for the vectors $x_1, \cdots, x_n$, etc., and $dx = dx_1 \cdots dx_n$, etc. We assume that ϕ vanishes outside a bounded set and is of class C_{n+2}. Then ϕ permits a Fourier integral representation of the form (see Vol. I, p. 80)

$$\phi(x) = \int \cdots \int_{-\infty}^{\infty} \psi(y) e^{i(xy)} \, dy,$$

where

(31) $$(2\pi)^n \psi(y) = \int \cdots \int_{-\infty}^{\infty} \phi(\xi) e^{-i(\xi y)} \, d\xi.$$

The function ϕ is of class C_{n+2}; it therefore follows by integration by parts that $r^{n+2} \psi(y)$ is bounded, and consequently that the integral

$$\int \cdots \int_{-\infty}^{\infty} r \, |\psi(y)| \, dy$$

converges. Then

$$(32) \qquad u(x, x_0) = \int \cdots \int_{-\infty}^{\infty} \psi(y)e^{i(xy)} Z(y, x_0) \, dy$$

is of class C_k and clearly satisfies conditions (17) and (24). Hence the Cauchy problem has been solved. The assumption that ϕ vanishes outside a bounded region is not a restriction, since it can be shown that u depends only on the initial values ϕ in a *bounded* region. (This follows, for example, from the uniqueness theorem; see p. 237.)

The expression (32) for u does not involve the initial function ϕ directly, only its Fourier transform ψ. Let us, therefore, substitute the expression for ψ from (31) in (32); we obtain

$$(33) \qquad u(x, x_0) = \int \cdots \int_{-\infty}^{\infty} \phi(\xi) \, K(x - \xi, x_0) \, d\xi,$$

where

$$(34) \qquad (2\pi)^n K(x, x_0) = \int \cdots \int_{-\infty}^{\infty} e^{i(xy)} Z(y, x_0) \, dy.$$

To interchange the order of integrations leading to (34) is certainly legitimate if the integral for K converges absolutely. Since $r^{k-1} Z$ is bounded, this is assured when the order k of the equation and the number $n + 1$ of the independent variables satisfy the inequality

$$(35) \qquad k \geq n + 2.$$

The representation (33) shows that in this case u depends *continuously* on the initial function ϕ. This no longer follows from the preceding considerations if the inequality (35) is violated (as in the case of the wave equation in higher dimensions). However a modified formula (33) can still be derived in these cases. We assume that there exists an integer κ for which

$$(36) \qquad n - 1 \geq 2\kappa \geq n + 2 - k.$$

Such a κ can certainly be found whenever the order k of the differential equation is at least 4. Denoting the Laplacian $\partial^2/\partial x_1^2 + \cdots + \partial^2/\partial x_n^2$ by Δ, we can write the solution u given by (32) in the form

$$u = \Delta^{\kappa} \int \cdots \int_{-\infty}^{\infty} \psi(y) r^{-2\kappa} e^{i(xy)} Z(y, x_0) \, dy,$$

where the integral converges absolutely at the origin because $2\kappa < n$. If we now set

$$(2\pi)^n K^*(x, x_0) = \int \cdots \int_{-\infty}^{\infty} r^{-2\kappa} e^{i(xy)} Z(y, x_0) \, dy,$$

the integral for K^* converges absolutely and we obtain the representation

$$u = \Delta^\kappa \int \cdots \int_{-\infty}^{\infty} \phi(\xi) K^*(x - \xi, x_0) \, d\xi.$$

Since this expression can also be written in the form

$$u = \int \cdots \int_{-\infty}^{\infty} \Delta^\kappa \phi(x + \xi) K^*(-\xi, x_0) \, d\xi,$$

we see that u depends continuously on ϕ and its derivatives of order $\leq 2\kappa$. (The solution in the case $k = 2$ omitted here can be found by other means; see e.g. Ch. VI, §12).

A similar theory can be established for systems with constant coefficients; however, we shall omit it here.

In §3, 6 we have constructed solutions of differential equations for which $P_\kappa = 0$ for all κ except $\kappa = k$ (i.e., for the case with no dispersion) by superposition of plane waves of arbitrary, not necessarily exponential, shape. Thus the use of Fourier analysis could be avoided (see Ch. VI, §§14, 15).

§6. Typical Problems in Differential Equations of Mathematical Physics[1]

1. Introductory Remarks. To find the "general solution", i.e., the totality of all solutions of a partial differential equation, is a problem which hardly ever occurs. The goal usually is to single out specific individual solutions by adding further conditions to the differential equations. For $n + 1$ independent variables these additional restrictions usually refer to n-dimensional manifolds, which sometimes appear as boundaries, sometimes as "initial manifolds", and sometimes as discontinuity surfaces of domains within which the solutions are to be found ('boundary", "initial", and "jump conditions"). In particular, *initial value problems*, or "Cauchy problems", were

---

[1] Compare Hadamard [2], particularly Chapter I, and Hadamard's comprehensive paper [1], in which other types of problems are also discussed.

discussed in §4: Along the plane $x_0 = t$ values for the solution u and, as the case requires, for its derivatives with respect to t, were prescribed as functions of the coordinates $x_1, x_2, \cdots, x_n$. We seek a solution u for $t \geq 0$ which at $t = 0$ represents the prescribed "initial state". Such solutions of initial value problems may be continued for $t < 0$, so that the manifold $t = 0$ lies within the domain of definition of the solution. In other words, the state at $t = 0$ may be interpreted as the result of a previous state whose continuation into the future is governed by the same laws. In the case of differential equations of first order, for which the solution of the initial value problem was constructed in Chapter II, a continuation of this kind is automatic. For problems of higher order a similar result is contained in Ch. I, §7 where analytic differential equations and analytic initial conditions are treated. However, to assume that either the differential equations or the initial conditions are analytic is in general an artificial restriction; moreover, even for analytic differential equations the analytic character of all the solutions is not evident *a priori*. It is therefore reasonable to consider initial or boundary value problems without regard to continuation of solutions beyond these boundaries.

A typical *boundary value problem*—one of the central problems of analysis—is to find a solution of the potential equation $\Delta u = 0$ which is regular in the interior of a given domain, i.e. continuous there together with its first and second derivatives, and which attains prescribed continuous, but not necessarily analytic, boundary values on the boundary of the domain. In the cases $n = 2$ and $n = 3$ this boundary value problem was solved explicitly for circular or spherical domains by means of Poisson's integral (cf. Ch. I, §3, 2 and Vol. I, Ch. VII, p. 513). In Chapter IV of this volume and in Volume III, we shall construct and discuss the solutions for arbitrary domains.

In other boundary value problems for the potential equation, some linear combination of the function and its normal derivative is given on the boundary. Problems formulated in this way were discussed in Vol. I, Ch. VI, from the standpoint of the calculus of variations; they will be solved in Volume III.

Potential theory gives a good illustration of the fact that the "general" solution may be useless for boundary value problems. For example, the well-known general solution of the Laplace equation for $n = 2$ is

$$u = f(x + iy) + g(x - iy),$$

where f and g are arbitrary analytic functions of a complex variable. However, this form of the solution is of no value for the treatment of the general boundary value problem.

Next, we mention a boundary value problem of nonlinear character which arises in connection with minimal surfaces. Suppose a space curve Γ is given in x, y, u-space whose projection C bounds a domain in the x,y-plane. In order to find a minimal surface $u(x, y)$ bounded by Γ, we pose the following *boundary value problem for the equation of minimal surfaces*

$$(1) \qquad (1 + u_y^2)u_{xx} - 2u_xu_yu_{xy} + (1 + u_x^2)u_{yy} = 0:$$

Find a solution $u(x, y)$ of equation (1) which is twice continuously differentiable in G and takes on prescribed values on C (Plateau's problem in unsymmetric form).

Further, the important class of "mixed problems" should be pointed out. We consider a fixed domain G in the space of the variables x_1, x_2, $\cdots$, x_n with boundary Γ which is assumed to have appropriate regularity properties. In the domain G and for $t \geq 0$ we seek a function $u(x_1, x_2, \cdots, x_n, t)$ or $u = u(x, t)$ which satisfies a given differential equation $L[u] = 0$, which attains on Γ, prescribed boundary conditions which may also depend on t, and which satisfies prescribed initial conditions for $t = 0$ in G. An example is given by a string stretched between the points $x = 0$ and $x = 1$ with the boundary conditions

$$u(0, t) = u(1, t) = 0$$

and the initial conditions.

$$u(x, 0) = \phi(x), \qquad u_t(x, 0) = \psi(x).$$

In many mixed problems the boundary conditions for the differential equations are nonhomogeneous.

If $L[u]$ is a linear, homogeneous operator, one may distinguish two types:

1. The boundary conditions are homogeneous; e.g., the function u vanishes at the boundary. Such conditions appear in *vibration*

problems of a bounded system extended over the domain G; the motion is assumed to begin at $t = 0$ with a given initial state.   In Volume I these vibration problems were thoroughly discussed, particularly on the basis of the theory of eigenfunctions.

2. The initial conditions are homogeneous; e.g., the function u and possibly some of its derivatives are required to vanish for $t = 0$. The boundary conditions, however, are nonhomogeneous.   Problems of this kind, e.g. *problems of transient response*, play an important role in many applications.

In principle, transient problems may be reduced to problems of type 1. We need only subtract from the unknown function in the differential equation a function which satisfies the prescribed initial and boundary conditions and otherwise is arbitrarily chosen. For the difference we then obtain a problem of the vibration type refer- ring to a nonhomogeneous differential equation, a problem which may therefore be attacked immediately by the method of eigen- functions according to Vol. I, Ch. V. Notwithstanding the possibility of such a reduction, an independent treatment of transient problems is desirable both from the theoretical and from the practical stand- point. Methods particularly adapted to applications will be dis- cussed in Appendix 2 to Chapter V.

Radiation problems which, in §4, 5 were treated as limiting cases of initial value problems for nonhomogeneous equations may in- stead be treated as limiting cases of transient problems; as such, they also belong to the class of mixed problems.

Finally, let us mention a number of typical examples which do not fall into the categories described above.   *Riemann's mapping problem* requires that a given domain G of the x, y-plane be mapped conformally on the unit disc $u^2 + v^2 < 1$. Analytically this leads to the following boundary value problem: Find, for the Cauchy-Riemann differential equations

$$u_x = v_y, \qquad u_y = -v_x,$$

a pair of solutions $u(x, y)$, $v(x, y)$, defined in the domain G with boundary Γ, continuous and once continuously differentiable in G; in addition, the functions u and v should possess continuous boundary values which satisfy the boundary condition $u^2 + v^2 = 1$ and map Γ on the unit circle in a one-to-one way.

In this formulation we are evidently no longer directly concerned with a simple boundary value problem of potential theory, even though solving this problem may be reduced to the solution of an ordinary boundary value problem (as was shown in Vol. I, p. 377 and as will be shown in the next chapter from another point of view).

More general is *Plateau's problem* in parametric form: to construct a minimal surface bounded by a given curve Γ in space. According to §1, 4, it can be formulated as the following problem in differential equations: Inside the unit circle $u^2 + v^2 < 1$, find three functions x, y, z of u, v which satisfy the equations

$$(2) \qquad \Delta x = \Delta y = \Delta z = 0,$$

the additional conditions

$$(3) \qquad x_u^2 + y_u^2 + z_u^2 = x_v^2 + y_v^2 + z_v^2, \quad x_u x_v + y_u y_v + z_u z_v = 0$$

and possess boundary values $x(s)$, $y(s)$, $z(s)$ which are continuous functions of the arc length s on the unit circle and which form a parametric representation of the given space curve Γ. Although Plateau's problem in the unsymmetric form (see eq. (1)) does not always possess a solution, it can always be solved in the form given here.[1]

Another typical example involving the potential equation is the "jet problem" of plane hydrodynamics. It differs essentially from the classical boundary value problem inasmuch as values are prescribed on *"free"* (i.e., not *a priori* given) *boundaries*. We consider the problem of two-dimensional irrotational flow of an incompressible fluid escaping from a symmetric nozzle. The flow is assumed symmetric; the axis of symmetry, say the x-axis, is a streamline and may be replaced by a rigid wall. Suppose that G is the infinite domain in the x, y-plane bounded below by the x-axis and above by a curve consisting of two parts, the nozzle boundary D and the jet boundary S. The given nozzle boundary D extends from a point A backward asymptotically approaching the horizontal line $y = b$. The unknown jet boundary S is to extend forward from A asymptotically approaching the horizontal line $y = 1$ (see Figure 7). One wants to find a "stream-function" $\psi(x, y)$, defining the desired flow, which satisfies the differential equation $\Delta\psi = 0$. Since the boundaries of G are streamlines, ψ is constant there; we may require $\psi = 0$ on the x-axis and $\psi = 1$ on $D + S$. On S the pressure is constant; hence by

[1] Cf. R. Courant [2].

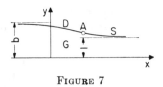

Bernoulli's theorem, $\partial \psi / \partial \nu$ = const. there (ν denotes the outward normal). Moreover, we require that $\partial \psi / \partial y$ should tend to 1 as x tends to $+\infty$ and to $1/b$ as x tends to $-\infty$. The problem is a "free" boundary value problem. The jet boundary S and the constant b occurring in the conditions on ψ are not prescribed in advance but are to be determined as part of the solution. Accordingly, an additional boundary condition is imposed besides those in an ordinary boundary value problem, i.e., the normal derivative is required to be constant on S.[1]

Finally a problem of yet another type should be mentioned, that of *scattering*: An "incoming" wave given *a priori*, e.g., a plane wave, must be modified; by an additional solution of the wave equation a "scattered wave" must be found to make the sum satisfy conditions imposed by obstacles. We shall discuss such problems in Ch. IV, §5.

2. Basic Principles. The types of differential equations enumerated in article 1 originate from situations in physics, mechanics, or geometry. The character of additional conditions such as boundary conditions or initial conditions is similarly motivated by physical reality. Nevertheless a motivation from a purely mathematical point of view should also be given. We do not, however, want to lose ourselves in an attempt at complete systematization; instead, in connection with some typical examples, we establish important guiding principles which will be supported in the further course of our studies.

The main principle is: *Boundary value problems are naturally associated with elliptic differential equations, while initial value problems, mixed problems and radiation problems arise in connection with hyperbolic and parabolic differential equations.*

---

[1] The problem of the free jet was first considered by Helmholtz in 1868. He and his successors obtained solutions for a variety of special nozzle shapes. The existence of solutions for general nozzle shapes was established by A. Weinstein in 1929. For a historical survey and bibliography of the theory of wakes and jets in two dimensions, see Weinstein [1].

A mathematical problem which is to correspond to physical reality should satisfy the following basic requirements:

(1) The solution must exist.

(2) The solution should be uniquely determined.

(3) The solution should depend continuously on the data (requirement of stability).

The first requirement expresses the logical condition that not too much, i.e., no mutually contradictory properties, is demanded of the solution.

The second requirement stipulates completeness of the problem —leeway or ambiguity should be excluded unless inherent in the physical situation.[1]

The third requirement, particularly incisive, is necessary if the mathematical formulation is to describe observable natural phenomena. Data in nature cannot possibly be conceived as rigidly fixed; the mere process of measuring them involves small errors. For example, prescribed values for space or time coordinates are always given within certain margins of precision. Therefore, a mathematical problem cannot be considered as realistically corresponding to physical phenomena unless a variation of the given data in a sufficiently small range leads to an arbitrary small change in the solution. This requirement of "stability" is not only essential for meaningful problems in mathematical physics, but also for approximation methods.

Any problem which satisfies our three requirements will be called a *properly posed* problem.

Another important point of view is suggested by examples of initial value and radiation problems. In many cases it is clear that the solutions do not depend on the totality of the given data; questions concerning the *domain of influence of the data* or *the domain of dependence of the solution* arise.

In support of the general thesis a few examples will serve for orientation, principally with respect to our third requirement.

First we consider the elliptic differential equation $\Delta u = 0$ in a domain G with the boundary Γ. The existence of the solution has already been demonstrated (cf. Vol. I, Ch. V, §15) at least in such special cases as the circle, the sphere, or the rectangle (a general proof

---

[1] Cases do occur in which uniqueness is not a proper requirement. For example, in the case of multiple eigenvalues whole families of solutions of the eigenvalue problem exist.

will be given later in Chapter IV). In any case the boundary value problem for an arbitrary piecewise smooth boundary satisfies our requirement of uniqueness. This follows immediately from the fact that every potential function which is regular in G reaches its maximum and minimum values on Γ (cf. Ch. IV, §1); therefore, it vanishes identically if its boundary values are zero. If we have two solutions belonging to the same boundary values, then their difference is a potential function with the boundary values zero; i.e., the difference vanishes identically. Hence the two solutions are the same.

The requirement that the solutions depend continuously on the boundary values is also satisfied. The difference of two solutions whose prescribed boundary data differ everywhere by an amount less than ϵ in absolute value is again a potential function and cannot have an absolute value greater than ϵ in the interior of G because it assumes its maximum and minimum on Γ. The boundary value problem of the potential equation is therefore properly posed according to our requirements.

We see moreover that the *domain of dependence* of the solution at every point in the region is the *whole boundary*; i.e., the value of the solution u at any closed subdomain G_1 depends on the boundary values on every part of the boundary Γ. If a part C of Γ had no influence on the value of u in the subdomain G_1, then we would obtain in this subregion the same solution u if the boundary values were altered on C but not on the rest of the boundary, C'. By forming the difference of the two sets of boundary values we would obtain a solution in G_1 hence also in G identically vanishing with nonidentically vanishing boundary values. This is obviously absurd since for such a solution *all* the boundary values must vanish.

Cauchy's initial value problem for the potential equation, unlike the boundary value problem, is not properly posed. We shall show that the first requirement (existence of a solution) and the third requirement (continuous dependence on the data) are violated.

Prescribe, for example, the initial conditions $u(x, 0) = 0$, $u_y(x, 0) = g(x)$ for the equation $\Delta u = 0$. According to a principle of potential theory (see Chapter IV), any solution in the upper half-plane, $y > 0$, can be continued into the lower half-plane by reflection and would automatically be analytic on the x-axis. Thus $g(x)$ must be an analytic function of x and cannot be arbitrarily given as, e.g., an everywhere twice continuously differentiable but nonanalytic function. (In the case of analytic initial values the solution was constructed in Ch. I, §7.)

The following example pointed out by Hadamard shows that the solution of such an initial value problem with analytic initial values does not depend continuously on the data: Consider a sequence of initial value problems for the equation $\Delta u = 0$; for the n-th problem $(n = 1, 2, \cdots)$ we use the analytic initial data

$$u(x, 0) = 0, \qquad u_y(x, 0) = g_n(x) = \frac{\sin nx}{n^2}$$

which approach the function $g(x) = 0$ uniformly with increasing n. The solution of the initial value problem for g_n is

$$u(x, y) = \frac{\sinh ny \sin nx}{n^2}.$$

As n increases, this solution does not approach the solution $u = 0$ which belongs to the initial values $g(x) = 0$. In other words, although the data change by an arbitrarily small amount, the change in the solution cannot be confined to a small range. Thus the initial value problem for the potential equation is certainly not properly posed. A more general theorem to this effect was given on p. 216.

On the other hand, the initial value problem for the simplest hyperbolic equation, the wave equation $u_{xx} - u_{tt} = 0$, satisfies all three requirements. The initial conditions

$$u(x, 0) = \phi(x), \qquad u_t(x, 0) = \psi(x)$$

yield the solution

$$2u(x, t) = \phi(x + t) + \phi(x - t) + \int_{x-t}^{x+t} \psi(\tau) \, d\tau$$

for $t > 0$. The solution of this hyperbolic problem exists, is uniquely determined, and, obviously, depends continuously on the given data $\phi(x)$ and $\psi(x)$. As to the domain of dependence for the solution: $u(x, t)$ depends only on those values $\phi(\xi)$ and $\psi(\xi)$ for which $x - t \leq \xi \leq x + t$.

For this hyperbolic differential equation, however, a boundary value problem would be meaningless. If, for instance, we replace the wave equation by the equivalent equation $u_{xy} = 0$ for a function $u(x, y)$, then we can no longer arbitrarily prescribe boundary values as in the case of a rectangle with sides parallel to the axes. Since the derivatives u_y must be equal at corresponding opposite points of the sides $x = $ const. of the rectangle and since a similar statement holds for u_x, u can be arbitrarily given on only two adjacent sides of the rectangle and hence a boundary value problem is impossible.

Considerations analogous to those for hyperbolic differential equations may also be applied to the parabolic case, e.g., to the equation of heat conduction.

The general thesis stated here, concerning the question of when problems in differential equations are properly posed, will be confirmed and amplified in subsequent parts of this book.

3. Remarks about "Improperly Posed" Problems.   The stipulations of article 2 about existence, uniqueness, and stability of solutions dominate classical mathematical physics. They are deeply inherent in the ideal of a unique, complete, and stable determination of physical events by appropriate conditions at the boundaries, at infinity, at time $t = 0$, or in the past.   Laplace's vision of the possibility of calculating the whole future of the physical world from complete data of the present state is an extreme expression of this attitude.   However, this rational ideal of causal-mathematical determination was gradually eroded by confrontation with physical reality.   Nonlinear phenomena, quantum theory, and the advent of powerful numerical methods have shown that "properly posed" problems are by far not the only ones which appropriately reflect real phenomena.   So far, unfortunately, little mathematical progress has been made in the important task of solving or even identifying and formulating such problems which are not "properly posed" but still are important and motivated by realistic situations.   The present book will have to be restricted essentially but not exclusively to the classical mathematical field of well-posed problems.

Nevertheless, we can indicate a few examples of such meaningful but not "properly posed" problems.   First we recall that the initial value problem for Laplace's equation is not "properly posed" according to the preceding article.   Yet it is of physical significance. For example, G. I. Taylor has shown that an important question of stability leads to such an improperly posed problem: Consider a system of two incompressible fluids separated by an interface, moving toward the lighter of the fluids.   This phenomenon can be described by means of a velocity potential.   This potential turns out to be a solution of an improper initial value problem for Laplace's equation.

Overdetermined problems form another type of meaningful "improper" problem. For example, we may seek a function, harmonic inside the unit circle, which has prescribed values in the concentric circle of radius $\frac{1}{2}$.

Improperly posed problems of great potential importance in numerical analysis so far have not been reached by the main stream of active research. Thus, for initial value problems where the initial data are only approximately known from observations, approximate methods, e.g., finite difference methods, are used to construct the solution. Then the question arises: How can later observations be used to improve and extend the computed solution?

Progress concerning such problems could be of the greatest value, for example, for mathematical weather prediction.[1]

4. General Remarks About Linear Problems. The analogy between problems of linear differential equations and systems of a finite number of linear algebraic equations for an equal number of unknowns was pointed out before (Vol. I, Ch. V, §1). Differential equations, for example, can be replaced by difference equations. Later, in Volume III, this idea (which of course requires a passage to the limit) will be carried out in detail.[2] We merely recall the following *alternative* for N linear equations in N unknowns: *Either the corresponding homogeneous problem possesses a nontrivial solution, or the general nonhomogeneous problem possesses a unique solution for arbitrarily given data.* Ambiguity or nonuniqueness in the solution of the general nonhomogeneous problem implies a nontrivial solution of the homogeneous problem. Moreover, the alternative can also be formulated in the following way: For N linear equations in N unknowns the existence and the uniqueness of the solution of the general nonhomogeneous problem are equivalent.

We may expect linear problems of mathematical physics which are correctly posed to behave like a system of N linear algebraic equations in N unknowns. Thus we arrive at the following heuristic principle: If for a correctly posed problem in linear differential equations the corresponding homogeneous problem possesses only the "trivial solution" zero, then a uniquely determined solution of the general nonhomogeneous problem exists. However, if the homogeneous problem has a nontrivial solution, the solvability of the non-

---

[1] Concerning not-well posed problems, compare F. John [2] and C. Pucci [2]. Also see Ch. VI, §17.

[2] See, for example, R. Courant, K. O. Friedrichs, H. Lewy [1].

homogeneous problem requires the fulfillment of certain additional conditions.

In the first volume we found the principle of this alternative well confirmed and we shall gain a deeper insight in the following chapters.

Appendix 1 to Chapter III

§1. *Sobolev's Lemma*

On several previous occasions we have made use of the significant fact that boundedness of positive definite quadratic integrals involving derivatives of a function may permit boundedness of the functions themselves. A general statement of this type is *Sobolev's lemma.* It consists of an estimate of a function by "L_2-bounds" of its derivatives (that means bounds for the integrals of their squares):

1) If $u(x)$ is defined in an open domain G (n dimensions) then for y in G

$$(1) \qquad |u(y)|^2 \leq C\sum_{j\leq\nu}R^{2j-n}\int_G |D^j u(x)|^2\, dx\,, \qquad \text{if} \quad \nu > \frac{n}{2}\,,$$

where R is the distance from y to the boundary of G; C is a constant depending only on ν and n. Here $D^j u$ represents any jth order derivative, and summation is to extend over all derivatives of orders shown.

In (1′) below we state a similar result for points y in the closure of G under some smoothness condition for the boundary of G.

Proof of 1) Let $h(t)$ be a function of class[1] C^∞ of one variable $t \geq 0$ which is identically one for $0 \leq t \leq \frac{1}{2}$ and vanishes for $t \geq 1$. Assuming that y is the origin and setting $|x| = r$ we define[2]

$$\zeta(r) = h\left(\frac{r}{R}\right).$$

Note that

$$(2) \qquad \left|\left(\frac{\partial}{\partial r}\right)^k \zeta\right| \leq R^{-k}C_k \qquad (k \geq 0)\,,$$

where C_k is constant depending only on the function $h(t)$. Since $h(0) = 1$ and $h(1) = 0$ we have

---

[1] The class C^N denotes functions with derivatives of order up to N, C^∞ the class of functions with derivatives of all order.

[2] We may choose $h(t) = e^{1/4}e^{-(t-1)^{-2}}(1 - e^{-(t-1/2)^{-2}})$ for $1/2 \leq t \leq 1$.

$$u(0) = - \int_0^R \frac{\partial}{\partial r} (\zeta u) \, dr \, .$$

Now, representing by $d\omega$ an element of the unit sphere about $y = 0$, and by ω the area of the $(n - 1)$-dimensional unit sphere, we obtain by integrating with respect to ω

$$\omega u(0) = - \iint \frac{\partial}{\partial r} (\zeta u) \, dr \, d\omega.$$

Integrating by parts $(\nu - 1)$ times with respect to r leads to

$$\omega u(0) = \frac{(-1)^\nu}{(\nu - 1)!} \iint r^{\nu-1} \frac{\partial^\nu}{\partial r^\nu} (\zeta u) \, dr d\omega \, .$$

Setting $r^{\nu-1} = r^{\nu-n} \cdot r^{n-1}$ we find by Schwarz' inequality

$$| u(0) | \leq \text{const.} \sqrt{\iint \left| \frac{\partial^\nu}{\partial r^\nu} (\zeta u) \right|^2 dV} \sqrt{\iint r^{2(\nu-n)} \, dV} \, ,$$

where $dV = r^{n-1} dr \, d\omega$ is the element of volume; the constant depends only on ν. Since $2\nu > n$ the last integral converges and we have

$$| u(0) |^2 \leq \text{const.} \, R^{2\nu-n} \iint \left| \frac{\partial^\nu}{\partial r^\nu} (\zeta u) \right|^2 dV.$$

Together with (2) the inequality

$$\left| \frac{\partial^j}{\partial r^j} u \right|^2 \leq C_j' \sum | D^j u |^2,$$

where C_j' depends only on j and n yields (1).

DEFINITION: G is said to satisfy the "cone condition" if every point y of $\bar{G}$ (the closure of G) is the vertex of a finite cone C_y (i.e., the intersection of a cone with a sphere of radius R about its vertex) which lies in $\bar{G}$ (the closure of G) and whose volume is greater than some constant V.

$1'$) *If G satisfies the "cone condition" then*

$$(1') \qquad | u(y) |^2 \leq \text{const.} \sum_{j \leq \nu} \int_G | D^j u(x) |^2 \, dx \qquad \left(\nu > \frac{n}{2} \right),$$

where the constant depends only on ν, n, and the values of R and V.

The proof of the sharper lemma $(1')$ is the same as that of (1) except that the integration $d\omega$ extends only over the interior solid angle of the cone C_y from y.

Integral inequalities such as (1) play a role in variational methods for the construction of solutions.[1]

§2. Adjoint Operators

1. Matrix Operators. A very important concept in the theory of linear operators is that of the operator L^* adjoint to an operator L.

For finite spaces the concept is a matter of formal linear algebra. We consider a linear operator

$$(1) \qquad w^i = \sum_{j=1}^{l} a_{ij} u^j \qquad (i = 1, \cdots, k),$$
$$w = L[u],$$

in matrix notation

$$w = Au,$$

which transforms an l-vector u with l components $u^1, \cdots, u^l$ into a k-vector[2] w with k components $w^1, \cdots, w^k$; if $l \neq k$ the transformation matrix A is not square.

Together with the operator L we consider the bilinear form (or inner product)

$$(2) \quad (v,w) = \sum_{i=1}^{k} v^i w^i = (v,Au) = (v,L[u]) = \sum_{i=1}^{k} \sum_{j=1}^{l} a_{ij} u^j v^i,$$

formed with a k-vector v. Then the adjoint operator L^* transforms a k-vector v into an l-vector z and is defined by the *transformation* with the transposed matrix A^*

$$(1') \qquad z_j = \sum_{i=1}^{k} a_{ij} v_i \qquad (j = 1, \cdots, l)$$

or

$$z = A^* v \quad \text{or} \quad z = L^*[v].$$

The adjoint operator is linked with the bilinear form (2):

$$(3) \qquad (v, w) = (v, L[u]) = (L^*[v], u).$$

No matter whether or not $k = l$ the equation

$$(4) \qquad vL[u] - uL^*[v] = 0$$

or

$$(v, Au) - (u, A^*v) = 0$$

---

[1] See Volume III and L. Nirenberg [2], Lecture II.

[2] Vectors may be considered as row vectors or column vectors, as the meaning of the matrix notation indicates.

characterizes the relationship between the two adjoint operators L and L^*.

From (4) we find immediately $(A^*)^* = A$; furthermore, for the product of two matrices AB

$$(AB)^* = B^*A^*,$$

as seen by replacing in (4) A by AB and then writing $(v, ABu) = (A^*v, Bu) = (B^*A^*v, u)$. If the matrix A is symmetric, then the corresponding operator is called *self-adjoint* which means identical with the adjoint operator.

In passing we recall a basic theorem of linear algebra: If $k < l$ then the system of l linear equations $L^*[v] = z$ for the components v and given values of the components z is overdetermined. It is solvable if and only if a number of, say, r, linear homogeneous compatibility conditions $Rz = 0$ are satisfied. The adjoint matrix $S = R^*$ leads to a transformation $u = Sh$ of a vector h with r components into a vector u with k components; it represents all the solutions of the underdetermined system of equations $L[u] = 0$ expressed by r independent arbitrary quantities $h_1, \cdots, h_r$.

2. Adjoint Differential Operators. We now consider differential operators $L[u]$ of any order. Again, the number k of the components of the vector L and the number l of components of the vector v are not necessarily assumed equal. As in article 1 we may form the inner product of $L[u]$ with any l-vector v and integrate over a domain G with piecewise smooth boundary Γ. Then we integrate $vL[u]$ by parts to remove the derivatives of u. Thus we obtain a relation of the form

$$(5) \qquad vL[u] - uL^*[v] = \sum_{i=1}^{n} P_{x_i}^i = \operatorname{div} P$$

or after integration

$$(6) \qquad \iint_G (vL[u] - uL^*[v])\, dx = \int P\xi\, dS,$$

where P is a vector with components P^i, dS is the element of Γ, and ξ denotes the outward normal unit vector on Γ. The components P^i depend bilinearly on the functions u, v as well as derivatives of u, v, and have coefficients which may depend on x. While L^* is uniquely determined (by (5)) the vector P is not, but merely within an arbitrary additive vector R whose divergence vanishes identically.

The operator L^* thus uniquely defined is called the adjoint to L. It transforms a k-vector into an l-vector.

Obviously the integral of $(v, L[u])$ corresponds to the bilinear form of article 1. If the function v and its derivatives, or u and its derivatives, vanish at the boundary then the boundary integral in (6) is zero and our definition corresponds to that in article 1 without modification.

We note some examples: The operator L of order zero:

$$L[u] = au$$

clearly is self-adjoint, i.e., $L^* = L$. Next we consider the first order operator

$$L[u] = D_j u,$$

where $D_j = \partial/\partial x_j$; clearly

$$L^*[u] = -D_j u.$$

The adjoint operator to a first order operator

$$L[u] = \sum A^\nu u_\nu + Bu$$

as in §2 is simply

$$L^*[v] = -(vA^\nu)_{x_\nu} + Bv$$

and the vector P has the components

$$P_\nu = (vA^\nu u).$$

For a scalar operator of second order

$$L[u] = \sum a^{ij} u_{ij} + \sum a^i u_i + bu$$

the adjoint operator is

$$L^*[v] = \sum \left[\frac{\partial^2 (va^{ij})}{\partial x^i \, \partial x^j} - \frac{\partial}{\partial x_i} (va^i) + vb \right],$$

and the components of P can be written

$$P^i(v, u) = \sum_j \left[va^{ij} \frac{\partial u}{\partial x_j} - \frac{\partial (va^{ij})}{\partial x_j} u + va^i u \right].$$

The coefficients a^{ij} may be matrices, i.e., the operator may be a system of single operators.

For a self-adjoint operator, a^{ij} must be symmetric matrices and

$$\sum_j \left(\frac{\partial a^{ij}}{\partial x^j} + \frac{\partial a^{ji}}{\partial x^j} \right) = 2\, a^i, \quad \sum_i \frac{\partial a^i}{\partial x^i} = 0.$$

Let L be any operator of, say, order m written in the form[1]:

$$L[u] = \sum_{|p| \leq m} a_p D^p u,$$

then

$$L^* = \sum (-1)^{|p|} D^p a_p u.$$

Incidentally, to form the adjoint operators and the vectors P one may make use of the easily established law

$$(LM)^* = M^* L^*$$

for the product of two operators.

Appendix 2 to Chapter III

The Uniqueness Theorem of Holmgren

Existence and uniqueness of the solution of Cauchy's problem in the small is guaranteed for any analytic differential equation provided that the data are analytic, that the initial surface is analytic and noncharacteristic and that the solution is assumed to be analytic (see Ch. I, §7). For nonanalytic data or equations the existence of a solution cannot be expected in general without further restrictive conditions such as the hyperbolic character of the differential equation and the space-like nature of the initial manifold and data. It is remarkable, however, that the *uniqueness* of the solution can still be proved in cases when existence theorems for *arbitrary* data are no longer valid. The most important theorem in this connection was given by Holmgren.[2] It concerns linear analytic[3] differential

---

[1] See for the notation Ch. VI, §3, 1.

[2] E. Holmgren [1].

[3] The argument of Cauchy and Kowalewski only shows that there cannot exist more than one *analytic* solution of an analytic Cauchy problem, but leaves open the possibility that other nonanalytic solutions exist. Holmgren's theorem denies this possibility. The uniqueness theorem for the solution of

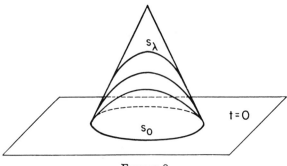

FIGURE 8

equations in any number of independent variables, single equations as well as systems:

If $L[u]$ is a linear differential operator with analytic coefficients and if the Cauchy initial data vanish on a smooth noncharacteristic surface S_0, then any solution u of $L[u] = 0$ with these initial data vanishes identically in a suitably small neighborhood of any closed subset of S_0.

Note that the theorem does assume, but does not assert the existence of, the solution u. Of course it asserts the uniqueness for arbitrary not necessarily analytic Cauchy data on S_0, since the difference of two such solutions has the Cauchy data zero, hence must vanish identically according to the preceding formulation of the theorem.

The proof is essentially very simple: We consider a lens-shaped domain G bounded by a suitably small portion of S_0 and a neighboring analytic surface S. In the vicinity of S we solve the backward Cauchy problem with initial Cauchy data zero for the inhomogeneous differential equation

$$L^*(v) = p(x),$$

where L^* is the operator adjoint to L and p any polynomial in the variables $x_1, \cdots, x_n$. This function v certainly can be constructed on the basis of Ch. I, §7 for a vicinity of S_0. Assume S can be chosen so near to S_0 that v exists in the same domain G for all polynomials

---

Cauchy problems for *nonanalytic* differential equations is still a challenge. It has been proved generally for the case of two independent variables by T. Carleman [2]. For higher dimensions it can be proved for many hyperbolic equations for space-like initial manifolds by methods similar to those described in this chapter. Compare for uniqueness theorem in more dimensions Claus Müller [1], A. P. Calderón [1], and L. Hörmander [5].

$p(x)$. Then Green's formula, that means integration of $uL^*(v) - vL(u) = up(x)$ over G, yields, because of the vanishing Cauchy data for v on S and for u on S_0,

$$\int \cdots \int_G up(x) \, dx = 0 .$$

Since the polynomials $p(x)$ form a complete set of functions in B it follows immediately that u is identically zero in G.

All that has to be proved is the possibility of choosing S as stipulated. This can be done without difficulties in line with the remarks in Ch. I, §7, 4, a reasoning which shall not be carried out here in detail.

As an additional remark it might be stated[1] that Holmgren's theorem can be extended from a statement in the small to a statement in the large referring to a lens G which is swept by a family of analytic surfaces S_λ for positive values of the parameter λ such that for all $\lambda > 0$ on S_λ the characteristic determinant is bounded away from zero uniformly in λ. For example, consider the wave equation $u_{xx} + u_{yy} + u_{zz} - u_{tt} = 0$ and take as the hypersurface S_0 the "spherical disk" $x^2 + y^2 + z^2 \leq 1$, $t = 0$. It is intuitively clear and easily verified that S_0 can be embedded into a family of surfaces S_λ with the same boundary which fill the whole characteristic hypercone having S_0 as base. This cone is indeed the exact domain in which u is determined uniquely by the Cauchy data on S_0. See Figure 8.

[1] See F. John [3].

CHAPTER IV

Potential Theory and Elliptic Differential Equations[1]

An exhaustive theory of elliptic differential equations is outside the scope of this work. We shall consider mainly, although not exclusively, linear differential equations of second order with particular emphasis on potential theory which is typical for the theory of more general differential equations and which is in itself an important branch of analysis.

In Volume I and in the preceding chapters numerous questions of potential theory have been discussed. In the first part of the present chapter a somewhat systematic treatment will be presented. The second, less elementary part, should provide readers a selective approach to farther-reaching theories concerned with more general types of elliptic problems.

A supplement, due to L. Bers, briefly develops the theory of pseudo-analytic functions, a significant recent contribution to the theory of differential equations of second order in two independent variables.

§1. *Basic Notions*

1. Equations of Laplace and Poisson, and Related Equations. We consider functions $u(x_1, x_2, \cdots, x_n) = u(x)$ of n variables $x_1, x_2, \cdots, x_n$, or the vector x, in a domain G of x-space with boundary Γ. The differential equation

$$(1) \qquad \Delta u = \sum_{\nu=1}^{n} \frac{\partial^2 u}{\partial x_\nu^2} = 0$$

is called *Laplace's equation* or the potential equation. Its solutions are called potential functions or *harmonic functions*. The associated

---

[1] We refer the reader to the following books: O. D. Kellogg [1], I. Petrovskii [1], C. Miranda [1], L. Lichtenstein [1], F. John [4] and L. Bers [2]. For more recent work the reader is referred to the bibliography of symposia and colloquia [1], [2], [3]; and E. Magenes and G. Stampacchia [1], L. Nirenberg [2], M. I. Visik, and O. A. Ladyzhenskaya [1], and L. Gårding [1].

nonhomogeneous equation is called *Poisson's equation*. We introduce, $\Gamma(n/2)$ denoting the Γ-function, the surface area ω_n of the unit sphere in n dimensions

$$(2) \qquad \omega_n = \frac{2(\sqrt{\pi})^n}{\Gamma\left(\dfrac{n}{2}\right)}$$

and write Poisson's equation in the form

$$(3) \qquad \Delta u = -\omega_n\,\mu(x_1, x_2, \cdots, x_n),$$

where $\mu(x_1, x_2, \cdots, x_n)$ is a given function of the point x.

Solutions of the potential equation which have continuous second derivatives in a domain G are called *regular in G*. (Later we shall see that analyticity of u is a consequence.) Here, as in the following, G denotes an *open, connected* and, unless expressly noted to the contrary, *bounded* region of space. We denote by $G + \Gamma$ the *closed* region obtained from G by adjunction of the boundary Γ. Similarly, in case μ is continuous in G, solutions of Poisson's equation (3) with continuous second derivatives are called regular. We shall consider principally the cases $n = 2$ and $n = 3$, denoting the coordinates x_1, x_2 and x_1, x_2, x_3 by x, y and x, y, z, respectively.

For $n = 2$ the "general solution" of the potential equation is the real part of any analytic function of the complex variable $x + iy$. For $n = 3$ one can also easily obtain solutions which depend on arbitrary functions. For example, let $f(w, t)$ be analytic in the complex variable w for fixed real t. Then, for arbitrary values of t, both the real and imaginary parts of the function

$$u = f(z + ix\cos t + iy\sin t, t)$$

of the real variables x, y, z are solutions of the equation $\Delta u = 0$. Further solutions may now be obtained by superposition:

$$(4) \qquad u = \int_a^b f(z + ix\cos t + iy\sin t, t)\, dt.$$

For example, if we set

$$f(w, t) = w^n e^{iht},$$

where n and h are integers, and integrate from $-\pi$ to $+\pi$, we get homogeneous polynomials

$$u = \int_{-\pi}^{\pi} (z + ix\cos t + iy\sin t)^n e^{iht}\, dt$$

in x, y, z.   Introducing polar coordinates $z = r \cos \theta$, $x = r \sin \theta \cos \phi$, $y = r \sin \theta \sin \phi$, we obtain

$$u = 2r^n e^{ih\phi} \int_0^\pi (\cos \theta + i \sin \theta \cos t)^n \cos ht \, dt,$$

i.e., except for a constant factor, the functions

$$u = r^n e^{ih\phi} P_{n,h}(\cos \theta)$$

where the $P_{n,h}(x)$ are the associated Legendre functions (cf. Vol. I, p. 505).

By *transformation to polar coordinates* r, ϕ for $n = 2$ or r, θ, ϕ for $n = 3$, i.e., by means of the transformation

$$\left.\begin{array}{l} x = r \cos \phi \\ y = r \sin \phi \end{array}\right\} \qquad \text{in the plane}$$

or

$$\left.\begin{array}{l} x = r \sin \theta \cos \phi \\ y = r \sin \theta \sin \phi \\ z = r \cos \theta \end{array}\right\} \qquad \text{in space,}$$

the Laplacian becomes

(5)
$$\Delta u = \frac{1}{r} \left[\frac{\partial}{\partial r}(r u_r) + \frac{\partial}{\partial \phi}\left(\frac{u_\phi}{r}\right) \right] \qquad (n=2),$$

$$\Delta u = \frac{1}{r^2 \sin \theta} \left[\frac{\partial}{\partial r}(r^2 u_r \sin \theta) + \frac{\partial}{\partial \theta}(u_\theta \sin \theta) + \frac{\partial}{\partial \phi}\left(\frac{u_\phi}{\sin \theta}\right) \right] (n=3),$$

(cf. Vol. I, p. 225).   From these formulas we deduce the following frequently applied theorem:

If $u(x, y)$ is a regular harmonic function in the plane domain G, the function

(6)
$$v(x, y) = u\left(\frac{x}{r^2}, \frac{y}{r^2}\right) \qquad (r^2 = x^2 + y^2).$$

also satisfies the potential equation and is regular in the region G' obtained from G by inversion with respect to the unit circle. The corresponding theorem holds in space, except that here we must set

(7)
$$v = \frac{1}{r} u\left(\frac{x}{r^2}, \frac{y}{r^2}, \frac{z}{r^2}\right) \qquad (r^2 = x^2 + y^2 + z^2).$$

For the proof we introduce polar coordinates and show: If $u(r, \phi)$

and $u(r, \theta, \phi)$ are harmonic then the functions $v(r, \phi) = u(1/r, \phi)$ and $v(r, \theta, \phi) = u(1/r, \theta, \phi)/r$ are also harmonic. This follows at once from the formulas (5), if we note that

$$r^4 \frac{1}{r} \frac{\partial}{\partial r} (rv_r) = \frac{1}{\rho} \frac{\partial}{\partial \rho} (\rho u_\rho) \qquad \text{for } n = 2,$$

$$r^5 \frac{1}{r^2 \sin \theta} \frac{\partial}{\partial r} (r^2 v_r \sin \theta) = \frac{1}{\rho^2 \sin \theta} \frac{\partial}{\partial \rho} (\rho^2 u_\rho \sin \theta) \qquad \text{for } n = 3,$$

where $\rho = 1/r$.

The reader may verify the analogous theorem in n dimensions for the function

$$(8) \qquad v = \frac{1}{r^{n-2}} u \left(\frac{x_1}{r^2}, \frac{x_2}{r^2}, \cdots, \frac{x_n}{r^2} \right).$$

Thus, *except for the factor r^{2-n}, the harmonic character of a function is invariant under inversions r with respect to spheres. Moreover, the harmonic property is retained completely under similarity transformations, translations, and simple reflections across planes.*

Let the function u be regular and harmonic in a bounded domain G. If we invert G with respect to a sphere of unit radius whose center, say the origin, lies in G, the interior of G is carried into the exterior G' of the inverted boundary surface Γ'. The harmonic function

$$v = \frac{1}{r^{n-2}} u \left(\frac{x_1}{r^2}, \frac{x_2}{r^2}, \cdots, \frac{x_n}{r^2} \right)$$

is then called regular in this exterior region G'. That is, we define *regularity in a domain G extending to infinity* in the following way: We invert G with respect to a sphere with center outside of G, and thus transform G into a bounded domain G'. *The harmonic function u is called regular in G if the above function v is regular in G'.* In particular, u is called *regular at infinity* if G contains a neighborhood of the point at infinity and a value is assigned to the function u at the point at infinity such that v is regular in G'. According to this definition, for example, the function $u = $ const. is regular at infinity in the plane, but not in spaces of three or more dimensions. In space, for arbitrary a, the functions

$$u = 1 - a + \frac{a}{r}$$

are harmonic outside the unit sphere and have the boundary values $u = 1$ on the sphere. But $u = 1/r$ is the only function of this family which is regular in the region exterior to the unit sphere.

For any number n of dimensions the only solutions of the potential equation (1) that depend merely on the distance r of the point x from a fixed point ξ, e.g., the origin, are (except for arbitrary multiplicative and additive constants) the functions

(9)
$$\gamma(r) = \frac{1}{(n-2)\omega_n} r^{2-n} \qquad (n > 2),$$

$$\gamma(r) = \frac{1}{2\pi} \log \frac{1}{r} \qquad (n = 2),$$

which exhibit the so-called *characteristic singularity* for $r = 0$.

Every solution of the potential equation in G of the form

$$\psi(x_1, x_2, \cdots, x_n ; \xi_1, \xi_2, \cdots, \xi_n) = \psi(x, \xi)$$

$$= \gamma(r) + w \qquad \left(r^2 = \sum_{\nu=1}^{n} (x_\nu - \xi_\nu)^2 \right)$$

for ξ inside G and w regular, is called a *fundamental solution* with a singularity at the parameter point ξ. (See, e.g., Ch. III, §2.)

Corresponding fundamental solutions can easily be obtained also for the more general differential equation

$$\Delta u + cu = 0,$$

where c is a constant. After introducing polar coordinates, we ask for solutions of the form $u = \psi(r)$ with $r^2 = \sum_{\nu=1}^{n} (x_\nu - \xi_\nu)^2$. For ψ we obtain the ordinary differential equation

(10)
$$\psi'' + \frac{n-1}{r} \psi' + c\psi = 0.$$

If we set $\psi(r) = r^{-(1/2)(n-2)} \phi(\sqrt{c}\, r)$, this equation becomes Bessel's equation

(11)
$$\phi'' + \frac{1}{\rho} \phi' + \phi - \left(\frac{n-2}{2} \right)^2 \frac{\phi}{\rho^2} = 0 \qquad (\rho = \sqrt{c}\, r).$$

The desired fundamental solution ψ is then simply the solution of equation (11) which is infinite at the origin. Thus, we have for odd n

(12)
$$\psi = r^{-(1/2)(n-2)} J_{-(1/2)(n-2)}(\sqrt{c}\, r)$$

and for even n

$$(13) \qquad \psi = r^{-(1/2)(n-2)} N_{(1/2)(n-2)}(\sqrt{cr}),$$

where N_ν is the νth Neumann function. (See Ch. III, §2.)

2. Potentials of Mass Distributions. For $n = 3$, the fundamental solution of Laplace's equation

$$\frac{1}{r} = \frac{1}{\sqrt{(x - \xi)^2 + (y - \eta)^2 + (z - \zeta)^2}}$$

corresponds physically to the gravitational potential at the point $P(x, y, z)$ of a unit mass concentrated at the point (ξ, η, ζ).[1]

Let $\mu(\xi, \eta, \zeta)$ be the density of a distribution of mass in ξ, η, ζ-space. We call the integral

$$(14) \qquad u(x, y, z) = \iiint_G \frac{\mu(\xi, \eta, \zeta)}{r} \, d\xi \, d\eta \, d\zeta$$

$$(r^2 = (x - \xi)^2 + (y - \eta)^2 + (z - \zeta)^2)$$

extended over the appropriate region G of the space, the *potential of a spatial mass distribution of density* μ in the region G. If the point P with coordinates x, y, z lies outside G, then it follows immediately by differentiating under the integral sign that u is a harmonic function. If the point P lies in the region G and if μ is piecewise continuously differentiable[2], then, as we saw earlier (cf. Vol. I, Ch. V),

[1] Here and elsewhere the word potential is meant in the physical sense. It denotes a quantity whose gradient furnishes a field of force. The concept of potential is usually connected with Laplace's equation, although for a precise mathematical terminology it may be desirable not to call solutions of Laplace's equation potential functions but rather to use the more specific term, harmonic functions.

[2] The following definition should be recalled: A surface is called *piecewise smooth* if it is composed of a finite number of parts, each of which is congruent to a surface represented by a function

$$x_n = z = f(x_1, x_2, \cdots, x_{n-1}),$$

where f is continuous and has continuous first derivatives in a corresponding domain including its boundary. If, in addition, each function f has continuous derivatives of second order, the surface is said to possess *piecewise continuous curvature*. Obviously similar definitions can be applied to curves. A function is said to be *piecewise continuous* in G if it is continuous in G except for jump discontinuities at isolated points and along piecewise smooth curves or surfaces, and if only a finite number of such discontinuities are present in any closed subdomain of G. If the first derivatives of a continuous function are piecewise continuous in G, the function is called *piecewise continuously differentiable* in G.

the potential u satisfies *Poisson's equation*

$$(15) \qquad \Delta u = -4\pi\mu.$$

Generally for n dimensions the integral formed with the fundamental solution $\gamma(r)$

$$u(x_1, x_2, \cdots, x_n) = u(x)$$

$$(14a)$$

$$= \int \cdots \int_G \mu(\xi_1, \xi_2, \cdots, \xi_n) \gamma(r)\, d\xi_1\, d\xi_2 \cdots d\xi_n,$$

where the domain of integration G in the ξ-space contains the point x, satisfies the Poisson equation

$$(15a) \qquad \Delta u = -\omega_n \mu(x)$$

if μ possesses continuous derivatives. Again this integral is called the potential of the mass in G with density μ. Presently we shall give a proof under slightly milder assumptions concerning μ and examine the solution u from other points of view.

First, however, we prove the following theorem:

Let $\mu(x, y, z)$ be a function which is bounded (in absolute value) *and integrable in the region G. Then the potential* (14) *and its first derivatives are everywhere uniformly continuous; these derivatives can be obtained by differentiation under the integral sign. If, in addition, μ is piecewise continuously differentiable in G, then the second derivatives of u are also continuous in the interior of G and Poisson's equation $\Delta u = -4\pi\mu$ is satisfied.*

To prove the first part of this theorem, we consider the function

$$(16) \qquad u_\delta(x, y, z) = \iiint_G \mu(\xi, \eta, \zeta) f_\delta(r)\, d\xi\, d\eta\, d\zeta,$$

where $f_\delta(r)$ is any auxiliary function differing from the fundamental solution $1/r$ only in a small sphere of radius δ about the point $r = 0$; in the interior of this sphere we let $f_\delta(r)$, unlike $1/r$, remain bounded. At the surface of the sphere we connect it with $1/r$ in a continuous and continuously differentiable way. For example, we define

$$(17) \qquad f_\delta(r) = \begin{cases} \dfrac{1}{2\delta}\left(3 - \dfrac{r^2}{\delta^2}\right), & r \le \delta \\[3mm] \dfrac{1}{r}, & r > \delta. \end{cases}$$

From the inequality

$$(18) \qquad |u_\delta - u| \leq 4\pi M \int_0^\delta \left(f_\delta + \frac{1}{r}\right) r^2 \, dr = \frac{18\pi}{5} M\delta^2,$$

where M denotes a bound for $|\mu|$, it follows at once that the sequence u_δ converges uniformly for all x, y, z to the potential u as $\delta \to 0$ and that u is uniformly continuous.

The differentiability of the function $f_\delta(r) = g(x - \xi, y - \eta, z - \zeta)$ carries over directly to the functions u_δ. In fact we have

$$\frac{\partial u_\delta}{\partial x} = \iiint_G \mu \frac{\partial}{\partial x} f_\delta(r) \, d\xi \, d\eta \, d\zeta.$$

Now let $\chi(x, y, z)$ be defined by the convergent integral

$$(19) \qquad \chi = \iiint_G \mu \frac{\partial}{\partial x} \frac{1}{r} \, d\xi \, d\eta \, d\zeta$$

which results from (14) by formal differentiation under the integral sign. Then

$$\frac{\partial u_\delta}{\partial x} - \chi = \iiint_{r<\delta} \mu \frac{\partial}{\partial x} \left(f_\delta - \frac{1}{r}\right) d\xi \, d\eta \, d\zeta$$

and hence

$$(20) \qquad \left|\frac{\partial u_\delta}{\partial x} - \chi\right| \leq 4\pi M \int_0^\delta \left(\left|\frac{\partial f_\delta}{\partial r}\right| + \frac{1}{r^2}\right) r^2 \, dr = 5\pi M\delta;$$

that is, the sequence $\partial u_\delta/\partial x$ converges uniformly in x, y, z to the function $\chi(x, y, z)$. From well-known theorems of analysis it then follows that χ is uniformly continuous and that $\chi = u_x$. The same facts follow in an analogous way for the derivatives u_y and u_z.

Without additional assumptions on the function μ we cannot prove that u has continuous second derivatives. However, if μ is once continuously differentiable then the integral

$$u_x = \iiint_G \mu \frac{\partial}{\partial x} \frac{1}{r} \, d\xi \, d\eta \, d\zeta = -\iiint_G \mu \frac{\partial}{\partial \xi} \frac{1}{r} \, d\xi \, d\eta \, d\zeta$$

may be integrated by parts and expressed as

$$-\iint_\Gamma \frac{\mu}{r} n_1 \, dS + \iiint_G \frac{\mu_\xi}{r} \, d\xi \, d\eta \, d\zeta,$$

where dS represents the element of area on the surface Γ and n_1 represents the cosine of the angle between the exterior normal on Γ and the ξ-axis. In view of the argument given above, we may differentiate this expression and obtain continuous expressions for the second derivative of u. For example, at a point $P: (x, y, z)$, we obtain

$$(21) \qquad u_{xx}(P) = -\iint_\Gamma \mu \frac{\partial}{\partial x} \frac{1}{r} n_1\, dS + \iiint_G \mu_\xi \frac{\partial}{\partial x} \frac{1}{r}\, d\xi\, d\eta\, d\zeta.$$

The second integral may also be written in the form

$$\iiint_G (\mu - \mu(P))_\xi \frac{\partial}{\partial x} \frac{1}{r}\, d\xi\, d\eta\, d\zeta$$

which, upon integration by parts, is equivalent to

$$\iint_\Gamma (\mu - \mu(P)) \frac{\partial}{\partial x} \frac{1}{r} n_1\, dS - \iiint_G (\mu - \mu(P)) \frac{\partial^2}{\partial\xi\partial x} \frac{1}{r}\, d\xi\, d\eta\, d\zeta$$

(the integration by parts is valid since $\mu - \mu(P)$ vanishes to first order at P). If now we substitute this expression into the expression for $u_{xx}(P)$ and collect terms we obtain

$$(21')\qquad
\begin{aligned}
u_{xx}(P) &= -\mu(P) \iint_\Gamma \frac{\partial}{\partial x} \frac{1}{r} n_1\, dS \\
&\qquad - \iiint_G (\mu - \mu(P)) \frac{\partial^2}{\partial\xi\partial x} \frac{1}{r}\, d\xi\, d\eta\, d\zeta \\
&= -\mu(P) \iint_\Gamma \frac{\partial}{\partial x} \frac{1}{r} n_1\, dS \\
&\qquad + \iiint_G (\mu - \mu(P)) \frac{\partial^2}{\partial x^2} \frac{1}{r}\, d\xi\, d\eta\, d\zeta.
\end{aligned}$$

From this and analogous formulas we deduce the last part of our theorem; that is, *for piecewise continuously differentiable μ the second derivatives of u are continuous and $\Delta u = -4\pi\mu$.*

The fact that the representation $(21')$ for u_{xx} does not involve the derivatives of μ suggests the search for a wider class of functions μ for which the integrals in the representation are convergent and for which the conclusions of the theorem hold. For this purpose we introduce the important class of Hölder continuous functions.

HÖLDER CONTINUITY: A function μ is called Hölder continuous in a region G or is said to *satisfy a Hölder condition* with exponent α,

$0 < \alpha < 1$, and coefficient K in G if for every pair of points P, Q in G the inequality

$$| \mu(P) - \mu(Q) | \leq K\overline{PQ}^{\alpha}$$

holds. Here $\overline{PQ}$ represents the distance from P to Q; α and K are called the constants of the Hölder inequality.

We observe that the last integral in (21′) is absolutely convergent if μ is Hölder continuous in G, so that the right side of (21) is a well-defined function $v(P)$; for then, with α, K as the constants of the Hölder condition, the integrand of the volume integral is bounded in absolute value by

(22) $$\frac{CK}{r^{3-\alpha}}$$

(with C a numerical constant), which is integrable.

We now state the following theorem (which is a sharper form of the theorem stated previously):

If $\mu(x, y, z)$ is a Hölder continuous function in G, then the potential (14) has uniformly continuous first derivatives which may be obtained by differentiation under the integral sign. Furthermore, u has continuous second derivatives which are given by (21) and analogous expressions, and u satisfies Poisson's equation $\Delta u = -4\pi\mu$.

In order to prove, say, that $u_{xx}(P)$ exists, we first assume that the function μ may be uniformly approximated by differentiable functions μ_n which satisfy a uniform Hölder condition with constants α and some K'. For these we define functions u_n by (14) which, from the above, have continuous second derivatives. It suffices to show that the $u_{n_{xx}}$ converge to v uniformly in any closed subdomain of G. From (21′) we have

$$| v(P) - u_{n_{xx}}(P) | \leq | \mu_n(P) - \mu(P) | \left| \iint_{\Gamma} \frac{\partial}{\partial x} \frac{1}{r} n_1 \, dS \right|$$

$$+ \iiint_G | \mu - \mu(P) - \mu_n + \mu_n(P) |$$

$$\cdot \left| \frac{\partial^2}{\partial\xi\partial x} \frac{1}{r} \right| d\xi \, d\eta \, d\zeta.$$

For P confined to a closed subdomain of G the first expression may be made uniformly small by choosing n large enough. In estimating the volume integral we divide G into two parts $G = G_1 + G_2$, where G_1 is a sphere about P with radius R. Because of (22) and the uniform Hölder continuity[1] the integrand is bounded in absolute value by $2CK'/r^{3-\alpha}$ so that the integral over G_1 is not greater than $C_1 K' R^\alpha/\alpha$, which can be made arbitrarily small by choosing R small enough (here C_1 is a numerical constant). In G_2 the integrand is bounded by $[|\,\mu - \mu_n\,| + |\,\mu(P) - \mu_n(P)|]C_2/R^3$; thus, once R is fixed, the second integral can be made small if n is large enough since the μ_n converge uniformly to μ.

To complete the proof we must show that μ can be uniformly approximated by uniformly Hölder continuous functions μ_n. We know, indeed, that μ can be approximated uniformly by polynomials of the form

$$\mu_n = P_n(x, y, z)$$

$$= \frac{1}{c^3} \iiint_{-1}^{1} \mu(\xi, \eta, \zeta)[1 - (\xi - x)^2]^n[1 - (\eta - y)^2]^n$$
$$\cdot [1 - (\zeta - z)^2]^n d\xi \, d\eta \, d\zeta$$

with

$$c = \int_{-1}^{1} (1 - t^2)^n dt$$

(cf. Vol. I, Ch. II, §4, 2, p. 68). Moreover, it can be verified that these μ_n satisfy a uniform Hölder condition with exponent α and some coefficient K', provided μ is Hölder continuous with exponent α and coefficient K.

We have shown that the solution u of $\Delta u = -4\pi\mu$ given by (14) has continuous second derivatives provided μ is Hölder continuous. One can show furthermore that the second derivatives of u are Hölder continuous (with the same exponent as μ) in any closed subdomain of G. This remark together with important consequences will be examined more closely in §7.

Moreover, an entirely analogous procedure establishes Poisson's equation (15a) in two dimensions for

$$u(x, y) = \iint_G \mu(\xi, \eta) \log \frac{1}{r} d\xi \, d\eta \quad (r^2 = (x - \xi)^2 + (y - \eta)^2)$$

and for (14a) in any number of dimensions.

[1] The uniform limit of functions satisfying a Hölder condition with constants α, K satisfies the same conditions.

There also occur, for $n = 3$, potentials of mass distributions and of so-called "double layers" on surfaces, as well as potentials of mass distributions along curves. For $n = 2$, we have potentials of single and double layers along curves. The corresponding concepts for $n > 3$ need not be discussed here.

The potential of a surface distribution of mass with surface density ρ on a surface F is defined by an integral of the form

$$(23) \qquad u = \iint_F \frac{\rho}{r}\, dS,$$

where dS is the element of area. Along a curve C with arc length s the potential of a distribution with density τ is defined by

$$(24) \qquad u = \int_C \frac{\tau}{r}\, ds$$

for $n = 3$, or by

$$(24') \qquad u = \int_C \tau \log \frac{1}{r}\, ds$$

for $n = 2$.

Potentials of double layers result from superposition of dipole potentials (cf. Vol. I, Ch. VII, p. 515). The potential of a single dipole is defined as

$$\frac{\partial}{\partial \nu} \frac{1}{r} = -\frac{\cos (\nu, r)}{r^2} \qquad\qquad (n = 3)$$

or

$$\frac{\partial}{\partial \nu} \log \frac{1}{r} = -\frac{\cos (\nu, r)}{r} \qquad\qquad (n = 2),$$

where $\partial/\partial \nu$ denotes differentiation in some direction of ξ, η, ζ-space (or of the ξ, η-plane) and (ν, r) is the angle between this direction and the radius vector from the point $P(x, y, z)$ to the point $Q(\xi, \eta, \zeta)$ (or the corresponding angle for $n = 2$).

The potential of a double layer of density σ over a surface F (for the case $n = 3$) or on a curve C (for the case $n = 2$) is then given by expressions of the form

$$(25) \qquad u(x, y, z) = \iint_F \sigma \frac{\partial}{\partial \nu} \frac{1}{r}\, dS$$

or

$$(25') \qquad u(x, y) = \int_C \sigma \frac{\partial}{\partial \nu} \log \frac{1}{r} \, ds,$$

respectively, where $\partial/\partial \nu$ now denotes differentiation in the positive direction of the normal to the surface or curve.

3. Green's Formulas and Applications. The most important tool for potential theory is provided by *Green's formulas*. In a three-dimensional bounded region G with volume-element $dg = dx \, dy \, dz$ and boundary Γ, which we assume to be piecewise smooth,[1] two functions u and v are related by the two Green's formulas

$$\iiint_G (u_x v_x + u_y v_y + u_z v_z) \, dg + \iiint_G v \, \Delta u \, dg$$

$$(26) \qquad\qquad\qquad\qquad = \iint_\Gamma v \frac{\partial u}{\partial \nu} \, dS,$$

$$\iiint_G (u \, \Delta v - v \, \Delta u) \, dg = \iint_\Gamma \left(u \frac{\partial v}{\partial \nu} - v \frac{\partial u}{\partial \nu} \right) dS.$$

In the first formula we assume continuity of u and v in the closed region $G + \Gamma$, continuity of the first derivatives of u and v in G, together with continuity of the first derivatives of u in $G + \Gamma$ and of the second derivatives of u in G. In the second formula we assume continuity of the first derivatives of both u and v in $G + \Gamma$ and continuity of the second derivatives in G. In addition, the existence of the integrals over G is assumed in both cases. Precisely analogous formulas hold in the plane.

If $\Delta u = 0$ and $v = 1$ we obtain *Gauss' integral theorem*

$$(27) \qquad\qquad\qquad \iint_\Gamma \frac{\partial u}{\partial \nu} \, dS = 0.$$

In other words *if a harmonic function is regular in a bounded region G and is continuously differentiable in $G + \Gamma$, then the surface integral of its normal derivative has the value zero.*

As a consequence of (27) we now state the following theorem on potentials of constant double layers:

Given a piece of surface F bounded by a curve C and a point P not in F, then in the case of the constant distribution $\sigma = 1$ the double-layer potential of F at P is equal in absolute value to the solid angle at P sub-

---

[1] Cf. footnote 2, p. 245.

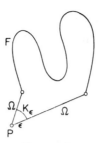

FIGURE 9

tended by C. In particular the double-layer potential of a surface enclosing a region G has the constant value -4π in G, and the value zero outside.

To prove the theorem, we construct the cone Ω generated by the lines joining P to points of the boundary curve C of F.[1] For simplicity, we assume that the surface F and the part of the surface of the cone between P and C enclose a simply connected region. We cut off the vertex of the cone by means of a sufficiently small sphere K_ϵ, of radius ϵ, and let G_ϵ denote the domain bounded by F, Ω, K_ϵ in which $u = 1/r$ is everywhere regular. We infer from (27) that

$$\iint_F \frac{\partial}{\partial \nu} \frac{1}{r} \, dS + \iint_\Omega \frac{\partial}{\partial \nu} \frac{1}{r} \, dS + \iint_{K_\epsilon} \frac{\partial}{\partial \nu} \frac{1}{r} \, dS = 0$$

and hence, since $\partial(1/r)/\partial \nu$ vanishes on Ω and has the constant value $-(1/\epsilon^2)$ on K_ϵ, our assertion is proved (cf. Figure 9).

For $n = 2$ we have the analogous theorem: *The double-layer potential of an arc C with the constant distribution $\sigma = 1$ at a point P not on C is equal to the angle at P subtended by the arc. In particular, $u = -2\pi$ inside and $u = 0$ outside any closed curve C.*

If we set $v = u$ in the first formula (26), we obtain the identity

$$(28) \qquad D[u] = \iiint_G (u_x^2 + u_y^2 + u_z^2) \, dg = \iint_\Gamma u \frac{\partial u}{\partial \nu} \, dS$$

[1] The sign of this solid angle is determined uniquely if a positive and negative side are associated with the surface: The portion of the surface F and the cone Ω determined by the point P and the curve C bound a closed domain, which we first assume to be simply connected. The sign of the solid angle is negative if the positive normal to F leads out of the domain, and is otherwise positive. In the general case, we suppose F to be composed of a finite number of surfaces for each of which the above assumption is satisfied. We then see that our double-layer potential is equal to the sum of the corresponding solid angles, each taken with its appropriate sign.

for every regular harmonic function u in a bounded region G whose first derivatives are continuous in $G + \Gamma$. This formula remains valid if the region G contains a full neighborhood of the point at infinity, provided that the harmonic function is regular there, that is, provided that the function

$$\frac{1}{r} u \left(\frac{x}{r^2}, \frac{y}{r^2}, \frac{z}{r^2} \right)$$

is regular at the origin. The integral $D[u]$, which is known as *Dirichlet's integral*, plays a particularly important role in potential theory. As we have already seen in Vol. I, p. 192, it links potential theory and the calculus of variations. In this connection it will be decisive for existence proofs. (See Volume III.)

Here we can already draw the following conclusion from identity (28):

Let u be a regular harmonic function in G which is continuous and continuously differentiable in $G + \Gamma$. Then a) if the function u vanishes on Γ, it vanishes identically in G, and b) if the normal derivative $\partial u / \partial \nu$ vanishes on the boundary, the function u is constant in G. In both cases it follows from (28) that $D[u] = 0$, hence $u = $ const. In the first case the constant must coincide with the boundary value zero.

Let G be a sphere of radius R, center P, and surface Γ. Let G_0 be a concentric sphere with radius $R_0 < R$ and surface Γ_0. Set $v = 1/r$ (where r is the distance from the center of the sphere) and apply the second formula (26) to the region lying between Γ and Γ_0. Then, in view of (27), we obtain for a harmonic function u

$$(29) \qquad \frac{1}{4\pi R_0^2} \iint_{\Gamma_0} u \, dS = \frac{1}{4\pi R^2} \iint_{\Gamma} u \, dS.$$

Letting R_0 converge to zero, we obtain the mean value relation

$$(30) \qquad u(P) = \frac{1}{4\pi R^2} \iint_{\Gamma} u \, dS.$$

In other words: *The value of a harmonic function at a point P is equal to the arithmetic mean of its values on any sphere with center at P provided that the function is regular inside and continuous in the closed region bounded by the sphere.*[1]

---

[1] Obviously this theorem need not require the harmonic function to be continuously differentiable in the *closed* region bounded by the sphere, although this requirement must be made for the application of (26) and (27) when proving the theorem.

From this mean value theorem we draw some important consequences. First, we state the

MAXIMUM PRINCIPLE: *Let the function u be regular and harmonic in a connected region G and continuous up to and on the boundary* Γ. *Then its maximum and minimum are always assumed on the boundary; maximum and minimum are assumed in the interior if and only if u is a constant.*

To prove the theorem, we consider the subset F of the closed domain $G + \Gamma$ consisting of the points at which u assumes its maximum value M in $G + \Gamma$. Since u is continuous in $G + \Gamma$, F is a closed set. Now if F should contain an interior point P_0 of G there would exist a family of spheres with center P_0 all contained in G, and the mean value of the function u on each one of these spheres would be $u(P_0) = M$. But since $u \leq M$, this is possible only if $u = M$ throughout every sphere of center P_0 contained entirely in G. Thus, F contains at least all points inside the largest sphere around P_0 which lies entirely in $G + \Gamma$. The same argument may now be repeated for any other interior point of F, hence F must coincide with $G + \Gamma$. But this would mean that u is constant (in fact, equal to M). Hence, for any u which is not a constant, F can consist only of boundary points. A corresponding argument shows that the minimum m is assumed on the boundary, and only for constant u also in the interior of G.

The following is an immediate corollary to the maximum principle: *If a regular harmonic function in G which is continuous in $G + \Gamma$ is constant on the boundary then it is constant in the whole region.*

In particular, this implies the

UNIQUENESS THEOREM: *Two harmonic functions in G which are continuous in $G + \Gamma$ and coincide on the boundary are identical throughout G.*

This is true because the difference of two such functions is itself a regular harmonic function which is continuous in $G + \Gamma$, and vanishes on the boundary; hence it is identically zero in G.

Green's formulas (26) undergo an important modification if for v we substitute a function having the characteristic singularity of the potential equation at a point P. Let P be an interior point of G with the coordinates x, y, z. We set

$$v(\xi, \eta, \zeta) = \frac{1}{r} + w(\xi, \eta, \zeta),$$

where $r^2 = (x - \xi)^2 + (y - \eta)^2 + (z - \zeta)^2$ and w is an arbitrary twice continuously differentiable function in G which is continuous and continuously differentiable in $G + \Gamma$. We apply Green's formulas (26) to a subregion $G - K_\epsilon$ of G obtained by deleting a small sphere K_ϵ of radius ϵ and center P. By applying Green's formulas to this region and taking the limit as $\epsilon \to 0$, in the familiar simple manner, we obtain

$$
\iiint_G (u_\xi v_\xi + u_\eta v_\eta + u_\zeta v_\zeta) \, dg + \iiint_G u \, \Delta v \, dg
$$

(31)

$$
= pu + \iint_\Gamma u \frac{\partial v}{\partial \nu} \, dS,
$$

(31') $\qquad \iiint_G (u \, \Delta w - v \, \Delta u) \, dg = pu + \iint_\Gamma \left(u \frac{\partial v}{\partial \nu} - v \frac{\partial u}{\partial \nu} \right) dS;$

here the dg integrations are in ξ, η, ζ-space and

$$
v = \frac{1}{r} + w, \qquad p = \begin{cases} 4\pi \text{ for } P \text{ in } G, \\ 2\pi \text{ for } P \text{ in } \Gamma, \\ 0 \text{ for } P \text{ outside } G; \end{cases}
$$

in case P lies on Γ, it is assumed that Γ has a continuous tangent plane at P.[1] For u and w we require, as in (26), existence of the integrals over G, continuity of u and w in $G + \Gamma$, continuity of the first and second order derivatives of u and w in G; for (31) we need continuity of the first derivatives of u in $G + \Gamma$, and for (31') continuity of the first derivatives of u and w in $G + \Gamma$.

Under analogous conditions we have an analogous system of formulas in the plane:

$$
\iint_G (u_\xi v_\xi + u_\eta v_\eta) \, dg + \iint_G u \, \Delta w \, dg
$$

(32)

$$
= pu + \int_\Gamma u \frac{\partial v}{\partial \nu} \, ds,
$$

(32') $\qquad \iint_G (u \, \Delta w - v \, \Delta u) \, dg = pu + \int_\Gamma \left(u \frac{\partial v}{\partial \nu} - v \frac{\partial u}{\partial \nu} \right) ds$

[1] E.g., if P is a conical vertex of the boundary, then at this point instead of 2π we have to set p equal to the vertex angle of this conical vertex.

with area element dg and $v = \log 1/r + w$,

$$p = \begin{cases} 2\pi \text{ for } P \text{ inside } G, \\ \pi \text{ for } P \text{ on } \Gamma, \\ 0 \text{ for } P \text{ outside } G. \end{cases}$$

In particular, if we set $w = 0$ in (31'), we obtain the representation

$$(33) \quad u = -\frac{1}{4\pi} \iiint_G \frac{\Delta u}{r} dg + \frac{1}{4\pi} \iint_\Gamma \frac{1}{r} \frac{\partial u}{\partial \nu} dS - \frac{1}{4\pi} \iint_\Gamma u \frac{\partial}{\partial \nu} \left(\frac{1}{r}\right) dS$$

for u valid in the interior of G. Thus:

Every function u continuously differentiable in $G + \Gamma$ and having continuous second derivatives in G can be considered as the potential of a distribution consisting of the spatial distribution of density $-\Delta u/4\pi$ in the region G, and of the (single-layer) surface distribution of density $(1/4\pi)\partial u/\partial\nu$ and the double-layer (dipole distribution) of density $-u/4\pi$ on the boundary surface Γ.

In particular, a harmonic function u satisfies the relation

$$(34) \quad u = \frac{1}{4\pi} \iint_\Gamma \frac{1}{r} \frac{\partial u}{\partial \nu} dS - \frac{1}{4\pi} \iint_\Gamma u \frac{\partial}{\partial \nu} \left(\frac{1}{r}\right) dS.$$

In other words, *every function u which is regular harmonic in G and continuously differentiable in $G + \Gamma$ can be represented as the potential of a distribution on the boundary surface consisting of the single layer of density $(1/4\pi)\partial u/\partial\nu$ and the double-layer (dipole distribution) of density $-u/4\pi$.*

The mean value theorem of potential theory can be deduced also from formula (31'). In fact, if we apply (31') to a sphere of radius R and set $w = -1/R = $ const., v vanishes on the surface of the sphere and relation (30) follows directly.

The problem of carrying over all these results to the case $n = 2$, and more generally to an arbitrary number n of dimensions, is left to the reader. For arbitrary n, and with the same assumptions as before for u, v and the region G, Green's formulas

$$(35) \quad \begin{aligned} \iint \cdots \int_G \left(\sum_{i=1}^n u_{x_i} v_{x_i} + v\, \Delta u\right) dg &= \int \cdots \int_\Gamma v \frac{\partial u}{\partial \nu} dS, \\ \iint \cdots \int_G (u\, \Delta v - v\, \Delta u)\, dg &= \int \cdots \int_\Gamma \left(u \frac{\partial v}{\partial \nu} - v \frac{\partial u}{\partial \nu}\right) dS \end{aligned}$$

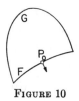

FIGURE 10

are satisfied, as well as the following formulas analogous to (31) and (31′):

$$\iint \cdots \int_{G} \left(\sum_{i=1}^{n} u_{\xi_i} v_{\xi_i} + u\,\Delta w \right) dg = pu + \int \cdots \int_{\Gamma} u\,\frac{\partial v}{\partial \nu}\,dS,$$

(36)

$$\iint \cdots \int_{G} (u\,\Delta w - v\,\Delta u)\,dg = pu + \int \cdots \int_{\Gamma} \left(u\,\frac{\partial v}{\partial \nu} - v\,\frac{\partial u}{\partial \nu} \right) dS.$$

Here, for $n > 2$, we set

$$v = \frac{1}{(n-2)r^{n-2}} + w, \qquad p = \begin{cases} \omega_n & \text{for P in G} \\ \tfrac{1}{2}\omega_n & \text{for P on } \Gamma \\ 0 & \text{for P outside G.} \end{cases}$$

4. Derivatives of Potentials of Mass Distributions. We saw in article 2 that the potential (14) of a spatial distribution is continuous and possesses continuous first derivatives if the density of the distribution is bounded and integrable. Now we shall study the continuity properties of surface potentials and their derivatives for single layers and double layers as the point $P(x, y, z)$ passes through the surface F. We make use of the following method:[1] As indicated in Figure 10, we consider a point P_0 on a portion of the surface F, regarding F as part of the boundary Γ of a spatial region G. We suppose the function ρ (or σ), which represents the density of the distribution on F, to be continued into this region G in a sufficiently continuous manner. We apply Green's formulas (31) and (31′) to the region G, using suitable functions u and v. Since it will only be of importance to study discontinuities of the function in question, it is expedient to omit from our formulas expressions which remain continuous as P passes through the surface F. The symbolic nota-

[1] Our method is related to a procedure of Erhard Schmidt. See E. Schmidt [1].

tion $\equiv 0$ ("congruent to zero") will be employed for such an expression; the region G, moreover, will be so chosen that the positive normal to F is directed outward.

We note first that the potential of a single layer is continuous as P passes through the surface F. Hence we have to study only the behavior of the potential of a double layer and its derivatives, along with the derivatives of the potential of a single layer.

We assume that in the neighborhood of the point P_0 the surface F has continuous curvature (cf. footnote 2, p. 245) and that the distribution density is twice continuously differentiable on this surface.[1] Then:

a) *Across the surface F, the value of the double-layer potential at the point P_0 has jump discontinuities given by*

$$
\lim_{P \to P_0^+} u(P) - u(P_0) = 2\pi\sigma(P_0),
$$

(37)

$$
\lim_{P \to P_0^-} u(P) - u(P_0) = -2\pi\sigma(P_0),
$$

where the symbol $P \to P_0^+$ denotes approach from the positive side of the surface and $P \to P_0^-$ approach from the negative side.

b) *The derivative of the double-layer potential $u(P)$ in the direction of the normal to the surface varies continuously when P penetrates the surface along the normal at P_0. However, the tangential derivatives $\partial u/\partial t$ (in other words, the derivatives perpendicular to the normal) vary discontinuously in accordance with the jump relations*

$$
\lim_{P \to P_0^+} \frac{\partial u(P)}{\partial t} - \frac{\partial u(P_0)}{\partial t} = 2\pi \frac{\partial \sigma(P_0)}{\partial t},
$$

(38)

$$
\lim_{P \to P_0^-} \frac{\partial u(P)}{\partial t} - \frac{\partial u(P_0)}{\partial t} = -2\pi \frac{\partial \sigma(P_0)}{\partial t}.
$$

c) *The potential of a single layer and its tangential derivatives vary continuously as we pass through P_0. However, the normal derivative has a jump of magnitude*

$$
(39) \qquad \left[\frac{\partial u}{\partial \nu} \right] = \frac{\partial u}{\partial \nu_+} + \frac{\partial u}{\partial \nu_-} = -4\pi\rho(P_0).
$$

[1] These assumptions can be weakened, even by using the method applied here.

Here $\partial/\partial\nu_+$ denotes differentiation in the direction of the positive surface normal at P_0, $\partial/\partial\nu_-$ differentiation in the direction of the negative normal.

We consider first a double layer of density σ, and suppose this density to be continued into $G + \Gamma$ in a continuously differentiable way as a function $\sigma(x, y, z)$; then, setting $u = \sigma$ and omitting all terms which are continuous across the boundary, we write Green's formula (31) in the form

$$p\sigma + \iint_F \sigma \frac{\partial}{\partial\nu}\frac{1}{r}\,dS \equiv \iiint_G \left(\sigma_\xi \frac{\partial}{\partial\xi}\frac{1}{r} + \sigma_\eta \frac{\partial}{\partial\eta}\frac{1}{r} + \sigma_\zeta \frac{\partial}{\partial\zeta}\frac{1}{r}\right) dg,$$

using the symbol $\equiv$ as defined above. Evidently all boundary integrals that do not depend on F are continuous with respect to the point P and have continuous derivatives of all orders. Since, according to article 2, the right-hand side of this expression is continuous, we have

$$p\sigma + \iint_F \sigma \frac{\partial}{\partial\nu}\frac{1}{r}\,dS \equiv 0,$$

which contains the statement about the behavior of the potential of the double layer.

In order to prove the assertion concerning the derivative of the double-layer potential, we suppose the boundary distribution σ continued into G as a twice continuously differentiable function in such a way that the normal derivative $\partial\sigma/\partial\nu$ vanishes. We then apply the second Green's formula (31'), which can now be written as

$$p\sigma + \iint_F \sigma \frac{\partial}{\partial\nu}\frac{1}{r}\,dS = -\iiint_G \frac{\Delta\sigma}{r}\,dg.$$

Since the right-hand side is continuous, we once more obtain the result concerning the discontinuity of the double-layer potential itself. In addition, since the right-hand side has continuous derivatives with respect to x, y, z, it follows that the derivatives of the double-layer potential have the same discontinuity properties as the derivatives of the expression $-p\sigma$, which is what we wished to prove.

Our assertion about the derivatives of the single-layer potential follows in precisely the same manner when we apply Green's formula (31'). However, we must now choose the function w in (31') in such a way that along the surface F this function is identically zero and the normal derivative $\partial w/\partial \nu$ equals the density ρ of the surface distribution.

That such a function w exists follows easily from continuity assumptions for F and the boundary distribution.

Analogous theorems and discontinuity relations may be proved for the potential in the plane, with the difference that in relations (37) and (38) the factor 2π is replaced by π, and in (39) the factor 4π is replaced by 2π.

§2. Poisson's Integral and Applications

1. The Boundary Value Problem and Green's Function. In Vol. I, Ch. V, §14, we have already discussed the representation of the solution of the boundary value problem by means of a *Green's function* which is independent of the particular boundary values and of the particular right-hand side of the differential equation.

We now pose the first boundary value problem of potential theory, also known as Dirichlet's problem, for a bounded region G in x-space with piecewise smooth[1] boundary Γ:

Let continuous boundary values be prescribed on Γ. In particular, let the boundary values be the values of a function $f(x_1, x_2, \cdots, x_n)$ or $f(x)$ which has continuous third derivatives in $G + \Gamma$. Find a solution of the differential equation $\Delta u = 0$ which is continuous in $G + \Gamma$ and regular in G, and coincides with f on Γ.—In §4 we shall see that the assumptions of differentiability of the boundary values may easily be removed by a simple limiting process.

The boundary value problem assumes a somewhat different form when we introduce the function $u - f = v$ in place of u. For the function v we then have the homogeneous boundary values zero on Γ but the nonhomogeneous differential equation

$$\Delta v = -\Delta f.$$

Now let $Q = (\xi_1, \xi_2, \cdots, \xi_n)$ or ξ be a fixed interior point, $P = x$ a variable point of G, and γ the function given by formula (9), §1. *Then we define Green's function $K(P, Q)$ of the differential expression $\Delta u = \sum_{\nu=1}^{n} \partial^2 u/\partial x_\nu^2$ for the region G as a specific fundamental solution of $\Delta u = 0$, depending on the parameter Q, of the form*

[1] Cf. footnote 2, p. 245.

$$K(P, Q) = K(x_1, x_2, \cdots, x_n; \xi_1, \xi_2, \cdots, \xi_n)$$

(1)
$$= K(x, \xi) = \gamma(r) + w,$$

$$r^2 = \sum_{i=1}^{n} (x_i - \xi_i)^2 = (x - \xi)^2$$

which vanishes for P on Γ and for which the component w is continuous in G + Γ and regular in G (so that K is regular in G except at the point P = Q). *This function is symmetric in the arguments P and Q:* $K(P, Q) = K(Q, P)$ (cf. Vol. I, p. 365).

Green's function vanishes on Γ and is positive on a small sphere about Q; it is therefore, according to the maximum principle, *positive in the interior of G.*

If v is a continuous and continuously differentiable function satisfying the equation $\Delta v = -\Delta f$ in G + Γ, then formula (36) of §1 at once yields the representations

(2)
$$v = \iint_G \cdots \int K(x, \xi) \Delta f \, d\xi,$$

and

(3)
$$u = f + \iint_G \cdots \int K(x, \xi) \, \Delta f \, d\xi$$

for the solution of the original boundry value problem.

In formulas (2) and (3) the solution u does not really depend on the values of the function f in the interior of G. For, by applying Green's formula, we easily obtain

$$\iint_G \cdots \int K \, \Delta f \, d\xi = -f - \int_\Gamma \cdots \int \frac{\partial K}{\partial \nu} f \, dS$$

and hence

(4)
$$u = - \int_\Gamma \cdots \int \frac{\partial K}{\partial \nu} f \, dS.$$

In this formula only the boundary values of f appear.

Often, however, it is desirable to retain the representation of the solution of the boundary value problem in the form (3). In particular, we shall use the form (3) in proving the following theorem, which is converse to the above results but does not contain such stringent assumptions on K and v:

If $K(P, Q)$ is Green's function for the bounded region G, then, for piecewise differentiable $g(x_1, x_2, \cdots, x_n) = g(x)$, the expression

$$v = \iint_G \cdots \int K(x, \xi)\, g(\xi)\, d\xi$$

represents a solution of Poisson's equation

$$\Delta v = -g,$$

continuous in $G + \Gamma$ and vanishing on the boundary Γ.

That v satisfies the differential equation in G follows immediately from the integral representation and from the piecewise continuous differentiability of g (cf. §1, 2, p. 246). To show that v vanishes on the boundary Γ, it is not enough to recall that Green's function vanishes there, since K does not tend to zero uniformly in Q as the point P approaches Γ. In order to circumvent this difficulty, we use the following lemma, whose proof will be given in article 2:

If B is a subregion of G with a diameter smaller than h, then

$$\iint_B \cdots \int K(x, \xi)\, d\xi < \epsilon(h),$$

where $\epsilon(h)$, depends only on h, not on the special choice of B, and tends to zero with h.

Let P in G approach the point $\bar{P}$ of the boundary Γ. Let B_h be that subregion of G which lies in the sphere of diameter h with center at $\bar{P}$, and let G' be the remainder of G. Then

$$v = \iint_{G'} \cdots \int Kg\, d\xi + \iint_{B_h} \cdots \int Kg\, d\xi.$$

The term depending on the region G' clearly converges to zero as P approaches $\bar{P}$. The term $v_h = \iint_{B_h} \cdots \int Kg\, d\xi$ can be estimated immediately by

$$|v_h| < M\epsilon(h)$$

if M is a bound for $|g|$. Thus, when P is sufficiently near $\bar{P}$, $|v| < M\epsilon(h)$. Since h was arbitrary, the assertion is proved.

By this theorem the solution u of the boundary value problem then follows from (2) and (3), provided K is known. In other words: *Solving the boundary value problem with general prescribed boundary values is equivalent to finding the Green's function*, which corresponds to a special boundary value problem depending on the point Q as parameter.

2. Green's Function for the Circle and Sphere. Poisson's Integral for the Sphere and Half-Space. In Vol. I, Ch. V, §15, 2 we constructed Green's function for circular and spherical fundamental domains. The considerations used there may be applied without difficulty also to the case of the n-dimensional potential expression. Let $\gamma = \psi(r)$ be the fundamental solution of the differential equation $\Delta u = 0$ in n dimensions:

$$(5) \qquad \psi(r) = \frac{1}{(n-2)\omega_n} r^{2-n} \qquad \text{(for } n > 2\text{)},$$

$$\psi(r) = \frac{1}{2\pi} \log \frac{1}{r} \qquad \text{(for } n = 2\text{)}.$$

Then Green's function for the sphere of radius R is given at once by means of the expression

$$(6) \qquad K(x, \xi) = \psi(r) - \psi\left(\frac{\kappa}{R} r_1\right),$$

where

$$\kappa^2 = \sum_{\nu=1}^{n} \xi_\nu^2, \quad r^2 = \sum_{\nu=1}^{n} (x_\nu - \xi_\nu)^2 = (x - \xi)^2, \quad r_1^2 = \sum_{\nu=1}^{n} \left(x_\nu - \frac{R^2}{\kappa^2} \xi_\nu\right)^2$$

r_1 denoting the distance of the point x from the reflected image

$$\left(\frac{R^2}{\kappa^2} \xi_1, \frac{R^2}{\kappa^2} \xi_2, \cdots, \frac{R^2}{\kappa^2} \xi_n\right) \qquad \text{or} \qquad \frac{R^2}{\kappa^2} \xi$$

of the point $(\xi_1, \xi_2, \cdots, \xi_n)$. This function clearly satisfies all the requirements; in particular, it vanishes on the sphere since $r = (\kappa/R)r_1$ whenever $(x_1, x_2, \cdots, x_n)$ lies on the sphere.

Green's function for the circle (or sphere) can be used as a *majorant for Green's function for an arbitrary bounded domain* G in the following manner. Let Q be any point in G and R a number so large that the circle (or sphere) of radius R around any of the points of G contains the domain $G + \Gamma$ entirely in its interior. If r denotes the distance of $(x_1, x_2, \cdots, x_n)$ from Q, then it can easily be seen that

$$\psi(r) - \psi(R) = \frac{1}{2\pi} \log \frac{R}{r} \qquad (n = 2),$$

$$\psi(r) - \psi(R) = \frac{1}{(n-2)\,\omega_n}\,(r^{2-n} - R^{2-n}) \qquad (n > 2)$$

represent Green's function for the circle and sphere, respectively, of radius R about Q with Q as singularity (that is, with the parameter point Q at the center of the circle or sphere). If $K(P, Q)$ is the Green's function of G with Q as singularity, the difference $K - (\psi(r) - \psi(R))$ is regular in G and nonpositive on Γ; thus everywhere in G

$$0 \le K \le \psi(r) - \psi(R).$$

This leads easily to the estimate, used in article 1, for subregions B of G with diameter less than h.

With the aid of (6) we shall now evaluate $\partial K/\partial \nu$ in the integral

$$u = -\iint_\Gamma \frac{\partial K}{\partial \nu} f \, dS$$

over the surface of the sphere and obtain the solution of the boundary value problem

$$\Delta u = 0, \qquad\qquad u = f \text{ on } \Gamma,$$

for the sphere. After an easy calculation we find

(7) $$\frac{\partial K}{\partial \nu} = \psi'(r)\,\frac{R^2 - \rho^2}{rR}, \qquad\qquad \rho^2 = \sum_{i=1}^n x_i^2,$$

which, when substituted in the integral formula, yields

(8) $$u = -\frac{R^2 - \rho^2}{R} \iint_\Gamma \frac{\psi'(r)}{r} f \, dS.$$

If we suppose f given as a function of the coordinates $\xi_1, \xi_2, \cdots, \xi_n$ on the unit sphere and substitute the values (5) for $\psi(r)$, we obtain Poisson's integral formula

(9) $$u(x) = \frac{R^{n-2}(R^2 - \rho^2)}{\omega_n} \iint \frac{f \, d\omega_n}{(\rho^2 + R^2 - 2\rho R \cos\theta)^{n/2}}.$$

In this expression the integral is taken over the surface of the n-dimensional unit sphere, $\rho^2 = x_1^2 + x_2^2 + \cdots + x_n^2 = x^2$, and θ is the angle between the vector ρ and the radius drawn to the variable point of integration ξ.

This formula has already been derived for $n = 2$ and $n = 3$ on p. 22, and in Vol. I, p. 514, respectively.

From the above remarks it follows that the solution of the boundary value problem for a sphere is given by (9) in case the boundary values $f(\Gamma)$ are determined by the values on Γ of a continuous function $f(x)$, which has continuous first and second, and piecewise continuous third derivatives in G. For example, these hypotheses are obviously satisfied for $f \equiv 1$. In this case, Poisson's integral formula, in combination with the uniqueness theorem of §1, implies that the kernel

$$(10) \qquad H(P,Q) = \frac{R^2 - \rho^2}{R\omega_n\, r^n}\,, \qquad r^2 = (x - \xi)^2, \quad \rho^2 = x^2,$$

which is always positive inside the sphere, yields, when integrated with respect to ξ over the surface of the sphere of radius R the value 1 with surface element $do = R^{n-1}d\omega$:

$$(11)\qquad \iint H(P,Q)\, do = \frac{R^{n-2}(R^2 - \rho^2)}{\omega_n}$$
$$\cdot \iint \frac{d\omega_n}{(\rho^2 + R^2 - 2\rho R \cos\theta)^{n/2}} = 1.$$

Moreover, for fixed $Q = \xi$, this kernel $H(P, Q)$ as a function of $P = x$ satisfies the potential equation in the interior of the sphere, as we recognize at once if we write it in the form

$$R\omega_n\, H = -\frac{1}{r^{n-2}} + \frac{2}{n-2}\sum_{\nu=1}^{n} \xi_\nu \frac{\partial}{\partial \xi_\nu} \frac{1}{r^{n-2}}$$

with

$$r^2 = \sum_{\nu=1}^{n} (x_\nu - \xi_\nu)^2 = (x - \xi)^2.$$

On the sphere itself, H vanishes everywhere except at the point $P = Q$, where it increases beyond all bounds as P approaches Q from within.

We now can easily free ourselves of the restrictive assumptions on the boundary values f. We show: *Poisson's integral formula solves the boundary value problem for the sphere even if all we require of the boundary values is continuity on the surface of the sphere.* For with this assumption we can differentiate (9) arbitrarily often under the integral sign provided P is an interior point of the sphere. Hence

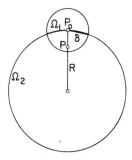

FIGURE 11

u also satisfies the potential equation and we have only to show that, on approaching the boundary, it tends to the prescribed boundary values.

Let P_1 be an arbitrary boundary point and P a nearby point in the interior (Figure 11). We assume P_0 is the point on the surface of the sphere determined by the radius going through P. Since the inequality

$$| u(P) - f(P_1) | \leq | u(P) - f(P_0) | + | f(P_0) - f(P_1) |$$

holds and since the boundary function is continuous, it is sufficient to show that u tends to the prescribed boundary values when Γ is approached radially, i.e., that, with the surface element dS_Q at Q,

$$(12) \qquad u(P) - f(P_0) = \iint_\Gamma H(P, Q) \, [f(Q) - f(P_0)] \, dS_Q$$

becomes arbitrarily small as P approaches P_0 along the radius through P_0.

For the proof we divide the surface of the sphere into two domains Ω_1 and Ω_2 by means of an arbitrarily small sphere of radius δ about P_0. We assume that P is already inside this small sphere and hence, that the distance h between P and P_0 is less than δ. Now if $|f| \leq M$ on this sphere, and $| f(Q) - f(P_0) | \leq \sigma(\delta)$ on Ω_1, where σ approaches zero with δ, then (12) leads to the estimate

$$| u(P) - f(P_0) | \leq 2M \iint_{\Omega_2} H \, dS_Q + \sigma(\delta) \iint_{\Omega_1} H \, dS_Q$$

$$< 2M \iint_{\Omega_2} H \, dS_Q + \sigma(\delta),$$

where the last inequality follows from (11). Now on Ω_2 the kernel H is smaller than the expression

$$\frac{1}{R}\frac{R^2 - \rho^2}{\omega_n(\delta/2)^n} < \frac{2Rh}{\omega_n(\delta/2)^n} .$$

Thus it follows that

$$| u(P) - f(P_0) | \leqq \frac{4MR^{n-1}}{(\delta/2)^n} h + \sigma(\delta).$$

If we choose δ so small that $\sigma(\delta) \leq \epsilon/2$, and select h in such a way that

$$\frac{4MR^{n-1}}{(\delta/2)^n} h < \frac{\epsilon}{2},$$

then $| u(P) - f(P_0) | < \epsilon$. This completes the proof.

A corresponding integral representation and corresponding results may be obtained if G is a half-space instead of a sphere. If Γ is the x,y-plane and G the domain $z > 0$, then *Poisson's integral for a half-space*

$$(9') \qquad u(x, y, z) = \frac{z}{2\pi} \iint_{-\infty}^{\infty} \frac{f(\xi, \eta)\, d\xi\, d\eta}{[(x - \xi)^2 + (y - \eta)^2 + z^2]^{3/2}}$$

solves the boundary value problem for G with arbitrary boundary values $f(x, y)$, under the condition that inversion of G with respect to a sphere lying outside G (cf. §1, p. 242) results in a boundary value problem with continuous boundary values for the bounded reflected domain G' and its boundary Γ'.

Under corresponding assumptions for an n-dimensional domain $G: x_n > 0$ with boundary $\Gamma: x_n = 0$, we have the integral formula

$$
\begin{aligned}
(9'') \quad & u(x_1, x_2, \cdots, x_n) = u(x) \\
& = \frac{2x_n}{\omega_n} \int_{-\infty}^{\infty} \cdots \int \frac{f(\xi_1, \xi_2, \cdots, \xi_{n-1})\, d\xi_1\, d\xi_2 \cdots d\xi_{n-1}}{[(x_1 - \xi_1)^2 + \cdots + (x_{n-1} - \xi_{n-1})^2 + x_n^2]^{n/2}} .
\end{aligned}
$$

3. Consequences of Poisson's Formula. The mean value theorem and hence also the maximum principle are immediate consequences of Poisson's representation. Moreover, uniqueness of the solution to the boundary value problem follows from the maximum principle

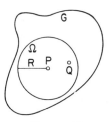

FIGURE 12

applied to the difference of any two solutions (cf. §1, 3). Thus, any two harmonic functions which agree on the boundary are identically equal throughout the region.

Poisson's formula enables us to prove the following important inequality: For $\rho < R$, the kernel H (see (10)) is always positive and lies between the values

$$\frac{1}{R\omega_n}\left(\frac{1}{R+\rho}\right)^{n-2}\frac{R-\rho}{R+\rho} \quad \text{and} \quad \frac{1}{R\omega_n}\left(\frac{1}{R-\rho}\right)^{n-2}\frac{R+\rho}{R-\rho}.$$

Let u be a non-negative regular harmonic function in a domain G (Figure 12); let Ω be a sphere entirely contained in G, R its radius, P its center, and Q an arbitrary interior point. Then Poisson's integral and the mean value theorem imply *Harnack's inequality*[1]

$$\left(\frac{R}{R+\rho}\right)^{n-2}\frac{R-\rho}{R+\rho}u(P) \leq u(Q) \leq \left(\frac{R}{R-\rho}\right)^{n-2}\frac{R+\rho}{R-\rho}u(P).$$

If u is regular in every bounded region of the space, and if P and Q are any two points, we can always choose an arbitrarily large sphere with center P containing Q. Considering Harnack's inequality and letting the radius R of the sphere go to infinity, we immediately obtain $u(P) = u(Q)$. Hence, *a harmonic function which is regular and positive in every bounded region of the space is a constant*.

Another important consequence of Poisson's integral formula is the *analyticity of harmonic functions*:

Every function u which is harmonic and regular in a domain G can be expanded in a power series in a neighborhood of every interior point P of G; that means it is analytic.

---

[1] Harnack's inequality has been extended to general second order linear elliptic equations. See J. Serrin [1] and J. Moser [2]. See also §§7 and 8 and the appendix in L. Bers and L. Nirenberg [2].

For the proof, we represent u by Poisson's integral formula for a sphere of sufficiently small radius R so that the sphere is interior to G. Taking P as the origin of the coordinate system, we shall expand the Poisson kernel as a power series in $x_1, x_2, \cdots, x_n$; $\xi_1, \xi_2, \cdots, \xi_n$ and then obtain a power series for $u(x_1, x_2, \cdots, x_n) = u(x)$ by a termwise integration in Poisson's integral.

We define a function $w(x, \xi)$ by

$$w = \frac{\rho^2 - 2\kappa\rho \cos\theta}{R^2},$$

where

$$\rho^2 = \sum_{i=1}^{n} x_i^2 = x^2, \quad \kappa^2 = \sum_{i=1}^{n} \xi_i^2 = \xi^2 \quad \text{and} \quad \kappa\rho \cos\theta = \sum_{i=1}^{n} x_i \xi_i = x\xi;$$

we do *not* require the point ξ to be on the sphere $\kappa = R$. In addition, we define $K(w)$ by

$$K(w) = \frac{1}{(1 + w)^{n/2}}$$

and note that

$$\left(1 - \frac{\rho^2}{R^2}\right) K(w(x, \xi))$$

agrees with the kernel

$$\frac{1 - \dfrac{\rho^2}{R^2}}{\left(1 - 2\dfrac{\rho}{R} \cos\theta + \dfrac{\rho^2}{R^2}\right)^{n/2}}$$

of Poisson's integral when $(\xi_1, \xi_2, \cdots, \xi_n)$ is on the sphere $\kappa = R$. We shall expand $K(w)$ in powers of $x_1, x_2, \cdots, x_n$; $\xi_1, \xi_2, \cdots, \xi_n$ in order to derive a power series representation for the Poisson kernel valid for $\kappa = R$.[1]

We observe that w is a polynomial in x and ξ, each of whose terms is non-negative if all the products $x_i \xi_i (i = 1, 2, \cdots, n)$ are non-positive. For arbitrary $\epsilon > 0$, we restrict $\xi = (\xi_1, \xi_2, \cdots, \xi_n)$ and $x = (x_1, x_2, \cdots, x_n)$ to the (solid) spheres

---

[1] The Poisson kernel as defined for $\kappa \leq R$ by (10) cannot be represented as a power series valid for $\kappa \leq R$ since the kernel becomes singular as ξ approaches x.

(13) $$\kappa < (1 + \epsilon)R$$

and

(14) $$\rho < \frac{R}{3 + 2\epsilon},$$

respectively. It follows that

(15) $$|w| \leq \frac{\rho^2 + 2\kappa\rho}{R^2} < \frac{7}{9} + \eta(\epsilon),$$

where η goes to zero with ϵ; hence $|w| < 1$ for sufficiently small ϵ.

Clearly, $K(w)$ can be expanded in an absolutely convergent power series in w valid for $|w| < 1$. Therefore, splitting the powers of w into their separate terms, we can obtain a power series in $x_1, x_2, \cdots, x_n$; $\xi_1, \xi_2, \cdots, \xi_n$ which, for sufficiently small ϵ, converges to $K(w)$ for $\xi_1, \xi_2, \cdots, \xi_n$ in the sphere (13) and $x_1, x_2, \cdots, x_n$ in the sphere (14); on the sphere $\kappa = R$, this power series multiplied by $1 - \rho^2/R^2$ will then represent the kernel of Poisson's integral.[1]

---

[1] It is sometimes useful to expand the Poisson kernel in powers of ρ/R obtaining a series

$$\sum_{\nu=0}^{\infty} \left(\frac{\rho}{R}\right)^\nu \psi_\nu (\cos\theta).$$

In the case $n = 2$,

$$\psi_\nu(\cos\theta) = 2^\nu T_\nu(\cos\theta)$$

where $T_\nu(x)$ is the νth *Tchebycheff polynomial* (cf. Vol. I, p. 88). Hence

$$\psi_\nu(\cos\theta) = 2\cos\nu\theta,$$

$$\psi_0(\cos\theta) = 1$$

and

$$\frac{1 - \dfrac{\rho^2}{R^2}}{1 - 2\dfrac{\rho}{R}\cos\theta + \dfrac{\rho^2}{R^2}} = 1 + 2\sum_{\nu=1}^{\infty} \left(\frac{\rho}{R}\right)^\nu \cos\nu\theta.$$

In the case $n = 3$,

$$\psi_\nu(\cos\theta) = (2\nu + 1)P_\nu(\cos\theta)$$

where $P_\nu(x)$ denotes the νth *Legendre polynomial* (cf. Vol. I, p. 83). Here

$$\frac{1 - \dfrac{\rho^2}{R^2}}{\left(1 - 2\dfrac{\rho}{R}\cos\theta + \dfrac{\rho^2}{R^2}\right)^{3/2}} = \sum_{\nu=0}^{\infty} (2\nu + 1)\left(\frac{\rho}{R}\right)^\nu P_\nu(\cos\theta).$$

(continued)

Finally, for the harmonic function u, Poisson's integral formula yields a power series in x_1, x_2, $\cdots$, x_n valid in the sphere (14).

Another consequence of Poisson's formula is the

REFLECTION PRINCIPLE: *If a function is harmonic in a domain and continuous up to the boundary, and if it vanishes on a spherical or plane part of the boundary, then it can be continued (analytically) by reflection as a harmonic function across that part of the boundary.*

It is sufficient to carry out the proof for a portion of a plane (or straight line) S which forms the plane part of the boundary of a hemispherical (or semicircular) domain H.[1]

Let u be harmonic in H and continuous in $H + \Gamma$ with values zero on S. Reflect H across S, thus obtaining a sphere (or circle) K.

We now assign boundary values on the boundary of K. To each point P on $\Gamma - S$ we assign the value $u(P)$ and to each point P on the boundary of K, which is the reflection (mirror image) of a point P' on $\Gamma - S$, we assign the value $-u(P')$. Then Poisson's integral yields a potential function U which is regular in K and has these prescribed boundary values. If, for any point P in K, we now denote its mirror image by P', it is clear that not only $U(P)$ but also $-U(P')$ (as a function of P) solves the boundary value problem for K. Hence, because of uniqueness of the solution to the boundary value problem, we conclude that $U(P) = -U(P')$ for all P in K and consequently that U vanishes on S. Finally, since $u(P)$ and $U(P)$ agree on the (total) boundary Γ of H, we infer from the uniqueness that $u \equiv U$ in H.

A similar reflection principle can be proved for a harmonic function in H whose normal derivative vanishes on S. In this case one continues u as an even function across S.

Using the associated Legendre functions $P_{\nu,h}(x)$ for $\nu > 0$, we can write the quantities $P_\nu(\cos \theta)$ in the form

$$(2\nu + 1)P_\nu(\cos \theta) = P_\nu(\cos \beta)P_\nu(\cos \beta')$$

$$+ 2 \sum_{h=1}^{\nu} \frac{(\nu - h)!}{(\nu + h)!} \cos h(\phi - \phi')P_{\nu,h}(\cos \beta)P_{\nu,h}(\cos \beta')$$

where we have set $\cos \theta = \cos \beta \cos \beta' + \sin \beta \sin \beta' \cos (\phi - \phi')$.

[1] If a domain G has a plane as part of its boundary, we use hemispheres entirely in G with centers on the plane.

Reflection across the surface of a sphere is just as simple as reflection across a plane.

From Poisson's integral we can also deduce the following important theorem.

WEIERSTRASS' CONVERGENCE THEOREM: *A sequence of potential functions u_n , which are regular in G and continuous in $G + \Gamma$, and have boundary values f_n converging uniformly on Γ, converges uniformly in G to a potential function with the boundary values $f = \lim f_n$* .

We restrict ourselves here to three or two dimensions without affecting the generality of our conclusions.

The uniform convergence claimed in Weierstrass' convergence theorem follows directly from the maximum principle. For if each potential function u_n is regular in G and continuous in $G + \Gamma$, the same is true of the difference $u_n - u_m$ for arbitrary n and m; therefore, this difference assumes its maximum (and minimum) on the boundary. Hence, for all points of the closed domain $G + \Gamma$ we have the inequality

$$| u_n - u_m | \leq \max | f_n - f_m |$$

from which the asserted uniform convergence in G and the fact that $u = \lim u_n$ has the boundary values $f = \lim f_n$ follow at once. From Poisson's integral it can be seen that the limit function u satisfies the potential equation in G. In fact, let K be an arbitrary sphere of radius R entirely contained in G and denote by $\bar{u}_n$ and $\bar{u}$ the respective boundary values of u_n and u on K. Then in the interior of K for every u_n , and hence also for u with the boundary values $\bar{u}$, we have the integral formula

$$u = \frac{R(R^2 - \rho^2)}{4\pi} \iint_K \frac{\bar{u}\, d\omega}{(\rho^2 + R^2 - 2\rho R \cos \theta)^{3/2}}$$

which immediately implies that u is a potential function in K.

Another consequence, also a convergence theorem, is

HARNACK'S THEOREM: *If a nondecreasing or nonincreasing sequence of regular harmonic functions in G converges at a single point of G, then it converges at all points of G, and the convergence is uniform in every closed interior subdomain.*

It will be sufficient to prove the theorem for a nondecreasing sequence. For arbitrary m and $n > m$ we consider the difference $\phi = u_n - u_m$, which is then non-negative, and draw a sphere K_a of radius a, still contained entirely in G, about the point of convergence P. If Q is any other point of this sphere and $\rho < a$ is its distance from the center P, Harnack's inequality yields

(16) $$0 < \phi(Q) < \frac{a + \rho}{a - \rho} \phi(P) \left(\frac{a}{a - \rho}\right)^{n-2}$$

This shows the uniform convergence of the sequence u_n in every sphere of radius $r \leq a - \delta$. Since every closed interior subdomain of G can be covered by a finite number of spheres of suitable radius $r < a$ lying entirely in G, we see that a repeated application of our argument proves the assertion. That the limit function is a potential function in G follows directly from Poisson's integral or from the theorem of Weierstrass.

We now state a theorem of fundamental importance:

If the set $\{u(P)\}$ of regular harmonic functions is uniformly bounded in G; i.e., if for all u of the set and all P in G

$$|u(P)| \leq M,$$

then in every closed interior subdomain of G the sets of derivatives $\{u_x\}$, $\{u_y\}$, and $\{u_z\}$ are also uniformly bounded.

We consider a sphere K of radius a lying in the interior of G, with center P and surface S. Since u_x is a potential function the mean value theorem holds:

$$u_x(P) = \frac{1}{4\pi a^2} \iint_S u_x \, dS.$$

The mean value theorem for the interior of K follows without difficulty:

$$u_x(P) = \frac{3}{4\pi a^3} \iiint_K u_x \, dg$$

(cf. §3).

Integrating by parts, we obtain

$$u_x(P) = \frac{3}{4\pi a^3} \iint_S u \frac{\partial x}{\partial \nu} \, dS.$$

Since $|\partial x/\partial \nu| \leq 1$ and $|u| \leq M$, this yields the estimate

(17) $$|u_x(P)| \leq \frac{3M}{a}.$$

Similarly, we have

(18) $$|u_y(P)| \leq \frac{3M}{a},$$

and

(19) $$| u_z(P) | \leq \frac{3M}{a}.$$

Now let G_a be a closed subdomain of G the distance of whose points from Γ is greater than a; then for all points P of G_a the above inequalities hold for all functions $u(P)$ of the set, which establishes our theorem.

A direct consequence of this result is the following selection theorem ("compactness" theorem): *From any uniformly bounded set of regular harmonic functions in G, a sequence $u_n(P)$ may be selected which converges to a harmonic function uniformly in every closed interior subdomain G' of G.*

Since the derivatives of u are also uniformly bounded in every fixed interior closed subdomain, the set $\{u(P)\}$ is equicontinuous there, which assures the possibility of selecting a uniformly convergent sequence (cf. Vol. I, p. 59). Again, from Weierstrass' convergence theorem we conclude that the limit function u is a potential function in G. In particular, our considerations enable us to state: *Any convergent sequence of uniformly bounded harmonic functions must converge uniformly in every closed subdomain, and hence possesses a harmonic limit function.*

§3. The Mean Value Theorem and Applications

1. **The Homogeneous and Nonhomogeneous Mean Value Equation.** We have already considered the mean value theorem of potential theory:

For every regular potential function u in a region G, the average value over the surface Ω_R of a sphere of radius R entirely contained in G is equal to the value u_0 of the function at the center of the sphere,

(1) $$u_0 = \frac{1}{4\pi R^2} \iint_{\Omega_R} u \, d\Omega_R.$$

A mean value theorem for the interior of the sphere follows immediately. Since (1) holds for all R for which the corresponding spheres remain inside G, we can multiply by R^2 and integrate between the limits 0 and a; we obtain

(2) $$u_0 = \frac{3}{4\pi a^3} \iiint_{K_a} u \, dg,$$

where K_a is a solid sphere of radius a. In other words, *the average value of u over the interior of a sphere lying entirely in G is equal to the value u_0 at the center of the sphere.*

A corresponding mean value theorem valid for spheres holds also for the solutions of the nonhomogeneous Poisson equation $\Delta u = -4\pi\mu$; it is obtained by a simple specialization of Green's formula (31′) derived in §1, 3. Applying this formula to a spherical domain K_R of radius R and center P, and setting

$$w = -\frac{1}{R}, \quad \text{i.e.} \quad v = \frac{1}{r} - \frac{1}{R},$$

we obtain the identity

(3) $$\frac{1}{4\pi R^2} \iint_{\Omega_R} u \, d\Omega_R = u_0 + \frac{1}{4\pi} \iiint_{K_R} \left(\frac{1}{r} - \frac{1}{R} \right) \Delta u \, dg,$$

valid for every continuous function $u(x, y, z)$ having continuous first and piecewise continuous second derivatives. For solutions of Poisson's equation we therefore have the following mean value theorem:

For an arbitrary sphere K_R entirely contained in G, every solution of the equation $\Delta u = -4\pi\mu$ regular in G satisfies the relation

(4) $$u_0 = \frac{1}{4\pi R^2} \iint_{\Omega_R} u \, d\Omega_R + \iiint_{K_R} \left(\frac{1}{r} - \frac{1}{R} \right) \mu \, dg.$$

As in the special case $\mu \equiv 0$, we get an equation involving the mean value over the interior of the sphere by multiplying by R^2 and integrating with respect to R. After an easy calculation we find

(5) $$u_0 = \frac{3}{4\pi R^3} \iiint_{K_R} u \, dg + \frac{1}{2R^3} \iiint_{K_R} \frac{(R - r)^2(2R + r)}{r} \mu \, dg.$$

Analogous theorems hold in the plane:

For an arbitrary circle K_R entirely contained in G, every solution of the equation $u_{xx} + u_{yy} = -2\pi\mu$ regular in G satisfies the relations

(6) $$u_0 = \frac{1}{2\pi R} \int_{\Gamma_R} u \, ds + \iint_{K_R} \mu \log \frac{R}{r} \, dg,$$

(7) $$u_0 = \frac{1}{\pi R^2} \iint_{K_R} u \, dg + \frac{1}{R^2} \iint_{K_R} \left(R^2 \log \frac{R}{r} - \frac{R^2 - r^2}{2} \right) \mu \, dg.$$

Generally, in n-dimensional space: *For an arbitrary sphere K_R*

entirely contained in G, every solution of the equation $\Delta u = -\omega_n \mu$ *which is regular in G satisfies the relations*

(6')
$$u_0 = \frac{1}{\omega_n R^{n-1}} \int \cdots \int_{\Omega_R} u \, d\Omega_R$$
$$+ \frac{1}{n-2} \iint \cdots \int_{K_R} \left(\frac{1}{r^{n-2}} - \frac{1}{R^{n-2}} \right) \mu \, dg,$$

(7')
$$u_0 = \frac{n}{\omega_n R^n} \iint \cdots \int_{K_R} u \, dg - \iint \cdots \int_{K_R} \Psi(r, R) \mu \, dg,$$

where

$$\Psi(r, R) = \frac{1}{\omega_n} \left[\frac{1}{n-2} \left(\frac{1}{R^{n-2}} - \frac{1}{r^{n-2}} \right) + \frac{1}{2R^{n-2}} \left(1 - \frac{r^2}{R^2} \right) \right].$$

It should be noted that equations (5), (7) and (7') can also be obtained from Green's formulas (32), (35), and (36) of §1 if we substitute for v the function

$$v = \gamma(r) - \gamma(R) + \frac{1}{2R^n} (r^2 - R^2)$$

with

$$\gamma(r) = \frac{1}{(n-2)r^{n-2}}$$

which vanishes, along with its normal derivatives, on the surface Ω_R of the sphere and satisfies the equation

$$\Delta v = \frac{n}{R^n}$$

inside K_R.

2. The Converse of the Mean Value Theorems. Remarkably, the mean value properties are characteristic for the solutions of the corresponding differential equations. We first prove the following *converse of the mean value theorem of potential theory*:

Suppose a function $u(x, y, z)$ *is continuous in a region G, and satisfies for every sphere* K_R *contained in G the mean value relation*

$$u_0 = \frac{1}{4\pi R^2} \iint_{\Omega_R} u \, d\Omega_R.$$

Then u is harmonic in G.

We shall give two proofs of this theorem.

Proof 1 (using Poisson's integral formula): Let G' be a sphere entirely contained in G and Γ' its boundary. Let v be the potential function given by Poisson's formula (9) in §2 as the solution of the boundary value problem $\Delta v = 0$ in G', $v = u$ on Γ'. Since v is harmonic, it satisfies the mean value theorem, consequently the difference $u - v$ must also have this property and thus satisfies the maximum (and minimum) principle (cf. §1, 3). But $u - v = 0$ on Γ', and therefore $u \equiv v$ in G'. Clearly, u is a solution of the equation $\Delta u = 0$ inside every sphere in G; hence u is harmonic everywhere in G.

Proof 2: It is easy to give a direct proof of the converse to the mean value theorem without using Poisson's integral formula. If we knew that u had continuous second derivatives in G, the assertion would follow immediately from the identity (3)

$$\frac{1}{4\pi R^2} \iint_{\Omega_R} u \, d\Omega_R - u_0 = \frac{1}{4\pi} \iiint_{K_R} \left(\frac{1}{r} - \frac{1}{R} \right) \Delta u \, dg$$

by dividing both sides by R^2 and letting R approach zero. Because of the continuity of Δu, the right side would converge to the value

$$\Delta u_0 \lim_{R \to 0} \frac{1}{4\pi R^2} \iiint_{K_R} \left(\frac{1}{r} - \frac{1}{R} \right) dg = \frac{1}{6} \Delta u_0$$

and the left side would vanish by hypothesis for all R, i.e., $\Delta u_0 = 0$. Thus the theorem will be proved if we can show that u is twice continuously differentiable in G.

Let G_a be a subdomain of G whose points are at a distance greater than a from the boundary of G. Then relation (1) certainly holds for all spheres K of radius $R \leq a$ with centers in G_a. Holding the center fixed, we multiply (1) by $R^2 f(R)$, where $f(R)$ is an arbitrary continuous function, and integrate with respect to R from 0 to a; we obtain

$$(8) \qquad\qquad C u_0 = \iiint_K f(r) u \, dg,$$

where $C = 4\pi \int_0^a r^2 f(r) \, dr$.

For convenience we extend f so that $f(R)$ vanishes identically for $R > a$. If we place the center of the sphere K at the point (x, y, z)

and set

$$L(x, y, z) = f(\sqrt{x^2 + y^2 + z^2}),$$

we can write (8) in the form

(9) $Cu(x, y, z) = \iiint_{-\infty}^{\infty} L(x - \xi, y - \eta, z - \zeta)u(\xi, \eta, \zeta) \, d\xi \, d\eta \, d\zeta.$

Since f can be chosen in such a way that L has derivatives of arbitrarily high order, (9) implies: A continuous function u having the mean value property (1) has continuous derivatives of all orders in the interior of G. This completes the second proof of the theorem.

So far we have assumed that the mean value property (1) is satisfied for *every* sphere entirely contained in G, where G can be an arbitrary finite or infinite domain. However, it turns out that if G is a finite domain it is not necessary to assume the mean value property for *every* sphere. Specifically, we have the following theorem:

Let G be a finite region of space for which the boundary value problem for the equation $\Delta w = 0$ is solvable for arbitrary continuous boundary values. Suppose that a function u is continuous in $G + \Gamma$ and has the mean value property,

(10) $$u_0 = \frac{1}{4\pi h^2} \iint_{\Omega_h} u \, d\Omega_h,$$

at every interior point P of G for at least one single sphere with center at P and radius $h(P) > 0$ such that the sphere is contained in $G + \Gamma$. Then u is a potential function in G.

We emphasize particularly that h may be a completely arbitrary function of x, y, z and thus, for example, may have arbitrary discontinuities.

The proof is obtained by an argument slightly different from the one used in the first proof of the preceding theorem. Assume v to be the potential function which coincides with u on Γ and, as a potential function, obviously satisfies the mean value equation (10) for any h in G. It follows that the difference $u - v$ satisfies condition (10) for $h(P)$. We now consider the closed set F of those points of G at which $u - v$ attains its maximum M. Let P_0 be the point of F which has minimal distance from Γ. If P_0 were an interior point of G, there would exist a sphere with radius $h(P_0) > 0$ and center P_0

lying entirely in G for which the mean value equation holds and on which consequently $u - v = M$. Hence, contrary to our assumption, there would exist points of F which are closer to Γ than P_0. Thus P_0 lies on Γ. Similarly, it can be seen that the minimum of $u - v$ is also assumed on the boundary. But $u - v$ vanishes on Γ. Hence we conclude that $u \equiv v$ in G.

In proving the converse of the mean value theorems we assumed that u is continuous; this assumption is not superfluous. It cannot in general be expected that the continuity of u follows from the mean value property, even when this property holds for every sphere. In fact, for $n = 1$, i.e., for the one-dimensional mean value equation

$$u(x) = \frac{1}{2h} \left(u(x + h) + u(x - h) \right),$$

nonlinear discontinuous functions u may be constructed which satisfy this equation for arbitrary x and h.[1]

Assuming the mean value property only for a single radius $h(P)$ at each point P one must require that G be a finite region. For an infinite domain, for example even in the special case $h = \text{const.} = l$, continuous nonharmonic functions can be found which satisfy the equation

$$(11) \qquad u_0 = \frac{1}{4\pi l^2} \iint_{\Omega_l} u \, d\Omega_l$$

at every point in G. If we assume that u depends only on x, it is easily seen that (11) goes over into the integral equation

$$(12) \qquad u(x) = \frac{1}{2l} \int_{x-l}^{x+l} u(\xi) \, d\xi.$$

We obtain special solutions of this equation by setting $u(x) = e^{i\gamma x}$, provided γ is a root of the transcendental equation

$$(13) \qquad \frac{\sin \gamma l}{\gamma l} = 1.$$

Apart from the solution $\gamma = 0$ this equation possesses infinitely many complex roots $\gamma = \alpha + i\beta$, represented by the points of intersection of the curves

$$\frac{\sin \alpha l}{\alpha l} = \frac{1}{\cosh \beta l}, \qquad \cos \alpha l = \frac{\beta l}{\sinh \beta l}$$

[1] See G. Hamel [1].

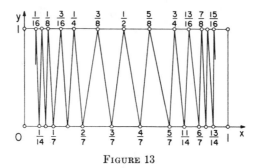

FIGURE 13

in the α,β-plane. The real functions $u = e^{-\beta x} \cos \alpha x$ and $u = e^{-\beta x} \sin \alpha x$ then also satisfy equation (12); a potential function is obtained only if $\alpha = \beta = 0$.

If the mean value property is assumed only for a single radius $h(P)$ at each point P, it is essential that u be continuous not only in the finite open domain G but also in the *closed* domain $G + \Gamma$. Figure 13 shows a simple example[1] of a function $u(x)$, continuous only in an open interval, which at every point x of the interval satisfies the one-dimensional mean value equation

$$(14) \qquad u(x) = \frac{1}{2h} \left(u(x + h) + u(x - h) \right)$$

for corresponding $h(x) > 0$ but which, nevertheless, is not linear.

This function is continuous and piecewise linear in the interval $0 < x < 1$, and zigzags back and forth between the lines $y = 0$ and $y = 1$. Clearly, it is not continuous at the end points $x = 0$ and $x = 1$. Let the peaks lying on the upper line have the abscissas a_ν, and those lying on the lower line the abscissas b_ν. In the neighborhood of every point of the interval which does not coincide with one of the points a_ν or b_ν, $u(x)$ is linear and hence certainly has the mean value property (14) for some definite $h(x)$, in fact, for infinitely many h. If we now choose the sequence a_ν in such a way that to every a_ν there exist two points $a_\alpha < a_\nu$ and $a_\beta > a_\nu$ for which $a_\nu = \frac{1}{2}(a_\alpha + a_\beta)$, then $u(x)$ satisfies relation (14) also at the upper peaks, and in fact $h(a_\nu) = a_\beta - a_\nu = a_\nu - a_\alpha$. If in addition we choose b_ν in the same way, but so that b_ν does not coincide with a_μ for any ν and μ, and so that in every interval between two adjacent

[1] This example was suggested by Max Shiffman.

a_ν precisely one b_μ occurs, then we obtain a function $u(x)$, which is continuous for $0 < x < 1$, has the property (14), but which is nevertheless not linear, as shown in Figure 13.

The construction of two sequences a_ν and b_ν of the type described is possible in many ways. For example, as in Figure 13, we can choose the points with the abscissas

$$a = \begin{cases} \dfrac{1}{2^k}, & \dfrac{3}{2^{k+2}}, \\[2ex] 1 - \dfrac{1}{2^k}, & 1 - \dfrac{3}{2^{k+2}} \end{cases} \qquad (k = 1, 2, \cdots)$$

on $y = 1$, and the points with the abscissas

$$b = \begin{cases} \dfrac{1}{7}\dfrac{1}{2^{k-1}}, & \dfrac{3}{7 \cdot 2^k}, \\[2ex] 1 - \dfrac{1}{7}\dfrac{1}{2^{k-1}}, & 1 - \dfrac{3}{7}\dfrac{1}{2^k} \end{cases} \qquad (k = 0, 1, 2, \cdots)$$

on $y = 0$, which are symmetric with respect to $x = \frac{1}{2}$.

The general mean value theorem for the nonhomogeneous Poisson equation also has a converse:

Let u be continuous and μ piecewise continuously differentiable in G. For every sphere K lying in G, let these functions satisfy the equation (4)

$$u_0 = \frac{1}{4\pi R^2} \iint_\Omega u \, d\Omega + \iiint_K \left(\frac{1}{r} - \frac{1}{R} \right) \mu \, dg$$

or the equivalent integrated equation (5). *Then u satisfies Poisson's equation $\Delta u = -4\pi\mu$ in G.*

We notice first that in case μ is only continuous in G, the triple integral in (4) divided by R^2 approaches the value

(15) $$\mu_0 \lim_{R \to 0} \frac{4\pi}{R^2} \int_0^R \left(\frac{1}{r} - \frac{1}{R} \right) r^2 \, dr = \frac{4\pi}{6} \mu_0$$

as $R \to 0$; consequently the limit

(16) $$\Theta(u_0) = \lim_{R \to 0} \frac{6}{R^2} \left\{ \frac{1}{4\pi R^2} \iint_{\Omega_R} u \, d\Omega_R - u_0 \right\}$$

also exists everywhere in G, and in fact we have

(17) $$\Theta(u_0) = -4\pi\mu_0 .$$

If u has continuous second derivatives, then as we remarked on page 278,

(18) $$\Theta(u) = \Delta u.$$

Our theorem is thus proved if we can show that u has continuous second derivatives. Indeed, we shall show more generally:

If u is continuous in G and μ is piecewise continuously differentiable in G, and if, for every sphere K lying in G, u satisfies the mean value relation (4) *or* (5), *then u has continuous second derivatives in G.*

If u satisfies (4) we proceed as in the proof for the special case $\mu \equiv 0$. We again choose a function $f(R)$ vanishing identically for $R > a$ and having derivatives of sufficiently high orders, multiply equation (4), which is valid for a fixed point P in the subdomain G_a, by $4\pi R^2 f(R)$ and integrate between the limits 0 and a. This leads to

(19) $$Cu_0 = \iiint_K uf(R)\, dg + \iiint_K \mu F(R)\, dg,$$

where

$$C = 4\pi \int_0^a R^2 f(R)\, dR$$

and

$$F(r) = 4\pi \int_r^a \left(\frac{1}{r} - \frac{1}{R}\right) R^2 f(R)\, dR$$

$$= \frac{C}{r} - \frac{4\pi}{r} \int_0^r R^2 f(R)\, dR - 4\pi \int_r^a Rf(R)\, dR.$$

If we now set

$$L(x, y, z) = f(\rho), \qquad \text{where} \qquad \rho = \sqrt{x^2 + y^2 + z^2}$$

and

$$H(x, y, z) = \frac{4\pi}{\rho} \int_0^\rho R^2 f(R)\, dR + 4\pi \int_\rho^a Rf(R)\, dR$$

then, by a suitable choice of $f(R)$, L and H will have continuous derivatives up to order N everywhere. With the center of the sphere K at (x, y, z), we then obtain the representation

$$Cu(x, y, z) = \iiint_{-\infty}^{\infty} L(x - \xi, y - \eta, z - \zeta)u(\xi, \eta, \zeta)\,d\xi\,d\eta\,d\zeta$$

(20)
$$- \iiint_G H(x - \xi, y - \eta, z - \zeta)\mu(\xi, \eta, \zeta)\,d\xi\,d\eta\,d\zeta$$

$$+ C \iiint_G \frac{\mu}{r}\,d\xi\,d\eta\,d\zeta$$

since $f(r)$ and consequently $F(r)$ vanish identically for $r > a$.

The first two integrals on the right have derivatives up to the Nth order. The third integral is the potential of the spatial distribution μ and, according to the results of §1, 2, is twice continuously differentiable in G. Hence the function u itself is twice continuously differentiable in G_a, and therefore satisfies the equation $\Delta u = -4\pi\mu$ in G_a. Since a may be chosen arbitrarily small, this result holds everywhere in the interior of G.

Corresponding theorems hold in n-dimensional space and follow immediately from the next theorem, whose proof is similar to the one just given: *If u is continuous in G and μ is piecewise continuously differentiable in G, and if, for every sphere K lying in G, u satisfies equations (6') or (7') (respectively (6) or (7) for $n = 2$), then u is twice continuously differentiable in G.*

3. Poisson's Equation for Potentials of Spatial Distributions. In §1, 2, we have proved that for $n = 3$ dimensions the potential u of a mass distribution μ in G satisfies Poisson's equation $\Delta u = -4\pi\mu$. A different proof will be given now making use of the converse to the mean value theorem, stated and proved on page 282 of the previous article.

We first let the function $\mu(x, y, z)$ be piecewise continuous in G, and consider the potential

$$(21) \qquad u(x, y, z) = \iiint_G \frac{\mu(\xi, \eta, \zeta)}{r}\,dg$$

of the mass distribution μ, where $dg = d\xi\,d\eta\,d\zeta$. For the moment we assume the distribution μ to be continued outside of G by $\mu = 0$. In addition, let P_0 be an arbitrary point of the space and K an arbitrary sphere of radius R with P_0 as center.

Since $u(x, y, z)$ is continuous everywhere, we can integrate both sides of (21) over the sphere's interior; as a matter of fact the right side can be integrated under the integral sign. We obtain

$$\iiint_K u(x, y, z)\, dx\, dy\, dz = \iiint_G \mu(\xi, \eta, \zeta) F(\xi, \eta, \zeta; P_0)\, dg,$$

where $F(\xi, \eta, \zeta; P_0) = \iiint_K dx\, dy\, dz/r$ is the potential at (ξ, η, ζ) of
the sphere K with a uniform distribution of unit mass. Thus

$$(22) \qquad F(\xi, \eta, \zeta; P_0) = \begin{cases} \dfrac{4\pi}{3}\dfrac{R^3}{r} & (r \geq R) \\[2mm] 2\pi\left(R^2 - \dfrac{r^3}{3}\right) & (r \leq R), \end{cases}$$

where r now denotes the distance of (ξ, η, ζ) from P_0.
Substitution of the values given by (22) leads to

$$(23) \qquad \iiint_K u(\xi, \eta, \zeta)\, dg = \frac{4\pi}{3} R^3 \iiint_{G^*} \frac{\mu}{r}\, dg + 2\pi \iiint_K \left(R^2 - \frac{r^2}{3}\right)\mu\, dg,$$

where G^* is the subdomain of G not covered by K. Since

$$\iiint_{G^*} \frac{\mu}{r}\, dg = u_0 - \iiint_K \frac{\mu}{r}\, dg,$$

we obtain the equation

$$\frac{4\pi}{3} R^3 u_0 = \iiint_K u\, dg - 2\pi \iiint_K \left(R^2 - \frac{r^2}{3} - \frac{2}{3}\frac{R^3}{r}\right)\mu\, dg$$

or

$$u_0 = \frac{3}{4\pi R^3}\iiint_K u\, dg + \frac{1}{2R^3}\iiint_K \frac{(R - r)^2(2R + r)}{r}\mu\, dg,$$

in other words, just the mean value equation (5).

*The potential of a piecewise continuous mass distribution satisfies the
mean value relation* (5), *and hence also the equivalent relation* (4), *for
every sphere.*

Taking into consideration the result obtained on page 282, we can
make the following statement:

If μ is continuous in G, then for the potential (21) *the expression*

$$\Theta(u) = \lim_{R \to 0} \frac{6}{R^2} \left\{ \frac{1}{4\pi R^2} \iint_{\Omega_R} u \, d\Omega_R - u_0 \right\}$$

exists everywhere in G, and in fact

(24) $\Theta(u) = -4\pi\mu.$

If μ is piecewise continuously differentiable in G, $\Theta(u) = \Delta u$ and hence $\Delta u = -4\pi\mu$.

4. Mean Value Theorems for Other Elliptic Differential Equations. The mean value theorems for the Laplace and Poisson equations may be read off immediately from the identity

(25) $\dfrac{1}{4\pi R^2} \iint_{\Omega_R} u \, d\Omega_R = u_0 + \dfrac{1}{4\pi} \iiint_{K_R} \left(\dfrac{1}{r} - \dfrac{1}{R} \right) \Delta u \, dg,$

which holds for continuous functions u having continuous first and piecewise continuous second derivatives. Instead of this identity we can easily obtain a more general one, leading to a Taylor expansion of the mean value

$$M(R) = \frac{1}{4\pi R^2} \iint_{\Omega_R} u \, d\Omega_R$$

as a function of R for a fixed center P_0.

We begin by considering Green's formula

(26) $Cu_0 = \iiint_K (u \, \Delta v - v \, \Delta u) \, dg,$

where K is an arbitrary sphere of radius R and center P_0; v has the form

(26′) $v(r) = \dfrac{C}{4\pi r} + w(r),$

where $w(r)$ is twice continuously differentiable for $r \leq R$, and, on the surface Ω_R of the sphere,

(26″) $v = \dfrac{\partial v}{\partial r} = 0.$

For u we can take any function with continuous second derivatives.

If we assume that u has continuous derivatives up to order $(2m + 2)$ in $K + \Omega_R$, then, for all $\nu \leq m$, we have the identity

(27) $C \, \Delta^\nu u_0 = \iiint_K (\Delta^\nu u \, \Delta v - v \, \Delta^{\nu+1} u) \, dg.$

By $\Delta^\nu u$ we understand the νth iteration of the Laplacian; e.g., $\Delta^1 u = \Delta u$, $\Delta^2 u = \Delta\Delta u$, etc.

In addition, let v_1, v_2, $\cdots$ be a sequence of functions of the type (26'), defined by the differential equations

$$(28) \qquad \Delta v_{\nu+1} = v''_{\nu+1} + \frac{2}{r} v'_{\nu+1} = v_\nu \qquad (\nu = 1, 2, \cdots),$$

the boundary conditions (26''), and the initial function

$$(29) \qquad v_0 = \frac{1}{4\pi}\left(\frac{1}{r} - \frac{1}{R}\right) = \frac{1}{4\pi}\frac{R-r}{Rr}.$$

One easily verifies that the solutions of this recursion system are given by the functions

$$(30) \qquad v_\nu = \frac{1}{4\pi(2\nu+1)!}\frac{(R-r)^{2\nu+1}}{Rr} \qquad (\nu = 0, 1, \cdots)$$

which all have the form (26') with $C_\nu = \dfrac{R^{2\nu}}{(2\nu+1)!}$.

Replacing v in (27) by v_ν yields

$$(31) \qquad C_\nu \Delta^\nu u_0 = \iiint_K (v_{\nu-1} \Delta^\nu u - v_\nu \Delta^{\nu+1} u)\, dg;$$

if we sum over all the equations corresponding to $\nu = 1, 2, \cdots, m$, we conclude from this expression:

$$\sum_{\nu=1}^m C_\nu \Delta^\nu u_0 = \iiint_K (v_0 \Delta u - v_m \Delta^{m+1} u)\, dg.$$

In view of (25) and (30), we easily obtain

$$(32) \qquad \begin{aligned} \frac{1}{4\pi R^2}\iint_{\Omega_R} u\, d\Omega_R &= \sum_{\nu=0}^m \frac{R^{2\nu}}{(2\nu+1)!}\Delta^\nu u_0 \\ &\quad + \frac{1}{4\pi(2m+1)!}\iiint_{K_R}\frac{(R-r)^{2m+1}}{Rr}\Delta^{m+1}u\, dg, \end{aligned}$$

where $\Delta^0 u_0$ is to be interpreted as u_0. This identity is valid for every $(2m+2)$-times continuously differentiable function u and every sphere K contained in G.[1]

If u is arbitrarily often differentiable in G and if the remainder

[1] In this connection see P. Pizetti [1].

term tends to zero with increasing m, (32) leads to the infinite series

$$(33) \qquad M(R) = \frac{1}{4\pi R^2} \iint_{\Omega_R} u \, d\Omega_R = \sum_{\nu=0}^{\infty} \frac{R^{2\nu}}{(2\nu + 1)!} \Delta^{\nu} u_0.$$

This is the case, for example, when $\Delta^m u$ vanishes identically in G from a certain m on. The expansion (33) then breaks off with $\nu = m - 1$ and we can state:

Every solution of the differential equation $\Delta^m u = 0$ regular in G, i.e., having continuous derivatives up to the $(2m)$th order, satisfies the mean value relation

$$(34) \qquad \frac{1}{4\pi R^2} \iint_{\Omega} u \, d\Omega = \sum_{\nu=0}^{m-1} \frac{R^{2\nu}}{(2\nu + 1)!} \Delta^{\nu} u_0$$

for an arbitrary sphere contained in G.

For instance, solutions of the equation $\Delta\Delta u = 0$ satisfy the mean value relation

$$(35) \qquad \frac{1}{4\pi R^2} \iint_{\Omega} u \, d\Omega = u_0 + \frac{R^2}{6} \Delta u_0.$$

Another example is given by the solutions of the differential equation $\Delta u + cu = 0$. Here

$$\Delta^{m+1} u = (-1)^{m+1} c^{m+1} u$$

and, since the remainder term converges to zero with increasing m, we have

$$M(R) = u_0 \sum_{\nu=0}^{\infty} \frac{(-1)^{\nu} c^{\nu}}{(2\nu + 1)!} R^{2\nu} = u_0 \frac{\sin R\sqrt{c}}{R\sqrt{c}}.$$

For every solution of the equation $\Delta u + cu = 0$ which is regular in G, the mean value relation

$$(36) \qquad \frac{1}{4\pi R^2} \iint_{\Omega} u \, d\Omega = u_0 \frac{\sin R\sqrt{c}}{R\sqrt{c}}$$

is valid for an arbitrary sphere in G.

Similar arguments hold in the plane and generally in n-space. Thus, in the plane we obtain the identity

$$(37) \quad M(R) = \frac{1}{2\pi R} \int_{\Omega} u \, ds = \sum_{\nu=0}^{m} \left(\frac{R}{2}\right)^{2\nu} \frac{\Delta^{\nu} u_0}{(\nu!)^2} + \iint_{K} v_m \Delta^{m+1} u \, dg,$$

where v_m is to be determined from the recursion formula

(37')
$$v_{\nu+1} = \int_r^R \rho v_\nu(\rho) \log \frac{\rho}{r} d\rho,$$

$$v_0 = \frac{1}{2\pi} \log \frac{R}{r}.$$

In n-dimensional space we have

(38)
$$M(R) = \frac{1}{\Omega} \iint_\Omega u \, d\Omega$$

$$= \Gamma\left(\frac{n}{2}\right) \sum_{\nu=0}^m \left(\frac{R}{2}\right)^{2\nu} \frac{\Delta^\nu u_0}{\nu! \, \Gamma\left(\nu + \frac{n}{2}\right)} + \iiint_K v_m \Delta^{m+1} u \, dg,$$

where the $v_m(r)$ are given by the recursion system

(38')
$$v_{\nu+1} = \frac{1}{(n-2)r^{n-2}} \int_r^R \rho v_\nu(\rho)(\rho^{n-2} - r^{n-2}) \, d\rho,$$

$$v_0 = \frac{1}{(n-2)\omega_n} \left(\frac{1}{r^{n-2}} - \frac{1}{R^{n-2}}\right).$$

As before, these formulas lead to the following theorem:

Every solution of the differential equation $\Delta u + cu = 0$ regular in G satisfies the mean value relation

(39)
$$\frac{1}{\Omega} \iint_\Omega u \, d\Omega = u_0 \Gamma\left(\frac{n}{2}\right) \sum_{\nu=0}^\infty \frac{(-1)^\nu \left(\frac{R\sqrt{c}}{2}\right)^{2\nu}}{\nu! \, \Gamma\left(\nu + \frac{n}{2}\right)}$$

$$= u_0 \frac{\Gamma\left(\frac{n}{2}\right) J_{(n-2)/2}(R\sqrt{c})}{\left(\frac{R\sqrt{c}}{2}\right)^{(n-2)/2}}$$

$$= u_0 p(R)$$

in an arbitrary sphere lying in G. Here, $J_\nu(x)$ is the νth Bessel function. In particular, in the plane we have

(40)
$$\frac{1}{2\pi R^2} \int_\Omega u \, ds = u_0 J_0(R\sqrt{c}).$$

For odd n the factor $p(R)$ of u_0 in (39) can be expressed in terms of the derivatives of the function $\sin R\sqrt{c}/R\sqrt{c}$. In fact (cf. Vol. I, Ch. VII, p. 488)

$$p(R) = \frac{(-1)^{(n-3)/2} 2^{n-2} \Gamma\left(\dfrac{n}{2}\right)}{\sqrt{\pi}} \frac{d^{(n-3)/2}}{d(R^2 c)^{(n-3)/2}} \frac{\sin R\sqrt{c}}{R\sqrt{c}}.$$

§4. The Boundary Value Problem

1. Preliminaries. Continuous Dependence on the Boundary Values and on the Domain. We return to the central problem of the theory, the *boundary value problem*. We have already proved the *uniqueness* of its solution, that is, the theorem:

There exists at most one function harmonic in G which takes on given continuous boundary values f on Γ.

Similarly, we have seen that the solution of the boundary value problem depends continuously on the boundary values: *If f_ν is a sequence of continuous functions converging uniformly to f on Γ, then the sequence of potential functions u_ν with boundary values f_ν converges in the interior of G to a potential function u with boundary values f.*

The proof follows immediately from the convergence theorems of §2, 3. It is, therefore, sufficient to demonstrate the solvability of the boundary value problem for special boundary values on Γ, say for polynomials in x_1, x_2, $\cdots$, x_n; then the solution of the problem for arbitrary continuous boundary values can be obtained by a limiting process. In accordance with the considerations of §2, 1, it is sufficient to construct the *Green's function K*.

In the construction we restrict ourselves to the dimensions $n = 2$ and $n = 3$ and obtain the following result:

In the case $n = 2$, Green's function can be constructed for any domain G whose boundary Γ consists of a finite number of continuous curves such that every point of Γ is the end point of a straight-line segment which otherwise lies entirely outside G.

In the case $n = 3$, Green's function can be constructed for every domain G whose boundary Γ consists of a finite number of continuous surfaces such that every point of Γ is the vertex of a tetrahedron otherwise lying entirely outside G.

In Volume III the boundary value problem will be treated once again, but under essentially more general assumptions and from different points of view.

The following discussion applies to a bounded domain G. For an unbounded domain G which does not cover the whole space we obtain Green's function by transforming the domain G into a bounded domain G' by inversion with respect to a suitable circle or sphere. On the basis of the theorem of §1, p. 242, Green's function for G is then obtained at once from Green's function for G'.

In order to simplify the construction of Green's function for more general domains, we shall first prove certain statements about the continuous dependence of Green's function on its domain. We consider a monotonic sequence of subdomains G_ν converging to G, such that each G_ν contains $G_{\nu-1}$ as a subdomain and, in addition, every fixed interior point of G is contained in all G_ν from a certain ν on. The following theorem then holds:

a) *If there exists a Green's function K_ν for every domain G_ν and a Green's function K for the limit-domain G, then the sequence K_ν converges uniformly to K in $G + \Gamma$.*[1]

Of greater significance, however, is a more general convergence theorem for the case when the existence of Green's function for the limit-domain G is not known in advance. This Green's function K can then be constructed by a corresponding limiting process. The justification for this is provided by the following refinement of theorem a):

b) *If the sequence of domains G_ν converges monotonically to G, and if for every G_ν the associated Green's function K_ν exists, then the sequence K_ν converges in G to a limit function*

$$K = \lim_{\nu \to \infty} K_\nu .$$

This limit function K is Green's function for the domain G if certain conditions are satisfied.

These conditions are: For two dimensions, *to every boundary point P of G there exists a finite straight-line segment which lies outside G and has an end point at P,*[2] and for three dimensions, *to every boundary*

----

[1] In the part of G not belonging to G_ν, one can suppose K_ν continued by, say, $K_\nu \equiv 0$.

[2] We remark that for $n = 2$ our considerations can easily be extended to the more general situation when P can be reached by a polygon consisting of a finite or a countable number of segments instead of one straight-line segment. In this case the boundary may be an arbitrary Jordan curve. However, since the boundary value problem for such a general boundary will be solved directly in a different way in Volume III, we omit this generalization here.

point P of G there exists a tetrahedron, which lies outside G and has a vertex at P.

To prove theorems a) and b) we first make the following remark. The functions K_ν as well as the functions $K_\nu - \gamma$, which are obtained by subtracting the fundamental solution γ and which are regular harmonic in G_ν, form monotonic sequences. For if we let ν be so large that the singularity point Q lies in G_ν then for $\mu > \nu$ the boundary values of the potential functions $K_\mu - K_\nu$, which are regular in G_ν, are non-negative on Γ_ν. Hence these functions are non-negative throughout G_ν.

Moreover, we have for the same reason, in case a), $K_\nu \leq K$, hence also $K_\nu - \gamma \leq K - \gamma$. In case b), there exists at least one sphere that lies entirely outside G. If $\hat{K}$ is Green's function for the exterior domain of this sphere then $K_\nu \leq \hat{K}$ and $K_\nu - \gamma \leq \hat{K} - \gamma$. That is, in both cases, the monotonically increasing sequence $K_\nu - \gamma$ is bounded, and hence convergent in G. The limit function $H = \lim_{\nu\to\infty} K_\nu$ is certainly non-negative throughout G and, by the convergence theorems of §2, 3, H is harmonic in G.

For theorem a), we conclude from $K_\nu \leq K$ that $H \leq K$. But since K has the boundary values zero, the same is true for the limit function H. In other words, $H = K$ is Green's function for the limit domain G.

In order to prove theorem b), we anticipate, as a result of the considerations of article 2, that we can construct Green's function for the exterior of a straight-line segment in the plane and for the exterior of a tetrahedron in space. Now let K^* be Green's function for such a segment (or for such a tetrahedron) which can be attached to a boundary point P_0 of G in the manner described in b). For every point of G our considerations give us immediately:

$$0 \leq H(P) \leq K^*(P);$$

since $K^*(P_0) = 0$, it follows just as before that the boundary value of H at P_0 is $H(P_0) = 0$. According to our assumptions, H thus has zero boundary values everywhere, and is, therefore, Green's function of the limit domain G.

We stress the fact that these discussions are by no means restricted to simply connected domains, but may be carried over essentially unchanged to *multiply connected domains*.

In the next section we shall show, using the *alternating procedure of*

H. A. Schwarz, how we can carry over the construction of Green's function K for relatively special domains $G_\nu = G$ to more general domains by approximating these monotonically by the G_ν.

2. Solution of the Boundary Value Problem by the Schwarz Alternating Procedure. A simple convergent process permits us to solve the boundary value problem for a domain B whenever this domain is the union of two overlapping domains G and G' (or of any finite number of such domains) for which we may assume the solvability of the boundary value problem with arbitrary continuous boundary values. Suppose that the boundaries Γ and Γ' of G and G' are composed of a finite number of parts which have continuous tangents or tangent planes. We assume further that the boundaries of G and G' intersect at an angle different from zero, and not at vertices or along edges of Γ or Γ'. Since, for example, the boundary value problem can be solved for the circle and half-plane, or for the sphere and half-space by means of Poisson's integral, the alternating procedure immediately permits the solution for domains which are unions of a finite number of overlapping circles and half-planes in the plane, or spheres and half-spaces in space. The construction of the union B is illustrated in Figures 14 and 15. The latter shows how doubly connected domains can be obtained from simply connected domains, and, accordingly, how the boundary value problem can also be solved for such domains.

Since, for this procedure, it makes no essential difference whether the overlapping takes place over one or several disjoint regions, we shall consider the case represented in Figure 14 where G and G' have only one single domain D in common. Let the boundary Γ of G consist of the parts a and b, where b lies in G'. Let a' and b' be defined in a corresponding way for the domain G'. We assume that

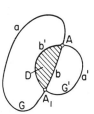

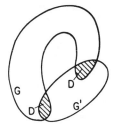

FIGURE 14   FIGURE 15

on the boundary $a + a' = \Lambda$ of B continuous boundary values are given which are bounded in absolute value by some constant M.

1) The alternating procedure consists of the following process: We supplement the boundary values assigned to a in a continuous manner by means of arbitrary values on b which are less than M in absolute value, and solve the boundary value problem for G with the continuous boundary values thus defined on Γ. The solution u_1 takes values on b' which, together with the values prescribed on a', furnish a continuous assignment of boundary values on Γ'. With these we solve the boundary value problem for G' and obtain a function u_1'. This function itself takes values on b which, with the boundary values on a, again result in a continuous boundary value assignment for G. Let the solution of the corresponding boundary value problem be u_2. Continuing alternately in this way we arrive at a sequence of potential functions $u_1, u_2, \cdots$, in G and a corresponding sequence $u_1', u_2', \cdots$ in G'. In general,

$$\text{on } b': \quad u_\nu' = u_\nu, \quad \text{hence} \quad u_{\nu+1} - u_\nu = u_{\nu+1}' - u_\nu',$$

$$\text{and on } b: \quad u_\nu' = u_{\nu+1}, \quad \text{hence} \quad u_{\nu+1} - u_\nu = u_\nu' - u_{\nu-1}'.$$

We now assert: The functions u_ν in G and u_ν' in G' converge uniformly to potential functions u and u', which are identical in the intersection D of G and G'. These limit functions define a regular potential function in the whole domain which solves the boundary value problem for B.

The proof depends on the following lemma:

Under the above assumptions on G and G', let the function v be regular harmonic in G, vanish on a, and satisfy the inequality

$$0 \leq |v| \leq 1$$

on b. Then there exists a positive constant $q < 1$, depending only on the configuration of the domains G and G', such that v satisfies the inequality

$$|v| \leq q$$

everywhere on b'. A corresponding statement holds, of course, for G'. Clearly, we can choose for q a constant applying to both domains.

We shall prove this lemma at the end of the article; however, we first use it to complete the proof of the convergence of the alternating procedure. By M_ν we denote the maximum of

$$| u_{\nu+1} - u_\nu | = | u'_{\nu+1} - u'_\nu | \text{ on } b',$$

and correspondingly by M'_ν the maximum of

$$| u_{\nu+1} - u_\nu | = | u'_\nu - u'_{\nu-1} | \text{ on } b.$$

If we identify the function v of our lemma for G with $(u_{\nu+1} - u_\nu)/M'_\nu$, we obtain at once

$$M_\nu \leq q M'_\nu.$$

Similarly we find

$$M'_\nu \leq q M_{\nu-1}$$

and hence

$$M_\nu \leq q^2 M_{\nu-1}.$$

Therefore the quantities M_ν and M'_ν converge to zero, and are in fact majorized by the terms of a geometric series with ratio $q^2 < 1$.

This immediately implies the uniform convergence of the series

$$u_1 + \sum_{\nu=1}^{\infty} (u_{\nu+1} - u_\nu) = \lim_{n \to \infty} u_n = u$$

in $G + \Gamma$, and of the corresponding series

$$u'_1 + \sum_{\nu=1}^{\infty} (u'_{\nu+1} - u'_\nu) = \lim_{n \to \infty} u'_n = u'$$

in $G' + \Gamma'$. Accordingly, u and u' are potential functions in G and G' which have the prescribed boundary values on a and a', respectively. For the intersection D, bounded by b and b', we have $u'_\nu - u_\nu = 0$ on b', while on b the difference $u'_\nu - u_\nu = u'_\nu - u'_{\nu-1}$ tends uniformly to zero. Hence the limit functions u and u' coincide in D and together define a regular harmonic function in B which solves the boundary value problem.

If we repeat our procedure a finite number of times, we are led to the following theorem: *If G is the union of a finite number of over-lapping domains G_1, G_2, $\cdots$, G_n whose piecewise smooth boundaries intersect at nonzero angles and never intersect at corners or edges, then the boundary value problem can be solved for G whenever it can be solved for each individual G_ν.*

In particular, we see that it is possible to solve the boundary value problem for every domain which can be covered by a finite number of circles and half-planes, or spheres and half-spaces. For example, if G consists of the entire plane with the exception of an interval $0 \leq x \leq 1$, we may consider this domain as the union of the four half-planes

$$x < 0, \qquad x > 1, \qquad y < 0, \qquad y > 0$$

and solve the boundary value problem individually for each of these half-planes by using Poisson's integral. The alternating procedure then immediately yields the solution for G. In three-dimensional space the same is true for the exterior of a tetrahedron which can be considered as the union of four half-spaces.

If we observe that an arbitrary domain can be regarded as the limit domain of a monotonically increasing sequence of domains G_ν, each of which consists of a finite number of circles or spheres, we can apply theorem b) of article 1 and obtain the general theorem:

In the plane, Green's function, and hence the solution of the boundary value problem, exists for every domain G such that each of its boundary points can be reached by a straight-line segment lying outside G. In space the same holds for every domain G for which each boundary point is the vertex of a tetrahedron otherwise lying outside G.[1]

2) Proof of the lemma: To prove the lemma we first consider the two-dimensional case (Figure 16). We form the potential of a double layer on the arc b with the density 1, i.e., the expression

$$w(P) = w(x, y) = \int_b \frac{\partial \log \frac{1}{r}}{\partial \nu} \, ds,$$

which represents the angle swept out by a radius vector from the point P in G to a point which runs along the arc b. This function, regular harmonic in the interior of G, has continuous boundary values

---

[1] We remark in passing that the alternating procedure we have described, carried out for several domains which are followed through cyclically as the process is continually repeated, is essentially identical with the famous *balayage method of Poincaré*. (Compare the numerous presentations in the literature.) The difference is that Poincaré at once assumed countably many circles or spheres which are followed through in a definite order as the process is repeated. The proof given here, however, differs from the usual proof of the balayage method.

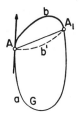

FIGURE 16

on the boundary arcs a and b. As a boundary point approaches the end point A along the arc a, the corresponding boundary values of w tend toward a limit R_{A-} which is equal to the angle between the secant AA_1 and the tangent at A directed toward b. The corresponding limit R_{A+} of the boundary values on b, on the other hand, equals the angle between AA_1 and the tangent of A now directed toward a. This leads to the relation

$$R_{A+} - R_{A-} = \pi.$$

If one approaches the boundary point A from the interior of the domain along any ray that makes an angle α with the tangent at A directed toward b, one obtains as a limiting value the linear combination

$$\frac{\alpha}{\pi} R_{A+} + \left(1 - \frac{\alpha}{\pi}\right) R_{A-}.$$

From this we see finally that for an arbitrary sequence of points P_ν in G converging to A the corresponding function values $w(P_\nu)$ can only have limit points which lie between R_{A+} and R_{A-}. Similar statements can, of course, be made for the other end point A_1 of the arc.

We now consider a function ρ defined on the boundary $\Gamma = a + b$ and having the constant values π on b and 0 on a. Let $\bar{w}$ be the boundary values of w; then the difference $\bar{w} - \rho$ is a continuous function on Γ. By hypothesis, there exists a regular harmonic function Ω in G having these boundary values.

We define the function

$$S(P) = \frac{w - \Omega}{\pi}$$

which is bounded in $G + \Gamma$ and regular harmonic in G. On a it has the boundary values 0 and on b the boundary values 1. But from the preceding remarks concerning w, we see that as we approach A or A_1 from the interior, S can have limiting values only between 0 and 1, and that as we approach them along a ray which makes an angle α with the tangent we obtain the boundary value α/π, hence a value smaller than 1.

In particular, if we approach the points A and A_1 along the boundary arc b' of G', then the limiting values of the angle α are the angles β and β_1 at which b' intersects Γ. Everywhere on b', S satisfies the inequality

$$S \leq q < 1.$$

For otherwise there would exist a sequence of points P_ν on b' such that $S(P_\nu) \to 1$. Just as in the case of the proof of the maximum and minimum theorem, this sequence cannot have a limit at any point in the interior of b'. But a limit point at A or A_1 is also impossible since, as we saw, the limits at these points are $\beta/\pi < 1$ and $\beta_1/\pi < 1$, respectively.

With the function v of our lemma we now form the differences $S - v = \Lambda$, whose boundary values vanish on a and are certainly non-negative on b. This function, regular harmonic in G, can neither become negative in the interior of G nor lead to negative limiting values as one approaches an interior point of the arcs a or b. Approaching the end points A or A_1, Λ has the same limit values as S which, therefore, must lie between zero and one. Hence $S - v \geq 0$ everywhere in the closed domain G, and, in particular,

$$v \leq S \leq q$$

on the arc b'. The same argument for the sum $S + v$ leads to $S + v \geq 0$ in G; hence, combining our results, we obtain

$$|v| \leq S \leq q < 1$$

on b'. This proves our lemma.

The proof has the advantage that it can be carried over immediately to three and more dimensions. To do this we take w to be the double layer potential of a surface distribution on the boundary with density 0 or 1.

3. The Method of Integral Equations for Plane Regions with

Sufficiently Smooth Boundaries. Another way of solving the boundary value problem in two dimensions for certain special regions, which differs essentially from the alternating procedure and the balayage method, is presented by Fredholm's method of integral equations. This is an extension of an older method due to C. Neumann, valid for convex regions. The boundary value problem will be reduced to a Fredholm equation of the second kind. We shall refrain from developing the method under the most general hypotheses, and assume that the boundary curve Γ can be represented parametrically by functions $x(t)$ and $y(t)$ which have continuous derivatives up to and including the fourth order. Here we assume that the parameter t is the arc length along the curve Γ.

We attempt to represent the desired potential function $u(x, y)$ in the form

$$(1) \qquad u(x, y) = \int_\Gamma \sigma(t) \frac{\partial \gamma}{\partial \nu} \, dt, \qquad \left(\gamma = \log \frac{1}{r} \right)$$

as the potential of a double layer of density $\sigma(s)$ on the boundary Γ. This integral has a meaning even when $P(x, y)$ is a point of the boundary Γ, say the point for which the value of the arc length is s, i.e., $P = (x(s), y(s))$. Then for any point $x(s)$, $y(s)$ on Γ, the above integral takes the form

$$(2) \qquad u(x(s), y(s)) = u(s) = \int_\Gamma \sigma(t) \frac{\partial \gamma(s, t)}{\partial \nu} \, dt;$$

here the expression

$$(3) \qquad K(s, t) = -\frac{1}{\pi} \frac{\partial \gamma(s, t)}{\partial \nu} = \frac{\cos \alpha}{\pi r} = \frac{1}{\pi} \frac{d\phi}{dt}$$

(cf. Figure 17 for the notations of the angles α and ϕ) converges as

FIGURE 17

$t \to s$ to the value $(1/2\pi)k(s)$, where $k(s)$ is the curvature of the boundary curve at the point s, and as such is twice continuously differentiable. In fact, it can be seen that the kernel $K(s, t)$ itself possesses continuous derivatives up to the second order.

We assume that $\sigma(t)$ is a continuously differentiable function of the arc length t. If P approaches the boundary point P_0 from the interior of the domain, then by the jump-discontinuity theorem[1] derived in §1, 4 the potential $u(P)$ converges to the boundary value $u_i(P_0) = u(P_0) - \pi\sigma(P_0)$, or, in virtue of (2), to the value

$$(4) \qquad u_i(s) = -\pi \int_\Gamma K(s, t)\sigma(t) \, dt - \pi\sigma(s).$$

It seems reasonable now to reverse our argument and determine the distribution $\sigma(s)$ from the integral equation

$$(5) \qquad \sigma(s) = -\int_\Gamma K(s, t)\sigma(t) \, dt - \frac{1}{\pi}f(s),$$

where $f(s) = u_i(P_0)$ represents the given boundary values. From article 1, we may assume without loss of generality that the boundary values $f(s)$ are continuously differentiable. If $\sigma(s)$ is a solution of this integral equation, then its potential

$$u = \int_\Gamma \sigma(t) \frac{\partial \gamma}{\partial \nu} \, dt$$

satisfies the potential equation in the interior of G. As a consequence of the differentiability of $f(s)$ and of $K(s, t)$, the distribution $\sigma(s)$ is also continuously differentiable. In other words, the conditions needed to apply the discontinuity theorem for the plane in §1, 4 are satisfied. Accordingly, upon approaching Γ, the potential u assumes the boundary values

$$-\pi \int_\Gamma K(s, t)\sigma(t) \, dt - \pi\sigma(s) = f(s).$$

Thus, we can solve the boundary value problem which we have posed if we can solve integral equation (5).

---

[1] Although it was assumed in §1, 4 that σ is twice continuously differentiable, it is evident upon inspecting the proof given there that it would be sufficient to assume that $\sigma(t)$ is merely once continuously differentiable for the particular result employed here (that is, for ascertaining the jump of the double-layer potential itself and not of its normal derivative).

Fredholm's theorems, proved in Vol. I, Ch. III, can now be applied to the integral equation (5). In terms of our problem, they state that to every continuously differentiable $f(s)$ there exists a uniquely determined continuously differentiable function $\sigma(s)$ which satisfies the integral equation (5), provided that the corresponding homogeneous integral equation

$$(6) \qquad \sigma(s) = -\int_\Gamma K(s, t)\sigma(t)\, dt$$

has only the trivial solution $\sigma \equiv 0$. In other words, the existence proof for our special domain is complete once we can show that the eigenvalue $\lambda = -1$ can never occur among the eigenvalues of the homogeneous equation

$$(7) \qquad \lambda v(s) = \int_\Gamma K(s, t)v(t)\, dt.$$

In the case of a convex boundary with continuous curvature and length L this is an immediate consequence of the relation

$$\int_\Gamma K(s, t)\, dt = \frac{1}{\pi}\int_0^L \frac{d\phi}{dt}\, dt = 1$$

and of the inequality

$$K(s, t) = \frac{\cos \alpha}{\pi r} \geq 0$$

which follows from the convexity. For if M is the maximum of $|v|$ on Γ, we have

$$|\lambda|\,|v| \leq M \int_\Gamma K(s, t)\, dt = M,$$

and hence for $|v| = M$

$$|\lambda|M \leq M.$$

Equality holds only when v is a constant. If $v \not\equiv 0$, then $M \neq 0$ and hence

$$|\lambda| \leq 1.$$

In fact $|\lambda| = 1$ only for constant v. But the eigenvalue corresponding to the eigenfunction $v = $ const. is $\lambda = +1$, so that we obtain the inequality $-1 < \lambda \leq +1$, which excludes the value $\lambda = -1$.

In the case of a boundary which is not convex, as we have already noted, the kernel $K(s, t)$ is, according to our assumptions, twice continuously differentiable. This implies the same property for all the eigenfunctions of the equation

$$\lambda v(s) = \int_\Gamma K(s, t)v(t) \, dt.$$

Now if $\sigma(s)$ is a solution of the equation (6) then, as a consequence of the jump relations of §1, 4, the potential

$$(8) \qquad\qquad u(x, y) = \int_\Gamma \sigma(t) \frac{\partial \gamma}{\partial \nu} \, dt$$

assumes the interior boundary values

$$u_i(s) = \int_\Gamma \sigma(t) \frac{\partial \gamma(s, t)}{\partial \nu} \, dt - \pi\sigma(s) = 0$$

on Γ, and by the uniqueness theorem vanishes identically in the interior of G. Thus the inner normal derivative of $u(x, y)$ on Γ also vanishes everywhere.

We now consider the potential (8) outside G. Since in our case the conditions for the discontinuity theorem for the plane (cf. §1, 4) are satisfied, we obtain on Γ the exterior boundary values

$$u_e(s) = \int_\Gamma \sigma(t) \frac{\partial \gamma(s, t)}{\partial \nu} \, dt + \pi\sigma = 2\pi\sigma(s)$$

and, since $\sigma(s)$ is twice continuously differentiable, the outer normal derivative $\partial u_e/\partial \nu = 0$. But at infinity ru is bounded, as easily seen from (1); hence u vanishes identically also outside G (this follows with the aid of the lemma on p. 294) and, in particular, assumes the exterior boundary values

$$u_e(s) = 2\pi\sigma(s) = 0$$

on Γ. Every solution of equation (6) thus vanishes identically: $\lambda = -1$ cannot be an eigenvalue of the homogeneous integral equation

$$\lambda v(s) = \int_\Gamma K(s, t)v(t) \, dt.$$

Thus we have carried out the existence proof for the solution of the boundary value problem for our special domain G.

A similar argument holds in space, although in order to be able to apply Fredholm's theory we must replace the kernel $\partial(1/r)/\partial\nu$, which is not square integrable, by the square integrable iterated kernel.

It is to be noted, however, that in spite of its elegance the method of integral equations is inferior to the previously developed procedure, since the occurrence of even an ordinary corner leads to singularities of the kernel K, so that the immediate application of Fredholm's theory is ruled out.

4. Remarks on Boundary Values. In the plane, the methods developed in articles 1 and 2 furnish the solution of the boundary value problem for every plane domain bounded by an arbitrary Jordan curve. In three or more dimensions, however, matters are more complicated since there exist domains for which the boundary value problem is no longer solvable in the strong sense; in other words, for prescribed continuous boundary values we cannot always expect that the boundary values are assumed at all the boundary points. This is illustrated by the following counter-example due to Lebesgue.

We first calculate the potential of a mass distribution concentrated on the segment of the x-axis between 0 and 1 with the linear density $\tau(x) = x$:

$$u(x, y, z) = \int_0^1 \frac{\xi \, d\xi}{\sqrt{(\xi - x)^2 + \rho^2}} = A(x, \rho) - 2x \log \rho,$$

where $\rho^2 = y^2 + z^2$ and

$$A(x, \rho) = \sqrt{(1 - x)^2 + \rho^2} - \sqrt{x^2 + \rho^2}$$
$$+ x \log | (1 - x + \sqrt{(1 - x)^2 + \rho^2})(x + \sqrt{x^2 + \rho^2}) |.$$

As the origin is approached through positive values of x, $A(x, \rho)$ tends to 1; however, the limit of the expression $-2x \log \rho$ depends essentially on the path of the approach. For example, if we approach the origin on the surface $\rho = |x|^n$, $-2x \log \rho$ converges to zero for every n, and thus u tends to the value 1. On the other hand, if we take the surface $\rho = e^{-c/2x}$ ($c > 0$, $x > 0$) which has an "infinitely sharp" peak at the origin, $-x \log \rho$ converges to c; hence the potential u tends to $1 + c$. This means that all the equipotential surfaces $u = 1 + c$ ($c > 0$) meet at the origin; moreover, all derivatives of the curve $\rho = f(x)$, from which these surfaces are formed by

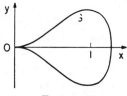

FIGURE 18

rotation about the x-axis, vanish at the origin. Such a surface $u = 1 + c$ is indicated in Figure 18.

If we now take as fundamental domain the region G bounded by an equipotential surface $u = 1 + c$ $(c > 0)$, and solve the exterior boundary value problem for G and the boundary values $u = 1 + c$, we find that the solution is furnished by the function $u(x, y, z)$ given above. But from our considerations it follows that, if we approach the origin in a suitable way, this solution can converge to any value between 1 and $1 + c$.

By inversion with respect to the sphere

$$(x - \tfrac{1}{2})^2 + y^2 + z^2 = \tfrac{1}{4}$$

and a suitable translation we can obtain from this example a corresponding example for an interior problem. G is mapped into a domain G' of ξ, η, ζ-space which has an infinitely sharp interior peak at the point $\xi = -\tfrac{1}{2}$, $\eta = 0$, $\zeta = 0$ (cf. Figure 19). The boundary values $1 + c$ go over into the boundary values

$$v = \frac{1 + c}{2r}, \qquad (r = \sqrt{\xi^2 + \eta^2 + \zeta^2}),$$

continuous on Γ'; the solution of the interior boundary value problem

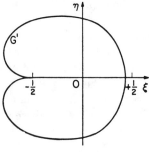

FIGURE 19

for these boundary values and the domain G' is given by the potential function

$$v(\xi, \eta, \zeta) = \frac{1}{2r}\, u\left(\frac{\xi}{4r^2} + \frac{1}{2}, \frac{\eta}{4r^2}, \frac{\zeta}{r^2}\right),$$

regular in G'. Approaching the point $\xi = -\frac{1}{2}$, $\eta = 0$, $\zeta = 0$ in an appropriate way, any value between 1 and $1 + c$ may again occur as limit value for v.

In Volume III we shall replace the requirement of assuming exact boundary values at every boundary point in three and more dimensions by the weaker requirement of *assuming boundary values in the mean* which is sufficient in itself to determine the solution uniquely. Only in the special case of two dimensions does this requirement assure the assumption of the boundary values at every boundary point. However, exact boundary values are assumed and derivatives at the boundary remain continuous if the boundary and boundary values are locally sufficiently smooth. (Compare the remarks on page 346 for more general elliptic equations.)

4a. Capacity[1] and Assumption of Boundary Values. N. Wiener [2] has observed that the question of assumption of boundary values is connected with the concept of capacity, a concept of independent significance.

To define capácity of a closed surface Γ in n dimensions (we may think of $n = 3$) we consider Γ as the inner boundary of a shell S with the outer boundary Γ^* and suppose the boundary value problem for S solved by a function u with the boundary values $u = 1$ on Γ and $u = 0$ on Γ^*. Whether or not derivatives of u on the boundary exist, the surface integral extended over any smooth surface Γ' inside the shell and homologous to Γ and Γ^*

$$\int_{\Gamma'} \frac{\partial u}{\partial n}\, ds = \kappa$$

exists; it is independent of the special choice of Γ', as seen easily by applying Green's formula to an internal shell bounded by two surfaces Γ'. This number κ then defines the "capacity" of S with respect to Γ, Γ^*; if Γ^* is the point at infinity then we call κ simply the capacity of Γ. This concept remains meaningful even for solutions of the boundary value problem in a generalized sense as indicated above; Γ may be in particular a finite piece of a surface which is made into a closed manifold by counting separately both sides of the surface.

[1] A modern exposition can be found in M. Brelot [1].

Physically κ represents the total electrical charge to be applied to a conductor Γ in order to generate the constant potential 1 on Γ while the potential at Γ^* is kept at the level zero.

Wiener's theorem characterizes those points P at the boundary of a domain G in which possibly the prescribed values are not always attained if the boundary value problem is solved in a generalized sense.[1]

Such "exceptional points" correspond in the example of article 4 to the end points of a sharp spine, or even simpler to points on a line pointing into the domain and artificially defined as part of the boundary.

We consider on a small sphere of radius λ about the boundary point P that part, H_λ, which is in G and bounds with a part of Γ a small subdomain of G containing P as boundary point. The capacity of H_λ in the sense defined above is called κ_λ. Then Wiener's theorem states: *The boundary point P is exceptional if and only if the capacity κ_λ tends sufficiently fast to zero with λ, precisely if the series*

$$\sum_{\nu=1}^{\infty} \kappa_\lambda^2 \qquad \text{for } \lambda = \left(\frac{1}{2}\right)^\nu$$

converges.

It should be recalled that §4, 1 states a sufficient condition for the regularity of a boundary point.

5. Perron's Method of Subharmonic Functions. In the following article we insert an account of an elegant method by Perron.[2] The method deemphasizes the constructive element and concentrates on the mere existence proof of the solution. It is connected with significant general concepts and has, in the form presented here, the advantage of being capable of generalization to other second order elliptic differential equations. (See remarks in §7.)

To find a function harmonic in a fixed bounded domain G, con-

---

[1] Such a generalized formulation of the boundary value problem could be given by considering G as a monotone limit of domains G_n with smooth boundaries consisting of regular points, such that G_{n+1} contains G_n and that any closed subdomain of G is contained in all but a finite number of the domains G_n. It then can be shown that the corresponding solutions u of the respective boundary value problems converge to a harmonic function u in G and that u is independent of the specific mode of approximation of G by G_n. Naturally u is then interpreted as the solution of the boundary value problem for G. See N. Wiener [1].

[2] See O. Perron [1].

tinuous in its closure $G + \Gamma$, and assuming prescribed continuous values ϕ on the boundary Γ Perron utilizes the notions of subharmonic and superharmonic functions, which are generalizations to higher dimensions of convex and concave functions of one variable, just as harmonic functions are generalizations of linear functions of one variable (i.e., solutions of the one-dimensional Laplace equation). We proceed to define these and some other basic concepts.

Let v be a continuous function in a domain D, and C a closed sphere (sometimes called a ball) in D. (C will represent a closed sphere in D throughout this article.) We shall denote by $M_C[v]$ the uniquely defined continuous function which is harmonic in the interior of C and equal to v in the remainder of D. We say that the function v is *subharmonic* (*superharmonic*) in D if v satisfies the inequality

$$v \leq M_C[v] \qquad (v \geq M_C[v])$$

for every sphere C in D. For example, any function w satisfying $\Delta w \geq 0$ is subharmonic, as follows from the maximum principle of §6, 4.

Subharmonic (superharmonic) functions have the following obvious properties: First, $v \geq 0$ implies $M_C[v] \geq 0$ (for the minimum of $M_C[v]$ in C occurs on the boundary of C); hence if $v \geq w$ then $M_C[v] - M_C[w] = M_C[v - w] \geq 0$. Secondly, if v is subharmonic (superharmonic) then $-v$ is superharmonic (subharmonic); and finally, any linear combination with non-negative constant coefficients of subharmonic (superharmonic) functions is also subharmonic (superharmonic).

We shall also need some additional properties which will be formulated only for subharmonic functions; since the negative of a superharmonic function is subharmonic, analogous properties hold for superharmonic functions.

1) The maximum principle for subharmonic functions: *If v is subharmonic in a domain D and has a maximum point in the interior of D then $v = $ const.*; it follows that if v is subharmonic in a bounded domain and continuous in its closure, then v assumes its maximum on the boundary.

To prove this we consider $w = M_C[v]$ where C is a sphere in D with center at a maximum point P. Denoting by V the maximum of v on the boundary of C, we have $V \leq v(P) \leq w(P)$. Thus w is harmonic in C with an interior maximum point, implying (as we have shown in §1, 3, p. 255) that $w = \text{const.}$ in C, so that $v = w = v(P)$ on the boundary of C. Since this argument holds also for all smaller spheres centered at P, we have $v = v(P)$ in all of C. This shows that the set of maximum points in D is open. By the continuity of v, this set is also closed in D and, since D is connected, it is identical with D.

As a consequence of the reasoning in 1) we have:

$1'$) *If v is continuous and subharmonic in a neighborhood of every point of a domain then v is subharmonic in the whole domain,* i.e., subharmonicity is a local property. To see this we first observe that, by selecting sufficiently small spheres C, we can immediately extend the proof of the maximum principle given in 1) to apply to a function v which is merely locally subharmonic. Now let C be any sphere in the domain, and set $w = M_C[v]$. Then $v - w$ is locally subharmonic in C and, by our remark, satisfies the maximum principle. Since $v - w = 0$ on the boundary of C, it follows that $v \leq w$ in C, showing that v is subharmonic.

2) *If v_1, v_2, $\cdots$, v_m are subharmonic in D, then the same holds for $v = max(v_1, v_2, \cdots, v_m)$.* For every sphere C in D we observe, by a previous remark, that

$$v_i \leq M_C[v_i] \leq M_C[v] \qquad (i = 1, 2, \cdots, m),$$

so that $v \leq M_C[v]$.

3) *If v is subharmonic in D, then $w = M_C[v]$ is also subharmonic in D.* Let C' be any closed sphere in D. We must show that

$$w = M_C[v] \leq M_{C'}[w].$$

This clearly holds if C' lies either completely inside or completely outside of C. Thus, we need only consider the case where C' is partly inside and partly outside of C. If P is a point in $C' - C$ (the shaded region of Figure 20) then at P

$$w = M_C[v] = v \leq M_{C'}[v] \leq M_{C'}[w]$$

since $v \leq w$. Suppose now that P belongs to the intersection of C and C'. In this region w and $M_{C'}[w]$ are harmonic and, on the boundary, $w - M_{C'}[w] \leq 0$ according to the result just obtained for the shaded region. By the maximum principle for harmonic functions the inequality holds throughout the region.

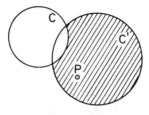

FIGURE 20

The boundary value problem will be solved with the aid of special subharmonic and superharmonic functions.

A function v continuous in $G + \Gamma$ is a *subfunction* (*superfunction*) if it is subharmonic (superharmonic) in G and if, on the boundary Γ of G, $v \leq \phi$ ($v \geq \phi$) where ϕ are the given boundary values. Clearly, the constant functions min ϕ and max ϕ are subfunctions and superfunctions, respectively.

The properties of subharmonic functions may be extended to subfunctions; we formulate the following

Lemma: Let F denote the class of all subfunctions. Then

a) *the functions in F are uniformly bounded from above (by* max ϕ*);*

b) *if v_1, v_2, $\cdots$, v_n belong to F then the function* max $(v_1, v_2, \cdots, v_n)$ *also belongs to F;*

c) *if v belongs to F and C is any closed sphere in G, then $M_C[v]$ also belongs to F.*

We further observe that, by the maximum principle, no subfunction can exceed any superfunction at any point of G.

The method of solving the boundary value problem is motivated by the following considerations. Suppose w is the solution. Then if v is any subfunction we have $v \leq w$ in all of $G + \Gamma$. This follows by noting that $v - w$ is subharmonic in G and nonpositive on Γ, and that the maximum principle holds for subharmonic functions. Moreover, w is itself a subfunction. Hence, if we define a function u which at any point in $G + \Gamma$ is equal to the least upper bound of the values of all subfunctions at this point, then $w = u$. Perron's method consists in showing that the function u so defined (which always exists) is harmonic in G, continuous, and equal to ϕ at "regular" boundary points (to be defined later).

Theorem: *The function* u *defined in* $G + \Gamma$ *by*

$$u(P) = \underset{v \in F}{\text{l.u.b. }} v(P)$$

is harmonic in G.

Proof:[1] Let K be a closed sphere in G and K_1 a sphere concentric to K with half the radius. We shall prove that u is harmonic in K_1.

We select a subset F^* of F whose functions are uniformly bounded from below and which has properties b) and c). (Every subset of F has property a).) For example, we can construct the subset F^* by choosing a fixed function v_1 in F as a lower bound and then prescribing that F^* should contain the function max (v, v_1) for every function v in F: Clearly

$$u(P) = \underset{v \in F^*}{\text{l.u.b. }} v(P).$$

Next, let Q_j be a sequence of points dense in K and let $v_{j,k}$ be functions in F^* such that

$$0 \leq u(Q_j) - v_{j,k}(Q_j) \leqq \frac{1}{k} \qquad\qquad (j, k = 1, 2, \cdots).$$

Moreover, we define for all points in $G + \Gamma$

$$v_k = M_k [\max(v_{1,k}, v_{2,k}, \cdots, v_{k,k})].$$

The v_k belong to F^* and $v_k \to u$ at all points Q_j. To show that u is harmonic in K_1, it suffices, by Harnack's theorem (cf. §2, 3) to show that the v_k converge to u uniformly in K_1. We note first: The uniform boundedness of the absolute value of the functions v_k, which are harmonic in the interior of K, implies by the theorem on p. 274 the uniform boundedness of their first derivatives in K_1 and hence the equicontinuity of the v_k in K_1. By the lemma stated below $v_k \to u$ uniformly in K_1.

Lemma: *If a sequence of equicontinuous functions defined on a bounded set* A *converges in a dense subset of* A, *then it converges uniformly in all of* A. To prove this we proceed as in the last step of the proof of Arzela's theorem (see Vol. I, p. 59).

Thus we have established that u is harmonic in G. We now want to find conditions for the boundary Γ of G which are sufficient to

---

[1] This version of the proof was suggested by Walter Littman.

insure that u be continuous in $G + \Gamma$ and assume the prescribed boundary values.

To this end we define for any boundary point Q of G a *barrier function* w_Q at Q as a function which is superharmonic in G, continuous in $G + \Gamma$, and positive in $G + \Gamma$ except at Q where it vanishes. A boundary point of G possessing such a barrier function is called *regular*.

We note that a boundary point Q is regular if there exists a "local barrier function" at Q, i.e., a function W_Q such that W_Q is continuous in the intersection $(G + \Gamma) \cap S$ of $G + \Gamma$ and some closed sphere S with center at Q, superharmonic in the interior of $G \cap S$, and positive in $(G + \Gamma) \cap S$ except at Q where it vanishes. To see this, let S_1 be a sphere concentric to S with half the radius; then W_Q attains a positive minimum m in $(G + \Gamma) \cap (S - S_1)$. Define

$$w_Q = \begin{cases} \min \{m, W_Q\} & \text{in } S \cap (G + \Gamma) \\ m & \text{in } (G + \Gamma) - S. \end{cases}$$

Clearly w_Q has all the properties of a barrier function except perhaps superharmonicity. However, it is easy to see that w_Q is superharmonic in two overlapping domains since it equals a constant m in $G - \bar{S}_1$ (where $\bar{S}_1$ is the closure of S_1) and is the minimum of two superharmonic functions in $S \cap G$. Thus w_Q is locally superharmonic; by 1') it follows that w_Q is superharmonic.

We shall prove that *at every regular point* Q, u *is continuous and* $u(Q) = \phi(Q)$. Since ϕ is continuous, one can find for any $\epsilon > 0$ a sufficiently large positive constant k such that the functions

$$v(P) = \phi(Q) - \epsilon - kw_Q(P)$$

and

$$V(P) = \phi(Q) + \epsilon + kw_Q(P)$$

are subfunctions and superfunctions, respectively. Since no subfunction can be greater than any superfunction, we have

$$\phi(Q) - \epsilon - kw_Q(P) \leq u(P) \leq \phi(Q) + \epsilon + kw_Q(P)$$

for all P in $G + \Gamma$, i.e.,

$$| u(P) - \phi(Q) | \leq \epsilon + kw_Q(P).$$

As P approaches Q, $w_Q(P) \to 0$; thus we have

$$| u(P) - \phi(Q) | < 2\epsilon$$

for P sufficiently close to Q.

To complete the solution of the boundary value problem in G we derive some conditions sufficient to insure the existence of appropriate barrier functions. Consider first the case of $n > 2$ dimensions. To any boundary point Q for which there exists a closed sphere S whose only common point with $G + \Gamma$ is Q, we may take as barrier function the harmonic function

$$w_Q(P) = \frac{1}{R^{n-2}} - \frac{1}{r^{n-2}},$$

where R is the radius of S and r represents the distance of P to the center of S. This function clearly satisfies all conditions.

For $n = 2$ we can construct a barrier function at each boundary point Q which is the initial point of an arc A lying, except for Q, outside of $G + \Gamma$, and having no double points. Let C be a circle with radius less than 1 centered at Q and so small that its circumference intersects A. Starting at Q and considering the points in A ordered by a parametrization, let z_0 be the first point of A to be on the circumference of C, and A' the part of A which consists of all points "preceding" z_0 in A. Then $C - A'$ is simply connected and the function

$$w_Q(P) = -\operatorname{Re} \frac{1}{\log z} = -\frac{\log \rho}{|\log \rho|^2 + \theta^2}$$

(ρ, θ are the polar coordinates of P about Q), which is single valued in $(G + \Gamma) \cap C$, is a local barrier there.

Thus *the Dirichlet problem can be solved for $n = 2$, if G has the property that each boundary point Q is the initial point of a non-selfintersecting arc lying, except for Q, outside of $G + \Gamma$* (in particular, the interior of every Jordan curve is such a domain G^1); *for $n > 2$ the problem is certainly solvable, if to every boundary point Q of G there exists a closed sphere having only the point Q in common with $G + \Gamma$.*

§5. The Reduced Wave Equation. Scattering

1. Background. We consider the "reduced wave equation" (see also Ch. III, §3, 2)

[1] See M. H. A. Newman [1], Ch. VI, §4.

(1) $$L[U] = \Delta U + \omega^2 U = 0$$

for a function $U(x_1, x_2, x_3)$ (we confine ourselves to three space variables x). (1) is of elliptic type and can be treated in the same way as Laplace's equation. It originates from the wave equation

(2) $$\Delta u - u_{tt} = 0$$

when we assume the wave $u(x, t)$ to be simple harmonic with frequency ω in the time t, i.e.,

$$u_{tt} = -\omega^2 u.$$

Then u is of the form

$$u = V_1(x) \cos \omega t + V_2(x) \sin \omega t,$$

where $L[V_1] = L[V_2] = 0$. Introducing

$$U = V_1 + iV_2,$$

we obtain the representation

$$u = \operatorname{Re}[U(x)e^{-i\omega t}] = \tfrac{1}{2}(Ue^{-i\omega t} + \bar{U}e^{i\omega t})$$

for the real wave u in terms of a complex "standing wave" solution, where U is a general complex valued solution of (1). We shall also consider the more general equations

(1a) $$\Delta U + \omega^2 U + gU = 0,$$

(2a) $$\Delta u - u_{tt} + gu = 0,$$

assuming that the coefficient g does not depend on t and vanishes outside a region R with smooth boundary B, so that outside R the equations (1) or (2) hold (see Figure 21).

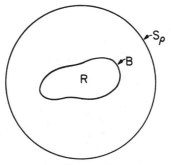

FIGURE 21

Equation (1) occurred in Volume I as the equation of vibrations. The present section concerns a quite different physical phenomenon governed by the same equation, namely *scattering* of progressing waves. The mathematical problem of scattering will be discussed in article 3 after some preparations in articles 1 and 2. For progressing waves the use of complex notation is somewhat essential, since they cannot be represented by a single term $U \cos \omega t$ or $U \sin \omega t$ with real U.

We introduce the distance $r = |x - x'|$ between the point x and a parameter point x' and recall from Chapter III the "fundamental solutions" (in slightly different notation)

$$K(x, x') = K(r) = -\frac{e^{i\omega r}}{4\pi r} \quad \text{and} \quad \bar{K}(r) = -\frac{e^{-i\omega r}}{4\pi r}.$$

K corresponds to an outgoing spherical wave and $\bar{K}$ to a converging one:

$$u = -\frac{e^{-i\omega(t-r)}}{4\pi r}, \quad u = -\frac{e^{-i\omega(t+r)}}{4\pi r}$$

with spherical phase surfaces about the point x' moving with speed 1 out toward infinity or in toward the center.

The main fact concerning the fundamental solutions is: For any differentiable function $f(x')$ the integrals

$$J(x) = \iiint f(x') K(r) \, dx' \quad \text{or} \quad \iiint f(x') \bar{K}(r) \, dx'$$

$$(dx' = dx'_1 \, dx'_2 \, dx'_3)$$

over a domain x' containing x satisfies the differential equation

$$L[J] = f(x).$$

The proof is the same as in the case $\omega = 0$, that is, for the harmonic equation, and may be omitted.

For large r the fundamental solutions K and $\bar{K}$ satisfy the relations

$$(4') \quad \left| \frac{\partial K}{\partial r} - i\omega K \right| = O\left(\frac{1}{r^2}\right), \quad \left| \frac{\partial \bar{K}}{\partial r} + i\omega \bar{K} \right| = O\left(\frac{1}{r^2}\right),$$

where as usual the symbol $O\left(\frac{1}{r^2}\right)$ means a function of order not exceeding $\frac{1}{r^2}$.

According to A. Sommerfeld, conditions of this type characterize outward and inward radiation.

2. Sommerfeld's Radiation Condition. Every complex-valued solution U of (1) which is regular outside a surface B can be decomposed in a unique manner into a sum

$$U = U_1 + U_2,$$

where U_1 is a regular solution of (1) in the whole space and where U_2 satisfies *Sommerfeld's condition for outward radiation* in the form[1]

$$(4) \qquad \lim_{\rho \to \infty} \iint_{r=|x-x'|=\rho} \left| \frac{\partial U_2}{\partial r} - i\omega U_2 \right|^2 dS_\rho = 0;$$

dS_ρ denotes the surface element of a large sphere S_ρ of radius ρ about a fixed point x'. (4) will be motivated presently. Correspondingly, of course, if in (4) i is replaced by $-i$ an *inward radiation* is characterized.

For the proof we consider a large sphere S_ρ of radius ρ about the point x' which contains B. For points x in the ring D_ρ between B and S_ρ the solution U can be represented in quite the same way as a harmonic function (see §1): We have

$$U(x) = \iint_{S_\rho} \left(U \frac{\partial K}{\partial n} - K \frac{\partial U}{\partial n} \right) dS_\rho$$

$$- \iint_{B} \left(U \frac{\partial K}{\partial n} - K \frac{\partial U}{\partial n} \right) dS = U_1 + U_2,$$

where now

$$K = -\frac{1}{4\pi R} e^{i\omega R}, \qquad R = |x - x''|,$$

x'' running over the surfaces S_ρ and B respectively and $\partial/\partial n$ denotes differentiation with respect to the outward normal in x''-space. Outside of B both integrands are solutions of (1) as functions of x, since K and $\partial K/\partial n$ are solutions. Obviously, as in the case of harmonic functions, U_1 is regular everywhere inside of S_ρ. Moreover U_1 does not depend on the radius ρ since the solution U under consideration and the component U_2 do not depend on ρ. Therefore U_1 is indeed regular in the whole space.

---

[1] This form, due to W. Magnus, is less demanding than the preceding formulation (4'); yet it suffices for the characterization of outward radiation.

To show that U_2 satisfies Sommerfeld's radiation condition (4) we observe that for x' fixed and x'' on B and for $x - x' = r\eta$, $x'' - x' = \sigma\zeta$, $|\eta| = |\zeta| = 1$

$$R = |x - x''| = \sqrt{r^2 - 2r\sigma\eta\cdot\zeta + \sigma^2} = r - \sigma\eta\cdot\zeta + O\left(\frac{1}{r}\right)$$

$$K = -\frac{e^{ir\omega}}{4\pi r}\, e^{-i\omega\eta\cdot(x''-x')} + O\left(\frac{1}{r^2}\right)$$

$$\frac{\partial K}{\partial n} = -i\omega K\eta\cdot\frac{\partial x''}{\partial n} + O\left(\frac{1}{r^2}\right).$$

As before the symbol $O(f(\lambda))$ means a quantity of order not greater than $f(\lambda)$ for large positive λ, that is, a quantity absolutely below $cf(\lambda)$ with constant c. Consequently

$$(5) \qquad U_2 = -\frac{e^{ir\omega}}{4\pi r}\,\psi(\eta) + O\left(\frac{1}{r^2}\right),$$

where the "form factor"

$$\psi(\eta) = \iint_B e^{-i\omega\eta\cdot(x''-x')}\left(\frac{\partial U}{\partial n} + i\omega\eta\cdot\frac{\partial x''}{\partial n}\right) dS$$

is a regular analytic function of the unit vector η. Analogous expressions, identical with those obtained by formal differentiation of (5) with respect to r, hold for $\partial K/\partial r$. Therefore Sommerfeld's condition is satisfied by U_2, even in the stronger form

$$\frac{\partial U_2}{\partial r} - i\omega U_2 = O\left(\frac{1}{r^2}\right)$$

uniformly in η.

To motivate Sommerfeld's condition (4) as characterizing outward radiation we show: Under the condition (4) a positive amount of energy flows in a period ω through a large sphere S_ρ in the outward direction.

The energy E contained at the time t in a domain $\sum$ with the boundary S is defined by

$$E = \frac{1}{2}\iiint_\Sigma (u_t^2 + u_{x_1}^2 + u_{x_2}^2 + u_{x_3}^2)\, dx_1\, dx_2\, dx_3.$$

Using Green's formula and (2) one obtains easily

$$\frac{dE}{dt} = \iint_S u_t \frac{\partial u}{\partial n} \, dS.$$

We take $\sum$ to be the ring between a large sphere S_0 and a larger sphere S_ρ ; then the boundary integral

$$F(t) = \iint_{S_\rho} u_t \frac{\partial u}{\partial n} \, dS$$

is interpreted as the *energy flux* through the outer boundary S_ρ , and

$$\int_t^{t+2\pi/\omega} F(t) \, dt = \Gamma$$ as the total flux through S_ρ during a period.
Writing $u = \frac{1}{2}(Ue^{-i\omega t} + \bar{U}e^{i\omega t})$ we find easily

$$\Gamma = \frac{\pi i}{2} \iint_{S_\rho} \left(\bar{U} \frac{\partial U}{\partial n} - U \frac{\partial \bar{U}}{\partial n} \right) dS.$$

This flux is independent of ρ, as follows immediately if we apply Green's formula for the ring between S_ρ and $S_{\rho'}$ to the two solutions U and $\bar{U}$ of the reduced wave equation.

Now we assume for U the Sommerfeld condition (4) expressed in the form

$$\iint_{S_\rho} (U_n \bar{U}_n + \omega^2 U\bar{U} + i\omega (\bar{U}U_n - U\bar{U}_n)) \, dS \to 0$$

or

$$\frac{2\omega}{\pi} \Gamma + \iint_{S_\rho} (|U_n|^2 + \omega^2 |U|^2) \, dS \to 0 \qquad \text{for} \quad \rho \to \infty,$$

where U_n means $\partial U/\partial n$. Since Γ does not depend on ρ we conclude that $\Gamma \leq 0$; that is, during a period energy is lost, or at least not gained, by outward flux through S_ρ. Thus indeed (4) indicates a phenomenon of outward radiation.

Finally we prove: *The decomposition of U into an everywhere regular solution and a solution satisfying Sommerfeld's condition is unique.* It is sufficient to prove that an everywhere regular solution U satisfying (4) vanishes identically. Let U be such a solution. Then $\Gamma = 0$, since U and $\bar{U}$ are regular solutions inside S_ρ. Hence

$$\iint_{S_\rho} (|U_n|^2 + \omega^2 |U|^2) \, dS \to 0 \qquad \text{for} \quad \rho \to \infty.$$

We have then for any x inside the sphere S_ρ about the point x'

$$|U(x)| = \left| \iint_{S_\rho} (UK_n - KU_n)\, dS \right|$$

$$\leq \left(\iint_{S_\rho} (|U|^2 + |U_n|^2)\, ds \right)^{\frac{1}{2}} \left(\iint_{S_\rho} |K_n|^2 + |K|^2)\, dS \right)^{\frac{1}{2}} \to 0 \text{ for } \rho \to \infty$$

since K and K_n are of order $1/\rho$. Hence $U(x) = 0$ as stated.

3. Scattering. The phenomenon of scattering may be described as follows: An "incoming" wave, which means an everywhere regular solution $U_1(x) = \chi(x)$ of (1), is given. (E.g., the incoming wave may be simply a plane wave $e^{i\omega(\alpha x)}$ or may be a wave bundle originating from a plane wave by integration with respect to the unit vector α within a certain solid angle.) This incoming wave U_1 is modified by an obstacle which produces a *scattered wave* U_2 such that a wave $U = U_1 + U_2$ results. The obstacle may be represented by either of two conditions: (a) On the boundary B a condition such as $U = 0$ or $\partial U/\partial n = 0$ is imposed and the solution is considered merely on and outside B; or (b) the obstacle is represented by the modifying term g in (2a) or (1a), a term which vanishes outside B. In the case (b) the solution is considered in the whole x-space.

In both cases the solution $U = U_1 + U_2$ differs from the originally given incoming wave $U_1 = \chi(x)$, assumed known, by an additional superimposed wave U_2 which accounts for the effect of the obstacle. Moreover, we stipulate: The compensating wave U_2 should be an outward scattered wave, that is, U_2 should satisfy Sommerfeld's condition (4). To find U_2 for a given incoming wave $U_1 = \chi$, is the mathematical problem of scattering.

We indicate the solution briefly, based on the representation of solutions of (1) or (1a) as described in article 1.

In case (a) we have to determine U_2 outside B from the boundary condition if $U_1 = \chi$ is given on B.

In case (b) we have to determine U_2 from (1a).

The problem reduces immediately to a Fredholm integral equation. In case (a) this integral equation concerns values of $U = U_2$ on the boundary B and the normal derivative $U_n = \partial U/\partial n$ on the boundary B. The integral equation is implied in the formula valid for x *on B*

$$\tfrac{1}{2} U_2 = \iint_B (U_2 K_n - KU_{2_n})\, dS,$$

where either the quantity

$$U_2 = U - U_1 = U - \chi$$

or the normal derivative

$$U_{2_n} = U_n - \chi_n$$

is given on B, and the other quantity is to be determined on B.

In case (b) the integral equation for U_2 originates even more directly:

$$U_2(x) = \chi(x) - \iint_R g(x')K(x' - x)U(x')\, dx',$$

or

$$U_2(x) = G(x) + \iint_R K^*(x, x')\, U(x')\, dx',$$

with the kernel K^*

$$K^*(x, x') = g(x')K(x - x'),$$

and with the known term

$$G(x) = \chi(x) + \iint_R \chi(x')\, g(x')\, K(x - x')\, dx'.$$

According to the theory in Vol. I, p. 205, this Fredholm integral equation for the scattered wave U_2 can be reduced to variational problems such as: To make stationary the functional of ϕ:

$$J(\phi) = \iint g\phi^2\, dx' + \iint g(x)\, g(x')\, K(x - x')\, \phi(x)\, \phi(x')\, dx\, dx'$$

$$- 2 \iint g(x)\, g(x')\, \phi(x)\, dx,$$

and equivalent problems. Such reductions have proved very useful for solving scattering problems numerically.[1]

Finally, we realize that we could just as well have treated a problem somewhat complementary to that of scattering of an incoming wave

---

[1] One could easily characterize the form factor $\tilde{\psi}(\eta)$ (see article 2) by a similar integral equation or variational problem.

For particular reference to these variational problems and numerous practical applications, see J. Schwinger [1] and [2] and B. Lippman and J. Schwinger [1].

by outward radiation: Let us suppose that a converging wave U_2' is given which satisfies Sommerfeld's condition for inward radiation. Then the problem is to find an everywhere regular outgoing wave U_1' resulting from an obstacle R, or from a modifying term q. Mathematically that means one should decompose a given solution U into a sum $U = U_2' + U_1'$, where the two components satisfy the conditions as stipulated.

Since a given solution U of (1) can be decomposed in either way, we are led to consider a converging wave U_2' and ask for the diverging wave U_2' into which it will be scattered.

§6. Boundary Value Problems for More General Elliptic Differential Equations. Uniqueness of the Solutions

Although the harmonic equation $\Delta u = 0$ is typical for elliptic differential equations, a more general theory, even with the restriction to equations of second order, would require much new discussion which would transcend the framework of this book; moreover, this theory is not yet completely developed. Therefore, we call attention to the literature,[1] and limit ourselves here to a brief presentation of some main points concerning the boundary value problem and the construction of particular solutions. In Volume III we shall resume the theory of linear elliptic problems from the more general point of view of the variational calculus. We now treat the uniqueness problem: Under which conditions is the solution of the boundary value problem uniquely determined?

1. Linear Differential Equations. Let $L[u] = 0$ be the elliptic differential equation

$$(1) \quad L[u] \equiv \sum_{i,k=1}^{n} a_{ik} u_{ik} + \sum_{i=1}^{n} b_i u_i + cu \equiv M[u] + cu = 0,$$

where $u_{ik} = \partial^2 u/\partial x_i\, \partial x_k$, $u_i = \partial u/\partial x_i$. Let the coefficients $a_{ik} = a_{ki}$, b_i, c be continuous functions of the variables x_1, x_2, $\cdots$, x_n in a bounded domain G of the n-dimensional space R_n. The quadratic form

$$\sum_{i,k=1}^{n} a_{ik}(x_1, x_2, \cdots, x_n)\xi_i \xi_k$$

---

[1] See the references in the footnote on page 240. In particular, the book by Miranda contains a detailed and extensive bibliography.

is supposed to be positive definite in the parameters ξ at all points x of G. Then we can state the following

UNIQUENESS THEOREM: *Under the assumption $c \leq 0$, i.e., for the equation*

$$(1') \quad L[u] \equiv M[u] + cu = \sum_{i,k=1}^{n} a_{ik} u_{ik} + \sum_{i=1}^{n} b_i u_i + cu = 0, \quad c \leq 0,$$

there exists at most one solution which has continuous derivatives up to the second order in G, is continuous in $G + \Gamma$, and assumes prescribed boundary values on the boundary of G.[1] In other words, a solution of equation $(1')$ which vanishes on Γ vanishes identically in G.

We first prove: *If a twice continuously differentiable function u has a maximum at an interior point P, then at this point $M[u] \leq 0$. It follows that if, in addition, $c(P) < 0$ and $u(P) > 0$, then at P, $L[u] < 0$.* (For a stronger formulation of this maximum principle, see article 4.)

For if u has a maximum at P then all the first derivatives u_i vanish there and the matrix of the second derivatives

$$u_{ik}(P) = b_{ik}$$

is the matrix of a nowhere positive quadratic form. Thus at the point P, $M[u]$ equals $S = \sum_{i,k=1}^{n} a_{ik} b_{ik}$, which is the trace of the product of the two matrices (a_{ik}) and (b_{ik}). This trace cannot be positive. For if we transform (a_{ik}) by an orthogonal transformation into the diagonal matrix (p_i) with $p_i > 0$, and (b_{ik}) by the same transformation into the matrix (β_{ik}), the value S remains invariant and we have

$$S = \sum_{i=1}^{n} p_i \beta_{ii}.$$

Since along with (b_{ik}) the matrix (β_{ik}) is also the matrix of a nowhere-positive quadratic form, it follows that $\beta_{ii} \leq 0$ and hence $S \leq 0$. This proves our assertion.

Suppose now that u is a solution of $L[u] = 0$ which vanishes on Γ, and that $c < 0$. Applying our result to u and to $-u$ we find that u

---

[1] Without the condition $c \leq 0$ we certainly cannot expect uniqueness in general as we see at once from the differential equation $\Delta u + cu = 0$ in case c is one of the positive eigenvalues corresponding to the boundary condition $u = 0$.

has no positive maximum or negative minimum in G; this, together with $u = 0$ on Γ, implies that u *vanishes identically in G.*

The case $c \leq 0$ may be reduced to the case $c < 0$ by the following device due to Picard. We set

$$u = z(x)v(x)$$

and obtain for v a differential equation of the form

$$(2) \quad z \sum_{i,k=1}^{n} a_{ik} v_{ik} + z \sum_{i=1}^{n} \beta_i v_i + v \left(cz + \sum_{i,k=1}^{n} a_{ik} z_{ik} + \sum_{i=1}^{n} b_i z_i \right) = 0,$$

where the β_i are certain point functions continuous in G. If for z we choose the function

$$z = C - e^{\mu x_1},$$

we obtain

$$(3) \qquad \sum_{i,k=1}^{n} a_{ik} v_{ik} + \sum_{i=1}^{n} \beta_i v_i + c^* v = 0$$

with

$$c^* = c - \frac{1}{z} \left(a_{11} \mu^2 + b_1 \mu \right) e^{\mu x_1}.$$

Since $a_{11} > 0$, we can choose the constants C and μ in such a way that everywhere in G we have $c^* < 0$ and $z > 1$. From our earlier results it then follows that v, and hence also $u = zv$, vanishes identically in G. This completes the proof of the uniqueness theorem.

2. Nonlinear Equations. The uniqueness theorem for solutions of equation $(1')$ may be applied to prove uniqueness for certain nonlinear equations of the second order

$$F(x_1, \cdots, x_n, u, u_1, \cdots, u_{nn}) = 0$$

if the following conditions are satisfied: For all $x_1, x_2, \cdots, x_n$ in G and *all values of the other arguments* of F the equation is elliptic; i.e., the quadratic form

$$\sum_{i,k=1}^{n} \frac{\partial F}{\partial u_{ik}} \xi_i \xi_k$$

is positive definite, and $\partial F / \partial u \leq 0$. (These conditions can be slightly weakened.) For if u and v are solutions having the same boundary values, then $w = u - v$ satisfies the equation

$$\sum_{i,j=1}^{n} \tilde{F}_{u_{ij}} w_{ij} + \sum_{i=1}^{n} \tilde{F}_{u_i} w_i + \tilde{F}_u w = 0$$

where in general $\tilde{\phi}$ represents a mean value

$$\tilde{\phi} = \int_0^1 \phi(x_1, \cdots, x_n, tu + (1 - t)v, tu_1 + (1 - t)v_1,$$
$$\cdots, tu_{nn} + (1 - t)v_{nn}) \, dt.$$

This equation for w is obtained by expressing

$$F(x_1, \cdots, x_n, u, \cdots, u_{nn}) - F(x_1, \cdots, x_n, v, v_1, \cdots, v_{nn})$$

as an integral. It has the form (1') with $c = \tilde{F}_u \leq 0$; thus we conclude that $w = 0$.

Let us now consider a quasi-linear equation

$$\sum_{i,k=1}^{n} a_{ik} u_{ik} + d = 0$$

and assume that the coefficients a_{ik} and d are functions of $x_1, x_2, \cdots,$ $x_n, u_1, u_2, \cdots, u_n$ alone (i.e., independent of u). Then we may prove uniqueness of the boundary value problem under weaker conditions.

If u is a solution of the equation such that the corresponding matrix (a_{ik}) is positive definite everywhere in G, then any solution v of the equation which agrees with u on the boundary equals u throughout G. Note that the equation need not be *assumed* elliptic beforehand for the function v.

To prove this, we again set $u - v = w$ and consider the identity

$$\sum_{i,k=1}^{n} a_{ik}[u]u_{ik} - \sum_{i,k=1}^{n} a_{ik}[v]v_{ik} + d[u] - d[v] = 0,$$

where we have used the abbreviations

$$a_{ik}[u] = a_{ik}(x_1, x_2, \cdots, x_n, u_1, u_2, \cdots, u_n),$$

etc. This identity may be written in the form

$$\sum_{i,k=1}^{n} a_{ik}[u]w_{ik} + \sum_{i,k=1}^{n} v_{ik}(a_{ik}[u] - a_{ik}[v]) + d[u] - d[v] = 0.$$

Applying the mean value theorem (of differential calculus) to the second sum and to the following term, we obtain

$$\sum_{i,k=1}^{n} a_{ik}[u]w_{ik} + \sum_{i=1}^{n} a_i w_i = 0,$$

where the $a_i = a_i[u, v]$ are certain functions of $x_1, x_2, \cdots, x_n$, $u_1, u_2, \cdots, u_n, v_1, v_2, \cdots, v_n$. If we now substitute the particular u and v under consideration into the $a_{ik}[u]$ and $a_i[u, v]$ and regard the result as a linear equation for w, then this equation is elliptic and of the form (1'). Thus $w = 0$.

3. Rellich's Theorem on the Monge-Ampère Differential Equation. Finally, as an example of the case of a nonlinear differential equation which does not satisfy the conditions of the preceding section, we insert a remark about the boundary value problem for the (nonlinear) *Monge-Ampère equation*:

$$(4) \quad L[u] = E(u_{xx}\, u_{yy} - u_{xy}^2) + Au_{xx} + 2Bu_{xy} + Cu_{yy} + D = 0.$$

Let the coefficients A, B, C, D, E be continuous functions of x and y in G which satisfy the inequality

$$(5) \quad AC - B^2 - DE > 0.$$

We can then state the uniqueness theorem[1]:

There exist at most two solutions of equation (4) *which assume the same boundary values on* Γ.

Proof: If u is a solution of (4), then by (4) and (5) we have the inequality

$$(6) \quad (Eu_{xx} + C)(Eu_{yy} + A) - (Eu_{xy} - B)^2 > 0.$$

From this it follows that the expression $(Eu_{xx} + C)(Eu_{yy} + A)$ must be greater than zero; hence, neither of the two factors can vanish in G. Both are either always greater than or always less than zero. Therefore, the above theorem is proved if we can show that there exists at most one solution of the boundary value problem for which

$$(7) \quad Eu_{xx} + C > 0 \quad \text{(hence also } Eu_{yy} + A > 0\text{)}$$

and at most one solution for which

$$(8) \quad Eu_{xx} + C < 0 \quad \text{(hence also } Eu_{yy} + A < 0\text{)}$$

holds everywhere in G. It is sufficient to consider case (7).

If we assume that two solutions u and v of the boundary value

---

[1] See F. Rellich [1].

problem exist, both of which satisfy the inequality (7), then the difference $w = u - v$ must satisfy the two equations

$$0 = L[\omega + v] - L[v] = E(\omega_{xx}\,\omega_{yy} - \omega_{xy}^2) + (Ev_{xx} + C)\omega_{yy}$$
$$+ (Ev_{yy} + A)\omega_{xx} - 2(Ev_{xy} - B)\omega_{xy},$$
$$0 = L[u] - L[u - \omega] = -E(\omega_{xx}\,\omega_{yy} - \omega_{xy}^2) + (Eu_{xx} + C)\omega_{yy}$$
$$+ (Eu_{yy} + A)\omega_{xx} - 2(Eu_{xy} - B)\omega_{xy},$$

from which the relation

$$(9) \qquad\qquad P\omega_{xx} - 2Q\omega_{xy} + R\omega_{yy} = 0$$

follows by addition. Here the coefficients

$$P = Ev_{yy} + A + Eu_{yy} + A,$$
$$Q = Ev_{xy} - B + Eu_{xy} - B,$$
$$R = Ev_{xx} + C + Eu_{xx} + C$$

are continuous point functions in G. The quadratic form

$$P\xi^2 - 2Q\xi\eta + R\eta^2,$$

being the sum of two positive definite quadratic forms by (6) and (7), is itself positive definite. Just as in article 1 and article 2, we conclude from (9) and the boundary condition $\omega = 0$ that ω vanishes identically in G, proving the statement of the uniqueness theorem.

From simple examples it can be seen that *in general* we really have to expect the *existence of two different solutions*. Thus, the boundary value problem for

$$u_{xx}\,u_{yy} - u_{xy}^2 = 4$$

with the boundary condition

$$(10) \qquad\qquad u = 0$$

on the unit circle has the solutions

$$u = x^2 + y^2 - 1 \qquad \text{and} \qquad v = 1 - x^2 - y^2.$$

For the first solution, $Eu_{xx} + C = 2$, for the second one, $Ev_{xx} + C = -2$.

On the other hand, if the function E vanishes at any point P of G, *then the boundary value problem can have at most one solution.* For at the point P we then have $Eu_{xx} + C = C(P)$ and since $Eu_{xx} + C$ does not change its sign in G, sign $(Eu_{xx} + C)$ equals sign $C(P)$ everywhere. That is, the sign of $Eu_{xx} + C$ is the same for all solutions u.[1]

4. The Maximum Principle and Applications. Returning to the linear equation (1) for $c \leq 0$ we now state the maximum principle in its strong form.

Maximum Principle:[2] *If a function u satisfies $M[u] \geq 0$ and has a maximum at an interior point then $u \equiv$ const.*

Consequently, the maximum of any function u which is continuous in $G + \Gamma$ and satisfies $M[u] \geq 0$ in G is assumed on the boundary Γ. (This is called the weak form of the maximum principle.) Clearly, an analogous minimum principle can be formulated.

The uniqueness theorem of article 1 follows also from the following corollary to the maximum principle.

Corollary: Let u satisfy equation $(1')$ in G; if u has a positive interior maximum then $u \equiv$ const. Consequently, $u \leq 0$ in G whenever u is continuous in $G + \Gamma$, nonpositive on Γ and satisfies $L[u] \geq 0$ in G.

For the proof of the corollary suppose u has a positive maximum at an interior point P. Since u is continuous it is positive in some neighborhood of P; but in this neighborhood $M[u] = L[u] - cu \geq 0$ because $c \leq 0$ and thus, by the maximum principle, $u \equiv$ const. there. Hence the set of maximum points is open in G. On the other hand, by the continuity of u, it is also closed in G and is therefore the whole of G. It follows that u is equal to a positive constant everywhere in G.

The proof of the maximum principle is based on the following

Lemma: Let S be an open sphere and P_0 a point on its boundary. Assume that the coefficients of $M[u]$ are bounded in S and that there

---

[1] It should be mentioned that the Monge-Ampère differential equation can be obtained from a simple *variational problem.* In doing this we neglect the additional term $Au_{xx} + 2Bu_{xy} + Cu_{yy}$ and consider the equation

$$u_{xx} u_{yy} - u_{xy}^2 = p(x, y).$$

As one can easily verify, this is Euler's equation for the variational expression

$$J[u] = \iint_G [u_x^2 u_{yy} - 2u_x u_y u_{xy} + u_y^2 u_{xx} + 6pu] \, dx \, dy.$$

[2] In this formulation the result is due to E. Hopf [2]. A slight modification of his proof is presented here.

exists a positive constant m such that

(11) $$\sum_{i,k=1}^{n} a_{ik}\,\xi_i\,\xi_k \geq m \sum_{i=1}^{n} \xi_i^2$$

holds for all values ξ and for all points x in S. Assume further that u is twice continuously differentiable in S, continuous in $S + P_0$, satisfies $M[u] \geq 0$ and $u < u(P_0)$ in S. Then the exterior normal derivative du/dn at P_0, understood as the lim. inf. of $\Delta u/\Delta n$, is positive.

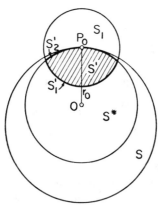

FIGURE 22

Proof: Let S^* be a smaller sphere internally tangent to S at P_0 (see Figure 22). Then P_0 is the only maximum point of u in $\bar{S}^*$, the closure of S^*. Choose the origin as the center of S^* and set $r^2 = \sum_{i=1}^{n} x_i^2$. Let r_0 denote the distance between P_0 and the origin. Denote by S' the intersection of $\bar{S}^*$ with a fixed closed sphere S_1 having P_0 as center and radius less than r_0. The boundary of S' consists of spherical caps of S_1 and $\bar{S}^*$ which we denote by S_1' and S_2', respectively.

We now introduce the auxiliary function

$$h = e^{-\alpha r^2} - e^{-\alpha r_0^2},$$

which is positive in S^* and vanishes on its boundary. With sufficiently large α we can make

$$M[h] = e^{-\alpha r^2}\left[4\alpha^2 \sum_{i,k=1}^{n} a_{ik}\,x_i\,x_k - 2\alpha \sum_{i=1}^{n} (a_{ii} + b_i\,x_i) \right]$$

positive in the interior of S'; for there r and hence $\sum_{i,k=1}^{n} a_{ik} x_i x_k$ is bounded away from zero by (11). On S_2' the function u is less than $u(P_0)$, in fact it is bounded away from $u(P_0)$. Thus for fixed sufficiently small ϵ the function

$$v = u + \epsilon h$$

is less than $u(P_0)$ on S_2'. Let us now consider the function v in S'. In the interior of S' we have $M[v] = M[u] + \epsilon M[h] > 0$. According to the statement on p. 321 in article 1, $\max_{S'} v$ occurs on the boundary of S'. But $v < u(P_0)$ on S_2' and $v = u < u(P_0)$ on S_1' (except at P_0); moreover, $v(P_0) = u(P_0)$. Thus $\max_{S'} v$ occurs at P_0. It follows then that at P_0

$$\frac{dv}{dn} = \frac{du}{dn} + \epsilon \frac{dh}{dn} \geq 0$$

and, since $dh/dn < 0$, that

$$\frac{du}{dn} > 0,$$

which proves the lemma.

The maximum principle now follows easily. Let u satisfy $M[u] \geq 0$ in G. If $u \not\equiv$ const. and has an interior maximum point one can find a closed sphere lying in G which has a maximum point of u on its boundary but none in its interior. According to the lemma, at this point $du/dn > 0$, which contradicts the fact that the first derivatives of u vanish at an interior maximum point.

As observed earlier, the maximum principle implies that any function satisfying $M[u] \geq 0$ in G assumes its maximum at a boundary point. If there exists an open sphere S in G containing the point P_0 on its boundary, and if in S the coefficients of M are bounded and satisfy (11), then the lemma and the maximum principle furnish the following useful result:[1] Either $u \equiv$ const. in G or the exterior normal derivative du/dn is positive at P_0. For such a region G this implies,

---

[1] See E. Hopf [1]. It should be mentioned that the results of the lemma and of the last statement hold true even if the lowest eigenvalue of the matrix (a_{ik}) is permitted to vanish at P_0, i.e., if the equation ceases to be elliptic at P_0, provided that the boundary of S at P_0 is not characteristic. In this case the expression $\sum_{i,k=1}^{n} a_{ik} x_i x_k$ occurring in the proof of the lemma is still bounded away from zero near P_0, hence the rest of the argument can be applied. For related results see also C. Pucci [1].

for example, that Green's function for the equation $M[u] = 0$ (which vanishes on the boundary of G and has a singularity at an interior point) satisfies $du/dn < 0$ at every boundary point, since the function has its minimum on the boundary.

This result provides us also with a simple proof for the uniqueness of solutions of the second boundary value or Neumann problem for the equation $M[u] = 0$ which can be easily extended to the equation (1'). The problem is best formulated in the following way: A continuous function ϕ is defined on the boundary Γ of G which is assumed to possess a continuously varying normal. It is required to find a solution u of $M[u] = 0$ which is continuous and has continuous first partial derivatives in $G + \Gamma$, takes on a prescribed value at some fixed point P, and whose exterior normal derivative on Γ equals ϕ within an additive constant,

$$\frac{du}{dn} = \phi + C.$$

To prove the uniqueness we must show that any such solution u of $M[u]$ satisfying $du/dn = \text{const.} = k$ on Γ and vanishing at P is identically zero, and hence $k = 0$. We assume that the coefficients of $M[u]$ are bounded in G and satisfy (11), and that to every point P_0 of Γ we may find an open sphere S lying entirely in G and having P_0 on its boundary. If u is a solution, then the maximum principle implies that it assumes its maximum or minimum at points of the boundary Γ. At these points, according to our previous argument, du/dn is positive or negative, respectively, unless u is identically constant. But $du/dn = k$, hence it follows that in fact $u = \text{const.}$ in G; moreover, u vanishes identically since it vanishes at P.

Besides furnishing a uniqueness proof for the solution u of (1') (and, consequently, also for the solution of $L[u] = f$) which has given boundary values $u = \phi$ on the boundary Γ of G, the maximum principle may also be used to estimate u.

We claim that if g is any function satisfying the conditions

$$-L[g] \geq \max |f| \qquad \text{in } G$$

and

$$g \geq \max |\phi| \qquad \text{on } \Gamma$$

then

$$|u| \leq g.$$

For the proof it suffices to show that $v = u - g$ is nonpositive. But this follows from the corollary to the maximum principle, since v satisfies the conditions

$$L[v] = L[u] - L[g] = f - L[g] \geq 0$$

and since, on the boundary, $v = \phi - g \leq 0$.

We now construct such a function g, assuming for convenience that the bounded domain G lies in the half-space $x_1 \geq 0$. We assume that there are positive constants m, b such that, throughout G,

$$a_{11} \geq m, \qquad -b_1 \leq b.$$

Let us set

$$g = \max |f| (e^{\alpha \bar{x}} - e^{\alpha x_1}) + \max |\phi|,$$

where $x_1 \leq \bar{x}$ in G and α is a positive constant to be chosen so that g satisfies the given conditions. Clearly $g \geq \max |\phi|$. Furthermore, for *sufficiently large* α

$$-L[g] = \max |f| [e^{\alpha x_1}(a_{11} \alpha^2 + b_1 \alpha) - c(e^{\alpha \bar{x}} - e^{\alpha x_1})] - c \max |\phi|$$

$$\geq \max |f| (a_{11} \alpha^2 + b_1 \alpha)$$

$$\geq \max |f|.$$

The choice of α depends only on m and b.

Thus we have obtained the following *a priori* estimate:

The solution u of equation (1') with boundary values $u = \phi$ is bounded by

$$(12) \qquad |u| \leq \max |\phi| + \max |f| (e^{\alpha \bar{x}} - 1),$$

where α is a constant depending only on b and m, and $\bar{x}$ is a constant such that $|x_1| \leq \bar{x}$ in G.

Even if we do not assume $c \leq 0$, it is still possible to obtain an estimate of the form

$$(13) \qquad |u| \leq k (\max |\phi| + \max |f|),$$

provided the domain G is sufficiently narrow in, say, the x_1 direction or, more precisely, provided that

$$(14) \qquad (\max c)(e^{\alpha \bar{x}} - 1) < 1.$$

(The constant k then depends on b, m, max c, and $\bar{x}$.)   For in this case, we write the equation as

$$M[u] + c^- u = (c^- - c)u + f = \tilde{f}$$

where $c^- = \min(c, 0)$, and we can apply the a priori estimate (12). We find

$$\max |u| \leq \max |\phi| + \max |\tilde{f}|(e^{a\bar{x}} - 1)$$

$$\leq \max |\phi| + (e^{a\bar{x}} - 1)(\max |f| + \max |u| \max c)$$

or

$$\max |u| \leq \frac{\max |\phi| + \max |f| \, (e^{a\bar{x}} - 1)}{1 - \max c \, (e^{a\bar{x}} - 1)} \, .$$

We observe that (13) *implies uniqueness of the solution.*

§7. *A Priori Estimates of Schauder and Their Applications*

Boundary value problems for nonlinear elliptic equations were first investigated in a systematic way by Serge Bernstein. In his fundamental papers it became evident that the essential and usually most difficult step in solving a nonlinear boundary value problem consists in proving sufficiently strong a priori estimates for the solution.

J. Schauder[1] has derived certain such a priori estimates for solutions $u(x_1, x_2, \cdots, x_n)$ of linear elliptic equations of the form

$$(1) \qquad L[u] \equiv \sum_{i,k=1}^{n} a_{ik} u_{ik} + \sum_{i=1}^{n} b_i u_i + cu = f$$

in a bounded domain G. Using them he solved the boundary value problem directly (for $c \leq 0$) without having to construct a fundamental solution of the equation.

The estimates hold for uniformly elliptic equations (1) with bounded Hölder continuous coefficients, i.e., for equations satisfying the following conditions: There exist positive constants m, M, and α

---

[1] See J. Schauder [2] and [1]. Simplified proofs have been given by C. Miranda [1], by A. Douglis, L. Nirenberg, and by others. S. Agmon, A. Douglis, and L. Nirenberg [1] have derived analogous estimates for general boundary value problems for elliptic equations of arbitrary order. See especially pp. 657–669. Other references are given there.

$(0 < \alpha < 1)$ such that in G

(a) for all real ξ_1, ξ_2, $\cdots$, ξ_n, the inequality

$$\sum_{i,k=1}^{n} a_{ik}\,\xi_i\,\xi_k \geq m \sum_{i=1}^{n} \xi_i^2$$

holds;

(b) $|\,a_{ik}\,|$, $|\,b_i\,|$, $|\,c\,| \leq M$, $(i,\,k\,=\,1,\,2,\,\cdots,\,n)$;

(c) the coefficients a_{ik}, b_i, c satisfy a Hölder condition (see §1, 2) with exponent α and coefficient M.

We shall merely describe the estimates without proof (see footnote on previous page) and present Schauder's method of solving the boundary value problem for (1). In addition, we shall give some applications of the estimates which may be used to derive, for solutions of (1), analogues of many properties of harmonic functions.

1. Schauder's Estimates. In order to give a succinct form to the estimates, which are expressed as bounds for the derivatives of solutions of (1), it is convenient to introduce classes of functions having derivatives of various orders and suitable "norms" expressing the "size" of functions in these classes. We define, therefore, C_m (m a non-negative integer) as the class of functions $u(x_1,\,x_2,\,\cdots,\,x_n)$ having partial derivatives up to order m which are continuous in $G + \Gamma$, and $C_{m+\alpha}$ (m a non-negative integer, and $0 < \alpha < 1$) as the class of functions u in C_m whose mth-order partial derivatives satisfy a Hölder condition in $G + \Gamma$ with exponent α (see §1, 2). C_0, sometimes called C, is the class of functions continuous in $G + \Gamma$.

Denoting any of the mth-order derivatives of u by $D^m u$, we introduce as "norm" in C_m

$$\|\,u\,\|_m = \max_{P \in G+\Gamma} |\,u(P)\,| + \max_{P \in G+\Gamma} |\,D^1 u(P)\,|$$
$$+ \cdots + \max_{P \in G+\Gamma} |\,D^m u(P)\,|,$$

where the maxima are also taken over all derivatives of the orders shown. Denoting by $H_\alpha[D^m u]$ the smallest constant K with the property that all the mth order derivatives of u satisfy a Hölder condition in $G + \Gamma$ with exponent α and coefficient K, we introduce as norm in $C_{m+\alpha}$

$$\|\,u\,\|_{m+\alpha} = \|\,u\,\|_m + H_\alpha[D^m u].$$

The class C_a is now defined for every number $a \geq 0$. It is clear that C_a is linear, i.e., that any finite linear combination with real coefficients of functions in C_a is itself in C_a. Furthermore, it is easily seen that the norm $\| u \|_a$ has the following properties:

$$\| u \|_a \geq 0 \qquad\qquad (\| u \|_a = 0 \text{ only if } u = 0),$$

$$\| cu \|_a = | c | \| u \|_a \qquad\qquad \text{for any real constant } c,$$

$$\| u + v \|_a \leq \| u \|_a + \| v \|_a \qquad \text{(the triangle inequality)}.$$

Therefore, the class C_a may be regarded as a linear space whose elements or "points" are the functions u, and on which the norm $\| \ \|_a$ defines a metric or notion of distance between any pair of functions u, v: $\| u - v \|_a$. In this space we define convergence of a sequence of functions $\{u_n\}$:

$$u_n \to u \quad \text{or} \quad \lim u_n = u \quad \text{if and only if} \quad \| u_n - u \|_a \to 0.$$

(Convergence with respect to the norm $\| \ \|_m$, m an integer, is equivalent to the uniform convergence of the function u and its derivatives up to order m in $G + \Gamma$.) Clearly every convergent sequence $\{u_n\}$ is a Cauchy sequence, i.e., has the property that

$$\| u_m - u_n \|_a \to 0 \quad \text{as} \quad m, n \to \infty.$$

In addition, the normed linear space C_a is complete; i.e., in C_a every Cauchy sequence converges. The proof of this statement is left to the reader. Thus the space C_a is a complete normed linear space, i.e., a Banach space.[1] For functions defined in any closed subdomain B of G we shall denote the corresponding classes and norms by C_a^B and $\| \ \|_a^B$.

Besides the spaces C_a we shall need spaces of continuously differentiable functions whose derivatives are permitted to become infinite at the boundary Γ in a prescribed way. Let d_P denote the distance from any point P in G to the boundary of G, and for any pair of points P, Q, set $d_{P,Q} = \min (d_P, d_Q)$. For any function u continuous in $G + \Gamma$ and having continuous derivatives up to order m in G define

$$\widehat{\| u \|}_m = \underset{P \in G}{\text{l.u.b.}} | u(P) | + \underset{P \in G}{\text{l.u.b.}} d_P \cdot | D^1 u(P) |$$

$$+ \cdots + \underset{P \in G}{\text{l.u.b.}} d_P^m \cdot | D^m u(P) |,$$

[1] See, e.g., S. Banach [1] for a study of these spaces. See also N. Dunford and J. T. Schwartz [1] and Supplement, §15, of this chapter.

where l.u.b., the least upper bounds, are taken over all derivatives of the orders shown; $\|\widehat{u}\|_m$ may be infinite. For the functions u whose mth-order derivatives satisfy a Hölder condition with exponent α in every closed subdomain of G we define

$$\hat{H}_\alpha [D^m u] = \underset{P,Q \in G}{l.\,u.\,b.}\ d_{P,Q}^{m+\alpha} \cdot \frac{|\ D^m u(P)\ -\ D^m u(Q)\ |}{|\ P - Q\ |^\alpha},$$

where the least upper bound is also taken over all derivatives of order m and $|\ P - Q\ |$ denotes the distance from P to Q; furthermore, we define

$$\widehat{\|\ u\ \|}_{m+\alpha} = \widehat{\|\ u\ \|}_m + \hat{H}_\alpha[D^m u].$$

Let $\hat{C}_m$ be the class of functions u continuous in $G + \Gamma$ which have continuous derivatives up to the mth order in G and for which $\|\widehat{u}\|_m$ is finite. For $0 < \alpha < 1$, let $\hat{C}_{m+\alpha}$ be the subclass of functions in $\hat{C}_m$ for which $\widehat{\|\ u\ \|}_{m+\alpha}$ is finite. The classes $\hat{C}_a$ and corresponding "norms" $\|\ \ \|_a$ are thus defined for all $a \geq 0$. Moreover, the norms $\widehat{\|\ \ \|}_a$ have the properties listed above for the norms $\|\ \ \|_a$, and $\hat{C}_a$ forms a linear space which under the norm $\widehat{\|\ \ \|}_a$ is easily seen to be *complete*. Thus $\hat{C}_a$ is a Banach space. Clearly, if u is in $\hat{C}_a$ then it belongs to the class of functions C_a with respect to any closed subdomain of G.

We are now in a position to formulate the *a priori* estimates. We observe first that conditions like (b) and (c) on the coefficients may be expressed more concisely in the present notation so that, instead of conditions (a), (b), and (c), we now have

$$(2) \qquad \sum_{i,k=1}^{n} a_{ik}\,\xi_i\,\xi_k \geq m \sum_{i=1}^{n} \xi_i^2$$

and

$$(3) \qquad \|\ a_{ik}\ \|_\alpha\,,\ \|\ b_i\ \|_\alpha\,,\ \|\ c\ \|_\alpha \leq 2M.$$

The estimates are of two kinds: "interior" estimates, in any closed subdomain B of G, and estimates "up to the boundary", in all of $G + \Gamma$. We shall require f to be in $\hat{C}_\alpha$ and C_α, respectively.

INTERIOR ESTIMATES: If u is a solution of (1) whose second derivatives are Hölder continuous (with exponent α) in every closed sub-domain of G, then $\widehat{\| u \|}_{2+\alpha}$ is finite and

(4) $$\widehat{\| u \|}_{2+\alpha} \leq K(\widehat{\| u \|}_0 + \widehat{\| f \|}_\alpha).$$

Here K is a constant depending only on m, α, M, and the diameter of G.

In formulating the estimates up to the boundary we require that the domain G and the boundary values ϕ of u be sufficiently *smooth*. We call a domain G smooth if its boundary is smooth; i.e., if we can cover Γ by a finite number of spheres which have the property that, singling out one coordinate, say x_n, one can express the part of the boundary contained in each of these spheres in the form

$$x_n = g(x_1, x_2, \cdots, x_{n-1}),$$

where g is assumed to have Hölder continuous second derivatives with exponent α. In addition, in terms of the local parameters $x_1, x_2, \cdots, x_{n-1}$, the boundary values ϕ are also assumed to be smooth, i.e., to have Hölder continuous second derivatives with exponent α. Using the fixed finite number of local parameter systems on the boundary and the norms $\| \phi \|_{2+\alpha}$ in each sphere, we may define a norm $\| \phi \|'_{2+\alpha}$ for the function ϕ as the maximum of the $\| \phi \|_{2+\alpha}$.

ESTIMATES UP TO THE BOUNDARY: Let u be a solution in $C_{2+\alpha}$ of (1) in a smooth domain G with smooth boundary values ϕ. Then

(5) $$\| u \|_{2+\alpha} \leq K_1(\| u \|_0 + \| f \|_\alpha + \| \phi \|'_{2+\alpha}),$$

where K_1 is a constant depending only on m, α, M, and the domain G.

The derivation of Schauder's estimates is too lengthy for presentation here. It is based on estimates for the second derivatives of solutions of a very special equation, namely, the Poisson equation $\Delta u = f$. One of these, which we will use in solving the boundary value problem, asserts that *if u is a solution of $\Delta u = f$ in a domain in which f is in C_α, $0 < \alpha < 1$, then u is in $C_{2+\alpha}$ in every closed subdomain.* The statements concerning the Poisson equation are derived in turn from the integral expressions for these second derivatives which were presented in (21') in §1 and which used the fundamental solution of the Laplace equation. (In §1, 2 we only showed that if f is in C_α then the function u, given by (14) in §1, is in C_2 in every closed subdomain.)

In solving the boundary value problem we also make use of a theorem for harmonic functions due to O. D. Kellogg:[1]

Let u be a harmonic function defined in a smooth domain G and having smooth boundary values. Then u is in $C_{2+\alpha}$.

Before solving the boundary value problem we remark that, if $c \leq 0$ in (1), the estimate (13) in §6 holds and the Schauder estimates (4) and (5) may be written in the form

$$(4') \qquad \widehat{\| u \|_{2+\alpha}} \leq K'(\widehat{\| f \|_{\alpha}} + \| \phi \|_0'),$$

$$(5') \qquad \| u \|_{2+\alpha} \leq K_1'(\| f \|_{\alpha} + \| \phi \|_{2+\alpha}'),$$

where K' and K_1' are constants depending only on m, α, M, and the diameter of G.

2. Solution of the Boundary Value Problem. We describe Schauder's method for solving the boundary value problem for (1) with $c \leq 0$ and coefficients satisfying conditions (a), (b), (c) or (2) and (3). Let us first assume that the bounded domain G and the boundary values ϕ are smooth and that f is in C_{α}. Then we have the following

THEOREM: *There exists a unique solution u of equation (1) in $C_{2+\alpha}$ with boundary values ϕ.*

In proving the theorem it suffices to assume that $\phi \equiv 0$ since the smoothness assumptions insure that there exists a function $\bar{\phi}$ in $C_{2+\alpha}$ with boundary values ϕ on Γ. The function $u - \bar{\phi}$ then vanishes on Γ and is a solution of the equation $L[u] = f - L[\bar{\phi}]$. Our aim is to show in a number of steps that the given operator L can be inverted.

For the solution we use the *continuity method*. We construct the one-parameter family of elliptic operators

$$L_t[u] \equiv tL[u] + (1 - t)\Delta u, \qquad 0 \leq t \leq 1,$$

where $L_0 = \Delta$ and $L_1 = L$. We shall show that all the L_t, and in particular L_1, can be inverted. Calling T the set of those values of t in the unit interval for which L_t has an inverse, we shall prove that T is the whole interval by showing:

1) T contains $t = 0$,

2) T is an open set in the unit interval; i.e., for every t_0 in T there exists an $\epsilon(t_0) > 0$ such that every t (in the unit interval) for which $| t - t_0 | < \epsilon(t_0)$ lies in T,

3) T is a closed set.

[1] See O. D. Kellogg [2].

The conclusion follows once these properties have been verified. For T is then a nonempty set in the interval which is both open and closed; hence it is the whole interval.

To prove 1) we have to show the existence of a solution of $\Delta u = f$ which is continuous in $G + \Gamma$, vanishes on Γ, and belongs to $C_{2+\alpha}$. Extend f to a sphere S containing $G + \Gamma$ in its interior in such a way that f belongs to C_α in S. In §1, 2, we constructed a particular solution v of $\Delta u = f$ in S given by the integral (14) in §1. By article 1 this solution is in $C_{2+\alpha}$ in every closed subdomain of S, in particular in $G + \Gamma$. It follows that its values on Γ are smooth (in the sense defined previously in this section). Let w be the harmonic function on G which equals v on Γ. By Kellogg's theorem mentioned in article 1 w is in $C_{2+\alpha}$ in G. It follows, therefore, that the function $u = v - w$ is the desired solution.

Before proving 2) and 3) we remark that, because of assumptions (a), (b), and (c), and because of the form of L_t, there exist positive constants m_1, M_1, α such that the coefficients of L_t satisfy conditions (a), (b), (c) for all t with the constants m, M replaced by m_1, M_1. It follows from the estimates $(5')$ that there exists a constant K_2 such that the required solution u_t of $L_t[u] = f$ (for any t) satisfies

$$(6) \qquad \| u_t \|_{2+\alpha} \leq K_2 \| f \|_\alpha .$$

We now continue with the proof of 2). Let t_0 be in T and let f be any function in C_α. We wish to solve the equation

$$L_t[u] = f$$

with $u = 0$ on the boundary for a function u in $C_{2+\alpha}$, provided t is close enough to t_0. This equation may be written in the form

$$L_{t_0}[u] = L_{t_0}[u] - L_t[u] + f$$

or

$$L_{t_0}[u] = (t - t_0)(\Delta u - L[u]) + f.$$

Substituting any function u which is in $C_{2+\alpha}$ into the expression on the right we obtain a function F in C_α. Since t_0 is in T there exists a function v in $C_{2+\alpha}$ satisfying

$$L_{t_0}[v] = (t - t_0)(\Delta u - L[u]) + f \equiv F,$$

with $v = 0$ on the boundary. We may consider v as defined by an inhomogeneous linear transformation on u,

$$v = A(u).$$

By an iteration procedure we seek a "fixed point" of this transformation, i.e., a function u such that

$$u = A(u).$$

An easy calculation which relies on assumptions (b) and (c) yields the inequality

$$\widehat{\| F \|}_\alpha \leq K_3 \mid t - t_0 \mid \| u \|_{2+\alpha} + \| f \|_\alpha,$$

where K_3 is a fixed constant independent of u. Applying the *a priori* estimate (6), we have

(7) $$\| v \|_{2+\alpha} \leq K_2 K_3 \mid t - t_0 \mid \| u \|_{2+\alpha} + K_2 \| f \|_\alpha.$$

It follows, therefore, that an inequality $\| u \|_{2+\alpha} \leq 2K_2 \| f \|_\alpha$ would imply

$$\| A(u) \|_{2+\alpha} = \| v \|_{2+\alpha} \leq 2K_2 \| f \|_\alpha$$

provided $2K_2 K_3 \mid t - t_0 \mid \leq 1$.

In addition, we note that if $v_1 = A(u_1)$ and $v_2 = A(u_2)$, then $v_1 - v_2$ is a solution of

$$L_{t_0}[v_1 - v_2] = (t - t_0)(\Delta(u_1 - u_2) - L[u_1 - u_2])$$

with $v_1 - v_2 = 0$ on the boundary. Using the estimate (7) we have

(8)
$$\| v_1 - v_2 \|_{2+\alpha} \leq K_2 K_3 \mid t - t_0 \mid \| u_1 - u_2 \|_{2+\alpha}$$
$$\leq \tfrac{1}{2} \| u_1 - u_2 \|_{2+\alpha}$$

for $2K_2 K_3 \mid t - t_0 \mid \leq 1$. Then, if we take t so close to t_0 that $2K_2 K_3 \mid t - t_0 \mid \leq 1$, the transformation $A(u)$ maps the set of functions satisfying $\| u \|_{2+\alpha} \leq 2K_2 \| f \|_\alpha$ into itself and is by (8) contracting.

If we now start with some function $u_{(0)}$ in this set and define

$$u_{(i+1)} = A(u_{(i)}) \qquad (i = 0, 1, \cdots),$$

then the $u_{(i)}$ converge uniformly to a function u in $C_{2+\alpha}$ satisfying

$$u = A(u).$$

The function u is then a solution of the equation $L_t[u] = f$ with $u = 0$ on the boundary. Thus the interval $\mid t - t_0 \mid \leq 1/2K_2 K_3$ belongs to T, and T is open.

In order to prove *3*), let t be a limit point of a sequence $\{t_i\}$ in T and consider any f in C_α. Corresponding to this sequence we have a sequence of solutions $\{u_i\}$ in $C_{2+\alpha}$ such that

$$L_{t_i}[u_{(i)}] = f$$

with $u_{(i)} = 0$ on Γ. Applying the estimate (6) we have

$$\| u_{(i)} \|_{2+\alpha} \leq K_2 \| f \|_\alpha .$$

Thus the $u_{(i)}$ and their first and second derivatives are equicontinuous in $G + \Gamma$. We therefore select a subsequence, denoted again by $u_{(i)}$, which converges to a function u while the first and second order derivatives of the $u_{(i)}$ converge to those of u, the convergence being uniform in every closed subdomain. Clearly, u is in $C_{2+\alpha}$ and satisfies the equation

$$L_t[u] = f$$

with $u = 0$ on Γ. Consequently u is the desired solution for L_t and t is in T. This completes the proof of the theorem.

Using this theorem and the interior estimates, (1) can be easily solved under weaker conditions as expressed in the following

THEOREM: *If the bounded domain G is smooth, f is in $\hat{C}_\alpha$ and the boundary values ϕ are continuous, then there exists a unique solution u of* (1) *and u is in $\hat{C}_{2+\alpha}$.*

For the proof we approximate the given functions ϕ and f uniformly by three times differentiable functions ϕ_n and differentiable functions f_n such that the norms $\widehat{\| f_n \|}_\alpha$ are uniformly bounded. By the previous theorem there exist solutions u_n in $C_{2+\alpha}$ of

$$L[u_n] = f_n$$

with $u_n = \phi_n$ on Γ. Applying the inequality (13) in §6 to differences $u_n - u_m$ we see that the functions u_n converge uniformly to a continuous function u. From the interior estimate (4') we know that the norms $\widehat{\|u_n\|}_{2+\alpha}$ are uniformly bounded. It follows, therefore, that u is in $\hat{C}_{2+\alpha}$ and that its derivatives are the limits (uniform in closed subdomains) of the derivatives of an appropriate subsequence of the u_n. The function u is then the desired solution.

With the aid of the interior estimates we may also solve the problem in a wide class of domains which are not smooth. Let G be a bounded domain which may be expressed as the union of a sequence G_n of smooth domains where each G_n is contained in the following G_{n+1} and has boundary Γ_n . Assume that to every point Q on the boundary Γ of G there exists a "strong barrier function", i.e., a function w_Q which is twice continuously differentiable in G and continuous and non-negative in $G + \Gamma$, vanishes only at Q, and satisfies

$$L[w_Q] \leq -1.$$

(In the next section we shall construct such functions w_Q for a certain class of domains.)   We can now state the following more general

THEOREM: *Assume that for a domain G with the properties just described f is in $\hat{C}_\alpha$ and that ϕ is a continuous function in $G + \Gamma$. Then there exists a unique solution u in $\hat{C}_{2+\alpha}$ of (1) which agrees with ϕ on Γ.*

Proof: By the previous theorem there exists a solution u_n of $L[u] = f$ in G_n which agrees with ϕ on Γ_n . Application of the interior estimate $(4')$ shows that the norms $\widehat{\|u_n\|}_{2+\alpha}$ of u_n in G_n are uniformly bounded. It follows that a subsequence of the functions u_n converges (uniformly in each closed subdomain of G) to a function u having finite norm $\widehat{\| u \|}_{2+\alpha}$ in G and satisfying there the differential equation $L[u] = f$. If we now define u to be ϕ on Γ, we have only to show that u is continuous in $G + \Gamma$.

For this purpose let Q be a point on Γ and w_Q the associated strong barrier function.   From the properties of w_Q it follows that for any $\epsilon > 0$ there exists a constant k, such that

$$| \phi - \phi(Q) | \leq \epsilon + kw_Q \qquad \text{in } G + \Gamma.$$

If we set

$$W = \epsilon + k_1 w_Q ,$$

where $k_1 = \max (k, \text{l.u.b.} | f - c\phi(Q) |)$, then

$$| \phi - \phi(Q) | \leq W \qquad \text{in } G + \Gamma.$$

We observe furthermore that

$$L[W] = c\epsilon + k_1 L[w_Q]$$
$$\leq - \text{l.u.b.} | f - c\phi(Q) |$$

since $c \leq 0$ and $L[w_Q] \leq -1$.

It follows from our construction of W that we have

$$W \pm (u_n - \phi(Q)) \geq 0 \qquad \text{on } \Gamma_n ,$$

and

$$L[W \pm (u_n - \phi(Q))] = L[W] \pm (f - c\phi(Q)) \leq 0 \quad \text{in } G_n.$$

Applying the maximum principle to $W \pm (u_n - \phi(Q))$ we conclude that

$$| u_n - \phi(Q) | \leq W \qquad \text{in } G_n,$$

and, going to the limit, that

$$| u - \phi(Q) | \leq W \qquad \text{in } G.$$

By the continuity of w_Q, however, we have $W \leq 2\epsilon$ in some neighborhood of Q; thus u is continuous at Q. Since Q was any boundary point, u is continuous in $G + \Gamma$, and the theorem is proved.

3. Strong Barrier Functions and Applications. To complete the discussion of the boundary value problem we show how to construct a strong barrier function w_Q for boundary points Q that have the following property: There exists a closed sphere S_Q whose intersection with $G + \Gamma$ is the single point Q. Let R be the radius of S_Q and let r denote the distance from the center of S_Q which we take as origin. The operator L given by (1) is assumed to be uniformly elliptic, i.e., to satisfy (a) and (b); furthermore, c is assumed to be nonpositive.

Let us set

$$w_Q = k_1(R^{-p} - r^{-p}),$$

where k_1 and p are positive constants. Clearly w_Q is non-negative in $G + \Gamma$ and zero only at Q. A direct calculation yields

$$L[w_Q] = k_1 \, pr^{-p-4}[-(p+2) \sum_{i,j=1}^{n} a_{ij} \, x_i \, x_j + r^2 \sum_{i=1}^{n} (a_{ii} + b_i \, x_i)] + cw_Q$$

$$\leq k_1 \, pr^{-p-2}[-(p+2)m + \sum_{i=1}^{n} (a_{ii} + b_i \, x_i)]$$

in virtue of (a) and the inequality $c \leq 0$. If we now choose p, and then k_1, sufficiently large we have

$$L[w_Q] \leq -1.$$

Thus with k_1 and p chosen appropriately (depending only on R, m, M, and G) the function w_Q has the desired properties of a strong barrier function.

We remark[1] that with the aid of the strong barrier functions we may extend Perron's method (cf. §4, 5) for the solution of Laplace's equation to the solution of Dirichlet's problem for the general homogeneous elliptic equation $(1')$ in §6, namely,

$$L[u] = \sum_{i,k=1}^{n} a_{ik}\, u_{ik} + \sum_{i=1}^{n} b_i\, u_i + cu = 0, \qquad c \le 0,$$

provided only that this problem can be solved in the small; i.e., provided that for sufficiently small spheres there exist solutions with arbitrary continuous boundary values. In this case, the analogues of subharmonic (superharmonic) functions can be easily defined: A continuous function u is a generalized subharmonic (superharmonic) function if, for every sufficiently small sphere C in G, $u \le\ (\ge)\ M_C[u]$. Here $M_C[u]$ equals u outside C and is, in C, that solution of $(1')$ in §6 which agrees with u on the boundary of C. The properties 1), 2), 3) given in §6, 4 for subharmonic functions are easily extended to generalized subharmonic functions with the aid of the maximum principle for equation $(1')$ in §6, an appropriate form of which is expressed in the corollary of §6, 4.

If we wish to find a solution u of $(1')$ in §6 with given boundary values ϕ, we define appropriate generalized subfunctions (superfunctions) as those generalized subharmonic (superharmonic) functions which are not greater (less) than ϕ on the boundary. As in the harmonic case the desired solution u is then given by the least upper bound of all subfunctions. In order to prove that u is a solution of $(1')$ in §6 we must be able to obtain, for a solution in a sphere, an estimate for the first derivatives in a concentric smaller sphere, such that the estimate depends only on the bound of the solution and on the spheres. We must also assert that a uniform limit of solutions of $(1')$ in §6 is a solution (cf. §4, 5). For equations with Hölder continuous coefficients these assertions follow from the interior estimates of Schauder. To show that u agrees with ϕ on the boundary we then use the strong barrier functions.

The strong barrier functions may also be used to estimate at boundary points the first derivatives of solutions of (1) which vanish on the boundary.

---

[1] This remark is due to P. D. Lax. Also see G. Tautz [1] and E. F. Beckenbach and L. K. Jackson [1]. N. Simonoff [1] has also applied a modification of the Perron method to solve the nonlinear elliptic equation of second order under the assumption that a solution with prescribed continuous boundary values may be found for small domains.

LEMMA: Let u be a solution of

$$(1') \qquad\qquad L[u] = f, \qquad\qquad c \leq 0,$$

where L is uniformly elliptic, i.e., satisfies (a) and (b). Assume that u and its first derivatives are continuous in $G + \Gamma$ and that u vanishes on Γ. Assume that the domain G is bounded, has a continuously turning tangent plane on the boundary and has the following property: There exists a positive constant R such that to every point Q on the boundary Γ there is a closed sphere of radius R whose intersection with $G + \Gamma$ is the single point Q. Then at every boundary point

$$(9) \qquad\qquad \left| \frac{\partial u}{\partial x_i} \right| \leq k_2 \, \mathrm{l. u. b.} \, |f| \qquad (i = 1, 2, \cdots, n),$$

where k_2 is a constant depending only on m, M, and G.

Proof: Let Q be any boundary point. Since u vanishes on Γ and since

$$\left| \frac{\partial u}{\partial x_i} \right| \leq \left| \frac{\partial u}{\partial n} \right|,$$

it suffices to estimate the interior normal derivative $\partial u / \partial n$ at Q. Let w_Q be the strong barrier function constructed above. From the construction it is clear that at Q

$$\left| \frac{\partial w_Q}{\partial n} \right| \leq k_2$$

for some constant k_2 which depends only on m, M, and G, and not on the choice of the boundary point Q. Set $v = w_Q \, \mathrm{l.u.b.} \, |f|$; clearly

$$L[v \pm u] \leq 0.$$

Since u vanishes on the boundary we find, by applying the maximum principle to $v \pm u$, that

$$|u| \leq v$$

in G. Both u and v vanish at Q, and hence

$$\left| \frac{\partial u}{\partial n} \right| \leq \frac{\partial v}{\partial n} \leq k_2 \, \mathrm{l.u.b.} \, |f|$$

at this point, which immediately implies inequality (9).

4. Some Properties of Solutions of $L[u] = f$. We assume here that f is in C_α and that conditions (a), (b), and (c) at the beginning of this section hold in G. For any statements regarding solutions in closed subdomains of G the same requirements must hold merely in every closed subdomain.

We first derive the following property: *If u is a twice continuously differentiable solution of equation* (1) *then the second derivatives of u are Hölder continuous in every closed subdomain of G.*

It suffices to establish this fact in spheres lying in G. In a closed sphere S in G, we write equation (1) in the form

$$\sum_{i,j=1}^{n} a_{ij} u_{ij} + \sum_{i=1}^{n} b_i u_i = f - cu.$$

By the theorem on page 340 there exists in S a solution v of the equation

$$\sum_{i,j=1}^{n} a_{ij} v_{ij} + \sum_{i=1}^{n} b_i v_i = f - cu$$

which agrees with u on the boundary. Furthermore, the second derivatives of v satisfy a Hölder condition with exponent α in any smaller concentric sphere. But the solution of the equation for v is unique; thus $u \equiv v$, and the Hölder continuity of the second derivatives of u follows. From now on we shall assume that all solutions have this property. Using the interior estimate (4), we have as an immediate consequence the analogue of the compactness theorem for harmonic functions of §2, 3.

Any uniformly bounded set of solutions of (1) *in G possesses a subsequence which converges uniformly to a solution in every closed subdomain B of G.*

In addition, *a uniform limit of solutions of* (1) *is a solution.*

We can infer even more about the solutions of equation (1'). In this case the interior estimate (4') is valid and therefore we have the analogue of the Weierstrass convergence theorem for harmonic functions (cf. §2, 3):

A sequence of solutions u_n of (1'), *continuous in the closure of G with uniformly converging boundary values ϕ_n , converges uniformly to a solution u with boundary values $\phi = \lim \phi_n$.*

This follows from the estimate (4') applied to differences $u_n - u_m$:

$$\| u_n - u_m \|_{2+\alpha} \leq K_0' \| \phi_n - \phi_m \|_0' \to 0.$$

The interior estimate (4) may be used also to extend one of the basic properties of harmonic functions to solutions of (1): *Every solution of equation (1) is infinitely differentiable provided f and the coefficients are infinitely differentiable.* In fact, the analyticity of f and of the coefficients implies the analyticity of any solution; however, we shall not prove this here (see the references in the next section).

We prove a precise form of the differentiability theorem:[1]

If f and the coefficients are in $C_{m+\alpha}$, where m is a non-negative integer and $0 < \alpha < 1$, then for every closed subdomain B, any twice differentiable solution u is in $C^B_{m+2+\alpha}$.

For $m = 0$ the proof has already been given. Now we present the proof for $m = 1$ (for $m > 1$ the argument may simply be repeated). Let B' and B'' be closed subdomains of G such that they contain in their interiors B and B', respectively. Let $h_0 > 0$ be so small that for $h \leq h_0$ and all points $P: (x_1, x_2, \cdots, x_n)$ in B' the points $P_h : (x_1 + h, x_2, \cdots, x_n)$ lie in B''. We subtract (1) at P from (1) at P_h and divide by h. Then, denoting the difference quotients with fixed h by

$$u^h = \frac{1}{h}(u(P_h) - u(P)), \qquad a^h_{ik} = \frac{1}{h}(a_{ik}(P_h) - a_{ik}(P)),$$

$$b^h_i = \frac{1}{h}(b_i(P_h) - b_i(P)), \qquad c^h = \frac{1}{h}(c(P_h) - c(P)),$$

$$f^h = \frac{1}{h}(f(P_h) - f(P)),$$

and the operator whose coefficients are the difference quotients of the original operator L by L^h, we may write the resulting equation in the form

(10) $$L[u^h(P)] = -L^h[u(P_h)] + f^h \equiv F_h.$$

Since f and the coefficients are in $C_{1+\alpha}$, their difference quotients may be expressed as integrals, e.g.,

$$a^h_{ik} = \int_0^1 \frac{\partial}{\partial x_1} a_{ik}(P_{th}) \, dt,$$

[1] See E. Hopf [4] where a proof of the theorem (and also a proof of analyticity of the solution for an analytic equation) is given without using the existence theory for the boundary value problem for the case $m = 0$. See also Appendix 5 in S. Agmon, A. Douglis, L. Nirenberg [1].

where P_{th} is the point $(x_1 + th, x_2, \cdots, x_n)$, and thus they are in $C_\alpha^{B'}$. In addition, since the solution u is in $C_{2+\alpha}^{B''}$, F_h (which involves derivatives of u at points in B'') is in $C_\alpha^{B'}$. Furthermore,

$$\| F_h \|_\alpha^{B'} \le K_4,$$

where K_4 is a constant *independent of h*. Regarding u^h as a solution of equation (10) we may apply the interior estimate (4), with B a closed subdomain of B', to obtain the estimate

$$\| u^h \|_{2+\alpha}^B \le K_5,$$

where K_5 is a constant independent of h.

From this estimate it follows that there exists a sequence $\{h_n\} \to 0$ for which $\{u^{h_n}\}$, $\{u_i^{h_n}\}$, $\{u_{ik}^{h_n}\}$ converge uniformly in B to a function v and its derivatives v_i, v_{ik}, respectively. In addition, the function v is in $C_{2+\alpha}^B$. But for $h \to 0$, u^h converges simply to u_{x_1}. Therefore, we have proved that u_{x_1} is in $C_{2+\alpha}^B$. Similarly, we can show that u_{x_i} is in $C_{2+\alpha}^B$ $(i = 1, 2, \cdots, n)$, i.e., that u is in $C_{3+\alpha}^B$.

We may easily apply this method to the general nonlinear second order equation

$$F(x_1, x_2, \cdots, x_n, u, u_1, u_2, \cdots, u_{nn}) = 0$$

which is elliptic, i.e., for which the quadratic form

$$\sum_{i,k=1}^n F_{u_{ik}} \xi_i \xi_k$$

is positive definite everywhere. In addition, if we use the theorem on the differentiability of solutions of linear elliptic equations (see page 345), we may easily show that any solution u in $C_{2+\alpha}$, $0 < \alpha < 1$, of the nonlinear equation is infinitely differentiable provided F possesses derivatives of all orders with respect to all its arguments. In fact, it has been shown that the statement is true even if u is in C_2.[1] Furthermore, the solution is analytic if F is analytic (see the next article).

5. Further Results on Elliptic Equations; Behavior at the Boundary. In the preceding article we have shown, with the aid of the Schauder inequalities, that a solution of second order elliptic equations possesses strong properties of smoothness provided that the

---

[1] Cf. L. Nirenberg [3] and [1]. See also C. B. Morrey [4].

coefficients in the equation are sufficiently smooth and provided that some smoothness of the solution is assumed at the outset.

The question of the differentiability and analyticity of solutions of differential equations has been considered by many mathematicians since the classical and fundamental work of S. Bernstein on second order equations.[1]   The paper by E. Hopf[2] has had much influence on later work.   In this paper Hopf makes use of a "parametrix" to obtain integral representations for the derivatives of the solution.   (A parametrix is a fundamental solution of an approximating differential equation having only second order terms and constant coefficients equal to the corresponding coefficients of the given equation at a fixed point.)   In the proof of analyticity of the solution of nonlinear equations, these representation formulas are extended to complex values of the independent variables.

C. B. Morrey[3] has proved analyticity of solutions of general nonlinear elliptic equations of arbitrary order using also integral formulas which are extended to the complex domain.   In addition, he proved analyticity at the boundary of a solution of the Dirichlet problem for such equations, assuming all the data to be analytic.   A. Friedman[4] has established similar results by estimating all derivatives of a solution u and showing that its Taylor series converges to u.

The proofs of analyticity mentioned above assume at the beginning some degree of smoothness of the solution.   For linear elliptic equations of arbitrary order, differentiability of the solutions at the boundary under very mild restrictions has been proved for a wide class of boundary conditions.   We particularly refer to Nirenberg[5] for further references.   For nonlinear equations there is still work to be done in reducing the differentiability hypotheses.   In the case of two independent variables much of the work in this direction has been initiated by C. B. Morrey.[6]   In higher dimensions there have recently appeared the papers by E. de Giorgi,[7] J. Nash[8] and J. Moser.[9] We mention also the interesting work on Monge-Ampère equations

---

[1] See S. Bernstein [1].
[2] See E. Hopf [4].
[3] See C. B. Morrey [2].
[4] See A. Friedman [1].
[5] See L. Nirenberg [2].
[6] See C. B. Morrey [3] and [1].
[7] See E. de Giorgi [1].
[8] See J. Nash [1].
[9] See J. Moser [1].

by E. Heinz[1] which extends results of H. Lewy (see references in Heinz' paper) to nonanalytic equations.

Considering equations of general type, L. Hörmander[2] has characterized all differential operators L, with constant coefficients, so that any solution u of $L[u] = f$ is infinitely differentiable whenever f is. For linear equations with variable coefficients both Hörmander[3] and B. Malgrange[4] have extended such a characterization.

We turn now to the question of the regularity at the boundary of solutions of differential equations satisfying differential boundary conditions. Hörmander[5] considered solutions of a differential equation $L[u] = f$ with constant coefficients in a half-space, which satisfy differential boundary conditions (also with constant coefficients) on the planar boundary. He characterized those operators and boundary conditions with the property that the solution is infinitely differentiable at the boundary provided the given functions (f, and the values of the boundary operators acting on u) are infinitely differentiable. For linear elliptic equations with variable coefficients the differentiability at the boundary of solutions satisfying boundary conditions has been proved for a wide class of boundary conditions. In the case of Dirichlet boundary conditions a clear exposition is given in L. Nirenberg.[6]

In certain so-called "free boundary" problems, part of the problem involves the determination of the domain in which the solution of some elliptic equation exists when given boundary conditions are imposed. These arise in gas dynamics and, of course, in the theory of water waves. One has then to prove not only that the solution is analytic at the unknown boundary, but that the boundary is also analytic. The fundamental work on these problems is due to H. Lewy.[7]

We conclude this article by showing how the differentiability theorems of the preceding article may be extended to solutions of the Dirichlet problem at the boundary. Since it is a question of local behavior we shall assume that a portion Γ_1 of the boundary lies

[1] See E. Heinz [2].
[2] See L. Hörmander [4].
[3] See L. Hörmander [2].
[4] See B. Malgrange [2].
[5] See L. Hörmander [3].
[6] See L. Nirenberg [6].
[7] See H. Lewy [2]. See also P. Garabedian, H. Lewy and M. Schiffer [1].

in a plane, say $x_n = 0$, and establish the differentiability of the solution on and near Γ_1, which we assume to be an open set in $x_n = 0$. For a "smooth" boundary this situation may be realized by a local transformation of variables. By subtraction of a suitable function we may assume that u vanishes on Γ_1.

Consider then a twice continuously differentiable solution of (1) in a domain G, with Γ_1 on its boundary. Suppose that the conditions (a), (b), (c) of the beginning of §7 are satisfied, and that u belongs to $C_{1+\alpha}$ in $G + \Gamma_1$ and vanishes on Γ_1. Then we have the theorem:

If f and the coefficients are in $C_{m+\alpha}$, for m a non-negative integer and $0 < \alpha < 1$, then for every bounded subdomain B with closure in $G + \Gamma_1$, u belongs to $C_{m+2+\alpha}$.

This theorem contains the analogous theorem on p. 345, as a special case.

We remark first in virtue of that case that it suffices to prove the theorem locally, i.e., for small B, for instance for B a small hemisphere with center in Γ_1. We may then also assume that the coefficient c vanishes; this may be achieved by dividing u by a positive solution of the homogeneous equation in B. There is a subdomain A of G, with smooth boundary, containing the closure of B in its interior and whose closure also lies in $G + \Gamma_1$. Let ζ be an infinitely differentiable function defined in A which equals one in B and vanishes, together with all its derivatives, at boundary points of A lying in G.

We shall first prove the theorem for $m = 0$. Setting $v = \zeta u$ we see easily that $L[v]$ belongs to C_α in the closure of A. Applying the basic existence and uniqueness theorems on pages 336 and 339 we see that v belongs to $C_{2+\alpha}^A$, so that u belongs to $C_{2+\alpha}^B$. Now we outline the proof for $m = 1$. Following the proof of the theorem on page 345 we consider the difference quotient u^h, taking differences in the directions parallel to the first $n - 1$ axes, i.e., parallel to the plane containing Γ_1. u^h satisfies

$$L[u^h(P)] = -L^h[u(P_h)] + f^h.$$

Setting $v_h = \zeta u^h$, we see easily that $\| L[v_h] \|_{2+\alpha}^A$ is bounded by a constant independent of h, and it follows as on page 346 that the functions u_{x_i}, $i = 1, \cdots, n - 1$ belong to $C_{2+\alpha}^B$. Thus all derivatives of u of the form $\partial^2 u/\partial x_i \partial x_j$, $i < n$, $j \le n$ belong to $C_{1+\alpha}^B$. With the aid of the differential equation we may express the derivative $\partial^2 u/\partial x_n^2$ in terms of these and lower order derivatives, and we infer that it belongs also to $C_{1+\alpha}^B$. Thus $u \in C_{3+\alpha}^B$. For $m > 1$ this argument also applies.

In a similar way we can prove differentiability, up to any order, of a solution u of the nonlinear equation $F = 0$ on page 346 which vanishes on the boundary if u belongs to $C_{2+\alpha}$ and the boundary and F are sufficiently smooth. For a general result in this direction we refer to the paper by Agmon, Douglis, and Nirenberg.[1]

§8. Solution of the Beltrami Equations

In Ch. III, §2, we have shown that the problem of locally reducing any elliptic equation

$$au_{xx} + 2bu_{xy} + cu_{yy} + \cdots = 0$$

to the form

$$u_{\sigma\sigma} + u_{\rho\rho} + \cdots = 0$$

by a transformation $\sigma(x, y)$, $\rho(x, y)$ is equivalent to that of finding a solution of the Beltrami equations

$$(1) \qquad \sigma_x = \frac{b\rho_x + c\rho_y}{\sqrt{ac - b^2}}, \qquad -\sigma_y = \frac{a\rho_x + b\rho_y}{\sqrt{ac - b^2}},$$

for which $\sigma_x \rho_y - \sigma_y \rho_x \neq 0$. In this section we construct such a solution under the assumption that the functions a, b, c satisfy a Hölder condition with exponent α, $0 < \alpha < 1$.[2]

For any differentiable complex-valued function w of x, y, which we denote by $w(z)$ $(z = x + iy)$, let us introduce the formal differentiation operations

$$\frac{\partial w}{\partial z} \equiv w_z = \frac{1}{2}\left(\frac{\partial w}{\partial x} - i\frac{\partial w}{\partial y}\right), \qquad \frac{\partial w}{\partial \bar{z}} \equiv w_{\bar{z}} = \frac{1}{2}\left(\frac{\partial w}{\partial x} + i\frac{\partial w}{\partial y}\right).$$

It is clear that these operations commute and that the usual rules for differentiation hold,

$$\frac{\partial}{\partial z}\left(u(z)w(z)\right) = uw_z + u_z w,$$

[1] See S. Agmon, A. Douglis, and L. Nirenberg [1].
[2] The result and the idea of the proof are due to A. Korn and L. Lichtenstein. The version given here was found, independently, by L. Bers and by S. S. Chern. The Beltrami equation (1) or (3) can be solved, in a generalized sense, if $\mu(z)$ is a measurable function, $|\mu| \leq \theta < 1$, as was shown by C. B. Morrey. For references see L. Ahlfors and L. Bers [1].

with a similar identity for $\partial/\partial\bar{z}$. If f is a differentiable complex-valued function of w, then

$$[f(w(z))]_z = f_w \, w_z + f_{\bar{w}} \, \bar{w}_z$$

and

$$[f(w(z))]_{\bar{z}} = f_w \, w_{\bar{z}} + f_{\bar{w}} \, \bar{w}_{\bar{z}} \, .$$

Green's identity takes the form

$$(2) \qquad \iint (u_z + w_{\bar{z}}) dx \, dy = \frac{i}{2} \oint (u d\bar{z} - w dz),$$

where $dz = dx + i dy$ and $d\bar{z} = dx - i dy$, and the Laplace operator Δw equals $4w_{z\bar{z}}$. The assertion that $w(z)$ is an analytic function of z may be expressed by the statement

$$w_{\bar{z}} = 0.$$

Now consider the Beltrami equations (1) and set

$$w = \sigma + i\rho.$$

We can immediately derive the equations

$$2w_{\bar{z}}\sqrt{ac - b^2} = (b - ia + i\sqrt{ac - b^2})\rho_x + (c - ib - \sqrt{ac - b^2})\rho_y \, ,$$

$$2w_z\sqrt{ac - b^2} = (b + ia + i\sqrt{ac - b^2})\rho_x + (c + ib + \sqrt{ac - b^2})\rho_y \, .$$

A simple calculation shows that the coefficients ρ_x and ρ_y on the right are proportional, and we find

$$\frac{w_{\bar{z}}}{w_z} = \frac{c - a - 2ib}{c + a + 2\sqrt{ac - b^2}}$$

or

$$(3) \qquad w_{\bar{z}} = \mu w_z , \qquad \mu = \frac{c - a - 2ib}{c + a + 2\sqrt{ac - b^2}} \, .$$

It is easily seen that (1) may in turn be derived from (3) so that (1) and (3) are equivalent. The function μ is Hölder continuous (see §1, 2) because of the conditions on the functions a, b, c, and its absolute value is less than 1.

To show that it suffices to solve equation (3) for a solution w with $w_z \neq 0$ at the points in question, we state the condition $\sigma_x \rho_y - \sigma_y \rho_x \neq 0$ in terms of $w = \sigma + i\rho$. Observing that

$$\sigma_x \rho_y - \sigma_y \rho_x = \frac{a\rho_x^2 + 2b\rho_x \rho_y + c\rho_y^2}{\sqrt{ac - b^2}},$$

we see that our condition is equivalent to $\sigma_x^2 + \sigma_y^2 + \rho_x^2 + \rho_y^2 \neq 0$ which, for a solution of (3), can be expressed as

$$w_z \neq 0.$$

The form (3) of the Beltrami equations implies that the system possesses very special properties. We note first: If f is an analytic function and w is a solution then $f(w(z))$ is also a solution, since

$$f_{\bar{z}} - \mu f_z = f_w(w_{\bar{z}} - \mu w_z) = 0.$$

Conversely, if w is a particular solution of (3) which maps a domain G in the z-plane in a one-to-one continuous way onto a domain G' in the w-plane, and for which $w_z \neq 0$, then any other solution v of (3) in G is an analytic function of w in G'. For we find from (3)

$$0 = v_{\bar{z}} - \mu v_z = v_w(w_{\bar{z}} - \mu w_z) + v_{\bar{w}}(\bar{w}_{\bar{z}} - \mu \bar{w}_z)$$
$$= v_{\bar{w}} \, \bar{w}_{\bar{z}}(1 - |\mu|^2),$$

so that $v_{\bar{w}} = 0$.

Thus, if we have a particular solution w which has the required mapping property, the problem of finding a solution satisfying other conditions (say on the boundary) may be reduced to that of finding an analytic function $f(w)$ satisfying certain analogues of these conditions. For example, the problem of finding a solution of (3) in a circle S in G having prescribed real part on the boundary and prescribed value at a particular point P is reduced to the problem of finding an analytic function on the image S' of S having prescribed real part on the boundary and prescribed value at P'. We may therefore consider this problem as solved.

In order to construct a particular solution of (3) which provides the required mapping, at least in a small neighborhood, we wish to establish the following theorem:

If $\mu(z)$ satisfies a Hölder condition with exponent α in a neighborhood of $z = 0$ and $|\mu(z)| < 1$, then equation (3) has a solution $w(z)$ in a neighborhood of the origin such that $w_z(0) \neq 0$ and the derivatives of $w(z)$ satisfy a Hölder condition with exponent α.

It suffices to prove the theorem under the assumption that

$$\mu(0) = 0.$$

For if we introduce as new variable $\zeta = z + \mu(0)\bar{z}$ then

$$w_z = w_\zeta + w_{\bar{\zeta}} \, \overline{\mu(0)}, \qquad w_z = w_\zeta \, \mu(0) + w_{\bar{\zeta}},$$

and equation (3) takes the form

$$w_{\bar{\zeta}} = \hat{\mu}(\zeta)w_{\zeta},$$

where

$$\hat{\mu}(\zeta) = \frac{\mu(z) - \mu(0)}{1 - \overline{\mu(0)}\mu(z)}.$$

We note that if $|\mu| < 1$ the transformation $z \rightarrow \zeta$ has a positive Jacobian and $|\hat{\mu}| < 1$. Moreover, $\hat{\mu}(\zeta)$ satisfies a Hölder condition with exponent α and $\hat{\mu}(0) = 0$.

In order to prove the theorem we first study the equation $u_{\bar{z}} = f$, where f is assumed to satisfy a Hölder condition with exponent α in a circle $S_r : |z| < r$. Setting

$$\pi v = \iint_{S_r} f(\xi, \eta) \log |\zeta - z| \, d\xi \, d\eta$$

and remembering $\Delta v = 4 \, (\partial/\partial \bar{z})(\partial/\partial z) \, v = f(\xi, \eta)$ we are led to the following lemmas.

LEMMA 1: *The function*

$$u(z) = -\frac{1}{\pi} \iint_{S_r} \frac{f(\zeta)}{\zeta - z} \, d\xi \, d\eta \qquad (\zeta = \xi + i\eta)$$

has continuous partial derivatives in S_r given by

$$u_{\bar{z}}(z) = f(z), \qquad u_z(z) = -\frac{1}{\pi} \iint_{S_r} \frac{f(\zeta) - f(z)}{(\zeta - z)^2} \, d\xi \, d\eta,$$

so that u is a particular solution of the equation $u_{\bar{z}} = f$.

We leave the proof of this lemma to the reader; it is similar to the proof of the second theorem of §1, 2. There the differentiability of a potential of a Hölder continuous mass distribution was demonstrated and expressions for the derivatives were obtained.

LEMMA 2: *In S_r let $f(z)$ satisfy a Hölder condition with exponent α and coefficient H, and set*

$$p(z) = -\frac{1}{\pi} \iint_{S_r} \frac{f(\zeta) - f(z)}{(\zeta - z)^2} \, d\xi \, d\eta.$$

Then there exists a constant C depending only on α such that

(4) $$|p(z)| \leq CHr^{\alpha},$$

and $p(z)$ satisfies a Hölder condition with exponent α and coefficient CH.

Proof: We derive (4) by taking absolute values under the integral sign in the expression for $p(z)$ and using the inequality

$$| f(\zeta) - f(z) | \leq H | \zeta - z |^{\alpha}.$$

In order to prove the Hölder continuity of $p(z)$ let z_1 and $z_3 \neq z_1$ be fixed in S_r and set $\delta = | z_1 - z_3 |$. Let z_2 be a point in S_r satisfying $| z_2 - z_1 | \leq \delta, | z_2 - z_3 | \leq \delta$ and $1 - | z_2 | \geq \delta/10$. We shall show that, for some constant $\tilde{C}_1, | p(z_i) - p(z_2) | \leq \tilde{C} H \delta^{\alpha}, i = 1,3$, from which the desired inequality $| p(z_1) - p(z_3) | \leq 2\tilde{C} H \delta^{\alpha}$ follows. We consider only the case $i = 1$:

$$(5) \qquad p(z_1) - p(z_3) = -\frac{1}{\pi} \iint_{S_r} g(\zeta) \, d\xi \, d\eta$$

where

$$g(\zeta) \equiv \frac{f(z_1) - f(\zeta)}{(z_1 - \zeta)^2} - \frac{f(z_2) - f(\zeta)}{(z_2 - \zeta)^2}.$$

Assume that Δ_1 (which has boundary $\dot{\Delta}_1$) is the intersection of S_r and the circle with center z_1 and radius $2\delta = 2 | z_1 - z_2 |$, and denote by Δ_2 the complement of Δ_1 in S_r. Then

$$\frac{1}{\pi} \iint_{\Delta_1} | g | \, d\xi \, d\eta \leq \frac{H}{\pi} \iint_{|\zeta - z_1| < 2\delta} \frac{d\xi \, d\eta}{| \zeta - z_1 |^{2-\alpha}}$$

$$+ \frac{H}{\pi} \iint_{|\zeta - z_2| < 3\delta} \frac{d\xi \, d\eta}{| \zeta - z_2 |^{2-\alpha}}$$

$$\leq C_1 H \delta^{\alpha},$$

where C_1 is a constant depending only on α. To estimate the integral of g over Δ_2 we observe first that

$$g(\zeta) = \frac{f(z_1) - f(z_2)}{(z_2 - \zeta)^2} + (f(z_1) - f(\zeta)) \left[\frac{1}{(z_1 - \zeta)^2} - \frac{1}{(z_2 - \zeta)^2} \right]$$

$$\equiv \frac{f(z_1) - f(z_2)}{(z_2 - \zeta)^2} + \frac{(f(z_1) - f(\zeta))(z_2 - z_1)}{(z_1 - \zeta)^3} \frac{z_1 - \zeta}{z_2 - \zeta} \left(1 + \frac{z_1 - \zeta}{z_2 - \zeta} \right)$$

$$\equiv g_1(\zeta) + g_2(\zeta).$$

Then

$$\frac{1}{\pi} \iint_{\Delta_2} g_1(\zeta) \, d\xi \, d\eta = \frac{1}{\pi} (f(z_1) - f(z_2)) \iint_{\Delta_2} \frac{d\xi \, d\eta}{(z_2 - \zeta)^2}.$$

By Green's identity (2) and setting $u = 1/(z_2 - \zeta)$, $w = 0$, we have

$$\iint_{\Delta_2} \frac{d\xi \, d\eta}{(z_2 - \zeta)^2} = \frac{i}{2} \oint_{|\zeta|=r} \frac{d\bar{\zeta}}{z_2 - \zeta} - \frac{i}{2} \oint_{\dot{\Delta}_1} \frac{d\bar{\zeta}}{z_2 - \zeta}$$

$$= -\frac{i}{2} \int_{\dot{\Delta}_1} \frac{d\bar{\zeta}}{z_2 - \zeta}$$

since the integral on $|\zeta| = r$ vanishes. Thus

(6) $$\left| \frac{1}{\pi} \iint_{\Delta_2} g_1 \, d\xi \, d\eta \right| \le \frac{H}{2\pi} \delta^\alpha \oint_{\dot{\Delta}_1} \frac{|d\bar{\zeta}|}{|z_2 - \zeta|} \le C_2 H \delta^\alpha,$$

where C_2 is an absolute constant, since $|z_2 - \zeta| \ge \delta/10$ for $\zeta \, \epsilon \, \dot{\Delta}_1$.

Finally we observe that in Δ_2

$$\left| \frac{z_1 - \zeta}{z_2 - \zeta} \right| \le 2,$$

so that

$$|g_2(\zeta)| \le \frac{6H\delta}{|z_1 - \zeta|^{3-\alpha}}$$

and

$$\frac{1}{\pi} \iint_{\Delta_2} |g_2(\zeta)| \, d\xi \, d\eta \le \frac{6H\delta}{\pi} \iint_{|\zeta-z_1|>2\delta} \frac{d\xi \, d\eta}{|z_1 - \zeta|^{3-\alpha}} \le C_3 H \delta^\alpha,$$

where C_3 is a constant depending only on α. This combined with (5) and (6) yields the required inequality

$$|p(z_1) - p(z_2)| \le (C_1 + C_2 + C_3)H|z_1 - z_2|^\alpha.$$

We are now in a position to prove the theorem. Suppose w is the desired solution; then, in virtue of Lemma 1, w_z differs from

$$-\frac{1}{\pi} \iint_{S_r} \frac{\mu(\zeta)w_z(\zeta) - \mu(z)w_z(z)}{(\zeta - z)^2} \, d\xi \, d\eta$$

by an analytic function. Taking as this analytic function the constant 1 we proceed to solve the following equation for w_z :

$$w_z = -\frac{1}{\pi} \iint_{S_r} \frac{\mu(\zeta)w_z(\zeta) - \mu(z)w_z(z)}{(\zeta - z)^2} \, d\xi \, d\eta + 1.$$

Then w is easily obtained and satisfies equation (3), as will be shown at the end of this section.

Let $C(r, \alpha)$ denote the set of complex-valued functions $\omega(z)$ defined in S_r and satisfying a Hölder condition with exponent $\alpha < 1$. For ω in $C(r, \alpha)$ we introduce the norm

$$\| \omega \|_r = \text{l.u.b.} \, | \omega(z) | + r^\alpha \, \underset{z_1 \neq z_2}{\text{l.u.b.}} \, \frac{| \omega(z_1) - \omega(z_2) |}{| z_1 - z_2 |^\alpha} \, ;$$

then $C(r, \alpha)$ is a Banach space. One can easily see that for ω and τ in $C(r, \alpha)$

$$\| \omega\tau \|_r \leq \| \omega \|_r \| \tau \|_r .$$

Now let $\mu(z)$ be the given function defined in S_r ; it satisfies a Hölder condition with exponent α, and moreover

$$\mu(0) = 0.$$

We define the operator T in $C(r, \alpha)$ by

$$T[\omega] = - \frac{1}{\pi} \iint_{S_r} \frac{\mu(\zeta)\omega(\zeta) - \mu(z)\omega(z)}{(\zeta - z)^2} \, d\xi \, d\eta.$$

Clearly, T is a linear operator and, in virtue of Lemma 2,

$$\| T[\omega] \|_r \leq Kr^\alpha r^{-\alpha} \| \mu\omega \|_r + r^\alpha Kr^{-\alpha} \| \mu\omega \|_r$$

$$\leq 2K \| \mu \|_r \| \omega \|_r .$$

Since $\mu(0) = 0$, the norm $\| \mu \|_r$ tends to zero as $r \to 0$. Therefore, for r sufficiently small, say $r < r_0$, we have

$$2K \| \mu \|_r < \tfrac{1}{2},$$

so that

(7) $$\| T[\omega] \|_r < \tfrac{1}{2} \| \omega \|_r \qquad\qquad \text{for } r < r_0 .$$

Now we wish to solve the equation

$$\omega = T[\omega] + 1 \equiv \hat{T}[\omega]$$

in a circle S_r with $r < r_0$. According to (7) the class of functions ω in $C(r, \alpha)$ satisfying

$$\| \omega \|_r < 2$$

is mapped into itself by the transformation $\hat{T}$ which is contracting since

$$\| \hat{T}[\omega_1] - \hat{T}[\omega_2] \|_r \leq \tfrac{1}{2} \| \omega_1 - \omega_2 \|_r .$$

By a well-known argument the functions ω_n which we define by iteration,

$$\omega_1 = 0, \qquad \omega_{n+1} = T[\omega_n] + 1 \qquad (n = 1, 2, \cdots),$$

converge in the norm $\| \ \|_r$ to a function ω satisfying

$$\omega = T[\omega] + 1$$

with $\| \omega \|_r < 2$. Moreover, we see that

$$| \omega(z) - 1 | < \tfrac{1}{2}\cdot 2 = 1$$

and therefore $\omega(z) \neq 0$ in S_r.

We complete the proof of the theorem by showing that the function

$$w(z) = -\frac{1}{\pi} \iint_{S_r} \frac{\mu(\zeta)\omega(\zeta)}{(\zeta - z)} \, d\xi \, d\eta + z$$

has the desired properties. By Lemma 1, w has continuous partial derivatives given by

$$w_{\bar{z}} = \mu\omega$$

and

$$w_z = T[\omega] + 1 = \omega;$$

consequently w satisfies the differential equation (3) and $w_z \neq 0$. Since ω is in $C(r, \alpha)$ the derivatives of w satisfy in S_r a Hölder condition with exponent α.

§9. The Boundary Value Problem for a Special Quasi-Linear Equation. Fixed Point Method of Leray and Schauder

In this section we shall illustrate the use of a topological method in proving existence theorems. This "fixed point method," related to ideas of Poincaré, was clearly envisioned by G. D. Birkhoff and Kellogg, and developed into a powerful tool by Schauder and Leray.[1]

Instead of the general theory due to Schauder and Leray, we shall use only a special result known as

SCHAUDER'S FIXED POINT THEOREM: *If T is a continuous mapping of a closed convex compact set of a Banach space into itself then T has a fixed point.*

---

[1] See G. D. Birkhoff and O. D. Kellogg [1] and J. Leray and J. Schauder [1].

The theorem is proved in §15 of the supplement to this chapter, where "convex compact sets in a Banach space" are defined. Consider a quasi-linear elliptic equation

$$(1) \qquad A(x, y, z)z_{xx} + 2B(x, y, z)z_{xy} + C(x, y, z)z_{yy} = 0$$

in a bounded domain G whose boundary consists of a finite number of distinct closed curves Γ. Suppose that G is a smooth domain (as in §7, 1, page 335) and that ϕ is a smooth function defined on the boundary Γ. Here the smoothness of G and ϕ may be described in terms of the arc length s on Γ as follows: The functions $x(s)$, $y(s)$, describing a boundary curve, and the function $\phi(s)$ have Hölder continuous first and second derivatives with exponent α, $0 < \alpha < 1$. For functions A, B, C possessing certain properties (to be specified later), we seek a solution $z(x, y)$ of (1) which equals ϕ on Γ. By the maximum principle any solution will satisfy the inequality

$$| z | \leq \max | \phi | = M_0.$$

Thus we merely consider A, B, C defined for (x, y) in $G + \Gamma$ and $| z | \leq M_0$, and stipulate the following properties:

 (a) $A\xi^2 + 2B\xi\eta + C\eta^2 \geq m(\xi^2 + \eta^2)$ for all real ξ, η,
 (b) $| A |, | B |, | C | \leq M$, and
 (c) A, B, C satisfy Hölder conditions with respect to x, y, z with exponent α and coefficient M. (Here m, M, and α are some fixed positive constants.)

Using the Schauder theory of linear elliptic equations of §7, we shall prove the following existence theorem:

Under conditions (a), (b), (c) *and the postulated conditions on G and ϕ, there exists a solution $z(x, y)$ of equation* (1) *which equals ϕ on Γ. Furthermore, z belongs to $C_{2+\alpha}$ (see §7, 1).*

The proof is based on an *a priori* estimate for solutions of a linear elliptic equation of the form

$$(2) \qquad L[u] \equiv a(x, y)u_{xx} + 2b(x, y)u_{xy} + c(x, y)u_{yy} = 0$$

in a bounded domain G whose boundary consists of a finite number of closed curves with continuously turning tangent. In addition, we assume that G has the following property: For a positive number R and every point Q on the boundary Γ, there exists a circle of radius R whose only common point with $G + \Gamma$ is Q. We also assume that the coefficients a, b, c are Hölder continuous in u and that the equation is uniformly elliptic; i.e., that for some positive constants m, M the inequalities

$$(3) \qquad a\xi^2 + 2b\xi\eta + c\eta^2 \geq m(\xi^2 + \eta^2), \qquad |a|, |b|, |c| \leq M$$

hold in G.

A PRIORI ESTIMATE: Let u be a solution of (2) which, together with its first order derivatives, is continuous in $G + \Gamma$. Assume that there exists a twice continuously differentiable function $\bar{u}$ in $G + \Gamma$ which equals u on Γ, which together with its first order derivatives is continuous in $G + \Gamma$, and whose first and second order derivatives are bounded in absolute value by a constant K. Then there exists a constant k depending only on m, M, and G (i.e., independent of the coefficients) such that

$$(4) \qquad |u_x|, |u_y| \leq kK$$

throughout G. We shall establish this *a priori* estimate later in this section.

In order to apply this general estimate to our nonlinear equation we observe first that, because of the assumed smoothness of G and ϕ, it is not difficult to construct a continuous function $\bar{z}$ in $G + \Gamma$ that equals ϕ on Γ and has first and second order derivatives which are bounded in absolute value by some constant $\bar{K}$ and are continuous in $G + \Gamma$ and G, respectively. If we then regard the desired solution z of (1) as a solution of a linear equation of the form (2) with $a(x, y) = A(x, y, z(x, y))$, etc., the estimate (4) yields

$$(5) \qquad |z_x|, |z_y| \leq k\bar{K}.$$

We observe further that if a function z has continuous first derivatives in $G + \Gamma$ bounded in absolute value by $k\bar{K}$, it satisfies a Lipschitz condition

$$|z(P) - z(Q)| \leq \kappa k\bar{K} |P - Q| \qquad \text{for all } P, Q \text{ in } G + \Gamma,$$

where κ is a constant depending only on the domain G. Let us now prove the existence theorem stated earlier. Consider the Banach space C_0 (see §7, 1) of continuous functions z in $G + \Gamma$, with

$$\| z \| = \max |z|.$$

Denote by S the subset of functions z in C_0 satisfying

$$|z| \leq \max |\phi|$$

and the Lipschitz condition

$$(6) \qquad \frac{|\, z(P) - z(Q)\,|}{|\, P - Q\,|} \leq \kappa k \bar{K} \qquad \text{for all } P, Q \text{ in } G + \Gamma.$$

It is easily seen that S is a compact, convex[1] subset of C_0.

After these preparations we proceed by a typical process of iterations: If in the coefficients $A(x, y, z)$, $B(x, y, z)$, and $C(x, y, z)$ in (1) we substitute for z a function $z(x, y)$ belonging to S, then A, B and C become functions a, b and c of x, y alone. From the Hölder continuity of A, B, C (condition (c)) and from (6) it follows that a, b, c satisfy a Hölder condition with exponent α and some fixed coefficient which is independent of the particular function $z(x, y)$.

Now with such a function $z(x, y)$ inserted, we solve the linear elliptic equation

$$A(x, y, z(x, y))u_{xx} + 2B(x, y, z(x, y))u_{xy} + C(x, y, z(x, y))u_{yy} = 0$$

for a function u which equals ϕ on Γ. It follows from the first theorem of §7, 2 that this is possible. We use here the postulated "smoothness" of G and ϕ. The solution u is unique, belongs to $C_{2+\alpha}$ and satisfies $|\, u\,| \leq \max |\, \phi\,|$. By the *a priori* estimate u satisfies (5), and hence (6), so that u lies in S. Therefore, the transformation

$$u = T[z]$$

defined as the "solution operator" of this linear problem is a transformation of S into itself.

If we can show that T is a continuous transformation then, by Schauder's fixed point theorem, T has a fixed point z. In this case z belongs to $C_{2+\alpha}$ and is the desired solution of equation (1). In order to show the continuity of T let $\{z^{(n)}\}$ be a sequence of functions in S converging uniformly to a function z in S, and set $u^{(n)} = T[z^{(n)}]$. The functions $u^{(n)}$ are solutions of equations whose coefficients satisfy (a), (b), and a uniform Hölder condition independent of n. Furthermore, $|\, u^{(n)}\,| \leq \max |\, \phi\,|$. Applying the Schauder estimates up to the boundary we infer that the norms $\| u^{(n)} \|_{2+\alpha}$ are uniformly bounded. Therefore, we may select a subsequence of the $u^{(n)}$ which converges together with the first and second order derivatives to a function u and its corresponding derivatives. But then u satisfies the limit differential equation

$$A(x, y, z(x, y)u_{xx} + 2B(x, y, z(x, y))u_{xy} + C(x, y, z(x, y))u_{yy} = 0$$

---

[1] For a definition of a compact convex set see Supplement, §15, p. 404.

and equals ϕ on Γ, that is, $u = T[z]$. Thus a subsequence of the $u^{(n)}$ converges to $T[z]$. From the uniqueness of the limit function it follows that the original sequence $\{u^{(n)}\}$ converges to $T[z]$, and the continuity of T is established.

In the remainder of this section we derive the *a priori* estimate (4). Our proof is divided into two parts. First we show that the functions u_x and u_y assume their maximum and minimum on the boundary. Then we establish the validity of inequality (4) at all boundary points which implies the validity of (4) throughout G.

We derive a maximum principle for u_x and u_y under the assumption that the coefficients of equation (2) are Hölder continuous. (Another derivation may be given which requires only the ellipticity (3) of the equation.) It suffices to consider the function u_x and to show that its maximum in any circle D in G occurs on the boundary. (It then follows that also the minimum of u_x occurs on the boundary.)

In case u has continuous third order derivatives and the coefficients a, b, c are once differentiable, we find, on dividing (2) by c and differentiating with respect to x, that $u_1 = u_x$ satisfies the elliptic differential equation

$$\left(\frac{a}{c} u_{1x} + \frac{2b}{c} u_{1y}\right)_x + u_{1yy} = 0,$$

to which the maximum principle of §6, 4 may be applied to give the desired result.

If the coefficients a, b, c are merely Hölder continuous in a circle D we may approximate them uniformly by twice differentiable functions $a^{(n)}, b^{(n)}, c^{(n)}$ satisfying a uniform Hölder condition. By the Schauder theory of linear equations, the equations

$$a^{(n)}u_{xx}^{(n)} + 2b^{(n)}u_{xy}^{(n)} + c^{(n)}u_{yy}^{(n)} = 0,$$

with $u^{(n)} = u$ on the boundary of D, will have solutions $u^{(n)}$ in D converging uniformly to u. Furthermore $u_x^{(n)} \to u_x$, $u_y^{(n)} \to u_y$. Since the $a^{(n)}, b^{(n)}, c^{(n)}$ are twice differentiable, the functions $u^{(n)}$ possess continuous third order derivatives (see §7, 4). Therefore, the previous paragraph applies and we may conclude that max $u_x^{(n)}$, and hence also max u_x, occurs on the boundary of D.

To show that (4) holds on the boundary we set $w = u - \bar{u}$. Since w vanishes on Γ and satisfies the differential equation

$$L[w] = -L[\bar{u}] = f,$$

and since, by the conditions on $\bar{u}$,

(7) $$|f| \leq 3M\bar{K},$$

the function w and the domain G satisfy all the conditions of the lemma at the end of §7, 3. We conclude that on Γ

$$|w_x|, |w_y| \leq 3k_2 M\bar{K},$$

where k_2 is a constant depending only on m, M, and G. Consequently

$$|u_x|, |u_y| \leq \bar{K} + 3k_2 M\bar{K}$$
$$\leq (1 + 3Mk_2)\bar{K} \qquad \text{on } \Gamma,$$

which is the desired inequality (4) on the boundary.

§10. Solution of Elliptic Differential Equations by Means of Integral Equations

The preceding sections are supplemented here by a brief account of a quite different approach to the solution of elliptic partial differential equations, generalizing and extending the method of integral equations of §4, 3.[1]

General considerations of functional analysis, as they appear explicitly and implicitly on various occasions in this work, suggest the use of integral equations.

We consider a linear elliptic differential operator $L[u] = f$ and seek to find the inverse operator

$$u = R[f] = L^{-1}[f],$$

where f is arbitrarily given in a suitably restricted function space (e.g., f continuous in $G + \Gamma$) and u is subject to appropriate boundary conditions (e.g., $u = 0$ on Γ in the case of a second order operator L).

Now a differential operator $L[u]$ transforms a function into another, $L[u] = f$, with generally a lower order of regularity, e.g., transforms a twice continuously differentiable function into a merely continuous function. More precisely, it transforms a certain function space S, in which it is applicable, into a wider function space $\tilde{S}$. Conversely, the inverse transformation $L^{-1}[f]$ transforms $\tilde{S}$ into a subspace S; in this sense (as in the stricter sense based on the concept of norm, see §7) it is a smoothing transformation. Such smoothing transforma-

---

[1] This method, based on the concept of a "parametrix", was discovered by E. E. Levi [4] and later by D. Hilbert [1], pp. 1–65. For a simplified development as well as complete results and proofs see F. John [1].

tions are in principle easy to handle.   In our case it is a matter of
analytic technique to represent the smoothing transformation L^{-1}
by means of integral operators which then may lead to the convergent
proçesses of Fredholm theory (cf. Vol. I, Ch. III) for finding the
solution u.

**1. Construction of Particular Solutions. Fundamental Solutions.
Parametrix.**   We first consider the case of a linear differential equation of second order for a function $u(x, y)$ of two independent variables and assume the differential equation in the form

(1) $$L[u] \equiv \Delta u + au_x + bu_y + cu = f$$

with a, b, c, f continuously differentiable in $G + \Gamma$.

For analytic coefficients a, b, c the question of existence of solutions
of (1) in sufficiently small domains G is answered in the affirmative
by the Cauchy-Kowalewski existence theorem of Ch. I, §7, 4.   Under
less restrictive assumptions for a, b, c, f, however, the proof of the
existence of even only a particular solution of (1) requires other
methods, as for example the following one introduced by E. E. Levi.

We consider the function

(2) $$\psi(x, y; \xi, \eta) = -\log \sqrt{(x - \xi)^2 + (y - \eta)^2} = -\log r,$$

called the "parametrix".   This is a function of x, y and of the parameter point (ξ, η) which possesses at $x = \xi$, $y = \eta$ the characteristic
singularity suggested by the principal part Δu of $L[u]$.   The parametrix does not satisfy the differential equation, but $L[\psi]$ has at $x = \xi$,
$y = \eta$ a singularity of only first order in $1/r$.

With arbitrary continuously differentiable $\rho(x, y)$ the integral

(3) $$u = \iint_G \psi(x, y; \xi, \eta)\rho(\xi, \eta) \, d\xi \, d\eta$$

as well as the more general expression

(4) $$u = \omega(x, y) + \iint_G \psi(x, y; \xi, \eta)\rho(\xi, \eta) \, d\xi \, d\eta$$

will not satisfy (1) if we choose $\omega(x, y)$ to be three times continuously
differentiable in G but otherwise arbitrary.   However, by a proper
choice of ρ for given ω we can construct a u which satisfies (1).

To prove this we substitute (4) into (1).   Because of the assumptions regarding the differentiability of ρ we have (see Vol. I, Ch. V, §14, 5)

$$\Delta u = \Delta \omega - 2\pi\rho,$$

and furthermore

$$L[u] = L[\omega] - 2\pi\rho + \iint_G (a\psi_x + b\psi_y + c\psi)\rho(\xi, \eta)\, d\xi\, d\eta.$$

If for brevity we set

(5)
$$K(x, y; \xi, \eta) = \frac{1}{2\pi} (a\psi_x + b\psi_y + c\psi)$$

$$= -\frac{1}{2\pi}\left[a(x, y)\, \frac{x - \xi}{r^2} + b(x, y)\, \frac{y - \eta}{r^2} + c(x, y)\log r \right]$$

and

$$g(x, y) = \frac{1}{2\pi} (L[\omega] - f)$$

then we obtain for ρ the *integral equation*

(6) $$\rho(x, y) = \iint_G K(x, y; \xi, \eta)\rho(\xi, \eta)\, d\xi\, d\eta + g(x, y).$$

Fredholm's theory cannot be applied directly to this integral equation since the kernel (5) at the point $x = \xi$, $y = \eta$ becomes infinite like $1/r$ and thus is not square integrable. But it is easy to see that the iterated kernel

$$K_2(x, y; \xi, \eta) = \iint_G K(x, y; s, t)K(s, t; \xi, \eta)\, ds\, dt$$

is square integrable. Instead of (6) we therefore first consider the iterated integral equation

(7) $$\rho(x, y) = \iint_G K_2(x, y; \xi, \eta)\rho(\xi, \eta)\, d\xi\, d\eta + h(x, y)$$

with

$$h = g + \iint_G K(x, y; \xi, \eta)g(\xi, \eta)\, d\xi\, d\eta.$$

To this equation Fredholm's theorems can be applied.

The homogeneous integral equation corresponding to (7),

$$(8) \qquad \rho(x, y) = \iint_G K_2(x, y; \xi, \eta)\rho(\xi, \eta) \, d\xi \, d\eta,$$

can have a solution ρ which does not vanish identically only if

$$\iint_G \left[\iint_G K_2^2(x, y; \xi, \eta) \, dx \, dy \right] d\xi \, d\eta \geq 1.$$

Therefore, if we choose the domain G sufficiently small, so that the value of the integral on the left becomes less than 1, (8) has only the solution $\rho \equiv 0$.

The function $g(x, y) = (1/2\pi)(L[\omega] - f)$ is continuous and continuously differentiable in G; hence the same is true of the function $h(x, y)$ since the theorem proved in §1, 2, p. 246 for three variables also holds for two variables.

Thus we see from Fredholm's theorems: *For sufficiently small domains G and arbitrary h, there exists a solution of the integral equation* (7). This solution is continuously differentiable and satisfies the original integral equation (6). In fact, if we set for the moment

$$(6') \qquad v = g + \iint_G K(x, y; \xi, \eta)\rho(\xi, \eta) \, d\xi \, d\eta,$$

then (7) means

$$\rho = \iint_G K(v - g) \, d\xi \, d\eta + h.$$

After multiplication by K and integration it follows that

$$v - g = \iint_G K_2(v - g) \, d\xi \, d\eta + \iint_G Kh \, d\xi \, d\eta$$

or

$$v = \iint_G K_2 \, v \, d\xi \, d\eta + h;$$

i.e., v also satisfies equation (7) and must, because of the uniqueness of the solution, agree with ρ. For $v = \rho$ however, (6') is identical with integral equation (6).

If we now use this $\rho(x, y)$ in the expression (4)

$$u = \omega + \iint_G \psi(x, y; \xi, \eta)\rho(\xi, \eta) \, d\xi \, d\eta,$$

then

$$L[u] = L[\omega] + 2\pi\left\{\iint_G K\rho \, d\xi \, d\eta - \rho\right\} = f;$$

i.e., u is a solution of (1) with continuous derivatives up to the second order in G, still depending on the arbitrary function ω. Thus we have shown the existence of solutions of our differential equation in a sufficiently small domain G.

In particular, if we set

$$\omega = -\log \sqrt{(x - x_0)^2 + (y - y_0)^2}$$

and choose for G a sufficiently small domain G^* around the point (x_0, y_0) from which, however, this point is excluded by a small circular disk of radius δ, we obtain, according to the above result, solutions in G^* of the form

$$u^*(x, y) = -\log \sqrt{(x - x_0)^2 + (y - y_0)^2}$$

$$+ \iint_{G*} \psi(x, y; \xi, \eta)\rho^*(\xi, \eta) \, d\xi \, d\eta.$$

It can easily be shown that in the limit $\delta \to 0$ the function ρ^* converges to a function ρ, such that

$$\iint_G \psi(x, y; \xi, \eta)\rho(\xi, \eta) \, d\xi \, d\eta$$

possesses continuous derivatives up to the second order in the limit domain $G = \lim G^*$. Then the function

$$\gamma(x, y; x_0, y_0) = -\log \sqrt{(x - x_0)^2 + (y - y_0)^2}$$

(9)
$$+ \iint_G \psi(x, y; \xi, \eta)\rho(\xi, \eta) \, d\xi \, d\eta$$

satisfies the equation $L[\gamma] = f$ everywhere in G except at the point

$x = x_0$, $y = y_0$; furthermore, since $\gamma - \log 1/r$ is regular everywhere in G, $\gamma(x, y; x_0, y_0)$ is the fundamental solution of equation (1).

2. Further Remarks. The specific difficulty of the method lies in the fact that the kernel of the integral equation obtained is singular. While in our case this difficulty could be overcome by passing to the iterated integral equation, it is preferable to modify the method[1] and thereby obtain one that is easier to apply. We can replace the above parametrix ψ by another parametrix $\psi'(x, y; \xi, \eta)$ for which $L[\psi']$ has at $x = \xi$, $y = \eta$ a singularity of lower order than $L[\psi]$ and then operate with ψ'.

The method of the generalized parametrix can be extended to any number of independent variables and to higher order linear differential equations as well as to systems.

For details and for further steps leading to the solution of the boundary value problem the reader is referred to the literature.[1]

Appendix to Chapter IV

Nonlinear Equations

Little is known concerning boundary value problems for general nonlinear differential equations (in more than two independent variables)[2]

$$(1) \qquad F(x_1, \cdots, x_n, u, u_1, \cdots, u_n, u_{11}, \cdots, u_{nn}) = 0,$$

where we suppose F to be twice continuously differentiable in its $p = \frac{1}{2}(n^2 + n) + 2n + 1$ arguments in a p-dimensional domain.

---

[1] See F. John [1].

[2] We refer the reader to the papers listed in article 5 of §4 and give here some additional references. For $n = 2$ a considerable amount of work has been done. We mention first the basic paper by C. B. Morrey [3]. For related work see L. Nirenberg [3] and [1] and C. B. Morrey [4], together with their bibliographies, and L. Bers and L. Nirenberg [1] and [2]. The article by E. Heinz [1] contains further references, in particular to the work of A. V. Pogorelov. For work on equations of the Monge-Ampère type we have already referred in §7, article 5, to a paper of E. Heinz [2], which contains reference to the work of H. Lewy.

We mention also, for such equations in higher dimensions, A. D. Aleksandrov [1].

In connection with equations of subsonic flow in gas dynamics see L. Bers [3], which contains an extensive bibliography.

For work in higher dimensions see, in addition, H. O. Cordes [2]. See also R. Finn and D. Gilbarg [1] and A. A. Kiselev and O. A. Ladyzhenskaya [1].

1. Perturbation Theory. Consider a solution u of equation (1) in a smooth domain G (as in Ch. IV, §7, 1) for which (1) is elliptic. Without restricting generality we can assume $u \equiv 0$, in which case the matrix

$$\left(\frac{\partial F}{\partial u_{ij}} (x_1, \cdots, x_n, 0, \cdots, 0) \right)$$

is positive definite at every point in G. Let

$$R(x_1, \cdots, x_n, u, u_1, \cdots, u_n, u_{11}, \cdots, u_{nn})$$

be a given twice continuously differentiable function. We ask the question: *Does there exist a solution of*

$$F(x_1, \cdots, x_n, u, u_1, \cdots, u_n, u_{11}, \cdots, u_{nn})$$

$$(2) \qquad = \epsilon R(x_1, \cdots, x_n, u, u_1, \cdots, u_n, u_{11}, \cdots, u_{nn}),$$

$$u = \phi \text{ on } \Gamma,$$

for sufficiently small ϵ with $\| \phi \|'_{2+\alpha} < \epsilon$ (see Ch. IV, §7, 1)?
In case

$$\frac{\partial F}{\partial u} (x_1, \cdots, x_n, 0, \cdots, 0) \leq 0,$$

the answer is in the affirmative.

To prove this statement we write the equation in the form

$$L[u] \equiv \sum_{i,j=1}^{n} \frac{\partial F}{\partial u_{ij}} (x_1, \cdots, x_n, 0, \cdots, 0) u_{ij}$$

$$+ \sum_{i=1}^{n} \frac{\partial F}{\partial u_i} (x_1, \cdots, x_n, 0, \cdots, 0) u_i$$

$$+ \frac{\partial F}{\partial u} (x_1, \cdots, x_n, 0, \cdots, 0) u$$

$$= L[u] - F(x_1, \cdots, x_n, u, u_1, \cdots, u_n, u_{11}, \cdots, u_{nn})$$

$$+ \epsilon R(x_1, \cdots, x_n, u, u_1, \cdots, u_n, u_{11}, \cdots, u_{nn})$$

$$= Q[u],$$

that is, we keep the linear, or first order, part of the expansion of F in u and its derivatives on the left-hand side of the equation and combine the nonlinear parts in the operator $Q[u]$. We now define recursively the sequence of functions u_m as solutions of

$$L[u_m] = Q[u_{m-1}],$$
$$u_m = \phi \text{ on } \Gamma \qquad (m = 1, 2, \cdots),$$
$$u_0 = 0.$$

That the solutions u_m exist follows from the Schauder theory of Ch. IV, §7, 2, and from our assumption $\partial F/\partial u \leq 0$.

With the aid of the Schauder estimates of Ch. IV, §7, 1, it is not difficult to show that for ϵ sufficiently small the functions u_m converge to a solution of (2).

A similar perturbation theorem for elliptic equations of arbitrary order is proved in the paper by Agmon, Douglis, and Nirenberg[1] as an application of the Schauder estimates for such equations.

2. The Equation $\Delta u = f(x, u)$. We consider the boundary value problem

(3) $$\Delta u = f(x_1, x_2, \cdots, x_n, u) = f(x, u), \qquad u = \phi \text{ on } \Gamma,$$

with f defined and having continuous first derivatives for all x in $G + \Gamma$ and for all u.

a) Assuming that f is bounded,

(4) $$|f(x, u)| \leq N,$$

we shall prove that (3) has a solution.

We shall assume that the boundary and the boundary values ϕ are smooth (as in Ch. IV, §7, 1), in fact that $\phi = 0$. Indeed, by setting $u = v + h$, where h is the harmonic function equalling ϕ on Γ, we have the following equivalent boundary value problem for v:

$$\Delta v = f(x, v + h), \qquad v = 0 \text{ on } \Gamma.$$

If the domain is not sufficiently small the solution of (3) need not be unique. Consider, for example, in the case of one independent variable x the equation $u_{xx} = f(u)$, on the interval $0 \leq x \leq 2\pi$, where $f(u)$ is a bounded function which equals $-u$ for $-1 \leq u \leq 1$. The functions $u = \lambda \sin x$, $|\lambda| \leq 1$, are solutions vanishing at the end points. Similar examples can be constructed in all dimensions.

Before proceeding to solve (3) we note a general inequality holding

---

[1] See S. Agmon, A. Douglis and L. Nirenberg [1].

for all continuous functions u in $G + \Gamma$ which vanish on Γ and have continuous first and second derivatives in G: *For every compact subdomain $\mathfrak{a}$ of G we have*

$$(5) \qquad \max_{\mathfrak{a}} \left| \frac{\partial u}{\partial x_i} \right| \leq c \ \mathrm{l.u.b.}_{\mathfrak{a}} \ | \Delta u |,$$

where the constant c depends on $\mathfrak{a}$ and G. (A stronger form of (5) holds, with $\mathfrak{a}$ replaced by G.) To verify (5) we observe that if $K(x - y)$ is the fundamental solution for the Laplace operator ($K(x) = \Omega \, | \, x \, |^{2-n}$ for $n > 2$, $K(x) = \Omega \log | \, x \, |$ for $n = 2$, with Ω a suitable constant depending on n) and if $\zeta(y)$ is a twice continuously differentiable function which is identically 1 in $\mathfrak{a}$ and vanishes on Γ and in a neighborhood of Γ, then

$$u(x) = \int_{G} \zeta(y) K(x - y) \Delta_y u(y) \, dy$$

$$- \int_{G} u(y) \Delta_y (\zeta(y) K(x - y)) \, dy, \qquad x \text{ in } \mathfrak{a}.$$

It may be shown, using the results of Ch. IV, §1, 2, that

$$\max_{\mathfrak{a}} \left| \frac{\partial u}{\partial x_i} \right| \leq c' \ \mathrm{l.u.b.}_{G} \ | \Delta u | + \mathrm{l.u.b.}_{G} \ | u |$$

with c' a suitable constant. Since, however, u vanishes on Γ we have, from inequality (13) of Ch. IV, §6, $| \, u \, | \leq K \, \mathrm{l.u.b.}_{G} \, | \, \Delta u \, |$ which, when inserted into the preceding estimate, yields (5).

A solution of (3) may be derived with the aid of Schauder's fixed point theorem (see Ch. IV, §7) but we shall present a proof using iterations. Let $v(x)$ be the solution of

$$\Delta v = -N, \qquad\qquad v = 0 \text{ on } \Gamma,$$

where N is taken from (4). According to the maximum principle of Ch. IV, §6, 4, it follows that $v \geq 0$. Set $k = \mathrm{l.u.b.} \ (\partial f(x, u)/\partial u)$ for all x in G and $-\max v \leq u \leq \max v$, so that

$$(6) \qquad \begin{aligned} f(x, u) - f(x, w) - k(u - w) &\geq 0 \\ \text{for} \quad -\max v \leq u &\leq w \leq \max v. \end{aligned}$$

Our solution of (3) will be obtained as the limit of the sequence of functions u_m defined by the equations

$$(7) \qquad \begin{aligned} L[u_m] &\equiv \Delta u_m - k u_m = f(x, u_{m-1}) - k u_{m-1}, \\ u_m &= 0 \text{ on } \Gamma, \qquad u_0 = v. \end{aligned}$$

By the Schauder theory of Ch. IV, §7, 2, these functions u_m exist. We observe first that

$$L[u_1] = f(x, v) - kv \geq -N - kv = Lv.$$

Applying the maximum principle we see that $u_1 \leq v$ and, from the resulting inequality

$$\Delta u_1 = k(u_1 - v) + f(x, v) \leq N = -\Delta v,$$

we see that $u_1 > -v$, so that

$$-v \leq u_1 \leq v.$$

We show by induction that

(8) $$-v \leq u_m \leq u_{m-1} \leq v \qquad (m = 1, 2, \cdots).$$

Having verified (8) for $m = 1$, assume it to hold for some m. Then

$$L[u_{m+1} - u_m] = f(x, u_m) - f(x, u_{m-1}) - k(u_m - u_{m-1}) \geq 0,$$

by (6); thus, by the maximum principle, $u_{m+1} \leq u_m$. Consequently

$$\Delta u_{m+1} = k(u_{m+1} - u_m) + f(x, u_m) \leq N = -\Delta v,$$

so that, again by the maximum principle, $-v \leq u_{m+1}$, verifying (8) for $m + 1$, and thus for all m.

The functions $| u_m |$ are uniformly bounded, and hence, by (7), so are the values of $| \Delta u_m |$. Inequality (5) then implies that the first derivatives of u_m, and hence, by (7), the first derivatives of Δu_m, are uniformly bounded in absolute value on compact subsets $\mathfrak{a}$ of G. By Schauder's interior estimates of Ch. IV, §7, 1, the second derivatives of the u_m are also uniformly bounded in absolute value, and equicontinuous, on every compact subdomain. By Arzela's theorem (and the usual diagonalization process) it follows that a subsequence $\{u_{m_i}\}$ converges (together with derivatives up to second order) to a function u in G (and its corresponding derivatives). Since, however, the sequence $\{u_m\}$ is monotonic, *the entire sequence* converges to u. Letting $m \to \infty$ in (8) we see that $-v \leq u \leq v$; hence u is continuous in $G + \Gamma$ if we set $u = 0$ on Γ. Letting $m \to \infty$ in (7) we conclude that u is a solution of (3).

It is easily seen that the solution u just constructed is the largest solution. For, if w is any other solution of (3), the maximum principle implies that $| w | \leq v$ and, by induction, that

$$w \leq u_m \qquad (m = 0, 1, \cdots).$$

Had we started our iteration process with $u_0 = -v$ we would have obtained the smallest solution.

b) Consider again the problem (3) assuming more generally that $f(x, u) = f_1(x, u) + f_2(x, u)$ where $|f_2|$ is bounded and $\partial f_1/\partial u \geq 0$. (An example which occurs in a number of mathematical and physical problems is $f = e^u$.) As before we may suppose $\phi = 0$. By the mean value theorem we may rewrite (3) in the form

$$L[u] = \Delta u - \frac{\partial f_1}{\partial u} (x, \tilde{u})u = f_1(x, 0) + f_2(x, u), \qquad u = 0 \text{ on } \Gamma,$$

where $\tilde{u}(x)$ lies between 0 and $u(x)$. Then the estimate

(9) $$|u| \leq K \max |f_1(x, 0) + f_2(x, u)| \leq M$$

is a consequence of inequality (13) of Ch. IV, §6, applied to the above $L[u]$.

We shall find the solution of (3) by solving a modified problem, for which we may use the result of a). Let $\hat{u}(u)$ be a continuously differentiable monotonically increasing function for $-\infty < u < \infty$, such that $\hat{u}(u) = u$ for $|u| \leq M$, and $|\hat{u}(u)| \leq 2M$ for all u. For $\hat{f}_1(x, u) = f_1(x, \hat{u}(u))$ consider the modified equation

(3′) $$\Delta u = \hat{f}_1(x, u) + f_2(x, u) = \hat{f}(x, u), \qquad u = 0 \text{ on } \Gamma.$$

Clearly $\partial \hat{f}_1/\partial u \geq 0$, and $\hat{f}_1(x, 0) = f_1(x, 0)$. From (9) it follows that a solution of (3′) is bounded in absolute value by M and is therefore a solution of (3). We observe now that

$$|\hat{f}(x, u)| \leq \underset{|u|<2M}{\text{l.u.b.}} |f_1(x, u) + f_2(x, u)| \leq N$$

for some constant N and conclude by a) that (3′), and hence (3), has a solution.

If $\partial f/\partial u \geq 0$ the solution of (3) is unique according to the general uniqueness theorem of Ch. IV, §6, 2. In this case one can also use the continuity method to solve (3) by considering, for $0 \leq t \leq 1$, the one-parameter family of problems

(3″) $$\Delta u = tf(x, u), \qquad u = 0 \text{ on } \Gamma.$$

For $t = 0$ the solution is $u = 0$. According to the discussion in a) the set T of those values t for which (3″) has a solution in $C_{2+\alpha}$ (see Ch. IV, §7, 2) is open. To complete the continuity method we need only show that T is closed; T, being nonempty and both open and closed, is then the entire interval.

To see that T is closed let u_m be a sequence of solutions of $(3'')$ corresponding to values t_m which converge to some $\bar{t}$; we must show that $(3'')$ for $\bar{t}$ admits a solution. By (9) the functions $| u_m |$ and hence $| f(x, u_m) |$ are uniformly bounded. By (5) (in its strong form) the first derivatives of the u_m, and hence of $f(x, u_m(x))$, are uniformly bounded in absolute value. Applying the Schauder estimates and Arzela's theorem we now see that a subsequence of the u_m converges to a solution u of $(3'')$; this completes the existence proof by the continuity method.

c) In general, the solvability of the boundary value problem for the differential equation

$$(10) \qquad \Delta u = f(x_1, \cdots, x_n, u, u_1, \cdots, u_n) = F[u]$$

can be demonstrated provided the *fundamental domain* G *is chosen sufficiently small.* We shall confine ourselves to a sphere G of radius R and to boundary values $u = 0$ on Γ, assuming f to be continuous and to have continuous first derivatives in all arguments.

We note first that from the strong form of (5) we obtain the inequality

$$(11) \qquad \max_G \left| \frac{\partial u}{\partial x_i} \right| \leq cR \text{ l.u.b.}_G \, | \Delta u |,$$

where the constant c depends only on the dimension n.

We shall show that for R sufficiently small the sequence of functions defined by

$$\Delta u_{(m)} = F[u_{(m-1)}],$$

$$u_{(m)} = 0 \text{ on } \Gamma \qquad\qquad (m = 1, 2, \cdots)$$

$$u_0 = 0$$

converges to a solution of (10).

Introducing the norm (see Ch. IV, §7, 1)

$$\| u \|_1 = \max | u | + \max \left| \frac{\partial u}{\partial x_i} \right|,$$

we denote by μ a positive constant which bounds $| f |$, $| \partial f / \partial u |$ and $| \partial f / \partial u_i |$ $(i = 1, 2, \cdots, n)$ for $\| u \|_1 \leq 1$. If $\| u_{(m)} \|_1 \leq 1$ we see, by (11), that

$$\left| \frac{\partial u_{(m+1)}}{\partial x_i} \right| \leq cR\mu$$

and hence, since $u_{(m+1)} = 0$ on Γ,

$$| u_{(m+1)} | \leq cR^2\mu.$$

If we now choose R so that $cR\mu + cR^2\mu \leq 1$ we find, by induction, that in general $\| u_{(j)} \|_1 \leq 1$. Using the mean value theorem we find furthermore, that

$$| \Delta(u_{(m+1)} - u_{(m)}) | = | F[u_{(m)}] - F[u_{(m-1)}] |$$

$$\leq n\mu \| u_{(m)} - u_{(m-1)} \|_1 .$$

By (11) we conclude that

$$\| u_{(m+1)} - u_{(m)} \|_1 \leq (cR + cR^2)n\mu \| u_{(m)} - u_{(m-1)} \|_1 .$$

Restricting R further by the requirement $(cR + cR^2)n\mu \leq \frac{1}{2}$ we have

$$\| u_{(m+1)} - u_{(m)} \|_1 \leq \tfrac{1}{2} \| u_{(m)} - u_{(m-1)} \|_1$$

and, therefore,

$$\| u_{(m+1)} - u_{(m)} \|_1 \leq 2^{-m} \| u_{(1)} - u_{(0)} \|_1 ,$$

from which it follows that the $u_{(m)}$ and their first derivatives converge uniformly to a function u and its first derivatives, respectively.

With the aid of Schauder's interior estimates we easily see that the second derivatives of the $u_{(m)}$ also converge uniformly in every compact subset of G, and therefore that u has continuous second derivatives in G and satisfies (10).

Supplement to Chapter IV

Function Theoretic Aspects of the Theory of Elliptic Partial Differential Equations[1]

The theory of Laplace's equation in the plane is essentially equivalent to the theory of complex analytic functions. One can associate a theory of generalized analytic functions, the so-called pseudo-analytic functions, with every linear elliptic partial differential equa-

[1] This supplement is a contribution of Lipman Bers.

tion of second order for an unknown function of two independent variables. We shall briefly describe some salient facts concerning this theory;[1] we shall also indicate other connections between function theory and elliptic equations.

§1. Definition of Pseudoanalytic Functions

We first recall how complex analytic functions are connected with Laplace's equation. Let $\Phi(x, y)$ be a harmonic function which solves Laplace's equation

$$\Delta\Phi \equiv \Phi_{xx} + \Phi_{yy} = 0.$$

Set $u = \Phi_x$, $v = -\Phi_y$. Then the functions $u(x, y)$, $v(x, y)$ satisfy the Cauchy-Riemann equations

$$u_x - v_y = 0, \qquad u_y + v_x = 0$$

so that the complex gradient $w = u + iv$ is an analytic function of the complex variable $z = x + iy$. Moreover, u and v themselves are also harmonic functions. On the other hand, one can find a conjugate harmonic function Ψ by the Cauchy-Riemann equations

$$\Phi_x - \Psi_y = 0,$$

$$\Phi_y + \Psi_x = 0.$$

The function Φ is the real part of the complex analytic function $\Omega = \Phi + i\Psi$. The functions Ω and w are connected by the relation $w(z) = d\Omega(z)/dz$.

Consider now a general linear partial differential equation of elliptic type

$$(1) \quad a_{11}\phi_{\xi\xi} + 2a_{12}\phi_{\xi\eta} + a_{22}\phi_{\eta\eta} + a_1\phi_\xi + a_2\phi_\eta + a_0\phi = 0.$$

Assume that in the domain considered the leading coefficients $a_{ij}(\xi, \eta)$ have Hölder continuous partial derivatives, and that the equation possesses a positive solution $\phi_0(\xi, \eta)$. If we introduce new independent variables $x = x(\xi, \eta)$, $y = y(\xi, \eta)$, such that the mapping $(\xi, \eta) \rightarrow (x, y)$ is a homeomorphism[2] satisfying the Beltrami equa-

---

[1] This theory is due to L. Bers; an independent development was given by I. Vekua. An exposition of the theory is given in Bers [1] together with a complete bibliography. See also I. Vekua [1].

[2] A homeomorphism is a topological mapping, that is, a mapping which is one-to-one and continuous together with its inverse.

tions (cf. Ch. III, §2 and Ch. IV, §8) associated with the metric

$$a_{22} \, d\xi^2 - 2a_{12} \, d\xi \, d\eta + a_{11} \, d\eta^2,$$

and if we also introduce the new unknown function $\Phi = \phi/\phi_0$, then equation (1) takes on the canonical form

(1') $$\Delta\Phi + \alpha(x, y)\Phi_x + \beta(x, y)\Phi_y = 0.$$

We shall henceforth consider only such equations.

As before, set $u = \Phi_x$, $v = -\Phi_y$. These functions satisfy the system of differential equations

(2)
$$u_x - v_y = -\alpha u + \beta v,$$
$$u_y + v_x = 0,$$

which is a special case of the elliptic system

(3)
$$u_x - v_y = \alpha_{11} u + \alpha_{12} v,$$
$$u_y + v_x = \alpha_{21} u + \alpha_{22} v,$$

first studied by Hilbert and by Carleman. Using the complex notation introduced in Ch. IV, §8 we may write (3) in the form

(4) $$w_{\bar{z}} = a(z)w + b(z)\bar{w},$$

where

$$4a = \alpha_{11} + \alpha_{22} + i\alpha_{21} - i\alpha_{12}, \qquad 4b = \alpha_{11} - \alpha_{22} + i\alpha_{21} + i\alpha_{12}.$$

A continuously differentiable solution of equation (4) will be called a *pseudoanalytic function of the first kind* defined by the system (3) or an $[a, b]$-pseudoanalytic function. The complex gradient of every solution of (1') is an $[a, \bar{a}]$-pseudoanalytic function, with $4a = -\alpha - i\beta$.

On the other hand, we shall show in §4 that given an equation of the form (1') we can find, under fairly general conditions, two real-valued functions $\tau(x, y)$ and $\sigma(x, y) > 0$, satisfying the equations

(5)
$$\sigma_x - \tau_y = \alpha\sigma,$$
$$\sigma_y + \tau_x = \beta\sigma$$

(a system which is also of the form (3)). These relations show that equation (1') is equivalent to the elliptic system

(6)
$$\sigma\phi_x + \tau\phi_y = \psi_y,$$
$$\sigma\phi_y - \tau\phi_x = -\psi_x,$$

from which it can be obtained by eliminating ψ. A function $\phi + i\psi$ corresponding to a solution of an elliptic system (6) with Hölder continuous coefficients is called a *pseudoanalytic function of the second kind* associated with this system. Thus every solution of equation (1′) is the real part of a pseudoanalytic function of the second kind.

Note that a linear combination of two pseudoanalytic functions, with real constant coefficients, is again pseudoanalytic, but that the product of two pseudoanalytic functions is, in general, not pseudoanalytic.

We shall show below that pseudoanalytic functions have significant properties in common with ordinary analytic functions, and that pseudoanalytic functions of the first and second kind are to be considered as two representations of the same mathematical entity.

§2. An Integral Equation

In studying solutions of equation (4), that is, $[a, b]$-pseudoanalytic functions of the first kind, we shall assume that the coefficients $a(z)$, $b(z)$ are defined everywhere, satisfy a Hölder condition, and vanish identically outside a large circle $|z| = R$. (These assumptions are made only for the sake of simplicity, and the results to be derived hold under much more general conditions.)

The treatment of equation (4) is based on certain properties of the complex-valued double integral

$$q(z) = -\frac{1}{\pi} \iint_D \frac{\rho(\zeta)}{\zeta - z} \, d\xi \, d\eta \, ,$$

where $\zeta = \xi + i\eta$. We assume that the complex-valued function $\rho(z)$ vanishes outside a circle $|z| = R$ and satisfies the inequality $|\rho| \leq M$. Then

$$|q(z)| \leq \frac{KM}{1 + |z|^\epsilon} \, , \qquad |q(z_1) - q(z_2)| \leq KM |z_1 - z_2|^\epsilon$$

for every ϵ, $0 < \epsilon < 1$, where the constant K depends only on ϵ and R. Also, if in the neighborhood of some point the function $\rho(z)$ satisfies a Hölder condition, then the function $q(z)$ has Hölder continuous partial derivatives in this neighborhood and is a solution of equation

$$(7) \qquad\qquad q_{\bar z} = \rho.$$

All these statements can be proved by the methods described in Ch. IV, §8.

We also note that relation (7) holds even if we do not assume the Hölder continuity of ρ, provided we know that there exists some continuously differentiable function $Q(z)$ with $Q_{\bar{z}} = \rho$. This can be derived from Green's identity as the reader may easily verify.

Let $w(z)$ be a bounded continuous function defined in a domain D. Then $w(z)$ is $[a, b]$-*pseudoanalytic if and only if the function* $f(z)$ *defined by*

$$(8) \qquad f(z) = w(z) + \frac{1}{\pi} \iint_D \frac{a(\zeta)\, w(\zeta) + b(\zeta)\, \overline{w(\zeta)}}{\zeta - z}\, d\xi\, d\eta$$

is analytic in D.

To prove this, observe that the double integral in (8) is a Hölder continuous function of z, so that the Hölder continuity of either of the functions w or f implies that of the other. If, in addition, w is Hölder continuous, the continuous differentiability of either of the functions w or f implies that of the other. By differentiating (8) and applying (7) we obtain for w the differential equation

$$f_{\bar{z}} = w_{\bar{z}} - aw - b\bar{w}$$

in D. Thus w satisfies (4) if and only if $f_{\bar{z}} \equiv 0$, that is, if and only if $f(z)$ is analytic.

As a consequence of this result we obtain the theorem on removable singularities: *If $w(z)$ is pseudoanalytic and bounded in the domain* $0 < |z - z_0| < r$, *it can be defined at z_0 in such a way that it is pseudo-analytic in the whole disk $|z - z_0| < r$.* The proof follows by observing that the corresponding theorem holds for analytic functions, and that a double integral is insensitive to the removal of a point from the domain of integration.

For a given function $f(z)$, equation (8) may be considered as a linear integral equation for the unknown function $w(z)$. This equation is always uniquely solvable, as has been proved by Vekua. We shall, however, not make use of this fact.

§3. Similarity Principle

We call two complex-valued functions $w(z)$ and $f(z)$, defined in a domain D, *similar* if the ratio w/f is bounded, bounded away from zero, and continuous on the closure of the domain. We shall prove four theorems asserting that every pseudoanalytic function is similar to some analytic function, and vice versa.

a) *Let $w(z)$ be $[a, b]$-pseudoanalytic in a domain D. Then there exists an analytic function $f(z)$ and a complex-valued continuous function $s(z)$ such that*

$$(9) \qquad w(z) = e^{s(z)}f(z).$$

Furthermore, $s(z)$ is continuous in the closure of D, and has a bound and a modulus of Hölder continuity depending only on the coefficients a, b.

The proof consists simply in exhibiting the function $s(z)$. If $w(z) \equiv 0$, there is nothing to prove. If w does not vanish identically, let D_0 denote the open subset of D in which $w(z) \neq 0$, and set

$$s(z) = -\frac{1}{\pi} \iint_{D_0} \left[a(\zeta) + b(\zeta) \frac{\bar{w}(\zeta)}{w(\zeta)} \right] \frac{d\xi \, d\eta}{\zeta - z}, \qquad f(z) = e^{-s(z)} w(z).$$

The function $s(z)$ is continuous everywhere; in D_0 it is continuously differentiable and satisfies the equation $s_{\bar{z}} = a + b\bar{w}/w$. Therefore, $f_{\bar{z}} \equiv 0$ in D_0, so that $f(z)$ is analytic in D_0. It follows from the theorem of removable singularities of analytic functions that $f(z)$ is also analytic at every isolated point of the complement $D - D_0$.

Now let z_0 be a nonisolated point of $D - D_0$. Then there exists a sequence of points $\{z_\nu\}$ such that $w(z_\nu) = 0$, $\nu = 1, 2, \cdots$, and $z_\nu \to z_0$. Selecting a subsequence we may assume that the argument of $z_\nu - z_0$ converges to an angle θ. A simple application of the mean value theorem shows that $w_x \cos \theta + w_y \sin \theta = 0$ at z_0. On the other hand, $w(z_0) = 0$ and, therefore, by (4), also $2w_{\bar{z}} = w_x + iw_y = 0$ at z_0. Hence $w_x(z_0) = w_y(z_0) = 0$, so that $\lim_{z \to z_0} [w(z)/(z - z_0)] = 0$ and also $\lim_{z \to z_0} [f(z)/(z - z_0)] = 0$. It follows that $f(z)$ has a complex derivative also at every nonisolated point of $D - D_0$, so that $f(z)$ is analytic in the whole domain D.

Since an analytic function which does not vanish identically has isolated zeros, we conclude *a posteriori* that $D - D_0$ consists of isolated points only, so that we may write

$$(10) \qquad s(z) = -\frac{1}{\pi} \iint_D \left[a(\zeta) + b(\zeta) \frac{\bar{w}(\zeta)}{w(\zeta)} \right] \frac{d\xi \, d\eta}{\zeta - z}.$$

That this function has the required properties follows from the statements made in §2 of this supplement.

b) *Under the hypothesis of a) assume that the boundary of D contains a simple closed twice continuously differentiable curve C and that D is* *located entirely inside or entirely outside of C. Then the function $s(z)$ may be chosen to be real on C and to vanish at a given point z_0 on C.*

We shall prove this theorem only for the case where C is the unit circle, and D a subdomain of the unit disk. The case of a general curve C may be treated similarly, or may be reduced to the case of the unit circle by a conformal transformation.

The proof is based on a classical theorem (due to A. Korn and I. I. Privaloff) on conjugate functions. *Let $g(z) = U + iV$ be analytic in the unit disk. If $V(x, y)$ is continuous on the unit circle and satisfies there a Hölder condition with constant H and exponent $\alpha < 1$, then the function $g(z)$ satisfies in the closed unit disk a Hölder condition with exponent α and constant kH, where k depends only on α.* The proof of Privaloff's theorem is given in §14 of this supplement.

In order to prove statement b), let s be defined by (10) and set

$$t(z) = \frac{1}{2\pi} \int_0^{2\pi} \frac{e^{i\theta} + z}{e^{i\theta} - z} \, \text{Im } s(e^{i\theta}) \, d\theta,$$

$$s_0(z) = s(z) - it(z) - s(z_0) + it(z_0).$$

It follows from Privaloff's theorem that the analytic function $t(z)$, the real part of which is equal to Im $s(z)$ on $|z| = 1$, is continuous in the closed unit disk and has a bound and a Hölder modulus of continuity depending only on the coefficients a, b. It is clear that the function $e^{-s_0(z)}w(z)$ is analytic. On the other hand, $s_0(z)$ is real on $|z| = 1$ and vanishes at z_0.

c) *Let $f(z)$ be an analytic function defined in a domain D. Then there exists a function $s(z)$ which is continuous in the closure of D and vanishes at a prescribed point z_0 such that $w(z) = e^{s(z)}f(z)$ is $[a, b]$-pseudoanalytic. The function $s(z)$ may be chosen so as to have a bound and a modulus of Hölder continuity depending only on a, b.*

In view of the theorem on removable singularities proved above we lose no generality by removing all zeros of the analytic function f from the domain D. Thus we may assume that $f(z) \neq 0$ in D. If we can find a function $s(z)$ satisfying the differential equation

$$(11) \qquad s_{\bar z} = a + b \frac{\bar f}{f} e^{\bar s - s},$$

the function $e^s f$ will be $[a, b]$-pseudoanalytic.

In order to find s we consider the operator T which transforms a bounded continuous function $s(z)$ in D into the function $\sigma(z) - \sigma(z_0)$, where

$$(12) \quad \sigma(z) = -\frac{1}{\pi} \iint_D \left[a(\zeta) + b(\zeta) \frac{\overline{f(\zeta)}}{f(\zeta)} e^{\bar{s}(\zeta)-s(\zeta)} \right] \frac{d\xi \, d\eta}{\zeta - z}.$$

The bounded continuous functions in D form a real vector space B, since a linear combination of two bounded continuous functions with real coefficients is again bounded and continuous. In this space one may introduce the norm $\| s \| = \text{l.u.b.} \, | s(z) |$. If $\{s_n\}$ is a Cauchy sequence in this norm, that is if $\lim_{m,n \to \infty} \| s_n - s_m \| = 0$, then there exists a bounded continuous function $s(z)$ such that $\| s_n - s \| \to 0$. Thus the space B is a complete normed vector space—a Banach space.[1] From the properties of the double integral discussed in §2 of this supplement, it follows that every function $\bar{\sigma} = Ts$ has a bound and a modulus of continuity depending only on a, b and, if D is unbounded, is uniformly $O(| z |^{-\epsilon})$ for $z \to \infty$. Let Λ denote the set of all functions having this bound and this modulus of continuity and, for unbounded D, this behavior at infinity. The set Λ is convex (see §15 of this supplement). Moreover, it follows from Arzela's theorem (cf. Vol. I, Ch. II, §2) that from every sequence of functions in Λ we may select a uniformly convergent subsequence, i.e., a subsequence which converges in the norm of B. Thus Λ is a compact subset of B. It is not difficult to see that T is a continuous mapping of B into Λ, and, in particular, a continuous mapping of Λ into itself.

We now make use of the Schauder fixed point theorem[2] (cf. §9, and §15 of this supplement) which asserts that *a continuous mapping of a compact convex set of a Banach space into itself has a fixed point.* The theorem implies that there exists a function $s(z)$ in Λ such that $s = Ts$; i.e.,

$$s(z) = -\frac{1}{\pi} \iint_D \left[a(\zeta) + b(\zeta) \frac{\overline{f(\zeta)}}{f(\zeta)} e^{\bar{s}(\zeta)-s(\zeta)} \right] \left[\frac{1}{\zeta - z} - \frac{1}{\zeta - z_0} \right] d\xi \, d\eta.$$

This function satisfies the desired differential equation (11) and vanishes at z_0.

d) *Let the domain D and the curve C be as in Theorem* b), *and let $f(z)$ be a given analytic function defined in D. Then the function*

---

[1] See §15 of this supplement.

[2] By a longer argument, the use of the Schauder fixed point theorem can be avoided.

$s(z)$, *whose existence is asserted in* c), *may be chosen so as to be real on C and to vanish at a given point z_0 of C.*

The proof is very similar to that of b). Again we consider only the case of the unit circle. In order to find the desired solution of equation (11) we consider the operator T_1 which transforms a bounded continuous function $s(z)$ in D into the function $\sigma_1(z) - \sigma_1(z_0)$ where

$$\sigma_1(z) = \sigma(z) - \frac{i}{2\pi} \int_0^{2\pi} \frac{e^{i\theta} + z}{e^{i\theta} - z} \operatorname{Im} \sigma(e^{i\theta})\, d\theta$$

(with $\sigma(z)$ defined by (12)), and apply the Schauder fixed point theorem to the equation $s = T_1 s$. The fact that T_1 maps B into a compact convex subset follows from Privaloff's theorem.

§4. Applications of the Similarity Principle

As a first application we state a theorem on local behavior of pseudo-analytic functions. *Let $w(z)$ be pseudoanalytic in the domain $0 < |z - z_0| < r$. Then either $w(z)$ comes arbitrarily close to every complex number as $z \to z_0$* (essential singularity), *or there exists a positive integer n and a complex number $\alpha \neq 0$ such that*

$$(13) \qquad w(z) \sim \frac{\alpha}{(z - z_0)^n}, \qquad z \to z_0$$

(*that means w has a pole of order n*), *or $w(z)$ has at z_0 a removable singularity. If $w(z)$ is regular at z_0 and $w(z_0) = 0$, $w \not\equiv 0$, then there exists a positive integer n and a complex number $\alpha \neq 0$ such that*

$$(14) \qquad w(z) \sim \alpha(z - z_0)^n, \qquad z \to z_0$$

(zero of order n).

This assertion follows at once from the similarity principle a) and the corresponding classical statements on analytic functions. As a corollary we obtain the well-known theorem by Carleman asserting that a solution of (4) which does not vanish identically has only isolated zeros and vanishes at most of finite order at any point. This theorem implies, in particular, that a pseudoanalytic function is uniquely determined by its values on any open set. The same statement must then hold for solutions of elliptic differential equations (1) with smooth nonanalytic coefficients. By different methods this unique continuation theorem can be extended also to elliptic equations of the form (1) with bounded measurable coefficients.

Aronszajn[1] has proved the unique continuation theorem for second order equations in n-space with sufficiently smooth coefficients. We have used above the part of the similarity principle which describes the structure of pseudoanalytic functions. Now we give an application of the existence statement of the principle. We shall show that an elliptic equation of the form $(1')$ possesses a *Green's function* with respect to an arbitrary domain D bounded by a simple closed twice continuously differentiable curve C. In order to construct this function, let $g(z_0, z)$ be the Green's function of Laplace's equation belonging to the domain considered, with the singularity at z_0, and set $4a = -\alpha - i\beta$. The similarity principle d) asserts that there exists an $[a, \bar{a}]$-pseudoanalytic function $w(z)$ in D such that the ratio $w/(g_x - ig_y)$ is uniformly continuous, different from zero and real on C. Set, for some point z_1 on C,

$$G(z_0, z) = \text{Re} \int_{z_1}^{z} w(z) \, dz.$$

This line integral is not changed if the path of integration is deformed so as to avoid the singular point z_0. In fact, if C is a closed curve in D which does not contain z_0 in its interior G, we have, by Green's theorem,

$$\text{Re} \oint_C w \, dz = \text{Re} \, 2i \iint_G w_{\bar{z}} \, dx \, dy = 0.$$

In order to prove that the function $G(z_0, z)$ is single-valued, we must also show that the line integral evaluated along some closed curve around the singularity is zero. But $g = 0$ on C and hence $g_x \, dx + g_y \, dy = 0$, so that

$$\text{Re}(w \, dz) = 0 \quad \text{on } C$$

and thus

$$\oint_C dG = \text{Re} \oint_C w \, dz = 0$$

[1] See N. Aronszajn [1]. Independently the same result was proved by H. O. Cordes [1]. A. P. Calderón [1] has proved a rather general uniqueness theorem for initial value problems. These papers extend the basic idea of T. Carleman [1]. Independently B. Malgrange [1] and P. D. Lax [1] have demonstrated the interesting relationship between the unique continuation property and the Runge property for elliptic equations. Recently A. Plis [3] and [1] and, independently, P. Cohen [1] gave examples of elliptic equations for which unique continuation does not hold.

which implies the assertion. The function $G(z_0, z)$ is a solution of equation $(1')$. It is constant on C and has, as can easily be seen, a logarithmic singularity at the point z_0. It is therefore the desired Green's function. Note that this method of construction gives at once the existence and continuity of the normal derivative of the Green's function on the boundary.

In a similar way we can construct a solution of equation $(1')$ which vanishes on an arc of C and equals 1 on the complementary arc. Once in possession of this function it is easy to solve the first boundary value problem for equation $(1')$.

Another application of the similarity principle is the following result: *There exists one and only one bounded $[a, b]$-pseudoanalytic function $w(z)$ defined in the whole plane which takes on the given value α at a given point z_0.* In fact, by c) there exists a pseudoanalytic function of the form $\alpha e^{s(z)}$ where $s(z)$ is bounded in the whole plane and $s(z_0) = 0$. The uniqueness statement follows from Theorem a), which implies that every bounded pseudoanalytic function defined in the whole plane is of the form $e^{s(z)}f(z)$ where $f(z)$ is a bounded entire analytic function and hence, by Liouville's theorem, a constant. Thus a bounded "entire" pseudoanalytic function either has no zeros, or vanishes identically.

We are now in a position to verify the statement made in §1 of this supplement that equation $(1')$ is equivalent to a system of the form (6). We assume that the coefficients α, β are Hölder continuous and vanish outside a large circle. By the theorem just proved there exists a bounded solution (σ_0, τ_0) of system (5) defined in the whole plane and such that $\sigma_0 = 1$, $\tau_0 = 0$ at $z = 0$. We claim that $\sigma_0 > 0$ everywhere. In fact, assume that $\sigma_0 = 0$ at some point z_0; then the bounded solution of (5), $\sigma = \sigma_0$, $\tau = \tau_0 - \tau_0(z_0)$, vanishes at z_0 and hence identically, which is absurd.

§5. Formal Powers

We assume that the coefficients of equation (4) satisfy the hypotheses made in §2 of this supplement so that the similarity principle is applicable to the whole plane. According to this principle there exists a pseudoanalytic function of the first kind which is similar to the analytic function $\alpha(z - z_0)^n$ where n is a positive or negative integer. We denote this function by $w(z) = Z^{(n)}(\alpha, z_0, z)$ and call it a formal power. It is asymptotic to $\alpha(z - z_0)^n$ as $z \to z_0$ and is $O(|z|^n)$ at infinity. Using the similarity principle, it is easy to see that these properties determine the functions $Z^{(n)}$ uniquely. It also follows from the similarity principle that there exists a constant K depending only on the equation considered such that

$$\frac{1}{K} \mid \alpha \mid \mid z - z_0 \mid^n \leq \mid Z^{(n)}(\alpha, z_0, z) \mid \leq K \mid \alpha \mid \mid z - z_0 \mid^n.$$

The formal powers satisfy the relation

$$Z^{(n)}(\lambda \alpha + \mu \beta, z_0, z) = \lambda Z^{(n)}(\alpha, z_0, z) + \mu Z^{(n)}(\beta, z_0, z),$$

where λ and μ are real. This is proved by verifying that the right-hand side has the properties characterizing uniquely the function on the left-hand side. It can also be shown that $Z^{(n)}(\alpha, z_0, z)$ is a continuous function of z_0.

With the aid of the formal powers one can give analytic expressions for arbitrary pseudoanalytic functions, similar to the classical results of function theory. We shall state these formulas without proof.

Let $w(z)$ be a pseudoanalytic function defined in the domain D interior to a simple closed smooth curve C, and assume that w is continuous on C. Then the "Cauchy formula"

$$(15) \qquad w(z) = \frac{1}{2\pi} \int_C Z^{(-1)}[iw(\zeta) \, d\zeta, \zeta, z]$$

holds for z in D. The integral is to be interpreted in the following sense: If the parametric representation of the curve C is $\zeta(s)$, $0 \leq s \leq L$, then, for any function χ defined on C,

$$\int_C Z^{(n)}[\chi(\zeta) \, d\zeta, \zeta, z] = \int_0^L Z^{(n)}\{\chi[\zeta(s)]\zeta'(s), \zeta(s), z\} \, ds.$$

The integral in (15) is zero if the point z lies outside of C.

Consider now a pseudoanalytic function defined in the domain $0 < \mid z - z_0 \mid < R$. Then $w(z)$ admits the unique expansion

$$w(z) = \sum_{n=-\infty}^{+\infty} Z^{(n)}(\alpha_n, z_0, z)$$

which converges in this domain. If infinitely many of the coefficients α_n with $n < 0$, are different from zero, the function has an essential singularity at z. If a finite number of the coefficients α_n with $n < 0$ are different from zero, the function has a pole, and if only coefficients α_n with positive n appear in the expansion, the function is regular at z_0.

These results can, of course, be restated as results on solutions of second order partial differential equations of elliptic type, in which no reference to pseudoanalytic functions is made.

§6. Differentiation and Integration of Pseudoanalytic Functions

Two solutions of equation (4), $F(z)$ and $G(z)$, are said to form a pair of *generators* if they have no zeros and if the ratio G/F has a positive imaginary part. Thus for the Cauchy-Riemann equations the two functions 1 and i form a pair of generators, and so do the functions e^z, ie^z. Under fairly general conditions one can show that generators always exist. For instance, under the hypothesis made in §2 of this supplement, the existence of generators follows from the last theorem stated in §4 of this supplement if we take as generators two bounded entire solutions of (4) which assume at the origin the values 1 and i.

Let (F, G) be a generating pair for equation (4). Every complex-valued function $w(z)$ may be written in a unique way in the form

$$(16) \qquad w(z) = \phi(z)F(z) + \psi(z)G(z),$$

where the functions ϕ, ψ are real. It will be convenient to associate with every function w the function $\omega(z) = \phi + i\psi$. If the function w is pseudoanalytic of the first kind, the function ω is pseudoanalytic of the second kind as will be shown below.

Since any two sufficiently smooth functions F, G which do not vanish and for which the imaginary part of G/F is positive form a generating pair for *some* equation of the form (4), we shall talk henceforth about (F, G)-pseudoanalytic functions.

The function (16) is (F, G)-pseudoanalytic if and only if

$$(17) \qquad \phi_{\bar{z}} F + \psi_{\bar{z}} G = 0.$$

In fact, by hypothesis, we have

$$F_{\bar{z}} = aF + b\bar{F},$$
$$G_{\bar{z}} = aG + b\bar{G},$$

so that

$$w_{\bar{z}} - aw - b\bar{w} = \phi_{\bar{z}} F + \psi_{\bar{z}} G.$$

Now define the real-valued functions $\tau(z)$ and $\sigma(z) > 0$ by the relation $\sigma - i\tau = iF/G$. Then equation (17) becomes identical with system (6) thus justifying our terminology.

If (16) is an (F, G)-pseudoanalytic function, its (F, G)-*derivative* is defined by the formula

$$(18) \qquad \frac{d_{(F,G)}w}{dz} = \dot{w} = \phi_z F + \psi_z G.$$

If the functions $A(z)$, $B(z)$ are determined by the equations

$$F_z = AF + B\bar{F},$$

$$G_z = AG + B\bar{G},$$

then (18) may also be written in the form

$$\dot{w} = w_z - Aw - B\bar{w}.$$

If $\dot{w}$ is known, the function w, or rather the corresponding pseudo-analytic function ω of the second kind can be obtained by integration. In fact, define the *dual* generating pair $(F, G)^* = (F^*, G^*)$ by the relations

$$FF^* - GG^* \equiv 2,$$

$$\bar{F}F^* - \bar{G}G^* \equiv 0.$$

Then, by (17) and (18), $2\phi_z = F^*\dot{w}$, $2\psi_z = -G^*\dot{w}$, so that

$$(19) \qquad \omega(z_2) - \omega(z_1) = \int_{z_1}^{z_2} [\mathrm{Re}\ (F^*\dot{w}\ dz) - i\ \mathrm{Re}\ (G^*\dot{w}\ dz)].$$

It is remarkable that the (F, G)-derivative of an (F, G)-pseudo-analytic function is itself pseudoanalytic, though, in general, with respect to another generating pair. To prove this we compute[1] $\dot{w}_{\bar{z}}$ and get

$$\dot{w}_{\bar{z}} = (\phi_z F + \psi_z G)_{\bar{z}}$$

$$= \phi_z(aF + b\bar{F}) + \psi_z(aG + b\bar{G})$$

$$+ (\phi_{\bar{z}} F + \psi_{\bar{z}} G)_z - \phi_{\bar{z}}(AF + B\bar{F}) - \psi_{\bar{z}}(AG + B\bar{G}).$$

Noting that $(\phi_{\bar{z}} F + \psi_{\bar{z}} G)_z \equiv 0$ and expressing ϕ_z, $\phi_{\bar{z}}$, ψ_z, $\psi_{\bar{z}}$ in terms of $\dot{w}$ and $\bar{\dot{w}}$ by means of (17), (18) we obtain the equation

$$\dot{w}_{\bar{z}} = a\dot{w} - B\bar{\dot{w}}$$

which shows that $\dot{w}$ is $[a, -B]$-pseudoanalytic.

---

[1] The fact that $\dot{w}$ is continuously differentiable can be proved using the existence theorems for equation (1′).

A pair of generators (F_1, G_1) belonging to the equation

$$w_{\bar{z}} = aw - B\bar{w}$$

is called a *successor* of (F, G). The original pair (F, G) is itself a successor of a pair (F_{-1}, G_{-1}) which can be obtained as follows: (F_{-1}, G_{-1}) is the dual of a successor of a dual of (F, G). The easy proof of this statement is left to the reader. Thus a given generating pair (F, G) can be embedded (in fact, in infinitely many ways) into a sequence of generating pairs

$$(20) \quad \cdots, (F_{-2}, G_{-2}), (F_{-1}, G_{-1}), (F_0, G_0), (F_1, G_1), (F_2, G_2), \cdots$$

such that (F_ν, G_ν) is a successor of $(F_{\nu-1}, G_{\nu-1})$. The sequence is called periodic with period n if $(F_n, G_n) = (F_0, G_0)$. The smallest n for which this can happen for any sequence in which (F_0, G_0) can be embedded is called the minimal period of (F_0, G_0), and (F_0, G_0) is said to have minimal period ∞ if it cannot be embedded into a periodic sequence. It has been shown by Protter[1] that there exist generators with given minimal periods.

With respect to a generating sequence (20) an (F_0, G_0)-pseudo-analytic function $w(z)$ has derivatives of all orders defined by the recurrence relations

$$w^{[0]} = w, \qquad w^{[n+1]} = \frac{d_{(F_n, G_n)} w^{[n]}}{dz} \qquad (n = 0, 1, \cdots).$$

One can show that, just as in the case of analytic functions, the sequence of numbers $\{w^{[n]}(z_0)\}$, for some fixed z_0, determines the function w uniquely.

Using a generating sequence (20) and the integration process described above one can construct by quadratures a sequence of particular (F_ν, G_ν)-pseudoanalytic functions called local formal powers and denoted by $Z_\nu^{(n)}(\alpha, z_0, z)$. These are defined by the recurrence relations

$$Z_\nu^{(0)}(\alpha, z_0, z) = \alpha,$$

$$\frac{d_{(F_\nu, G_\nu)} Z_\nu^{(n)}(\alpha, z_0, z)}{dz} = Z_{\nu+1}^{(n-1)}(n\alpha, z_0, z) \qquad (n = 1, 2, \cdots),$$

where α and z_0 are complex constants. The name "powers" is justi-

[1] See N. H. Protter [1].

fied by the remark that for $(F_\nu, G_\nu) = (1, i)$ $(\nu = 0, \pm 1, \cdots)$ we have $Z_\nu^{(n)}(\alpha, z_0, z) = \alpha(z - z_0)^n$. The global formal powers described in §5 of this supplement are a special case of local formal powers.

§7. Example. Equations of Mixed Type

A particularly simple class of pseudoanalytic functions is obtained by taking as generators the functions $F \equiv 1$, $G \equiv i\beta(y)$, $\beta(y)$ being a positive function. By (17), the function $\phi + i\beta\psi$ is $(1, i\beta)$-pseudoanalytic if and only if ϕ and ψ satisfy the equations[1]

$$(21) \qquad \begin{aligned} \phi_x &= \beta(y)\psi_y, \\ \phi_y &= -\beta(y)\psi_x. \end{aligned}$$

Without referring to the general theory of (F, G)-differentiation and integration one verifies at once that if (ϕ, ψ) is a solution of (21) then (ϕ', ψ'), with

$$\phi' = \phi_x, \qquad \psi' = \psi_x,$$

as well as (Φ, Ψ), where Φ and Ψ are defined by the path-independent integrals

$$(22) \qquad \begin{aligned} \Phi &= \int (\phi\, dx - \beta\psi\, dy), \\ \Psi &= \int \left(\psi\, dx + \frac{\phi}{\beta}\, dy\right), \end{aligned}$$

are also solutions of (21). The generating pair $(1, i\beta(y))$ is its own successor, $\phi' + i\beta\psi'$ is the $(1, i\beta)$-derivative of $\phi + i\beta\psi$, and the latter function is the $(1, i\beta)$-derivative of $\Phi + i\beta\Psi$.

The local formal powers defined above can be written down explicitly. For the sake of simplicity we consider only the powers $Z^{(n)}(a, 0, z)$. Set

$$(23)$$

$$Y^{(0)}(y) = 1, \qquad Y^{(1)}(y) = \int_0^y \frac{d\eta}{\beta(\eta)}, \qquad Y^{(2)}(y) = 2! \int_0^y \beta(\eta) Y^{(1)}(\eta)\, d\eta,$$

$$Y^{(3)}(y) = 3! \int_0^y \frac{Y^{(2)}(\eta)}{\beta(\eta)}\, d\eta, \cdots,$$

---

[1] See L. Bers and A. Gelbart [1].

$$\bar{Y}^{(0)}(y) = 1, \qquad \bar{Y}^{(1)}(y) = \int_0^y \beta(\eta)\, d\eta, \qquad \bar{Y}^{(2)}(y) = 2! \int_0^y \frac{\bar{Y}^{(1)}(\eta)}{\beta(\eta)}\, d\eta,$$

$$\bar{Y}^{(3)}(y) = 3! \int_0^y \beta(\eta)\, \bar{Y}^{(2)}(\eta)\, d\eta, \cdots,$$

Then, for real λ and μ and $n = 1, 2, \cdots$,

(24)
$$Z^{(n)}(\lambda + i\mu, 0, x + iy) = \lambda \sum_{j=0}^{n} \binom{n}{j} x^{n-j} i^j Y^{(j)}(y)$$
$$+ i\mu \sum_{j=0}^{n} \binom{n}{j} x^{n-j} i^j \bar{Y}^{(j)}(y).$$

It is important to observe that the same formalism works for systems of the form

(25)
$$\phi_x = \beta_1(y)\psi_y,$$
$$\phi_y = -\beta_2(y)\psi_x,$$

where we do not assume the real-valued functions β_1 and β_2 to be positive. One has merely to replace in formulas (22) through (24) β by β_2 and $1/\beta$ by $1/\beta_1$. System (25) may now be either elliptic, or hyperbolic, or of mixed type. As an example of a system of mixed type consider

$$\phi_x = \psi_y,$$
$$\phi_y = -y\psi_x.$$

Elimination of ϕ leads to the so-called Tricomi equation[1]

$$y\psi_{xx} + \psi_{yy} = 0,$$

which is important in transonic gas dynamics. Our formalism furnishes a series of polynomials satisfying this equation.

In a similar elementary way, one can treat the system

(26)
$$\alpha_1(x)\phi_x = \beta_1(y)\psi_y,$$
$$\alpha_2(x)\phi_y = -\beta_2(y)\psi_x.$$

---

[1] The theory of this equation was initiated in the celebrated memoir by Tricomi [1]. An extensive bibliography on equations of mixed type may be found in L. Bers [3].

The differentiation and integration formulas

$$\phi' = \alpha_1 \phi_x, \qquad\qquad \psi' = \psi_x/\alpha_2,$$

$$\Phi = \int (\alpha_2 \phi \, dx - \beta_2 \psi \, dy), \qquad \Psi = \int [(\psi/\alpha_1) \, dx + (\phi/\beta_1) \, dy]$$

lead from a solution (ϕ, ψ) of (26) to solutions of the system

$$(27) \qquad\qquad \frac{\Phi_x}{\alpha_2} = \beta_1 \Psi_y, \qquad \frac{\Phi_y}{\alpha_1} = -\beta_2 \Psi_x.$$

If $\alpha_1 = \alpha_2 = \alpha > 0$, $\beta_1 = \beta_2 = \beta > 0$, system (26) means that the function $\phi + i\psi$ is pseudoanalytic of the second kind with respect to the generators $F = (\alpha/\beta)^{1/2}$, $G = i(\beta/\alpha)^{1/2}$. The (F, G)-derivative of $\phi F + \psi G$ is easily computed to be

$$(\alpha/\beta)^{1/2} \phi_x + i(\beta/\alpha)^{1/2} \psi_x = F_1 \phi' + G_1 \psi',$$

where

$$F_1 = (\alpha\beta)^{-1/2}, \qquad G_1 = i(\alpha\beta)^{1/2}.$$

This derivative is (F_1, G_1)-pseudoanalytic, since ϕ' and ψ' satisfy (27). Thus (F_1, G_1) is a successor of (F, G) and, by the same token, (F, G) is a successor of (F_1, G_1). The minimum period of our generating pair is therefore 2 unless α is a constant.

§8. General Definition of Pseudoanalytic Functions

We return to the general theory of pseudoanalytic functions and observe that the (F, G)-derivative defined in §6 of this supplement can be obtained by a limiting process which is an extension of the process of ordinary complex differentiation.

More precisely, let $w(z)$ be represented in the form (16) where $F(z)$ and $G(z)$ are fixed continuous complex-valued functions with Im $(G/F) > 0$ and where ϕ and ψ are real-valued. We form the "difference quotient"

$$(28) \quad \frac{1}{h} [\phi(z + h)F(z) + \psi(z + h)G(z) - \phi(z)F(z) - \psi(z)G(z)]$$

and ask whether it has a limit as the complex number h approaches 0 in all possible ways. If this limit exists, it will in particular exist for $h = \delta \to 0$ and for $h = i\delta \to 0$, where δ is a real variable. The two limits in question are, of course,

$$\frac{\partial \phi}{\partial x} F + \frac{\partial \psi}{\partial x} G \quad \text{and} \quad \frac{1}{i} \left(\frac{\partial \phi}{\partial y} F + \frac{\partial \psi}{\partial y} G \right).$$

The condition that the two limits be equal yields at once equation (17) and, if we denote their common value by $\dot{w}$, we obtain relation (18).

In the general theory of pseudoanalytic functions one starts with two generators $F(z)$, $G(z)$ which are not assumed to be differentiable, and calls a function (16) (F, G)-pseudoanalytic in a domain if at every point z of this domain the ratio (28) has a finite limit for $h \to 0$. Most of the theorems stated above remain valid if F and G are Hölder continuous. Pseudoanalytic functions can still be characterized by the differential equation (6); however they will not, in general, satisfy an equation of the form (4) and the similarity principle may cease to be valid.

§9. *Quasiconformality*[1] *and a General Representation Theorem*

In studying geometric properties of pseudoanalytic functions it is convenient to work with functions of the second kind. By virtue of a differential inequality, which is a consequence of the differential equation (6), it turns out that these functions share, in a certain sense, many of the geometric properties of analytic functions.

An analytic function of a complex variable considered as a mapping of the plane is conformal at every point at which the derivative does not vanish. This means that at such points the mapping is in the small a similarity transformation: It takes infinitesimal circles into infinitesimal circles. It is natural, and it turns out to be very useful, to consider mappings which take infinitesimal circles into infinitesimal ellipses of uniformly bounded eccentricity. Such mappings are called *quasiconformal*.

For a mapping $w(z) = u(x, y) + iv(x, y)$ having continuous partial derivatives and a nonvanishing Jacobian the geometric condition just stated can be expressed by either of the three equivalent differential inequalities

$$(29) \qquad \max_{0 \leq \theta \leq 2\pi} | w_x \cos \theta + w_y \sin \theta |^2 \leq Q(u_x v_y - u_y v_x),$$

[1] The theory of quasiconformal mappings has been initiated by Grötzsch and developed by Ahlfors, Lavrentieff, Morrey, Teichmüller, and others. An extensive bibliography may be found in H. P. Künzi [1].

$$(29') \qquad\qquad u_x^2 + u_y^2 + v_x^2 + v_y^2 \le 2K(u_x v_y - u_y v_x),$$

$$(29'') \qquad\qquad |w_z| \le k\,|w_{\bar z}|.$$

Here $Q \ge 1$, $K \ge 1$, $0 \le k < 1$, and the constants Q, K, k are connected by the relations

$$K = \frac{1}{2}\left(Q + \frac{1}{Q}\right), \qquad k = \frac{Q-1}{Q+1}.$$

It is easy to verify that every pseudoanalytic function of the second kind $\omega = \phi + i\psi$, or more generally every solution $\omega = \phi + i\psi$ of an elliptic system

$$u_x = \alpha_{11} v_x + \alpha_{12} v_y,$$

$$-u_y = \alpha_{21} v_x + \alpha_{22} v_y,$$

will satisfy these differential inequalities if the system is uniformly elliptic, that is, if $\alpha_{12} > 0$ and

$$0 < \frac{(\alpha_{12} + \alpha_{21})^2}{4\alpha_{12}\alpha_{21} - (\alpha_{11} + \alpha_{22})^2} < \text{const.};$$

the constant Q of quasiconformality depends only on the constant in this inequality.

It is useful to define quasiconformality in a more general way. Instead of requiring that the partial derivatives occurring in the inequalities (29) should be continuous, we shall assume only that these derivatives exist, satisfy the inequalities almost everywhere and are locally square integrable, and that the continuous functions $u(x, y)$, $v(x, y)$ are absolutely continuous in one variable for almost all fixed values of the other.

The main property of quasiconformal functions can be expressed by the following theorem, which we state here without proof (and not in its most general form):

Let $w(z)$ be a quasiconformal function defined in the unit disk. Then w admits the representation

$$(30) \qquad\qquad w(z) = f[\chi(z)],$$

where $\zeta = \chi(z)$ is a homeomorphism (one-to-one and bicontinuous mapping) of $|z| \le 1$ onto $|\zeta| \le 1$ with $\chi(0) = 0$, $\chi(1) = 1$, which satisfies together with its inverse χ^{-1} a uniform Hölder condition depending only on the constant Q in (29), and where $f(\zeta)$ is an analytic function of the complex variable ζ, $|\zeta| < 1$.

In this theorem an arbitrary quasiconformal function is resolved into an analytic function and a function for which a Hölder condition is known. In §3 of this supplement we derived a somewhat similar decomposition (equation (9)) for functions satisfying the differential equation (4). It is easy to see that the same decomposition (9) holds for functions satisfying the differential inequality

$$(31) \qquad | w_{\bar{z}} | \leq k' | w |$$

for some constant k'. Both results, the similarity principle for solutions of the differential inequality (31) and the decomposition theorem (30) for quasiconformal functions, are special cases of a more general representation theorem which we now state, again without proof.[1]

Let $w(z)$ be a function defined in the unit disk and satisfying there the differential inequality

$$(32) \qquad | w_{\bar{z}} | \leq k | w_z | + k' | w | + k''.$$

(Here $k < 1$ and the partial derivatives are subject to the same conditions as before.) *Then the function w can be represented by*

$$(33) \qquad w(z) = e^{s(z)} f[\chi(z)] + s_0(z),$$

where $\zeta = \chi(z)$ is a homeomorphism of the unit disk onto itself with $\chi(0) = 0$ and $\chi(1) = 1$; $s(z)$ and $s_0(z)$ are continuous on the closed unit disk, real on the circumference and vanish at $z = 1$, and $f(\zeta)$ is an analytic function of the complex variable ζ. The functions s, s_0, χ and the inverse homeomorphism χ^{-1} have bounds and moduli of Hölder continuity depending only on the constants in (32). (In particular, if $k'' = 0$, then $s_0 \equiv 0$.)

The importance of the representation theorem lies in the fact that every solution of a uniformly elliptic system with bounded coefficients

$$(34) \qquad \begin{aligned} u_x &= \alpha_{11} v_x + \alpha_{12} v_y + \beta_{11} u + \beta_{12} v + \gamma_1 , \\ -u_y &= \alpha_{21} v_x + \alpha_{22} v_y + \beta_{21} u + \beta_{22} v + \gamma_2 \end{aligned}$$

necessarily satisfies a differential inequality of the form (32). In particular, if the system is homogeneous ($\gamma_1 \equiv \gamma_2 \equiv 0$), it follows from the representation theorem that a solution has the *unique con-*

---

[1] The representation (30) is due to C. B. Morrey [3] and the representation (33) is due to L. Bers and L. Nirenberg [1]. See also B. V. Boyarskii [1].

tinuation property: It cannot vanish in an open set without vanishing identically. This result is an extension (but not a direct consequence) of Carleman's theorem discussed in §4 of this supplement.

§10. *A Nonlinear Boundary Value Problem*

The representation theorems stated in the preceding section can be used in order to obtain *a priori* estimates and existence theorems for boundary value problems for nonlinear elliptic equations. We illustrate this here by considering a relatively simple example.

We want to solve the Neumann problem for a quasi-linear equation of the form

$$(35) \quad \begin{aligned} a(x, y, \phi, \phi_x, \phi_y)\phi_{xx} + 2b(x, y, \phi, \phi_x, \phi_y)\phi_{xy} \\ + c(x, y, \phi, \phi_x, \phi_y)\phi_{yy} = 0. \end{aligned}$$

We assume that the coefficients a, b, c are defined for $x^2 + y^2 < 1$ and for all values of ϕ, ϕ_x, ϕ_y and satisfy a uniform Hölder condition, and that the equation is uniformly elliptic, i.e., that[1]

$$(36) \qquad a > 0, \qquad ac - b^2 \equiv 1, \qquad a + c \leq \text{const.}$$

Let $\tau(z)$ be a real-valued function defined on the unit circle and satisfying a uniform Hölder condition. The Neumann problem consists in finding a solution ϕ of this equation defined and continuously differentiable in the unit disk and satisfying on its boundary the conditions

$$(37) \qquad \begin{aligned} \frac{\partial \phi}{\partial n} &= x\phi_x + y\phi_y = \tau + k \qquad \text{for } |z| = 1, \\ \phi &= 0 \qquad \text{at } z = 1, \end{aligned}$$

where k is a constant to be determined.

We shall show that this problem always has a solution. The proof is based on the following

A PRIORI ESTIMATE: *Assume that a solution of the Neumann problem is given. Then the function ϕ and its first derivatives ϕ_x, ϕ_y have a bound, and the first derivatives satisfy Hölder conditions, the bound and the Hölder conditions depending only on the constant in* (36) *and on the given boundary function τ.*

Proof: We remark first that a bound on the first derivatives will

---

[1] This example is taken from L. Bers and L. Nirenberg [2].

imply a bound on the function. Next we estimate the constant k. Since a nonconstant solution of equation (35) cannot achieve its maximum or minimum at an interior point, the normal derivative must change sign on the boundary. Hence k cannot exceed the maximum of $|\tau|$. Thus we know a bound and a Hölder modulus of continuity for the function $\tau(z) + k$. At some point on the boundary the tangential derivative of ϕ must vanish. At this point

$$w \equiv \phi_x - i\phi_y = \bar{z}(\tau + k).$$

Hence we need only find a Hölder modulus of continuity for w. This function can also be written in the form

$$w = u + iv$$

where we set $u = \phi_x$, $v = -\phi_y$. Since equation (35) is equivalent to

$$u_x = \frac{2b}{a} v_x + \frac{c}{a} v_y,$$

$$u_y = -v_x,$$

the function $w(z)$ will be quasiconformal with a certain constant depending only on the constant in (36). It will, therefore, have a representation of the form (30). The boundary condition (37) can now be written in the form

$$\operatorname{Re}[zw(z)] = \tau(z) + k, \qquad |z| = 1,$$

or

$$\operatorname{Re}[\chi^{-1}(\zeta)f(\zeta)] = \tau[\chi^{-1}(\zeta)] + k, \qquad |\zeta| = 1.$$

Since we know the Hölder continuity of the mappings $\zeta = \chi(z)$, $z = \chi^{-1}(\zeta)$, we may write this in the form

$$\operatorname{Re}[\zeta e^{i\lambda(\zeta)} f(\zeta)] = \sigma(\zeta), \qquad |\zeta| = 1,$$

where λ is real and the functions λ and σ have known bounds and Hölder moduli of continuity. An easy extension of Privaloff's theorem (stated and proved later, cf. §14 of this supplement) yields a Hölder condition for $f(\zeta)$ in $|\zeta| \leq 1$. Since $w(z) = f[\chi(z)]$, we obtain a Hölder condition on w.

We consider now our Neumann problem for a linear equation of the form (35), that is for an equation in which the coefficients a, b, c depend only on x and y, and state first the

EXISTENCE AND UNIQUENESS THEOREM: *For a linear uniformly elliptic equation of the form* (35) *the Neumann problem has one and only one solution.*

The uniqueness proof follows at once from the observation, already made, that for a nonconstant solution the normal derivative must change sign on the unit circle. The existence statement, on the other hand, is a consequence of the following

CONTINUITY LEMMA: *Consider a sequence of equations*

$$(38) \qquad a^{(n)}(x, y)\phi_{xx} + 2b^{(n)}(x, y)\phi_{xy} + c^{(n)}(x, y)\phi_{yy} = 0.$$

We assume that all equations are uniformly elliptic with the same constant and that their coefficients satisfy the same Hölder condition. We assume also that for $n \to \infty$ the coefficients converge toward the coefficients of an equation

$$(39) \qquad a(x, y)\phi_{xx} + 2b(x, y)\phi_{xy} + c(x, y)\phi_{yy} = 0$$

at all points within the unit disk. Let $\phi^{(n)}$ be, for every n, a solution of equation (38) *satisfying the boundary conditions of our Neumann problem. Then the functions $\phi^{(n)}$ converge uniformly in the closed unit disk, together with their first derivatives, toward a solution of the Neumann problem for equation* (39).

Proof: In view of the *a priori* estimate all functions $\phi^{(n)}$, $\phi_x^{(n)}$, $\phi_y^{(n)}$ are uniformly bounded and equicontinuous. Hence, by Arzela's theorem, we may select a subsequence $\{\phi^{(n_j)}\}$ which converges uniformly, together with its first derivatives, toward a function $\phi(x, y)$. It is clear that this function satisfies the boundary conditions of the Neumann problem. On the other hand, it follows from the Schauder estimates (see Ch. IV, §7) that in every closed subdomain of the unit disk the second derivatives of the functions $\phi^{(n_j)}$ converge uniformly. Hence, the limit function satisfies equation (39). In view of the uniqueness statement already proved we conclude *a posteriori* that the selection of the subsequence was unnecessary and that $\phi^{(n)} \to \phi$.

The continuity lemma implies that the Neumann problem can be solved for a linear equation (39) if the coefficients of this equation can be approximated by coefficients of equations for which the Neumann problem has already been solved. We know that for equations with very smooth coefficients this can be done, say, by the method of integral equations (cf. Ch. IV, §10). Thus the existence of a solution is proved in general. (We remark that the same method leads to the existence of a solution for uniformly elliptic equations with coefficients which not only fail to be Hölder continuous, but which even are discontinuous.)

We return now to the nonlinear equation (35). Let $\mathfrak{B}$ denote the Banach space of continuously differentiable functions Φ on the closed unit disk with the norm

$$\| \Phi \| = \max | \Phi | + \max | \Phi_x - i\Phi_y |.$$

Let Λ denote the subset of $\mathfrak{B}$ consisting of those functions which satisfy the boundary conditions of our Neumann problem and the *a priori* estimates derived for solutions of this problem. It is easy to see that Λ is a convex compact subset of $\mathfrak{B}$. Let Φ be any function belonging to Λ. With the aid of this function we form the linear equation

$$a(x, y, \Phi(x, y), \Phi_x(x, y), \Phi_y(x, y))\phi_{xx}$$
$$+ 2b(x, y, \Phi(x, y), \Phi_x(x, y), \Phi_y(x, y))\phi_{xy}$$
$$+ c(x, y, \Phi(x, y), \Phi_x(x, y), \Phi_y(x, y))\phi_{yy} = 0$$

and denote by ϕ the uniquely determined solution of the Neumann problem for this linear equation. If we set

$$(40) \qquad\qquad \phi = T(\Phi),$$

then T is a mapping of Λ into Λ. That this mapping is continuous can be easily shown using the continuity lemma stated above. By the Schauder fixed point theorem the mapping T must have a fixed point. In other words, there exists a function ϕ such that $\phi = T(\phi)$. This is the desired solution of the Neumann problem.

The same method works in more general situations. For instance, we could prove the solvability of the Neumann problem for a nonlinear equation of the form

$$a(x, y, \phi, \phi_x, \phi_y)\varphi_{xx} + 2b(x, y, \phi, \phi_x, \phi_y)\phi_{xy}$$
$$+ c(x, y, \phi, \phi_x, \phi_y)\phi_{yy} = d(x, y, \phi, \phi_x, \phi_y)$$

which is uniformly elliptic and in which the right-hand side satisfies an inequality of the form

$$| d(x, y, \phi, \phi_x, \phi_y) | \leq k'(| \phi_x | + | \phi_y |) + k''.$$

In this case we would derive the appropriate estimates using not the representation (30) but the more general representation (33).

§11. *An Extension of Riemann's Mapping Theorem*

The concept of quasiconformality of a mapping leads in a natural way to a far-reaching extension of Riemann's mapping theorem. The latter theorem asserts that a given simply connected domain, say—for the sake of simplicity—a domain bounded by a Jordan curve, can be mapped onto another given Jordan domain conformally, i.e., in such a way that infinitesimal circles go into infinitesimal circles. The mapping may be chosen so that three given boundary points of one domain go into three given boundary points of the other.

It is natural to ask whether one can map a given Jordan domain onto another so that at every point the following condition is satisfied: An infinitesimal ellipse with given eccentricity e and given slope θ of the major axis is taken into an infinitesimal ellipse of given eccentricity e' and given slope θ' of the major axis. We may require that the numbers e, e', θ, θ' should depend both on the point considered and on its image point. It is not difficult to verify that under proper continuity assumptions the geometric condition for such a mapping can be expressed analytically by requiring that the mapping function $w(z) = u + iv$ satisfy a quasi-linear system of partial differential equations of elliptic type, of the form

$$
\begin{aligned}
u_x &= \alpha_{11}(x, y, u, v)v_x + \alpha_{12}(x, y, u, v)v_y \,, \\
-u_y &= \alpha_{21}(x, y, u, v)v_x + \alpha_{22}(x, y, u, v)v_y \,.
\end{aligned}
$$

(41)

Let us assume that the coefficients of the system are Hölder continuous functions and that the system is uniformly elliptic. Under these conditions the following result (first proved by Miss Z. Schapiro[1]) holds: *A given Jordan domain D in the z-plane can be mapped onto a given Jordan domain D' in the w-plane by a pair of functions* $u(x, y)$, $v(x, y)$ *which satisfy equation* (41). The mapping can be chosen so that three given boundary points of D go into three given boundary points of D'.

This theorem can be proved by the method described in the preceding section. The representation formula (30) for quasiconformal functions yields the necessary *a priori* estimates.

---

[1] See Z. Schapiro [1]. This problem and generalizations have been treated by a number of authors; we refer to the bibliography in B. V. Boyarskii [1].

§12. *Two Theorems on Minimal Surfaces*

In our previous discussion we emphasized the similarities which exist between analytic functions of a complex variable and solutions of elliptic partial differential equations. In the case of nonlinear equations, however, new phenomena arise, which have no counterpart in the theory of analytic functions. We illustrate this by mentioning two theorems on the solutions of one of the simplest and most important nonlinear equations, the equation of minimal surfaces (cf. Ch. I, §6):

$$(42) \qquad (1 + \phi_y^2)\phi_{xx} - 2\phi_x\,\phi_y\,\phi_{xy} + (1 + \phi_x^2)\phi_{yy} = 0.$$

The first theorem (due to S. Bernstein) states that *every solution of equation* (42) *defined in the whole plane is a linear function.* Note that for the solution of the Laplace equation the assertion of the theorem would be true only if we knew in advance that the solution in question has bounded derivatives, in which case the assertion would follow from the well-known theorem of Liouville on entire analytic functions.

In a similar way Riemann's theorem on removable singularities, which, as we have seen earlier, is applicable also to linear elliptic equations, holds for the equation of minimal surfaces in a much stronger form (L. Bers).

A single-valued solution of equation (42) *defined in a deleted neighborhood of a point has at this point a removable singularity.* In other words, the solution can be defined at the point in such a way as to be regular there. Note that we do not require *a priori* that the solution be bounded.

We shall not give here the proofs of these theorems, but mention only that one way of proving them involves again the application of complex function theory.[1]

§13. *Equations with Analytic Coefficients*

Up to now we have discussed primarily the application of function theoretical methods to linear partial differential equations of elliptic type under very mild restrictions on the coefficients. In the case where the coefficients are themselves analytic functions of two variables there exists a quite different possibility of utilizing complex function theory for the study of the solutions.

Consider the linear partial differential equation

$$(43) \quad \phi_{xx} + \phi_{yy} + \alpha(x, y)\phi_x + \beta(x, y)\phi_y + \gamma(x, y)\phi = 0$$

---

[1] Bibliographical references on these theorems and generalizations are given in L. Bers [4]. See also R. Finn [1] and R. Osserman [1]. Particularly simple proofs were given by Johannes Nitsche [1] and [2].

in which the coefficients are real analytic functions and can therefore be defined also for complex values of the independent variables. We assume, for the sake of simplicity, that the coefficients are entire analytic functions of x and y. We treat the complex variables $z = x + iy$ and $\bar{z} = x - iy$ as independent variables; writing (43) in terms of these new variables, we obtain an equation of the form

$$(44) \qquad \phi_{z\bar{z}} + A\phi_z + B\phi_{\bar{z}} + C\phi = 0$$

which is *formally* a hyperbolic equation in canonical form. Note that in doing so we must necessarily consider also complex-valued solutions. We do not lose any solutions in the process since, as was pointed out earlier (p. 345), all solutions of linear elliptic equations with analytic coefficients are themselves analytic functions.

It is well known that one can express all solutions of hyperbolic equations in canonical form in the real domain by means of an integral formula involving the so-called Riemann function of the equation and arbitrary functions of one variable (see Chapter V). It is also possible to carry out the same formalism in the complex domain. Then one obtains integral operators that transform an arbitrary function of one complex variable, which must of course be assumed analytic, into a solution of the elliptic differential equation (44). In the case of the Laplace equation this operator is not an integral operator; it merely implies that the real part of an analytic function should be taken.

The method of integral operators, which we have described only in very general terms, has been developed and applied to many particular cases by S. Bergman, by I. Vekua, and by their followers. For details and the actual formulas, as well as for extensions to differential equations of higher order and to differential equations for more than two independent variables, we refer to the literature.[1]

§14. *Proof of Privaloff's Theorem*[2]

In this section we give a proof of Privaloff's theorem formulated in §3 of this supplement.

Consider an analytic function $g(z) = U + iV$ defined for $|z| < 1$. The imaginary part $V(z)$ is continuous for $|z| \leq 1$, and, for all real

---

[1] See S. Bergman .[1]. An extensive bibliography is contained in M. Z. Krzywoblocki [1] and I. Vekua [2].

[2] See J. Privaloff [1]. The theorem had already occurred in A. Korn [1]. The proof given here is due to Bers. A generalization of the theorem is given in Section 3 of the paper by S. Agmon, A. Douglis, and L. Nirenberg [1].

θ, θ' and some α, $0 < \alpha < 1$,

$$| V(e^{i\theta}) - V(e^{i\theta'}) | \leq H | e^{i\theta} - e^{i\theta'} |^{\alpha}.$$

For some fixed θ' denote by $\Phi(z)$ the harmonic function (single-valued in the unit disk) $\Phi(z) = \text{Re}\,[(1 - ze^{-i\theta'})^{\alpha}]$. This function may be defined so that $\Phi(0) = 1$. Then $\Phi(z) = |z - e^{i\theta'}|^{\alpha} \cos \alpha\nu$, where ν is the angle between the straight lines issuing from $e^{i\theta'}$ to the points 0 and z. Hence the inequality

$$-\frac{H\Phi(z)}{\cos \frac{1}{2}\alpha\pi} \leq V(z) - V(e^{i\theta'}) \leq \frac{H\Phi(z)}{\cos \frac{1}{2}\alpha\pi}$$

holds on the unit circle, and, by the maximum principle for harmonic functions, also everywhere on the unit disk. In particular, the harmonic function $V(z) - V(e^{i\theta'})$ considered in the circular disk $|z - re^{i\theta'}| < 1 - r$ does not exceed in modulus

$$(H/\cos \tfrac{1}{2}\alpha\pi)[2(1 - r)]^{\alpha}.$$

At the center of this disk the absolute values of the derivatives V_x, V_y are therefore not greater than $(H/\cos (1/2)\alpha\pi)2^{\alpha} (1 - r)^{\alpha-1}$ But $g'(z) = V_y + iV_x$ and hence, since θ' was arbitrary, we have

$$| g'(z) | \leq \frac{4H}{\cos \frac{1}{2}\alpha\pi} \frac{1}{(1 - | z |)^{1-\alpha}}.$$

Using this inequality it is easy to show that

$$| g(z_1) - g(z_2) | = \left| \int_{z_1}^{z_2} g'(\zeta)\, d\zeta \right| \leq kH \left| z_1 - z_2 \right|^{\alpha},$$

where k depends only on α.

We state now a slight *extension of Privaloff's theorem* which is useful in applications (cf. §10 of this supplement): Let $\lambda(z)$ be a real-valued ·function defined on the unit circle $|z| = 1$ for which a modulus of Hölder continuity *with an exponent less than one* is known. Let $g(z)$ be an analytic function defined in the unit disk and continuous on the circumference. Assume that the function $\sigma(z) = \text{Re}\,[g(z)ze^{i\lambda(z)}]$ satisfies on the unit circle a uniform Hölder condition. Then $g(z)$ *satisfies a uniform Hölder condition on the closed unit disk, which depends only on the bounds and the Hölder conditions for λ and σ.*

In order to prove this theorem let $h(z)$ be an analytic function defined in the unit disk whose imaginary part coincides with $\lambda(z)$ on the unit circle. Such a function exists since the Dirichlet problem for the harmonic equation can be solved. If we require that $\operatorname{Re} h$ should vanish at the origin, this function h will be uniquely determined and, by virtue of Privaloff's theorem, a bound and Hölder condition for it will be known. The analytic function

$$g_1(z) = g(z)ze^{h(z)}$$

satisfies on the unit circle the boundary condition $\operatorname{Re} g_1(z) = e^{\operatorname{Re} h}\sigma$. Thus, by Privaloff's theorem, we can find a Hölder modulus of continuity for the function $g(z)$ on the unit circle, and then, by another application of the same theorem, in the whole unit disk.

§15. Proof of the Schauder Fixed Point Theorem

The Schauder fixed point theorem is an extension to infinite dimensional spaces of a celebrated theorem due to Brouwer. The latter theorem asserts that a continuous mapping of a closed bounded convex set in Euclidean n-space into itself has a fixed point. The proof of the Brouwer fixed point theorem can be found in most texts on topology.

Since the Schauder fixed point theorem deals with mappings in a Banach space, let us first recall the meaning of this term:

A linear vector space (over the real numbers) is a set of elements (called also points) which can be added and multiplied by real numbers in such a way that the usual laws of arithmetic hold. More precisely, if x, y, z $\cdots$ denote elements of the space and λ, μ, ν, $\cdots$ real numbers, we require: $1)$ $x + y = y + x$; $2)$ $x + (y + z) = (x + y) + z$; $3)$ there exists an element 0 such that $x + 0 = x$; $4)$ the equation $x + y = 0$ has, for every y, a unique solution (denoted by $-y$); $5)$ $\lambda(x + y) = \lambda x + \lambda y$; $6)$ $(\lambda + \mu)x = \lambda x + \mu x$; $7)$ $\lambda(\mu x) = (\lambda\mu)x$; $8)$ $1x = x$. A linear space is normed if to every element x a real number $\| x \|$ is assigned such that $\| x \| \geq 0$, $\| x \| = 0$ if and only if $x = 0$; $\| \lambda x \| = | \lambda | \, \| x \|$; and $\| x + y \| \leq \| x \| + \| y \|$. In a normed space the distance between two elements x and y is defined as $\| x - y \|$. A sequence of elements $\{x_n\}$ is said to converge to an element x if $\| x_n - x \| \to 0$. A set Λ in the space is called closed if, for every sequence of elements $\{x_n\}$ in Λ such that $\| x_n - x \| \to 0$, x belongs to Λ.

A normed vector space is called complete, or a Banach space, if every Cauchy sequence converges; i.e., if to every sequence of elements

$\{x_n\}$ such that $\| x_n - x_m \| \to 0$ for $n, m \to \infty$, there exists an element x such that $\| x_n - x \| \to 0$.

A function or mapping from a Banach space $\mathfrak{B}$ into another Banach space or into the real numbers is called continuous if sufficiently small changes in the argument produce arbitrarily small changes in the function values.

A set of elements Λ in a Banach space is called convex if it contains the straight segment joining any two points belonging to it. More precisely, if x and y are elements in Λ, all elements of the form $\lambda x + \mu y$ belong to Λ, if $\lambda \geq 0, \mu \geq 0$ are real numbers such that $\lambda + \mu = 1$.

A set Λ of elements in a Banach space is called compact if every infinite sequence of elements in Λ contains a subsequence converging to an element of Λ.

Let Λ be any set of elements in a Banach space. It is contained in some closed convex set, for instance in the whole space. Since the intersection of any number of closed convex sets is easily seen to be closed and convex, there exists a smallest closed convex set $\hat{\Lambda}$ containing Λ; it is called the convex hull of Λ. In particular, if Λ consists of a finite number of points, the convex hull of Λ lies in a finite dimensional subspace of the Banach space, and may be considered as a bounded closed convex set in Euclidean space.

Now let Λ be a convex compact set of a Banach space $\mathfrak{B}$, and let T be a not necessarily linear continuous mapping of Λ into itself. The Schauder fixed point theorem asserts that this mapping has a fixed point.

In order to prove this theorem we first construct an auxiliary mapping S which maps Λ continuously into a finite-dimensional closed convex subset[1] of Λ and has the additional property that

$$\| S(x) - x \| < \epsilon$$

for every x in Λ, where ϵ is a preassigned positive number.

This construction proceeds as follows: We can find a finite sequence of points

(45) $x_1, x_2, \cdots, x_N$

in Λ such that for every x in Λ, $\| x - x_j \| < \frac{1}{2}\epsilon$ for some j. In fact, let x_1 be any point in Λ. If all other points lie within a distance

---

[1] By a finite-dimensional set in a Banach space, we mean here a set contained in a finite-dimensional subspace.

$\frac{1}{2}\epsilon$ from x_1 our task is fulfilled. Otherwise there will be a point x_2 such that $\| x_1 - x_2 \| \geq \frac{1}{2}\epsilon$. If not all points of Λ have distance less than $\frac{1}{2}\epsilon$ from either x_1 or x_2, then there will be an x_3 in Λ with $\| x_1 - x_3 \| \geq \frac{1}{2}\epsilon$, $\| x_2 - x_3 \| \geq \frac{1}{2}\epsilon$. Continuing in this way we construct the sequence (45). Note that the process must break off after a finite number of steps, since otherwise we would obtain an infinite sequence of points Λ such that any two of them have distance exceeding $\frac{1}{2}\epsilon$. Such a sequence could not contain a convergent subsequence, and this would contradict the compactness of Λ.

Having constructed the sequence (45), we define for x in Λ,

$$\mu_j(x) = \begin{cases} \| x - x_j \| & \text{if} \quad \| x - x_j \| < \frac{1}{2}\epsilon \\ \epsilon - \| x - x_j \| & \text{if} \quad \frac{1}{2}\epsilon < \| x - x_j \| < \epsilon, \\ 0 & \text{if} \quad \| x - x_j \| > \epsilon \end{cases}$$

$$\lambda_j(x) = \frac{\mu_j(x)}{\sum_{\nu=1}^{N} \mu_\nu(x)},$$

$$S(x) = \sum_{j=1}^{N} \lambda_j(x) x_j.$$

It is easy to see that the mapping $S(x)$ is continuous. It maps Λ into the convex hull Λ_0 of the points (45). Furthermore, since $\lambda_j \geq 0$, $\sum_{j=1}^{N} \lambda_j = 1$ and $\lambda_j(x) = 0$ if $\| x - x_j \| \geq \epsilon$, we see that

$$\| x - S(x) \| = \| \sum_{j=1}^{N} \lambda_j(x) x - S(x) \| \leq \sum_{j=1}^{N} \lambda_j(x) \| x - x_j \| < \epsilon.$$

Consider next the combined mapping ST. It maps Λ_0 (the convex hull of the points (45)) continuously into itself. By the Brouwer theorem this mapping has a fixed point y. Thus we have $S[T(y)] = y$ and $\| T(y) - y \| < \epsilon$.

It follows from the preceding result that we can find for every n a point y_n in Λ such that $\| T(y_n) - y_n \| < 1/n$. Since Λ is compact there will be a subsequence $\{y_{n_i}\}$ which converges to a point y in Λ. By the continuity of the mapping T we have

$$T(y) = \lim T(y_{n_i}) = y.$$

This point y is the desired fixed point.

There exists also a more general form of the Schauder fixed point theorem which asserts that a continuous mapping of any closed convex set of a Banach space into a compact subset of itself has a fixed point. One can prove this result from the preceding one by verifying that the convex hull of any compact set is compact. A more general result was proved by Tychonoff.[1]

---

[1] See N. Dunford and J. Schwartz [1], p. 456.

CHAPTER V

Hyperbolic Differential Equations in Two Independent Variables

Introduction

The following two chapters are concerned with hyperbolic differential equations of wave propagation. In this chapter problems for two independent variables x, y or x, t are treated (in the later sections we shall frequently write t instead of y to emphasize t as the time variable); Chapter VI will deal with more than two independent variables. Some modified repetition of material already touched upon in Chapter III will be necessary for a unified presentation.

Single hyperbolic differential equations, in particular of second order, will be discussed in the beginning of this chapter, in accordance with the historic development of the subject.[1] But the main emphasis will later be directed toward hyperbolic systems of differential equations, specifically systems of first order equations. This not only leads to greater generality and simplicity, but also corresponds directly to many physical problems, since these problems often occur in terms of such systems.

As a main result hyperbolic initial value problems in two independent variables will be solved with essentially the same degree of completeness as problems of ordinary differential equations; for that matter, the solutions can be constructed by iterative methods quite similar to those used for ordinary differential equations.

The concept of characteristics, already developed in Chapters I, II, and III, plays a decisive role in the treatment of hyperbolic problems not only in two but also in more independent variables (see Chapter VI). We shall first review and supplement our previous discussion of characteristics for two independent variables and then apply this theory to the solution of the basic initial value problems.

Characteristic curves C have the following properties, each of which can be used as a definition (see also Chapters I, II, III):

1) Along a characteristic curve the differential equation (or,

[1] See R. Sauer [1].

407

for systems, a linear combination of the equations) represents an interior differential equation.

1a) Initial data on a characteristic curve cannot be prescribed freely, but must satisfy a compatibility condition if these data are to be extended into "integral strips".

2) Discontinuities (of a nature to be specified later) of a solution cannot occur except along characteristics.

3) Characteristics are the only possible "branch lines" of solutions, i.e., lines for which the same initial value problem may have several solutions.

For systems of quasi-linear differential equations of first order the initial data or "Cauchy data" are simply the values of the unknown functions on the initial curve. The first property is inherent in the basic fact: A direction is characteristic at a point P if there exists a linear combination of the differential equations for which all the unknowns are differentiated at P only in this direction. (A system is hyperbolic if it can be replaced by a linearly equivalent one in which each differential equation contains at every point differentiation only in one "characteristic" direction.) The second and third properties for hyperbolic systems can also be read off from this special form of the equations.

As we shall see, initial value problems for single differential equations of higher order are always reducible to problems for systems of first order equations with specially chosen initial conditions (see also Ch. I, §7). Nevertheless, we shall first briefly discuss single differential equations, mainly equations of second order, without such a reduction. (The reader primarily interested in a systematic presentation may omit many of the details of §1.)

§1. Characteristics for Differential Equations Mainly of Second Order

1. Basic Notions. Quasi-Linear Equations. We consider the quasi-linear differential operator of second order

$$(1) \qquad L[u] \equiv ar + bs + ct$$

and the differential equation

$$(2) \qquad L[u] + d \equiv ar + bs + ct + d = 0,$$

where

$$r = u_{xx}, \qquad s = u_{xy}, \qquad t = u_{yy}$$

and where a, b, c, d are given functions of the quantities x, y, u, $p = u_x$, $q = u_y$ in the region under consideration. (It is always assumed that all functions and derivatives which occur are continuous unless the contrary is expressly stated.)

As in Chapter II, we start with the initial value problem, to extend an initial strip into an integral strip. We define a strip C_1 of first order as follows: Two functions $x = X(\lambda)$ and $y = Y(\lambda)$ of a parameter λ determine a curve C_0 in the x,y-plane; together with a function $u = U(\lambda)$ they determine a curve $\bar{C}_0$ "above C_0" in x,y,u-space. Planes tangent to the curve $\bar{C}_0$ are defined by two additional functions $P(\lambda)$ and $Q(\lambda)$ (with normal given by the direction numbers P, Q, -1) subject to the "strip relation"

$$(3) \qquad \dot{U} = P\dot{X} + Q\dot{Y}$$

which expresses that the curve $\bar{C}_0$ and the tangent plane are parallel (here the dot denotes differentiation with respect to the parameter λ). Throughout we assume

$$\dot{X}^2 + \dot{Y}^2 \neq 0.$$

A given surface $u(x, y)$ induces such a strip C_1 over a base curve C_0 if we identify U, P, and Q of $\bar{C}_0$ with the values of u, p, and q on C_0, that is, $U(\lambda) = u(X(\lambda), Y(\lambda))$ and, in addition, $P(\lambda) = u_x(X(\lambda), Y(\lambda))$ and $Q(\lambda) = u_y(X(\lambda), Y(\lambda))$.[1] For a strip on a surface $u(x, y)$ we shall write u, p, q (and x, y) instead of U, P, Q (and X, Y) whenever the meaning is clear.

Often it is useful to represent the base curve C_0 in the x,y-plane by a relation $\phi(x, y) = 0$. We assume that the curve $\phi = 0$ in the x,y-plane and also the curve $\bar{C}_0$ on the surface $u = u(x, y)$ separate a region $\phi < 0$ from a region $\phi > 0$. We also assume that $\phi = 0$ is a regular curve, i.e., that ϕ_x and ϕ_y do not vanish simultaneously.

Further we define a strip C_2 of second order by considering three additional functions $R(\lambda)$, $S(\lambda)$, $T(\lambda)$ corresponding to the second derivatives r, s, t of a surface $u(x, y)$ through C_1 and satisfying the strip relations

$$\dot{P} = R\dot{x} + S\dot{y}, \qquad \dot{Q} = S\dot{x} + T\dot{y}.$$

---

[1] The strip relation (3) expresses the fact that the integral surface $u(x, y)$, contains the strip.

The basic initial value problem[1] for the differential equation (2) can be stated as follows: Given a strip C_1 of first order, find a solution $u(x, y)$ of (2) such that the surface $u(x, y)$ contains the strip. Naturally we assume that the functions of λ defining C_1 possess continuous derivatives of first and, if needed, of higher order.

Instead of aiming at once at this initial value problem we ask the following less incisive question: Do the relations (3) and (2) permit us to extend a given strip C_1 uniquely into a strip C_2 satisfying (2)? Such a strip C_2 we shall call an *integral strip*.

Writing the relations for an integral strip C_2 in the form

$$\dot{p} = r\dot{x} + s\dot{y}, \qquad \dot{q} = s\dot{x} + t\dot{y},$$

we obtain the system

$$ar + bs + ct = -d,$$

(4)
$$\dot{x}r + \dot{y}s = \dot{p}$$

$$\dot{x}s + \dot{y}t = \dot{q}$$

of three linear equations for r, s, t along C_1. As a result, we have the following *alternative* for every point P of C_1 : Either for every point P of $\bar{C}_0$

$$Q \equiv \begin{vmatrix} a & b & c \\ \dot{x} & \dot{y} & 0 \\ 0 & \dot{x} & \dot{y} \end{vmatrix} \equiv a\dot{y}^2 - b\dot{x}\dot{y} + c\dot{x}^2 \neq 0,$$

in which case C_1 is called a *free strip* and the second derivatives r, s, t along C_0 are uniquely determined by the strip C_1 and the differential equation; or at some point P of $\bar{C}_0$

(5)
$$Q \equiv a\dot{y}^2 - b\dot{x}\dot{y} + c\dot{x}^2 = 0.$$

At such a point P, i.e., at a point where this "characteristic condition" is satisfied, the strip functions are said to form a *characteristic element*.

---

[1] For two independent variables x, y it is indeed useful to visualize solutions, strips, and characteristics in the three-dimensional x,y,u-space, as we did in Chapter II. However, often we shall concentrate on the plane of the independent variables x, y, and consider strips or curves in the x,y,u-space as curves in the x,y-plane carrying values of $u, p, q, \cdots$. When the context is clear, we shall take the liberty of using either of these definitions according to convenience.

In the following it is assumed that either the entire strip considered is free, or that it consists only of characteristic elements. In the second case the *strip* is called *characteristic* (see Ch. III, §2).

If C_1 is free, i.e., if $Q \neq 0$ everywhere along $\bar{C}_0$, then the integral strip C_2 is uniquely determined as an extension of C_1. Moreover, differentiating (4) we see that integral strips of higher order along $\bar{C}_0$ are also uniquely determined. For the third derivatives r_x, s_x, t_x, for example, we obtain the three equations

$$ar_x + bs_x + ct_x = -a_x r - b_x s - c_x t - d_x,$$

$$\dot{x}r_x + \dot{y}s_x = \dot{r},$$

$$\dot{x}s_x + \dot{y}t_x = \dot{s}$$

whose right sides are known and whose left sides have the non-vanishing determinant Q.

If $Q \equiv 0$ along $\bar{C}_0$, i.e., if a strip C_1 of first order consists entirely of characteristic elements, then the vanishing of the determinant (5) implies a linear relation between the left sides, and hence also between the right sides of equations (4), with coefficients depending only on x, y, u, p, q. Along C_1, this relation represents a new condition on p, q beyond the strip relation (3) which must be satisfied if C_1 can be extended to a second-order integral strip C_2. We call such a strip C_1 of first order a *characteristic strip*; its carrier $\bar{C}_0$ will be called a *characteristic curve* in x,y,u-space and its projection C_0 a *characteristic base curve*, or merely a characteristic curve in the x,y-plane.

Along a characteristic strip C_1 imbedded in a second-order integral strip, the second derivatives r, s, t are not determined uniquely, but only up to an arbitrary additive solution of the homogeneous equations associated with (4).

Summing up: *Either the strip C_1 is free, in which case the differential equation determines the second and higher derivatives of u on C_1 uniquely if u, p, q are given; or it contains points which satisfy the characteristic condition (5). If C_1 consists entirely of such points, it can be extended to an integral strip C_2 only if an additional condition is fulfilled. Further, the extension is no longer unique. The strip C_1 is then called characteristic.*

Consider, for example, the linear differential equation $u_{xy} = 0$ and the strip C_1 given by $x = \lambda$, $y = 0$, $u = k\lambda$ (for constant k), $p = k$,

$q = f(\lambda)$. All points of this strip satisfy (5), and, along the strip, equations (4) imply $\dot{q} = 0$. The strip must therefore be subjected to the further restriction $q = $ const. if it is capable of extension to an integral strip. In other words, among the strips whose elements all satisfy (5) only the plane strips are characteristic.

The characteristic condition is also obtained by the following argument (which can be generalized to n independent variables) referring to the base manifold $\phi(x, y) = 0$ ($\phi_x^2 + \phi_y^2 \neq 0$) of the strip C_1. (See Ch. I, App., §1 and Ch. III, §2.) We call a differential operator of second order applied to u an *interior differential operator* along C_1, or an operator within C_1, if it can be expressed on the curve $\bar{C}_0$ in terms of the quantities describing the strip C_1. For example, u_{xy} is such an interior differential expression in the strip $x = \lambda$, $y = 0$, $u = 0$, $p = 0$, $q = f(\lambda)$, since $u_{xy} = \dot{q}$.

We now pose the question: What conditions must the strip C_1 satisfy in order that the quasi-linear differential operator (1) be an interior operator along C_1? The answer is: The necessary and sufficient condition along C_1 is the *characteristic condition*

$$(6) \qquad Q(\phi, \phi) \equiv a\phi_x^2 + b\phi_x \phi_y + c\phi_y^2 = 0;$$

here $Q(\phi, \phi)$ is known as the "characteristic form".

Proof: In place of x and y we introduce new coordinates $\eta = \phi(x, y)$ and $\lambda = \psi(x, y)$ in such a way that λ (or ψ) is identical with the parameter previously introduced along C_1 while ϕ is a variable "leading out" of C_1. Then, for any function $u(x, y)$,

$$u_{xx} = u_{\phi\phi}\, \phi_x^2 + 2u_{\phi\psi}\, \phi_x\, \psi_x + u_{\psi\psi}\, \psi_x^2 + u_\phi\, \phi_{xx} + u_\psi\, \psi_{xx},$$

$$u_{xy} = u_{\phi\phi}\, \phi_x\, \phi_y + u_{\phi\psi}(\phi_x\, \psi_y + \phi_y\, \psi_x) + u_{\psi\psi}\, \psi_x\, \psi_y + u_\phi\, \phi_{xy} + u_\psi\, \psi_{xy},$$

$$u_{yy} = u_{\phi\phi}\, \phi_y^2 + 2u_{\phi\psi}\, \phi_y\psi_y + u_{\psi\psi}\, \psi_y^2 + u_\phi\, \phi_{yy} + u_\psi\, \psi_{yy},$$

and hence, if $Q(\phi, \psi)$ denotes the polar form (cf. Vol. I, Ch. I, §1, 4) of the quadratic form Q,

$$L[u] = u_{\phi\phi}\, Q(\phi, \phi) + 2u_{\phi\psi}\, Q(\phi, \psi) + u_{\psi\psi}\, Q(\psi, \psi)$$

$$(7)$$

$$+ u_\phi\, L[\phi] + u_\psi\, L[\psi].$$

Along C_1, differentiation with respect to $\psi = \lambda$ denotes interior differentiation, while differentiation with respect to ϕ is differentiation leading out of C_1 (cf. Ch. II, App. I, §1). Along C_1, the function u and the first derivatives of u are known, as well as all second-order derivatives obtained from those of first order by differentiating with respect to $\lambda = \psi$. Thus the only term in $L[u]$ which contains second derivatives not lying in C_1 is $u_{\phi\phi} Q(\phi, \phi)$. The condition $Q(\phi, \phi) = 0$ *on $\phi = 0$ is, therefore, necessary and sufficient to assure that $L[u]$ is an interior operator.

We now consider the differential equation (2)

$$L[u] + d = 0.$$

Again we immediately recognize the *alternative*: Either $Q(\phi, \phi) \neq 0$ for every point of C_1, in which case the outward derivative $u_{\phi\phi}$ and with it all the second derivatives of u are uniquely determined along C_1; or else $Q(\phi, \phi) = 0$ at a point P of C_1, and then the differential equation (2) represents at this point of C_1 an aditional condition for the strip quantities. If we assume that the strip C_1 is given (so that the first derivatives of u are known along the curve C_0) and if $Q(\phi, \phi) = 0$ everywhere on C_1, then this new condition has the form of an ordinary differential equation for the quantity $u_\phi = \kappa$ as a function of $\psi = \lambda$, namely,

(8) $$2\kappa_\lambda Q(\phi, \psi) + \kappa L[\phi] + \cdots = 0,$$

where the dots stand for terms which are known on the strip.

Clearly, the characteristic conditions (6) and (5) are equivalent. As a matter of fact, since

$$\phi_x \dot{x} + \phi_y \dot{y} = 0$$

i.e.,

$$\frac{\phi_x}{\phi_y} = -\frac{\dot{y}}{\dot{x}},$$

the left side of (6) agrees with the left side of (5), except for a factor. Characteristic strips C_1 can exist only in regions where

$$4ac - b^2 \leq 0;$$

otherwise equation (5) (or (6)) cannot be satisfied by real ratios $\dot{x}:\dot{y}$ (or $\phi_x:\phi_y$).

We recall the following definitions: The differential operator $au_{xx} + bu_{xy} + cu_{yy}$ is *hyperbolic at a point* $P:(x, y, u, p, q)$ of the five-dimensional x,y,u,p,q-space if

$$(9) \qquad\qquad 4ac - b^2 < 0$$

at that point (cf. Ch. III, §2, 1). Similarly, it is called *hyperbolic on a surface* $u = u(x, y)$ if condition (9) is satisfied at every point of the surface. Naturally, if condition (9) is satisfied for a point P of the five-dimensional x, y, u, p, q-space, it remains valid for a suitably chosen neighborhood of P.

In the following we shall always suppose that the differential operators are hyperbolic at the points considered.

If the differential operator is linear, then the hyperbolic character depends only on x, y and not on u, p, q. In particular, the characteristic base curves C_0 are then determined by the differential operator independently of u, p, q.

Finally, we emphasize the following important fact: *The characteristic conditions for a differential equation* (2) *are invariant under arbitrary transformations of the independent variables* x, y.

This follows immediately from the fact that the characteristic condition is necessary and sufficient for $L[u]$ to be an interior operator on C_1. To obtain a formal proof we transform x, y into ξ, η. Thus

$$a u_{xx} + b u_{xy} + c u_{yy} + d = \alpha u_{\xi\xi} + \beta u_{\xi\eta} + \gamma u_{\eta\eta} + \delta,$$

where the coefficients α, β, γ, δ on the right are functions of ξ, η, u, u_ξ, u_η. Then, as one easily verifies,

$$a\phi_x^2 + b\phi_x \phi_y + c\phi_y^2 = \alpha\phi_\xi^2 + \beta\phi_\xi \phi_\eta + \gamma\phi_\eta^2,$$

from which the assertion follows.

2. Characteristics on Integral Surfaces. So far we have restricted ourselves to the discussion of the variables along a strip; now we consider the total extent of a surface $J : u = u(x, y)$ which we assume to be an integral surface of equation (2). Along such an integral surface, not only u but also $p = u_x$ and $q = u_y$, and hence also the coefficients a, b, c, d, are known functions of x and y. We assume that along the entire surface in question the condition (9) is satisfied, i.e., that (2) is hyperbolic. The characteristic condition

$$a\dot{y}^2 - b\dot{x}\dot{y} + c\dot{x}^2 = 0$$

then defines two distinct real values ζ_1 and ζ_2 for the ratio $\dot{y}/\dot{x}$, and thus a set of *two distinct* one-parameter *families of characteristic curves on the integral surface,*[1] as solutions of the ordinary differential equations

$$\frac{dy}{dx} = \zeta_1, \qquad \frac{dy}{dx} = \zeta_2 .$$

As will be seen, introducing these two families as parameter curves greatly simplifies the problem of solving the differential equation.

In the elliptic case, when $4ac - b^2 > 0$, no such characteristics exist. In the parabolic limiting case, when $4ac - b^2 = 0$, the two characteristic families coincide.

Characteristic strips are the only possible *branch strips* of an integral surface. By a branch strip we mean a strip along which two distinct integral surfaces touch, so that they agree in the quantities u, p, q, but differ in some derivatives of higher order. On a noncharacteristic strip, all second derivatives (and similarly all higher ones as far as they exist and are continuous) are uniquely determined. Therefore, branch strips are necessarily characteristic.

If a differential equation is elliptic, the integral surface can carry no branch strips. If, in addition, we assume that the differential equation is ˙analytic, and that accordingly the derivatives of any order are uniquely determined along every strip on the integral surface, we are naturally led to the conjecture that every such solution of the elliptic type must be an *analytic* function. A proof of this will be given in Appendix 1 of this chapter (see also Ch. IV, §7). Here we merely remark that the presence of branch strips implies the existence of nonanalytic solutions of the equation.

We consider once more the characteristic condition (6), $Q = a\phi_x^2 + b\phi_x\phi_y + c\phi_x^2 = 0$, where a, b, c are known functions of x and y, on the integral surface J. Let ϕ be a solution of (6) considered as a partial differential equation; we have seen that then $\phi = 0$ represents a characteristic curve on J. Clearly, if ϕ is a solution of (6), so is ϕ minus a constant, and hence all curves of the family $\phi = $ const. are characteristics on J. Conversely, if $\phi = $ const. represents such a family of characteristics, the function ϕ satisfies equation (6) considered as a partial differential equation.

An illustration is given by the differential equation $u_{xy} = 0$.

---

[1] Again we take the liberty of using the name "characteristic curves" for the space curves on J as well as for their projection on the x,y-plane. In particular, for linear differential equations, we shall mostly talk about the latter since they are fixed and do not depend on the solution.

The characteristic condition is $\phi_x \phi_y = 0$. It is satisfied by any ϕ which is either a function of x alone or a function of y alone. This leads us to the lines $x = $ const. and the lines $y = $ const. as characteristic base curves.

3. Characteristics as Curves of Discontinuity. Wave Fronts. Propagation of Discontinuities. We have introduced characteristics as possible branch curves of integral surfaces along which certain derivatives of u have discontinuities. The present article concerns a remarkable fact which will be discussed later (e.g., Ch. VI, §4) in greater generality: The magnitude of such jump discontinuities is controlled by an ordinary differential equation of first order along the characteristic curves.

We describe this phenomenon of propagation here for the case of discontinuities of second order derivatives.

Let C be a curve given by $\phi(x, y) = 0$ and separating the regions $\phi \geq 0$, and $\phi \leq 0$; let $u(x, y)$ be a solution of the differential equation (2) in each of these regions and such that u and its first derivatives are continuous across C but "exterior" second derivatives of u suffer a jump across C. Interior derivatives, however, remain continuous along $\phi = 0$ in the following sense: If, as before, $\lambda = \psi$ and $\eta = \phi$ are coordinates on the integral surface J in a neighborhood of C, where λ is the parameter on C, then all derivatives of u, p, q with respect to λ are to remain continuous across C.

We denote by $[f]$ the jump of a function f across C in the direction of increasing values of ϕ. By assumption, u_x is continuous, and so is its interior derivative (cf. Ch. II, §1), given by $u_{xx} \phi_y - u_{xy} \phi_x$, as well as the interior derivative of u_y, given by $u_{xy} \phi_y - u_{yy} \phi_x$. For the jumps we thus have the two relations

$$[u_{xx}]\phi_y - [u_{xy}]\phi_x = 0,$$

$$[u_{xy}]\phi_y - [u_{yy}]\phi_x = 0,$$

which at once yield

$$[u_{xx}] = \kappa\phi_x^2, \qquad [u_{xy}] = \kappa\phi_x \phi_y, \qquad [u_{yy}] = \kappa\phi_y^2$$

for some factor of proportionality κ. (Incidentally, it is immediately seen that $\kappa = [u_{\phi\phi}]$.)

We now consider the differential equation (2) at two points P_1 and P_2 on different sides of the curve C on J, subtract the two resulting equations from each other, and allow P_1 and P_2 to converge to a point P of C (Figure 23). The continuous terms vanish and we have

$$a[u_{xx}] + b[u_{xy}] + c[u_{yy}] = 0,$$

or, according to our above result, cancelling the factor κ,

$$a\phi_x^2 + b\phi_x \phi_y + c\phi_y^2 = Q(\phi, \phi) = 0,$$

which is simply the characteristic relation. Thus it is confirmed that discontinuities of the indicated type can occur only along a characteristic.

For a physical interpretation we consider $y = t$ as time and think of the solution $u(x, t)$ as a "wave", or simply a quantity that varies in x-space with the time t. If this wave has a discontinuity along the characteristic $\phi(x, t) = 0$, we imagine $\phi = 0$ solved in the form $x = x(t)$, with

$$\frac{dx}{dt} = -\frac{\phi_t}{\phi_x}.$$

Then we can interpret the curves of discontinuity $\phi = 0$ in the x,t-plane as the paths of points of discontinuity x which move along the x-axis with the time t and the velocity dx/dt.

The factor of proportionality κ is a measure of the discontinuity. It has the following remarkable property: *Along the characteristic C, the factor κ satisfies an ordinary homogeneous linear differential equation,* namely,

(10) $$\alpha\kappa_\lambda + \beta\kappa = 0,$$

where the coefficients α and β are given by

$$\alpha = 2Q(\phi, \psi), \qquad \beta = L[\phi] + Q_\phi(\phi, \phi).$$

For the proof, we differentiate the equation $L[u] + d = 0$, where L is given by (7), with respect to ϕ and form the expression for the

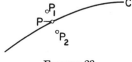

FIGURE 23

jump. Only terms containing derivatives with respect to ϕ of second or higher order contribute. Recalling that $Q(\phi, \phi) = 0$ on C, we are led to the differential equation (10) for κ.

This result implies that the factor of discontinuity either does not vanish anywhere or vanishes everywhere along the portion of the x-axis considered.

The discontinuities discussed above occur in the second or higher derivatives. The meaning of the differential equation excludes discontinuities in the first derivatives. Nevertheless, in §9 we shall generalize the concept of solution so that such discontinuities and others are admissible. In fact, discontinuities in the first derivatives can be produced along an arbitrary free curve C. We can extend C in different ways to a strip C_1. Then (see the second part of this chapter) we can solve the corresponding initial value problems and, by joining one solution on one side of C to any other solution on the other side, we obtain a generalized solution u which is continuous and has discontinuous first derivatives across C.

Discontinuities in the first derivatives are excluded by the meaning of the differential equation. It would be a pointless generalization to admit generalized "solutions" u which, while remaining continuous, have discontinuous derivatives across a (free) curve C. Such functions u always could be formed by considering on different sides of C two different solutions u_1 and u_2 for which the initial strips C_1 and C_2 are different strips through C; for example $u = 0$ for $y > 0$ and $u = y$ for $y < 0$ in the case of the equation $u_{yy} - u_{xx} = 0$ and the line $y = 0$ as the curve C.

Yet, in §9 we shall give a nontrivial extension to the concept of solution by defining "weak solutions". Such solutions may have discontinuities of the first derivatives, and it will turn out that here again the characteristics are distinguished as the only possible carriers of discontinuities.

In the case of linear differential equations this distinct property of characteristics as lines of discontinuity will be seen to prevail even for discontinuities of u itself. In all these cases we shall find ordinary differential equations of the form (10) for the propagation of these discontinuities.

4. General Differential Equations of Second Order. It is simple to extend the preceding results to a general second order differential equation

$$(11) \qquad F(x, y, u, p, q, r, s, t) = 0.$$

Once again we consider a curve C, in x,y,u-space, assumed to be supplemented so as to form a first-order strip C_1 or a second order strip C_2. We assume that C_2 is an integral strip, i.e., that the corresponding quantities x, y, u, p, q, r, s, t satisfy the equation $F = 0$.

The first step toward solving the initial value problem is to extend the initial integral strip C_2 into an integral strip C_3 of third order carrying quantities which correspond to the third derivatives along C of a function $u(x, y)$ satisfying (10).

We can obtain the characteristic condition as follows: Let λ be the parameter along the curve C whose projection C_0 on the x,y-plane is given by the equation $\phi(x, y) = 0$. We introduce λ and ϕ as new coordinates in the neighborhood of the base curve C_0. Then $u(x, y)$ becomes a function $u(\lambda, \phi)$ and we write

$$F(x, y, u, p, q, r, s, t) = G(\lambda, \phi, u, u_\lambda, u_\phi, u_{\lambda\lambda}, u_{\lambda\phi}, u_{\phi\phi}).$$

The initial integral strip C_2 is called *characteristic* if not all the higher derivatives of u, in particular if not the third derivatives, are determined along it by the differential equation. If the second outward derivative $u_{\phi\phi}$ can be calculated from the differential equation $G = 0$ along the strip, i.e., if the differential equation can be written in the form

$$u_{\phi\phi} = g(\lambda, \phi, u, u_\lambda, u_\phi, u_{\lambda\lambda}, u_{\lambda\phi}),$$

then by differentiation we obtain the third outward derivative $u_{\phi\phi\phi}$, and clearly with it all the other third derivatives along the strip C_2. In other words, if $G_{u_{\phi\phi}} \neq 0$, all higher derivatives along C_2 can be determined. The strip C_2 then is free, i.e., not characteristic. Hence the condition $G_{u_{\phi\phi}} = 0$ must be satisfied for a characteristic strip C_2. This "characteristic condition" may easily be put into the form

(12) $$F_r \phi_x^2 + F_s \phi_x \phi_y + F_t \phi_y^2 = 0.$$

In a somewhat different way, more in line with article 1, we can derive this characteristic condition by calculating the third derivatives, say r_x, s_x, t_x along the initial strip C_2. We differentiate the equation $F = 0$ with respect to x, use the abbreviation

$$\{F\}_x = F_p r + F_q s + F_u p + F_x,$$

employ the strip relations, and arrive at the linear system

$$F_r r_x + F_s s_x + F_t t_x = -\{F\}_x ,$$

$$\dot{x} r_x + \dot{y} s_x = \dot{r},$$

$$\dot{x} s_x + \dot{y} t_x = \dot{s}.$$

Precisely the same considerations apply to this system as to system (4) in article 1: *If the determinant*

$$Q = \begin{vmatrix} F_r & F_s & F_t \\ \dot{x} & \dot{y} & 0 \\ 0 & \dot{x} & \dot{y} \end{vmatrix} = F_r \dot{y}^2 - F_s \dot{x} \dot{y} + F_t \dot{x}^2$$

is different from 0, then the integral strip C_2 is free and the higher derivatives along C_2 are uniquely determined. If the determinant vanishes everywhere along the strip, the strip is called characteristic. In this case, the quantities x, y, u, p, q, r, s, t determining the characteristic strip must satisfy two additional conditions so that the strip can be extended to an integral surface (or merely to an integral strip of third order). The first condition expresses the fact that the right-hand sides of the above equations satisfy the same linear dependence as the left-hand sides; the second one is obtained similarly when we differentiate the equation $F = 0$ with respect to y.

If the base curve of the strip is given by $\phi(x, y) = 0$, the characteristic condition

$$(12') \qquad F_r \dot{y}^2 - F_s \dot{x} \dot{y} + F_t \dot{x}^2 = 0$$

differs from condition (12) only by a factor. Condition (12) or (12') can be satisfied at a point of a strip C_2 by a real ratio $-\phi_y : \phi_x$ or $\dot{x} : \dot{y}$ only if at that point

$$4 F_r F_t - F_s^2 \leq 0.$$

As before, we call the differential operator F *hyperbolic* at a point of the eight-dimensional x, y, u, p, q, r, s, t-space if at that point the strong inequality

$$(13) \qquad 4 F_r F_t - F_s^2 < 0$$

holds. The differential operator F is called hyperbolic on a strip of second order or on a surface $u(x, y)$ if (13) holds at every point.

As in the special case of quasi-linear differential equations, it must be noted that the characteristic condition on a given integral surface can be considered as a first order partial differential equation for ϕ, and the characteristic curves consist of the entire family of solutions $\phi = $ const.

5. Differential Equations of Higher Order. In discussing an nth-order differential equation for the function $u(x, y)$, we shall introduce the following abbreviation:

$$p_\nu = \frac{\partial^n u}{\partial x^\nu \partial y^{n-\nu}} \qquad (\nu = 0, 1, \cdots, n).$$

The nth order differential equation then has the form

$$(14) \qquad F(x, y, u, \cdots, p_0, \cdots, p_n) = 0$$

in which the derivatives of order less than n are not written out explicitly. We obtain the concept of characteristics and the characteristic condition briefly in the following way: Suppose that $u = u(x, y)$ is an integral surface, and that on it a curve C together with the associated strip C_n of nth order is given with the base curve $\phi(x, y) = 0$ separating a region $\phi > 0$ from a region $\phi < 0$. Let λ be a parameter in this strip. On the integral surface $u(x, y)$ we again introduce λ and ϕ as independent variables instead of x and y. We write $u(\lambda, \phi)$ for the function into which $u(x, y)$ is transformed, and denote by

$$\omega = u_{\phi \cdots \phi} = \frac{\partial^n u}{\partial \phi^n}$$

the nth derivative of u leading out of the strip C_n. Then we have

$$p_\nu = \omega \phi_x^\nu \phi_y^{n-\nu} + \cdots,$$

where the dots stand for terms in which the nth order derivative ω does not occur. The given differential equation (14) is now regarded as a differential equation for $u(\lambda, \phi)$. If along C equation (14) can be put into the form

$$\omega = f(\lambda, \phi, u, \cdots),$$

where ω itself no longer appears explicitly on the right-hand side, we can, by differentiation with respect to ϕ, uniquely determine the $(n + 1)$st outward derivative ω_ϕ and all other derivatives of u of order $n + 1$. This is always possible if $F_\omega \neq 0$. If $F_\omega = 0$ for all λ, we shall say that C_n is a *characteristic strip*. Written in the old variables, this relation becomes the fundamental *characteristic condition*

$$(15) \qquad F_{p_n}\phi_x^n + F_{p_{n-1}}\phi_x^{n-1}\phi_y + \cdots + F_{p_0}\phi_y^n = 0.$$

We define an nth order strip with the base curve $\phi = 0$ as a characteristic strip if (1) relation (15) is satisfied along it, and (2) if it is an integral strip.

The question whether a characteristic strip defined in this way can always be imbedded in an integral surface, i.e., the question of the existence of a solution of the initial value problem, is here left open.

We now return to the consideration of characteristic strips on a given integral surface J. Relation (15) holds on $J : u = u(x, y)$ if we insert the function $u(x, y)$ and the values of its derivatives everywhere and restrict x, y by the supplementary condition $\phi = 0$.

If the curve $\phi = 0$ is written in the form $y = y(x)$, then we have for the slope $y' = -\phi_x : \phi_y$ and the characteristic relation goes over into

$$(16) \qquad F_{p_n}y'^n - F_{p_{n-1}}y'^{n-1} + \cdots = 0,$$

an *ordinary first order differential equation for the characteristic curve* C on J. Conversely, every solution of this ordinary differential equation furnishes a characteristic curve on J.

If one interprets the characteristic condition (15) as a partial differential equation for the function $\phi(x, y)$ of two independent variables, then every solution of the equation will give not only a single characteristic curve C but actually an entire family of characteristic curves $\phi(x, y) = c$ with the parameter c on J.

The nature of the roots of the algebraic equation in ζ,

$$(17) \qquad F_{p_n}\zeta^n - F_{p_{n-1}}\zeta^{n-1} + \cdots + (-1)^n F_{p_0} = 0,$$

determines the type of the differential equation (14) in the neighborhood of the integral strip. If the algebraic equation has n distinct real solutions, the differential equation is called *totally hyperbolic*, or merely *hyperbolic*, at a point of the $x, y, u, \cdots, p_0, \cdots, p_n$-space, or along a strip C_n, or on a surface $u = u(x, y)$. If all the solutions of the algebraic equation are imaginary, the differential equation is called *elliptic*. If some real solutions coincide, the differential equation is often called *parabolic*. (However, we shall later modify this classification by classing as hyperbolic some types of equations with multiple characteristics for which Cauchy's initial value problem remains solvable.)

6. Invariance of Characteristics under Point Transformations. The characteristics of partial differential equations are invariant with respect to point transformations. In other words, under a point transformation the characteristics go over into the corresponding characteristics of the transformed differential equation. The proof is sufficiently clear from a consideration of the simple case of the second order differential equation $F(u_{xx}, u_{xy}, u_{yy}, \cdots) = 0$ which becomes $G(u_{\xi\xi}, u_{\xi\eta}, u_{\eta\eta}, \cdots) = 0$ when transformed to the new independent variables ξ, η. Once again the invariance follows either immediately from the intuitive meaning of the characteristic strips or from the identity

$$F_{u_{xx}}\phi_x^2 + F_{u_{xy}}\phi_x\,\phi_y + F_{u_{yy}}\phi_y^2 = G_{u_{\xi\xi}}\phi_\xi^2 + G_{u_{\xi\eta}}\phi_\xi\,\phi_\eta + G_{u_{\eta\eta}}\phi_\eta^2,$$

which follows by elementary calculations.

The invariance of characteristics has the following consequences: Hyperbolic problems are *time-reversible*; i.e., the hyperbolic character is preserved under the transformation $x' = x$, $y' = -y$, y being interpreted as the time.

Since every nowhere characteristic curve can be transformed into an arbitrary nowhere characteristic straight line, it suffices to solve Cauchy's initial value problem merely for the x-axis as the initial line.

7. Reduction to Quasi-Linear Systems of First Order. Without loss of generality one may restrict the discussion of initial value problems to problems for quasi-linear systems of first order (see also Ch. 1, §7, 2). Here we repeat the usual procedure for such a reduction: If the differential equation (12),

$$F(x, y, u, p, q, r, s, t) = 0,$$

is not quasi-linear then one obtains by differentiating, e.g., with respect to y, the quasi-linear differential equation

$$(F)_y = F_r\,r_y + F_s\,s_y + F_t\,t_y + F_p p_y$$

(18)

$$+ F_q\,q_y + F_u\,q + F_y = 0.$$

This, together with

$$u_y = q, \qquad p_y = s, \qquad q_y = t, \qquad r_y = s_x, \qquad s_y = t_x$$

represents a quasi-linear system of first order for the unknown functions u, p, q, r, s, t.

If initial values (at $y = 0$) $u = f(x)$, $u_y = g(x)$ are given, then the initial values for the system are $u = f$, $p = f'$, $q = g$, $r = f''$, $s = g'$, and $t(x, 0)$ calculated from $F = 0$. A solution u, p, q, r, s, t of the system furnishes a solution u of the original equation. To see this we note that our system of equations yields

$$s_y = t_x = q_{yx} = q_{xy}, \qquad (s - q_x)_y = 0.$$

Since $s = q_x$ at $y = 0$ we have $s \equiv q_x \equiv u_{xy}$. Similarly, one can identify the other unknowns in the system with derivatives of u. Finally, one verifies that $dF/dy = 0$; since initially $F = 0$, it follows that u solves the original equation.

Such a reduction to quasi-linear systems of first order can always be carried out under our over-all assumption of sufficient differentiability of the functions involved (see also Ch. I, §7, 2). Accordingly, from now on we shall concentrate on quasi-linear systems of first order, reviewing and supplementing our discussion of Ch. III, §1.

§2. Characteristic Normal Forms for Hyperbolic Systems of First Order

1. Linear, Semilinear and Quasi-Linear Systems. A quasi-linear system of first order is written in the form[1]

$$(1) \qquad L^\kappa[u] \equiv \sum_{i=1}^{k} (a^{\kappa i} u_x^i + b^{\kappa i} u_y^i) = c^\kappa \qquad (\kappa = 1, 2, \cdots, k),$$

or in obvious matrix notation with matrices A, B and the vector c

$$(1') \qquad L[u] = Au_x + Bu_y = c.$$

We assume that in the region considered the matrix B is nonsingular, specifically that the determinant $\| B \| = \| b^{\kappa i} \|$ does not vanish.

If A, B do not depend on the unknown vector u and if c depends linearly on u, then the system is *linear*; if A and B are independent

---

[1] We consider only "determined systems" (cf. Ch. I, §2, 3), where the number k of equations equals the number of unknowns.

of u, and c depends on u, but not linearly, the system is called *semi-linear* and can be treated almost as a linear system. Otherwise, if A, B depend on u the system is *quasi-linear*. As we saw, any nonlinear system if differentiated becomes essentially equivalent to a quasi-linear one.

Because of $\| B \| \neq 0$ we may solve (1) with respect to u_y and write it in the equivalent form

$$(1'') \qquad u_y + A u_x = c,$$

with $B = I$ the unit matrix, which implies that the lines $y = $ const. are noncharacteristic or free.

We supplement basic facts concerning systems $(1'')$, considering first the linear or semilinear case (cf. Ch. III, §2, 1). Characteristic curves $C: \phi(x, y) = 0$ are given by differential equations

$$(2) \qquad dx:dy = \tau$$

or

$$(2') \qquad \tau\phi_x + \phi_y = 0,$$

with τ defined as a root of the algebraic equation

$$(3) \qquad Q = \| A - \tau I \| = 0.$$

These equations state that the matrix $A - \tau I$ is singular; hence for a characteristic curve there exists an eigenvector l with components $l_1, \cdots, l_k$ for which

$$(4) \qquad lA = \tau l.$$

In agreement with Ch. III, §2, hyperbolicity of L is defined by the requirement that all roots $\tau_1, \cdots, \tau_k$ of $Q = 0$ are real and that in addition there exist k linearly independent eigenvectors $l^1, l^2, \cdots,$ l^k. In particular, this is the case if all the roots τ and therefore all characteristics are distinct.

The hyperbolic system can be replaced by a linearly equivalent system obtained if one combines the equations $(1'')$ with the components of the eigenvectors $l^1, \cdots, l^k$, respectively or multiplies (1) or $(1')$ or $(1'')$ with the left eigenvectors l, observing (4). From $(1'')$ the equivalent system for $l = l^i$, $\tau = \tau_i$ results in:

$$(5) \qquad l^i L[u] = \Lambda^i[u] = l^i(u_y + \tau_i u_x) = l^i c \ (i = 1, \cdots, k).$$

Now under the assumption that the system is linear or semilinear we can simplify (5) further by introducing instead of the unknowns u a new linearly equivalent unknown vector U with the k components

$$(6) \qquad\qquad U^\kappa = l^\kappa u \qquad\qquad (\kappa = 1, \cdots, k).$$

If along the characteristic C_i we introduce the directional differentiation

$$(7) \qquad\qquad D_i = \frac{\partial}{\partial y} + \tau_i \frac{\partial}{\partial x}$$

and write

$$g^i = l^i c + u(l_y^i + \tau l_x^i) \qquad\qquad (\tau = \tau_i),$$

then the system becomes

$$(8) \qquad\qquad D^i U^i = g^i.$$

Denoting differentiation with y as the independent variable along the characteristic C by d/dy, (8) can be written in the form

$$(9) \qquad\qquad \frac{d}{dy} U^i = g^i(x, y, U) \qquad\qquad \text{on } C_i,$$

or briefly in matrix form

$$(9') \qquad\qquad U_y + TU_x = g,$$

where T is the diagonal matrix with the diagonal elements τ_i. These equations (5), (7), (9), (9') are the *normal form of the system*. They exhibit the fact: A semilinear hyperbolic system is equivalent to a *symmetric system* with coefficients A and B, one of whose matrices, e.g. that of A, is positive definite.[1]   In particular, *we may always assume A in* (1'') *is a symmetric matrix.*

[1] In matrix notation the transformation of the differential equation (1'') to a symmetric (diagonal) form can be described as follows: We set $U = Hu$, where the rows of the matrix H are the eigenvectors l^i, and substitute

$$u = H^{-1}U, \qquad u_x = H^{-1}U_x + H_x^{-1}U, \qquad u_y = H^{-1}U_y + H_y^{-1}U$$

into (1''). Then we obtain

$$AH^{-1}U_x + H^{-1}U_y = c - (AH_x^{-1} + H_y^{-1})U.$$

Now, multiplying by H from the left, we obtain the diagonal form

$$U_y + HAH^{-1}U_x = G,$$

where $G = H(c - AH_x^{-1}U - H_y^{-1}U)$ is a known vector, depending linearly on U but not depending on the derivatives of U.

For quasi-linear systems the introduction of the new variables U would not immediately lead to a simplification such as (7) or (9) since the derivatives of the coefficients l again contain the unknowns. We must therefore be satisfied with the normal form (5). Nevertheless, by the simple device of differentiating the equations (5) with respect to x and y one could obtain for the derivatives of u simplified equations for which the nonlinear quasi-linear character is de-emphasized so that introduction of those linear combinations then becomes effective.

Restricting ourselves again to linear operators L we can immediately read off from the normal form the basic properties of characteristic curves. In particular the normal form exhibits the fact that along a characteristic curve the linear combination lL is an internal differential operator.

Furthermore, the normal form suggests that beyond the solutions with discontinuous derivations considered in §1 we may admit generalized *solutions which* themselves *have discontinuities*, suffer in particular a jump discontinuity across a curve C but are otherwise regular, and for which C is characteristic. In Ch. VI, §4 we shall discuss discontinuities and their transmission along characteristics generally. Here merely the following fact should be pointed out, which is evident from (8).

All components of U except U^κ are continuous across C^κ. The jump $[U^\kappa] = \omega$ propagates along C according to the ordinary linear differential equation[1]

$$\frac{d\omega}{dy} = g(x, y, \omega).$$

2. The Case $k = 2$. Linearization by the Hodograph Transformation.[2] For hyperbolic systems of two equations, just as for a single differential equation of second order, one may (as in §1) introduce the two families of characteristic curves as coordinate curves. This means one introduces two "characteristic parameters" α and β such that

$$\alpha = \phi(x, y) = \text{const.} \quad \text{and} \quad \beta = \psi(x, y) = \text{const.}$$

represent the two distinct characteristic families which in the quasi-

---

[1] In the quasi-linear case the equation for the jump was investigated by J. Nitsche [3].

[2] See Ch. III, §2, 5.

linear case still depend on the individual solution u^1, u^2 under consideration, but in the linear case are independent of u and are *a priori* known. The introduction of characteristic parameters α, β is particularly convenient in the treatment of the nonlinear Monge-Ampère equation (see App. I, §2).

We may consider u^1, u^2, x, y as four functions of the two independent variables α, β and obtain an equivalent system of four equations in characteristic independent parameters

(11a) $\qquad h^{11}u_\alpha^1 + h^{12}u_\alpha^2 = \gamma^1, \qquad h^{21}u_\beta^1 + h^{22}u_\beta^2 = \gamma^2,$

(11b) $\qquad\qquad x_\alpha = \tau^1 y_\alpha, \qquad x_\beta = \tau^2 y_\beta,$

where the $h^{i\kappa}$, γ^κ, τ^κ are known functions of u^1, u^2, x, y, x_α, $\ldots$, y_β.

In the linear case the equations (11b) are independent of the equations (11a); in the quasi-linear case they are coupled. At any rate, together they replace the original system of two equations by a simpler system of four equations in which, incidentally, the independent variables do not occur explicitly.

In an important special case a quasi-linear system in two unknowns can immediately be reduced to a linear system by simply interchanging the role of dependent and independent variables, i.e., by considering x and y as functions of $u = u^1$ and $v = u^2$. This reduction is possible if the system consists of differential equations which are homogeneous in the derivatives and have the form

$$a^1 u_x + b^1 u_y + c^1 v_x + d^1 v_y = 0,$$

$$a^2 u_x + b^2 u_y + c^2 v_x + d^2 v_y = 0,$$

where $a^1, \cdots, d^2$ are functions of u, v alone, and if the Jacobian $J = u_x v_y - u_y v_x$ does not vanish. We have

$$u_x = J y_v, \qquad u_y = -J x_v,$$

$$v_x = -J y_u, \qquad v_y = J x_u$$

and our system is indeed transformed into the linear system

$$a^1 y_v - b^1 x_v - c^1 y_u + d^1 x_u = 0,$$

$$a^2 y_v - b^2 x_v - c^2 y_u + d^2 x_u = 0$$

for $x(u, v)$ and $y(u, v)$.

This transformation plays an important role in fluid dynamics, where u, v denote the velocity components of a steady two-dimensional flow. It is called hodograph transformation, because the image in the u,v-plane of a particle path in the x,y-plane is the "hodograph" of the particle, that is, the path described by the velocity vector.

§3. Applications to Dynamics of Compressible Fluids

The motion of compressible fluids offers instructive and significant illustrations of the concept of characteristics.[1] In Chapter VI we shall consider such flows depending on three space variables and time. Here we restrict ourselves to cases which can be described by only two independent variables.

1. One-Dimensional Isentropic Flow. The differential equations for the velocity $u(x, t)$ and density $\rho(x, t)$ of a one-dimensional isentropic flow are

$$L^{1}[u, \rho] \equiv \rho u_{x} + u\rho_{x} + \rho_{t} = 0,$$

$$L^{2}[u, \rho] \equiv \rho u u_{x} + \rho u_{t} + c^{2}\rho_{x} = 0.$$

The "sound speed" $c(\rho)$ is given by $c = \sqrt{f'(\rho)} > 0$, where $p = f(\rho)$ is a given monotonic function expressing the pressure p in terms of ρ. (The condition $f'(\rho) > 0$ expresses hyperbolicity of the system.) For this first-order quasi-linear system, the characteristic directions $\tau = dx:dt$ are defined by

$$\begin{vmatrix} 1 & u - \tau \\ u - \tau & c^{2} \end{vmatrix} = 0,$$

hence $\tau = u \pm c$. The system is hyperbolic since there are two distinct real roots τ. For each of these values of τ, there exists a nontrivial solution $(\lambda^{1}, \lambda^{2})$ of the system

$$\lambda^{1} + (u - \tau)\lambda^{2} = 0,$$

$$(u - \tau)\lambda^{1} + c^{2}\lambda^{2} = 0.$$

When $\tau = u + c$, $\lambda^{1}:\lambda^{2} = c$; when $\tau = u - c$, $\lambda^{1}:\lambda^{2} = -c$. With

---

[1] To facilitate comparison with the literature on hydrodynamics, we use in the present section a notation slightly deviating from that in §2. For example, we write for y the letter t, indicating time.

Further details about the examples we shall give here can be found in R. Courant and K. O. Friedrichs [1].

each of these values of $\lambda^1:\lambda^2$, the linear combination $\lambda^1 L^1 + \lambda^2 L^2$ involves differentiation of u and of · in the same direction (the corresponding characteristic direction). Accordingly, we have the system of differential equations

$$\Lambda^1[u, \rho] \equiv \rho[u_t + (u + c)u_x] + c[\rho_t + (u + c)\rho_x] = 0,$$

$$\Lambda^2[u, \rho] \equiv \rho[u_t + (u - c)u_x] - c[\rho_t + (u - c)\rho_x] = 0.$$

We may introduce characteristic parameters $\alpha = \phi(x, t)$, $\beta = \psi(x, t)$ where

$$\psi(x, t) = \text{const.}$$

and

$$\phi(x, t) = \text{const.}$$

are the characteristic curves defined by

$$-\psi_t:\psi_x = dx:dt = u + c$$

and

$$-\phi_t:\phi_x = dx:dt = u - c.$$

Since $-\psi_t:\psi_x = x_\alpha:t_\alpha$ and $-\phi_t:\phi_x = x_\beta:t_\beta$, we obtain the equations

(1a)
$$\rho u_\alpha + c\rho_\alpha = 0,$$
$$\rho u_\beta - c\rho_\beta = 0,$$

(1b)
$$x_\alpha = (u + c)t_\alpha,$$
$$x_\beta = (u - c)t_\beta$$

for the four functions u, ρ, x, t of the characteristic parameters α, β (cf. §2, 3).

The characteristic curves in this example can be interpreted as representing "sound waves" (small disturbances) whose velocity

$$\frac{dx}{dt} = u \pm c,$$

sometimes called "characteristic speed", differs from the "stream velocity" u by $\pm c$.

We point out that equations (1a) lead to the so-called "*Riemann*

invariants", that is, to functions which are constant along characteristics, in the following way.

We wish to write (1a) in the form

$$\frac{dr}{d\alpha} = 0, \qquad \frac{ds}{d\beta} = 0$$

with suitable functions $r(u, \rho)$, $s(u, \rho)$, in order to deduce that
$r = $ const. along $\beta = $ const. and $s = $ const. along $\alpha = $ const. Clearly,
we must find functions g and l such that $r_u = g\rho$, $r_\rho = gc$; $s_u = l\rho$,
$s_\rho = -lc$, where the "integrating factors" $g(u, \rho)$ and $l(u, \rho)$ satisfy
the compatibility conditions $(g\rho)_\rho = (gc)_u$, $(l\rho)_\rho = (-lc)_u$. One
easily finds that

$$g = \frac{1}{2\rho}, \qquad l = -\frac{1}{2\rho}$$

are solutions leading to the Riemann invariants

$$2r = u + \int_{\rho_0}^{\rho} \frac{c(\rho')}{\rho'}\, d\rho' = \text{const.} \qquad \text{along } \beta = \text{const.,}$$

$$2s = -u + \int_{\rho_0}^{\rho} \frac{c(\rho')}{\rho'}\, d\rho' = \text{const.} \qquad \text{along } \alpha = \text{const.}$$

In the important case of ideal "polytropic" gases, the pressure is
given by

$$p = f(\rho) = A\rho^\gamma,$$

where A and γ are constants, $\gamma > 1$. In this case,

$$c^2 = f'(\rho) = A\gamma\rho^{\gamma-1}$$

and if we take $\rho_0 = 0$, the Riemann invariants are

$$r = \frac{u}{2} + \frac{c}{\gamma - 1}, \qquad s = -\frac{u}{2} + \frac{c}{\gamma - 1}.$$

These expressions can be used to solve the initial value problem.

Finally, we note that the nonlinear system $L^1[u, \rho] = 0$, $L^2[u, \rho] = 0$
is of the type which admits linearization by the hodograph transformation of §2. Thus the following linearized equations are obtained for x and t as functions of u and ρ:

$$\rho t_\rho - ut_u + x_u = 0,$$

$$\rho ut_\rho - \rho x_\rho - c^2 t_u = 0$$

They are valid as long as the Jacobian $u_x \rho_t - u_t \rho_x$ differs from zero. (For a detailed analysis, see R. Courant and K. O. Friedrichs [1].)

2. Spherically Symmetric Flow. The differential equations

$$L^1[u, \rho] \equiv \rho u_x + u\rho_x + \rho_t = -\frac{2\rho u}{x}$$

$$L^2[u, \rho] \equiv \rho u u_x + \rho u_t + c^2 \rho_x = 0$$

for spherical isentropic flow can be treated in the same way. Since the operators L^1 and L^2 are the same as in the one-dimensional flow (see article 1), the characteristic directions τ are given by $\tau = u \pm c$. Likewise the values of λ^1/λ^2 are again given, except for a factor of proportionality, by $\lambda^1/\lambda^2 = \pm c$, yielding the system of differential equations

$$\Lambda^1[u, \rho] = -\frac{2c\rho u}{x},$$

$$\Lambda^2[u, \rho] = \frac{2c\rho u}{x}$$

with the same operators Λ^1, Λ^2 as before. The equations analogous to (1a) are

$$\rho u_\alpha + c\rho_\alpha = -\frac{2c\rho u}{x} t_\alpha,$$

(2a)

$$\rho u_\beta - c\rho_\beta = \frac{2c\rho u}{x} t_\beta$$

and equations (1b) are unchanged.

3. Steady Irrotational Flow. In a steady two-dimensional irrotational isentropic flow, the velocity components u, v satisfy the system

$$L^1[u, v] \equiv u_y - v_x = 0,$$

$$L^2[u, v] \equiv (c^2 - u^2)u_x - uv(u_y + v_x) + (c^2 - v^2)v_y = 0,$$

where the sound speed c is a given function of $u^2 + v^2$. In this case, the characteristic directions τ satisfy

$$\begin{vmatrix} -\tau & c^2 - u^2 + uv\tau \\ -1 & -uv - (c^2 - v^2)\tau \end{vmatrix} = (c^2 - v^2)\tau^2 + 2uv\tau + c^2 - u^2 = 0.$$

At all points where $u^2 + v^2 > c^2$ (that is, where the flow is supersonic), this equation has two distinct real roots τ^1, τ^2 and the system is hyperbolic. For each of these values of τ, the corresponding value of $\lambda^1:\lambda^2$ is $-uv - (c^2 - v^2)\tau$. By taking the linear combination $\lambda^1 L^1 + \lambda^2 L^2$ in each case, we arrive at the system

$$\Lambda^i[u, v] \equiv (c^2 - u^2)(u_y + \tau^i u_x)$$
$$+ (c^2 - v^2)\tau^i(v_y + \tau^i v_x) = 0 \quad (i = 1, 2).$$

The equations for u, v, x, y as functions of α, β are

(3a)
$$(c^2 - u^2)u_\alpha + (c^2 - v^2)\tau^1 v_\alpha = 0,$$
$$(c^2 - u^2)u_\beta + (c^2 - v^2)\tau^2 v_\beta = 0,$$

and

(3b)
$$x_\alpha = \tau^1 y_\alpha,$$
$$x_\beta = \tau^2 y_\beta.$$

Equations (3a) can be written in the equivalent form

(3a′)
$$\tau^2 u_\alpha + v_\alpha = 0,$$
$$\tau^1 u_\beta + v_\beta = 0.$$

For the differential equations

$$L^1[u, v] \equiv u_y - v_x = 0,$$

$$L^2[u, v] \equiv (c^2 - u^2)u_x - uv(u_y + v_x) + (c^2 - v^2)v_y = -\frac{c^2 v}{y}$$

of steady three-dimensional irrotational isentropic flow with cylindrical symmetry, the operators L^1, L^2 are the same as in the previous example so that the characteristic directions τ^1, τ^2 again satisfy

$$(c^2 - v^2)\tau^2 + 2uv\tau + c^2 - u^2 = 0.$$

The equations analogous to (3) in this case are

(4a)
$$u_\alpha + \frac{1}{\tau^2}v_\alpha = -\frac{c^2 v}{(c^2 - u^2)y}x_\alpha,$$
$$u_\beta + \frac{1}{\tau^1}v_\beta = -\frac{c^2 v}{(c^2 - u^2)y}x_\beta$$

and equations (3b).

Again, the equations of steady irrotational flow allow linearization by the hodograph transformation of §2. Provided that

$$u_x v_y - v_x u_y \neq 0,$$

we obtain the linear system

$$x_v - y_u = 0$$

$$(c^2 - u^2)y_v + uv(x_v + y_u) + (c^2 - v^2)x_u = 0$$

for x, y as functions of u and v.

4. Systems of Three Equations for Nonisentropic Flow. In a (nonsteady) one-dimensional nonisentropic flow, the velocity $u(x, t)$, pressure $p(x, t)$ and specific entropy $S(x, t)$ satisfy the system

$$L^1[u, p, S] \equiv \rho c^2 u_x + u p_x + p_t = 0,$$

$$L^2[u, p, S] \equiv \rho u u_x + \rho u_t + p_x = 0,$$

$$L^3[u, p, S] \equiv u S_x + S_t = 0,$$

where ρ and $c \neq 0$ are given functions of p and S. The characteristic directions τ are then defined by

$$\begin{vmatrix} c^2 & u - \tau & 0 \\ u - \tau & 1 & 0 \\ 0 & 0 & u - \tau \end{vmatrix} = 0 ;$$

hence, they are

$$\tau^1 = u + c,$$

$$\tau^2 = u - c,$$

$$\tau^3 = u,$$

and the system is hyperbolic. For each of the characteristic directions τ^1, τ^2, τ^3 there exists a nontrivial solution λ^1, λ^2, λ^3 of

$$c^2\lambda^1 + (u - \tau)\lambda^2 = 0,$$

$$(u - \tau)\lambda^1 + \lambda^2 = 0,$$

$$(u - \tau)\lambda^3 = 0,$$

i.e., except for a factor of proportionality,

$$(\lambda^1, \lambda^2, \lambda^3) = \begin{cases} (1, c, 0) & \text{for } \tau = \tau^1, \\ (-1, c, 0) & \text{for } \tau = \tau^2, \\ (0, 0, 1) & \text{for } \tau = \tau^3. \end{cases}$$

By forming the linear combination $\lambda^1 L^1 + \lambda^2 L^2 + \lambda^3 L^3$, in each case, we obtain the system

$$\Lambda^1[u, p, S] \equiv \rho c[u_t + (u + c)u_x] + p_t + (u + c)p_x = 0,$$

$$\Lambda^2[u, p, S] \equiv \rho c[u_t + (u - c)u_x] - [p_t + (u - c)p_x] = 0,$$

$$\Lambda^3[u, p, S] \equiv S_t + uS_x = 0.$$

Although we do not introduce characteristic parameters instead of the variables x, t as we did when there were two equations for two unknowns, we can, nevertheless, write these equations in a more concise form by letting s_1, s_2, s_3 be curve parameters on the three families of characteristic curves. These equations then take the form

$$\rho c \frac{\partial u}{\partial s_1} + \frac{\partial p}{\partial s_1} = 0,$$

$$\rho c \frac{\partial u}{\partial s_2} - \frac{\partial p}{\partial s_2} = 0,$$

$$\frac{\partial S}{\partial s_3} = 0.$$

In the present example, the characteristic curves determined by $dx : dt = u + c$ represent sound waves (as in the isentropic case) and the characteristic curves $dx : dt = u$ represent trajectories of particles. As another example, we consider the differential equations

$$\rho u u_x + \rho v u_y + c^2 \rho_x = 0,$$

$$\rho u v_x + \rho v v_y + c^2 \rho_y = 0,$$

$$\rho(u_x + v_y) + u\rho_x + v\rho_y = 0$$

(where the sound speed $c = c(\rho) \neq 0$ is a known function) for the velocity components $u(x, y)$, $v(x, y)$ and the density $\rho(x, y)$ of steady two-dimensional rotational isentropic flow. The characteristic directions τ satisfy

$$\begin{vmatrix} u - \tau v & 0 & 1 \\ 0 & u - \tau v & -\tau \\ c^2 & -\tau c^2 & u - \tau v \end{vmatrix} \equiv (u - \tau v)[(u - \tau v)^2 - (1 + \tau^2)c^2] = 0.$$

Two families of characteristic curves $dx : dy$ are determined by the roots of the quadratic factor

$$(1 + \tau^2)c^2 - (u - \tau v)^2 \equiv (c^2 - v^2)\tau^2 + 2uv\tau + c^2 - u^2 = 0,$$

an equation for τ already considered in our example for the irrotational case. These two families of characteristics will be real and distinct at points where $u^2 + v^2 > c^2$ (i.e., where the flow is supersonic). Sometimes the characteristic curves of the two families are referred to as *Mach lines*. The third family of characteristic curves is determined by $dx:dy = u:v$ (i.e., $u - \tau v = 0$) and consists of the *streamlines*, namely, the curves tangent to the velocity vectors. It is easy to see that a streamline cannot have the same direction as a Mach line at any point, so the system is totally hyperbolic for $u^2 + v^2 > c^2$.

For the first two characteristic directions, we obtain, except for a factor of proportionality,

$$(\lambda^1, \lambda^2, \lambda^3) = (-1, \tau^i, u - \tau_v^i) \qquad (i = 1, 2)$$

and for the third characteristic direction

$$(\lambda^1, \lambda^2, \lambda^3) = (u, v, 0).$$

Taking the linear combination $\lambda^1 L^1 + \lambda^2 L^2 + \lambda^3 L^3$ in each case and introducing curve parameters s_1, s_2, s_3 on the three families of characteristic curves, we arrive at the system

$$\rho v \frac{\partial u}{\partial s_i} - \rho u \frac{\partial v}{\partial s_i} - [uv + \tau^i(c^2 - v^2)]\frac{\partial \rho}{\partial s_i} = 0 \qquad (i = 1, 2),$$

$$\rho u \frac{\partial u}{\partial s_3} + \rho v \frac{\partial v}{\partial s_3} + c^2 \frac{\partial \rho}{\partial s_3} = 0.$$

5. Linearized Equations. The nonlinear equations of fluid dynamics can be approximated by linear equations if we assume an underlying constant state from which the actual flow deviates by "small" quantities such that second order terms in these quantities and their derivatives may be neglected. An example may suffice: We form a "linearized" approximation to the last case treated, i.e., steady two-dimensional rotational isentropic flow. We assume that the velocity differs only slightly from a constant velocity $\tilde{u}$ parallel to the x-axis and that the density ρ likewise is only slightly different from a constant density $\tilde{\rho}$. We set

$$u = \tilde{u} + \omega, \qquad v = \lambda, \qquad \rho = \tilde{\rho} + \sigma,$$

where ω, λ, σ are small quantities. We assume further that the motion of the fluid can be represented with sufficient precision if, in the differential equations, we omit products and higher powers of the quantities ω, λ, σ and their derivatives. Then the original system of differential equations is replaced by the following system of linear differential equations with constant coefficients:

$$\tilde{u}\tilde{\rho}\omega_x + c^2\sigma_x = 0,$$

$$\tilde{u}\tilde{\rho}\lambda_x + c^2\sigma_y = 0,$$

$$\tilde{\rho}(\omega_x + \lambda_y) + \tilde{u}\sigma_x = 0,$$

where $c = c(\tilde{\rho})$. We obtain at once the relation

$$\omega_{xy} = \lambda_{xx},$$

i.e.,

$$\omega_y - \lambda_x = F(y)$$

for some function $F(y)$. This expresses the so-called *vortex theorem*. Setting $\tilde{u}/c = k$, we obtain further the system

$$k\tilde{\rho}\lambda_x + c\sigma_y = 0,$$

$$k\tilde{\rho}\lambda_y - (1 - k^2)c\sigma_x = 0,$$

which leads by elimination to the two differential equations

$$(1 - k^2)\lambda_{xx} + \lambda_{yy} = 0,$$

$$(1 - k^2)\sigma_{xx} + \sigma_{yy} = 0.$$

From these equations we see at once that we are dealing with the hyperbolic case if $k > 1$, i.e., if $\tilde{u} > c$. The characteristics are then straight lines which form an angle, α, called the *Mach angle*, with the x-axis for which $|\sin\alpha| = 1/k = c/\tilde{u}$.

The two sets of characteristics can be visualized experimentally in the motion of a fluid or gas with the base velocity $\tilde{u}$ parallel to a plane wall. We then introduce a small roughness or a bump along a small interval AB of the wall which produces along the wall a small vertical component λ of the velocity. Under the assumption that this motion is represented by our approximated system of equations, this roughness is continued into the domain of the fluid along two parallel strips of characteristics starting at AB and making an angle α with the wall.

§4. *Uniqueness. Domain of Dependence*

In subsequent sections we shall frequently emphasize the variable y as time and write t instead of y.

The proofs of existence of a solution of Cauchy's initial value problem for a hyperbolic equation or a system

$$L[u] = f(x, t)$$

with prescribed initial values

$$u(x, 0) = \psi(x)$$

will be given in §§6, 7, 8. Here we assume existence of a solution and prove that the solution is uniquely determined by the data, i.e., by $f(x, t)$ and $\psi(x)$. Moreover, at a point P the solution is determined by data in merely a portion of the space Γ. This dependence of u on the data exhibits the finiteness of the speed of propagation of phenomena governed by hyperbolic equations.

1. Domains of Dependence, Influence, and Determinacy. The solution $u(P)$ of Cauchy's problem at a point P with coordinates x, t ($t > 0$) depends only on the data in a finite "domain of dependence" Γ_P of P which consists of the point set in the upper half-plane bounded by the outer characteristics[1] through P (see Figure 24). If the differential equation is homogeneous, i.e., if $f = 0$, then $u(P)$ depends only on the initial data $\psi(x)$ on the segment γ_P of the x-axis cut out by the outer characteristics.[2] This fact can also be expressed as follows: $u(P)$ does not depend on the data outside γ_P ; or, if $\psi(x) = \psi^*(x)$ in γ_P and $f = f^*$ in Γ_P , then the solutions $u(P)$ and $u^*(P)$ of $L[u] = f$ and $L[u] = f^*$ with initial values $\psi(x)$ and $\psi^*(x)$, respectively, coincide at P. Sometimes one alludes to this situation by saying that the quantity $u(P)$ does not "notice" the data outside Γ_P .

---

[1] If an outer characteristic crosses another, the outer characteristic is to be continued as an arc of the other.

[2] For simple visualization one likes to assume as in Figure 24 that in the domain of the x,t-plane under consideration the characteristics through a point P do form a bundle of k curves which do not cross each other and which join P with points on the x-axis. Figure 25 relating to the equation $u_{tt} - t^2 u_{xx} = 0$ with the characteristics $(t - \tau_0)^2 - (x - \xi)^2 = 0$ illustrates the possibility that the two characteristics through a point, on $t = \tau_0$, touch each other so that the outer characteristics do not consist of one analytic piece. However, such crossing of characteristics is actually not excluded by the subsequent reasoning.

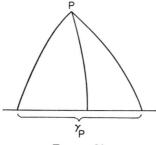

FIGURE 24

From now on we restrict ourselves to homogeneous hyperbolic equations $L[u] = 0$ (as we know, this entails no specialization), and we call the segment γ_P of the x-axis the *domain of dependence* of P.

Correspondingly, we define as the *domain of influence* of a set $\mathfrak{S}$ on the x-axis that region of the upper half-plane whose points P have domains of dependence with points in $\mathfrak{S}$. If $\mathfrak{S}$ is an interval, then its domain of influence is bounded by the outside characteristics[1] pointing from the end points of $\mathfrak{S}$ into the upper half-plane. This domain of influence is unbounded if we can extend the characteristics for arbitrarily large values of t.

Finally, the *domain of determinacy* of an interval $\mathfrak{S}$ on the x-axis is the set of those points P with $t > 0$ whose domains of dependence γ_P are entirely in $\mathfrak{S}$. Obviously, this domain of determinacy is bounded by $\mathfrak{S}$ and the inner characteristics pointing from the end points of $\mathfrak{S}$ into the upper half-plane; $\mathfrak{S}$ is the domain of dependence of the point of intersection of these characteristics.

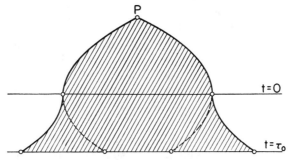

FIGURE 25. Domain of dependence for $(\partial/\partial t + t\partial/\partial x)(\partial/\partial t - t\partial/\partial x) u = 0$

[1] See preceding footnote.

These closely interrelated concepts acquire their meaning from the uniqueness proofs given in the next articles.

In passing, we remark that these concepts and the subsequent uniqueness theorems point merely to a negative fact, namely, to the nondependence of $u(P)$ on data outside of γ_P. In the case of two independent variables one can, however, show that γ_P is indeed the domain of dependence in a strict sense: For each point Q in γ_P, there exist data ψ such that $u(P)$ actually depends on the values of ψ at Q and in the vicinity of Q. For more than two independent variables, we shall see in Chapter VI that the corresponding question concerning the domain of dependence in the strict sense presents great difficulties.

If the admissible data $\psi(x)$ for Cauchy's problem were restricted to analytic functions, then the values of ψ on an arbitrarily small open set of the initial line would determine ψ on the whole line. Hence, the concepts of domain of dependence, influence, and determinacy would be pointless. To understand these important notions, a wider class of data must be admitted. Thus it is in the nature of hyperbolic problems to construct the solutions by methods allowing for non-analytic data instead of relying on the Cauchy-Kowalewsky existence theorem of Ch. I, §7 in which the analytic character of the differential equation and of the initial conditions was assumed.

2. Uniqueness Proofs for Linear Differential Equations of Second Order. In the present section we shall establish the following facts: For linear second order hyperbolic equations, the domain of dependence of a point P is enclosed by the two characteristic curves through P backward in time toward the initial curve. If the equation is homogeneous, the base line γ of the resulting triangular region is the desired domain of dependence of P (this situation has already been discussed for the wave equation $u_{tt} - u_{xx} = 0$ in Ch. III, §7).

The domain of determinacy of the segment γ is the whole triangular region, since it contains all points whose domains of dependence are contained in γ. The two other ("outer") characteristics through the end points of γ bound the domain of influence of this segment which, in fact, consists of those points whose domains of dependence have points in common with γ.

To show that the base γ of our triangular region is the domain of dependence γ_P of the vertex P, it suffices to prove uniqueness by demonstrating that the solution of the homogeneous equation $L[u] = 0$

vanishes at the point P if the initial values are zero on γ; if the initial data of two solutions of $L[u] = f$ differ only outside γ, their difference is a solution of the homogeneous equation with initial values zero on γ.

The following uniqueness proof is based on certain *energy integrals* associated with the differential equation. We have already met an example of such an integral in connection with the uniqueness proof for the parabolic equation of heat conduction (see Ch. III, §6). For hyperbolic equations, a similar idea is effective provided the domains of integration are bounded by characteristics.[1]

The uniqueness theorem for linear problems asserts: *If, on a portion σ of the initial curve, the initial values are zero, the homogeneous equation has only the trivial identically vanishing solution in the triangular region cut out by the inner characteristics through the end points of σ.*

The idea of the proof is illustrated by the wave equation

$$(1) \qquad u_{xx} - u_{tt} = 0.$$

Let Γ be a triangular region of the x,t-plane bounded by an initial curve AB which is nowhere characteristic, and by the two characteristic lines PA and PB (Figure 26). Our object is to show: If u and the derivative u_t, and hence also u_x, vanish on AB, then u vanishes identically in the entire region Γ. To this end we cut off the vertex of our triangle by means of a straight line $t = \text{const.}$ which intersects PA and PB in the points C and D, and consider the remaining region Γ' with boundary γ'. Over this region we integrate the expression

$$-2u_t(u_{xx} - u_{tt}) = (u_x^2)_t + (u_t^2)_t - 2(u_x u_t)_x$$

which is a divergence. Taking into account the differential equation, we obtain by integration over Γ'

$$0 = \iint_{\Gamma'} [(u_x^2)_t + (u_t^2)_t - 2(u_x u_t)_x]\, dx\, dt$$

$$= \int_{\gamma'} [(u_x^2 + u_t^2)t_\nu - 2(u_x u_t)x_\nu]\, ds,$$

where x_ν, t_ν denote the direction cosines of the outward normals to β' (i.e., $AB + BD + DC + CA$) and where s is arc length. On the characteristic edges CA and DB, $x_\nu^2 = t_\nu^2 = \frac{1}{2}$. The corresponding part

---

[1] For references to the literature and, for the general case of $n + 1$ independent variables, see Ch. VI, §8.

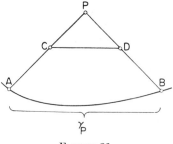

of the boundary integral can, therefore, be written in the form

$$\int_{AC+BD} \frac{1}{t_\nu} \left(u_x\, t_\nu - u_t\, x_\nu \right)^2 ds$$

and hence is certainly non-negative. On AB the integrand is zero in view of the vanishing initial data, and on CD we have $t_\nu = 1$, $x_\nu = 0$, $ds = dx$. Thus we obtain

$$\int_{CD} (u_x^2 + u_t^2)\, dx = 0$$

which implies that $u_x^2 + u_t^2 = 0$ everywhere on CD. Since the distance of CD from the x-axis was arbitrary, we conclude that u_x and u_t vanish at all points of Γ belonging to a neighborhood of the vertex P. But every point of Γ can be considered as the vertex of a corresponding smaller triangular region which is contained in Γ and whose base curve is contained in AB; therefore u_x and u_t vanish at every point of Γ. Hence u is constant throughout the entire region Γ and since u vanishes initially, u is identically zero, as asserted.

A similar argument can be used for the linear second order hyperbolic differential equation

$$L[u] = u_{tt} - u_{xx} - \alpha u_t - \beta u_x - \delta u = 0,^1$$

where α, β, δ are continuous functions of x and t. For the sake of brevity we confine ourselves to the initial value problem with initial values given on the line $t = 0$. Then the uniqueness theorem states: *If u and u_t vanish on the base AB of the triangle Γ* (i.e., the triangle ABP having the characteristic sides $x + t = $ const., $x - t = $ const.

---

[1] The general form of a linear second order hyperbolic differential equation is obtained from this by a transformation of coordinates.

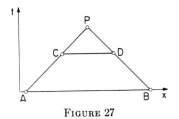

FIGURE 27

and the base AB on the line $t = 0$, see Figure 27), *then u vanishes throughout the triangle.*

We first note that for any point (x, t) in Γ,

$$u(x, t) = \int_0^t u_\tau(x, \tau) \, d\tau \, ;$$

hence, by Schwarz's inequality,

$$u^2(x, t) \le t \int_0^t u_\tau^2(x, \tau) \, d\tau \, .$$

Again, we cut off the vertex of our triangle by means of a line $t = $ const., obtaining a smaller triangle whose base we denote by H_t, and a trapezoid Γ_t with parallel sides AB and H_t. Then we have

$$\int_{H_t} u^2 \, dx \le t \iint_{\Gamma_t} u_\tau^2(x, \tau) \, dx \, d\tau$$

for any trapezoidal part (with base AB) of our triangle. Integrating this expression from $t = 0$ to $t = h$, we have

$$\iint_{\Gamma_h} u^2 \, dx \, dt \le \int_0^h \left[t \iint_{\Gamma_t} u_\tau^2 \, (x, \tau) \, dx \, d\tau \right] dt$$

$$\le \int_0^h \left[h \iint_{\Gamma_h} u_\tau^2(x, \tau) \, dx \, d\tau \right] dt$$

$$\le h^2 \iint_{\Gamma_h} (u_\tau^2 + u_x^2) \, dx \, d\tau \, .$$

Now we define an "energy integral"

$$E(h) = \int_{H_h} (u_x^2 + u_t^2) \, dx$$

and integrate the identity

$$0 = 2u_t L[u] = (u_x^2 + u_t^2)_t - 2(u_x u_t)_x - 2\alpha u_t^2 - 2\beta u_x u_t - 2\delta u u_t$$

over Γ_h. Since H_h corresponds to CD in Figure 27, we see (by the same reasoning as before) that

$$0 \leq \int_{AC+BD} \frac{1}{t_\nu} (u_x t_\nu - u_t x_\nu)^2 \, ds + E(h)$$

$$= 2 \iint_{\Gamma_h} (\alpha u_t^2 + \beta u_t u_x + \delta u u_t) \, dx \, dt$$

$$\equiv R,$$

from which we conclude that

$$E(h) \leq R.$$

We estimate the right-hand side, observing that

$$2|u_x u_t| \leq u_t^2 + u_x^2, \qquad 2|u u_t| \leq u_t^2 + u^2.$$

If M denotes an upper bound for the absolute values of the continuous functions α, β, δ, we obtain

$$R \leq 4M \iint_{\Gamma_h} (u_t^2 + u_x^2 + u^2) \, dx \, dt$$

and therefore, by the inequality

$$\iint_{\Gamma_h} u^2 \, dx \, dt \leq h^2 \iint_{\Gamma_h} (u_t^2 + u_x^2) \, dx \, dt$$

proved above, we have

$$R \leq 4M(1 + h^2) \iint_{\Gamma_h} (u_t^2 + u_x^2) \, dx \, dt \leq C \int_0^h E(\alpha) \, d\alpha,$$

where $C = 4M(1 + h^2)$ and h is the altitude of the trapezoid Γ_h. If l is any value such that $l > h$, we have

$$E(h) \leq C \int_0^h E(\alpha) \, d\alpha \leq C \int_0^l E(\alpha) \, d\alpha.$$

Integrating this relation with respect to h between the limits 0 and l, we have

$$\int_0^l E(h) \, dh \leq Cl \int_0^l E(h) \, dh.$$

If $E(h)$ were different from zero anywhere in the interval $0 \leq h \leq l$, then it would follow that

$$1 \leq Cl,$$

which is clearly impossible if we choose $l < 1/C$. Hence in the interval $0 \leq h \leq l$ we certainly have $E \equiv 0$. Repeating the procedure with $t = l$, $t = 2l$, $\cdots$ as initial lines, we see after a finite number of steps that E vanishes in the whole triangle Γ; therefore u is constant and, in fact, zero, as we wished to prove.

3. General Uniqueness Theorem for Linear Systems of First Order. The method of "energy integrals" permits a transparent uniqueness proof for hyperbolic systems of first order. We write the inhomogeneous system in the vector and matrix form

$$(2) \qquad u_t + A u_x + B u + c = 0,$$

where u is the unknown vector function, $A(x, t)$ and $B(x, t)$ are given matrices, and $c(x, t)$ is a given vector.

Without restriction of generality we assume for the proof that $A = (a_{ik}(x, t))$ is a *symmetric* matrix and recall that linear hyperbolic systems can be transformed into a symmetric form (e.g., into characteristic normal form, see §2, 2). We assume, moreover, that A has continuous derivatives and that B and c are continuous.

Through a point P with coordinates ξ, τ, we draw the characteristics C_1, C_2, $\cdots$, C_k backward, obtaining the points P_1, P_2, $\cdots$, P_k of intersection with the initial line $t = 0$[1] (see Figure 28). We can now state the following uniqueness theorem: *If $c = 0$ and $u(x, 0) = 0$ in a closed interval of the line $t = 0$ including all the points P_κ, then $u(\xi, \tau) = 0$.*

The smallest such interval, i.e., the interval cut out by the outer characteristics through P, contains the domain of dependence[2] of P. As a matter of fact, if $u(x, 0) = 0$ on $P_1 P_k$, then $u(x, t) = 0$ in the triangular region Γ_P cut out by the outer characteristics through P and having the interval $P_1 P_k$ of the line $t = 0$ as base.

For the proof, we use *Green's identity* which is equivalent to

$$(u, A u)_x = (u_x, A u) + (u, A_x u) + (u, A u_x);$$

[1] Some of the characteristics may be multiple throughout the region considered.

[2] The interval $P_1 P_k$ actually is the domain of dependence of P; the remarks made in article 2 apply here also.

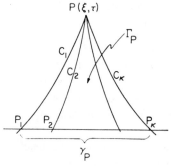

FIGURE 28

since A is symmetric, $(u, Au_x) = (Au, u_x)$, this identity reduces to

$$2(u, Au_x) = (u, Au)_x - (u, A_x u).$$

Taking the inner product of the differential equation (2) with the vector u and using the preceding relation we have (for $c = 0$)

$$\tfrac{1}{2}(u, u)_t + \tfrac{1}{2}(u, Au)_x - \tfrac{1}{2}(u, A_x u) + (u, Bu) = 0.$$

As a convenient device we introduce with a constant μ instead of u the unknown vector

$$v = e^{-\mu t}u$$

which leads to a differential equation

$$(2')\qquad\qquad v_t + Av_x + B^*v = 0$$

and the initial condition $v(x, 0) = 0$ where, with the unit matrix I,

$$B^* = B + \mu I.$$

The preceding quadratic identity becomes, if we write u again instead of v,

$$\tfrac{1}{2}(u, u)_t + \tfrac{1}{2}(u, Au)_x = (u, \hat{B}u).$$

The quadratic form on the right-hand side is formed with the matrix

$$\hat{B} = -B^* - \tfrac{1}{2}A_x,$$

which is the aim of the device. By choosing μ sufficiently large, we can insure that $\hat{B}$ is negative definite, that is, that $(u, \hat{B}u) \leq 0$.

Now we integrate over the trapezoid-like domain Γ_h which has the boundary $\beta_h = P_1 P_k A_k A_1$ (see Figure 29), denoting the components of the outward unit normal vector by x_ν, t_ν, and obtain Green's formula

$$0 \geq \frac{1}{2} \iint_{\Gamma_h} [(u, u)_t + (u, Au)_x]\, dx\, dt$$

$$= \frac{1}{2} \int_{\Gamma_h} [(u, u)t_\nu + (u, Au)x_\nu]\, ds$$

$$= \frac{1}{2} \int_{A_1 A_k + P_k P_1} (u, u)\, dx + \frac{1}{2} \int_{C_1 + C_k} x_\nu \left(u, \left[A + \frac{t_\nu}{x_\nu} I\right] u\right) ds.$$

With the notation

$$E(h) = \frac{1}{2} \int_{A_1}^{A_k} (u, u)\, dx,$$

this means

$$(3) \quad E(h) - E(0) \leq -\frac{1}{2} \int_{C_1 + C_k} x_\nu \left(u, \left[A + \frac{t_\nu}{x_\nu} I\right] u\right) ds.$$

We show that the right-hand side is nonpositive. For this purpose we recall that C_1 and C_k are characteristic curves $\phi^1(x, t) = 0$, $\phi^k(x, t) = 0$ and that, therefore, they satisfy the differential equations

$$-\frac{\phi_t^1}{\phi_x^1} = \tau^1, \qquad -\frac{\phi_t^k}{\phi_x^k} = \tau^k,$$

where τ^1, τ^k are eigenvalues of the matrix A, i.e., values for which $A - \tau I$ is singular (see §2).

Along $C_1 = \phi^1$, the outward normal has components proportional to ϕ_x^1, ϕ_t^1; hence $-t_\nu/x_\nu = \tau^1$. Since C_1 is the outer characteristic at

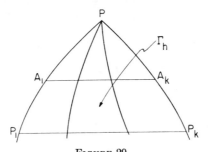

FIGURE 29

the left, τ^1 is the largest eigenvalue of A. Similarly, along $C_k = \phi^k$, the components of the normal are proportional to ϕ_x^k, ϕ_t^k and $-t_\nu/x_\nu = \tau^k$ is the smallest eigenvalue of A. We recall that, for a symmetric matrix A, the extreme eigenvalues can be characterized by

$$\tau^1 = \max \frac{(u, Au)}{(u, u)}, \qquad \tau^k = \min \frac{(u, Au)}{(u, u)}$$

(see Vol. I, Ch. I, §4, 1). Consequently

$$(u, u)\tau^1 \geq (u, Au), \qquad (u, u)\tau^k \leq (u, Au)$$

or

$$(u, [A - \tau^1 I]u) \leq 0, \qquad (u, [A - \tau^k I]u) \geq 0.$$

Since $x_\nu < 0$ on C_1 and $x_\nu > 0$ on C_k, we have

$$-\frac{1}{2} \int_{C_i} x_\nu \left(u, \left[A + \frac{t_\nu}{x_\nu} I \right] u \right) ds \leq 0 \qquad \text{for } i = 1, k.$$

This shows that the second term in (3) is nonpositive and thus we can write

(4) $$E(h) \leq E(0).$$

Since by assumption $E(0) = 0$, we have $0 \leq E(h) \leq 0$; hence $E(h) = 0$ for $h > 0$; therefore $u = 0$ for $t = h$.

4. Uniqueness for Quasi-Linear Systems. The preceding uniqueness theorem remains valid for quasi-linear systems of first order:

(5) $$u_t + A(x, t, u)u_x + B(x, t, u) = 0$$

even though the characteristics C_κ of (5) depend on the solution u.

We assume here that the matrices A and B possess continuous derivatives with respect to x, t, and u in the region under consideration.

Let u and v be two solutions of (5) defined in a domain D and having the same initial values on the initial interval. Then the function

$$z(x, t) = u - v$$

has the initial values zero.

Subtracting the differential equation for v from that for u we have, omitting explicit reference to x and t,

(6) $$z_t + A(v)z_x + [A(u) - A(v)]u_x + B(u) - B(v) = 0.$$

Because of the continuity and differentiability properties of the matrices A and B we may apply the mean value theorem:

$$A(u) - A(v) = H(u, v)z; \qquad B(u) - B(v) = K(u, v)z,$$

where H, K are continuous functions. We now consider u, v, u_x as known expressions in x, t and substitute these expressions in H and K as well as in $A(v)$; thereby (6) becomes a linear homogeneous differential equation for z of the form

$$z_t + az_x + bz = 0$$

with initial values zero. The theorem of article 3 asserts now that z is identically zero, and thus the uniqueness theorem is proved also for quasi-linear equations.

5. Energy Inequalities. Inequality (4) of article 3 asserts that the value of $E(h)$ for positive h is bounded by a fixed multiple of $E(0)$. In many equations governing physical phenomena, $E(h)$ can be interpreted as the energy contained in a portion of the physical system at time h. Inequality (4) in such instances (and by analogy in general) is therefore called an *energy inequality*.

In Ch. VI, §9 such energy inequalities will be used to prove the existence of solutions of initial value problems for n independent variables. For two independent variables, however, in §6 of the present chapter, solutions will be constructed by a direct method which at the same time furnishes a different proof for the uniqueness theorems just presented.

§5. *Riemann's Representation of Solutions*

The modern theory of hyperbolic partial differential equations was initiated by Bernhard Riemann's representation of the solution of the initial value problem for an equation of second order.[1] His paper gives neither a general existence proof, nor a construction for a solution, except in explicitly solvable examples. Assuming the existence of a solution u, it merely offers an elegant explicit integral representation of u in a form analogous to the representation of solutions of boundary value problems for elliptic equations by Green's function, and in fact, typical of a great variety of other such formulas expressing "linear functionals".[2] The elementary presentation of Riemann's theory will subsequently be greatly generalized.

---

[1] Riemann developed his theory in an appendix to his classical paper on the dynamics of compressible gases (see Riemann [1]).

[2] See footnote on p. 453 and further discussions, in particular Ch. VI, §15 and Appendix.

1. The Initial Value Problem.   Any second order linear hyperbolic differential equation can be written, after an appropriate transformation, in one of two forms (cf. Ch. III, §2, 1):

(1) $L[u] \equiv u_{xy} + au_x + bu_y + cu = f,$

(1') $L[u] \equiv u_{yy} - u_{xx} + au_x + bu_y + cu = f,$

where a, b, c, f are given (continuously differentiable) functions of the independent variables x, y.

The initial curve C is assumed "free", that is, nowhere tangent to a characteristic direction; the characteristics in equation (1) are the straight lines parallel to the coordinate axes; in equation (1') they are the lines $x + y =$ const. and $x - y =$ const.

Our aim is to represent a solution u at a point P in terms of f and the initial data, i.e., the values of u and one "outgoing" derivative of u on C. (Both $u_x = p$ and $u_y = q$ are known on C from such initial data.)   From §4 we infer that the domain of dependence of a point P: (ξ, η) is the region cut out by the two characteristics through P and the curve C. In the case of equation (1), it is the triangular region ◁ indicated in Figure 30.

If the initial curve degenerates into a right angle formed by the characteristics $x = \alpha$, $y = \beta$, we can no longer prescribe two conditions on C; instead we pose the "characteristic initial value problem" in which merely the values of the one quantity u on $x = \alpha$ and $y = \beta$ are prescribed.

2. Riemann's Function.   Riemann motivated the derivation of his representation formula by the analogy between a differential equation and a finite system of linear equations.   Such a system can be solved for unknowns u^i in the following way: Multiply the left sides by as yet undetermined quantities v^i, add these products, rearrange the resulting bilinear form in u^i and v^i by factoring out the unknowns u^i,

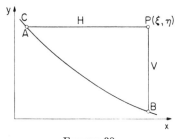

FIGURE 30

and finally determine the v^i so that the coefficients of all the u^i vanish except, say, u^1, whose coefficient is 1.

We thus obtain a representation for u^1 and similarly for u^2, etc. The corresponding quantities v^i depend on the coefficients but not on the right-hand sides of the equations.

The value of the solution of our differential equation at P can be represented in a similar way: One multiplies the differential equation by a function v, integrates over the region $\mathfrak{d}$, transforms the integral by Green's formula so that u appears as a factor of the integrand; then one tries to determine v in such a way that the required representation is obtained.

We have already introduced the concept of the adjoint L^* of an operator L (Ch. III, App. 1, §2), by the requirement that $vL[u] - uL^*[v]$ be a divergence expression. For equation (1), L^* is given by the formula

$$L^*[v] = v_{xy} - (av)_x - (bv)_y + cv$$

and we have

$$vL[u] - uL^*[v] = (vu_x + buv)_y - (uv_y - auv)_x .$$

Integrating over a domain G with boundary Γ, one obtains by Gauss' formula

$$-\iint_G (vL[u] - uL^*[v])\, dx\, dy$$

$$= \int_\Gamma [(vu_x + buv)\, dx + (uv_y - auv)\, dy].$$

We apply this formula to the domain of dependence of P. Using (1) we find

$$-\iint_G (fv - uL^*[v])\, dx\, dy$$

$$= \int_{AB+BP+PA} [v(u_x + bu)\, dx + u(v_y - av)\, dy]$$

$$= \int_{AB} [v(u_x + bu)\, dx + u(v_y - av)\, dy]$$

$$+ \int_{BP} u(v_y - av)\, dy - \int_{AP} v(u_x + bu)\, dx,$$

and since

$$\int_{AP} v(u_x + bu) \, dx = v(P)u(P) - v(A)u(A) - \int_{AP} u(v_x - bv) \, dx,$$

we have

$$u(P)v(P) = u(A)v(A) + \int_{AP} u(v_x - bv) \, dx + \int_{BP} u(v_y - av) \, dy$$

$$+ \int_{AB} [v(u_x + bu) \, dx + u(v_y - av) \, dy]$$

$$+ \iint_G (fv - uL^*[v]) \, dx \, dy.$$

To obtain a representation for $u(P) = u(\xi, \eta)$ we choose for v a function $R(x, y; \xi, \eta)$ subject to the following conditions:

a) As a function of x and y, R satisfies the adjoint equation

$$(2) \qquad\qquad L^*_{(x,y)} [R] = 0.$$

b) $R_x = bR$ on AP, $R_y = aR$ on BP, or, written out,

$$R_x(x, y; \xi, \eta) = b(x, \eta)R(x, y; \xi, \eta) \qquad \text{on } y = \eta,$$

$$R_y(x, y; \xi, \eta) = a(\xi, y)R(x, y; \xi, \eta) \qquad \text{on } x = \xi.$$

c) $R(\xi, \eta; \xi, \eta) = 1.$

Conditions b) are ordinary differential equations for R along the characteristics; integrating them and using c) to determine the value of the constant of integration, we get

$$(3) \qquad\qquad R(x, \eta; \xi, \eta) = \exp\left\{ \int_\xi^x b(\lambda, \eta) \, d\lambda \right\},$$

$$(3') \qquad\qquad R(\xi, y; \xi, \eta) = \exp\left\{ \int_\eta^y a(\lambda, \xi) \, d\lambda \right\},$$

for the value of R along the characteristics through P. The problem of finding a solution R of equation (2) with data (3), (3') is a "characteristic initial value problem". The existence of its solution will be discussed at the beginning of §6. The function R is called the Riemann function associated with the operator L; it provides

Riemann's representation formula

$$u(P) = u(A)R(A; \xi, \eta) + \int_{AB} [R(u_x + bu) \, dx + u(R_y - aR)dy]$$

$$+ \iint_G Rf \, dx \, dy.$$

In order to obtain a more symmetric expression, we add the identity

$$0 = \frac{1}{2} [u(B)R(B) - u(A)R(A)]$$

$$- \int_{AB} \left[\frac{1}{2} (u_x R + uR_x) \, dx + \frac{1}{2} (u_y R + uR_y) \, dy \right],$$

which follows by integrating uR by parts along C between A and B:

$$u(P) = \frac{1}{2} [u(A)R(A) + u(B)R(B)]$$

(4)
$$+ \int_{AB} \left(\left[\frac{1}{2} Ru_x + \left(bR - \frac{1}{2} R_x \right) u \right] dx \right.$$

$$\left. - \left[\frac{1}{2} Ru_y + \left(aR_y - \frac{1}{2} R_y \right) u \right] dy \right)$$

$$+ \iint_G Rf \, dx \, dy.$$

Formula (4) represents the solution of equation (1) for arbitrary initial values given along an arbitrary noncharacteristic curve C, by means of a solution R of the adjoint equation which depends on x, y and two parameters ξ, η.[1]

---

[1] Riemann's formula (4) is a special case of the following illuminating general principle: A "continuous linear functional" $u(P) = L[f]$, i.e., a quantity depending linearly and continuously on a function $f(Q)$, can be represented in the form

$$u(P) = \int K(Q, P)f(Q) \, dQ,$$

where the kernel K is a function of Q and P, and the integration is extended over the range of the variable Q. Naturally, such a general representation formula (due to F. Riesz) is valid only under suitable regularity conditions. But at any rate it constitutes a suggestive motivation for a great many procedures. In our case the solution u is a linear functional of $f(x, y)$ in G and of the initial values on C, and depends on ξ, η as parameters. As we shall see later in Chapter VI, the desirability of such integral representations motivates far-reaching generalizations.

In deriving the representation (4), we have *assumed* the existence of a solution u of $L[u] = f$, with initial data along AB, and of the Riemann function R. The following fact is easily verified: If Riemann's function exists and possesses sufficiently many derivatives with respect to its arguments, then the function $u(P)$ given by (4) is indeed a solution of $L(u) = f$ with the prescribed data, provided that C is nowhere characteristic. Still, this observation does not really simplify the general existence proof because Riemann's function, defined as the solution of a characteristic initial value problem (with seemingly *special* initial values), is in general not easier to construct than a solution of an *arbitrary* free initial value problem. In some special cases, however, it is possible to construct Riemann's function explicitly (see article 5, e.g., example 2).[1]

Assuming the existence of R, formula (4) shows that the solution u of (1) with prescribed data is unique. In view of Holmgren's theorem (cf. Ch. III, App. 2) this is not unexpected.

3. Symmetry of Riemann's Function. Suppose u is a solution of $L[u] = u_{xy} + au_x + bu_y + cu = 0$ defined in the rectangle $PADB$ (see Figure 31) whose sides AD and DB are characteristics into which the initial line AB degenerates. Then Riemann's representation formula (4) reads

$$u(P) = \frac{1}{2}\left[u(A)R(A) + u(B)R(B)\right]$$

$$-\frac{1}{2}\int_D^A (R_y\,u - u_y\,R - 2auR)\,dy - \frac{1}{2}\int_D^B (R_x\,u - u_x\,R - 2buR)\,dx.$$

Using the identity

$$R_z\,u - u_z\,R - 2guR = (uR)_z - 2R(u_z + gu)$$

and integrating the first term, we have

$$u(P) = u(D)R(D) + \int_D^A R(u_y + au)\,dy$$

(5)

$$+ \int_D^B R(u_x + bu)\,dx.$$

---

[1] Riemann's representation formula was first generalized to higher order linear equations in two independent variables by P. Burgatti [1] and F. Rellich [2]. E. Holmgren extended Riemann's method to systems of first order equations with two independent variables (Holmgren [2]). See also article 4 and Ch. VI, §15.

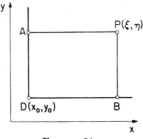

FIGURE 31

We now choose u to be the Riemann function of the adjoint equation $L^*[v] = v_{xy} - av_x - bv_y + dv$ (where $d = c - a_x - b_y$) with respect to the point D:

$$u = R^*(x, y; D).$$

The properties of the Riemann function (see a), b), c) in article 2) imply

$$L^{**}[R^*] = L[R^*] = 0,$$

$$R_y^* + aR^* = 0 \qquad \text{on } DA,$$

$$R_x^* + bR^* = 0 \qquad \text{on } DB,$$

and $R^*(D, D) = 1$. Therefore (5) yields

(6) $$R^*(P; D) = R(D; P).$$

In other words: Riemann's function of an operator L becomes Riemann's function of its adjoint operator L^* when one interchanges the parameters ξ, η with the variables x, y.[1] This "reciprocity" property of Riemann's function implies that, as a function of the parameters, R satisfies the equation

(7) $$L_{(\xi,\eta)} [R(x, y; \xi, \eta)] = 0.$$

While R originally was defined as a solution of the adjoint equation, it now appears also as a two-parameter family of solutions of $L[u] = 0$.

4. Riemann's Function and Radiation from a Point. Generalization to Higher Order Problems. We consider equation (1') (identifying

---

[1] In particular, it follows that if L is self-adjoint then R is symmetric in the parameters and arguments.

the variable y with the time t)

$$L[u] = u_{tt} - u_{xx} + au_x + bu_t + cu = f.$$

Let $\{f_k\}$ be a sequence of functions with the properties:

(i) $f_k \geq 0$ is different from zero only in a neighborhood N_k of a fixed point $Q = (\alpha, \beta)$;

(ii) $\iint_{N_k} f_k \, dx \, dt = 1$ for each k;

(iii) the neighborhoods N_k shrink to the point Q as $k \to \infty$.[1]

We denote by u_k the solution of $L[u] = f_k$ which vanishes together with its first derivatives along the initial curve C. Riemann's representation then shows easily that $\lim_{k \to \infty} u_k = u$ exists and

$$u(P) = R(Q; P).$$

The solution $u(x, t) = R(\alpha, \beta; x, t)$ can thus be interpreted physically as the intensity, at (x, t), of a unit radiation emitted from the point Q in space-time. In concise mathematical formulation, $u(x, t)$ is thus the solution for $t > 0$ of

$$L[u] = \delta(x - \xi, t - \tau)$$

with the initial condition $u(x, 0) = 0$.

For arbitrary f, Riemann's formula represents the superposition of radiation effects from all those points (α, β) above C which lie in the domain of dependence of the point (x, t). This suggests the proper generalization of Riemann's function to higher order problems and to more space variables (see Ch. VI, §15). A brief indication may suffice here since in Ch. VI, §15 the complete theory will be implicitly developed.

We consider for brevity an operator $L[u]$ for a vector u of k components and assume that L consists of k linear operators of first order. Then the Riemann tensor is defined in terms of Dirac's δ-function as the solution of the adjoint equation

$$L^*(R(x, t; \xi, \tau)) = 0 \qquad (t < \tau)$$

which satisfies for $t = \tau$ the end condition

$$R(x, \tau; \xi, \tau) = \delta(x - \xi)I$$

[1] Then $\lim_{k \to \infty} f_k$ corresponds to Dirac's δ-function.

with the unit matrix I.   We refer to Ch. VI, §15 for the actual construction and detailed explanation of the Riemann tensor R, which can essentially be constructed by iterated integration along the characteristic curves issuing from the point P toward $t = 0$ as shown in Ch. VI, §4.

5. Examples.

1) For the simplest wave equation

$$(8) \qquad\qquad u_{xy} = 0$$

the Riemann function $R(x, y; \xi, \eta) \equiv 1$, and hence the solution is given by

$$(9) \quad u(P) = \frac{1}{2}[u(A) + u(B)] + \frac{1}{2}\int_{AB}(u_x\, dx - u_y\, dy).$$

Introducing new coordinates

$$x + y = X,$$
$$x - y = T,$$

we obtain for equation (8)

$$(8') \qquad\qquad u_{TT} - u_{XX} = 0,$$

and for (9)

$$(9') \quad
\begin{aligned}
u(P) = {}& \frac{1}{2}[u(A) + u(B)] \\
& + \frac{1}{2}\int_{AB}\left[(u_X + u_T)\frac{1}{2}(dX + dT)\right. \\
& \qquad\qquad\qquad \left. - (u_X - u_T)\frac{1}{2}(dX - dT)\right] \\
= {}& \frac{1}{2}[u(A) + u(B)] + \frac{1}{2}\int_{AB}(u_T\, dX + u_X\, dT).
\end{aligned}$$

When the initial curve is the line $T = 0$, i.e., when A, B are the points $(X - T, 0)$, $(X + T, 0)$, (9') yields

$$(10) \quad
\begin{aligned}
u(X, T) = {}& \frac{1}{2}[u(X - T, 0) + u(X + T, 0)] \\
& + \frac{1}{2}\int_{X-T}^{X+T} u_T(\lambda, 0)\, d\lambda.
\end{aligned}$$

The general solution of (8') can be written in several forms, the simplest of which is

$$u(X, T) = f(X - T) + g(X + T),$$

where f and g are arbitrary functions. To solve an initial value problem one has to determine f and g from the initial data. This, of course, leads back to formula (10), which is usually obtained in exactly this way.

2) Next we consider the equation

$$L[u] = u_{xy} + cu = g(x, y),$$

with constant c, which is equivalent to the *telegraph equation* (see Ch. III, §4). This equation is self-adjoint. Since the operator L has constant coefficients, $R(P; Q)$ depends only on the relative position of the points P and Q. Moreover, letting Q be the origin, we observe that if

$$v(x, y) = R(x, y; 0, 0)$$

satisfies the conditions imposed on Riemann's function (see a), b), c) of article 2), then

$$w(x, y) = v(\alpha x, \alpha^{-1}y)$$

also satisfies them. Clearly $L^*[v(x, y)] = 0$ implies $L^*[w] = 0$, so that (a) is satisfied. Condition b) means (since $a = b = 0$) that $v = R$ remains constant along the coordinate axes. If v has this property, then w has it also; and finally, $v(0, 0) = 1$ implies $w(0, 0) = 1$. Since these conditions determine Riemann's function uniquely (see §6), it follows that $w(x, y) \equiv v(x, y)$ is a function of xy. More generally, for $Q = (\xi, \eta)$, Riemann's function $R(P; Q)$ has the form

$$R(x, y; \xi, \eta) = f(z),$$

where

$$z = (x - \xi)(y - \eta).$$

The equation $L^*[R] = 0$ then yields the equation $zf'' + f' + cf = 0$ for f which, if we set $\lambda = \sqrt{4cz}$, becomes Bessel's equation

$$\frac{d^2f}{d\lambda^2} + \frac{1}{\lambda}\frac{df}{d\lambda} + f = 0.$$

The solution, which is regular at the origin, is the Bessel function

$$f = J_0(\lambda)$$

(see Vol. I, Ch. VII, §2). Indeed,

$$R(x, y; \xi, \eta) = J_0\left(\sqrt{4c(x - \xi)(y - \eta)}\right)$$

is the required Riemann function because it satisfies the prescribed conditions on $x = \xi$, $y = \eta$, as one sees immediately.

3) In §3, 1 we discussed *one-dimensional isentropic flow*; we introduced the Riemann invariants r and s and brought the equations of motion into the form

$$\left(\frac{\partial}{\partial t} + (u + c)\frac{\partial}{\partial x}\right)r = 0,$$

$$\left(\frac{\partial}{\partial t} + (u - c)\frac{\partial}{\partial x}\right)s = 0.$$

This system can be linearized by the hodograph transformation (see §2, 3) and leads to

$$x_s - (u + c)t_s = 0,$$

$$-x_r + (u - c)t_r = 0.$$

We eliminate x and obtain the second order equation

$$-2ct_{rs} + (u - c)_s t_r - (u + c)_r t_s = 0.$$

For a polytropic gas, the Riemann invariants were found to be (see §3, 1)

$$r = \frac{u}{2} + \frac{c}{\gamma - 1}, \qquad s = -\frac{u}{2} + \frac{c}{\gamma - 1},$$

so that our second order equation becomes

$$(r + s)t_{rs} + \frac{\gamma + 1}{2(\gamma - 1)}(t_r + t_s) = 0.$$

We now change the notation to conform with the present section. Let $r = x$, $s = y$, $t = u$, and $(\gamma + 1)/2(\gamma - 1) = -n$. Then our second order equation becomes

$$(11) \qquad L[u] \equiv u_{xy} - \frac{n}{x + y}(u_x + u_y) = 0.$$

Its Riemann function, according to article 2, must satisfy the equation

$$(12) \qquad L^*[R] \equiv R_{xy} + \frac{n}{x+y}(R_x + R_y) - \frac{2n}{(x+y)^2}R = 0$$

with

$$R_x(x, \eta; \xi, \eta) = -\frac{n}{x+\eta}R(x, \eta; \xi, \eta),$$

$$R_y(\xi, y; \xi, \eta) = -\frac{n}{\xi+y}R(\xi, y; \xi, \eta)$$

and $R(\xi, \eta; \xi, \eta) = 1$. When these conditions are integrated along the characteristics, they furnish the data

$$(13) \qquad R(x, \eta; \xi, \eta) = \left(\frac{x+\eta}{\xi+\eta}\right)^{-n},$$

$$(13') \qquad R(\xi, y; \xi, \eta) = \left(\frac{\xi+y}{\xi+\eta}\right)^{-n}$$

for (12) (cf. equations (2), (3), and (3')).

Equation (11) has some particularly simple solutions when n is a positive integer. Multiplying (11) by $x + y$ and differentiating the result n times with respect to x we obtain, by Leibniz's rule

$$(14) \qquad (x+y)D_x^{n+1}u_y - nD_x^{n+1}u = 0,$$

where D_x and D_y denote $\partial/\partial x$ and $\partial/\partial y$.

We recall from the reciprocity of Riemann's function that R, as a function of ξ, η, satisfies

$$(12') \qquad L_{(\xi,\eta)}[R] = R_{\xi\eta} - \frac{n}{\xi+\eta}(R_\xi + R_\eta) = 0$$

(cf. equation (7)). Hence, we also have

$$(15) \qquad (\xi+\eta)D_\xi^{n+1}R_\eta - nD_\xi^{n+1}R = 0.$$

Equations (14) and (15) are ordinary differential equations for $D_x^{n+1}u$ and $D_\xi^{n+1}R$, respectively, and have the solutions

$$D_x^{n+1}u = A(x; \xi, \eta)(x+y)^n$$

and

$$D_\xi^{n+1}R = B(\xi; x, y)(\xi+\eta)^n.$$

We see, using equation (13), that for $\eta = y$, $D_\xi^{n+1}R(x, y; \xi, \eta) = 0$. Hence $B(\xi; x, \eta) = 0$ and, therefore, $B(\xi; x, y) = 0$. From this and the corresponding argument in case of the $(n + 1)$st η-derivative, we conclude that R is a polynomial in ξ, η of degree at most n.

As a solution of (12'), we try a polynomial with the correct boundary values for $\xi = x$, $\eta = y$:

(16) $$R(\xi, \eta; x, y) = \frac{(\xi + \eta)^n}{(x + y)^n} \, \psi(w),$$

where

$$w = -\frac{(\xi - x)(\eta - y)}{(\xi + \eta)(x + y)}$$

and

$$\psi(w) = 1 + a_1 w + \cdots + a_n w^n.$$

Substituting (16) into (12'), we find, after calculating $\psi_\xi = \psi' w_\xi$, $\psi_\eta = \psi' w_\eta$, $\psi_{\xi\eta} = \psi'' w_\xi w_\eta + \psi' w_{\xi\eta}$, that ψ satisfies the equation

(17) $$w(w - 1)\psi''(w) + (2w - 1)\psi'(w) - n(n + 1)\psi(w) = 0.$$

The only solution[1] of this equation with $\psi(0) = 1$ is the hypergeometric series

$$\psi(z) = \psi(1 + n, -n, 1; z).$$

Thus, we obtain R in the form

(18) $$R(x, y; \xi, \eta) = \left(\frac{\xi + \eta}{x + y}\right)^n \psi\left(1 + n, -n, 1; -\frac{(\xi - x)(\eta - y)}{(x + y)(\xi + \eta)}\right)$$

which, for positive integers n, is a polynomial of degree $n - 1$. Our argument shows that (18) represents Riemann's function for arbitrary n.[2]

§6. Solution of Hyperbolic Linear and Semilinear Initial Value Problems by Iteration

In the next sections, constructive existence proofs are given for the solutions of Cauchy's initial value problem. They are based on the

[1] See Magnus and Oberhettinger [1], Ch. II, §1.

[2] Copson [1] has made a survey of equations for which the Riemann function is known in closed form. The reader will also find there an interesting discussion of methods for obtaining explicit solutions.

same methods of iteration which are familiar for ordinary differential equations.

As said before it is possible to reduce Cauchy's initial value problem to one for which the initial data, or "Cauchy data", vanish: We subtract from the unknown function u a suitable[1] fixed function ω and consider the differential equation for the new unknown. Moreover, we can transform the independent variables so that a given nowhere characteristic initial curve C becomes the straight line $y = 0$. Without restricting the generality, we shall if convenient make use of such simplifications, again writing u for the unknown.

1. Construction of the Solution for a Second Order Equation. First we treat briefly the characteristic initial value problem, in particular with a view to proving the existence of Riemann's function of the preceding section.

We consider first the differential equation of second order

$$(1) \qquad L[u] = u_{xy} + au_x + bu_y + cu = f(x, y)$$

which, incidentally, can be written as a system

$$u_y = v - au$$

$$v_x = -bv + (a_x + ab - c)u + f$$

of two linear equations of first order for u and $v = u_y + au$.

Instead of solving Cauchy's problem for (1) it is just as simple and formally even more transparent to solve the more general semilinear differential equation

$$(1') \qquad u_{xy} = F(x, y, u, p, q),$$

where, as before, the abbreviations

$$u_x = p, \qquad u_y = q$$

are used. Again this differential equation could be replaced by a semilinear system of three equations such as

$$u_x = p, \qquad p_y = F(x, y, u, p, q), \qquad q_x = F(x, y, u, p, q)$$

for u, p, q.

---

[1] If the initial curve C is given by $y = y(x)$ and if on C the nonhomogeneous Cauchy data are given by $u = u(x)$, $u_x = p(x)$, $u_y = q(x)$ subject to the strip condition $u' = p + qy'$, then such a function is provided, for example, by $\omega(x, y) = u(x) + (y - y(x))q(x)$, as is easily verified.

We aim at constructing a solution u which along a smooth non-characteristic curve C has the initial values $u = p = q = 0$.

The function F is assumed to have continuous derivatives with respect to all its arguments x, y, u, p, q. The solution u is assumed to have continuous derivatives of first order and a mixed derivative $s = u_{xy}$, which by $(1')$ then is necessarily continuous.[1]

The problem remains meaningful, and the subsequent construction provides the solution also for the *characteristic initial value problem*. This is the limiting case in which the curve C consists of two characteristic lines AD and BD (Figure 32). For this problem, initial values of u only are admissible along C, and, as seen before, the differential equation (1) becomes an ordinary differential equation for p and q along the lines AD and BD, respectively. This equation, together with the condition $p = q = 0$ at D, determines p and q along C. Otherwise there is no further need for distinguishing in the following between Cauchy's problem for free or characteristic curves C.

The solution is constructed in a suitably small triangular portion $\sqcap$ of the x,y-plane, which contains or is adjacent to a given segment of the initial curve C and consists of points P connected by two characteristic arcs PP_1 and PP_2 with the initial curve C so that the points are on the given segment. $\sqcap$ is extended to a domain $\bar{G}$ of the (x,y,u,p,q)-space by inequalities $|u| \leq \mu$, $|p| \leq \mu$, $|q| \leq \mu$ with a suitable constant μ. In this domain $\bar{G}$ we assume $|F| < \lambda$, $|F_u| \leq \lambda$, $|F_p| \leq \lambda$, $|F_q| \leq \lambda$, with a constant λ. With the initial values $u = p = q = 0$ (or in the case of the characteristic initial curve, merely $u = 0$) equation $(1')$ can be written in the integrated form

$$(2) \qquad u(P) = \iint_{\sqcap} F(x, y, u, p, q) \, dx \, dy$$

for a point P in $\sqcap$ (see Ch. I, §1, 1).

Now we define an *integral transformation* of a function $v(x, y)$ into a function $v'(x, y)$ by

$$(2') \qquad v'(P) = Tv = \iint_{\sqcap} F(x, y, v, v_x, v_y) \, dx \, dy.$$

[1] No stipulations need be made for the second derivatives u_{xx}, u_{yy}. However, if F and the initial data possess continuous second derivatives, then the following construction also insures continuous derivatives $r = u_{xx}$, $t = u_{yy}$ and even continuous third derivatives p_{xy}, q_{xy}.

FIGURE 32

The desired solution then is a "fixed point" of this transformation T in function space, and this fixed element is obtained by a process of iteration. It is the limit for $n \to \infty$ of a sequence $u_n : u_{n+1} = Tu_n$, with an arbitrary function u_0 which satisfies the initial conditions, say $u = 0$, $p = 0$, $q = 0$.

For sufficiently small domains ⊓ this iteration indeed converges to the solution. To avoid duplication the proof will not be carried out in this article. As stated above, the present problem is equivalent to an initial value problem for linear, or semilinear, systems of first order; we shall presently solve such systems by iterations.

2. Notations and Results for Linear and Semilinear[1] Systems of First Order. In the following articles we shall write t instead of y. As in §2, we consider a system of k differential equations of first order for a function vector $u(x, t)$ with the components $u^1, u^2, \cdots, u^k$:

$$(3) \qquad u_t + Au_x + B = 0,$$

where the $k \times k$ matrix $A(x, t)$ and the vector $B(x, t, u)$ have continuous first derivatives and where B may depend on the variables u in a linear or nonlinear manner. Without restricting generality we may assume the x-axis $t = 0$ as the initial line, prescribing initial Cauchy data

$$u(x, 0) = \psi(x),$$

where the vector $\psi(x)$ possesses a continuous first derivative. The system (3) is assumed hyperbolic in the sense of §2; that is, the

---

[1] The results on quasi-linear systems which will be obtained from the considerations of this and the next section are due to J. Schauder [1]. Subsequent proofs have been given by M. Cinquini-Cibrario [1], K. O. Friedrichs [1], and by R. Courant and P. D. Lax [1]. More refined results have been found by A. Douglis [1], P. Hartman and A. Wintner [1], and P. D. Lax [5].

An existence theorem with stronger assumptions about differentiability was given for a single nonlinear equation of nth order by E. E. Levi many years earlier [2]. Levi's work on hyperbolic equations remained forgotten until nearly all his results were independently rediscovered. His remarks on nonlinear equations with multiple characteristics (see Levi [1]) have not yet been followed up, however.

matrix A has k real eigenvalues τ^1, τ^2, $\cdots$, τ^k and k linearly independent left eigenvectors l^1, l^2, $\cdots$, l^k which form a matrix Λ with determinant 1. The eigenvectors l^κ are assumed to have continuous derivatives with respect to x and t. Finally, for the first derivatives of the coefficients and consequently for the first derivatives of the eigenelements we assume a specific modulus of continuity, say *Lipschitz continuity*. The characteristic curves are defined by the ordinary differential equations $dx/dt = \tau^\kappa$ and the characteristics C_κ issuing from a point P with coordinates ξ, τ can be represented by functions

$$C_\kappa: x = x^\kappa(t; \xi, \tau)$$

which have continuous derivatives with respect to t as well as the parameters ξ, τ. In view of the eigenvalue relation $l^\kappa A = \tau^\kappa l^\kappa$ we obtain by multiplying (3) with the eigenvector l^κ the characteristic system of equations

(3′) $$l^\kappa D^\kappa u + l^\kappa B = 0,$$

or briefly

$$lDu + lB = 0,$$

where D^κ is the differential operator $D^\kappa = \partial/\partial t + \tau^\kappa(\partial/\partial x)$. The operator D^κ can be regarded as differentiation d/dt along C_κ.

In keeping with our assumptions we suppose that the eigenvalues τ^κ and the eigenvectors l^κ are functions of x and t with continuous first derivatives.

We shall occasionally omit explicit reference to a specific characteristic by writing D, l, instead of D^κ, l^κ and we note the useful formula $D(ab) = aD(b) + bD(a)$ for any pair of functions a and b.

The form (3′) of the equations suggests introducing new unknowns $U^\kappa = l^\kappa u$, or briefly $U = \Lambda u$ or $U = lu$. Because of $lD(u) = D(lu) - uDl$ the form

(3″) $$DU = -lB - (Dl)u = -b - (Dl)u$$

of the differential equations then results. Since u is given in terms of U by the relation $u = \Lambda'U$ where Λ' is the matrix reciprocal to Λ one can always express (3″) in the form

(3‴) $$DU = F(x, t; U),$$

where the right-hand side is a continuous function vector of the variables x, t, U and has continuous derivatives with respect to the variables U. Of course, D here stands for a diagonal matrix with the components D_κ.

We note that the initial values $\psi(x)$ of u go over into initial values $\Psi(x)$ for U given as the values for $t = 0$ of $\Psi(x, t) = \Lambda\psi$. Our assumptions insure the equivalence of the Cauchy problems for u and U.

In the x, t-plane we consider a closed domain G so that all characteristics C_κ followed from a point P in G backwards towards decreasing values of t, meet a given section $\mathscr{s}$ of the x-axis in points P_κ with coordinates $x_\kappa = x(0; \xi, \tau)$ so that $\mathscr{s}$ contains the domain of dependence for all points P of G. The strip $0 < t < h$ within G is called G_h (see Figure 33).

The main results of the next article will be: *Cauchy's problem with initial values ψ on $\mathscr{s}$ possesses a unique solution u with continuous derivatives in the strip G_h for sufficiently small h.*

According to §4 the solution is uniquely determined. We also shall see: *The solution can be extended into the whole domain G as long as the coefficients retain their continuity properties.* Moreover, if A and B have continuous derivatives of order up to n then the solution u likewise possesses such derivatives. And finally, if the coefficients and the data ψ depend on a parameter continuously and differentiably then the same holds for the solution.

3. Construction of the Solution. To construct the solution in terms of the equation $(3''')$ and the initial values $\Psi(x)$ we replace U on the right-hand side by a function vector $V = \Lambda v$ and consider the set S or "space" of functions in G having continuous derivatives and the initial values $\Psi(x)$. Integration of $(3'')$ or $(3''')$ along the characteristics C_κ from P_κ to P then suggests the study of the integral transformation, where $x^\kappa(t)$ is substituted for x in $F^\kappa(x, t, V(x, t))$,

$$
\begin{aligned}
W^\kappa(\xi, \tau) &= \Psi(x_\kappa) + \int_0^\tau F^\kappa(x, t, V)\, dt \\
&= \Psi^\kappa(x_\kappa) - \int_0^\tau (b^\kappa(x, t, V) + (D^\kappa l^\kappa)v)\, dt,
\end{aligned}
$$

(4)

or symbolically,

$$W = TV$$

of a function V in S into a function W with the same initial values. We then seek the solution of Cauchy's problem as an element which remains fixed under this transformation.

Of course, in the transformation T each component of $W(\xi, \tau)$ at

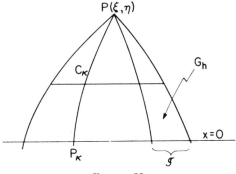

FIGURE 33

the point P is expressed by a corresponding integral over the characteristic C_κ; that means, in G^κ we have to replace x in $F^\kappa(x, t, V(x, t))$ by $x_\kappa(t; \xi, \tau)$ and then simply to integrate with respect to t from 0 to τ.

For a suitably narrow strip G_h the desired fixed element will be constructed by iterations as the uniform limit for $n \to \infty$ of

$$U_{n+1} = TU_n$$

starting with an arbitrary admissible $U_0 = \Psi(x, t)$, whose initial values are $\psi(x)$.

To this end we use the *maximum norm* $\| f \|$ or $\| f \|_0$ of a continuous function vector f, that is, the largest value of f attained in the closed domain G for any component of f in G. Now we set $N = \| \Psi(x, t) \|$ and restrict "admissible functions" in S by stipulating $\| V \| < 2N$. For these admissible functions there exists a common upper bound μ such that in G and all the more in G_h

$$\| F^\kappa \| < \mu, \qquad \| F^\kappa_V \| < \mu, \qquad \| F^\kappa_x \| < \mu, \qquad \| F^\kappa_t \| < \mu,$$

where F_V means the gradient of F with respect to the variables V. Also for the absolute values of the integrands in (4) we may then assume μ as an upper bound.

Then for h chosen sufficiently small the transformation T takes an admissible function V into an admissible function W. Indeed we observe that (4) implies $\| W \| \leq N + h\mu$; hence if we choose h so small that $h\mu < N$ then $\| W \| \leq 2N$. That W has continuous derivatives will be seen presently.

To prove uniform convergence of the iterations U_n we form the difference $Z_n = U_{n+1} - U_n$ for which by (4)

$$Z_n^\kappa(\xi, \tau) = \int_0^\tau (F^\kappa(x, t, U_{n+1}) - F^\kappa(x, t, U_n)) \, dt \qquad (x = x_\kappa(t; \xi, \eta)).$$

Since $F(x, t, V)$ has continuous derivatives with respect to V we may apply the mean value theorem under the integral sign and obtain with suitable intermediate values $\tilde{U}$ the relation

$$Z_n^\kappa(\xi, \tau) = \int_0^\tau F_U^\kappa(x, t; \tilde{U}^\kappa) \, Z_{n-1}(x, t) \, dt.$$

Since $\| F_U^\kappa \| < \mu$ we find immediately as before

$$\| Z_n \| < hk\mu \| Z_{n-1} \| \, .$$

Therefore, by choosing h small enough, such that $hk\mu = \theta < 1$, e.g., $\theta = 1/2$, is satisfied we assure

$$\| Z_n \| \leq \theta \| Z_{n-1} \| \, .$$

We may say: The transformation T is therefore *contracting in the maximum norm* and Z_n tends uniformly to zero in G_h as $n \to \infty$; hence the functions U_n indeed converge uniformly to a continuous function U in G.

The limit vector U obviously has the initial values $\Psi(x)$, is a fixed element under the transformation T, and solves the system of integral equations $U = TU$. Therefore, since the directional differential operator D^κ on these integrals produces the integrand, U also solves the differential equations $(3''')$ in the characteristic normal form $(3''')$.[1]

However, the existence of the directional derivatives does not immediately imply existence—or continuity—of the derivatives U_x, U_t of the limit function U. That these derivatives exist and are continuous will now be proved by showing that the first derivatives of U converge uniformly as the functions U do. Then according to elementary calculus, the limits of U have these limits as derivatives.

We may confine ourselves to the x-derivative (or the ξ-derivative), since the known directional derivatives D^κ then allow us to express the t-derivative. The term $(Dl)v$ would introduce second derivatives of the eigenvectors l if we differentiated (4) with respect to ξ under the integral sign while our assumptions imply only the existence of

---

[1] The above reasoning shows the uniqueness of the solution. In fact for a solution U all the functions U_n are equal to U if $U_0 = U$; then the difference of two solutions reproduces itself by the transformation T; because of the contracting property of T this implies the difference vanishes identically.

continuous first derivatives. We may easily overcome this obstacle by first assuming continuous second derivatives and after differentiation eliminate this assumption by a limit process. However we could use the following more direct device[1] to differentiate with respect to a parameter ξ an integral of the form $K(\xi) = \int_0^\tau P(t, \xi)DQ(t, \xi)\, dt$, where D denotes differentiation with respect to the variable t—in our case differentiation along a characteristic. Instead of $K(\xi)$ we consider a function of two parameters α and β: $H(\alpha, \beta) = \int_0^\tau P(t, \alpha)\, DQ(t, \beta)$, so that $K(\xi) = H(\xi, \xi)$ and $K_\xi = H_\alpha + H_\beta$ for $\alpha = \beta = \xi$. Then the term H_α is obtained by direct differentiation under the integral sign: $H_\alpha = \int_0^\tau P_\alpha(t, \alpha)DQ(t, \beta)\, dt$. To express the second term we integrate first by parts, and afterwards differentiate directly with respect to β; finally we set again $\alpha = \beta = \xi$. Adding the results we obtain

$$K_\xi = \int_0^\tau [P_\xi(t, \xi)DQ(t, \xi) - P(t, \xi)Q_\xi(t, \xi)]\, dt$$

$$+ P(\tau, \xi)Q_\xi(\tau, \xi) - P(0, \xi)Q_\xi(0, \xi).$$

This formula immediately allows us to differentiate the integral of $vD(l)$: Combining the various expressions and cancelling a boundary term we finally can write briefly for $W = TV$

$$(5) \quad W_\xi(\xi, \tau) = - \int_0^\tau x_\xi[b_x + b_V V_x + l_x Dv - v_x Dl]\, dt$$

$$+ l_x v \big|_{t=\tau} - l\psi' \big|_{t=0}.$$

Whether or not we express v under the integral sign in terms of V, this formula represents the derivative W_ξ in terms of V and the first derivatives of V without referring to second derivatives of l. It shows that W possesses continuous first derivatives. Furthermore, (5) exhibits the fact: The derivatives of W are bounded provided $\| V \|$ and $\| V \|_1$ remain bounded; here $\| V \|_1$ is the *maximum norm of first order*; that means the largest value which is attained in the domain by any of the quantities $| V_x^\kappa |$ or $| V_t^\kappa |$.[2]

More specifically one sees: For any given bound M of $\| V \|$ one can determine a bound M_1 for $\| V \|_1$ so that with sufficiently small

---

[1] Suggested by P. Ungar.

[2] One could also define $\| V \|_1$ slightly differently as the maximum attained for a component $| V^\kappa |$ or $| V_x^\kappa |$ or $| V_t^\kappa |$.

h, the quantities $\| W \|$ and $\| W \|_1$ are bounded by the same numbers M and M_1 respectively.

Moreover, one concludes exactly as above: For sufficiently small h the transformation TV contracts the first derivatives. Hence not only U_n but also the first derivatives of the iterates U_n converge in a suitably narrow strip G_h uniformly. Necessarily these limits are the corresponding derivatives of the limit function U. The function U therefore indeed is the solution of Cauchy's problem.

A remark may be added: From (5) it follows easily that a suitable "modulus of continuity" for V_ξ likewise becomes a modulus of continuity for W_ξ. In particular, if V_ξ (and V_τ) are "Lipschitz continuous" with a Lipschitz constant σ, i.e., if the difference quotients are absolutely bounded by σ, then the same holds for the derivatives of TV.

Hence the derivatives of the iterates are equicontinuous. By Arzela's theorem they therefore form a compact set, since they are bounded, and by the uniqueness theorem they then can define only one limit element; i.e., they converge uniformly. Thus a slightly different variant of the existence proof is obtained.

Finally we add an important remark for use in §7: Even if at each step the transformation (5) $T = T_n$ changes, equicontinuity of the derivatives, such as a Lipschitz condition, persists under the transformation T as long as the term $D^\epsilon l^\kappa$ remains uniformly bounded.

Moreover, the preceding reasoning extends without change to higher derivatives. If the initial data and the coefficients of the differential equation possess continuous derivatives up to the order n, then the solution has the same smoothness properties. We say: The smoothness properties of u are *continuable*, i.e., differentiability of any order is preserved in the transition from $t = 0$ to $t = h$. In Ch. VI, §10 we shall discuss the significance of the concept of *continuable initial conditions*.

The theorem stated at the beginning of article 2 holds in a larger region. To show this we use the line $t = h$ as the new initial line and the value of the solution $U(x, h)$ as the new initial data, and solve the same problem for the strip $h \leq t \leq 2h$. We continue stepwise in this way and thus construct the solution for arbitrarily large t as long as the assumptions of continuity and boundedness remain satisfied.

4. Remarks. Dependence of Solutions on Parameters. An important consequence of the estimates used to prove convergence of iterations is easily ascertained: If the coefficients of the differential operator or the initial data ϕ depend continuously on a parameter ϵ and possess continuous derivatives up to order s with respect to ϵ, then the same is true of the solution u as function of ϵ.

Finally it should be stated: The preceding constructions applied to initial values $\psi(x)$ lead to a generalized solution even if ψ has discontinuities, e.g., jump discontinuities at a certain point, say the point $x = 0$. Such discontinuities then propagate from this initial discontinuity point P^* along each of the characteristics issuing from P^*. A detailed analysis of discontinuous solutions in greater generality will be contained in Ch. VI, §4.

5. Mixed Initial and Boundary Value Problems.[1] Most physical phenomena of interest take place in a portion of space limited by boundaries stationary or moving. These boundaries are expressed mathematically by relations among the variables describing the physical system. Examples of such boundary conditions are: a) The normal component of the velocity of an ideal fluid at a moving wall must be equal to the normal component of the velocity of the wall. b) The displacement of a vibrating string fixed at the end points is zero at the end points. c) The normal component of the gradient of the amplitude of sound waves at a perfectly reflecting wall is zero.

We assume that the system is confined to the positive x-axis, $x \geq 0$ and aim at solving a linear or semilinear system in the characteristic normal form $(3'')$: $D^\kappa u^\kappa = F^\kappa(x, t, u^1, \cdots, u^n)$; the solution is to be constructed for a small domain adjacent to the positive x-axis $x > 0$ and to the time axis $t > 0$. The conditions imposed along the x-axis and the t-axis should establish a properly posed mixed problem. We assume that r is the number of positive eigenvalues τ:

$$\tau_1 \geq \tau_2 \geq \cdots \geq \tau_r > 0$$

while $\tau_{r+1}, \cdots, \tau_k$ are negative. By assumption no eigenvalue τ should vanish. Or: r is the number of characteristics through the origin O which point upward into the first quadrant of the x, t-plane. Then the first r characteristics through a point on $t > 0$ near enough to O also point into the first quadrant. Moreover, we visualize the characteristics as not crossing each other. Then the characteristic C_r through O separates the quadrant adjacent to O into two regions

---

[1] See also Appendix 2 of this chapter, also Ch. VI, §8, 4.

such that all the r characteristics drawn from a point P in the region left from C_r towards decreasing t intersect the positive t-axis, provided we restrict ourselves to a sufficiently small region adjacent to O (see Figure 34).

Along the positive x-axis we prescribe Cauchy initial data, i.e., the k values $u(x, 0) = \psi(x)$.

In addition to these Cauchy data on $x > 0$ we impose on $x = 0$ the r boundary conditions

$$(8) \qquad u^\rho = \chi^\rho(t),$$

with prescribed functions $\chi^\rho(t)$, $\rho \leq r$; or more generally

$$(9) \qquad u^\rho - \sum_{j=r+1}^{k} m^{j,\rho} u^j = \chi^\rho(t) \qquad (\rho = 1, \cdots, r).$$

Here the $(k - r)r$ quantities $m^{j,\rho}$ are given and may be chosen so that

$$(10) \qquad \sum_{j=r+1}^{k} | m^{j,\rho} | < 1.$$

The condition (8) insures that the r functions $u^1, \cdots, u^r$ can be linearly expressed by the remaining ones $u^{r+1}, \cdots, u^k$. The condition (10) is a mere technical convenience for the convergence proof below. It can always be satisfied by introducing instead of $u^{r+1}, \cdots, u^k$ the new dependent variables $\mu u^{r+1}, \cdots, \mu u^k$ with a sufficiently large positive factor μ.

Finally we note: Along the characteristics through O the solution will have discontinuities unless "consistency conditions" at O for the data are stipulated for $x = 0$, $t = 0$. In particular first order consistency, i.e., continuity of u, depends on the conditions

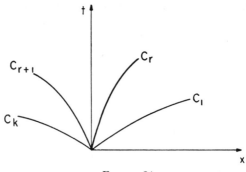

FIGURE 34

$$\chi^{\rho}(0) + \sum_{j=r+1}^{k} m^{j,\rho} \psi^{j}(0) = \psi^{\rho}(0) \quad (\rho = 1, \cdots, r).$$

Similar conditions are obtained by differentiation for consistency of derivatives.

We state: Under the conditions prescribed *the differential equation* (3) *has a unique solution in the quadrant* $0 \leq x$, $0 \leq t$. If the coefficients of the differential equation and the data possess continuous derivatives of order s and if the data satisfy the conditions of consistency up to order s, then the solution possesses continuous derivatives of order s. If consistency conditions are satisfied only up to the order p, then the solution has discontinuities of the derivatives of order greater than p along the characteristics C_1, $\cdots$, C_r issuing from the origin.

The construction of the solution proceeds by iterations almost exactly as in article 3. Again denoting by P_κ the intersections of C_κ with $t = 0$ or $x = 0$, we have in obvious notation

$$(11) \qquad u^{\kappa}(P) = u^{\kappa}(P_\kappa) + \int_{P_\kappa}^{P} F^{\kappa}(x, t, u^1, \cdots, u^k) \, dt.$$

If P_κ is on the t-axis, which can happen only for $k \leq r$, we substitute for $u^\kappa(P_\kappa)$ the values from (9), and if P_κ is on the x-axis, we substitute the Cauchy data $\psi^\kappa(P_\kappa)$. Then we define an integral transformation

$$v_{n+1} = Tv_n$$

as before, by defining the right side of (11) as Tu. To prove the convergence of the sequence $\{v_n\}$, we again consider the series of differences $w_{n+1} = v_{n+1} - v_n$. If P_κ is on the t-axis we have the recursion relations

$$w_{n+1}^{\kappa} = \sum_{i=r+1}^{k} m^{i,\kappa} w_n^i - \int_{P_\kappa}^{P} \sum_{i=w_n}^{k} b^{\kappa i} w_n^i \, dt.$$

The contracting character of the transformation linking w_{n+1} to w_n in the maximum norm as well as in the s norm then follows for a sufficiently small strip exactly as in article 3. Thus the solution is again obtained as a fixed element of the transformation T.

If a physical system is confined by two walls instead of one, say by the lines $x = 0$ and $x = a$, then boundary conditions have to be prescribed also on the line $x = a$ according to the same principles.

In other words, we need as many boundary conditions as there are characteristics issuing into the region from the point $(a, 0)$, and those conditions must satisfy the requirements of consistency and linear independence analogous to those stated before for the boundary $x = 0$. In this situation we obtain a unique solution in a half-strip.

A domain moving in time is described mathematically by giving its position at $t = 0$ as some interval (a, b) on the x-axis, and by giving the motion of the end points as a pair of curves issuing from a and b in the positive t-direction. If these curves are sufficiently smooth, the problem for such a region can be reduced to the one treated above by a transformation of independent variables.

So far we have assumed that the boundaries are nowhere characteristic; but the same method for constructing solutions can be used if one (or both) of the boundary curves is characteristic. In this case the number of data to be prescribed on the boundary is the same as the number of characteristic curves which enter the region *not including* the boundary itself. This is consistent with the observations in article 1 concerning characteristic initial value problems.

Examples: 1) The motion of a taut string in a plane is governed by the equation

$$(12) \qquad u_{tt} - c^2 u_{xx} = 0,$$

where u denotes displacement of the string and c is a constant depending on the mass density and tension of the string. The initial position and velocity of the string are given (Cauchy data) and the end points $(a, 0)$ and $(b, 0)$ are kept fixed, i.e., the boundary conditions are

$$u(a, t) = u(b, t) = 0.$$

Convert the second order equation (12) into a first order diagonalized system by introducing two new unknown functions $v = cu_x + u_t$, $w = cu_x - u_t$. Then we have

$$v_t - cv_x = 0,$$

$$w_t + cw_x = 0.$$

The characteristic speeds are $\pm c$. The region under consideration is the half-strip $a \le x \le b, t \ge 0$; exactly one of the characteristics enters this region from each corner. Therefore, $r = 1$ at both boundaries. Indeed we have one boundary condition on each boundary:

$$v(a, t) - w(a, t) = v(b, t) - w(b, t) = 0.$$

Furthermore, it is easily verified that the condition for linear independence is satisfied in this case.

2) The motion of a compressible gas in a tube closed off by moving pistons may be described in terms of the flow velocity u and the density ρ. The equations of continuity and conservation of momentum are

$$\rho_t + u\rho_x + \rho u_x = 0$$

and

$$u_t + uu_x + \frac{1}{\rho} p_x = 0.$$

Here p denotes pressure; the equation of state of the gas gives p as a known[1] (and monotonically increasing) function of ρ.

These equations are nonlinear. The theory of the initial value problem for such equations is quite parallel to the theory in the linear case; it will be developed in the next section. The similarity extends also to the theory of mixed initial boundary value problems as the example given below illustrates. Naturally, for nonlinear systems one must remain aware of the fact that statements are always meant to be valid merely for sufficiently small regions.

Initially u and ρ are prescribed as given functions of x ($a < x < b$). On the curves $x = a(t)$ and $x = b(t)$, described by the two pistons enclosing the tube, we require that the flow velocity be equal to the piston velocity:

$$u(a(t), t) = \frac{da}{dt},$$

$$u(b(t), t) = \frac{db}{dt}.$$

The characteristic speeds are $u + c$ and $u - c$, where $c^2 = dp/d\rho$. The region under consideration is $a(t) \leq x \leq b(t)$, $t \geq 0$. Again r is 1 at both boundaries; this is the number of conditions on each boundary.

In the theory developed for the linear case the number of conditions had to be equal to the number of characteristics entering the region. A similar condition must be satisfied in the nonlinear case. Such a condition is actually satisfied in our example, as immediately seen.

---

[1] We assume hereby, as in §3, 1, that the flow is isentropic.

§7. Cauchy's Problem for Quasi-Linear Systems

We turn now to a strictly quasi-linear system

(1) $$u_t + A(x, t; u)u_x + B(x, t; u) = 0$$

and briefly indicate how Cauchy's problem for (1) can be solved on the basis of the results of §6 by a slightly different iteration process. The result is exactly the same as that for linear or semilinear systems. In particular, if the coefficients A, B, and the initial values $\psi(x)$ possess Lipschitz-continuous first derivatives with respect to x, t, u, then in a suitable neighborhood $0 \leq t \leq h$ of a portion $\mathfrak{s}$ of the x-axis a uniquely determined solution with Lipschitz-continuous first derivatives exists, provided that the system is hyperbolic for the given initial values $u(x, 0) = \psi(x)$. As before, hyperbolicity means the existence of k independent left eigenvectors $(l^1, \cdots, l^k)$, which we may assume so normalized that the matrix Λ they form has determinant 1. Characteristics C_κ, eigenvalues τ^κ and eigenvectors now depend on the specific functions u. Generally we consider functions v, not necessarily solutions, with the fixed initial values $v(x, 0) = \psi(x)$, with Lipschitz-continuous first derivatives and satisfying inequalities

$$\| v \| \leq M , \qquad \| v \|_1 \leq M_1$$

with fixed M and M_1. We assume that the matrix $A(x, t; v)$ has k real eigenvalues $\tau^\kappa(x, t; v)$, briefly $\tau(v)$, and a matrix $\Lambda(v)$ of independent eigenvectors l^κ with Lipschitz-continuous derivatives with respect to all arguments[1] in a fixed domain G_h to be described presently provided we substitute for v any "admissible" function restricted by the above conditions. The solutions of the ordinary differential equations $dx/dt = \tau^\kappa$ are called the characteristics C^κ of the v-field. For admissible functions their slopes are uniformly bounded: $| \tau^\kappa | < \mu$, and we now specify that a closed domain G and in it the strip G_h consists of points such that the v-characteristics C^κ followed backwards towards $t = 0$ remain in G and intersect the portion $\mathfrak{s}$ of the x-axis. The set or "space" of all such functions v is again called S_h. The eigenvectors l and eigenvalues as well as their derivatives depend on the variables x, t—and v—and are Lipschitz continuous. As before, we define the characteristic differential operators in the v-field by $D = \partial/\partial t + \tau(\partial/\partial x)$.

Now we proceed by a natural iteration scheme. (Compare,

---

[1] For distinct eigenvalues these properties of the eigenelements follow from the assumptions about A.

incidentally, Ch. IV, §7, for a similar scheme for elliptic equations, and Ch. VI, §10, 5 for hyperbolic equations in more than two variables.)

To allow enough leeway for v, we may set $M = N + 1$ assuming a bound $\| \psi(x) \| < N$. Then, after substituting an admissible function $v(x, t)$ in A and B, equation (1) becomes linear as we treated in in §6. The solution u of Cauchy's problem with the given initial values $u(x, 0) = \psi(x)$ is called $u = Tv$, and we obtain the solution u of (1) as a fixed element of the transformation T, specifically as the uniform limit for $n \to \infty$ of iterations $Tu_n = u_{n+1}$ with $u_0 = \psi(x)$, where $T = T_n$ depends on n.

The construction proceeds as follows:[1] First we obtain $u = TV$ according to the procedure in §6, based on the introduction of a new function $U = \Lambda u$, $V = \Lambda v$, $\Psi = \Lambda \psi$, etc. Then in the basic formulas (5) and (6) of §6 we have merely to observe that the differentiations with respect to x must account for the dependence of $l, \cdots$, on v; that is, we must interpret $(d/dx)l = l_x + l_v v_x$, $\cdots$, thus introducing derivatives of v. This leads quite directly to the lemma: For $M = N + 1$ we can choose a sufficiently large bound M_1 and sufficiently small h such that any function v of S_h is transformed into another function $w = Tv$ of S_h.

Furthermore, the iterates u_n converge uniformly in G_h to a limit function u. This function is uniquely determined and obviously satisfies the differential equation in the characteristic form $lDu + lB = 0$. The contracting property and uniqueness follow directly as in §6 by considering for the difference $z = u - u^*$ of two admissible functions $u = Tv$ and $u^* = Tv^*$ with the difference $\zeta = v - v^*$, the differential equation

$$z_t + A(v)z_x + (A(v) - A(v^*))z_x^* + B(v) - B(v^*) = 0.$$

Using the mean value theorem and referring to (4) of §6, we have

$$z_t + A(v)z_x + \zeta K = 0,$$

where K is a bounded function of x, t, and obtain as in §6 an inequality of the form

$$\| z \| < M_3 hk \| K \| .$$

With sufficiently small h we assure the inequality $\| z \| \leq \frac{1}{2} \| \zeta \|$, and hence the contracting character of the transformation T. Therefore u_n converges uniformly to a function u in G_h .

[1] See R. Courant [3].

To show that u possesses Lipschitz continuous first derivatives and solves (1) in the strict sense, we rely on the remark at the end of §6; we may proceed by showing from §6, (5) that the transformation Tv preserves the Lipschitz continuity of the first derivatives. Therefore the first derivatives of the iterates u_n are equicontinuous, form a compact set and the limit function enjoys Lipschitz-continuous derivatives, as stated.

One could also use a slightly different argument: First we approximate A, B and ψ by suitably chosen functions with continuous and bounded second derivatives. Then the corresponding solutions u_n will have uniform bounds of the second derivatives. These bounds, after passing to the limit, will simply become Lipschitz bounds for the first derivatives of u. At any rate, the result for nonlinear systems remains the same as for linear ones.

Incidentally, it is not difficult to replace the assumption of Lipschitz continuity by stipulating any (concave) modulus of continuity.[1]

§8. Cauchy's Problem for Single Hyperbolic Differential Equations of Higher Order

Single differential equations of higher order present features sufficiently different from those of systems of first order to warrant a brief separate discussion.

A linear differential equation of order k for a single function $u(x, t)$ has been written before in the symbolic form

$$(1) \qquad L[u] \equiv (P^k + P^{k-1} + \cdots + P^0)u + f = 0,$$

where

$$(1a) \qquad P^\kappa = P^\kappa\left(\frac{\partial}{\partial t}, \frac{\partial}{\partial x}\right) = a_0^\kappa \frac{\partial^\kappa}{\partial t^\kappa} + a_1^\kappa \frac{\partial^\kappa}{\partial t^{\kappa-1}\partial x} + \cdots + a_k^\kappa \frac{\partial^\kappa}{\partial x^\kappa}$$

$$(\kappa = 0, 1, \cdots, k)$$

is a homogeneous polynomial of degree κ in $\partial/\partial t$, $\partial/\partial x$, with coefficients a_i^κ depending on x, t.

Cauchy's problem for the initial line $t = 0$ is to determine u from (1) if the values of u and of its partial derivatives u_t, u_{tt}, $\cdots$, $u_{t^{k-1}}$ are prescribed on $t = 0$. Without essentially restricting the generality we assume that the Cauchy initial data are zero, so that u and all the derivatives of u of order up to $k - 1$ vanish identically for $t = 0$.

We assume: In the region $t \geq 0$ under consideration, (1) is hyper-

[1] See A. Douglis [1].

bolic[1] in the sense that at each point there are k characteristic directions (see §2) with distinct characteristic derivatives

$$D_i = \frac{\partial}{\partial t} + \tau_i(x, t) \frac{\partial}{\partial x} \qquad (i = 1, \cdots, k).$$

As seen in §1, the coefficients $\tau_i = -\phi_t^i/\phi_x^i$ correspond to the family $\phi^i(t, x) = $ const. of characteristic curves which satisfy the characteristic equation

$$P^k(\phi_t, \phi_x) = 0.$$

We assume that the lines $t = $ const. are not characteristic, hence $\phi_x \neq 0$ and $a_0^k \neq 0$. Dividing by a_0^k, we may then replace a_0^k by 1 and the characteristic equation by

$$P^k(\tau, -1) = 0.$$

In the present section, we shall indicate several methods for solving Cauchy's problem for (1) in this hyperbolic case. We consider also a wider class of cases, in which Cauchy's problem is solvable even though multiple characteristics may occur.

1. Reduction to a Characteristic System of First Order. Cauchy's problem can be reduced to an initial value problem for a linear system of first order in the diagonalized form.

We shall first carry out such a reduction by introducing the derivatives of u as new unknown functions (as we did before). The resulting first order system will have the same k distinct characteristics as (1) and, in addition, the lines $x = $ const. as trivial multiple characteristics. This system can be diagonalized into the normal form treated in §6, 2. In article 3, we shall discuss more elegant and general reductions.

Writing a_i instead of a_i^k and setting $a_0 = 1$ we replace (1) by a system of differential equations for $\frac{1}{2}k(k - 1)$ functions

$$p^{i,j}(x, t) \qquad\qquad (i + j \leq k - 1);$$

these functions are to be identified with the derivatives $\partial^{i+j}u/\partial t^i \partial x^j$. The principal equation is

---

[1] For constant coefficients the more inclusive definition of hyperbolicity given by L. Gårding [2] includes equations with real multiple characteristics. Some of the methods in this section apply also to such equations. Again it should be said that the aim of such general definitions is to provide verifiable criteria which insure that the Cauchy problem is properly posed.

$$(2) \quad p_t^{k-1,0} + (a_1 p_x^{k-1,0} + a_2 p_x^{k-2,1} + \cdots + a_k p_x^{0,k-1}) + H = 0,$$

where all kth order derivatives in (1a), except the kth derivative with respect to t, are replaced by x-derivatives of the quantities $p^{i,j}$ for $i + j = k - 1$. H is a linear expression in the quantities $p^{i,j}$ with $i + j < k - 1$, and does not contain derivatives. Equation (2) is now supplemented by the equations

$$(2') \quad p_t^{i,j} - p^{i+1,j} = 0 \qquad \text{for } i + j = 0, 1, \cdots, k - 2,$$

$$(2'') \quad p_t^{i,j} - p_x^{i+1,j-1} = 0 \qquad \text{for } i + j = k - 1 \text{ and } i \neq k - 1.$$

As initial conditions we choose those given or induced by the Cauchy data; that is, we choose for $t = 0$ the functions $p^{0,i}$ equal to $\partial^i u(x, t)/\partial t^i$, $i = 0, \cdots, k - 1$. Then $(2')$ and $(2'')$ imply vanishing Cauchy data for $p^{i,j}$ and moreover show that everywhere $p^{i,j} = \partial^{i+j}(p^{0,0})/\partial x^i \partial t^j$ for $i + j < k$.

Now we denote by $U(x, t)$ the column vector with the components $p^{i,j}$, arranged in the order of decreasing $i + j$ and increasing j, as in (2), $(2')$, and $(2'')$. Then our system has the form

$$(2''') \qquad U_t + A U_x + Bu + c = 0,$$

where the matrix A is

$$\begin{pmatrix}
a_1 & a_2 & \cdots & a_{k-1} & a_k & 0 & \cdots & 0 \\
-1 & 0 & \cdots & 0 & 0 & 0 & \cdots & 0 \\
0 & -1 & \cdots & 0 & 0 & 0 & \cdots & 0 \\
\multicolumn{8}{c}{\cdots\cdots\cdots\cdots\cdots\cdots\cdots\cdots} \\
0 & 0 & \cdots & -1 & 0 & 0 & \cdots & 0 \\
\multicolumn{8}{c}{\cdots\cdots\cdots\cdots\cdots\cdots\cdots\cdots} \\
0 & 0 & \cdots & 0 & 0 & 0 & \cdots & 0
\end{pmatrix}$$

and the characteristic equation $\| I\tau + A \| = 0$ is easily found to be

$$\| I\tau + A \| = \tau^N P^k(-\tau, 1) = 0,$$

with $N = \frac{1}{2}k(k - 1)$. The factor τ^N corresponds to the trivial multiple characteristics $x = $ const. of our system.

Thus the system $(2''')$ falls into the category treated in §6, 2. Cauchy's problem with the data given above therefore possesses a unique solution which at the same time solves the original problem for equation (1).

2. Characteristic Representation of $L[u]$. A more general method of solving Cauchy's problem for a hyperbolic equation (1) is based on a representation of the operator $L[u]$ in terms of characteristic directional derivatives.

As a preparation we consider r not necessarily characteristic but distinct directional derivatives

$$D_i = \frac{\partial}{\partial t} + \tau_i(x, t) \frac{\partial}{\partial x} \qquad (i = 1, 2, \cdots, r),$$

with $\tau_i \neq \tau_j$ for $i \neq j$. The functions $\tau_i(x, t)$ are assumed to be sufficiently differentiable. Then, the following lemmas are easily verified:

Lemma a): $D_i(\alpha D_j) = \alpha D_i D_j + \beta D_j$, $\beta = D_i(\alpha)$.

Lemma b): $D_i D_j - D_j D_i = a(x, t) \dfrac{\partial}{\partial x} = \dfrac{a}{\tau_i - \tau_j} (D_i - D_j)$,

where the coefficient a is given by

$$a = -\frac{\partial \tau_i}{\partial t} + \frac{\partial \tau_j}{\partial t} + \tau_i \frac{\partial \tau_j}{\partial x} - \tau_j \frac{\partial \tau_i}{\partial x} = -D_j \tau_i + D_i \tau_j.$$

(*If τ_i, τ_j are constant, a vanishes and the operators D_i, D_j commute.*)

Lemma b′): *Permutation of the factors D_i in a product of k distinct directional derivatives leaves the product unchanged except for an additional linear differential operator of order less than k.* (*In case of constant coefficients this additional operator is zero.*)

By induction now the following lemma can be easily established:

Lemma c): *Any linear differential operator N_r of order $r < k$ can be represented as a sum of the form*

$$(3) \qquad N_r = \sum_{i \leq r+1} a_i U^{r,i} + N_{r-1},$$

where $U^{r,i}$ is the product of r of the $r + 1$ operators $D_1 \cdots D_{r+1}$; the omitted operator is D_i, e.g., $U^{r,r+1} = D_r \cdots D_1$ or

$$U^{r,1} = D_{r+1} D_r \cdots D_2.$$

The additional term N_{r-1} is an operator of order not exceeding $r - 1$.

By applying Lemma c) to N_{r-1} and proceeding we obtain

Lemma c′): *Any operator of order r can be written in the form*

$$(4) \qquad N_r = \sum a_i^s U^{s,i} \qquad (i \leq s + 1, s \leq r).$$

The proof of Lemma c) proceeds simply by induction: We have $\partial/\partial x = [1/(\tau_2 - \tau_1)](D_2 - D_1)$ and

$$\partial/\partial t = [1/(\tau_2 - \tau_1)](\tau_1 D_2 - \tau_2 D_1).$$

This establishes c) for $s = r = 1$. Now assuming c) to hold for $r = s \leq k - 1$ we show that $\partial(N_r)/\partial x$ and $\partial(N_r)/\partial t$ have the form stipulated by (4). Consider the term $U^{s,i}$. Both $\partial/\partial x$ and $\partial/\partial t$ have the form $pD_i + qD_{s+2}$. By Lemmas b) and b') we find $pD_i U^{s,i} = pU^{s+1,s+2} + N_s$ and $qD_{s+2} U^{s,i} = qU^{s+1,i}$. If we apply this reasoning to all the terms in N_s, the proof of Lemma c) is completed by induction.

The process terminates for $s = k - 1$, if altogether k derivatives D_i are considered. In fact, if k distinct derivatives D_i are given, it is in general not possible to express a linear differential operator of order $r = k$ in the form (3). We see immediately, e.g., from article 1: If the derivatives D_i are the k characteristic derivatives of an hyperbolic operator $L[u]$, then the decomposition of the characteristic polynomial P^k into linear factors yields the basic decomposition

$$(5) \qquad L[u] = Mu + N_{k-1} u,$$

where

$$(6) \qquad Mu = D_k D_{k-1} \cdots D_1 u,$$

and N_{k-1} is an operator of order not exceeding $k - 1$. Accounting for terms in N_{k-1} not containing derivatives of u, we can therefore write, by Lemma c'), the differential equation (1) in the normal form

$$(7) \quad L[u] = Mu + N_{k-1} u = Mu + \sum_{\substack{s<k \\ i \leq s+1}} a_s^i U^{s,i} u + au = -f(x, t).$$

3. Solution of Cauchy's Problem. Cauchy's problem for (1) with the initial conditions $u = 0$, $U^{s,i}u = 0$ for $t = 0$ can now be immediately solved by reduction to the diagonalized normal form.

Slightly modifying the notation we write $u = U^0$ and replace $U^{s,i}u$ simply by the symbol $U^{s,i}$. All the quantities U, $U^{s,i}$ are considered as unknowns forming a vector U with $k(k + 1)/2$ components.

Denoting by $l(U)$ a linear expression in the components of U we obviously have, by article 2 and by the equation (4),

$$(8) \qquad D_i U^{s,i} = l_{s,i}(U) \qquad\qquad (s < k - 1, i < s + 1),$$

and

$$(9) \qquad D_i\, U^{k-1,i} = Mu + l_{k-1,i}(U) = l^*_{k-1,i}(U) - f,$$

where $l^*_{k-1,i}(U)$ again is a linear expression in the components of U.
The system (8), (9) is precisely a characteristic diagonalized system as solved in §6.

Thus Cauchy's problem is solved by reduction to the characteristic normal form.

The preceding solution can be at once extended to the case of multiple characteristics under a natural condition.

To motivate this condition we observe that the form (5) for $L[u]$ remains valid even if some of the characteristic differentiations coincide. What may not be true now is the representation (7) which might fail for the term N_{k-1}. For example, for $k = 4$ if $D_1 = D_2$, $D_3 = D_4$ the operator $L[u] = D_4\, D_3\, D_2\, D_1 + D_4\, D_3\, D_3$ has the form (5) but not (7). However we conceive of multiple characteristics originating by a continuous variation of parameters which makes some originally distinct derivatives D_i coincide. The expression N_{k-1} also takes on the form given in (7) in which merely the respective factors D_i coincide. We now turn this plausibility observation into a "condition A": This condition simply stipulates that $L[u]$ can be put into the form (7), in which no term contains a power of a derivative D_i higher than contained in the principal term M. Then the reduction of (1) to the normal diagonalized form remains literally the same. Hence: Under "condition A", Cauchy's problem remains uniquely solvable and "well posed" also for multiple characteristics.[1]

4. Other Variants for the Solution.   A Theorem by P. Ungar.   In the preceding article, we have avoided extraneous characteristics, but have introduced a great number of unknowns, which implies that the characteristics appear with a high multiplicity.

a) In reducing a single equation to a diagonal system, we can avoid the introduction of extraneous characteristics as well as of super-

---

[1] Concerning the preceding articles and the next one, reference is made to the work of E. E. Levi [3]. The results were found independently by Anneli Lax [1]. In this paper it is shown for constant coefficients that for multiple characteristics "condition A" is not only sufficient but also necessary to make Cauchy's problem a properly posed one. For equations in n variables with constant coefficients Lars Gårding [2] has given necessary and sufficient criteria referring to the behavior of the roots of the polynomial associated with the operator.

fluous equations, as P. Ungar has shown, by using the following remarkable theorem.[1]

If L is a kth order operator satisfying "condition A", then there exists a reduction of Cauchy's problem for L[u] = f to one for a diagonal system of precisely k first order equations.

Write (5) with slightly modified order of the factors:

$$L = D_1 D_2 \cdots D_k + N_{k-1},$$

where equal D_j are consecutively numbered. There exists a chain of operators

$$L_0 = 1,$$

$$L_1 = D_1 L_0 + R_0,$$

$$\cdots\cdots\cdots\cdots\cdots\cdots,$$

$$\cdots\cdots\cdots\cdots\cdots\cdots,$$

$$L_{k-1} = D_{k-1} L_{k-2} + R_{k-2},$$

$$L = L_k = D_k L_{k-1} + R_{k-1},$$

where $R_\nu = \sum_{i=0}^{\nu} a_i^\nu L_i$. Introducing the new variables $u_0 = u$, $u_i = L_i u$ $(i = 1, 2, \cdots, k - 1)$, we obtain a diagonal system of equations

$$
\begin{aligned}
(10) \quad & u_\nu = D_\nu u_{\nu-1} + \sum_{i=0}^{\nu-1} a_i^{\nu-1} u_i \qquad (\nu = 1, 2, \cdots, k - 1), \\
& f = D_k u_{k-1} + \sum_{i=0}^{k-1} a_i^{k-1} u_i.
\end{aligned}
$$

As before, the prescribed Cauchy data for u induce corresponding initial values for the functions u_i. In particular, these latter are zero if the Cauchy data for u are to vanish.

b) Generalizing directly the solution of $u_{xy} = -au_x - bu_y + f$ given in §6, 1 we obtain another variant of the solution. We briefly sketch it as follows:

Write the differential equation $L[u] = f$ again in the form

$$L[u] = Mu + N_{k-1} u = -f$$

---

[1] See P. Ungar [1].

or

$$Mu = -Nu - f.$$

The equation

(11) $$Mu = -Nv - f$$

defines, together with the initial conditions, a transformation

$$u = Tv.$$

We have to show that the iteration

$$v^n = Tv^{n-1}$$

converges; in particular we show that

$$\| Tv^n - Tv^{n-1} \| \leq \tfrac{1}{2} \| v^n - v^{n-1} \| \, ,$$

where $\| w \|$ is an appropriate norm. It is convenient to use

$$\| w \| = \max_{0 \leq t \leq h} | Mw |$$

because in a sufficiently narrow strip $0 \leq t \leq h$, the estimate

(12) $$\max_{0 \leq t \leq h} | Nw | \leq \tfrac{1}{2} \max_{0 \leq t \leq h} | Mw |$$

for a function w with vanishing Cauchy data is easily proved by repeated integrations along characteristics and by repeated applications of Lemmas a) and b) of article 2.

Thus T can be shown to be a contracting transformation, and in a sufficiently narrow region adjacent to the initial line the iterations v^j converge uniformly to a function u for which

$$u = Tu,$$

i.e.,

$$Mu + Nu = -f.$$

It is not difficult to show that u not only possesses all relevant directional derivatives, but satisfies the initial conditions and thus solves our initial value problem.

5. Remarks. As in the case of hyperbolic systems, the preceding solutions of Cauchy's problem are not only unique but also depend continuously on the data. Thus Cauchy's problem is "properly

posed" in the sense of Ch. III, §6. This is inherent in the construction of our solutions.

Violation of "condition A" may vitiate the continuous dependence
of the solution of Cauchy's problem on the initial conditions. This
is shown by the example:

$$L[u] = u_{tt} + u_x = 0, \qquad u(x, 0) = 0, \qquad u_t(x, 0) = \frac{1}{n} e^{-n^2 x},$$

whose solution is

$$u = \frac{1}{n^2} \frac{e^{nt} - e^{-nt}}{2} e^{-n^2 x}.$$

Obviously, if $n \to \infty$, we have for $t = 0$ uniformly $u = 0$, $u_t \to 0$
while $u(x, t)$ diverges in every region $t > 0$. The initial value problem for vanishing initial values of u and u_t, on the other hand, has
the identically vanishing solution. Thus the continuity condition
is violated and Cauchy's problem is not properly posed for this
operator L.

§9. Discontinuities of Solutions. Shocks

Wave propagation phenomena are represented by solutions of
hyperbolic equations with prescribed initial and boundary data. If
the given data are discontinuous (as is the case, e.g., for wave motion initiated by an impulse), the solution also is discontinuous.

Our aim is a precise definition of "discontinuous solution". For
example: The function $u = f(x + t) + g(x - t)$ is a genuine solution of the wave equation $u_{tt} - u_{xx} = 0$, provided f and g are twice
differentiable. If f and g are no longer differentiable, u ought to be
regarded as a "solution in the generalized sense". We shall give
now an intrinsic characterization of such concepts.

1. Generalized Solutions. Weak Solutions.[1] As in §3, let $L[u] = 0$
be the linear system

$$L[u] \equiv Au_x + Bu_t + Cu = 0$$

for the unknown function vector u^1, u^2, $\cdots$, u^k. We define the
operator L^* adjoint to L by the condition that $\zeta L[u] - uL^*[\zeta]$ is a
divergence expression; i.e.,

---

[1] For a somewhat different and more thorough approach see Ch. VI, §4,
and Appendix.

$$L^*[\zeta] = -(A\zeta)_x - (B\zeta)_t + C\zeta$$

so that

(1) $$\zeta L[u] - uL^*[\zeta] = (\zeta Au)_x + (\zeta Bu)_t.$$

In the domain G in which u is considered we now introduce "test functions" ζ which vanish identically outside a subdomain R of G.[1] Integrating (1) over R, we obtain by Gauss's theorem

(2) $$\iint_R (\zeta L[u] - uL^*[\zeta])\ dx\ dt = 0.$$

If $L[u] = 0$, then

(3) $$\iint_R uL^*[\zeta]\ dx\ dt = 0.$$

Conversely, if (3) holds for a function u with continuous derivatives and for all admissible test functions ζ in all subdomains R of G, then (2) yields

(4) $$\iint_R \zeta L[u]\ dx\ dt = 0.$$

Hence, by the "fundamental lemma of the variational calculus" (cf. Vol. I, Ch. IV, §3, 1), we conclude that $L[u] = 0$.

We now give the generalization of the concept of solution: We allow the vector function u and its derivatives to be piecewise continuous, i.e., to possess jump discontinuities along piecewise smooth curves C. Such a function u is called a *weak solution* of the equation $L[u] = 0$ in G if

$$\iint_R uL^*[\zeta]\ dx\ dt = 0$$

for all admissible test functions ζ and all subdomains R of G.[2]

Assuming the discontinuous solution u of $L[u] = 0$ to be regular in all domains not containing C, we shall show that the lines C of discontinuity are necessarily characteristics. Suppose C divides a domain R into two parts R_1 and R_2. Integrate (2) by parts sepa-

---

[1] Such functions sometimes are called functions of compact support; the domain in which a function does not vanish is its "support".

[2] Of course, the definition could be generalized by demanding merely the integrability of u, but this would not be particularly useful for our purposes.

rately in R_1 and R_2. Since in both these domains $L[u] = 0$, and since $\zeta = 0$ on the boundary of R, we obtain (denoting by $[u]$ the jump of u across C)

$$\int_C \zeta(A[u]\phi_x + B[u]\phi_t) \, ds.$$

Here ϕ_x, ϕ_t denote the direction cosines of the normal to C, and ds is the arc length on C. But ζ is arbitrary on C, and hence, by the fundamental lemma of the calculus of variations, we conclude that

(5) $$(\phi_x A + \phi_t B)[u] = 0.$$

Under the assumption that the jump $[u]$ is different from zero, this linear homogeneous equation implies that the matrix $\phi_x A + \phi_t B$ is singular, i.e. (see §2), that C is a characteristic curve. As an example the reader will easily verify that $u = f(x + t) + g(x - t)$ is a weak solution of the wave equation even if f and g are discontinuous.

A similar definition of weak solutions can be given for equations of higher order.

Also generalized solutions u can be introduced which become infinite along characteristics. Take, for example, a weak solution u defined as a derivative $u = v_t$ of a function v which may suffer jump discontinuities, and stipulate

$$\iint_R v \frac{\partial}{\partial t} L^*[\zeta] \, dx \, dt = 0$$

for all smooth test functions ζ. As before, we find that discontinuities of v can occur only across characteristics, and that $[v]$ satisfies the same relation as $[u]$ in the preceding case, i.e., that

$$(A\phi_x + B\phi_t)[v] = 0.$$

2. Discontinuities for Quasi-Linear Systems Expressing Conservation Laws. Shocks. A similar theory of discontinuous solutions, highly important in the dynamics of compressible fluids, can be developed for quasi-linear systems, provided they have the form of divergence equations or "conservation laws"[1]

(6) $$L[u] = p_t(x, t, u) + q_x(x, t, u) + n(x, t, u) = 0,$$

[1] See remark (v) at the end of this article, which points out that such conservation laws are not uniquely determined by the differential equation alone, but need an additional physical motivation.

where p, q, n are twice continuously differentiable function vectors of their arguments x, t, in G and of u in some given domain.[1] Among others, differential equations which arise in connection with Hamilton's principle appear in this form.

To define weak solutions for systems of conservation laws, we again consider arbitrary smooth test functions ζ in $R \subset G$ vanishing outside R. We multiply (6) by ζ, integrate over R, and obtain

$$\iint_R \zeta L[u] \, dx \, dt = 0.$$

For smooth u, Gauss's theorem yields

$$(7) \qquad \iint_R (p\zeta_t + q\zeta_x - n\zeta) \, dx \, dt = 0.$$

Conversely, if (7) holds for a function u with continuous first derivatives and for all admissible test functions ζ, then we obtain by another application of Gauss's theorem

$$(8) \qquad \iint_R \zeta L[u] \, dx \, dt = 0.$$

As before, we can conclude that $L[u] = 0$. We call a function u a *weak solution* if it is piecewise continuous with piecewise continuous first derivatives and if (7) is satisfied for all admissible test functions ζ in all subdomains R of G.

Again, we shall derive relations for the jumps of a discontinuous weak solution u. Let C be a curve of discontinuity which divides R into two domains. Applying Gauss's theorem separately to each domain and taking into consideration that $\zeta = 0$ outside R and $L[u] = 0$ outside C, we obtain

$$\int_C \zeta(\phi_t[p] + \phi_x[q]) \, ds = 0,$$

and hence we have also the jump conditions

$$(9) \qquad \phi_t[p] + \phi_x[q] = 0,$$

across C. Here ϕ_t, ϕ_x again denote the direction cosines of the normal to C, and $[p]$, $[q]$ the jumps of p and q across C.

There are a number of incisive differences between this and the linear case:

---

[1] Linear systems can always be written in such a form.

(i) The relations for the discontinuities and for the slope of the curve C are not separated but interlocked. Discontinuity curves C, or "shocks", are *not* characteristics.

(ii) In the linear case, discontinuous solutions can be obtained as limits of genuine solutions; in fact, one may define weak solutions in this way. For nonlinear conservation laws, however, discontinuous weak solutions cannot be obtained as limits of smooth solutions.

(iii) The jump conditions (5) for the linear case suffice to determine a unique discontinuous solution to a given discontinuous initial (or mixed initial and boundary) value problem. The jump conditions (9) for the nonlinear case have to be supplemented, e.g., by a so-called "entropy" condition, in order to make them yield a unique solution of the corresponding problem.

(iv) Solutions of linear equations are discontinuous only if the prescribed data are discontinuous. In contrast, solutions of nonlinear equations with smooth (even analytic) initial data can develop discontinuities after a finite time interval has elapsed.

(v) Different systems of conservation laws may be equivalent as systems of differential equations, i.e., smooth solutions of one system are also smooth solutions of the other. But a discontinuous solution of one system need not (and in general will not) be a solution of the other. A striking example is the system consisting of the equations of conservation of mass, momentum, and energy on one hand, and the system of the equations of conservation of mass, momentum, and entropy on the other hand.

Discontinuities have been treated by Riemann, Hugoniot, Rankine, and others in connection with compressible flow.[1] A general theory of discontinuities in solutions of systems of conservation laws was developed by P. D. Lax.[2]

Appendix 1 to Chapter V

Application of Characteristics as Coordinates

§1. *Additional Remarks on General Nonlinear Equations of Second Order*

Cauchy's problem for nonlinear differential equations was reduced to one for quasi-linear systems of first order by a general pro-

---

[1] See R. Courant and K. O. Friedrichs [1], O. Oleinik [4], and I. M. Gelfand [1].

[2] P. D. Lax [2].

cedure. For the case of equations of second order it is of interest to treat the problem somewhat more directly[1] by introducing a characteristic coordinate system α, β, and deriving a system of equations of first order for the eight quantities x, y, u, p, q, r, s, t in terms of these characteristic parameters (see also §2, 3).

1. The Quasi-Linear Differential Equation. We consider first the quasi-linear equation

$$(1) \qquad L[u] \equiv au_{xx} + bu_{xy} + cu_{yy} + d = 0,$$

where a, b, c, d are given functions of x, y, u, $p = u_x$, $q = u_y$ with continuous second derivatives in the underlying domain of x,y,u,p,q-space.

A family of curves $\phi(x, y) = $ const. is characteristic (see §1, 1 and 2) if

$$(2) \qquad a\phi_x^2 + b\phi_x\phi_y + c\phi_y^2 = 0;$$

we require that

$$b^2 - 4ac > 0,$$

i.e., that (1) be hyperbolic in the part of x,y,u,p,q-space considered, and without restriction of generality, we may assume $a \neq 0$, $b \neq 0$, which means that the lines $x = $ const., $y = $ const. are not characteristic.

As in §2, 3, the two families of characteristic curves for a solution $u(x, y)$ are given by $\phi = \alpha(x, y) = $ const. and $\phi = \beta(x, y) = $ const. and differentiation along a characteristic curve C with respect to its parameter will be indicated by a dot. Using the abbreviations

$$u_{xx} = r, \qquad u_{xy} = s, \qquad u_{yy} = t,$$

we now write the equation (1) and the strip conditions along a characteristic C:

$$
\begin{aligned}
& ar + bs + ct + d = 0, \\
(3) \qquad & \dot{x}r + \dot{y}s \qquad - \dot{p} = 0, \\
& \dot{x}s + \dot{y}t - \dot{q} = 0.
\end{aligned}
$$

---

[1] For details see Hans Lewy [7] and the presentation in Hadamard's book [2]. Hans Lewy's solution of Cauchy's problem for a nonlinear equation of second order was a decisive step in the theory discussed in the present chapter. The case of second order is singled out by the fact that the two characteristic parameters can be introduced as independent variables.

On the one hand, C being characteristic, r, s, t cannot be determined uniquely from (3); on the other hand, the equations (3) must be compatible if a solution $u(x, y)$ is to exist. Therefore, the matrix

$$\begin{pmatrix} a & b & c & d \\ \dot{x} & \dot{y} & 0 & -\dot{p} \\ 0 & \dot{x} & \dot{y} & -\dot{q} \end{pmatrix}$$

has the rank 2, at most. This yields again the characteristic condition

$$(2')\qquad \begin{vmatrix} a & b & c \\ \dot{x} & \dot{y} & 0 \\ 0 & \dot{x} & \dot{y} \end{vmatrix} \equiv a\dot{y}^2 - b\dot{x}\dot{y} + c\dot{x}^2 = 0,$$

and

$$(4)\qquad \begin{vmatrix} a & c & d \\ \dot{x} & 0 & -\dot{p} \\ 0 & \dot{y} & -\dot{q} \end{vmatrix} \equiv d\dot{x}\dot{y} + a\dot{y}\dot{p} + c\dot{x}\dot{q} = 0.$$

Since $b^2 - 4ac > 0$, and $a \neq 0$, $c \neq 0$, we may write $(2')$ as

$$a\left(\frac{\dot{y}}{\dot{x}}\right)^2 - b\frac{\dot{y}}{\dot{x}} + c \equiv a\left(\tau^1 - \frac{\dot{y}}{\dot{x}}\right)\left(\tau^2 - \frac{\dot{y}}{\dot{x}}\right) = 0,$$

where $\tau^1(x, y, u, p, q)$, $\tau^2(x, y, u, p, q)$ are two distinct real functions and define the two families of characteristics by the differential equations

$$\tau^1 x_\alpha - y_\alpha = 0,$$

$$\tau^2 x_\beta - y_\beta = 0.$$

In order that α, β may be used as coordinates, the Jacobian

$$\frac{\partial(x, y)}{\partial(\alpha, \beta)} = x_\alpha y_\beta - x_\beta y_\alpha = (\tau^2 - \tau^1) x_\alpha x_\beta$$

must be different from zero, i.e., $x_\alpha^2 + x_\beta^2 \neq 0$, which may be assumed without losing generality.

We can now write a system of six differential equations for the five quantities x, y, u, p, q:

$$\text{(a)} \quad \tau^1 x_\alpha - y_\alpha \qquad\qquad = 0,$$

$$\text{(b)} \quad \tau^2 x_\beta - y_\beta \qquad\qquad = 0,$$

$$\text{(c)} \quad d\tau^1 x_\alpha \qquad + a\tau^1 p_\alpha + c q_\alpha = 0,$$

$$\text{(5)} \qquad \text{(d)} \quad d\tau^2 x_\beta \qquad + a\tau^2 p_\beta + c q_\beta = 0,$$

$$\text{(e)} \quad -p x_\alpha - q y_\alpha + u_\alpha \qquad = 0,$$

$$\text{(f)} \quad -p x_\beta - q y_\beta + u_\beta \qquad = 0.$$

The first two originate from $(2')$, the next two from (4), and the last two are the strip relations for u. One of these equations is superfluous; it is a consequence of the other five.[1] Thus we have derived a system of five first order equations for five quantities.

These equations form a hyperbolic system of first order in the characteristic parameters α, β of the type discussed in §7. If the initial values u, u_y are prescribed for the original equation on some nowhere characteristic curve, then initial values for x, y, u, p, q are immediately induced on that curve (now considered in the α, β coordinates).

Existence and uniqueness of the solution of (5) is assured by §7. Moreover, the solution x, y, u, p, q of the initial value problem for (5) with Cauchy data induced by the original Cauchy data for (1) yields a solution $u(x, y)$ of the original Cauchy problem for equation (1); this is shown as follows:

[1] In fact, to obtain equation (5f) as a consequence of the others, we differentiate the expression

$$B = u_\beta - p x_\beta - q y_\beta$$

with respect to α, and equation (5e) with respect to β. By subtraction we obtain

$$B_\alpha = p_\beta x_\alpha - p_\alpha x_\beta + q_\beta y_\alpha - q_\alpha y_\beta .$$

We express y_α, y_β in terms of x_α, x_β by means of equations (5a, b); then we combine the last two terms using equations (5c, d). This yields

$$B_\alpha = \left(\frac{a}{c} \tau^1 \tau^2 - 1 \right)(p_\alpha x_\beta - p_\beta x_\alpha).$$

But from the characteristic condition $(2')$, we have $\tau^1 \tau^2 = c/a$; hence

$$B_\alpha = 0.$$

This implies that $B = \text{const.}$ along each curve $\beta = \text{const.}$ Since the data are assumed to satisfy our system initially, it follows that $B \equiv 0$ throughout.

First, since the Jacobian $x_\alpha y_\beta - x_\beta y_\alpha$ does not vanish, x, y may be introduced as independent variables and u, p, q are continuously differentiable functions of x, y. To prove $u_x = p$, $u_y = q$, we consider the equivalent relations

$$A \equiv u_\alpha - px_\alpha - qy_\alpha = 0, \qquad B \equiv u_\beta - px_\beta - qy_\beta = 0.$$

The relation $A = 0$ is satisfied because of (5e); the relation $B = 0$, as seen above, is a consequence of (5a through e) and the fact that $B = 0$ initially. Finally, we have to verify that the quantities u, p, q, $u_{xx} = r$, $u_{xy} = s$, $u_{yy} = t$, obtained from (5) satisfy the differential equation (1). Indeed, (5) yields, because of

$$p_\alpha = rx_\alpha + sy_\alpha, \qquad q_\alpha = sx_\alpha + ty_\alpha,$$

the relation

$$0 = d\tau^1 x_\alpha + a\tau^1(rx_\alpha + sy_\alpha) + c(sx_\alpha + ty_\alpha)$$

$$= \tau^1 x_\alpha \left[d + ar + ct + s\left(a\tau^1 + \frac{c}{\tau^1} \right) \right].$$

But since $\tau^1 x_\alpha \neq 0$ and, because of the quadratic equation $a\tau^1 + (c/\tau^1) = b$, we have

$$0 = ar + bs + ct + d,$$

as we set out to prove.

Thus Cauchy's problem for the quasi-linear equation is solved and, implicitly, also the uniqueness of the solution is proved.

The assumption under which the solution has been established, may be summarized as follows: The initial strip, nowhere characteristic and everywhere smooth, carries initial values u, p, q which have continuous derivatives; the coefficients a, b, c, d have continuous derivatives up to second order. The uniqueness and existence of the solution is assured and the domain of dependence for a point P is bounded by the two characteristics through P and the arc of the initial curve between them.[1]

It also should be stated that the characteristic initial value problem can be formulated and solved in the same way.

2. The General Nonlinear Equation. The method of the previ-

---

[1] The assumption may be made somewhat weaker as shown by A. Douglis [1] (see also Hartman and Wintner [1]).

ous article applies almost literally to quasi-linear systems with the same principal part, i.e., systems of the form

$$(1') \qquad au^j_{xx} + bu^j_{xy} + cu^j_{yy} + d^j = 0 \qquad (j = 1, 2, \cdots, m),$$

where a, b, c, d^j are given functions of x, y, u^j, p^j, q^j.

The general nonlinear hyperbolic equation

$$(6) \qquad F(x, y, u, p, q, r, s, t) = 0$$

can then again be reduced to an equivalent system of quasi-linear equations with the same principal part by differentiating equation (6) with respect to x and y.

In general one obtains an equivalent first order canonical system in characteristic parameters—not five equations as before, but eight differential equations for the eight quantities x, y, u, p, q, r, s, t as functions of α, β.[1] (A special situation for the Monge-Ampère equation will be examined in the next section.)

The same procedure as in article 1 then produces twelve equations for the eight quantities x, y, u, p, q, r, s, t and it can be shown: a) Four of these twelve equations are consequences of the others and only eight of them are independent. b) If an initial value problem is posed for equation (6), corresponding initial values are induced for the system of these eight first order equations. The solution x, y, u, p, q, r, s, t of the system is again constructed by iterations and yields the solution

$$u(x, y) = u(\alpha(x, y), \beta(x, y))$$

for equation (6). c) The solution u of the system has continuous derivatives up to the third order if the initial values for x, y, u, p, q, r, s, t are continuously differentiable and if F has continuous derivatives with respect to x, y, u, p, q, r, s, t up to the third order.

§2. The Exceptional Character of the Monge-Ampère Equation
The Monge-Ampère equation

$$(1) \qquad F = Ar + Bs + Ct + D(rt - s^2) + E = 0,$$

of great interest in various fields such as differential geometry, is genuinely nonlinear—it is quadratic in r, s, t. But in contrast to the general nonlinear equation, which is reduced to a system of eight

[1] For details see H. Lewy [7] and J. Hadamard [2].

differential equations, the initial value problem for (1) can be reduced to one for a system of only five quasi-linear first order equations, just as in the case of a second order quasi-linear equation. This fact has interesting consequences; for instance, the class of initial data admissible for the Monge-Ampère equation is wider (i.e., is subject to less stringent smoothness requirements) than that for the general nonlinear equation.

Consider the equation (1), where A, B, C, D, E are smooth functions of x, y, u, p, q. The characteristic condition (9) of §1 becomes

$$(2) \qquad (A + Dt)\dot{y}^2 - (B - 2Ds)\dot{y}\dot{x} + (C + Dr)\dot{x}^2 = 0.$$

We may assume $A + Dt \neq 0$, $C + Dr \neq 0$. Equation (2) has real distinct roots τ^1, τ^2 if the equation (1) is hyperbolic, i.e., if the discriminant

$$(3) \qquad \Delta^2 = F_s^2 - 4F_r F_t = B^2 - 4AC + 4ED > 0.$$

Significantly, the second derivatives r, s, t do not enter the expression for the discriminant.

Moreover, the following identity is a consequence of the equation (1) and of the form of the discriminant Δ^2:

$$0 = D\{Ar + Bs + Ct + D(rt - s^2) + E\}$$
$$= (A + Dt)(C + Dr) - \tfrac{1}{4}(B - 2Ds)^2 + \tfrac{1}{4}\Delta^2,$$

or

$$(4) \qquad \frac{A + Dt}{\tfrac{1}{2}(B - 2Ds - \Delta)} = \frac{\tfrac{1}{2}(B - 2Ds + \Delta)}{C + Dr}.$$

We solve (2) for the ratio $\dot{y}/\dot{x}$ and obtain the roots

$$\tau^1 = \frac{B - 2Ds + \Delta}{2(A + Dt)}, \qquad \tau^2 = \frac{B - 2Ds - \Delta}{2(A + Dt)};$$

they yield, in terms of the characteristic independent variables, the equations

$$(5) \qquad \begin{aligned} (A + Dt)y_\alpha - \tfrac{1}{2}(B - 2Ds + \Delta)x_\alpha &= 0, \\ (A + Dt)y_\beta - \tfrac{1}{2}(B - 2Ds - \Delta)x_\beta &= 0 \end{aligned}$$

or

$$D(ty_\alpha + sx_\alpha) + Ay_\alpha - \tfrac{1}{2}(B + \Delta)x_\alpha = 0,$$

(5')

$$D(ty_\beta + sx_\beta) + Ay_\beta - \tfrac{1}{2}(B - \Delta)x_\beta = 0.$$

By the strip relation

$$\dot{q} = s\dot{x} + t\dot{y},$$

we have

$$\tfrac{1}{2}(B + \Delta)x_\alpha - Ay_\alpha - Dq_\alpha = 0,$$

(5'')

$$\tfrac{1}{2}(B - \Delta)x_\beta - Ay_\beta - Dq_\beta = 0.$$

Using the identity (4), we obtain from (5) the two additional equations

$$(C + Dr)x_\alpha - \tfrac{1}{2}(B - 2Ds - \Delta)y_\alpha = 0,$$

(6)

$$(C + Dr)x_\beta - \tfrac{1}{2}(B - 2Ds + \Delta)y_\beta = 0$$

or

$$D(rx_\alpha + sy_\alpha) + Cx_\alpha - \tfrac{1}{2}(B - \Delta)y_\alpha = 0,$$

(6')

$$D(rx_\beta + sy_\beta) + Cx_\beta - \tfrac{1}{2}(B + \Delta)y_\beta = 0,$$

and, by means of the strip condition

$$\dot{p} = r\dot{x} + s\dot{y},$$

we have

$$\tfrac{1}{2}(B - \Delta)y_\sigma - Cx_\alpha - Dp_\alpha = 0,$$

(6'')

$$\tfrac{1}{2}(B + \Delta)y_\beta - Cx_\beta - Dp_\beta = 0.$$

We observe: The system consisting of the five equations (5'') and (6'') and the strip condition

(7) $$u_\alpha - px_\alpha - qy_\alpha = 0$$

is equivalent to the original Monge-Ampère equation (1) in the following sense: If x, y, u, p, q is a solution of this system with initial values induced by the given initial values for the solution of (1), and if the Jacobian $\partial(x, y)/\partial(\alpha, \beta)$ does not vanish, then

$$u(\alpha(x, y), \beta(x, y)) = u(x, y)$$

is the solution of the initial value problem for (1).

To ascertain this fact, we first calculate the three quantities r, s, t from the four strip conditions

(8)
$$x_\alpha r + y_\alpha s \qquad - p_\alpha = 0,$$
$$x_\beta r + y_\beta s \qquad - p_\beta = 0,$$
$$x_\alpha s + y_\alpha t - q_\alpha = 0,$$
$$x_\beta s + y_\beta t - q_\beta = 0.$$

We eliminate r from the first two and t from the next two equations, obtaining

$$s = \frac{x_\alpha p_\beta - x_\beta p_\alpha}{\dfrac{\partial(x, y)}{\partial(\alpha, \beta)}} \qquad \text{and} \qquad s = \frac{y_\beta q_\alpha - y_\alpha q_\beta}{\dfrac{\partial(x, y)}{\partial(\alpha, \beta)}}.$$

The relations of the type $q_\alpha = q_x x_\alpha + q_y y_\alpha$ show that the equations (8) are compatible if and only if $q_x = p_y$. The functions r, s, t obtained from (8) clearly satisfy equations (5') and (6'), and hence also (5) and (6). From the first equations in (5) and (6), we obtain the identity

$$D\{Ar + Bs + Ct - D(rt - s^2) + E\} = 0,$$

or, since $D \neq 0$ (otherwise (1) would be quasi-linear),

$$Ar + Bs + Ct + D(rt - s^2) + E = 0.$$

Thus, the solution of the system leads to a solution of (1), and it is immediately verified that this solution has the correct initial values.

It suffices to require that the data for the Monge-Ampère equation have the smoothness properties: $u(x, 0)$ is twice, $u_y(x, 0)$ is once differentiable. In other words, the requirements are only as stringent as those in the quasi-linear case, and not so stringent as in the general nonlinear case.

Another remark also illuminates the exceptional character of the Monge-Ampère equation. It concerns the initial value problem: For a differential equation which is quadratic in the second derivatives

(9) $Ar^2 + Bs^2 + Ct^2 + Drs + Ert + Fst + Gr + Hs + It + K = 0,$

in which A, $\cdots$, K are functions of x, y, u, p, q, we consider the initial value problem along a curve $x(\lambda)$, $y(\lambda)$ by prescribing $u(\lambda)$,

$p(\lambda)$, $q(\lambda)$ in such a way that the strip relation $\dot{u} = p\dot{x} + q\dot{y}$ is satisfied. We must then extend this first order strip to an integral strip of second order by calculating the initial values of r, s, t from (9) and the strip relations

$$r\dot{x} + s\dot{y} = \dot{p}, \qquad s\dot{x} + t\dot{y} = \dot{q}.$$

Because of the quadratic nature of (9) this extension is, in general, possible in two ways. However, we can show: Among all equations of the form (9), the Monge-Ampère equation alone permits, in only one way, the completion of an arbitrary initial strip of first order to an integral strip.

To prove this, we set $\dot{y}/\dot{x} = -\alpha$ in the above strip relations, obtaining

$$s = \alpha t + \cdots, \qquad r = \alpha^2 t + \cdots,$$

where the dots stand for quantities which are known along the first order strip. Inserting these expressions for r and s into (9), we obtain for the coefficient of t^2

$$A\alpha^4 + D\alpha^3 + (E + B)\alpha^2 + F\alpha + C.$$

The vanishing of this expression for all α is equivalent to the equations

$$A = D = F = C = 0, \qquad E + B = 0,$$

which proves our assertion.

This result for the initial value problem seems particularly remarkable since, for the boundary value problem in case of the elliptic Monge-Ampère equations in Ch. IV, §5, 3, it was shown that an ambiguity is possible.

§3. Transition from the Elliptic to the Hyperbolic Case Through Complex Domains

Everywhere in this book all the variables were assumed essentially to be real; complex quantities were on occasion introduced merely in a formal way. In the following two sections, however, we shall briefly report about a much deeper use of complex variables initiated by H. Lewy[1] and carried further by H. Lewy, by P. Garabedian[2] and others. Many of the considerations of Chapter V hold almost without

[1] See H. Lewy [6].
[2] See P. Garabedian and H. M. Lieberstein [1].

modification if the functions f and the coefficients $a_{\nu\mu}$ are complex-valued functions of the real independent variables x, y. We may split the solution $u = u_1 + iu_2$ into their real and imaginary parts; thus, instead of n equations with complex-valued coefficients, we obtain $2n$ real equations of the same type for the functions u_1 and u_2. The theory of integration, uniqueness theorems, and previous results on continuous and differentiable dependence of the solutions on parameters remain unchanged.

Moreover, if the left-hand side of a real differential equation $F(x, y, u, \cdots) = 0$ is an analytic function of all its arguments and if we know, in addition, that the solution $u(x, y)$ depends analytically on x, y, then we can continue the differential equation and its solution analytically into the complex domain by considering $x = x_1 + ix_2$ and $y = y_1 + iy_2$ as complex variables. When we do this the distinction between types disappears so that a transition from the elliptic to the hyperbolic type becomes possible in principle.

The simplest typical example is offered by the differential equation

$$(1) \qquad \Delta u \equiv u_{xx} + u_{yy} = f(x, y, u, u_x, u_y),$$

which is elliptic (in the real domain). We assume that the right-hand side is an analytic function of its five arguments. If the solution u depends analytically on x and y, we can consider u as a function of the complex variables $x = x_1 + ix_2$, $y = y_1 + iy_2$ or as a complex function of the four real variables x_1, x_2, y_1, y_2. Then, in the real domain, the differential equation has the form

$$(2) \qquad u_{x_1x_1} + u_{y_1y_1} = f(x, y, u, u_{x_1}, u_{y_1}).$$

But since in the complex plane we can differentiate with respect to iy_2 as well as with respect to y_1, the complex analytic function u, considered as a function of the four variables x_1, x_2, y_1, y_2, also satisfies the equation

$$(3) \qquad u_{x_1x_1} - u_{y_2y_2} = f(x, y, u, u_{x_1}, -iu_{y_2})$$

which exhibits a hyperbolic character. The justification of such a transformation is bound up with the so-far-assumed analytic nature of the solution u, i.e., with the fact that the derivative of the function in the complex domain is independent of the direction of the differentiation.

We can now reverse our reasoning, i.e., start with a real solution of the original equation and attempt to continue this solution into the complex domain in such a way that the continuation satisfies the hyperbolic equation (3), or corresponding systems, and then demonstrate the analyticity of the complex-valued function so obtained. This is the basic idea used first by Hans Lewy in his method to prove the analyticity of the solutions of elliptic differential equations.[1]

§4. The Analyticity of the Solutions in the Elliptic Case

1. Function-Theoretic Remark. A complex-valued function $w(x_1, x_2, y_1, y_2) = w_1 + iw_2$ with continuous partial derivatives of first order is called an analytic function of the two complex variables $x = x_1 + ix_2, y = y_1 + iy_2$ in a region B of four-dimensional x_1, x_2, y_1, y_2-space if the Cauchy-Riemann equations

$$(1) \qquad \nabla w \equiv w_{x_1} + iw_{x_2} = 0, \qquad \Lambda w \equiv w_{y_1} + iw_{y_2} = 0$$

are satisfied there. An equivalent definition is: the function w is analytic in a neighborhood of a point $x = 0, y = 0$ if there exists a positive number M such that w can be expanded in a power series

$$(2) \qquad w = \sum_{\nu,\mu=0}^{\infty} a_{\nu\mu} x^\nu y^\mu$$

for $|x| \leq M, |y| \leq M$;[2] it is called analytic in a region B if it is analytic in a neighborhood of every point of B.

[1] Hans Lewy [6].

[2] That this property follows, for example, from the Cauchy-Riemann definition is seen with the help of a repeated application of Cauchy's integral representation for a complex variable: Let the Cauchy-Riemann relations (1) hold in the region B, defined by $|x| < M, |y| < M$. For every pair of numbers ξ_1, ξ_2 in $|\xi| \leq M/2$, the circle $K_x : |x - \xi| = M/2$ is contained entirely in B; so are all points x such that $|x - \xi| \leq M/2$. Hence, if for the moment we consider y_1, y_2 as parameters, w is represented in K_x according to Cauchy's integral formula by

$$w(x_1, x_2; y_1, y_2) = \frac{1}{2\pi i} \int_{K_x} \frac{w(\xi_1, \xi_2; y_1, y_2)}{(\xi_1 + i\xi_2) - (x_1 + ix_2)} (d\xi_1 + i\,d\xi_2).$$

Similarly, the circle $K_y : |y - \eta| = M/2$ and its interior lie in B if η_1 and η_2 are restricted by the relation $|\eta| \leq M/2$. Hence w can be represented once again in the form

$$w(\xi_1, \xi_2; y_1, y_2) = \frac{1}{2\pi i} \int_{K_y} \frac{w(\xi_1, \xi_2; \eta_1, \eta_2)}{(\eta_1 + i\eta_2) - (y_1 + iy_2)} (d\eta_1 + i\,d\eta_2).$$

By substitution we obtain Cauchy's double integral representation

$$w(x_1, x_2; y_1, y_2) = \frac{1}{4\pi^2} \int_{K_x} \int_{K_y} w(\xi_1, \xi_2; \eta_1, \eta_2) \frac{(d\xi_1 + i\,d\xi_2)(d\eta_1 + i\,d\eta_2)}{(\xi_1 + i\xi_2 - x)(\eta_1 + i\eta_2 - y)}.$$

The fraction in the integrand is now expanded in a power series in terms of x and y, as in the case of a single variable, and the resulting expression integrated term by term. For w we then obtain the desired power series representation.

2. Analyticity of the Solutions of $\Delta u = f(x, y, u, p, q)$. We assume that in the differential equation

$$(3) \qquad \Delta u = f(x, y, u, p, q)$$

the function f is a (real) analytic function of its five arguments and that $u(x, y)$ is a given twice continuously differentiable solution of the differential equation in a (real) neighborhood of $x = 0$, $y = 0$. The assumed analyticity of f is supposed to hold for this neighborhood and for the domain of values of u, p, q determined by the solution u considered. We assert that the solution u in question is not only twice continuously differentiable, but is even analytic.

We shall carry out the proof by means of *continuation into the complex domain*, extending u continuously to a complex-valued twice continuously differentiable function of x_1, x_2, y_1, y_2 which satisfies conditions (1).[1] By introducing complex variables $x = x_1 + ix_2$, $y = y_1 + iy_2$, we attempt to construct a complex function $u(x_1, x_2, y_1, y_2)$—which will afterward be proved analytic in x and y—that reduces for $x_2 = y_2 = 0$ to the given function $u(x, y) = u(x_1, y_1)$.

The continuation is carried out step by step; for fixed x_1 we first extend the original function $u(x_1, y_1)$ to a complex-valued function $u(x_1, x_2, y_1)$, and then extend the latter to a complex function $u(x_1, x_2, y_1, y_2)$. We continue f as an analytic function of complex arguments; it is then automatically continuously differentiable with respect to these arguments.

Our first step consists in considering x_1 as a parameter and trying to determine the new function $u(x_1, x_2, y_1)$ by means of the differential equation

$$(4) \qquad u_{y_1 y_1} - u_{x_2 x_2} = f(x_1 + ix_2, y_1, u, -iu_{x_2}, u_{y_1}),$$

which arises formally from the original equation (3) when we replace x by $x_1 + ix_2$. Here x_1 is regarded as a fixed parameter, while y_1 and x_2 are the two real independent variables in the complex-valued

---

[1] For our differential equation the proof could be obtained just as simply, in principle, by the application of methods of potential theory. However, the method of Hans Lewy presented here is interesting in itself, and opens the way to the solution of further problems (cf. Hans Lewy [5], [3], and [4]). Similar ideas of extension into a more-dimensional space have proved successful in different, but related, subjects (see Lewy [2]).

differential equation. We consider an initial value problem for this equation with the initial curve $x_2 = 0$.

On this curve one initial condition has the form

$$(5) \qquad u(x_1, 0, y_1) = u(x_1, y_1),$$

where the right-hand side is the original real solution of (3).

For a second initial condition, we determine u_{x_2} by making the initial requirement

$$(6) \qquad \nabla u \equiv u_{x_1} + i u_{x_2} = 0 \qquad \text{for} \quad x_2 = 0,$$

which states that the Cauchy-Riemann condition is to be satisfied on the initial curve. Thus by our earlier theory, $u(x_1, y_1)$ can be extended uniquely to $u(x_1, x_2, y_1)$ in a suitable neighborhood of the initial curve. Since, furthermore, $u(x_1, y_1)$ depends in a continuously differentiable way on the parameter x_1 in a certain interval (see Ch. V, §5), $u(x_1, x_2, y_1)$ is defined and continuously differentiable with respect to x_1 in a solid rectangular neighborhood of the point $x_1 = 0$, $x_2 = 0$, $y_1 = 0$. Similarly, the derivative u_{x_2} is continuously differentiable with respect to x_1.

Differentiating the second initial condition (6) with respect to the parameter x_1, we obtain $\partial(\nabla u)/\partial x_1 = u_{x_1 x_1} + i u_{x_2 x_1} = 0$. For $x_2 = 0$, we subtract equation (4) from (3) and obtain

$$u_{x_1 x_1} + u_{x_2 x_2} = 0 \qquad \text{for} \quad x_2 = 0$$

or, using the above relation,

$$u_{x_2 x_2} - i u_{x_2 x_1} = 0$$

or

$$(7) \qquad \frac{\partial}{\partial x_2} \nabla u \equiv \frac{\partial}{\partial x_2} (u_{x_1} + i u_{x_2}) = 0.$$

To equation (4) we now apply the Cauchy operator ∇. Setting for brevity $\nabla u = \omega$, we have upon differentiating formally,

$$\omega_{y_1 y_1} - \omega_{x_2 x_2} = f_x \nabla x + f_u \nabla u - i f_p \nabla u_{x_2} + f_q \nabla u_{y_1}.$$

Since $\nabla x = 0$, we finally obtain

$$\omega_{y_1 y_1} - \omega_{x_2 x_2} = f_u \omega - i f_p \omega_{x_2} + f_q \omega_{y_1}.$$

The coefficients on the right are known complex-valued functions of y_1 and x_2. Hence our equation is a linear homogeneous hyperbolic differential equation for ω, for which, by our earlier results, the solution of the initial value problem is uniquely determined. But since, in virtue of (6) and (7), the initial values of ω and $\partial\omega/\partial x_2$ vanish, $\omega = 0$ identically in a three-dimensional domain Q about the origin.

We now have to carry out the second step in the continuation, namely to continue u into a four-dimensional x_1, x_2, y_1, y_2 -domain. To this end, we consider any two values x_2 and y_1 in Q, and undertake the continuation of u with respect to the new variable, on the basis of the hyperbolic differential equation

$$(8) \qquad u_{x_1 x_1} - u_{y_2 y_2} = f(x, y, u, u_{x_1}, -iu_{y_2}).$$

On the line $y_2 = 0$ in the x_1, y_2 -plane we prescribe the initial condition

$$u(x_1, x_2, y_1, 0) = u(x_1, x_2, y_1),$$

and as a second initial condition we require that

$$(9) \qquad \Lambda u \equiv \left(\frac{\partial}{\partial y_1} + i\frac{\partial}{\partial y_2}\right) u = 0 \qquad \text{for} \quad y_2 = 0.$$

The function $u(x_1, x_2, y_1, y_2)$ is thus uniquely defined. Because of the continuity properties of the solution of our partial differential equation (8) with respect to the parameters x_2, y_1, we see that this solution is indeed defined and continuously differentiable with respect to the parameters in a four-dimensional neighborhood B of the origin.

Now, to prove the analyticity of u, we have only to show that everywhere in B the relations $\nabla u = 0$ and $\Lambda u = 0$ are satisfied. The relation $\Lambda u = 0$ for $y_2 = 0$ is just our initial condition (9). Furthermore, for $y_2 = 0$ both equations (8) and (4) are satisfied and, by subtraction, we obtain

$$u_{x_1 x_1} - u_{y_2 y_2} + u_{x_2 x_2} - u_{y_1 y_1} = 0 \qquad \text{for} \quad y_2 = 0.$$

But since on $y_2 = 0$ from the preceding discussion we also have $\nabla u \equiv u_{x_1} + iu_{x_2} = 0$, it follows by differentiating with respect to x_1 and x_2 and subtracting that $u_{x_1 x_1} + u_{x_2 x_2} = 0$; hence we obtain

$$(10) \qquad u_{y_1 y_1} + u_{y_2 y_2} = 0 \qquad \text{for} \quad y_2 = 0.$$

On the other hand, (9) also holds there, and thus differentiating with respect to the parameter y_1 we get $u_{y_1 y_1} + iu_{y_1 y_2} = 0$, and using (10),

$$\frac{\partial}{\partial y_2} \Lambda u = 0 \qquad \text{on} \quad y_2 = 0.$$

It now follows from the uniqueness theorem for the initial value problem in the same way as before that $\Lambda u = 0$ everywhere in the four-dimensional region B. The relation $\nabla u = 0$ in B is proved analogously.

Since u is thus seen to be an analytic function in a complex neighborhood of $x = x_1$, $y = y_1$, the *analyticity of the original solution* $u(x, y)$ *of the elliptic differential equation* $\Delta u = f$ *has been proved.*

3. Remark on the General Differential Equation $F(x, y, u, p, q, r, s, t)$ $= 0$. Lewy's idea can be used also in the general case of an analytic differential equation of second order in two variables. The following theorem holds: *If* $u(x, y)$ *is a three times continuously differentiable solution of an elliptic differential equation, which is assumed analytic in all its arguments, then* $u(x, y)$ *itself is necessarily an analytic function of the two variables* x *and* y.

For the details of the proof the reader is referred to the literature.[1] The basic idea is the following: As before, the differential equation is replaced by a quasi-linear system of differential equations. Now, however, because of the elliptic character, real characteristic parameters α, β cannot be introduced. Nevertheless, it is possible to carry out the reduction to a system of differential equations of the form

$$v_{\alpha\alpha}^{\nu} + v_{\beta\beta}^{\nu} = f_{\nu} (\alpha, \beta, v^1, v^2, \cdots ; v_{\alpha}^1 v_{\alpha}^2, \cdots ; v_{\beta}^1, v_{\beta}^2, \cdots)$$

for unknown functions v^1, v^2, $\cdots$. For such a system the theory of §2 can now be applied almost unchanged, and on this basis it is then possible to carry out the proof of the analytic character of the solution.

§5. Use of Complex Quantities for the Continuation of Solutions

Let u be a solution of an analytic elliptic equation which satisfies an analytic boundary condition on part of the boundary of the domain of definition.

By extension into the complex domain H. Lewy[2] as well as P.

---

[1] See H. Lewy [6] and a presentation of Lewy's proof by J. Hadamard [2].
[2] See H. Lewy [2].

Garabedian[1] has achieved the analytic continuation of solutions of elliptic equations across boundaries. Significant applications of the method to the continuation of minimal surfaces, to the formulation of a generalized reflection principle, to problems with free boundaries and to problems of shock-computations will be found in the literature quoted.[2]

In this section we merely indicate the method by treating a very elementary example.

We have seen in §4 that if u is a solution of an analytic elliptic equation in a real domain it can be continued into a complex domain by solving *initial value problems* for certain hyperbolic equations. If u satisfies an analytic boundary condition on an analytic segment of the boundary we can use the complex extension of the boundary condition to obtain *mixed initial-boundary value problems* for the hyperbolic equations which provide an analytical extension of u to a larger domain. This extended domain includes an extension across the boundary segment of the originally given domain of existence.

To obtain this result in some generality involves considerable geometrical and analytical complications. The most extensive results so far are due to Garabedian where references to earlier work of Lewy and Hadamard are given.

We illustrate the method in the easiest nontrivial case by considering a solution of equation (3) of §4 in a rectangle R: $-\sigma \le x \le \sigma$, $-\tau \le y \le 0$, in the lower half plane. Let $u = 0$ on $y = 0$ and assume u has continuous third derivatives in the closed region R. We will continue u into the half plane $y > 0$.

We introduce x_1 and x_2 as before. Equation (4) with initial conditions (5) and (6) of the preceding section and boundary condition

$$u = 0 \quad \text{on} \quad y = 0$$

constitute a hyperbolic semilinear boundary value problem which we know can be solved in some sufficiently small triangle bounded by characteristics on one side and the line $y = 0$ with the prescribed boundary values on the other (Figure 35).

Explicitly such a triangle can be described as

$$(1) \qquad |y| + |x_2| \le \rho, \qquad y \le 0.$$

x_1 enters the initial value problem only as a parameter. While the largest domain in the x_2, y-plane in which (4) (of §4) has a solution

[1] See P. Garabedian [2].
[2] See also a forthcoming book by P. Garabedian [1].

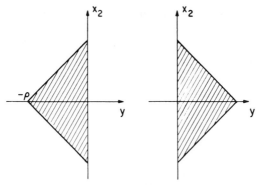

FIGURE 35

depends on x_1; for a sufficiently small value of ρ a solution will exist for all $|\,x_1\,| \leq \sigma$ and x_2, y satisfying (1).

Now we know u, u_y on $y = 0$, $|\,x_2\,| \leq \rho$, $|\,x_1\,| \leq \sigma$. These data enable us to find u by solving a Cauchy problem for (4) (§4) in a domain in $y > 0$ which includes part of the real half-plane $y > 0$, $x_2 = 0$.

One still has to verify that the continuation we obtained is a solution of the original equation. This can be done by showing as we did above that u satisfies the appropriate equation in the four-dimensional complex domain, and will not be carried out here.

It should be noted that for linear equations of the form (4) (§4) the region into which u can be continued depends only on R and the location of the singularities of the coefficients and not on the particular solution u itself. In particular, if the coefficients have no singularity then u can be continued into the whole reflected rectangle.[1]

Appendix[2] 2 to Chapter V
Transient Problems and Heaviside
Operational Calculus

Transient problems, or mixed problems (see Ch. III, §8), play a very important part in applications, e.g., in electrical engineering. They are the subject of an extensive literature in which emphasis is

---

[1] As Garabedian has observed, the reflection principle of Lewy (see H. Lewv [2]) can be obtained on this basis.

[2] This appendix, although closely connected with topics discussed in Chapters III, V, and VI, should be considered as an insert differing somewhat in style and emphasis from other parts of the book.

placed on Heaviside's symbolic *method of operators*. This method attacks problems in a strikingly direct way and often provides explicit answers not otherwise obtainable as simply. It was originally presented without a rigorous justification of the symbolic procedure; in fact, Heaviside displayed even a touch of contempt for the misgivings of professional mathematicians. Still the overwhelming success of Heaviside's method has forced a mathematical clarification which has led to a complete vindication and altogether greatly stimulated the development of symbolic methods.

An exhaustive discussion would exceed the scope of this work.[1] However, we shall present a theory of transient problems of at least the simplest type and give a few examples.

§1. *Solution of Transient Problems by Integral Representation*

1. Explicit Example. The Wave Equation. We seek the solution $u(x, t)$ of the wave equation

$$(1) \qquad u_{tt} - u_{xx} = 0$$

in an interval $0 \leq x \leq l$ satisfying the initial conditions

$$u(x, 0) = 0, \qquad u_t(x, 0) = 0$$

and the boundary conditions

$$(\alpha) \qquad u(0, t) = f(t), \qquad u(l, t) = 0$$

or

$$(\beta) \qquad u(0, t) = f(t), \qquad u_x(l, t) = 0,$$

where the "force" $f(t)$ is a given function of t.

The first problem (α) may refer to a string at rest at the time $t = 0$, fastened at the end point $x = l$, and subject at the other end point $x = 0$ to a motion prescribed by the forcing function $f(t)$ which causes a deflection u.

In the second problem (β), the string is subjected to the same motion at $x = 0$, but the end point $x = l$ of the string may now slide freely along a line perpendicular to the x-axis (position of rest). The second problem may also be interpreted as the problem of determining the voltage $u(x, t)$ in an ideal transmission line where the current u_x vanishes at the end point. In both problems we assume

---

[1] The reader will find a detailed exposition including many recent developments in the book of Mikusínski [1].

$$f(t) = 0 \qquad\qquad \text{for} \quad t \leq 0.$$

It is easy to solve this problem explicitly by adapting the functions $\phi(\lambda)$ and $\psi(\lambda)$ in the general solution of the wave equation

$$(2) \qquad u(x, t) = \phi(t + x) + \psi(t - x)$$

to the initial and boundary conditions. We subdivide the λ-axis into intervals $J_\nu : \nu l \leq \lambda < (\nu + 1)l$ and then determine ϕ and ψ successively in the different intervals.

First we consider the second problem, which corresponds to reflection at the end point. From the initial condition at $t = 0$ we obtain, for the desired functions, the relations

$$(3) \qquad \begin{aligned} \phi(x) + \psi(-x) &= 0, \\ \phi'(x) + \psi'(-x) &= 0. \end{aligned}$$

From the first of these we obtain by differentiation

$$(3') \qquad \phi'(x) - \psi'(-x) = 0.$$

Here x lies in the interval J_0 and $-x$ in the interval J_{-1}. It follows that $\phi(\lambda)$ and $\psi(\lambda)$ are constant for λ in these intervals; we may (by properly adjusting the arbitrary additive constant in ϕ and $-\psi$) assume that these constants are zero. From the boundary condition for $x = 0$, it follows that

$$(4) \qquad \phi(t) + \psi(t) = f(t)$$

holds for all intervals J_ν. The reflection condition at the other end point yields

$$(5) \qquad \lim_{x \to l} [\phi'(t + x) - \psi'(t - x)] = 0.$$

Assuming that $\phi(t + l) - \psi(t - l)$ is everywhere continuous we integrate and obtain

$$(6) \qquad \phi(t + l) - \psi(t - l) = 0,$$

which yields the recursion formula

$$(7) \qquad \phi(t + 2l) = \psi(t)$$

valid for all t in the intervals $J_{-1}, J_0, \cdots$. From (4) and (7), the functions ϕ and ψ and hence also $u(x, t)$ are uniquely determined by recursion for all values of $t > 0$. As is easily verified, the solution may be written explicitly in the following way:

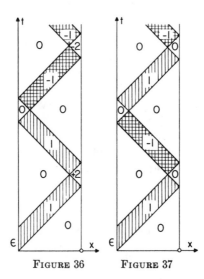

FIGURE 36   FIGURE 37

(8)
$$u(x, t) = f(t - x)$$
$$+ \sum_{\nu=1}^{\infty} (-1)^{\nu}[f(t - x - 2\nu l) - f(t + x - 2\nu l)].$$

Here the right side has the form of an infinite sum, and yet for every instant t, only a finite number of terms are different from zero. We may think of this series as representing wave trains; the solution consists of a superposition of such trains with wave form $f(\lambda)$ or $f(-\lambda)$, moving along the string $-\infty < x < \infty$ in both directions.

One aspect of the solution should be noted. Suppose that $f(t)$ represents an impulse,[1] i.e., $f(t) = 1$ in a small interval $0 \leq t \leq \epsilon$, and that $f(t) = 0$ everywhere else. Then, at the end point $x = l$, the function u will have the value 2 in the interval of time $l < t < l+\epsilon$, (cf. Figure 36). In electrical engineering applications where u represents a voltage, this means that the applied voltage can be doubled in a transmission line with infinite resistance at the end.

The problem (α) with fixed end point may be explicitly solved in an equally simple way:

(9) $u(x, t) = f(t - x) + \sum_{\nu=1}^{\infty} [f(t - x - 2\nu l) - f(t + x - 2\nu l)].$

[1] The word "impulse" will be used to describe instantaneous phenomena.

This also may be visualized by a superposition of waves which are similar in form. Figure 37 corresponds to an impulse with the forcing function $f(t)$ having the value 1 only in the interval $0 \leq t \leq \epsilon$ and the value zero otherwise. In our diagram the strip considered in the x,t-plane is divided into regions in which the function u has the values $+1$, -1, or zero.

2. General Formulation of the Problem. In investigating transient problems from a more general point of view, we restrict ourselves to the case of one space dimension x and the time coordinate t; the case of several space dimensions can be treated in an analogous way. We shall be concerned with the following problem.

Problem I: Suppose the differential equation

(10) $$a u_{tt} + b u_t = L[u] \equiv p u_{xx} + q u_x + r u$$

is given, where a, b, p, q, r are continuous functions of x alone in the interval $0 \leq x \leq l$ and satisfy, in this interval, the following conditions:

(α) $$p > 0,$$

(β) $$\begin{cases} a > 0 \text{ in the hyperbolic case} \\ a = 0, \, b > 0 \text{ in the parabolic case.} \end{cases}$$

A solution $u(x, t)$ of the differential equation (10) is sought in the interval $0 \leq x \leq l$ for the time $t > 0$ which satisfies the *initial conditions*

(11) $$u(x, 0) = \phi(x),$$
$$u_t(x, 0) = \psi(x) \text{ (in the hyperbolic case)}$$

and the *boundary conditions*

(12) $$u(0, t) = f(t),$$
$$p u_x(l, t) + \lambda u_t(l, t) = \sigma u(l, t).$$

Here $\phi(x)$, $\psi(x)$, and $f(t)$ are prescribed functions and ρ, λ, and σ are prescribed constants.

In particular, we consider the most important case, where $\phi = \psi = 0$, that is, where the state of rest prevails[1] at the time $t = 0$ (transient problem in the proper sense).

---

[1] As we saw in Ch. III, §8, the general case can always be reduced to this case in a formal way.

3. The Integral of Duhamel. For this choice of initial conditions, i.e., for $u(x, 0) = u_t(x, 0) = 0$ in the hyperbolic case or $u(x, 0) = 0$ in the parabolic case, the general problem I may be easily reduced to a problem with special functions $f(t)$ (see Ch. III, §4). We observe that since the coefficients are assumed to be independent of t, the derivatives u_t, u_{tt}, $\cdots$ (if they exist or can be properly defined) are solutions corresponding to the "forces" $f'(t), f''(t), \cdots$.

We introduce a solution $U(x, t)$ with the discontinuous boundary condition

$$U(0, t) = f(t) = \begin{cases} 1 & \text{for } t \geq 0 \\ 0 & \text{for } t < 0 \end{cases}$$

and the given boundary condition for $x = l$

$$U(x, 0) = 0,$$

and assume that every bounded region of the strip $0 \leq x \leq l, t \geq 0$ may be divided into a finite number of closed subregions in which U is continuous together with its derivatives up to the second order.[1] This function U may be defined either as the second time-derivative of a solution $U_2(x, t)$ with the boundary condition

$$f(t) = \begin{cases} \dfrac{t^2}{2} & \text{for } t \geq 0 \\ 0 & \text{for } t < 0 \end{cases}$$

or U may be defined simply by referring to the concept of generalized functions or distributions (see Ch. VI, §3 and Appendix).

Then the *theorem of Duhamel* holds: *If $f(t)$ and its derivative $f'(t)$ are piecewise continuous for $t > 0$, then the Duhamel integral*

$$(13) \qquad u(x, t) = \frac{\partial}{\partial t} \int_0^t U(x, t - \tau) f(\tau) \, d\tau$$

is the solution of problem I with the boundary conditions (12).

The function $U(x, t)$ need not be everywhere continuous as examples will show. In fact, we must expect discontinuities for U since the boundary condition $U(0, t) = 1$ together with the initial condition $U(x, 0) = 0$ means that at the point $x = 0$ and the time

---

[1] For the justification, see the discussion of propagation of discontinuities in Ch. V, §9, and later in Ch. VI, §4.

$t = 0$ an impulse occurs which increases the value $U(0, 0) = 0$ instantaneously to the value 1. This immediately gives us an intuitive meaning for the Duhamel integral (13).

We imagine the effect of the "force" $f(t)$ (at the left end of the interval) to consist of individual impulses occurring at the instants $\tau_0 = 0, \tau_1, \tau_2, \cdots, \tau_n$. Each impulse makes the value $u(0, \tau_{\nu-1})$ jump by the increment $f(\tau_\nu) - f(\tau_{\nu-1})$. If $U(x, t)$ is the special solution defined above, then the solution $u(x, t)$, corresponding to these impulses, can be written additively in the form

$$u(x, t) = \sum_{\nu=0}^{n} U(x, t - \tau_\nu)[f(\tau_{\nu+1}) - f(\tau_\nu)] + U(x, t)f(0)$$

$$(\tau_{n+1} = t).$$

If we assume that $f(t)$ is continuously differentiable for $t > 0$ but that $f(0)$ may differ from zero—this corresponds to a finite jump at the time $t = 0$—and if we carry out the passage to the limit by letting the time intervals go to zero, then we obtain the solution

$$u(x, t) = U(x, t) f(0) + \int_0^t U(x, t - \tau) f'(\tau) \, d\tau$$

(13′)

$$= \frac{\partial}{\partial t} \int_0^t U(x, t - \tau) f(\tau) \, d\tau$$

in accordance with assertion (13).

This heuristic argument can easily be replaced by a straightforward verification based on the identity

$$\frac{\partial}{\partial t} \int_0^t U(x, t - \tau) f(\tau) \, d\tau = U(x, t) f(0)$$

$$+ U_1(x, t) f'(0) + U_2(x, t) f''(0) + \int_0^t U_2(x, t - \tau) f'''(\tau) \, d\tau,$$

which holds since $U_2(x, 0) = U_1(x, 0) = 0$, where $U_1(x, t)$ is the solution with the boundary condition

$$f(t) = \begin{cases} t & \text{for} \quad t \geq 0 \\ 0 & \text{for} \quad t < 0. \end{cases}$$

This identity shows that the differential equation and the boundary condition at $x = l$ for U, U_1, and U_2 are satisfied by $u(x, t)$. Moreover, the initial conditions for u follow from the observation that the differential equation and the conditions $U_2(x, 0) = U_1(x, 0) = 0$ im ply $U(x, 0) = U_t(x, 0) = 0$. Finally, since $U(0, t) = 1$, (13′) im plies $u(0, t) = f(0) + \int_0^t f'(\tau) \, d\tau = f(t)$. Therefore, the first of the boundary conditions (12) holds.

Duhamel's integral formula (13) can be regarded as a representa tion of the linear operator T which transforms the given boundary values $f(t)$ into the solution $u(x, t)$; i.e., (13) represents $u = Tf$. However, Duhamel's integral formula remains valid not only for the solution operator T of the differential equation problem considered here but for all linear operators T which satisfy the following con ditions[1]:

a) The operator T is defined for all functions $f(t)$ which vanish for $t < 0$, and transforms $f(t)$ into a function $Tf(t)$ which also vanishes for $t < 0$. ($Tf(t)$ possibly depends on other variables[2] $x, \cdots$ in addition to t.)

b) $T \int_{\tau_1}^{\tau_2} f(t, \tau) \, d\tau = \int_{\tau_1}^{\tau_2} Tf(t, \tau) \, d\tau$; here τ is a parameter.

c) If $f(0) = 0$ and if $f(t)$ is differentiable, then

$$\frac{d}{dt} (Tf(t)) = T \frac{df(t)}{dt} .$$

d) If $Tf(t) = \phi(t)$, then for any $\tau > 0$

$$Tf(t - \tau) = \phi(t - \tau).$$

To prove equation (13) it suffices to represent $f(t)$ in the form

$$f(t) = \frac{d}{dt} \int_0^\infty \eta(t - \tau) f(\tau) \, d\tau ,$$

where

$$\eta(t) = \begin{cases} 1 & \text{for } t \geq 0 \\ 0 & \text{for } t < 0. \end{cases}$$

From this, by conditions a), b), c), and d), it follows that

---

[1] This remark stems from a note in the Russian translation of the original German edition.

[2] The dependence of functions on these other variables will not be shown explicitly.

$$Tf(t) = \frac{d}{dt} \int_0^\infty T\eta(t - \tau) f(\tau) \, d\tau = \frac{d}{dt} \int_0^t U(t - \tau) f(\tau) \, d\tau,$$

where

$$U = T\eta.$$

It can be seen that the conditions a), b), c), and d) essentially characterize the class of linear operators which are represented by Duhamel's integral.

4. Method of Superposition of Experimental Solutions. In discussing the initial value problem in Ch. III, §7, 1, we used the method of the Fourier integral, i.e., the superposition of solutions which are represented by means of exponential functions; this method can also be used, with appropriate changes, for the solution of transient problems. Here again we shall limit ourselves to a heuristic argument which will be completed by an existence theorem in §3. We consider the special problem for $u(0, t) = 1$, $u(x, 0) = u_t(x, 0) = 0$, and seek the function $U(x, t)$ discussed in the preceding section. First we construct special solutions of the differential equation (10)

$$au_{tt} + bu_t = L[u]$$

of the form

$$u = e^{\gamma t} v(x, \gamma) \qquad\qquad (\gamma = \alpha + i\beta).$$

For v the ordinary differential equation

(14) $$L[v] = (a\gamma^2 + b\gamma)v$$

follows, where γ now appears as a parameter. If at the end point $x = l$, we impose the boundary condition

(15) $$\rho v_x = (\sigma - \lambda\gamma)v$$

for v, then

$$u = v(x, \gamma)e^{\gamma t}$$

clearly satisfies the second of the given boundary conditions (12)

$$\rho u_x + \lambda u_t = \sigma u;$$

the same is true for any linear combination of solutions of this kind with various parameters γ. By superposition of such solutions we now try to satisfy the boundary condition $u(0, t) = 1$ for $t > 0$, and the initial conditions $u(x, 0) = 0$, $u_t(x, 0) = 0$.

For this purpose we assume that v and its relevant derivatives are analytic functions of the *complex* parameter $\gamma = \alpha + i\beta$ in the half-plane $\alpha > 0$. Integrating over a path L in the right half-plane of the complex variable γ, we now obtain new solutions of the differential equation (10) satisfying the second of the boundary conditions (12). These solutions take the form

$$(16) \qquad u(x, t) = \frac{1}{2\pi i} \int_L \frac{v(x, \gamma)}{\gamma} e^{\gamma t} d\gamma.$$

At $x = 0$, we obtain

$$u(0, t) = \frac{1}{2\pi i} \int_L \frac{v(0, \gamma)}{\gamma} e^{\gamma t} d\gamma,$$

and the problem is to choose the path of integration L and the available boundary value $v(0, \gamma)$ in such a way that

$$u(0, t) = 1 \qquad \text{for} \quad t > 0,$$
$$u(0, t) = 0 \qquad \text{for} \quad t < 0.$$

The boundary condition at $x = 0$ is satisfied if we set

$$(15') \qquad v(0, \gamma) = 1$$

and if we assume that L is an arbitrary parallel to the imaginary axis of the γ-plane in the right half-plane $\alpha > 0$. For these choices, we obtain the integral

$$(17) \qquad u(0, t) = \frac{1}{2\pi i} \int_{\alpha - i\infty}^{\alpha + i\infty} \frac{e^{\gamma t}}{\gamma} d\gamma = \frac{e^{\alpha t}}{2\pi} \int_{-\infty}^{\infty} \frac{e^{i\beta t}}{\alpha + i\beta} d\beta,$$

which by elementary theorems converges for all $\alpha \neq 0$, $t \neq 0$. Since the integral $\int_{l_1 + i\beta}^{l_2 + i\beta} \frac{e^{\gamma t}}{\gamma} d\gamma$, extended over any straight line segment parallel to the real α-axis with the finite length $l_2 - l_1$, tends to zero as $|\beta|$ increases, the integral (17) does not depend on α. This follows in the usual way from the Cauchy integral theorem (cf. Figure 38). Hence, if we allow α to increase without limit, it follows immediately that the equation $u(0, t) = 0$ holds for $t < 0$. In the case $t > 0$, using Cauchy's integral theorem and considering the residue for $\gamma = 0$, we have the equation

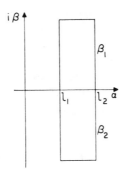

FIGURE 38

$$u(0, t) = 1 + \frac{1}{2\pi i \alpha} \int_{a-i\infty}^{a+i\infty} \frac{e^{\gamma t}}{\gamma} \, d\gamma,$$

where α is negative but otherwise arbitrary. Passing to the limit $\alpha \to -\infty$, we obtain $u(0, t) = 1$ for $t > 0$.

It is therefore plausible that the expression

$$(18) \qquad\qquad U(x, t) = \frac{1}{2\pi i} \int_L \frac{v(x, \gamma)}{\gamma} e^{\gamma t} \, d\gamma$$

is the desired solution of equation (10). Incidentally, we cannot always expect this integral to converge absolutely, since, for given x, $U(x, t)$ will in general possess discontinuities in t. This shortcoming can be met by the following method: For sufficiently large n we consider the expression

$$U_n(x, t) = \frac{1}{2\pi i} \frac{1}{n!} \int_L \frac{v(x, \gamma)}{\gamma^{n+1}} e^{\gamma t} \, d\gamma \, ,$$

which corresponds to the initial function $f(t) = t^n/n!$ for $t > 0$ and $f(t) = 0$ for $t < 0$. This representation for $U_n(x, t)$ converges well for large n and can easily be verified as a solution of the transient problem. Then U can be obtained by differentiation with respect to t. These preliminary considerations, at present heuristic, will be taken up again and carried out from a somewhat different point of view in §3.

§2. The Heaviside Method of Operators

In practice *Heaviside's symbolic method* may have great advantages compared to the procedure of §1, 4. It can be rigorously justified on the basis of the concepts given in §§1 and 3. Its advantage is that the formal calculation of the solution is set apart from the conceptual mathematical content. This separation makes it possible to postpone the substantiation of the result obtained symbolically. Moreover the realization of symbolic operators can be anticipated in the form of tables. Thus, in many applications, difficulties arising from the nonformal mathematical content may be largely avoided.

The basic idea of this formal calculus is to consider the linear functional operation which attaches the solution u to the prescribed boundary function $f(t)$ rather than to seek solutions $u(x, t)$ of the differential equation of §1, 3 directly. Again we restrict ourselves to the proper transient problem for which the state of rest prevails at the time $t = 0$.

1. The Simplest Operators. The method is based on the introduction of the *differentiation* and *integration operators* p and p^{-1} representing inverse operations. We consider functions of the time variable t for $t > 0$ (which we may continue as identically zero for $t < 0$) and define the integration operator p^{-1} by

$$(1) \qquad p^{-1} f(t) = g(t) = \int_0^t f(\tau) \, d\tau.$$

If we denote the differentiation operator by p so that

$$(2) \qquad pg(t) = f(t) = \frac{dg}{dt},$$

then the property

$$(3) \qquad pp^{-1} = p^{-1}p = 1$$

is the basis for constructing a calculus with rules corresponding to those of algebra. To insure the validity of this "inverse" relation we must impose a strongly restrictive condition: *The operator p may be applied only to functions $g(t)$ for which $g(0) = 0$.* Otherwise, we would have

$$p^{-1}pg = \int_0^t \frac{dg(\tau)}{d\tau} \, d\tau = g(t) - g(0),$$

$$pp^{-1}g = \frac{d}{dt} \int_0^t g(\tau) \, d\tau = g(t)$$

and therefore

$$p^{-1}pg \neq pp^{-1}g.$$

Unlike the operator p, however, the operator p^{-1} may be applied to arbitrary continuous functions. Now, if

$$Q(\lambda) = a_0 + a_1\lambda + \cdots + a_m\lambda^m$$

is any polynomial of degree m, we can define the *integral rational* operator $Q(p^{-1})$ in an obvious way and apply it to an arbitrary function f. The corresponding operator $Q(p)$ is defined as a linear differential operator of order m but may operate only on functions f which vanish at $t = 0$ together with their derivatives up to order $m - 1$.

If

$$P(\lambda) = b_0 + b_1 \lambda + \cdots + b_n \lambda^n$$

is another polynomial of degree n and if $Q(0) = a_0 \neq 0$, then

$$(4) \qquad R(p^{-1}) = \frac{P(p^{-1})}{Q(p^{-1})}$$

is called a *fractional rational regular operator.*

This operator,

$$R(p^{-1})f(t) = g(t),$$

can be defined in various ways:

First, for a given function f which is piecewise continuous for $t > 0$, one can characterize g as the solution of the differential equation

$$(5) \qquad a_0 g^{(m)} + a_1 g^{(m-1)} + \cdots + a_m g = \phi^{(m)},$$

where

$$\phi = P(p^{-1})f,$$

which is uniquely determined by the requirement that it satisfy the initial conditions

$$a_0 g(0) = \phi(0),$$

$$a_0 g'(0) + a_1 g(0) = \phi'(0),$$

$$(6) \qquad a_0 g''(0) + a_1 g'(0) + a_2 g(0) = \phi''(0),$$

$$\cdots\cdots\cdots\cdots\cdots\cdots\cdots\cdots\cdots\cdots$$

$$a_0 g^{(m-1)}(0) + a_1 g^{(m-2)}(0) + \cdots + a_{m-1} g(0) = \phi^{(m-1)}(0).$$

Secondly, g can be defined if we develop the rational function

$$P(\lambda)/Q(\lambda) = R(\lambda)$$

in the neighborhood of the origin in a power series in λ

$$R(\lambda) = \sum_{\nu=0}^{\infty} \alpha_\nu \lambda^\nu.$$

It is then easy to see that the corresponding series

$$R(p^{-1})f = \sum_{\nu=0}^{\infty} \alpha_\nu \, p^{-\nu} f(t)$$

converges for all positive values of t and agrees with the function $g(t)$ defined above.

If the coefficient a_0 is equal to 0, then the rational operator is called *irregular* and may be represented in the form

$$p^k R(p^{-1}),$$

where R is regular. To apply such an operator to the function f we must assume the conditions $f(0) = f'(0) = \cdots = f^{(k-1)}(0) = 0$.

Clearly, one may perform rational calculations with these operators according to the rules of algebra. For these operators, Duhamel's representation of §1 asserts: *If, for Heaviside's "unit function" $\eta(t)$ defined by*

$$\eta(t) = 1 \qquad\qquad \text{for}\quad t \geq 0,$$
$$\eta(t) = 0 \qquad\qquad \text{for}\quad t < 0,$$

the relation[1]

(7) $$T\eta(t) = H(t)$$

holds for an operator T, then we have

(8) $$Tf(t) = \frac{d}{dt} \int_0^t H(t - \tau) f(\tau)\, d\tau$$

for an arbitrary function $f(t)$.

To make the operational calculus a useful tool it is essential to find

---

[1] In the literature the unit function $\eta(t)$ is often not written explicitly. The operator symbol T then simply denotes the function $T\eta$. Incidentally, if we want to express that a function $f(t)$ defined for $t > 0$ is to be continued as zero for $t < 0$ then we shall sometimes write $f(t)\eta(t)$ instead of $f(t)$.

an interpretation not only for rational but also for more general functions of p in such a way that the rules of algebra, the Duhamel principle, and certain other rules (to be introduced presently) will hold in the extended domain of operators.

2. Examples of Operators and Applications. In this article we discuss a number of simple operators and indicate how they may be applied to solve differential equations.

1) For the operator

$$T = \frac{1}{1 + \alpha p^{-1}},$$

we have

$$T\eta = e^{-\alpha t}, \qquad Tf(t) = \frac{d}{dt} \int_0^t e^{-\alpha(t-\tau)} f(\tau) \, d\tau = g(t).$$

Here g is the solution of the differential equation

$$g' + \alpha g = f' \quad \text{with} \quad g(0) = f(0).$$

2) For the operator

$$T = \frac{1}{1 + \nu^2 p^{-2}},$$

we have

$$T\eta = \cos \nu t, \qquad Tf = g(t) = \frac{d}{dt} \int_0^t \cos \nu(t - \tau) f(\tau) \, d\tau.$$

Here g is the solution of the differential equation

$$g'' + \nu^2 g = f''$$

with the initial conditions

$$g(0) = f(0), \qquad g'(0) = f'(0).$$

3) We consider the nonhomogeneous linear differential equation with constant coefficients

$$(9) \qquad a_0 u^{(m)} + a_1 u^{(m-1)} + \cdots + a_m u = f(t)$$

under the initial conditions

$$(9)' \qquad u(0) = u'(0) = \cdots = u^{(m-1)}(0) = 0.$$

With these initial conditions we can write the differential equation symbolically in the form

$$Q(p)u = f(t) \qquad\qquad (Q(\lambda) = a_0 \lambda^m + \cdots + a_m);$$

the solution is obtained symbolically in the form

$$(10) \qquad\qquad u(t) = \frac{1}{Q(p)} f(t).$$

If the algebraic equation

$$Q(\lambda) = 0$$

possesses the m distinct roots $\alpha_1, \alpha_2, \cdots, \alpha_m \neq 0$, we can interpret the solution in a very elegant way by means of partial fractions. From

$$\frac{1}{pQ(p)} = \frac{c_0}{p_{\scriptscriptstyle\bullet}} + \sum_{\nu=1}^{m} \frac{c_\nu}{p - \alpha_\nu},$$

we have

$$u(t) = \frac{1}{Q(p)} f(t) = c_0 f(t) + \sum_{\nu=1}^{m} c_\nu \frac{p}{p - \alpha_\nu} f(t).$$

In the special case

$$f(t) = e^{i\omega t}\eta(t) \qquad\qquad (i\omega \neq \alpha_\nu),$$

we can, instead of using Duhamel's integral, realize our operator by the following even more elegant procedure.

We write

$$f(t) = \frac{p}{p - i\omega} \eta,$$

and obtain

$$u = \frac{p}{L[p]} \eta,$$

where, for brevity, we have set

$$L[p] = (p - i\omega)Q = a_0(p - \alpha_1) \cdots (p - \alpha_m)(p - i\omega).$$

We can write this rational operator in the form

$$\frac{p}{L[p]} = \frac{d_0\, p}{p - i\omega} + \sum_{\nu=1}^{m} \frac{d_\nu\, p}{p - \alpha_\nu},$$

where the coefficients of the expansion into partial fractions are given by

$$d_0 = \frac{1}{Q(i\omega)}, \qquad d_\nu = \frac{1}{a_\nu - i\omega} \frac{1}{Q'(\alpha_\nu)}.$$

Remembering our first example we now obtain immediately the desired solution

$$(11) \qquad u = d_0 e^{i\omega t} + \sum_{\nu=1}^{m} d_\nu e^{\alpha_\nu t}.$$

The coefficient d_0 is, of course, the most important for applications.

4) As another example we consider the "singular operators"

$$(12) \qquad p^{1/2}, \qquad p^{-1/2}$$

which must be defined in agreement with our earlier rules if we are to enlarge our store of operators in a consistent way. The theory of differentiation and integration of fractional order suggests the definition

$$(13) \qquad \begin{aligned} p^{-1/2}\eta &= 2\sqrt{t/\pi}; \quad p^{-1/2}f(t) = (2/\sqrt{\pi}) \frac{d}{dt} \int_0^t \sqrt{t - \tau}\, f(\tau)\, d\tau, \\ p^{1/2}\eta &= 1/\sqrt{\pi t}; \quad p^{1/2}f(t) = (1/\sqrt{\pi}) \frac{d}{dt} \int_0^t \frac{f(\tau)}{\sqrt{t - \tau}}\, d\tau. \end{aligned}$$

As can easily be verified, this definition is actually consistent with the requirement

$$p^{-1/2} p^{-1/2} \eta = p^{-1} \eta.$$

The same definition is implied by the following observation: The equation $p^{-n}\eta = ct^n$, where $c = 1/n!$ is a constant, holds. Therefore it is plausible to set

$$p^{-1/2}\eta = c\sqrt{t},$$

where the constant c must appropriately be determined. If we require that Duhamel's principle and the relation

$$p^{-1/2} p^{-1/2} \eta = p^{-1}\eta = t$$

be valid, then we have

$$t = c^2 \frac{d}{dt} \int_0^t \sqrt{t - \tau}\, \sqrt{\tau}\, d\tau = 2c^2 t \int_0^1 \sqrt{1 - \tau}\, \sqrt{\tau}\, d\tau = c^2 \frac{\pi}{4} t,$$

and hence

$$c = \frac{2}{\sqrt{\pi}}.$$

Thus the constant c is determined in agreement with (13).

5) A very important operator which is not rational, the *exponential operator*, is introduced for constant h by the definition

(14) $$e^{-hp}f(t) = f(t - h),$$

which is suggested by the Taylor series for e^{-hp} and for $f(t - h)$. However, this plausibility consideration cannot provide a satisfactory justification, because the definition should not depend on the analytic character of the function $f(t)$. Our definition is justified rather by the fact that it implies the relations

$$e^{-hp}e^{-kp}f(t) = e^{-(h+k)p}f(t)$$

and

$$\frac{d}{dh} e^{-hp}f(t) = -p e^{-hp}f(t);$$

the latter is equivalent to the equation

$$\frac{d}{dh} f(t - h) = -\frac{d}{dt} f(t - h).$$

6) Now we consider, for $h < 0$, the operator

(15) $$e^{-h\sqrt{p}},$$

which will be interpreted in article 5. Here we merely remark that, assuming we may differentiate our operator with respect to the parameter h, we obtain from $e^{-h\sqrt{p}}f = g$ the equation

$$\frac{\partial g}{\partial h} = -\sqrt{p}\, e^{-h\sqrt{p}}\, f.$$

For the parameter value $h = 0$, we can expect the value of the function $\partial g/\partial h$ to be given by

(16) $$\left.\frac{\partial g}{\partial h}\right|_{h=0} = -\sqrt{p}\,f = -\frac{1}{\sqrt{\pi}} \frac{d}{dt} \int_0^t \frac{f(\tau)}{\sqrt{t - \tau}}\, d\tau.$$

This illustrates the directness of the method of operators.

7) As a last example in this connection, we discuss the so-called *translation principle* of Heaviside:

If $T = \Phi(p)$ *is an operator and* k *is a constant, then the operator* $\Phi(p + k)$ *is given by*

$$(17) \qquad \Phi(p + k) = e^{-kt}\Phi(p)e^{kt}.$$

This can be proved immediately for all rational regular operators. First, by induction from n to $n + 1$, we show that the principle of displacement is valid for the operator $1/p^n$. It is then valid for all rational regular operators, since they can be expressed as power series in $1/p$. We may now introduce the displacement principle for irregular operators as a plausible postulate. For example, we define

$$(18) \qquad \sqrt{p + \alpha^2}\, f(t) = e^{-\alpha^2 t}\, \sqrt{p}\; e^{\alpha^2 t} f(t) = \frac{e^{-\alpha^2 t}}{\sqrt{\pi}} \frac{d}{dt} \int_0^t \frac{e^{\alpha^2 \tau} f(\tau)}{\sqrt{t - \tau}}\, d\tau$$

and, in particular,

$$(19) \qquad \begin{aligned} \sqrt{p + \alpha^2}\, \eta(t) &= \frac{e^{-\alpha^2 t}}{\sqrt{\pi}} \frac{d}{dt} \int_0^t \frac{e^{\alpha^2 \tau}}{\sqrt{t - \tau}}\, d\tau \\ &= \frac{2e^{-\alpha^2 t}}{\alpha \sqrt{\pi}} \frac{d}{dt} \left(e^{\alpha^2 t} \int_0^{\alpha \sqrt{t}} e^{-\tau^2}\, d\tau \right). \end{aligned}$$

Satisfactory justification for these irregular operators requires proving that their introduction is consistent with the elementary algebraic rules of calculation. This justification will be given in article 5.

3. Applications to Heat Conduction. We shall apply the method of operators to a few typical examples.

We consider in a semi-infinite interval *the heat equation*

$$(20) \qquad u_t - u_{xx} = 0.$$

Let the initial and boundary conditions be

$$u(x, 0) = 0,$$
$$u(0, t) = f(t), \qquad u(\infty, t) = 0.$$

We seek the operator $T = T(x)$ (depending on x as a parameter) which transforms the given function $f(t)$ into the desired solution $u(x, t)$. We write the differential equation in the form

$$(21) \qquad (T_{xx} - pT)f = 0$$

and the boundary conditions in the form

$$T(0) = 1, \qquad T(\infty) = 0.$$

Assuming our functions as zero for negative t, we have already accounted for the initial condition by writing

$$u_t = pTf.$$

Treating the differential equation

$$T_{xx} - pT = 0$$

as if p were a parameter, we immediately obtain the solution

(22) $$T = e^{-x\sqrt{p}}.$$

The question arises: How can this symbolic expression be interpreted; what functions are represented by the expressions

$$e^{-x\sqrt{p}}\eta, \qquad e^{-x\sqrt{p}}f(t)?$$

We can at least give a partial answer which covers the question directly relevant for applications: We can find the rate at which heat is transferred at the end point $x = 0$; i.e., we can find the expression $u_x(0, t)$ explicitly, by operating with T according to the ordinary rules. Applying the result of article 2, example 6, we obtain

$$u_x(0, t) = T_x(0)f = -\sqrt{p}\,f = -\frac{1}{\sqrt{\pi}}\frac{d}{dt}\int_0^t \frac{f(\tau)}{\sqrt{t - \tau}}\,d\tau.$$

One of the principal advantages of the symbolic calculus lies precisely in the possibility of obtaining partial results such as the one just given without necessarily being able to interpret the operators completely by easily expressible functions.

In a similar way we can treat the more *general equation of heat conduction* $u_t - u_{xx} + \alpha^2 u = 0$ for the interval $0 \leq x < \infty$ under the same initial and boundary conditions as before. For the operator $T(x)$ which produces the solution $u(x, t) = T(x)f(t)$, the symbolic differential equation

(23) $$T_{xx} = (p + \alpha^2)T$$

holds. The symbolic solution, under the boundary conditions

$$T(0) = 1, \qquad T(\infty) = 0,$$

is

$$(24) \qquad\qquad T = e^{-x\sqrt{p+\alpha^2}}.$$

We are even less able to interpret this operator fully than in the previous example. However, we can again answer an important part of the problem; we find the relation

$$u_x(0, t) = T_x(0)f = -\sqrt{p + \alpha^2}\, f = -\frac{e^{-\alpha^2 t}}{\sqrt{\pi}} \frac{d}{dt} \int_0^t \frac{e^{\alpha^2 \tau} f(\tau)}{\sqrt{t - \tau}}\, d\tau$$

for $u_x(0, t)$, where the expression on the right is taken from article 2, example 4.

In the special case of the unit function $f = \eta$ we have

$$(25) \qquad\qquad u_x(0, t) = -\frac{2e^{-\alpha^2 t}}{\alpha\sqrt{\pi}} \frac{d}{dt} \left(e^{\alpha^2 t} \int_0^{\alpha\sqrt{t}} e^{-\tau^2}\, d\tau \right).$$

4. Wave Equation. The simple transient problems discussed in §1, 1 also are illuminated by the operational calculus. We consider, for the interval $0 \leq x \leq l$, the problem of solving the differential equation

$$(26) \qquad\qquad u_{tt} - u_{xx} = 0$$

with the initial and boundary conditions

$$u(x, 0) = u_t(x, 0) = 0,$$

$$u(0, t) = f(t), \qquad u_x(l, t) = 0.$$

We set

$$u(x, t) = T(x)f.$$

For the operator T we obtain the differential equation

$$(27) \qquad\qquad T_{xx} - p^2 T = 0$$

with the conditions

$$T(0) = 1, \qquad T_x(l) = 0.$$

Thus we are led to the symbolic expression

$$(28) \qquad T(x) = \frac{\cosh p(l - x)}{\cosh pl} = \frac{e^{-px} + e^{-p(2l-x)}}{1 + e^{-2pl}}$$

or, if the expansion is written out, to

$$T(x) = e^{-px} + \sum_{\nu=1}^{\infty} (-1)^{\nu} [e^{-p(x+2\nu l)} - e^{-p(2\nu l - x)}].$$

On the basis of the examples in article 2, the interpretation is now very simple. We have

$$
\begin{aligned}
u(x, t) &= f(t - x) \\
&+ \sum_{\nu=1}^{\infty} (-1)^{\nu} [f(t - x - 2\nu l) - f(t + x - 2\nu l)]
\end{aligned}
$$

(29)

in agreement with the result obtained in §1, 1.

The other problem discussed there (problem (α) with fixed end point) can, of course, also be solved in a similar way. For the corresponding operator we set

(30)
$$T(x) = \frac{\sinh p(l - x)}{\sinh pl}$$

or

$$T = e^{-px} + \sum_{\nu=1}^{\infty} [e^{-p(x+2\nu l)} - e^{-p(2\nu l - x)}].$$

Hence

(31) $$u(x, t) = f(t - x) + \sum_{\nu=1}^{\infty} [f(t - x - 2\,\nu l) - f(t + x - 2\nu l)].$$

5. Justification of the Operational Calculus. Interpretation of Further Operators. The operational calculus can be justified rigorously. We first give a general definition of our operators and then, on the basis of this definition, we verify that the stipulated rules for calculation, the translation theorem, and Duhamel's principle are all valid. We also verify that this definition is consistent with those given previously.

The considerations of §1, 4 motivate the following definition:

Let $F(\gamma)$ be a regular analytic function of the variable $\gamma = \alpha + i\beta$ defined in the half-plane $\alpha > \bar{\alpha}_0$. Let L be an arbitrary line parallel to the imaginary axis in the half-plane $\alpha > \alpha_0$, or, if $\alpha_0 < 0$, L becomes an indented path of the form illustrated in Figure 39. Then, if the integral $(1/2\pi i) \int_L [F(\gamma)/\gamma] e^{\gamma t} d\gamma$ exists and is independent of the special choice of L for all $t > 0$, we define

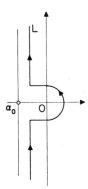

FIGURE 39

$$
F(p)\eta = \frac{1}{2\pi i} \int_L \frac{F(\gamma)}{\gamma} e^{\gamma t} \, d\gamma,
$$
(32)
$$
F(p)f = \frac{d}{dt} \int_0^t f(\tau) \, d\tau \, \frac{1}{2\pi i} \int_L \frac{F(\gamma)}{\gamma} e^{\gamma(t-\tau)} \, d\gamma.
$$

A sufficient condition for the existence of the integral (32) is, for example, the existence of a positive function $\Phi(\rho)$, for which

$$
\int_0^\infty \Phi(\rho) \, d\rho
$$

converges such that, for all $\gamma = \alpha + i\beta$ with $\alpha \geq \alpha_0 + \delta \ (\delta > 0)$, the inequality

$$
\left| \frac{F(\gamma)}{\gamma} \right| \leq \Phi(|\beta|)
$$

is satisfied. Then in the second integral (32) we may carry out the integration with respect to τ under the integral sign; accordingly, if we set

$$
D(\gamma, t) = \int_0^t f(\tau) e^{-\gamma\tau} \, d\tau
$$

we have

$$
F(P)f = \frac{1}{2\pi i} \frac{d}{dt} \int_L \frac{F(\gamma)}{\gamma} D(\gamma, t) \, e^{\gamma t} \, d\gamma.
$$

The validity of the laws of calculation for the operators defined in this way is insured by the *multiplication theorem*

$$(33) \qquad F(p)G(p) = FG(p),$$

or, in words: *The result of the successive application of two operators F and G can also be obtained by applying the operator which corresponds to the product function FG.*

It is sufficient to prove this theorem for the unit function $\eta(t)$. For the proof we make the following additional assumptions about the functions F and G:[1] In every half-plane $\alpha \geq \alpha_0 + \delta$ there exists a positive function $\psi(\rho)$ for which the integral $\int_0^\infty \psi^2 \, d\rho$ converges such that everywhere in this half-plane

$$|F| \leq \psi(|\beta|), \qquad |G| \leq \psi(|\beta|).$$

Then the integrals

$$\int_1^\infty \frac{\psi}{\rho} \, d\rho \qquad \text{and} \qquad \int_1^\infty \frac{\psi^2}{\rho} \, d\rho$$

also exist—the first because the Schwarz inequality

$$\int_1^\infty \frac{\psi}{\rho} \, d\rho \leq \sqrt{\int_1^\infty \psi^2 \, d\rho} \sqrt{\int_1^\infty \frac{d\rho}{\rho^2}}$$

holds. Hence the integrals (32) corresponding to the functions F, G, FG converge absolutely. Now, if we set

$$f(t) = G(p)\eta = \frac{1}{2\pi i} \int_L \frac{G(\delta)}{\delta} e^{\delta t} \, d\delta$$

and

$$D(\gamma, t) = \int_0^t f(\tau)e^{-\gamma\tau} \, d\tau = \frac{1}{2\pi i} \int_L \frac{G(\delta)}{\delta} \frac{e^{(\delta-\gamma)t} - 1}{\delta - \gamma} \, d\delta,$$

we obtain

$$FG\eta = \frac{1}{2\pi i} \frac{d}{dt} \int_L \frac{F(\gamma)}{\gamma} D(\gamma, t)e^{\gamma t} \, d\gamma.$$

On the basis of our assumptions it can easily be seen that differentiation under the integral sign always yields integrals which converge

---

[1] In fact, these assumptions are too restrictive for some important cases. However the theorem can be proved under much weaker assumptions. See, for example, W. von Koppenfels [1].

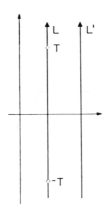

FIGURE 40

uniformly in a domain $t_1 \leq t \leq t_2$ with $t_1 > 0$; therefore we have

$$FG\eta = \frac{1}{(2\pi i)^2} \int_L \frac{F(\gamma)}{\gamma} d\gamma \int_{L'} \frac{G(\delta)}{\delta} \frac{\delta e^{\delta t} - \gamma e^{\gamma t}}{\delta - \gamma} d\delta.$$

Here L' is chosen as a straight line lying to the right of and parallel to L (see Figure 40). As we see from the estimate

$$\left| \int_{L'} \frac{G(\delta)}{\delta(\delta - \gamma)} d\delta \right| \leq \frac{1}{\alpha' - \alpha} \int_{L'} \left| \frac{G(\delta)}{\delta} \right| d\beta',$$

the second term of the inner integral becomes arbitrarily small as α' increases; therefore the integral vanishes altogether. Hence

$$FG\eta = \frac{1}{(2\pi i)^2} \int_L \frac{F(\gamma)}{\gamma} d\gamma \int_{L'} \frac{G(\delta)}{\delta - \gamma} e^{\delta t} d\delta.$$

It follows from our assumptions that we can interchange the order of integration[1] in this double integral. We obtain

[1] Let L_1 be a finite interval of the straight line L between the ordinates $-T$ and $+T$ (see Fig. 40); then we have

$$FG\eta = \lim_{L_1 \to L} \frac{1}{(2\pi i)^2} \int_{L'} G(\delta)e^{\delta t} d\delta \int_{L_1} \frac{F(\gamma)}{\gamma(\delta - \gamma)} d\gamma.$$

The assertion follows immediately from the estimate

$$\left| \int_{L-L_1} \frac{F(\gamma)}{\gamma(\delta - \gamma)} d\gamma \right| \leq \frac{1}{|\delta|} \sqrt{2 \int_T^\infty \psi^2 \, d\rho} \sqrt{2 \int_L \left(\frac{1}{|\gamma|^2} + \frac{1}{|\gamma - \delta|^2} \right) d\beta}$$

and from the convergence of the integral $\int_0^\infty \psi^2 \, d\rho$.

$$FG\eta = \frac{1}{(2\pi i)^2} \int_{L'} G(\delta) e^{\delta t} \, d\delta \int_L \frac{F(\gamma)}{\gamma(\delta - \gamma)} \, d\gamma.$$

Thus only the relation

$$\frac{1}{2\pi i} \int_L \frac{F(\gamma)}{\gamma(\delta - \gamma)} \, d\gamma = \frac{F(\delta)}{\delta}$$

remains to be proved.   But this relation is a consequence of Cauchy's integral theorem.

The justification of the *translation theorem* also follows immediately from the complex integral representation.   If $F(p)$ is a given operator and therefore

$$F(p)\eta = \frac{1}{2\pi i} \int_L \frac{F(\gamma)}{\gamma} e^{\gamma t} \, d\gamma,$$

then we have

$$F(p + k)\eta = \frac{1}{2\pi i} \int_L \frac{F(\gamma + k) e^{\gamma t}}{\gamma} \, d\gamma = \frac{e^{-kt}}{2\pi i} \int_L \frac{F(\gamma)}{\gamma - k} e^{\gamma t} \, d\gamma.$$

This expression means

$$F(p + k)\eta = e^{-kt} F(p) \frac{p}{p - k} \eta.$$

Since $\dfrac{p}{p - k} \eta = e^{kt}\eta$, it follows that

$$F(p + k)\eta = e^{-kt} F(p) e^{kt} \eta.$$

It is also easy to establish the equivalence of the definitions used in our earlier examples and the present integral definition.   We have:[1]

1)
$$\frac{1}{p^n} = \frac{1}{2\pi i} \int_L \frac{e^{\gamma t}}{\gamma^{n+1}} \, d\gamma,$$

where n is a positive integer.   The path of integration can be deformed into an arbitrary curve surrounding the origin; therefore

$$\frac{1}{p^n} = \frac{1}{n!} \left[\frac{d^n}{d\gamma^n} e^{\gamma t} \right]_{\gamma=0} = \frac{t^n}{n!}.$$

[1] For the sake of brevity, we shall write $F(p)$ for the expression $F(p)\eta.$

2)
$$\frac{p}{p+\alpha} = \frac{1}{2\pi i} \int_L \frac{e^{\gamma t}}{\gamma + \alpha}\, d\gamma = e^{-\alpha t}.$$

3)
$$\frac{p}{(p+\alpha)^{n+1}} = \frac{1}{2\pi i} \int_L \frac{e^{\gamma t}}{(\alpha + \gamma)^{n+1}}\, d\gamma = e^{-\alpha t}\frac{t^n}{n!}.$$

4)
$$\sqrt{p} = \frac{1}{2\pi i} \int_L \frac{e^{\gamma t}}{\sqrt{\gamma}}\, d\gamma = \frac{1}{\sqrt{t}}\frac{1}{2\pi i} \int_L \frac{e^{\gamma}}{\sqrt{\gamma}}\, d\gamma.$$

If we replace the variable of integration γ by $\kappa = \sqrt{\gamma}$, it follows that

$$\frac{1}{2\pi i} \int_L \frac{e^{\gamma}}{\sqrt{\gamma}}\, d\gamma = \frac{1}{\pi i} \int_{L'} e^{\kappa^2}\, d\kappa.$$

The path of integration L' in the κ-plane is (cf. §3, 3) the right branch of an arbitrary equilateral hyperbola and is equivalent to the imaginary axis. Thus we have

$$\sqrt{p} = \frac{1}{\sqrt{t}}\frac{1}{\pi} \int_{-\infty}^{\infty} e^{-\beta^2}\, d\beta = \frac{1}{\sqrt{\pi t}}.$$

5)
$$p^s = \frac{1}{2\pi i} \int_L \frac{e^{\gamma t}}{\gamma^{1-s}}\, d\gamma = t^{-s}\frac{1}{2\pi i} \int_L \frac{e^{\gamma}}{\gamma^{1-s}}\, d\gamma.$$

For the value of the integral one can find

$$\frac{1}{2\pi i} \int_L \frac{e^{\gamma}}{\gamma^{1-s}}\, d\gamma = \frac{1}{\Gamma(1-s)}$$

and therefore

$$p^s = \frac{t^{-s}}{\Gamma(1-s)} \qquad\qquad (s < 1).$$

6)
$$e^{-kp} = \frac{1}{2\pi i} \int_L \frac{e^{(t-k)\gamma}}{\gamma}\, d\gamma = \begin{cases} 0 & \text{for } t < k \\ 1 & \text{for } t > k. \end{cases}$$

7)
$$\frac{p}{p^2 + a^2} = \frac{1}{2\pi i} \int_L \frac{e^{\gamma t}}{\gamma^2 + a^2}\, d\gamma.$$

If we deform L into a curve surrounding the points $\pm ia$, we obtain

$$\frac{p}{p^2 + a^2} = \frac{1}{2\pi i} \oint e^{\gamma t}\frac{1}{2ia}\left(\frac{1}{\gamma - ia} - \frac{1}{\gamma + ia}\right) d\gamma = \frac{\sin at}{a}.$$

8) As an example for the interpretation of further operators we consider

$$(34) \qquad \sqrt{p}\, e^{-x\sqrt{p}} = \frac{1}{2\pi i} \int_L \frac{e^{-x\sqrt{\gamma}+\gamma t}}{\sqrt{\gamma}}\, d\gamma.$$

The integral on the right may easily be evaluated; we have

$$(35) \qquad \sqrt{p}\, e^{-x\sqrt{p}} = \frac{1}{\sqrt{\pi t}}\, e^{-x^2/4t}.$$

For the operator $e^{-x\sqrt{p}}$, instead of directly evaluating the integral

$$e^{-x\sqrt{p}} = \frac{1}{2\pi i} \int_L \frac{e^{-x\sqrt{\gamma}}}{\gamma}\, e^{\gamma t}\, d\gamma,$$

we obtain the interpretation as follows:

$$e^{-x\sqrt{p}} = \sqrt{p}\, e^{-x\sqrt{p}}\, \frac{1}{\sqrt{p}} = \sqrt{p}\, e^{-x\sqrt{p}}\, \frac{2}{\sqrt{\pi}}\, \sqrt{t}.$$

Therefore

$$e^{-x\sqrt{p}} = \frac{d}{dt} \int_0^t \frac{e^{-x^2/4\tau}}{\sqrt{\pi\tau}}\, \frac{2}{\sqrt{\pi}}\, \sqrt{t-\tau}\, d\tau = \frac{1}{\pi} \int_0^t \frac{e^{-x^2/4\tau}}{\sqrt{\tau(t-\tau)}}\, d\tau.$$

9) The formula

$$(36) \qquad \frac{p}{\sqrt{p^2+a^2}} = \frac{1}{2\pi i} \int_L \frac{e^{\gamma t}}{\sqrt{\gamma^2+a^2}}\, d\gamma = J_0(at)$$

leads to an interesting application of the multiplication theorem. If we split the operator $p/(p^2+a^2)$ into the product

$$\frac{p}{p^2+a^2} = \frac{1}{p}\, \frac{p}{\sqrt{p^2+a^2}}\, \frac{p}{\sqrt{p^2+a^2}},$$

then Duhamel's principle yields

$$\frac{p}{p^2+a^2} = \int_0^t J_0(a(t-\tau))\, J_0(a\tau)\, d\tau.$$

On the other hand, according to example 7, we have

$$\frac{p}{p^2+a^2} = \frac{\sin at}{a}.$$

Thus we obtain the following *integral theorem for Bessel functions:*

$$(37) \qquad \int_0^t J_0(a(t-\tau))\, J_0(a\tau)\, d\tau = \frac{\sin at}{a}.$$

10) Finally we consider (cf. Vol. I, p. 158) the Abel integral equation

$$(38) \qquad f(t) = \int_0^t \frac{\phi(\tau)}{(t - \tau)^\alpha} d\tau \qquad (0 < \alpha < 1),$$

i.e., the operator equation

$$pf = \Gamma(1 - \alpha)p^\alpha\phi.$$

Its solution is

$$\phi = \frac{1}{\Gamma(1 - \alpha)} p^{1-\alpha}f$$

or

$$(39) \qquad \phi = \frac{1}{\Gamma(\alpha)\Gamma(1 - \alpha)} \frac{d}{dt} \int_0^t \frac{f(\tau)}{(t - \tau)^{1-\alpha}} d\tau.$$

Then, since $\Gamma(\alpha)\Gamma(1 - \alpha) = \pi/\sin \pi\alpha$, it follows that

$$(40) \qquad \phi(t) = \frac{\sin \pi\alpha}{\pi} \frac{d}{dt} \int_0^t \frac{f(\tau)}{(t - \tau)^{1-\alpha}} d\tau$$

in agreement with the result obtained in Vol. I, p. 159.

§3. General Theory of Transient Problems

The preceding section contains applications but not a complete justification for the method of operators. It is questionable whether it is worthwhile to formulate general theorems from which these methods can be justified deductively. The fascination of the method of operators lies in the ease and naturalness with which it applies to problems of quite different types. To subsume all these possibilities under one comprehensive theorem would require, at best, a rather clumsy formulation. No such attempt will be carried out here in a complete way; but we shall take a step in this direction by following up §1, 4. We shall not only show how the method can be justified but also formulate a theorem which governs some relatively complicated examples. The basis for this discussion will be the Laplace transformation, which has often been used for similar purposes particularly by G. Doetsch.[1]

1. The Laplace Transformation. The Laplace transformation is immediately obtained if in the two theorems on the Mellin integral

[1] See G. Doetsch [2] and [1].

formulas (cf. Vol. 1, pp. 103–104) we replace the variable x by e^{-x} and the function $g(x)$ by $g(e^{-x}) = \phi(x)$. However, we prove the *Laplace inversion formulas* again independently (from the Fourier integral theorem) and under somewhat broader assumptions.

Theorem 1: *Let the function $\phi(s)$ of the complex variable $s = \sigma + i\tau$ be regular analytic in a strip $\alpha < \sigma < \beta$ of the plane. In any narrower strip $\alpha + \delta \leq \sigma \leq \beta - \delta(\delta > 0$, arbitrarily fixed), let a positive function $\Phi(\rho)$ be given such that $\int_0^\infty \Phi(\rho)\, d\rho$ exists and such that*

$$(1) \qquad\qquad |\phi(s)| \leq \Phi(|\tau|) \qquad\qquad (s = \sigma + i\tau)$$

everywhere in this strip. Then, for real x and fixed σ,

$$(2) \qquad\qquad \psi(x) = \frac{1}{2\pi i} \int_{\sigma-i\infty}^{\sigma+i\infty} \phi(s)e^{xs}\, ds$$

exists and, in the strip $\alpha < \sigma < \beta$, the equation

$$(3) \qquad\qquad \phi(s) = \int_{-\infty}^{\infty} \psi(x)e^{-xs}\, dx$$

is valid.

Theorem 2: *If $\psi(x)$ is a piecewise smooth function for real x and if the integral $\int_{-\infty}^{\infty} \psi(x)e^{-\sigma x}\, dx$ converges absolutely for $\alpha < \sigma < \beta$, then the inversion (2) follows from (3).*

Corollary. *If $\beta = \infty$, and therefore $\phi(s)$ is regular in the whole half-plane $\sigma > \alpha$, and if $\phi(s)$ is subject to the additional assumptions given above,[1] then we have $\psi(x) = 0$ for $x < 0$. Thus, in this case, the reciprocal formulas*

$$
\begin{aligned}
\psi(x) &= \frac{1}{2\pi i} \int_{\sigma-i\infty}^{\sigma+i\infty} \phi(s)e^{xs}\, ds, \\
(4) \\
\phi(s) &= \int_0^\infty \psi(x)e^{-xs}\, dx
\end{aligned}
$$

hold.

First we prove Theorem 2. Let

[1] We refer in particular to the assumption that there exists an appropriate $\Phi(\rho)$ for all $\sigma \geq \alpha + \delta$.

$$\psi_T(x) = \frac{1}{2\pi i} \int_{\sigma-iT}^{\sigma+iT} \phi(s)e^{xs} \, ds = \frac{e^{x\sigma}}{2\pi} \int_{-T}^{T} \phi(\sigma + i\tau)e^{ix\tau} \, d\tau.$$

Substituting for $\phi(\sigma + i\tau)$ the expression

$$\phi = \int_{-\infty}^{\infty} \psi(\xi)e^{-\xi(\sigma+i\tau)} \, d\xi$$

we obtain

$$\psi_T(x) = \frac{e^{x\sigma}}{2\pi} \int_{-T}^{T} d\tau \int_{-\infty}^{\infty} \psi(\xi)e^{-\sigma\xi}e^{-i(\xi-x)\tau} \, d\xi.$$

Since $\psi(x)e^{-x\sigma}$ is piecewise smooth and $\int_{-\infty}^{\infty} |\psi(x)| e^{-x\sigma} \, dx$ converges for every fixed σ in the interval $\alpha < \sigma < \beta$, it follows by the Fourier integral theorem (cf. Vol. I, p. 78) that the integral

$$\frac{1}{2\pi} \int_{-T}^{T} d\tau \int_{-\infty}^{\infty} \psi(\xi)e^{-\sigma\xi}e^{-i(\xi-x)\tau} \, d\xi$$

tends to the value $\psi(x)e^{-\sigma x}$ as T increases; therefore $\psi_T(x)$ tends to $\psi(x)$ as asserted.

In order to prove Theorem 1, we form the integral

$$\psi(x) = \frac{e^{x\sigma}}{2\pi} \int_{-\infty}^{\infty} \phi(\sigma + i\tau)e^{i\tau x} \, d\tau$$

which, under our assumptions, converges absolutely in the interval $\alpha < \sigma < \beta$.

We now show that this integral does not depend on σ. Consider the integral

$$J = \int_{\sigma_1}^{\sigma_2} \phi(\sigma + iT)e^{x(\sigma+iT)} \, d\sigma$$

over a segment of a straight line, parallel to the real axis; this segment has the fixed length $\sigma_2 - \sigma_1 > 0$ and is wholly contained in the strip $\alpha + \delta \leq \sigma \leq \beta - \delta$. Using Cauchy's integral theorem, we see that indeed $\psi(x)$ does not depend on σ if J tends to zero as $|T|$ runs through an appropriate sequence $|T_1|, |T_2|, \cdots$ which increases without limit (see Figure 41). And the integral J does indeed tend to zero because of the estimate

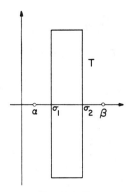

FIGURE 41

$$| J | \leq e^{|x|\sigma_2} \int_{\sigma_1}^{\sigma_2} | \phi(\sigma + iT) | \, d\sigma \leq e^{|x|\sigma_2} \Phi(| T |)(\sigma_2 - \sigma_1);$$

since the integral $\int_0^\infty \Phi(\rho) \, d\rho$ exists, there must be a sequence of values $T_1, T_2, \cdots$ for which $\Phi(| T |)$ tends to zero.

From the equation

$$\psi(x)e^{-x\sigma} = \frac{1}{2\pi} \int_{-\infty}^\infty \phi(\sigma - i\tau)e^{-i\tau x} \, d\tau,$$

it follows that $\psi(x)e^{-x\sigma}$ is the Fourier transform of the function $\phi(\sigma - i\tau)$, which is certainly piecewise smooth in τ. Thus, since the integral $\int_{-\infty}^\infty | \phi(\sigma - i\tau) | \, d\tau$ converges, we obtain by Fourier's inversion theorem

$$\phi(\sigma - i\tau) = \int_{-\infty}^\infty \psi(x)e^{-x(\sigma-i\tau)} \, dx,$$

$$\phi(s) = \int_{-\infty}^\infty \psi(x)e^{-xs} \, dx,$$

as asserted in Theorem 1.

To prove the corollary we note that under its hypotheses the estimate

$$(5) \qquad\qquad | \psi(x) | < e^{x\sigma} \frac{1}{\pi} \int_0^\infty \Phi(\rho) \, d\rho$$

is valid for all $\sigma \geq \alpha + \delta$ and for all x.   If x is negative the right side becomes arbitrarily small for sufficiently large σ, and hence

$$\phi(x) \equiv 0 \qquad\qquad \text{for} \quad x < 0,$$

as asserted in the corollary.

2. Solution of Transient Problems by the Laplace Transformation. We are now in a position to solve the transient problem I of §1 in a more general form than before (cf. §1, 4), even under the assumption that the initial state is not the state of rest.   The method of solution depends on the fact that problem I can be reduced to a different problem II with one less independent variable.   The two problems are equivalent on the basis of the Laplace transformation and its inverse, but in many cases the second problem can be treated simply and explicitly while the first cannot.   It is desirable to make the assumptions under which the transformation is applied broad enough to include the important practical applications.

We impose the following requirement on the solution $u(x, t)$ of problem I: There exists a real number α_0 such that with increasing t the functions

$$(6) \qquad u(x, t)e^{-\alpha_0 t}, \qquad u_x(x, t)e^{-\alpha_0 t}, \qquad u_{xx}(x, t)e^{-\alpha_0 t}$$

remain uniformly bounded in x if t increases beyond all bounds. Under this condition the associated Laplace transform, conveniently written in the form

$$\frac{v(x, \gamma)}{\gamma} = \frac{v}{\gamma} = \int_0^\infty u(x, t)e^{-\gamma t}\, dt,$$

exists for $\operatorname{Re} \gamma = \alpha > \alpha_0$ and is a regular analytic function of

$$\gamma = \alpha + i\beta$$

in the half-plane $\alpha > \alpha_0$.   On the basis of our assumptions, we obtain for the derivatives of this function

$$\frac{v_x}{\gamma} = \int_0^\infty u_x e^{-\gamma t}\, dt,$$

$$\frac{v_{xx}}{\gamma} = \int_0^\infty u_{xx} e^{-\gamma t}\, dt.$$

We deduce from these conditions that the Laplace transforms of the functions u_t and u_{tt} also exist. We have

$$\int_0^T u_t(x, t)e^{-\gamma t}\, dt = u(x, T)e^{-\gamma T} - \phi(x) + \gamma \int_0^T u(x, t)e^{-\gamma t}\, dt.$$

Since the right side converges for $T \to \infty$ and $\mathrm{Re}\, \gamma > \alpha_0$, the left side also converges, i.e., we have

$$\int_0^\infty u_t e^{-\gamma t}\, dt = v(x, \gamma) - \phi(x).$$

In the hyperbolic case $(a > 0)$ the existence of the integral

$$\int_0^\infty u_{tt} e^{-\gamma t}\, dt$$

also follows. We obtain

$$\int_0^\infty u_{tt} e^{-\gamma t}\, dt = (v - \phi)\gamma - \psi.$$

Moreover, if we multiply the differential equation (10) in §1 by $e^{-\gamma t}$ and integrate with respect to t from 0 to ∞, we obtain for v the nonhomogeneous ordinary differential equation

$$L[v] + (a\gamma^2 + b\gamma)\phi + a\gamma\psi = (a\gamma^2 + b\gamma)v.$$

This equation becomes homogeneous if the state of rest prevails initially. In an analogous way, we obtain the boundary conditions

$$v(0, \gamma) = \gamma \int_0^\infty f(t)e^{-\gamma t}\, dt,$$

$$\rho v_x = (\sigma - \lambda\gamma)v + \lambda\gamma\phi(l) \quad \text{for} \quad x = l.$$

Thus the following boundary value problem for an ordinary differential equation arises for the function v of the independent variable x and the complex parameter γ:

Problem II:

(7) $$L[v] + (a\gamma^2 + b\gamma)\phi + a\gamma\psi = (a\gamma^2 + b\gamma)v,$$

(7′)
$$v(0, \gamma) = \gamma \int_0^\infty f(t)e^{-\gamma t}\, dt$$
$$\rho v_x = (\sigma - \lambda\gamma)v + \lambda\gamma\phi(l) \quad \text{for} \quad x = l.$$

Let $f(t)$ be piecewise smooth for $t \geq 0$, let the integral $\int_0^\infty f(t)e^{-\alpha t}\, dt$

converge absolutely for $\alpha > \alpha_0$, and let $\phi(x)$ and $\psi(x)$ be continuous functions for $0 \leq x \leq l$.

We immediately deduce: *If problem II possesses a uniquely determined solution for every* $\gamma = \alpha + i\beta$ *with* $\alpha > \alpha_0$, *then there is at most one solution of the corresponding problem I which satisfies the requirements (6).* This is true because, under our assumptions, the Laplace transformation has a unique inverse, and therefore two different solutions of problem I would correspond to two different solutions of problem II.

It is even more important, however, to establish that by means of the Laplace inversion formula one can obtain the solution of problem I from the solution of problem II. We prove the following theorem: *Let* $v(x, \gamma)$ *be a solution of problem II which is continuous and possesses continuous derivatives with respect to x up to the second order in the interval* $0 \leq x \leq l$. *For every fixed x of this interval, let* $v(x, \gamma)$ *be regular everywhere in the half-plane* Re $\gamma > \alpha_0$ *of the complex γ-plane. Furthermore, in every half-plane* Re $\gamma \geq \alpha_0 + \delta$—*from which, in case* $\alpha_0 < 0$, *the origin is excluded by an arbitrarily small fixed circle—and in every fixed subinterval* $\epsilon \leq x \leq l$, *let an inequality of the form*

$$(8) \qquad \left| \frac{v(x, \gamma)}{\gamma} \right|_{L} \leq \Phi(|\beta|)$$

hold, where $\displaystyle\int_0^\infty \Phi(\rho)\, d\rho$ *exists. Now, if the function u defined by the integral*

$$(9) \qquad u(x, t) = \frac{1}{2\pi i} \int_L \frac{v(x, \gamma)}{\gamma} e^{\gamma t}\, d\gamma$$

is continuous in the domain $0 \leq x \leq l$, $t \geq 0$, $t^2 + x^2 \geq \epsilon > 0$ (ϵ *arbitrarily small) and has continuous first and second derivatives in the domain* $0 < x \leq l$, $t \geq 0$, $t^2 + x^2 \geq \epsilon$, *then u is the solution of the corresponding problem I.* Here the path of integration L is an arbitrary parallel to the imaginary axis lying in the half-plane $\alpha > \alpha_0$ or, in case $\alpha_0 < 0$, it is an indented path of the form illustrated in Figure 39 §2, 5.[1]

We begin the proof by showing that $u(x, t)$ satisfies the differential equation. As in the subsequent verification of the boundary and initial conditions, we shall carry out the necessary differentiations

[1] It is easy to prove that this result is identical with the representation of the solution by the Duhamel integral of §1, 3.

by using a device mentioned several times in §1.   We first form the
auxiliary function

$$(10) \qquad w(x, t) = \frac{1}{2\pi i} \int_L \frac{v(x, \gamma)}{\gamma^3} e^{\gamma t} d\gamma.$$

Because of assumption (8), this function can be twice differentiated
with respect to the time t under the integral sign, for $\epsilon \leq x \leq l, t \geq 0$.
In particular, we have

$$w_{tt} = u(x, t).$$

On the other hand, because of the differential equation (7), the integral

$$\frac{1}{2\pi i} \int_L \frac{L[v]}{\gamma^3} e^{\gamma t} d\gamma$$

converges uniformly in the domain $\epsilon \leq x \leq l, \epsilon \leq t \leq T$.   From
this we deduce that

$$L[w] = \frac{1}{2\pi i} \int_L \frac{L[v]}{\gamma^3} e^{\gamma t} d\gamma$$

holds.[1]   It then follows that

$$aw_{tt} + bw_t - L[w] = \frac{1}{2\pi i} \int_L \frac{e^{\gamma t}}{\gamma^3} [(a\gamma^2 + b\gamma)v - L[v]] d\gamma.$$

[1] It is sufficient to verify that

$$\frac{1}{2\pi i} \int_L \frac{pv_{xx} + qv_x}{\gamma^3} e^{\gamma t} d\gamma = pw_{xx} + qw_x$$

or, since $p > 0$, that

$$\Omega \equiv \frac{1}{2\pi i} \int_L \frac{(Pv_x)_x}{\gamma^3} e^{\gamma t} d\gamma = (Pw_x)_x$$

holds, where we have set $P(x) = \exp\left\{ \int_0^x \frac{q}{p} dx' \right\}$.

Integrating Ω as defined above by an integral, we see that

$$\int_\epsilon^x \frac{dx'}{P(x')} \int_\epsilon^{x'} \Omega \, dx'' = w(x, t) - w(\epsilon, t) - A(t) \int_\epsilon^x \frac{dx'}{P(x')},$$

where $A(t)$ does not depend on x.   Differentiation of the last equation yields
directly $\Omega = (Pw_x)_x$.

Thus, from the differential equation (7), we have

$$aw_{tt} + bw_t - L[w] = \frac{1}{2\pi i} \int_L \frac{e^{\gamma t}}{\gamma^3} [(a\gamma^2 + b\gamma)\phi + a\gamma\psi]\, d\gamma$$

or

(11) $$aw_{tt} + bw_t - L[w] = a\phi + (b\phi + a\psi)t.$$

If we differentiate twice with respect to t—this is possible since $u(x, t)$ is, by assumption, twice continuously differentiable—then for u we have the differential equation $au_{tt} + bu_t = L[u]$, which holds for all x, t with $0 < x \leq l, t > 0$.

The fact that u satisfies the boundary condition $u(0, t) = f(t)$ at the point $x = 0$ follows immediately from the inversion theorem and from the continuity of $u(x, t)$ which was assumed for $t > 0$, $0 \leq x \leq l$.

For the auxiliary function w, at the point $x = l$, the condition

$$\rho w_x + \lambda w_t - \sigma w = \frac{1}{2\pi i} \int_L \frac{e^{\gamma t}}{\gamma^3} [\rho v_x + (\lambda\gamma - \sigma)v]\, d\gamma$$

$$= \frac{\lambda\phi(l)}{2\pi i} \int_L \frac{e^{\gamma t}}{\gamma^2}\, d\gamma = \lambda t \phi(l)$$

follows. Hence, differentiating twice with respect to t, we have

$$\rho u_x + \lambda u_t - \sigma u = 0.$$

To verify the initial conditions we observe first that both $w(x, 0)$ and $w_t(x, 0)$ vanish for $x > 0$ because of

$$w(x, 0) = \frac{1}{2\pi i} \int_L \frac{v}{\gamma^3}\, d\gamma$$

and

$$w_t(x, 0) = \frac{1}{2\pi i} \int_L \frac{v}{\gamma^2}\, d\gamma$$

and because of assumption (8). ⸱In fact, the estimates

$$| w(x, 0) | \leq \frac{1}{\pi\alpha^2} \int_0^\infty \Phi(\rho)\, d\rho,$$

$$| w_t(x, 0) | \leq \frac{1}{\pi\alpha} \int_0^\infty \Phi(\rho)\, d\rho$$

are valid and yield this assertion as $\alpha \to \infty$. From our assumptions concerning $u(x, t)$, it follows that w and w_t are continuous together with their derivatives up to the second order in the domain

$$0 < x \leq l, \qquad t \geq 0,$$

and hence both $L[w]$ and $L[w_t]$ vanish for $t \to 0$. Equation (11) goes over into

(12) $a[w_{tt}(x, 0) - \phi(x)] = a[u(x, 0) - \phi(x)] = 0.$

Differentiating (11) we have, for $t = 0$,

$$a(w_{ttt} - \psi) + b(w_{tt} - \phi) = 0$$

or

(13) $a[u_t(x, 0) - \psi(x)] + b[u(x, 0) - \phi(x)] = 0.$

In the case $a \neq 0$ it follows that

$$u(x, 0) = \phi(x), \qquad u_t(x, 0) = \psi(x),$$

and in the case $a = 0, b \neq 0$, that

$$u(x, 0) = \phi(x),$$

which concludes the proof.

Finally we remark: Because of (8), the estimate for u,

$$|u(x, t)| \leq \frac{1}{\pi} e^{\alpha t} \int_0^\infty \Phi(\rho) \, d\rho,$$

is valid for arbitrary $\alpha \geq \alpha_0$. For $t < 0$ it follows that

(14) $u(x, t) = 0$

and for $t > 0$,

(15) $$|u(x, t)| \leq \frac{1}{\pi} e^{\alpha_0 t} \int_0^\infty \Phi(\rho) \, d\rho.$$

This shows that the assumption (6) initially made for the function u is actually satisfied by the solution just constructed.

3. Example. The Wave and Telegraph Equations. An illustration of the method is given by the *telegraph equation*

(16) $u_{tt} = u_{xx} - r^2 u$ $(r = \text{constant})$

with the initial conditions

(16′) $$u(x, 0) = u_t(x, 0) = 0$$

and the boundary conditions[1]

(16″)
$$u(0, t) = \frac{t^3}{3!},$$
$$\rho u_x + \lambda u_t = \sigma u \qquad \text{for} \quad x = l.$$

We have the corresponding problem II:

(17) $$v_{xx} = \kappa^2 v,$$

(17′)
$$v(0, \gamma) = \frac{1}{\gamma^3}$$
$$\rho v_x = (\sigma - \lambda\gamma)v \qquad \text{for} \quad x = l,$$

where we must set

$$\kappa^2 = \gamma^2 + r^2.$$

The solution is given by

(18)
$$v(x, \gamma) = \frac{\rho\kappa \cosh \kappa(l - x) + (\lambda\gamma - \sigma) \sinh \kappa(l - x)}{\rho\kappa \cosh \kappa l + (\lambda\gamma - \sigma) \sinh \kappa l} \frac{1}{\gamma^3}$$
$$= \frac{e^{-\kappa x} - \epsilon(\kappa)e^{\kappa(x-2l)}}{1 - \epsilon(\kappa)e^{-2\kappa l}} \frac{1}{\gamma^3},$$

where now

(19)
$$\epsilon(\kappa) = \frac{\lambda \sqrt{\kappa^2 - r^2} - \rho\kappa - \sigma}{\lambda \sqrt{\kappa^2 - r^2} + \rho\kappa - \sigma}.$$

As before, we see that an $\alpha_0 > 0$ exists such that the denominator in (18) possesses no zeros in the half-plane Re $\gamma > \alpha_0$; therefore v is regular everywhere in this half-plane. On any parallel L to the imaginary axis lying in the half-plane Re $\gamma \geq \alpha_0 + \delta$, the inequality

[1] Instead of the impulse function $U(x, t)$ we construct the function $U_3(x, t)$ (cf. §1, 3) in order to make certain, on the basis of the theorem of article 2, that the corresponding integral $(1/2\pi i)\int_L [v(x,\gamma)/\gamma]e^{\gamma t} \, d\gamma$ represents the desired solution. The assumptions of this theorem would not be satisfied for the integral belonging to $U(x, t)$. We can, however, obtain U from U_3 afterward in the form $U(x, t) = \partial^3 U_3(x, t)/\partial t^3$.

$$\left| \frac{v(x, \gamma)}{\gamma} \right| \leq \frac{A'}{(B + |\beta|)^4}$$

holds, where $A > 0$ and $B > 0$ are constants independent of x and γ. From this and from the corresponding estimates for v_x/γ and v_{xx}/γ, it is clear that

$$(20) \qquad U_3(x, t) = \frac{1}{2\pi i} \int_L \frac{v(x, \gamma)}{\gamma} e^{\gamma t} \, d\gamma$$

represents a function which is continuous and has continuous derivatives of the first and second orders in the domain $0 \leq x \leq l, t \geq 0$. This function is therefore the solution of (16).

Again we consider the special cases[1]

$$(21) \qquad \begin{aligned} u_x(l, t) &= 0 & \epsilon(\kappa) &= -1 \\ u(l, t) &= 0 & \epsilon(\kappa) &= 1. \end{aligned}$$

As before, we develop v into the series

$$(22) \qquad \gamma^3 v(x, \gamma) = \sum_{\nu=0}^{\infty} \epsilon^\nu e^{-\kappa(x+2\nu l)} - \sum_{\nu=1}^{\infty} \epsilon^\nu e^{\kappa(x-2\nu l)}$$

and substitute this series into (20). Since termwise integration is clearly permissible, we obtain a series of the form

$$(23) \qquad U_3(x, t) = S(x, t) + \sum_{\nu=1}^{\infty} \epsilon^\nu [S(2\nu l + x, t) - S(2\nu l - x, t)],$$

where $S(x, t)$ is defined by the integral

$$(24) \qquad S(x, t) = \frac{1}{2\pi i} \int_L \exp\{-x\sqrt{\gamma^2 + r^2} + \gamma t\} \frac{d\gamma}{\gamma^4}.$$

For $r = 0$, we immediately have

$$S(x, t) = \frac{1}{2\pi i} \int_L e^{\gamma(t-x)} \frac{d\gamma}{\gamma^4}.$$

Thus

$$(25) \qquad S(x, t) = S(t - x) = \begin{cases} \dfrac{(t - x)^3}{3!} & \text{for } t > x \\ 0 & \text{for } t < x, \end{cases}$$

[1] Another important case is the so-called "matching" $\epsilon = 0$ which can occur, however, only for $r = 0$, $\sigma = 0$, and $\lambda = \rho$. In this case no "reflected" waves occur.

and finally

$$U_3(x, t) = S(t - x)$$

(26)
$$+ \sum_{\nu=1}^{\infty} \epsilon^{\nu}[S(t - x - 2\nu l) - S(t + x - 2\nu l)],$$

in agreement with our results of §1, 1 for the special case

$$f(t) = \frac{t^3}{3!}, \qquad\qquad t > 0.$$

Only a finite number of terms of the series (26) do not vanish identically, and each of these terms possesses continuous derivatives up to the second order and piecewise continuous derivatives of the third order. Therefore we immediately obtain the function U for $r = 0$ by differentiation:

$$U(x, t) \overset{\cdot}{=} \eta(t - x) + \sum_{\nu=1}^{\infty} \epsilon^{\nu}[\eta(t - x - 2\nu l) - \eta(t + x - 2\nu l)],$$

where

$$\eta(t) = \begin{cases} 1 & \text{for} \quad t > 0 \\ 0 & \text{for} \quad t < 0. \end{cases}$$

To calculate the integral (24) in the case $r \neq 0$ we introduce instead of γ the quantity $\phi = \sigma + i\tau$ as the variable of integration, by means of the equation

$$\gamma = ir \cos \phi.$$

It follows that

(27) $\quad S(x, t) = -\dfrac{1}{2\pi r^3} \displaystyle\int_{L'} \exp\left\{ir(ix \sin \phi + t \cos \phi)\right\} \dfrac{\sin \phi}{\cos^4 \phi}\, d\phi$.

Here L' is the image in the ϕ-plane of the straight line L, that is, L' is the curve

$$\text{Re } (ir \cos \phi) = \text{constant} > \alpha_0$$

represented in Figure 42. If $t < x$, then the real part of the exponent

$$ir(ix \sin \phi + t \cos \phi),$$

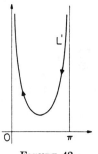

FIGURE 42

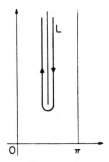

FIGURE 43

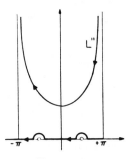

FIGURE 44

that is, the expression

(28) $r \sin \sigma(t \sinh \tau - x \cosh \tau)$,

becomes negatively infinite as $\tau \to \infty$ in the domain $0 < \sigma < \pi$, $\tau > 0$.

Thus L' can be contracted to a double straight line running through this domain (cf. Figure 43) and therefore

$$(29) \qquad\qquad S(x, t) \equiv 0 \qquad\qquad (t < x).$$

If $t > x$, then (28) becomes negatively infinite as $\tau \to \infty$ in the domain $-\pi < \sigma < 0$, $\tau > 0$ and L can be stretched into a curve L'' around the edge of the whole strip $-\pi < \sigma < \pi$ (cf. Figure 44). Since the integrand in (27) is periodic, we obtain

$$(30) \quad S(x, t) = \frac{1}{2\pi r^3} \int_{-\pi}^{\pi} \exp\{ir(ix \sin \phi + t \cos \phi)\} \frac{\sin \phi}{\cos^4 \phi} \, d\phi \, ,$$

where the points $\phi = +\pi/2$ are excluded by using small semicircles open at the bottom.

It is sufficient to calculate the function

$$(31) \quad f(x, t) = \frac{1}{2\pi r^4} \int_{-\pi}^{\pi} \exp\{ir(ix \sin \phi + t \cos \phi)\} \frac{d\phi}{\cos^4 \phi}$$

with $S = -f_x$. The fourth derivative of (31) with respect to t immediately turns out to be

$$\frac{\partial^4 f(x, t)}{\partial t^4} = \frac{1}{2\pi} \int_{-\pi}^{\pi} \exp\{ir(ix \sin \phi + t \cos \phi)\} \, d\phi = J_0(r\sqrt{t^2 - x^2}).$$

For $t = x$, $f(x, t)$ vanishes along with its derivatives with respect to t up to the third order. Actually, if we set $t = x$ in (31) and introduce $z = e^{i\phi}$ as the variable of integration, we obtain

$$f(x, x) = \frac{8}{i\pi r^4} \oint \frac{z^3}{(1 + z^2)^4} e^{irxz} \, dz \, ,$$

where the unit circle of the z-plane is to be chosen as the path of integration and the points $z = \pm i$ are to be excluded by indenting the path. We have immediately $f(x, x) = 0$; correspondingly we derive the vanishing of the derivatives for $t = x$. Thus we obtain

$$f(x, t) = \frac{1}{3!} \int_{x}^{t} (t - \tau)^3 J_0(r\sqrt{\tau^2 - x^2}) \, d\tau$$

and therefore, as the end result,

$$(32) \quad S(x, t) = \begin{cases} -\dfrac{\partial}{\partial x} \displaystyle\int_{x}^{t} \frac{(t - \tau)^3}{3!} J_0(r\sqrt{\tau^2 - x^2}) \, d\tau & \text{for} \quad t > x \\[4mm] 0 & \text{for} \quad t < x \, . \end{cases}$$

For the special case $r = 0$ we again obtain the earlier result

$$S(x, t) = \begin{cases} \dfrac{(t - x)^3}{3!} & \text{for } t > x \\ 0 & \text{for } t < x. \end{cases}$$

In the case $r \neq 0$ the series (23) possesses, according to (32), only a finite number of terms which do not vanish identically, and each of which has continuous derivatives up to the second order and piecewise continuous third derivatives. Consequently, by differentiation, we obtain for the impulse function $U = \partial^3 U_3 / \partial t^3$ the expression

$$(23') \quad U(x, t) = S(x, t) + \sum_{\nu=1}^{\infty} \epsilon^{\nu}[S(2\nu l + x, t) - S(2\nu l - x, t)],$$

where we have set

$$(24') \qquad S(x, t) = \begin{cases} -\dfrac{\partial}{\partial x} \displaystyle\int_x^t J_0(r\sqrt{\tau^2 - x^2})\, d\tau & \text{for } t > x \\ 0 & \text{for } t < x. \end{cases}$$

This function is a solution of problem (16) for the boundary condition $U(0, t) = 1$.

We refer to the literature for a great wealth of other significant examples.

CHAPTER VI

Hyperbolic Differential Equations in More Than Two Independent Variables

Introduction

The theory of hyperbolic differential equations of mathematical physics in more than two independent variables covers a field of such variety that a complete treatment is impossible in this book, even though we shall largely restrict ourselves to linear problems. The present chapter will discuss the subject from the selective viewpoint of wave propagation. Characteristic surfaces and bicharacteristic rays (along which disturbances are propagated) will play a central role in the discussion of Cauchy's initial value problem and of the problem of radiation.

The construction of solutions for more than two independent variables is no longer feasible by the simple iterative methods of Chapter V. Nevertheless, we shall be able to describe the general structure of solutions and to give a detailed analysis in relevant special cases.

Primarily we shall deal with Cauchy's problem for a single equation of arbitrary order and with systems of such equations for several unknown functions. In the rather elementary first two sections, special attention will be given to single equations of second order; later sections will emphasize *symmetric hyperbolic* systems of first order. These systems can be treated somewhat more simply than general equations or systems of higher order. Significantly, almost all equations of mathematical physics occur in the form of first order symmetric hyperbolic systems.—Some material from Chapter III will have to be repeated in a modified form.

The first part of this chapter is devoted to the questions of uniqueness, existence, construction, and geometry of solutions; the second part is concerned mainly with the representation of solutions in terms of the data and related questions.

The construction in the first part will clearly exhibit the basic fact of the existence of a finite domain of dependence of the solution on Cauchy data (as already seen in Chapter V for two independent variables). In contrast, the explicit representations of the solutions by "plane waves" in the second part are achieved only at the expense of obscuring the feature of a finite initial domain of dependence.

One should always keep in mind that, as a rule, statements and results refer to situations merely "in the small"; yet, whenever possible, we shall extend our results to larger domains appropriate to the underlying problems.

While we shall concentrate on *linear problems*, we shall indicate in sufficient detail generalizations to quasi-linear systems.

Finally we recall (Ch. III, §4, see also §10, 1 and §12, 5) that, in the case of linear hyperbolic problems, Duhamel's representation expresses the solution of inhomogeneous problems in terms of solutions of homogeneous problems. We may therefore restrict ourselves mainly to a discussion of homogeneous initial value problems.

We shall consider $n + 1$ independent variables x_0, x_1, $\cdots$, x_n, denoted by the vector x. Often, however, we shall distinguish the variable x_0 as the time t, and then write our independent variables as t, x. We shall also use conventional abbreviations such as

$$a_i b_i = \sum_{i=0}^{n} a_i b_i$$

when convenient.

PART I

Uniqueness, Construction, and Geometry of Solutions

§1. *Differential Equations of Second Order. Geometry of Characteristics*

1. **Quasi-Linear Differential Equations of Second Order.** We consider a second order quasi-linear differential equation

$$(1) \qquad L[u] + d = \sum_{i,k=0}^{n} a_{ik} u_{ik} + d = 0.$$

Here the abbreviations $u_i = \partial u/\partial x_i$ and $u_{ik} = \partial^2 u/\partial x_i \, \partial x_k$ are used. d and the coefficients $a_{ik} = a_{ki}$ are given functions of the independent variables x_0, x_1, $\cdots$, x_n, of the function u and of the derivatives u_i.[1] We assume, as always, that the quantities which occur are continuous in the region considered, unless the contrary is expressly stated.

As already described in Chapters I and III, the concept of characteristics originates from the problem of extending initial values on a surface C defined by $\phi(x_0, \cdots, x_n) = 0$, with grad $\phi \neq 0$, to a solution of (1).[2] These initial data on C consist of the values of u (which determine the interior derivatives of u) and the values of one exterior, or outward, derivative, say $u_\phi = \sum_{i=0}^{n} u_i \phi_i$. All first derivatives of u on C are then determined.[3]

While Cauchy's initial value problem aims at a solution of (1) in a full neighborhood of C, the differential equation considered merely along C presents the much simpler problem: Extend the initial data to an *integral strip*; i.e. find functions u, u_i, u_{ik} which satisfy (1) along C. (See Ch. III, §2.) It is this simple problem which leads immediately to the concept of characteristics.[4]

Instead of x_0, $\cdots$, x_n we introduce new independent variables λ_0, $\cdots$, λ_n one of which is $\lambda_n = \phi$ while λ_0, $\cdots$, λ_{n-1} are "interior variables" on C; thus the problem is more precisely formulated as follows: Given on C the quantities

$$(1') \qquad u(\lambda_0, \cdots, \lambda_{n-1}), \qquad u_\phi(\lambda_0, \cdots, \lambda_{n-1}),$$

construct a function $u(x_0, \cdots, x_n)$ so that on C this function and its derivative with respect to ϕ coincide with the given functions $(1')$ of λ_0, $\cdots$, λ_{n-1}, and so that u satisfies (1) on C.

Clearly, all the second derivatives of $u(x_0, \cdots, x_n)$ except $u_{\phi\phi}$ are uniquely determined from the data on C by differentiating the quantities $(1')$ with respect to interior variables. We now investigate whether the differential equation (1) and the initial data also deter-

[1] In this chapter, subscripts on functions subject to differential operators will denote partial derivatives, as for example, u_i, ϕ_i, ψ_i, ω_i, while subscripts on given coefficients, such as a_{ik}, b_i, c_i, are merely indices.

[2] Later in this chapter we shall modify the notation by reserving the letters C and ϕ for characteristic surfaces.

[3] For equations of order m, the appropriate *Cauchy data* consist of the values of u and of outward derivatives up to order $m - 1$.

[4] Compare throughout the discussions in Chapter III.

mine the quantity $u_{\phi\phi}$ along C. In terms of the variables λ_0, λ_1, $\cdots$, $\lambda_n = \phi$, equation (1) takes the form

$$(2) \qquad u_{\phi\phi}\, Q(\phi_i) + \cdots = 0$$

with the *characteristic form*

$$(3) \qquad Q(\phi_i) = Q(\phi_i\,, \phi_k) = \sum_{i,k=0}^{n} a_{ik}\, \phi_i\, \phi_k\,.$$

The dots in (2) stand for expressions which are all known from the initial data, i.e., which contain only interior derivatives of u and of the first derivatives of u. The second derivative $u_{\phi\phi}$ is therefore uniquely determined by the data at every point P on C where the coefficient of $u_{\phi\phi}$ in (2), i.e., the quadratic form $Q(\phi_i)$, does not vanish. Thus the following *alternatives* for every point P on C obtain: Either the exterior second derivative $u_{\phi\phi}$, and with it all the second derivatives, are uniquely determined by the data and the differential equation, or the differential equation represents an additional restriction on the data.

Subsequently it will be assumed that one alternative holds on the *entire* surface C for the data prescribed in (1'). In the first case we call the initial surface *free* (see Ch. III, §2), and in the second, *characteristic*. In the second case the *characteristic condition*

$$(4) \qquad Q(\phi_i\,, \phi_k) = Q(\phi_i) = \sum_{i,k=0}^{n} a_{ik}\, \phi_i\, \phi_k \equiv 0$$

is satisfied on C with the values of u and u_i, given by the data, substituted in the coefficients a_{ik}.

Although the characteristic condition (4) has the form of a first order partial differential equation for ϕ, the function

$$\phi(x_0\,, x_1\,, \cdots\,, x_n)$$

need not satisfy this differential equation identically; by definition, it must satisfy (4) only on C, that is, for $\phi = 0$. However, if C is given in the form

$$x_0 = \psi(x_1\,, x_2\,, \cdots\,, x_n),$$

then (4) does represent a partial differential equation for a function ψ of only n variables:

$$(5) \qquad \sum_{i,k=1}^{n} a_{ik}\, \psi_i\, \psi_k - 2\sum_{i=1}^{n} a_{i0}\, \psi_i + a_{00} = 0,$$

where expressions in x_1, x_2, $\cdots$, x_n have to be substituted for x_0 and u and u_i in the coefficients.

If in particular $a_{00} = -1$, $a_{i,0} = 0$ for $1 \leq i \leq n$, and if a_{ik} is independent of $x_0 = t$, then (1) becomes

$$(5') \qquad u_{tt} - \sum_{i,k=1}^{n} a_{ik} u_{ik} + \cdots = 0,$$

and the characteristic partial differential equation for ψ is

$$\sum_{i,k=1}^{n} a_{ik} \psi_i \psi_k = 1.$$

This case often occurs in physics.

For a given solution $u = u(x_0, x_1, \cdots, x_n)$ of the differential equation (1), the quantities u and the derivatives u_i are known functions of $x_0, x_1, \cdots, x_n$. Replace u and u_i wherever they occur in a_{ij} by these known functions; then *the characteristic condition* (4) *defines characteristic surfaces for the given solution* u. *If* (4) *is satisfied not only for* $\phi = 0$ *but identically in* $x_0, x_1, \cdots, x_n$, *then* $\phi = c = const.$ *furnishes a one-parameter family of characteristic surfaces depending on* c. *Conversely: If* $\phi = c$ *is such a family of characteristic surfaces, then* ϕ *satisfies the characteristic condition* (4) *interpreted as a first order partial differential equation.*

Furthermore, equation (2) shows that for a characteristic surface C: $Q(\phi_i, \phi_k) = 0$ the second order differential operator $L[u]$ is an *interior differential operator* in the following sense: If along C the values of u and the derivatives u_i are given, then $L[u]$ is known. In fact, from the data all the interior derivatives of u_i and also of u_ϕ are known. The second outward derivative $u_{\phi\phi}$ does not occur, since it appears in the form $u_{\phi\phi} Q$ and $Q = 0$. Hence $L[u] = 0$ may be considered on C as a partial differential equation of first order for the outward first derivative u_ϕ.

As already pointed out in Chapter III, characteristic surfaces can exist only if the condition (4) is satisfied by real functions ϕ; then the *quadratic form* $Q(\phi_i)$ must be indefinite. Equations for which the quadratic form Q in $n + 1$ variables can be reduced to a similar form in fewer variables by a linear transformation (parabolic equations) will not be considered in this chapter.

We now assume that the form Q is not merely *indefinite*, but, more specifically, that *the index of inertia is one*; i.e., at each point under consideration, Q can be brought into the form

$$Q = \Phi_1^2 + \Phi_2^2 + \cdots + \Phi_n^2 - \Phi_0^2$$

by means of a suitable local linear transformation of the independent variables x_i at a point P.

Then the differential equation (1) is called *hyperbolic*. Its principal part in the new variables at the point P is simply

$$u_{11} + u_{22} + \cdots + u_{nn} - u_{00} ,$$

the same as that of the *wave equation*. The n-dimensional surface element $\Phi_0 = 0$ is called space-like, and the direction of the Φ_0-axis may be considered as the time axis at P. Later in §3, we shall give a more general analysis of the concept of hyperbolicity (see also Ch. III, §2), not based on the quadratic character of Q.

The simplest instance is the wave equation

$$u_{11} + u_{22} + \cdots + u_{nn} - u_{00} = 0$$

as given above, with the corresponding characteristic condition

$$\phi_1^2 + \phi_2^2 + \cdots + \phi_n^2 - \phi_0^2 = 0.$$

Different indices of inertia for the quadratic form Q of course are also possible. We shall return to such "ultrahyperbolic" cases later in §16 of this chapter. The typical example is the differential equation

$$u_{11} + u_{22} - u_{33} - u_{44} = 0$$

with the characteristic condition

$$\phi_1^2 + \phi_2^2 - \phi_3^2 - \phi_4^2 = 0.$$

2. Linear Differential Equations. The general situation outlined in article 1 is simplified for a linear differential equation

(6) $$L[u] + d = 0 ,$$

where

(7) $$L[u] \equiv \sum_{i,k=0}^{n} a_{ik}\, u_{ik} + \sum_{i=0}^{n} a_i\, u_i + au,$$

and where the coefficients a_{ik}, a_i, a are given functions merely of the $n + 1$ independent variables $x_0, x_1, \cdots, x_n$. The characteristic condition (4) (or (5)) then depends only on the surface C, and not on the data; hence it is independent of the particular solution under consideration. The same is true if independence of u and its first derivatives is assumed merely for the coefficients of the principal part, i.e., for a_{ik}. Such equations are called *semilinear*. A surface $\phi = 0$ in x-space satisfying the relation (4), i.e., $Q(\phi_i) = 0$, is called a *characteristic surface of the linear differential equation* (6). Obviously hyperbolicity is in this case a property of the differential equation itself and does not depend on the given Cauchy data.

The relation between the characteristic condition (4) and the partial differential equation (5)

$$\sum_{i,k=1}^{n} a_{ik}\,\psi_i\,\psi_k - 2\sum_{i=1}^{n} a_{i0}\,\psi_i + a_{00} = 0,$$

can be described as follows: Suppose $\phi = \phi(x_0, x_1, \cdots, x_n)$ is a solution of equation (4), interpreted as a partial differential equation. If the equation $\phi = c = $ const. is solved for x_0:

$$x_0 = \psi(x_1, x_2, \cdots, x_n, c),$$

then a one-parameter family of solutions $x_0 = \psi(x_1, x_2, \cdots, x_n, c)$ of equation (5) is obtained. Conversely, if

$$x_0 = \psi(x_1, x_2, \cdots, x_n, c)$$

is a one-parameter family of solutions of the partial differential equation (5), and is solved for c in the form

$$c = \phi(x_0, x_1, \cdots, x_n),$$

then ϕ is a solution of the partial differential equation (4).

Now if $\phi = 0$ is an arbitrary characteristic surface satisfying equation (4) for $\phi = 0$, then the corresponding function

$$x_0 = \psi(x_1, x_2, \cdots, x_n)$$

is a solution of the partial differential equation (5). Every sufficiently smooth solution of such a first order partial differential equation can be embedded in a one-parameter family of solutions $x_0 = \psi(x_1, x_2, \cdots, x_n, c)$.[1] By solving for c we obtain a corresponding solution of the partial differential equation (4). Hence:

---

[1] For example, according to Chapter II, we can determine the solutions of the partial differential equation (5) by prescribing initial values which depend on a parameter c.

*Every characteristic surface $\phi = 0$ can be embedded in a one-param-
eter family of charcteristic surface $\phi = c$.* We may therefore assume such embedding without loss of generality, and shall always
do so unless the contrary is stated. Then ϕ is a solution of (4)
interpreted as a partial differential equation.

As an example of such an embedding, we consider, for $n = 2$,
the differential equation $u_{tt} - u_{xx} - u_{yy} = 0$ and the characteristic
cone $\chi \equiv t^2 - x^2 - y^2 = 0$. This function χ satisfies the differential equation

$$\chi_t^2 - \chi_x^2 - \chi_y^2 = 4\chi.$$

Hence the cone $\chi = 0$ is characteristic, but the surfaces $\chi = c$
are not characteristic for $c \neq 0$. On the other hand, if we embed
the original characteristic cone in the family of cones

$$\phi = t - \sqrt{x^2 + y^2} = c$$

we have

$$\phi_t^2 - \phi_x^2 - \phi_y^2 = 0,$$

so that the surfaces $\phi = c$ are characteristic for every constant c.
Corresponding statements hold for the wave equation in any number
of dimensions.

3. Rays or Bicharacteristics. According to the theory of first
order equations in Chapter II, every characteristic surface $\phi = 0$ or
$\phi = $ const. is generated by a family of *bicharacteristic curves* or
rays, which are closely linked with the differential equation (1) of
second order. These rays are defined in terms of a suitable curve
parameter s by the system of $n + 1$ ordinary differential equations

$$(8) \qquad \dot{x}_i = \tfrac{1}{2}Q_{\phi_i} \equiv \sum_{k=0}^{n} a_{ik}\phi_k \qquad (i = 0, 1, \cdots, n)$$

if the coefficients a_{ik} in (1) are independent of u and $u_1, \cdots, u_n$,
that is, in the case of a linear or "semilinear" equation (1).[1] For a
quasi-linear equation (1) we consider a fixed solution u and insert
in the coefficients a_{ik} the corresponding values $u(x)$ and $u_i(x)$; then
the rays are defined by $\dot{x}_i = \tfrac{1}{2}Q_{\phi_i}$. In any case one may supplement the bicharacteristic rays by introducing the "bicharacteristic
strip" $x_i(s)$, $p_i(s)$, where $p_i = \phi_i$, and define the strip quantities as

[1] In this case only the second order terms are required to be linear.

solutions of the canonical system (see Ch. II, §8)

$$(8')\qquad \dot{x}_i = \frac{1}{2}\frac{\partial Q}{\partial p_i}, \qquad \dot{p}_i = -\frac{1}{2}\frac{\partial Q}{\partial x_i} \qquad (i = 0, 1, \cdots, n).$$

The system $(8')$ then yields *all possible characteristic strips* for the solution u.

It should be remembered that these ordinary differential equations $(8')$ yield solutions along each of which $Q = \text{const.} = c$. To single out those genuinely belonging to (1) we must impose the further condition $Q = 0$ at one point of each of the rays, which then implies that $Q = 0$ along the whole ray.[1]

The integral curves of the system (8) subject to the condition $Q = 0$ are the *characteristic rays or bicharacteristics of the given second order differential equation* (1); they generate all members of families of characteristic surfaces $\phi = \text{const.}$

One also may recall from Chapter II: *If two distinct characteristic surfaces $\psi = t$ and $\chi = t$ are tangent at the time $t = 0$, then at every later time they have a common point of tangency which moves along a ray common to the two wave fronts.* This statement is equivalent to the theorem that two integral surfaces of a partial differential equation of first order (here the characteristic equation) which have a surface element in common have also an entire characteristic strip in common.

If the coefficients a_{ik} of the partial differential equation (1) *are constant, all the characteristic rays are straight lines.* We see this immediately from $(8')$. It follows also because complete integrals ϕ of (4) can be obtained as families of linear functions; by forming envelopes of one-parameter subfamilies, we immediately get straight lines as curves of contact.

The simplest example is again provided by the wave equation

$$u_{tt} - u_{x_1 x_1} - \cdots - u_{x_n x_n} = 0,$$

---

[1] In a slightly more general way we may associate with any (not necessarily characteristic) family of surfaces $\phi = \text{const.}$ a family of "transversal" curves defined by (8). Thus the planes tangent to these surfaces and the corresponding transversal directions are conjugate with respect to the quadratic surface $\sum_{i,k=0}^{n} a_{ik}\,\xi_i\,\xi_k = 0$.

The surface $\phi = \text{const.}$ is characteristic if and only if at every point the transverse direction is tangent to the surface. Then transverse differentiation is interior differentiation. In fact, the characteristic condition can immediately be written in the form $\sum_{i=0}^{n} \dot{x}_i\,\phi_i = 0$.

where we have set $x_0 = t$. The characteristic relation is

$$\phi_t^2 - \phi_{x_1}^2 - \phi_{x_2}^2 - \cdots - \phi_{x_n}^2 = 0,$$

and the rays are all the lines of x,t-space of the form $x_i = a_i + \alpha_i t$, where $\sum_{i=1}^n \alpha_i^2 = 1$.

The bicharacteristics, interpreted not in an $(n + 1)$-dimensional x,t-space but in an n-dimensional x-space, depending on the time parameter t, are arbitrary straight lines on which the point x moves with unit velocity. The characteristic surfaces, given by an equation $t = \psi(x_1, x_2, \cdots, x_n)$, satisfy the partial differential equation

$$\sum_{i=1}^{n} \psi_i^2 = 1.$$

Thus the characteristics for the wave equation are defined by a family of parallel surfaces $\psi = t$ (see Ch. II, §6) generated from an initial surface by a parallel motion with unit velocity along the normals. The rays are the associated orthogonal trajectories.

4. Characteristics as Wave Fronts. Characteristic surfaces play a role as "wave fronts", that is, surfaces across which solutions of (1) might suffer discontinuities, e.g., discontinuities of the second derivatives. Such discontinuities would imply different values of the second derivatives on both sides of the surface. Since on free surfaces the second derivatives are uniquely determined by the Cauchy data, such ambiguity is possible only along characteristics.

Such "wave fronts" occur, e.g., as frontiers beyond which there is no excitation at time t. The solution representing the disturbance vanishes identically on one side of this frontier but not on the other.

On many occasions we shall return to the significant concept of wave fronts. (See in particular §2.) Here the following remarks may suffice:

Let us now assume specifically that (1) is linear, and again set $x_0 = t$ and $\phi = \psi(x_1, x_2, \cdots, x_n) - t$. We interpret t as the time and u as a function in the n-dimensional x-space R_n with the time as a parameter. Then we deal with a solution $u(x_1, x_2, \cdots, x_n, t)$ of (1) with a surface of discontinuity

$$\psi(x_1, x_2, \cdots, x_n) = t,$$

which depends on the time t and moves through the x-space.

Let us, for convenience, assume the form (5′) of the differential equation. Along the rays, $dt/ds = 1$. Thus the curve parameter introduced above is identical with the time t, and the equations for the rays are

$$(9) \qquad \frac{dx_i}{dt} = \sum_{k=1}^{n} a_{ik}\,\psi_k \qquad (i = 1, 2, \cdots, n).$$

In the n-dimensional x-space R_n these rays traverse the wave fronts $\psi = t$ and we have

$$\sum_{i=1}^{n} \psi_i\,\dot{x}_i = \sum_{i,k=1}^{n} a_{ik}\,\psi_i\,\psi_k = 1.$$

The vector in R_n with components $\dot{x}_i$ is called the *ray transverse to the wave front* $\psi = t$.

By stipulating that the n-rowed matrix (a_{ik}) be positive definite, we insure the hyperbolicity of equation $(5')$. In this case, the *ray direction and the tangent plane to the wave front are conjugate with respect to the ellipsoid*

$$\sum_{i,k=1}^{n} a_{ik}\,\xi_i\,\xi_k = 1.$$

Besides the *ray velocity vector* with components $\dot{x}_i = v_i$ one may consider the *normal velocity vector* or *wave velocity vector of the progressing wave front*; this velocity is measured by following in time a point of $\psi = t$ on the orthogonal trajectories of the family $\psi = t = \text{const.}$ Its components η_i are proportional to the derivatives ψ_i, and are given by

$$\eta_i = \frac{\psi_i}{(\operatorname{grad}\psi)^2} \qquad (i = 1, 2, \cdots, n).$$

The normal and ray velocities are related by the equations

$$(10) \qquad v_i = \sum_{k=1}^{n} a_{ik}\,\eta_k\,(\operatorname{grad}\psi)^2 \qquad (i = 1, 2, \cdots, n).$$

For a more detailed discussion see §3.

5. Invariance of Characteristics. A few simple *invariance properties* are important.

Under a transformation

$$\xi_\nu = \xi_\nu(x_0, x_1, \cdots, x_n) \qquad (\nu = 0, 1, \cdots, n)$$

$u(x)$ may become $\omega(\xi)$; we write (see equation (7))

$$L[u] \equiv L'[u] + cu$$

$$= \sum_{\mu,\nu=0}^{n} \alpha_{\mu\nu}\, \omega_{\mu\nu} + \sum_{\mu=0}^{n} \beta_{\mu}\, \omega_{\mu} + c\omega$$

$$\equiv \Lambda[\omega] \equiv \Lambda'[\omega] + c\omega.$$

Then we have not only $L[u] = \Lambda[\omega]$, but also $L'[u] = \Lambda'[\omega]$.

We assert: *Characteristics are invariant with respect to arbitrary transformations of the independent variables.*

This is obvious from the conceptual meaning of the characteristic condition. To prove it by formal calculation, we set $\tau_{ji} = \partial\xi_j/\partial x_i$ and verify immediately $\alpha_{ik} = \sum_{j,l=0}^{n} a_{jl}\,\tau_{ij}\,\tau_{kl}$. Now, if

$$\phi(x_0, x_1, \cdots, x_n) = \psi(\xi_0, \xi_1, \cdots, \xi_n),$$

then by $\phi_\nu = \sum_{\mu=0}^{n} \psi_\mu\,\tau_{\mu\nu}$, we have the identity

$$\sum_{i,k=0}^{n} a_{ik}\,\phi_i\,\phi_k = \sum_{i,k=0}^{n} \alpha_{ik}\,\psi_i\,\psi_k,$$

showing that the characteristic form is invariant.

Occasionally, one may use this invariance by transforming a characteristic surface into the coordinate plane $x_n = 0$. We note that *the plane $x_n = 0$ is a characteristic surface if and only if*

(11) $$a_{nn}(x_0, x_1, \cdots, x_{n-1}, 0) = 0.$$

A necessary and sufficient condition for $x_n = $ const. to represent a family of characteristic surfaces is that the coefficient

$$a_{nn}(x_1, x_2, \cdots, x_n)$$

vanish identically.

Likewise the bicharacteristic rays are invariant; this means that the bicharacteristic direction coefficients $\dot{\xi}_i = \tfrac{1}{2}Q_{\psi_i}$ and $\dot{x}_\nu = \tfrac{1}{2}Q_{\phi_\nu}$ represent the same vector, i.e., are related by

$$\dot{\xi}_i = \sum_{\nu=0}^{n} \dot{x}_\nu\,\tau_{i\nu}.$$

Indeed, this equation follows immediately from the preceding formulas.[1]

---

[1] Likewise, *transverse differentiation is invariant with respect to arbitrary transformations of the independent variables* (see article 4), since for arbitrary functions χ the bilinear form $\dot{\chi} = \partial\chi/\partial s = \sum_{i,k=0}^{n} a_{ik}\,\phi_k\,\chi_i$ associated with the quadratic form Q is invariant.

6. Ray Cone, Normal Cone, and Ray Conoid. The directions of the rays through a point P form the quadratic "local ray cone" (or the Monge cone in the sense of Ch. II, §3) for the characteristic differential equation (4). This differential equation (4) itself is a condition for the direction coefficients $\xi_i = \phi_i$ not of the rays but of the normals of characteristic surface elements. Consider these normals as vectors ξ, emanating from the origin of a rectangular $\xi_0, \cdots, \xi_n$-space (which we may represent in the same coordinate system as $x_0, x_1, \cdots, x_n$). Then their end points lie on the quadratic "normal cone", or "dual" cone $\sum_{i,k=0}^{n} a_{ik}\xi_i\xi_k = 0$.

The ray directions are given by the equation

$$(12) \qquad\qquad \dot{x}_i = a_{ik}\,\xi_k,$$

where ξ_k satisfies the equation

$$(13) \qquad\qquad Q(\xi, \xi) = 0.$$

If A_{ik} is the inverse matrix to the matrix a_{ik}, then we have, from (12),

$$\xi_k = A_{ki}\,\dot{x}_i\,.$$

Substituting in (13), we obtain easily for the ray directions the equation

$$(14) \qquad\qquad A_{ik}\,\dot{x}_i\,\dot{x}_k = 0.$$

Conversely, if $\dot{x}_i$ satisfies (14), then the vector $\dot{x}_i$ points in a bicharacteristic direction.

By definition, the ray cone is the envelope of all the characteristic surface elements passing through P. In a dual way the normal cone is the envelope of all of the planes normal to generating rays passing through P.

The duality of the two quadratic cones can be described as follows: We define a collineation or duality transformation which associates with each ray through P the polar plane (i.e., the orthogonal plane) with respect to the imaginary cone $\sum_{i=0}^{n} y_i^2 = 0$ (*reciprocal transformation*). Then each of the two cones is the envelope of the polar planes of the rays of the other.

For example, for the differential equation

$$u_{tt} - u_{x_1 x_1} - u_{x_2 x_2} - \cdots - u_{x_n x_n} = 0,$$

both cones coincide if we identify the y-space and the x-space.

On the other hand, for the equation

$$u_{tt} - u_{x_1 x_1} - 2u_{x_2 x_2} = 0,$$

the equation of the normal cone is

$$\xi_1^2 + 2\xi_2^2 = \xi_0^2$$

and the equation of the ray cone is

$$x_1^2 + \tfrac{1}{2}x_2^2 = t^2.$$

If the coefficients a_{ik} of the differential equation are not constant, the situation remains essentially unchanged. We have merely to consider at each point the normal cone and the *local* ray cone of the characteristic directions.

For constant or nonconstant coefficients the *ray conoid* is defined as consisting of all rays through the point P and tangent at P to the local ray cone (see Chapter II). This conoid is a characteristic surface or wave front with P as "center of disturbance"; it is called a *spherical wave front*[1] about P.

In §3 we shall consider the relationship of these cones in much greater generality. Here we merely point out that the two parts of the ray cone issuing from P in time t are often distinguished as the *forward* ray cone, pointing towards increasing time, "into the future", and the *backward* cone, pointing "into the past".

7. Connection with a Riemann Metric. The following remarks may be inserted, even though they will not be used immediately: We introduce a metric with the line element

$$(15) \qquad d\sigma^2 = \sum_{i,k=0}^{n} A_{ik}\, dx_i\, dx_k$$

in the $(n + 1)$-dimensional space R_{n+1}. Then the generating rays of the conoid, in virtue of the characteristic relation, become "null rays", i.e., curves along which $d\sigma = 0$, or curves along which the distance between any two points vanishes. Conversely, all null curves for this metric are characteristic rays of the differential equation $L[u] = 0$.

These relations are easily visualized if we again distinguish $t = x_0$ as a time coordinate and consider a differential equation of the special type

$$(16) \qquad u_{tt} - \sum_{i,k=1}^{n} a_{ik}\, u_{ik} = 0,$$

[1] See article 7.

where we assume that the matrix (a_{ik}) is positive definite and that the coefficients a_{ik} do not depend on the time t. The characteristics $\psi(x_1, x_2, \cdots, x_n) - t = 0$ then satisfy the partial differential equation $\sum_{i,k=1}^{n} a_{ik}\, \psi_i\, \psi_k = 1$.

The system of all rays passing through a fixed point of the x-space forms the associated ray conoid. We can represent it in the form:

$$\omega(x_1, x_2, \cdots, x_n\,;\, x_1^0, x_2^0, \cdots, x_n^0) \equiv \omega(x; x^0) = t,$$

where x^0 is the vertex of the ray conoid and has the coordinates x_i^0.

Spherical wave fronts represented by the conoid are given by $\omega = t$. If (A_{ik}) again denotes the matrix reciprocal to (a_{ik}), we have

$$(17) \qquad \sum_{i,k=1}^{n} A_{ik}\, \dot{x}_i\, \dot{x}_k = \sum_{i,k=1}^{n} a_{ik}\, \omega_i\, \omega_k = 1$$

along the rays. If in the n-dimensional x-space we now introduce the metric with the line element

$$(18) \qquad d\rho^2 = \sum_{i,k=1}^{n} A_{ik}\, dx_i\, dx_k$$

then t stands for length along these rays, and the surface $\omega = t$, in terms of this metric, is actually a sphere of radius t about the midpoint x^0 provided one measures distances along the rays. (Comparison with Ch. II, §9 shows that these rays are nothing but the geodetic curves of the integrand in

$$\int \sqrt{\sum_{i,k=1}^{n} A_{ik}\, \dot{x}_i\, \dot{x}_k\, dt}.)$$

The square of the geodetic distance Γ from a point x, t to a parameter point ξ, τ satisfies the partial differential equation

$$\sum A_{ik}\Gamma_i\Gamma_k = 4\Gamma,$$

as was pointed out before for the case of the wave equation.

A direction dx_i is called *time-like* if

$$d\sigma^2 = dt^2 - d\rho^2 = dt^2 - \sum_{i,k=1}^{n} A_{ik}\, dx_i\, dx_k > 0$$

holds, and an element of the surface $\phi(x_0, x_1, \cdots, x_n) = 0$ is called *space-like*, in accordance with the definition in article 1, if

$$\phi_t^2 - \sum_{i,k=1}^{n} a_{ik}\, \phi_i\, \phi_k > 0$$

holds (cf. the end of article 2). Thus, in particular, the time-axis $dx_i = 0$ is indeed time-like and the surface $\phi \equiv t = 0$ space-like. A general discussion of the concepts of "time-like" and "space-like" will be given in §3.

The example of the wave equation

$$u_{tt} - \Delta u = 0$$

immediately illustrates our general concepts. The corresponding line elements are $d\rho^2 = \sum_{i=1}^{n} dx_i^2$ and $d\sigma^2 = dt^2 - \sum_{i=1}^{n} dx_i^2$.

8. Reciprocal Transformations. With a view to §3 we add a further remark about the reciprocal transformation, irrespective of the application to differential equations of second order.

As defined in article 6 the reciprocal transformation in the bundle through a fixed point is the linear transformation which associates with each ray represented by a vector ξ the plane

$$(19) \qquad\qquad \xi x = 0$$

in running coordinates x, i.e., the plane orthogonal to the ray ξ. Conversely the symmetric relation (19) associates with each plane through P the direction normal to the plane.

This linear transformation of rays into planes is extended into a transformation of a cone N generated by rays ξ into a cone S generated by the planes (19): If the ray ξ describes a conical surface

$$(20) \qquad\qquad N(\xi_0 , \xi_1 , \cdots , \xi_n) = N(\xi) = 0,$$

N denoting a homogeneous function of degree k, then by (19) we associate a conical surface S' with N, which is generated by the planes reciprocal to the rays of N. Precisely, we may consider S as the cone enveloped or rather supported by these planes.

The reciprocal transformation of a cone of *second order* again yields a cone of second order; but for cones N of higher order k the transformation will lead in general to surfaces of order different from k.

Since the relationship (20) is symmetric, we may state: The planes supporting one cone are the polar planes of the generators of the other reciprocal cone—polar planes with respect to the imaginary cone $\sum_{j=0}^{n} \xi_j^2 = 0$, the origin P being the vertex. We may express the situation also by singling out the coordinate x_0 and considering in the n-dimensional ξ-space the surface N^* in which N intersects the "plane" $\xi_0 = -1$. Likewise we may consider the surface S^*

in the n-dimensional $(x_1, \cdots, x_n)$-space which is the intersection of S with the plane $x_0 = 1$. Then the reciprocity between N and S or N^* and S^* is expressed by

$$(20a) \qquad x\xi = \sum_{i=1}^{n} x_i\, \xi_i = 1,$$

which states that for fixed x the point ξ ranges over the polar plane[1] of x with respect to the unit sphere, and vice versa.

To carry out analytically the transformation of a given surface $N^*(\xi_1, \cdots, \xi_n) = 0$ into S^* we would have to form the envelope of the x-planes (20a) whose normals ξ are restricted by $N^*(\xi) = 0$. Not only does this process in general lead to an algebraic expression for S^* of higher order than the order k of N,[2] but, as we shall confirm by significant examples in §§3 and 3a, the formation of envelopes may also lead to singularities such as isolated points or edges of regression. However, the geometrical definition of the transformation mitigates this complication: We consider not merely tangent planes of a surface but more generally *supporting planes* at a point, that is, planes through a point P of the surface which leave the surface on one side (at least in a neighborhood of P). Then the reciprocal transformation associates with the points of a surface N^* the locus S^* of the poles of the supporting planes of N^*, and vice versa, it associates with the points of S^* the locus N^* of the poles of the supporting planes of S^*.

Portions of N^* which are smoothly curved, of course, are mapped into points of the regular envelope of the planes polar to these points. Singularities such as cusps of the envelope must be expected at points corresponding to points where the curvature of N^* vanishes.[3] If a portion of a surface N^* has a singular point, carrying a bundle of supporting planes, then the corresponding poles form a portion of a linear x-manifold which at the boundary is tangent to this envelope.

The complications inherent in the occurrence of singularities seem to elude general simple geometrical descriptions. This is in keeping

[1] Had we defined N^* by the intersection of N with $\xi_0 = 1$ the polarity would be with respect to the sphere with imaginary radius i, which would in fact mean no significant difference.

[2] There are, however, cases, important in mathematical physics, when both surfaces are of the same order; see for example the discussion of crystal optics in §3a.

[3] Examples will be discussed in §§3 and 3a.

with the fact that our knowledge concerning algebraic surfaces in the real space is unsatisfactory.

However, for the transformation of a closed surface N' which is convex (N' may be a part of N^*), the situation is transparent and not obscured by complicating singularities. Then we show easily: *The surface S' reciprocal to a convex surface N' is again convex; it is the convex hull* of the point set obtained by forming the ordinary envelope of the polar planes of N'.

Indeed: Let us define by $\sum x_i \xi_i - 1 \leq 0$ the "positive" side of the plane $(x\xi) - 1 = 0$ polar to the point ξ. Assuming that the point $\xi = 0$ is inside N', we now define the reciprocal transformation of the closed interior $\overline{N'}$ of N' as the intersection, that is the common point set, of all half spaces $(x\xi) - 1 \leq 0$ for all points ξ in $\overline{N'}$. This point set $\overline{S'}$, as a domain common to a set of half spaces, is convex, and so is its boundary S', which obviously is the image of N'. We could also have defined the reciprocal image of the convex domain $\overline{N'}$ as the set of all the poles of all planes in the ξ-space which do not intersect $\overline{N'}$. The reader will easily recognize the equivalence of these various definitions.

At any rate, the character of S' as the convex hull of the envelope of the polar planes of the points on the smooth part of N' is now obvious.

The role of the reciprocal surfaces for differential equations of higher order will be further clarified in §3.

9. Huyghens' Construction of Wave Fronts. The theory of the complete integral, and the corresponding construction of envelopes of the solutions of the initial value problem for first order differential equations, leads immediately to the following important construction of wave fronts (cf. Ch. II, §§4 and 8, and §3 of the present chapter).

We consider a possible wave front $\psi(x_1, x_2, \cdots, x_n) = t$ satisfying the equation $a_{ik} \psi_i \psi_k = 1$. The spherical waves about the point P_0 may be denoted by $\omega(x_1, x_2, \cdots, x_n, P_0) = t$. If the front at $t = 0$ coincides with a prescribed surface W_0, then Huyghens' construction produces a wave front at the time t as follows: Around each point P_0 of W_0 we consider the spherical wave front $t = \omega(x, P_0)$, and for a fixed positive t we form the envelope of all these spheres in $x_1, x_2, \cdots, x_n$-space, letting P_0 range over the surface W_0. This leads to a surface $\psi(x_1, x_2, \cdots, x_n) = t$ containing the desired wave front. In other words: *At the time t the wave front is given by the en-*

velope of the spheres of radius t, defined in the sense of the above metric, whose centers are the points of the wave front at t = 0.[1]

10. Space-Like Surfaces. Time-Like Directions. We shall now further illuminate the concept of a "space-like" surface for second order hyperbolic equations. Its significance (see §§8, 9) is that the initial value problem can be solved if the initial surface is space-like.

If the second order operator $L[u]$ is hyperbolic, i.e., if the matrix a_{ik} has one positive and n negative eigenvalues, then all directions ξ satisfying $Q = 0$ form the *normal* cone in ξ-space. A surface element through P is called *space-like* if its normal ξ points into the interior of the cone, e.g., if $Q(\xi) > 0$. It is called *characteristic* if $Q(\xi) = 0$; and *non-space-like* if $Q(\xi) < 0$. A *space-like surface* is one which is space-like at every point.

As easily seen, the following definition is equivalent to the previous one: A surface element through P is space-like if it intersects the local ray cone through P only at P, i.e., if it separates the two parts of the cone.

A direction is *time-like* if it points into the interior of the local ray cone.

If the distance along a time-like direction is identified with the time t, then P is said to separate the "forward" from the "backward" ray cone, or the ray cone of the *future* from that of the *past*.

In §3 we shall see how these concepts are generalized, clarified, and refined for higher order problems.

§2. Second Order Equations. The Role of Characteristics

Instead of defining characteristics as surfaces on which the Cauchy data are not free (as we did in §1), we may use the following equivalent property which stresses a slightly different aspect of the concept: In a characteristic surface C the differential operator is an

---

[1] We call attention to what seems at first sight to be a paradox: Suppose $u(x_1, x_2, \cdots, x_n, t)$ is a solution of the differential equation $L[u] = 0$ with the wave front $\psi = t$. Let this wave front consist of a single surface W_t advancing in time through the space R_n. If we start out with the wave front W_0 at the time $t = 0$, Huyghens' wave construction—as in the example of the wave equation—can lead to two different "parallel surfaces" W_t and W_t', both of which satisfy the characteristic differential equation. However, by hypothesis only one of them is the carrier of the discontinuity of u at the time t, namely the one that actually corresponds to this time t, while the other surface would correspond to the time $-t$.

A characteristic surface can, but need not, contain discontinuities of the solution u, and the envelope construction can also lead, without contradicting our theory, to surfaces on which the wave is not discontinuous at the time t.

interior operator in a sense to be presently specified. This property was seen in Chapter V to be decisive for the construction of the solution of Cauchy's problem for two independent variables. For more independent variables it will not in general lead to a similar direct construction of the solution, except (see article 4) for special differential equations. However, main features of the solution can be analyzed in general by using the interior character of the differential operator along characteristics to study discontinuities of the solution. Thereby at least a skeleton of the solution can be constructed by solving only ordinary differential equations. Under suitable assumptions one can go even further towards constructing the complete solution of Cauchy's problem. The relevant fact, of great importance for wave propagation, is: Physically meaningful discontinuities of solutions occur only across characteristic surfaces (hence in this context such discontinuities are called wave fronts) and are propagated in these characteristics along bicharacteristic rays. This propagation is governed by a simple ordinary differential equation.

In this section we shall briefly indicate the situation for linear (and quasi-linear) differential equations of second order with a view to a more detailed general discussion of the subject in §§4 and 5.

1. Discontinuities of Second Order. We consider a surface $C: \phi(x_0, \cdots, x_n) = 0$ along which the first derivatives of a solution u of (1) of §1 are continuous[1] and so are all the tangential, or interior, derivatives of these first derivatives. The second derivatives u_{jk} (if they are not such interior derivatives) may suffer jump discontinuities across C.

As in Ch. V, §1, 3, for any function f which has a jump discontinuity across the surface $C: \phi = 0$, we denote the jump by (f). The expression $u_{ik}\phi_j - u_{ij}\phi_k$ is an interior derivative of u_i in C (cf. Ch. II, App. §1), and is therefore continuous across C. The same holds for $u_{ij}\phi_l - u_{jl}\phi_i$. The linear combination $u_{ik}\phi_j\phi_l - u_{jl}\phi_i\phi_k$ of these two continuous expressions is, therefore, also continuous across C. Hence for the jump quantities we obtain the relation

$$(u_{ik})\phi_j\phi_l = (u_{jl})\phi_i\phi_k,$$

and thus[2]

$$(u_{ik}) = \lambda\phi_i\phi_k,$$

[1] This equation may here be assumed as linear or quasi-linear.
[2] By assumption not all derivatives of ϕ vanish at the same point on $\phi = 0$.

where the factor λ of proportionality is a function defined on C which cannot vanish at points where any one of the second derivatives of u is discontinuous. Incidentally, as is easily seen,

$$\lambda = (u_{\phi\phi}).$$

Obviously, C must be characteristic, since otherwise all the second derivatives of u along C would be uniquely determined by u and u_i on C and hence could not suffer jumps. This fact is also seen immediately by considering the jump relation $(L[u]) = 0$ along C and using the preceding formulas; we obtain

$$0 = \sum_{i,k=0}^{n} a_{ik}(u_{ik}) = \lambda \sum_{i,k=0}^{n} a_{ik}\,\phi_i\,\phi_k \, .^{1}$$

Discontinuities in the first derivatives of an artificially "generalized solution" of u would be compatible with the differential equation along noncharacteristic surfaces (cf. Ch. V, §1, and later §3). However, in physically relevant generalized solutions[2] such jumps are in fact restricted to the characteristic surfaces, as we shall show in §3. The same is true for jumps of the function u itself and for other types of discontinuities.

Proofs of existence of the solutions analyzed in this article will be found in §4 and §10.

2. The Differential Equation along a Characteristic Surface. For brevity we shall confine ourselves to linear differential equations. We analyze the information expressed by the equation

$$(1) \qquad L[u] \equiv \sum_{i,k=0}^{n} a_{ik}\,u_{ik} + \sum_{i=0}^{n} a_i\,u_i + au = 0,$$

along a characteristic surface $\phi = 0$. Without loss of generality we may assume that the family of surfaces $\phi = c = $ const. is transformed into the family of coordinate planes $x_n = c = $ const. The results

---

[1] A similar analysis can be made for discontinuities across C which occur only in derivatives of order $2 + r$; we have merely to differentiate the differential equation r times and apply the same argument to the differentiated equation. The result is again: C is characteristic; and moreover, if the jump of the $(r + 2)-th$ outward derivative is $\lambda = (u_\phi{}^{r+2})$ then the jumps of the derivatives $D_0^{\alpha_0} \cdots D_n^{\alpha_n} u$ are equal to

$$\lambda\phi_0^{\alpha_0} \cdots \phi_n^{\alpha_n} \qquad (\alpha_0 + \cdots + \alpha_n = r + 2),$$

where D_i denotes $\partial/\partial x_i$.

[2] "Admissible" solutions will be characterized later in §4.

can then be formulated for arbitrary surfaces $\phi = c$ on the basis of the invariance properties proved in §1, 5.

We write (1) in the form

$$
\text{(2)} \quad L[u] \equiv \sum_{i,k=0}^{n-1} a_{ik} u_{ik} + \sum_{i=0}^{n-1} a_i u_i + au + a_{nn} u_{nn} \\
+ 2 \sum_{i=0}^{n-1} a_{in} u_{in} + a_n u_n = 0.
$$

Now we combine the terms which contain only interior differentiation on the surface C: $x_n = 0$, i.e., differentiation with respect to x_0, $x_1, \cdots, x_{n-1}$, and denote their sum by J. We have

$$
\text{(3)} \quad L[u] \equiv J + a_{nn} u_{nn} + a_n u_n + 2 \sum_{i=0}^{n-1} a_{in} u_{in} = 0.
$$

The assumption that the surfaces $\phi = x_n =$ const. are characteristic immediately implies $a_{nn} = 0$ and vice versa. Therefore on these characteristic surfaces we have

$$
\text{(4)} \quad L[u] = J + a_n u_n + 2 \sum_{i<n} a_{in} u_{in} = 0.
$$

Now, for $\phi = x_n$ the derivatives of the characteristic form $Q(\phi_i, \phi_k)$ are

$$
\frac{\partial Q}{\partial \phi_i} = 2a_{in} \quad i < n, \quad \frac{\partial Q}{\partial \phi_n} = 0.
$$

According to §1 the vector Q_{ϕ_i} is tangential to the surface $\phi =$ const. and points in the direction of the bicharacteristic rays. Introducing the outward derivative $u_\phi = v$ and a suitable parameter s on the rays in $\phi =$ const., we may therefore write (4) in the form

$$
\text{(5)} \quad J + \frac{\partial v}{\partial s} + a_n v = 0
$$

with $\partial/\partial s = 2 \sum_{i<n} a_{in} \partial/\partial x_i$.

According to §1, the characteristic differentiation is invariant[1] under the coordinate transformation which transforms the planes $x_n = c$ into other families of characteristic surfaces $\phi = c =$ const. Moreover, by definition $L[\phi - c]$ is invariant on $\phi = c$ under coordinate transformations (this invariance defines the transformation of the coeffi-

---

[1] Except for an arbitrary internal parameter along the bicharacteristic curves.

cients of L). Since for $\phi = x_n = 0$ we have $L[\phi] = a_n$, and, more generally, for $\phi = x_n = c$ we have $L[\phi - c] = a_n$, we can now write (5) along the characteristic surface $C_c : \phi = c$ in the form

$$(6) \qquad J + \frac{\partial v}{\partial s} + L[\phi - c]v = 0,$$

where $L[\phi - c]$ is a known function along C_c and J is likewise known on C_c if u alone is known. This remarkable form of the original differential equation will later reappear in more general contexts. It shows that indeed on C the Cauchy data cannot be chosen freely. We shall now use it to study the propagation of discontinuities.

3. Propagation of Discontinuities along Rays. Equation (6) represents a linear ordinary differential equation of first order for the outward derivative $v = u_\phi$. Such an ordinary differential equation is satisfied along each of the rays which generate the characteristic surface $\phi = c$.

Now we return for the moment to the assumption that the whole family of characteristics consists of the planes $\phi = x_n = c$. Then we have identically $a_{nn} = 0$.

We use (6) and the differential equation obtained from (1) by differentiation with respect to $x_n = \phi$, assume $a_{nn} = 0$, and conclude

$$\sum_{i,k<n} a_{ik} u_{ikn} + 2 \sum_{i<n} a_{in} u_{nni} + \sum a_i u_{ni} + a u_n + (a)_n u + a_n u_{nn}$$

$$+ \sum_{i,k<n} (a_{ik})_n u_{ik} + 2 \sum (a_{in})_n u_{ni} + \sum (a_i)_n u_i = 0.$$

On $\phi = 0$ the function u and the first derivatives of u, and hence also their interior derivatives, are assumed known. Lumping them together as an interior expression J^*, we have

$$J^* + 2 \sum_i a_{in} w_i + a_n w = 0 \qquad (w = u_{\phi\phi}).$$

If, as in (6), we denote differentiation along a ray in C by $\partial/\partial s$, we may write

$$(5a) \qquad J^* + \frac{\partial w}{\partial s} + a_n w = 0.$$

Let us now assume as in article 1 that $u_{\phi\phi}$ suffers a discontinuity across C

$$(u_{\phi\phi}) = (w) = \lambda,$$

while u and the first derivatives of u and their interior derivatives on C remain continuous. Then because of the continuity of J^*, (5a) yields immediately[1] the jump relation on C: $x_n = 0$

$$(7) \qquad\qquad \frac{\partial \lambda}{\partial s} + P\lambda = 0,$$

where $\partial/\partial s$ denotes differentiation along the bicharacteristic rays in C and $P = L[\phi] = L[x_n]$ is known on C.

Equation (7) is the law governing the *transmission of discontinuities* along rays in the characteristic discontinuity surface. It has the form of an ordinary differential equation and shows, incidentally, that a discontinuity cannot vanish at any point of a ray on which it is somewhere different from zero.

Because of the invariance of characteristics and characteristic differentiation (§1, 5) the relation (7) with $P = L[\phi]$ is valid for arbitrary, not necessarily plane, characteristics $\phi = 0$. To establish this fact, we should notice that if we transform x_n into ϕ, the outward derivative u_{nn} transforms into $u_{\phi\phi} \phi_{x_n}^2 + \cdots$. The dots denote terms which remain continuous across C and thus cancel out if we form the difference of $L[u]$ for both sides of C_i. The factor $\phi_{x_n}^2$ merely modifies the parameter s on the ray.

If the discontinuities occur across a characteristic surface $\phi = c$, the law (7) is to be modified to

$$\lambda_s + L[\phi - c]\lambda = 0.$$

We shall find that a similar law prevails for all types of singularities of solutions of linear differential equations, e.g., for discontinuities occurring in the first derivatives or even in the function u itself in a sense to be justified later (see §4, 3).

Finally, we call attention to the specific characteristic surface, the "ray conoid" introduced in §1. Its rays carry local discontinuities from the vertex into the x-space.

4. Illustration. Solution of Cauchy's Problem for the Wave Equation in Three Space Dimensions. As stated in the introduction,

---

[1] This is seen by considering (5a) for two points on opposite sides near a point on C, taking the difference, and passing to the limit for these two points coinciding on C.

there are cases in which the solution can actually be completed directly by the use of integration along rays.

The important example is the equation

$$(8) \qquad L[u] \equiv u_{tt} - \Delta u \equiv u_{tt} - u_{xx} - u_{yy} - u_{zz} = 0.$$

Using the ideas developed above, we shall solve the initial value problem for this equation, which has already been treated in Ch. III, §5 and which will be discussed in greater detail in §12. For $t = 0$ we prescribe the initial values

$$u(x, y, z, 0) = \phi(x, y, z), \qquad u_t(x, y, z, 0) = \psi(x, y, z).$$

We introduce the following differentiation symbols:

$$A_1 = y \frac{\partial}{\partial z} - z \frac{\partial}{\partial y}, \qquad A_2 = z \frac{\partial}{\partial x} - x \frac{\partial}{\partial z}, \qquad A_3 = x \frac{\partial}{\partial y} - y \frac{\partial}{\partial x}$$

and (slightly at variance with the notation of article 3)

$$\frac{\partial}{\partial s} = \frac{1}{t - \tau} \left[x \frac{\partial}{\partial x} + y \frac{\partial}{\partial y} + z \frac{\partial}{\partial z} + (t - \tau) \frac{\partial}{\partial t} \right],$$

$$\frac{\partial}{\partial \nu} = \frac{1}{t - \tau} \left[x \frac{\partial}{\partial x} + y \frac{\partial}{\partial y} + z \frac{\partial}{\partial z} - (t - \tau) \frac{\partial}{\partial t} \right].$$

Then on the characteristic cone K through the point $(0, 0, 0, \tau)$

$$K\colon (t - \tau)^2 - x^2 - y^2 - z^2 = 0,$$

the differentiations A_1, A_2, A_3 and the characteristic derivative $\partial/\partial s$ are interior differentiations, while $\partial/\partial \nu$ denotes the normal derivative. The occurrence of interior derivatives on K is expressed by the identity

$$(9) \quad \begin{aligned} \Psi[u] &= -(t - \tau)^2 L[u] - (t - \tau) \frac{\partial u}{\partial s} - (t - \tau) \frac{\partial}{\partial s} \left[(t - \tau) \frac{\partial u}{\partial \nu} \right] \\ &= (A_1^2 + A_2^2 + A_3^2)u, \end{aligned}$$

which holds for $(t - \tau)^2 = x^2 + y^2 + z^2$. Now on the three-dimensional sphere $x^2 + y^2 + z^2 = $ const., the surface integral of $A_1[v]$ is equal to zero for an arbitrary function v since the integral over a circle formed by intersection with $z = $ const. vanishes. The same is true, of course, for corresponding surface integrals of $A_2[v]$ and

$A_3[v]$, and hence also for $(A_1^2 + A_2^2 + A_3^2)v = \Psi[v]$.   Using the expression $\Psi[u]/(t - \tau)$ we now form the integral

$$\iiint_K \frac{\Psi[u]}{t - \tau}\, ds\, d\omega$$

over the surface of the characteristic half-cone K for $t > 0$.   Since $L[u] = 0$, this yields

$$\iiint_K \left\{\frac{\partial u}{\partial s} - \frac{\partial}{\partial s}\left[(t - \tau)\frac{\partial u}{\partial \nu}\right]\right\} ds\, d\omega = 0$$

and then, after integration with respect to s, we have

(10) $$4\pi\tau^2 u(P) - \iint u\, d\omega - \tau \iint \frac{\partial u}{\partial \nu}\, d\omega = 0,$$

where the integrals are to be taken over the surface of the sphere $x^2 + y^2 + z^2 = \tau^2$.   This is the solution already found in Ch. III, §5.

In the same way we obtain for the nonhomogeneous wave equation the solution given in Ch. III, §5.

The method given here, which is essentially due to *Beltrami*, depends on the fact that with the aid of interior differentiations we can write the data furnished by the differential equation along the characteristic cone in a particularly simple manner.[1]   By integrating only over the surface of the cone it is possible to find explicitly the solution u at the vertex in terms of initial values along the base curve.   This method reveals the "Huyghens character" of the wave equation.

One cannot develop a strictly analogous theory for arbitrary linear differential equations of second order, giving solutions by means of integration processes in the interior of the characteristic conoid along rays; such a theory would entail the general validity of Huyghens' principle, which certainly does not obtain.   It remains an interesting problem to find sufficient conditions for the validity of Huyghens' principle by considerations similar to those of this section.[2]

---

[1] This method has been extended for all odd values of n by M. Riesz [2].

[2] See L. Asgeirsson [1], K. L. Stellmacher [1], and A. Douglis [2].   See also, e.g., §18.

§3. *Geometry of Characteristics for Higher Order Operators*

For higher order equations and for systems of equations an essential generalization of the theories of §§1 and 2 is necessary (see also Chapters III and V) and will be given in the present section.

1. Notation. The following notation[1] will be used:
Differentiation with respect to x_κ will again be denoted by D_κ or D^κ:

$$D^\kappa = D_\kappa = \frac{\partial}{\partial x_\kappa} \qquad (\kappa = 0, \cdots, n).$$

Instead of $x_0, \cdots, x_n$ we shall sometimes write x; however, when $x_0 = t$ is to be singled out as the time variable, we shall write x for the n-dimensional space vector $x_1, \cdots, x_n$. The gradient operator will be denoted by D:

$$D = (D_0, \cdots, D_n).$$

Let $p = (p_0, \cdots, p_n)$ be any vector with $n + 1$ non-negative integer components. Let $\xi = (\xi_0, \cdots, \xi_n)$ be any vector with $n + 1$ components. Then we define

$$\xi^p = \xi_0^{p_0} \cdots \xi_n^{p_n}.$$

The components of ξ may be numbers or operators; in particular, for $\xi = D$, the symbol D^p denotes the differential operator

$$D^p = D_0^{p_0} \cdots D_n^{p_n}.$$

The order of the differential operator D^p is denoted by $|p|$:

$$|p| = p_0 + \cdots + p_n.$$

[1] Incidentally, one of the advantages of this notation (proposed by Laurent Schwartz) is that it allows a concise expression of Leibnitz' rule and Taylor's theorem. If we define

$$p! = p_1! p_2! \cdots p_n!,$$

we have

$$D^p(uv) = \sum_{q+r=p} \frac{p!}{q! \, r!} D^q u \, D^r v$$

and

$$f(x + \xi) = \sum_p \frac{1}{p!} \xi^p D^p f(x).$$

578 VI. HYPERBOLIC DIFFERENTIAL EQUATIONS

With this notation a differential equation of order m can be written in the following form:

$$L[u] = \sum_{|p| \leq m} a^p D^p u = f.$$

The coefficients a^p as well as the right-hand side f may be constant, or they may be functions of the independent variables x, or depend on x and partial derivatives of u up to order $m - 1$.

Most equations of mathematical physics appear as systems of k equations involving k unknown functions $u^1, \cdots, u^k$, which we consider as components of a single column vector u; the equations then are written in matrix form

(1) $$L[u] = \sum_{|p| \leq m} A^p D^p u = f.$$

Here the differential operator D^p acts on each component of the vector-valued function u, and the coefficients A^p are $k \times k$ square matrices and f is a column vector with k components.

Just as before, the coefficients A^p may be constants or they may depend on x; for quasi-linear equations they depend on u and its partial derivatives of order up to $m - 1$.

The terms of highest order constitute the *principal part* of the operator:

$$\sum_{|p|=m} A^p D^p.$$

With a view to concrete cases particular attention will be paid to three categories.

Case a). Systems of first order:

(1a) $$L[u] = \sum_{i=0}^{n} A^i D_i u + Bu = f.$$

Case b). Single equations of order m:

(1b) $$L[u] = \sum_{|p| \leq m} a^p D^p u = f.$$

The general case (1) of systems of equations of order m will be referred to as case c).

Of particular importance for applications are second order systems which may be written more explicitly in the form

$$L[u] = \sum_{i,j=0}^{n} A^{ij} D_i D_j u + \sum_{i=0}^{n} A^i D_i u + Bu = f,$$

with $k \times k$ matrices B, A^i, A^{ij}, where $A^{ij} = A^{ji}$.

2. Characteristic Surfaces, Forms, and Matrices. Cauchy's initial value problem is to determine solutions of $L[u] = f$ if "Cauchy data" are prescribed on a given surface[1] C. For an equation or system of equations of order m these data are the values of u and its partial derivatives up to order $m - 1$ with respect to some outgoing variable. (We assume that all the equations of the system have the same order m.) In particular, for first order systems (case a), the values of u themselves constitute the Cauchy data.

From the Cauchy data we can determine by interior (tangential) differentiation on C the values of all partial derivatives of u up to order $m - 1$, as well as the values of those partial derivatives of order m or higher which are obtained by further interior differentiation. A differential operator whose value on C is thus determined by the Cauchy data on C is called an *interior differential operator* of order m (see Ch. III, §2).

In this article we shall not attempt to construct solutions of the differential equations (1) but shall deal with the following *preliminary question* referring merely to the initial surface C:

For what surfaces C do there exist functions u with arbitrarily given Cauchy data which satisfy the equation $L[u] = f$ along C?

If one can always find such a function u, the surface is called *free* for the operator L; if it is nowhere free it is called a *characteristic surface* of L.

The following algebraic criterion in terms of the coefficients of L and the normal ξ to the surface C: $\phi(x) = 0$ decides whether the surface C is characteristic:[2] A homogeneous "characteristic form" $Q(\xi_0, \xi_1, \cdots, \xi_n)$ of order mk in the components of the vector $\xi = D\phi$ normal to C will presently be defined such that

$$(2) \qquad\qquad Q(\xi_0, \cdots, \xi_n) = Q(\xi) = 0$$

[1] By surface we mean as before a smooth n-dimensional hypersurface $\phi(x) = 0$ with $|D\phi| \neq 0$.

[2] See also Chapter V and Ch. III, §2. The present discussion partly repeats previous ones.

is necessary and sufficient for C to be characteristic.[1] For a single equation (1b) of order m the characteristic form Q is given by

$$(3) \qquad Q(\xi) = \sum_{|p|=m} a^p \xi^p;$$

for a system (1c) Q is given as a determinant by

$$(4) \qquad Q(\xi) = \| \sum_{|p|=m} A^p \xi^p \| .$$

In the particularly important case of a system of equations of first order (1a) we have

$$(5) \qquad Q(\xi_0, \cdots, \xi_n) = \| \sum A^i \xi_i \| .$$

For systems of second order,

$$Q = \| \sum A^{ij} \xi_i \xi_j \| .$$

For systems of equations of any order m the characteristic form is the determinant of the characteristic matrix

$$(6) \qquad A = \sum_{|p|=m} A^p \xi^p.$$

In the case $m = 1$ the characteristic matrix is

$$(7) \qquad A = \sum_{i=0}^{n} A^i \xi_i$$

and in the case $m = 2$

$$(8) \qquad A = \sum_{i,j=0}^{n} A^{ij} \xi_i \xi_j .$$

That $Q = 0$ is the characteristic condition follows from our previous reasoning: Introducing interior variables $\lambda_1, \cdots, \lambda_n$ in C and the outgoing variable $\phi = \lambda_0$, we merely express the condition that on C the equation $L[u] = 0$ does not determine all the quantities $(\partial/\partial\phi)^m u$ from arbitrary Cauchy data for u. The linear algebraic equations for these k quantities have the matrix A and the determinant Q. Naturally the characteristic forms Q and matrices A depend on the point x unless the principal part of L has constant coefficients.

In the matrix case (1a), (1c) the condition $Q = 0$ states that *the*

---

[1] Of course, the condition may be applied merely to a surface element of C if we want to stress its local character.

characteristic matrix A is singular; hence there exist *left nullvectors l* and *right nullvectors r* of k components such that

$$lA = Ar = 0.$$

If the rank of A is $k - 1$ then these nullvectors are uniquely determined within an arbitrary scalar factor for a given characteristic surface C at each point.

A single operator L is *interior* on every characteristic surface C.

If $L[u]$ is a matrix operator on the vector u, then the characteristic determinant condition $Q = 0$ states: The single operator

$$lL[u]$$

acting on the vector u is an interior operator along the characteristic surface C.

3. Interpretation of the Characteristic Condition in Time and Space. Normal Cone and Normal Surface. Characteristic Nullvectors and Eigenvalues. With $x_0 = t$ singled out as the time variable we may represent the characteristic surfaces, as before in §1, by $\phi(t, x) = \psi(x) - t = 0$, that is, as a surface $C: \psi(x) = \text{const.} = t$ moving in the n-dimensional x-space. We have $\phi_t = -1$, $\phi_{x_i} = \psi_{x_i} = \xi_i$, and consider the vector ξ in the n-dimensional space. The *normal velocity* of the moving surface $\psi = t$ is a vector v orthogonal to the surface such that $|v|$ measures the rate of change of ψ along this normal. We have therefore, with the n-dimensional normal vector $\xi = D\psi$ and a scalar λ, the relations:

$$v = \lambda\xi, \qquad 1 = \sum_{i=1}^{n} \xi_i v_i = \xi v, \qquad \lambda\xi^2 = 1.$$

Hence

$$\xi = \frac{v}{v^2}, \qquad v = \frac{\xi}{\xi^2}.$$

The vectors ξ and v are inverse or "dual" with respect to the unit sphere, and the characteristic condition (2) can be written

$$(9) \qquad Q(-1, \xi_1, \cdots, \xi_n) = 0 \qquad (\xi = D\psi).$$

Introducing in the n-dimensional x-space the unit vector

$$\alpha = \frac{\xi}{|\xi|}$$

normal to C, we obtain from (9), because of the homogeneity of $Q(\xi_0, \xi_1, \cdots, \xi_n)$, the equivalent equation

(9a) $$Q\left(-\mid v \mid, \alpha_1, \cdots, \alpha_n\right) = 0.$$

This is an algebraic equation of degree mk for the normal speed $\mid v \mid$ of that characteristic surface whose normal points in the direction α. Of course, for differential operators with nonconstant coefficients these relations refer to a fixed point in the x-space and to a particular time t.

Instead of (9a) we may also write

(9b) $$Q\left(-1, \frac{v_i}{\mid v \mid^2}\right) = 0.$$

It is important to note that for a system of first order

(1a) $$L[u] = u_t + \sum_{\nu=1}^{n} A^{\nu} u_{\nu} + \cdots = 0,$$

where $A^0 = I$ is the unit matrix, the nullvectors l or r for a characteristic surface $\phi = \psi(x) - t = 0$ are simply left or right eigenvectors of the matrix $\sum_{\nu=1}^{n} \xi_{\nu} A^{\nu}$ with $\psi_{x_\nu} = \xi_{\nu}$, and the *normal speeds* $\mid v \mid$ are the *eigenvalues* of the matrix

$$\hat{A} = \sum_{i=1}^{n} \alpha_i A^i.$$

This follows because $Q = \parallel A \parallel = 0$ means

$$\parallel -vI + \hat{A} \parallel = 0.$$

Without necessarily emphasizing the time t, we now define the *characteristic normal cone* at a point O of the $x_0, x_1, \cdots, x_n$-space as the cone generated by the normals to all characteristic surface elements at O. If $\xi_0, \cdots, \xi_n$ denote the running coordinates on the cone measured from the vertex O, then the homogeneous algebraic equation of degree mk

(10) $$Q(\xi_0, \xi_1, \cdots, \xi_n) = 0$$

represents the normal cone through O.

The normal cone is directly given by the differential operator L.

If we again single out $x_0 = t$, then the characteristic condition (9) in the $\xi_1, \cdots, \xi_n$-space is geometrically represented as a surface, called the *normal surface*, i.e., the intersection of the normal cone with the plane $\xi_0 = -1$: $Q(-1, \xi_1, \cdots, \xi_n) = 0$.

Likewise, we may interpret the form (9b) of the characteristic condition in the velocity space of the n-components $v_1, \cdots, v_n$ as a surface, which we call the *normal velocity surface*, or the *reciprocal normal surface*.[1] According to (9a) this surface is the locus of the points P whose distance from O in the α-direction is equal to the speed $|v|$. The two normal surfaces correspond to each other by inversion with respect to the unit sphere $|\xi|^2 = 1$ in ξ-space.[2]

4. Construction of Characteristic Surfaces or Fronts. Rays, Ray Cone, Ray Conoid. In the physical x-space the characteristic condition $Q(\xi) = 0$ for the normals $\xi = \operatorname{grad} \phi = D\phi$ must be interpreted in terms of characteristic surfaces $\phi(x_0, x_1, \cdots, x_n) = \text{const.}$ As in §1, we regard $Q = 0$ as a partial differential equation of first order for $\phi(x_0, \cdots, x_n)$; then *all* members of the family $\phi = \text{const.} = c$ are characteristic surfaces C_c.[3] They are constructed according to the theory of partial differential equations of first order developed in Chapter II.

Every solution of the characteristic partial differential equation $Q(D\phi) = 0$ is generated by an n-parameter family of characteristic curves belonging to the first order equation $Q = 0$. These "bicharacteristics", or *rays* associated with the operator L, are supplemented to *bicharacteristic strips* by attaching to them the values of

$$\xi_i = D_i \phi.$$

(We consider the values ξ_i along such a curve as functions of a curve parameter λ, and we shall denote by a dot differentiation with respect to λ.) Then, as in §1, the bicharacteristic strips satisfy the system of $2n + 2$ canonical ordinary differential equations

(11) $$\dot{x}_i = Q_{\xi_i},$$

(11a) $$\dot{\xi}_i = -Q_{x_i},$$

for which $Q(x, \xi)$ is an integral. We impose the condition $Q = 0$ at

---

[1] Both of these closely related normal surfaces are useful tools for the visualization of propagation phenomena. Incidentally, in the literature occasionally the nomenclature is the opposite of that used here.

[2] With or without singling out t we may interpret the geometrical relationships in projective n-dimensional space.

[3] If we consider not a family, but a single surface $\phi = 0$ and represent this surface in the form $\phi = -t + \psi(x_1, \cdots, x_n) = 0$, then the function ψ of the n variables x satisfies the differential equation $Q(-1, \psi_{x_1}, \cdots, \psi_{x_n}) = 0$, where t has to be replaced by ψ in the coefficients.

one point of each ray; then (11) and (11a) together with $Q = 0$ define a $2n$-parameter family of characteristic strips which is determined independently of the specific characteristic surfaces $\phi =$ const.

If $x_0 = t$ is singled out and we solve $Q(\xi) = 0$ in the form $\xi_0 - \chi(\xi_1, \cdots, \xi_n) = 0$, then again $\dot{x}_0 = 1$, i.e., λ can be identified with the time: $\lambda = t = x_0$.

If $\phi(x)$ is assumed known, then of course the system (11) alone determines the bicharacteristics on the characteristic surfaces C_c: $\phi =$ const. $= c$, after we substitute for ξ_i in Q the values ϕ_{x_i}, which depend on $x_0, \cdots, x_n$.

To construct the characteristic surface $\phi(x_0, x_1, \cdots, x_n) = 0$ through a prescribed $(n - 1)$-dimensional manifold $\mathcal{S}'$ in an n-dimensional initial surface $\mathcal{S}$, we may assume $\mathcal{S}$ in the form $x_0 = 0$, since any n-dimensional initial surface $\mathcal{S}$ can be transformed into $x_0 = 0$ and characteristic elements remain invariant. Now we prescribe in $\mathcal{S}$ initial values $\phi(0, x_1, \cdots, x_n) = \omega(x_1, \cdots, x_n)$ so that $\mathcal{S}'$ is defined by $\omega = 0$. Then we determine the bicharacteristic rays (and strips) through $\mathcal{S}$ by computing the initial values of ϕ_i on $\mathcal{S}$. For $t = x_0 = 0$ we have the relations $\phi(0, x_1, \cdots, x_n) = \omega(x_1, \cdots, x_n)$; hence $\phi_i = \omega_{x_i}$ for $i = 1, \cdots, n$. At $t = 0$ therefore $Q(\phi_0, \omega_{x_i}, \cdots, \omega_{x_n}) = 0$ is an algebraic equation of order mk for the initial value of ϕ_0 and determines mk (or fewer) real initial values of ϕ_0. Then (11), (11a) yield the same number of bicharacteristic strips, and these generate the same number of families of characteristic manifolds $\phi = 0$ fanning out from the initial manifold $\mathcal{S}'$.

Of particular significance is the case when the initial manifold $\mathcal{S}'$ is shrunk to a point O. We define:

The bicharacteristic rays through a fixed point O form the *ray conoid* through O; the bicharacteristic directions through O form the *local ray cone* or *Monge cone* through O. If the principal part of L has *constant coefficients*, and hence the form Q has constant coefficients, then the *rays* are *straight lines* and the local ray cone is identical with the ray conoid.

As pointed out in §1, 8 the local ray cone is dual to the normal cone. Even if the latter is a relatively simple algebraic cone of degree mk, the ray cone may have singularities, or isolated rays, and need not consist of separate smooth conical shells[1] (see article 5 and examples

---

[1] Sometimes it may be more appropriate to consider the local ray cone as being generated by its supporting planes rather than by its rays.

in §3a). For instance, the single differential operator of order 3 in three dimensions $L = D_0 D_1 D_2$ leads to

$$Q(\xi) = \xi_0 \xi_1 \xi_2 .$$

The normal cone through a point O in the ξ-space consists of the three planes parallel to the coordinate planes, and the rays are simply the lines parallel to the axes in x-space. Hence the ray cone degenerates into three lines, while through a nondegenerate initial manifold $\mathcal{g}'$ in $\mathcal{g}$ the three regular cylinders through $\mathcal{g}'$ parallel to the coordinate axes form the characteristic surfaces.

Finally we recall that the theory of the complete integral from Chapter II allows us to construct solutions of the first order differential equation $Q = 0$ as envelopes of families of other solutions; we shall presently interpret this theory as Huyghens' construction of the wave front.

5. Wave Fronts and Huyghens' Construction. Ray Surface and Normal Surfaces. Just as described in §2, 1, characteristic surfaces are of significance as possible carriers of discontinuities of solutions u of $L[u] = 0$. As before, one sees immediately that jump discontinuities occurring merely in the derivatives of order m cannot occur across a "free" surface C, where $Q \neq 0$. On such a surface arbitrary Cauchy data determine the derivatives of order m uniquely, and so they cannot be discontinuous across C. In other words, only characteristic surfaces can carry such discontinuities. For other types of discontinuities, the role of characteristics as the only possible carriers of admissible discontinuities will be analyzed in §4.[1]

The characteristic conoids are important because they represent the propagation of a disturbance initially concentrated at a point O. Such disturbances are called "spherical" wave fronts about the center O (see Ch. II, §9, and Ch. VI, §18.

In the case of *constant coefficients*, all rays are straight and neither the normal cone nor the ray cone depends on the vertex O. The ray cone through O which consists of straight rays is enveloped, or rather supported,[2] by the planes through O orthogonal to the normal cone $Q(\tau, \xi_1, \cdots, \xi_n) = 0$ with $\xi_0 = \tau$. Here we represent ξ, τ in the x-t-space, singling out $t = x_0$ as the time.

---

[1] The construction of §§4, 9, 10 will in fact show that for any characteristic surface there exist such discontinuous solutions u.

[2] As stated before, not all supporting planes are necessarily tangential to the (local) ray cone, which may have concave parts or isolated rays.

The intersection of the forward ray cone with the plane $t = 1$ is called the *ray surface*. It is an $(n - 1)$-dimensional surface in the n-dimensional x-space. The ray surface represents the locus of points reached at $t = 1$ by a disturbance u that started at $t = 0$ at the origin O. In this sense it may be called a *spherical wave front*, although it may consist of separate shells or parts.

To illuminate the situation we assume that for an arbitrary n-dimensional unit vector α the characteristic equation

$$Q\left(-v, \alpha_1, \cdots, \alpha_n\right) = 0$$

yields k real roots $v = v^\kappa$ such that the functions

$$(12) \qquad \phi(t, x) = vt - \alpha_1 x_1 - \cdots - \alpha_n x_n = vt - (\alpha x) = 0$$

represent "plane wave fronts" moving with the normal speed v in the α-direction (see Ch. III, §3) of the x-space.[1]

The plane wave fronts (12) passing through the origin at $t = 0$ do not necessarily support a whole smooth ray cone, but they may support its hull. The hull consists of parts of the ray cone connected by parts of planes (12) which are tangent to the ray cone along two bicharacteristics and which bridge the sector of the ray cone between these bicharacteristics (see §1, 8). Except for the outer hull, complicated geometrical shapes may arise in some cases. To give an intuitive and yet general description seems an elusive task.

This makes the following fact, implied in §1, 8, all the more significant: We consider for a given direction α the plane wave given by (12) which has the largest speed $v(\alpha)$. Now varying α, these wave fronts at the time t support the convex hull Γ of the ray cone. Thus the hull Γ represents the outer "spherical front" of the initial disturbance at O. We may say that the "inner" parts of the ray cone not on Γ represent slower "modes of propagation". The discussion will be resumed in article 7, and examples in §3a will illustrate the variety of geometrical possibilities which may arise.

For nonconstant coefficients the corresponding spherical wave fronts are no longer given by envelopes of plane wave fronts. Instead we consider more generally *plane-like wave fronts*, that is, solutions $\phi(t, x)$ of the characteristic partial differential equation $Q(\phi_{x_0}, \phi_{x_1}, \cdots, \phi_{x_n}) = 0$, which for $t = 0$ have the initial values

[1] This is essentially the assumption of hyperbolicity. (See Ch. III, §2 and article 7 of the present section.)

$(\alpha x) = \alpha_1 x_1 + \cdots + \alpha_n x_n$ with an arbitrary unit vector α. The plane-like wave fronts $\phi = 0$ moving through the x-space in time, and possibly losing their initial plane form, then lead to the same construction of the ray conoid and the hull Γ as the plane wave fronts for constant coefficients.

The following Huyghens' construction is a useful variant of the procedure, and yields the ray surface rather than the ray cone: For brevity we assume again constant coefficients and consider at the time $t = 1$ the plane wave fronts $v - (\alpha x) = 0$, where

$$Q(-v, \alpha_1, \cdots, \alpha_n) = 0$$

and α is a unit vector. As α varies over the unit sphere, the vector $(v\alpha_1, \cdots, v\alpha_n)$ traverses the normal velocity surface (9b) and the planes $v - (\alpha x) = 0$ envelop the ray surface.

Since the reciprocal normal surface has no physical meaning in itself and, moreover, is generally of higher algebraic order than the normal surface, it is preferable to express the connection between the two surfaces as follows: *The ray surface is the locus of the poles with respect to the unit sphere of the tangent planes or supporting planes of the normal surface* (9). This is easily seen, since the planes (12) are the polars of the points on the normal surface. (See §1, 8.)

In examples of §3a we shall see that for higher order equations this construction of the ray surface exhibits the possibility not only of degenerations of the ray surface (e.g., isolated points) but also of singularities such as sharp edges of regression. In keeping with the remarks of §1, 8 it should be pointed out that it is preferable to consider supporting planes rather than merely tangent planes, and to consider the concept of envelope in reference to supporting planes. Then the relationship between ray surface and normal surface is symmetric; each is the locus of the polars of the supporting planes of the other. Such a geometric definition may be applied separately to separate sheets of these surfaces.

Incidentally, to an isolated ray of the ray surface corresponds a plane sheet of the normal surface. (For the reciprocal normal surface or velocity surface the corresponding sheet is spherical.)

Finally we point out: For nonconstant coefficients the concepts of normal surface and ray surface retain their meaning; they refer to the local situation at a point and their relationship remains exactly the same as described above.

5a. Example. An example is the following equation of third order:

$$\left[\left(\frac{\partial}{\partial t} - \frac{\partial}{\partial y}\right)\left(\frac{\partial}{\partial t} + \frac{\partial}{\partial y}\right)^2 - \frac{\partial^2}{\partial x^2}\left(2\frac{\partial}{\partial t} + \frac{\partial}{\partial y}\right)\right] u = 0.$$

The corresponding characteristic equation is

$$(\phi_t - \phi_y)(\phi_t + \phi_y)^2 = \phi_x^2(2\phi_t + \phi_y),$$

and the equation of the normal cone in τ, ξ, η-space is

$$(\tau - \eta)(\tau + \eta)^2 = \xi^2(2\tau + \eta).$$

Thus the normal surface is the curve of third order (Folium of Descartes) in the ξ, η-plane:

$$(-1 - \eta)(-1 + \eta)^2 = \xi^2(-2 + \eta),$$

as in Figure 45, which illustrates how the normal cone may fail to consist of separate sheets. Here one sheet is an oval, and the other sheet touches the oval and extends to infinity. At the point of contact both sheets have corners, across which they connect analytically, so that they form together one connected algebraic curve with self-intersection.

The ray surface is shown in Figure 46. The segment between the points $x = \pm 1/\sqrt{2}$, $y = 1$ is the image of the double point of the normal surface. The cusped portion of the ray surface between these points is the image of the sheet of the normal surface which extends to infinity. The remaining convex portion of the ray surface is the image of the oval part of the normal surface.

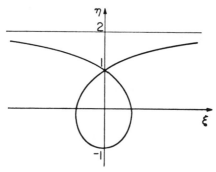

FIGURE 45

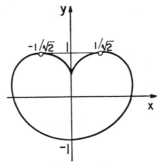

FIGURE 46

This example, although somewhat artificial, illustrates two possible peculiarities in the geometrical structure of the ray cone and ray surface. If the normal surface has double points, then the ray surface and the ray cone may not be convex, and a lid must be added to form the convex hull. If the normal surface has a sheet extending to infinity, or a point of inflection, then the ray surface will have a cusp at an isolated point. In §3a we shall meet examples of physical significance which exhibit such behavior.

6. Invariance Properties. Characteristic forms, characteristic matrices, and characteristic rays remain invariant under coordinate transformations. This fact follows immediately, as said before, from the conceptual definitions, in particular from the invariant character of interior differentiation and from the fact that bicharacteristics are curves along which characteristic surfaces touch each other. The analytic confirmation of the invariance properties exactly parallels §1, 5.

7. Hyperbolicity. Space-Like Manifolds, Time-Like Directions.[1] So far, no particular assumption concerning the reality of characteristic surfaces or of the normal cone has been made. Yet the main objective of the present chapter, i.e., the solution of Cauchy's problem, requires more than existence of real branches of these surfaces. A more stringent condition of *hyperbolicity* must be imposed to insure the solvability of Cauchy's problem. One could equate this solvability with the concept of hyperbolicity. However, for mathematical physics it is more suitable to stipulate conditions which can be verified by algebraic or geometric criteria, and which are sufficient

[1] Compare Ch. III, §2.

to assure the existence of a solution for as wide a class of problems as possible. All variants of such definitions[1] of hyperbolicity amount to requiring a maximum degree of reality of the algebraic normal cone.

Definition: At a point[2] O the operator $L[u]$ is called hyperbolic if there exist vectors ζ through O such that every two-dimensional plane π through ζ intersects the normal cone $Q(\xi) = 0$ in mk real and distinct lines.

Algebraically the stipulation of our definition is: If θ is an arbitrary vector (not parallel to ζ) then the line of points $\xi = \lambda\zeta + \theta$, λ being a parameter, must intersect the normal cone in mk distinct real points; i.e., the equation for λ

$$Q(\lambda\zeta + \theta) = 0$$

must have mk real distinct roots. Space elements at O orthogonal to the vectors ζ are called space-like, and ζ is called a space-like normal. Space-like surface elements separate the "forward" part of the ray cone from the "backward" part.[3]

It is useful to give a *second equivalent definition of hyperbolicity,* stressing the concept of space-like surfaces.

We first realize that we may decompose any vector θ into the sum of vectors parallel and orthogonal respectively to ζ and we may combine the former with ζ. Secondly, invoking the invariance properties of article 6, we may assume ζ to have the components $1, 0, 0, \cdots, 0$; hence θ has the first component 0. Then $Q(\lambda\zeta + \theta) = 0$ becomes $Q(\lambda, \theta_1, \cdots, \theta_n) = 0$.

This remark leads immediately to the following alternative definition:

An n-dimensional manifold $\mathcal{J}$ (or an element of it), which we may by a suitable coordinate transformation write as $x_0 = 0$, is called *space-like,* if for each point in $\mathcal{J}$ and arbitrary real values of $\xi_1, \cdots, \xi_n$, the equation $Q(\xi_0, \xi_1, \cdots, \xi_n) = 0$ possesses mk real distinct

[1] See also Ch. III, §2.

[2] It should be emphasized again that hyperbolicity is a local property of L, or Q, and that one may restrict oneself to surface elements through the P under consideration.

For quasi-linear operators hyperbolicity also depends on the local Cauchy data.

[3] Note that these parts are connected at infinity if interpreted in the projective space.

roots ξ_0. This means, according to article 4, from each $(n - 1)$-dimensional initial manifold of the x-space $\mathcal{S}$ there fan out mk different characteristic sheets.

$L[u]$ is called *hyperbolic* at O if such space-like manifolds (or elements) through O exist.

The space-like normals ζ at O form the inner "core" of the normal cone, bounded by the *inner sheet* of the cone.

The core of the normal cone is convex. This important theorem follows almost immediately from the definition: If it were false, there would exist vectors ζ in the core carrying planes which intersect the inner sheet more than twice so that they would have more than mk intersections with the cone. Geometrically, then, the normal cone may be visualized as consisting of the closed inner sheet bounding the inner "core" into which ζ points, and of further sheets, which form subsequent shells around the core. These sheets may be closed or they may extend to infinity; in any case they are such that all the *planes* π through ζ intersect in mk distinct *lines*.

The cone supported by the planes orthogonal to the generators of the convex inner sheet of the normal cone is the *convex hull* Γ *of the local ray cone*; specifically the hull of its *outer shell*[1] (for the proof see §1, 8). We define: Every direction from O into this outer shell is called *time-like*. A *curve* in the x-space is called *time-like* if its direction is everywhere time-like.[2]

Obviously, the concepts "space-like" and "time-like" do not depend on the coordinate system.

For single differential equations of second order, in which there is only one shell of the normal cone and of the local ray cone (see §1), the situation is easily seen to fit into the scheme discussed here.

It may be useful at this point to summarize observations and definitions concerning the characteristic cones made at various occasions in the preceding sections: If the algebraic condition of order k for the normal cone defines a maximal number of real and separate sheets, then, by the reciprocal transformation, to each of the sheets corresponds a distinct sheet of the local ray cone, or a distinct "mode of

[1] Again we stress the fact that nothing need be said here about the inner parts of the ray cone. It may consist of closed sheets corresponding to sheets of the normal cone which wind around the core, but it might have quite a different structure.

[2] For a different definition, not equivalent to ours, see F. John [4], p. 157.

propagation of a wave front". The convex inner core of the normal cone is transformed into the outer shell of the local ray cone which is automatically convex. If the normal cone consists of nested convex sheets, then the same is true of the ray cone. Otherwise the latter may contain isolated rays or singularities.[1]

In general, the preceding assumptions about separateness of sheets are not satisfied. For mathematical physics the condition of distinctness of the roots λ is too restrictive, since in many relevant cases multiple roots λ do occur, inasmuch as several sheets of the algebraic normal cone may touch each other, or intersect, or even completely coincide.[2] In article 9 we shall see that the definitions 1 or 2 could be easily widened so that they cover the case of characteristic surfaces with uniform multiplicity.[3] But even differential equations as simple as $u_{x_0 x_1 x_2} = 0$ do not fall under our definitions, although Cauchy's problem is explicitly solvable.

For equations with not entirely distinct sheets of the normal cone a useful generalization of the concept of hyperbolicity is difficult,[4] and the investigation of Cauchy's problem requires a more delicate analysis. Yet fortunately in mathematical physics difficulties arising from multiple intersections between the planes π and the normal cone do not in general interfere with uniqueness and existence proofs, as long as all roots λ of $Q(\lambda \zeta + \theta) = 0$ are real. The reason is that the differential equations of mathematical physics are in general *symmetric*.[5] In particular the *symmetric* hyperbolic systems of equations of first order have emerged as a central and most fruitful topic; in the context of mathematical physics nonsymmetric systems are of secondary importance.

---

[1] It may be remarked that the present state of the theory of algebraic surfaces does not permit entirely satisfactory applications to the questions of reality of geometric structures which confront us here.

[2] Such multiplicities still pose serious obstacles for the theory. For many systems of first order, e.g. Maxwell's equations, they always occur. Obviously single differential equations of order k with arbitrarily prescribed algebraic normal cones always can be constructed.

[3] For constant coefficients (and also to an extent for nonconstant coefficients) pertinent extensions of the concept of hyperbolicity have been achieved by Gårding and others by accounting also for the influence of lower order terms in the differential equations. (See L. Gårding [2] and A. Lax [1].)

[4] Compare also Ch. V, §8.

[5] The symmetry, incidentally, is closely related to the fact that these differential equations usually are Euler equations of a quadratic variational problem.

8. Symmetric Hyperbolic Operators. We consider linear (or quasi-linear) symmetric systems of first order, i.e., systems of the form

$$(14) \qquad L[u] = \sum_0^n A^i u_i + Bu,$$

where the matrices A^i are symmetric, but the matrix B need not be.

A symmetric system (14) is called *symmetric hyperbolic* (at O) if one of the matrices A^i or a linear combination

$$\sum \xi^i A^i$$

is definite, say positive definite. The n-dimensional manifolds $\mathcal{J}$ orthogonal to such vectors ξ are called *space-like*. Since a linear combination, with positive weights, of positive definite matrices is positive definite, it follows again that the set of such vectors ξ orthogonal to space-like elements forms a *convex cone*.

If for example A^0 is positive definite, then the n-dimensional spaces $x_0 = t = $ const. are space-like.

The close relationship of the concept of symmetric hyperbolic operators to the definition of article 7 is seen as follows, if for simplicity, we assume A^0 positive definite: For $\xi_0 = 1, \xi_1 = \cdots = \xi_n = 0$, the equation

$$Q(\lambda \xi + \theta) = Q(\lambda, \theta_1, \cdots, \theta_n) = 0$$

has indeed k real roots, since it is simply the symmetric determinant equation

$$(15) \qquad \| \lambda A^0 + \sum_1^n \theta_i A^i \| = 0,$$

with definite A^0. The roots λ are the eigenvalues of the matrix $\sum \theta_i A^i$ with respect to the positive definite matrix A^0. The essential difference is that in our definition of symmetric hyperbolicity no assumption is made about distinctness of the roots λ of (15). Indeed, as we shall see in §§8, 10, for symmetric systems Cauchy's problem is always solvable.

The following observation is relevant for many examples in physics: If A^0 is positive definite, we may linearly transform the system into

$$(16) \qquad L[u] = \tilde{L}[v] = v_t + \sum_{i=1}^n \tilde{A}^i v_i + \tilde{B} v,$$

where $\tilde{A}^i$, $\tilde{B}$, v denote the transformed quantities and where the matrices $\tilde{A}^i$ are still symmetric.[1]

Even in the case of multiple roots, the characteristic matrix has a complete set of independent k left eigenvectors l ($l = r$) such that $lL[u]$ is an interior operator in the respective characteristic surface, whether or not the surface is a multiple characteristic, i.e., allows several linearly independent nullvectors l.

9. Symmetric Hyperbolic Equations of Higher Order. A final remark should be made concerning *single equations or systems of higher order.* If they are the result of elimination from a symmetric hyperbolic system of first order, then they also should be called *symmetric hyperbolic* and are covered by the theory of such first order systems.

For example

$$u_{x_0 x_1 x_2} = 0$$

originates in this way from the symmetric system

$$u^1_{x_0} = u^2, \qquad u^2_{x_1} = u^3, \qquad u^3_{x_2} = 0.$$

A remarkable fact is: *Every single hyperbolic differential equation of second order can be reduced to a symmetric hyperbolic first order system.* We write (see §1) the differential equation for u in the self-explanatory form

$$(17) \qquad L[u] = u_{tt} + 2\sum_{i=1}^{n} a^i u_{it} - \sum_{i,k=1}^{n} a^{ik} u_{ik} + \cdots = 0,$$

where the dots denote terms containing at most first derivatives of u, and where $a^{ik} = a^{ki}$.

We assume (17) is hyperbolic and $t = $ const. a space-like surface; then the quadratic form $H(\xi_i, \xi_k) = \sum_{i,k=1}^{n} a^{ik}\xi_i\xi_k$ is positive definite

---

[1] The transformation is

$$\tilde{L}[v] = TLT(v),$$

so that

$$\tilde{A}^j = TA^jT.$$

Since A^0 is positive definite, it has a square root $C: A^0 = C^2$. We choose now $T = C^{-1}$. Clearly, $\tilde{A}^0 = I$, and since T is symmetric, the matrices $\tilde{A}^j$ are also symmetric.

and the point $\xi_1 = \xi_2 = \cdots = \xi_n = 0$ lies inside the ellipsoid

$$1 + 2\sum_{i=1}^{n} a_i \xi_i - H(\xi_i, \xi_k) = 0;$$

indeed, a surface element with components $\tau, \xi_1, \xi_2, \cdots, \xi_n$ of the normal is space-like if the equation

$$(\lambda - \tau)^2 + 2(\lambda - \tau)\sum_{i=1}^{n} a_i \xi_i - H(\xi_i, \xi_k) = 0$$

has two real roots λ with opposite signs.

To accomplish the reduction we simply replace in (17) the first derivatives u_t, u_{x_i} by new unknowns, v^0, v^i, respectively. Then we replace (17) by the system

$$v_t^0 + 2\sum_{i=1}^{n} a^i v_{x_i}^0 - \sum_{i,k=1}^{n} a^{ik} v_{x_k}^i + \cdots = 0,$$

$$\sum_{i=1}^{n} a^{ik}(v_t^i - v_{x_i}^0) = 0 \qquad (k = 1, 2, \cdots, n),$$

for the vector $v = (v_0, \cdots, v_n)$. This system is symmetric and hyperbolic.[1] If, as before, the initial values of v are restricted by the condition that they be obtained by identification with the given initial values of u_i and u_t, then the two problems are equivalent as is easily ascertained.

10. Multiple Characteristic Sheets and Reducibility. A brief remark may be added about the general case of first order systems that are not necessarily symmetric. The fan of k distinct characteristics C^κ described in the second definition of article 7 corresponds to k linearly independent nullvectors $l^1, \cdots, l^k$ such that $l^\kappa L[u]$ is an internal differential operator in the characteristic C^κ. Now, let us assume that s of these characteristic surfaces coincide, say $C^1, \cdots, C^s$, but that for the surface $\phi(x)$ we still have s linearly independent nullvectors $l^1, \cdots, l^s$ of the matrix A, and that this is true along the

---

[1] In fact, it has the form $A^0 v_t + \sum_{\nu=1}^{n} A_\nu^\nu v_\nu + \ldots = 0$, where

$$A^0 = \begin{pmatrix} 1 & 0 & \cdots & 0 \\ 0 & a^{11} & \cdots & a^{1n} \\ \cdots\cdots\cdots\cdots\cdots \\ 0 & a^{n1} & \cdots & a^{nn} \end{pmatrix}, A^\nu = \begin{pmatrix} 2a^\nu & -a^{\nu 1} \cdots & -a^{\nu n} \\ -a^{\nu 1} & 0 & \cdots & 0 \\ \cdots\cdots\cdots\cdots\cdots\cdots \\ -a^{\nu n} & 0 & \cdots & 0 \end{pmatrix}$$

whole characteristic surface.[1] Then the whole surface C is called s-fold, and all the s differential operators $l^i L$ for $i = 1, \cdots, s$ are interior in C.

We may call such operators $L[u]$ hyperbolic (in the generalized sense), even though A has the rank $k - s$ on C, as assumed.

This phenomenon of multiple sheets is closely related to the occurrence of *reducibility* of the characteristic form $Q(\xi_0, \cdots, \xi_n)$ into irreducible factors of lower degree[2]

$$Q = Q_1 Q_2 \cdots Q_j \cdots$$

and in particular, to the occurrence of s identical factors Q_i.

Characteristic surfaces must satisfy one of the equations $Q_j = 0$; moreover if we assume that the equation is hyperbolic, i.e., that there are k linearly independent nullvectors l^k, then each of the equations $Q_j = 0$ leads to exactly k_j independent nullvectors l^i if k_j is the degree of the polynomial Q_j. If s of the factors Q_j are identical, then the corresponding characteristic surface is s-fold and for it we have s linearly independent nullvectors.

Conversely, as a simple algebraic reasoning shows, occurrence of sheets s-fold according to our definition implies reducibility of Q in the form $Q = Q_1^s \cdot Q_2$.

It should be emphasized that in the case of reducibility also the rays should be defined (and for multiple sheets *must* be defined) with reference not to Q but to the respective irreducible factors Q_j.

As stated before, reducibility and multiple sheets occur in many differential equations of physics, such as the Maxwell equations, the linearized equations of hydromagnetic waves, the equations of elastic waves (see §3a and §13a). Also contacts or intersections of sheets are frequent.[3] However, since in these cases the system is symmetric hyperbolic, no difficulties arise from such multiplicities for the

---

[1] We do not discuss here the case when two or more sheets C^κ merely intersect or touch. (See M. Yamaguti and K. Kasahara [1].)

[2] In particular, such a situation arises if the system $L[u] = 0$ consists of blocks of equations each of which contains the derivatives of only some of the unknowns, such that the coupling of these blocks is effected only by lower order terms (weak coupling).

[3] It may be conjectured that such multiplicities must occur for systems of differential equations. In particular, every system containing an odd number of equations can be seen to have multiple roots. To clarify the conjecture in general would seem an interesting algebraic problem.

proofs of uniqueness and existence (while special attention is necessary for the study of propagation of singularities and related subjects).

11. Lemma on Bicharacteristic Directions. The following remarks are added to this section for use in §4.

Lemma: *At a fixed point x we consider the characteristic matrix A as a function of the variables $\xi_i = \partial\phi/\partial x_i$, $i = 0, \cdots, n$, subject to the condition $\| A \| = Q(\xi_0, \xi_2, \cdots, \xi_n) = 0$. If A has the rank $k - 1$, the differentiation along the bicharacteristic rays is given by the relation*

$$\dot{x}_i = lA_{\xi_i} r \qquad (i = 0, \cdots, n),$$

where the dot denotes differentiation along the ray with respect to a suitable curve-parameter, and l and r denote the left and right nullvectors of the matrix A.

The proof can be given by explicit calculation.[1] However, we proceed by the following implicit reasoning:

At a fixed point O we consider all the characteristic surface elements; they are orthogonal to the sheets of the normal cone and defined by the set of parameters $\xi_0, \cdots, \xi_n$, which in turn are restricted by $Q(\xi_0, \cdots, \xi_n) = 0$ but are otherwise considered as independent variables. The vectors r and l are functions of ξ.

By taking differentials of the equation $Ar = 0$ we obtain

$$A \, dr + \sum_{i=0}^{n} A_{\xi_i} r \, d\xi_i = 0.$$

[1] In the case of a system of first order $A = \sum A'\xi_\nu$, we denote the elements of A' by a_ν^{ij} and observe $Q_{\xi_\nu} = \sum_{i,j=1}^{k} a_\nu^{ij} \, \alpha^{ij}$, where α^{ij} is the minor to $a^{ij} = \sum_{i=0}^{n} a_\nu^{ij} \xi_\nu$ in the determinant $\|A\|$. By assumption, the nullvectors r and l are uniquely determined within a scalar factor as the minors of a row or column, respectively, of α. We have $\alpha^{11} : \alpha^{1j} = \alpha^{i1} : \alpha^{ij}$ $(i, j = 1, \ldots, k)$, or

$$\frac{1}{\alpha} \, \alpha^{ij} = \alpha^{1j}\alpha^{i1} = l_i r_j,$$

where $l_i = \alpha^{i1}$, $r_j = \alpha^{1j}$ are the components of the vectors l and r, respectively, and $1/\alpha = \alpha^{11}$. (We may assume $\alpha^{11} \neq 0$.)

Now

$$lA'r = \sum_{i,j=1}^{k} l_i a_\nu^{ij} r_j = \sum_{i,j=1}^{k} \frac{1}{\alpha} a_\nu^{ij}\alpha^{ij},$$

and hence

$$\alpha lA'r = Q_{\xi_\nu}.$$

After multiplying by l and observing that $lA = 0$, this becomes

$$\sum_0^n lA_{\xi_i} r \, d\xi_i = 0.$$

The differentials $d\xi_i$ are independent except for the linear relation

$$dQ = \sum_0^n Q_{\xi_i} \, d\xi_i = 0$$

since the variables ξ_i are restricted by $Q(\xi_0, \cdots, \xi_n) = 0$.

From these relations it follows immediately that the quantities $\partial Q / \partial \xi_i$ and $lA^i r$ are proportional, since n components of $d\xi_i$ can be chosen arbitrarily. This proves the lemma.

This result can be generalized to multiple characteristic sheets with a rank $k - s$ for A throughout the region under consideration. Such a sheet of the characteristic surface satisfies an equation of the form

$$\xi_0 = f(\xi_1, \cdots, \xi_n).$$

A possesses linearly independent right nullvectors r^1, r^2, $\cdots$ and left nullvectors l^1, l^2, $\cdots$. Then we have for arbitrary i and j

$$l^j A^\nu r^i = -l^j A^0 r^i f_\nu .$$

Thus independently of i and j, the proportion

$$l^j A^\nu r^i : l^j A^0 r^i = -f_\nu \qquad\qquad (\nu > 0)$$

is valid.

The proof is exactly as above: We consider, at a fixed point x, the equation

$$Ar^i = \sum_{\nu=0}^n A^\nu \xi_\nu r^i = 0$$

with $\xi_1, \cdots, \xi_n$ as independent parameters. Differentiating and multiplying on the left by l^j one obtains, as before,

$$\sum_{i=0}^n l^j A^\nu r^i \, d\xi_\nu = 0,$$

where for the sheet under consideration, $\xi_0 = f(\xi_1, \cdots, \xi_n)$ and

$$d\xi_0 = \sum_{\nu=1}^{n} f_\nu \, d\xi_\nu$$

is the only condition for the parameters $d\xi_\nu$. Therefore

$$l^j A^\nu r^i + l^j A^0 r^i f_\nu = 0 \qquad (\nu > 0).$$

§3a. Examples. Hydrodynamics, Crystal Optics, Magneto-hydrodynamics

1. Introduction. In this section we shall consider three examples[1] to illuminate the general theories of §3.

The (quasi-linear) equations of hydrodynamics show the physical significance of characteristics, rays and the ray conoid.

The second example, the equations of crystal optics, exhibits the important feature of *anisotropy*: The velocity of propagation of a wave depends upon the direction of propagation; in crystal optics the normal surface and the ray surface are surfaces of the fourth degree, and consequently are more complicated than in the case of hydrodynamics.

The third example, the equations of magnetohydrodynamics, illustrates the complicated structure which the normal cone and in particular the ray conoid may have in physically meaningful situations.

The still more complicated example of the equations of anisotropic elastic waves is treated by G. F. D. Duff.[2] Some examples exhibiting extreme degeneration in the normal surface, ray surface and domain of dependence are given by L. Gårding.[3]

In the case of symmetric hyperbolic systems, there is no special difficulty in proving the existence and uniqueness of the solution of the Cauchy problem (see §8, 10). Yet, for nonsymmetric systems or for single equations of order higher than two, possible complications in the geometrical structure of the normal cone, in particular multiple elements, introduce difficulties which have not been completely resolved.[4] Even in the case of symmetric hyperbolic systems,

---

[1] Complying with usage in physics, in this section we shall occasionally use *ad hoc* notations. The reader should fit them into the general framework of the notations of §3.

[2] See G. F. D. Duff [2]. See also V. T. Buchwald [1].

[3] See L. Gårding [4].

[4] For a discussion of single equations with multiple characteristics, see L. Gårding [2] and A. Lax [1].

the detailed analysis of the structure of solutions, specifically the treatment of propagation of singularities, is impeded by complications in the geometry of characteristic surfaces. Thus, the analysis in §4 applies immediately only where the multiplicity of the roots of the characteristic equation does not change.

2. The Differential Equation System of Hydrodynamics. As an example of a nonlinear problem we consider the system of *differential equations describing the motion of compressible fluids* in the x,y-plane. (The case of stationary flow was already treated in Ch. V, §3, 3.) If the components of the velocity of the fluid are denoted by $u(x, y, t)$, $v(x, y, t)$, and the density by $\rho(x, y, t)$ and if, as before, $p(\rho)$ denotes pressure as function of density, with $p'(\rho) > 0$, the quasilinear *Euler equations of motion* are

$$\rho u_t + \rho u u_x + \rho v u_y + p' \rho_x = 0,$$

(1) $$\rho v_t + \rho u v_x + \rho v v_y + p' \rho_y = 0,$$

$$\rho_t + u \rho_x + v \rho_y + \rho(u_x + v_y) = 0.$$

Let $\phi(x, y, t) = 0$ be a manifold along which u, v, ρ are given. Then all the derivatives of u, v, ρ, and in particular the outward derivatives $u_\phi, v_\phi, \rho_\phi$, are in general uniquely determined by their values along this manifold. However, this is not the case if on $\phi = 0$, or on the whole family $\phi = $ const., the characteristic condition is satisfied:

$$(2) \quad Q \equiv \begin{vmatrix} \rho(\phi_t + u\phi_x + v\phi_y) & 0 & p'\phi_x \\ 0 & \rho(\phi_t + u\phi_x + v\phi_y) & p'\phi_y \\ \rho\phi_x & \rho\phi_y & \phi_t + u\phi_x + v\phi_y \end{vmatrix} = 0.$$

Expanding the determinant, we obtain

$$(3) \quad Q \equiv \rho^2(\phi_t + u\phi_x + v\phi_y)[(\phi_t + u\phi_x + v\phi_y)^2 - p'(\phi_x^2 + \phi_y^2)] = 0,$$

and then omitting the factor ρ^2 and setting $\phi_t = \tau, \phi_x = \xi, \phi_y = \eta$, we have[1]

$$(3') \quad (\tau + u\xi + v\eta)[(\tau + u\xi + v\eta)^2 - p'(\xi^2 + \eta^2)] = 0.$$

These characteristic surfaces $\phi = 0$ in x,y,t-space, or the corresponding families of curves $t = \psi(x, y)$ in the x,y-plane, which we obtain by

---

[1] The term $\tau + u\xi + v\eta = \dot{\phi}$ is the rate of change of ϕ in time at a moving particle of the fluid.

setting $\phi = t - \psi(x, y)$, once again represent the possible manifolds of discontinuity or wave fronts associated with the motion of the fluid. Thus, corresponding to the first factor of $(3')$, we obtain for the characteristics the surface

$$(4) \qquad \tau + u\xi + v\eta = 0.$$

The corresponding sheet of the normal cone[1] in the ξ,η,τ-space is a plane. The projection of the associated ray on the x,y-plane is nothing but the stream line of the flow; the rays themselves, given in the three-dimensional x,y,t-space by $dx/dt = u$, $dy/dt = v$, represent the stream lines together with the velocity of flow.

Corresponding to the second factor of $(3')$, we have the characteristic manifold

$$(5) \qquad (\tau + u\xi + v\eta)^2 - p'(\xi^2 + \eta^2) = 0$$

which represents the quadratic sheet of the normal cone. The directions of the rays or bicharacteristics, given by the ratio $dt\!:\!dx\!:\!dy$, again represent the "propagation velocities" or ray velocities for the discontinuities. The rays with t as parameter satisfy the "Monge equation" (see Ch. II, §5)

$$(6) \qquad \left(\frac{dx}{dt} - u\right)^2 + \left(\frac{dy}{dt} - v\right)^2 = p',$$

as is easily verified.

In acoustics and hydrodynamics, $\sqrt{p'}$ is the *sound velocity*. Equation (6) therefore states: *The relative velocity of the propagation of discontinuities is equal to the sound velocity.*

These results and their relation to the stationary case mentioned earlier (Ch. V, §3, 3) can be further illuminated by the following geometric argument:

For given u and v, the local ray cone or *Monge cone of the characteristic differential equation* in x,y,t-space, whose vertex we assume to be at the origin $x = y = t = 0$, is represented by

$$\left(\frac{x}{t} - u\right)^2 + \left(\frac{y}{t} - v\right)^2 = p'(\rho).$$

---

[1] The normal cone is to be considered locally for a fixed vertex, say $t_0 = 0$, $x_0 = y_0 = 0$ and for fixed values of u, v, ρ.

It is therefore obtained by projection of the circle

$$(x - u)^2 + (y - v)^2 = p'$$

in the plane $t = 1$ from the origin. According to whether this circle encloses the origin $x = 0$, $y = 0$ or not, i.e., according to whether $u^2 + v^2 < p'$ or $u^2 + v^2 > p'$, the circular cone in question either contains the t-axis or slants so much that the t-axis lies outside of it. See Figure 47.

The transition to the stationary case is performed by setting all derivatives with respect to t equal to zero. Then we consider only tangent planes to the Monge cone for which $\phi_t = 0$, in other words, tangent planes perpendicular to the x,y-plane or containing the t-axis. Their lines of contact with the cone furnish the two characteristic directions in the stationary case. Now the two tangent planes to the Monge cone through the t-axis are real and distinct if and only if the t-axis lies outside of the cone, i.e., if the flow velocity $\sqrt{u^2 + v^2}$ is greater than the sound velocity $\sqrt{p'}$. The earlier result of Ch. V, §3 concerning stationary fluid motion is thus confirmed and illuminated.

3. Crystal Optics. The characteristic condition for *Maxwell's equations* for a vacuum was derived in Ch. III, §2, starting out from a somewhat different point of view. We will consider here the generalization of Maxwell's equations to the *case of crystal optics*. Maxwell's general equations relating the magnetic vector $\mathfrak{H}$, the electric vector $\mathfrak{E}$, the electric displacement $\mathfrak{D}$ and the magnetic displace-

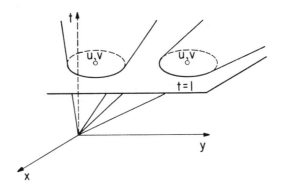

FIGURE 47

ment $\mathfrak{B}$ are given by

(7) $$\operatorname{curl} \mathfrak{H} = \frac{1}{c} \dot{\mathfrak{D}}, \qquad \operatorname{curl} \mathfrak{E} = -\frac{1}{c} \dot{\mathfrak{B}};$$

here c is the velocity of light, the dot indicates differentiation with respect to the time t, $\mu \mathfrak{H} = \mathfrak{B}$ (μ is the constant of permeability, usually assumed as 1). The components u_1, u_2, u_3 of the electric vector $\mathfrak{E}$ are related to the electric displacement $\mathfrak{D}$ by $\mathfrak{D} = (\epsilon_1 u_1, \epsilon_2 u_2, \epsilon_3 u_3)$, where ϵ_1, ϵ_2, ϵ_3 are three dielectric constants in the three axial directions. The occurrence of different constants ϵ characterizes the medium as a crystal.

From (7) one finds immediately $(\operatorname{div} \mathfrak{D})^{\cdot} = 0$ and $(\operatorname{div} \mathfrak{B})^{\cdot} = 0$. Assuming that initially $\operatorname{div} \mathfrak{D} = \operatorname{div} \mathfrak{B} = 0$ we have everywhere

(7a) $$\operatorname{div} \mathfrak{D} = 0, \qquad \operatorname{div} \mathfrak{B} = \operatorname{div} \mathfrak{H} = 0;$$

we shall always assume that (7a) is satisfied.

Elimination of the vector $\mathfrak{H}$ from (7) leads to three linear differential equations of second order for the electric vector:

$$
\begin{aligned}
\sigma_1 \ddot{u}_1 &= \Delta u_1 - \frac{\partial}{\partial x} \operatorname{div} \mathfrak{E} = \frac{\partial^2 u_1}{\partial y^2} + \frac{\partial^2 u_1}{\partial z^2} - \frac{\partial^2 u_2}{\partial x \partial y} - \frac{\partial^2 u_3}{\partial z \partial x}\cdot \\
(8) \quad \sigma_2 \ddot{u}_2 &= \Delta u_2 - \frac{\partial}{\partial y} \operatorname{div} \mathfrak{E} = \frac{\partial^2 u_2}{\partial z^2} + \frac{\partial^2 u_2}{\partial x^2} - \frac{\partial^2 u_3}{\partial y \partial z} - \frac{\partial^2 u_1}{\partial x \partial y}, \\
\sigma_3 \ddot{u}_3 &= \Delta u_3 - \frac{\partial}{\partial z} \operatorname{div} \mathfrak{E} = \frac{\partial^2 u_3}{\partial x^2} + \frac{\partial^2 u_3}{\partial y^2} - \frac{\partial^2 u_1}{\partial z \partial x} - \frac{\partial^2 u_2}{\partial y \partial z},
\end{aligned}
$$

with $\sigma_i = (\mu/c^2)\epsilon_i$.[1] We want to carry out briefly the algebraic calculations which result in equations for the normal cone, the ray cone, and the normal surface. Writing x_1, x_2, x_3 instead of x, y, z, we have the following equation for a characteristic manifold $\phi = t - \psi(x) = 0$, with $\xi_i = \phi_{x_i}$, $\rho^2 = |\xi|^2 = \xi_1^2 + \xi_2^2 + \xi_3^2$:

$$
(9) \quad H(\xi) = \begin{vmatrix} \rho^2 - \xi_1^2 - \sigma_1 & -\xi_1 \xi_2 & -\xi_1 \xi_3 \\ -\xi_2 \xi_1 & \rho^2 - \xi_2^2 - \sigma_2 & -\xi_2 \xi_3 \\ -\xi_3 \xi_1 & -\xi_3 \xi_2 & \rho^2 - \xi_3^2 - \sigma_3 \end{vmatrix} = 0,
$$

or after a short calculation

[1] See also the separate equations in §14a, 1 for each component of the vectors.

$$H(\xi) = (\rho^2 - \sigma_1)(\rho^2 - \sigma_2)(\rho^2 - \sigma_3)$$

(9a)
$$\cdot \left(1 - \frac{\xi_1^2}{\rho^2 - \sigma_1} - \frac{\xi_2^2}{\rho^2 - \sigma_2} - \frac{\xi_3^2}{\rho^2 - \sigma_3}\right) = 0.$$

The equation of the normal cone in the τ,ξ-space is

$$Q(\tau, \xi) = (\rho^2 - \sigma_1\tau^2)(\rho^2 - \sigma_2\tau^2)(\rho^2 - \sigma_3\tau^2)$$

(9b)
$$\cdot \left(1 - \frac{\xi_1^2}{\rho^2 - \sigma_1\tau^2} - \frac{\xi_2^2}{\rho^2 - \sigma_2\tau^2} - \frac{\xi_3^2}{\rho^2 - \sigma_3\tau^2}\right) = 0,$$

so that

$$H(\xi) = Q(-1, \xi).$$

Hence the normal surface is given by

(10)
$$F(\xi) = \sum_{i=1}^{3} \frac{\xi_i^2}{\rho^2 - \sigma_i} = 1.$$

The equation of the reciprocal normal surface (see §3, 3) is obtained if we replace ξ_i by $(1/\rho^2)\xi_i$:

(11)
$$G(\xi) = \sum \frac{\xi_i^2}{1 - \sigma_i\rho^2} = 1.$$

Another form of the equations of the normal surface is

(10a)
$$\sum \frac{\sigma_i\xi_i^2}{\rho^2 - \sigma_i} = 0,$$

and the equation for the reciprocal normal surface is

(11a)
$$\sum \frac{\sigma_i\xi_i^2}{1 - \sigma_i\rho^2} = 0.$$

(10a) can be found by straightforward calculations from (10) by using the identity

$$1 = \sum \frac{\xi_i^2}{\rho^2} = \sum \frac{(\rho^2 - \sigma_i)\xi_i^2}{(\rho^2 - \sigma_i)\rho^2} = \sum \frac{\xi_i^2}{\rho^2 - \sigma_i} - \frac{1}{\rho^2}\sum \frac{\sigma_i\xi_i^2}{\rho^2 - \sigma_i} ;$$

(11a) then follows directly.

The ray surface is either formed as the envelope[1] of normal

---

[1] Note that we obtain by this analytic construction the ray surface proper, and not directly its hull.

planes of the reciprocal normal surface, or equivalently as the locus of the poles with respect to the unit sphere of the planes tangent to the normal surface (10).

The tangent plane at the point ξ is represented in running coordinates ζ_i by $\sum_i F_{\xi_i}(\xi)(\zeta_i - \xi_i) = 0$ or $\sum F_{\xi_i}(\xi)\zeta_i = \sum \xi_\kappa F_{\xi_\kappa}$. The pole of this tangent plane has the coordinates

$$\eta_i = F_{\xi_i}(\xi) \bigg/ \sum_{\kappa=1}^{3} \xi_\kappa F_{\xi_\kappa}(\xi).$$

Elimination of the coordinates ξ_i from these algebraic equations and (10a) leads after some calculation to the *equation for the ray surface*:

(12) $$\sum_i \frac{\eta_i^2}{R^2 - \dfrac{1}{\sigma_i}} = 1 \qquad (R^2 = \sum \eta_i^2).$$

Both the normal surface and the ray surface are algebraic surfaces of order 4, the "surfaces of Fresnel". They are related to each other by changing σ_i into $1/\sigma_i$, ξ into η and ρ into R. As we shall analyze further in article 4, these surfaces consist of two closed sheets, the interior one being convex. Their projections from the origin into a four-dimensional space are the normal cone and ray cone respectively. The convex core of the one corresponds by the reciprocal transformation to the convex hull of the other.

4. The Shapes of the Normal and Ray Surfaces. The normal surface (10) is a surface of fourth degree, symmetric with respect to the origin. This surface is intersected by each line through the origin in four real points; it consists of two closed "mantles" or sheets which are separate except for four points at which they touch each other. Assuming that $\sigma_1 > \sigma_2 > \sigma_3$, these four double points, which define the principal axes of the bi-axial crystal, lie in the ξ_1, ξ_3-plane on the straight lines

$$\xi_1 \sqrt{\frac{1}{\sigma_3} - \frac{1}{\sigma_2}} \pm \xi_3 \sqrt{\frac{1}{\sigma_2} - \frac{1}{\sigma_1}} = 0.$$

These facts will be demonstrated by means of a calculation starting from equation (10). The formal similarity between equations (10) and (12) then shows that the corresponding statements are true for the ray surface, provided that σ_1, σ_2 and σ_3 are replaced by their reciprocals.

Multiplying (10) by $(\rho^2 - \sigma_1)(\rho^2 - \sigma_2)(\rho^2 - \sigma_3)$ and collecting terms of the same degree in ξ, we find that the normal surface has the equation

$$(13) \qquad -\sigma_1\sigma_2\sigma_3(1 - \Psi(\xi) + \rho^2\Phi(\xi)) = 0,$$

where

$$(13') \qquad \begin{cases} \rho^2 = \xi_1^2 + \xi_2^2 + \xi_3^2, \\[2mm] \Psi(\xi) = \dfrac{\rho^2 - \xi_1^2}{\sigma_1} + \dfrac{\rho^2 - \xi_2^2}{\sigma_2} + \dfrac{\rho^2 - \xi_3^2}{\sigma_3}, \\[2mm] \Phi(\xi) = \dfrac{\xi_1^2}{\sigma_2\sigma_3} + \dfrac{\xi_2^2}{\sigma_3\sigma_1} + \dfrac{\xi_3^2}{\sigma_1\sigma_2}. \end{cases}$$

If α is an arbitrary unit vector, then the straight line through the origin with direction α consists of points of the form $\xi = \rho\alpha$, where ρ is a parameter. The intersections of this line and the normal surface are given by the roots of the quadratic equation for ρ^2:

$$(14) \qquad \rho^4\Phi(\alpha) - \rho^2\Psi(\alpha) + 1 = 0.$$

The discriminant of this equation is

$$(15) \qquad X(\alpha) = \Psi^2(\alpha) - 4\Phi(\alpha).$$

Now if we set

$$A_1^2 = \frac{1}{\sigma_3} - \frac{1}{\sigma_2}, \qquad A_2^2 = \frac{1}{\sigma_1} - \frac{1}{\sigma_3}, \qquad A_3^2 = \frac{1}{\sigma_2} - \frac{1}{\sigma_1},$$

an easy calculation shows that for $\sum \alpha_i^2 = 1$

$$(16) \qquad X(\alpha) = \prod (\alpha_1 A_1 \pm \alpha_2 A_2 \pm \alpha_3 A_3),$$

where the product is to be extended over all four possible combinations of signs. Since A_1 and A_3 are real, and A_2 is pure imaginary, the factors of X can be paired into complex conjugates. Hence $X \geq 0$. In fact, $X = 0$ only if $\alpha_2 = 0$ and $\alpha_1 A_1 \pm \alpha_3 A_3 = 0$. Hence the four roots of (14) are real and distinct except if $\alpha_1 A_1 \pm \alpha_3 A_3 = 0$.

Thus the normal surface consists of two mantles, which have the equations

$$\Psi(\alpha) - \sqrt{X(\alpha)} = \frac{2}{\rho^2}, \qquad \Psi(\alpha) + \sqrt{X(\alpha)} = \frac{2}{\rho^2}.$$

Using the homogeneity of Ψ and Φ, we may define $X(\xi)$ by means of

$$(17) \qquad X(\xi) = \Psi(^2\xi) - 4\rho^2\Phi(\xi).$$

Then the inner and outer mantles have the equations

$$(18) \qquad \begin{cases} \Psi(\xi) + \left| \sqrt{X(\xi)} \right| = 2, \\ \Psi(\xi) - \left| \sqrt{X(\xi)} \right| = 2, \end{cases}$$

respectively. These mantles touch only at four points, which lie on the lines

$$\xi_1 A_1 \pm \xi_3 A_3 = 0,$$

in the ξ_1, ξ_3-plane (Figure 48).

In order to visualize the normal surface, we intersect it with the coordinate planes. The intersection with the plane $\xi_2 = 0$ is a circle and an ellipse with four points of intersection:

$$\xi_1^2 + \xi_3^2 - \sigma_2 = 0,$$

and

$$\frac{\xi_1^2}{\sigma_3} + \frac{\xi_3^2}{\sigma_1} - 1 = 0.$$

Of course, these points of intersection are the points where the inner and outer mantles come together. In the other two coordinate planes the intersections are also a circle and an elipse; in these cases the circle and ellipse do not intersect. In the plane $\xi_1 = 0$, we have

$$\xi_2^2 + \xi_3^2 - \sigma_1 = 0,$$

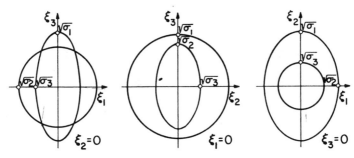

FIGURE 48. Intersections of the normal surface with the coordinate planes.

and

$$\frac{\xi_2^2}{\sigma_3} + \frac{\xi_3^2}{\sigma_2} - 1 = 0;$$

similarly in the plane $\xi_3 = 0$ we have

$$\xi_1^2 + \xi_2^2 - \sigma_3 = 0,$$

and

$$\frac{\xi_1^2}{\sigma_2} + \frac{\xi_2^2}{\sigma_1} - 1 = 0.$$

See Figure 48.

From the general theory of §3, the inner mantle of the normal surface must be convex. The outer mantle is not convex since it has four conical vertices directed inwards, while the corresponding vertices of the inner mantle are directed outwards.

As was remarked above, exactly the same facts are true for the ray surface if the parameters σ_1, σ_2 and σ_3 are replaced by their reciprocals. One should, however, pay attention to the fact noted in §3, 5 that the notion of reciprocal transformation requires modification when dealing with the conical vertices. The image of a convex surface enclosing the origin must be another convex surface. The convex inner mantle of the normal surface is associated with the convex hull of the ray surface. (Both are indicated by thick lines in Figure 49.) The conical vertices of the normal surface are transformed

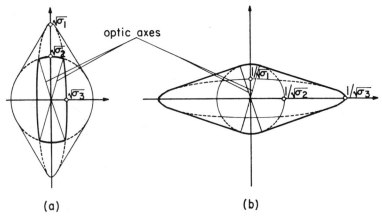

(a)   (b)

FIGURE 49   (a) Normal surface.   (b) Ray surface.

into four plane "lids" which, when added to part of the ray surface, form its convex hull. These lids are not formed as the envelope of plane characteristics, but they are parts of supporting planes of the ray surface proper (which may be represented as the envelope of plane characteristics).

Since the relationship between the normal and ray surfaces is reciprocal, the convex hull of the normal surface is mapped onto the inner mantle of the ray surface. The part of the outer mantle of the normal surface which is not contained in its convex hull is mapped onto the corresponding portion of the ray surface. (Both are indicated by dotted lines in Figure 49.) Just as the conical vertices of the normal surface correspond to the lids of the ray surface, the lids of the normal surface are mapped onto the conical vertices of the ray surface. The boundaries of the lids are parabolic curves, i.e., curves where one of the principal curvatures of the normal surface vanishes. See Figure 49.

In fact, the boundaries of the lids of the ray surface are circles: Since each of the boundaries is a curve of contact of a plane and a surface of fourth degree, each of the boundaries must be a curve of fourth degree with all points counted double, that is, a conic section. To see that the boundary of the lid is a circle we write the equation of the ray surface in homogeneous coordinates η, τ:

$$\tau^4 - \tau^2 \Psi(\eta) + R^2 \Phi(\eta) = 0,$$

where the parameters σ_1, σ_2 and σ_3 are replaced by their reciprocals in the expressions for Ψ and Φ. It follows that the ray surface contains the absolute circle of projective space, which is given by the equations

$$\tau = 0, \qquad R^2 = 0.$$

Hence every plane intersection of the ray surface contains the two absolute points of the plane. If the plane curve of intersection degenerates into two real curves of second order, then one of the two curves contains the absolute points, and hence is a circle. In particular this is true of the curve of contact of the lid with the ray surface. (Incidentally, this reasoning confirms that each intersection of the ray surface with a coordinate plane must contain a circle.)

5. Cauchy's Problem for Crystal Optics. Cauchy's problem for the system of first order equations of crystal optics, and also for other differential equations with constant coefficients such as the equations of magnetohydrodynamics, may be reformulated in terms of a system of second order or a single equation of higher order, by means of the process of elimination (see Ch. I, §2). We shall briefly demonstrate the equivalence of the corresponding Cauchy problems, restricting ourselves to the case of crystal optics.

By elimination of the magnetic vector from the original system of Maxwell's equations for the electric and magnetic vectors (7), we obtained for the electric vector the system of three equations of second order:

$$
\begin{aligned}
\sigma_1 \ddot{u}_1 &= \frac{\partial^2 u_1}{\partial x_2^2} + \frac{\partial^2 u_1}{\partial x_3^2} - \frac{\partial^2 u_2}{\partial x_1 \partial x_2} - \frac{\partial^2 u_3}{\partial x_1 \partial x_3} \\[2mm]
\sigma_2 \ddot{u}_2 &= \frac{\partial^2 u_2}{\partial x_3^2} + \frac{\partial^2 u_2}{\partial x_1^2} - \frac{\partial^2 u_3}{\partial x_2 \partial x_3} - \frac{\partial^2 u_1}{\partial x_2 \partial x_1} \\[2mm]
\sigma_3 \ddot{u}_3 &= \frac{\partial^2 u_3}{\partial x_1^2} + \frac{\partial^2 u_3}{\partial x_2^2} - \frac{\partial^2 u_1}{\partial x_3 \partial x_1} - \frac{\partial^2 u_2}{\partial x_3 \partial x_2}.
\end{aligned}
\tag{19}
$$

By the algorithm of elimination, we easily see that all the components w of the vector u satisfy the same differential equation of order six:

$$
D(\xi, \tau) w = 0,
\tag{20}
$$

where τ denotes $\dfrac{\partial}{\partial t}$, ξ_i denotes $\dfrac{\partial}{\partial x_i}$, and

$$
D(\xi, \tau) =
\begin{vmatrix}
\rho^2 - \xi_1^2 - \sigma_1 \tau^2 & -\xi_1 \xi_2 & -\xi_1 \xi_3 \\[2mm]
-\xi_2 \xi_1 & \rho^2 - \xi_2^2 - \sigma_2 \tau^2 & -\xi_2 \xi_3 \\[2mm]
-\xi_3 \xi_1 & -\xi_3 \xi_2 & \rho^2 - \xi_3^2 - \sigma_3 \tau^2
\end{vmatrix}
\tag{20'}
$$

From (13), we infer

$$
D(\xi, \tau) = -\sigma_1 \sigma_2 \sigma_3 (\tau^6 - \Psi(\xi) \tau^4 + \rho^2 \Phi(\xi) \tau^2).
\tag{21}
$$

Equation (20) may be reduced to an equation of order four, since τ^2 can be factored out of each term of D. If we set $v = w_{tt}$, then

$$
F(\xi, \tau) v = (\tau^4 - \Psi(\xi) \tau^2 + \rho^2 \Phi(\xi)) v = 0.
\tag{22}
$$

The equation (7a) allows us by an easy calculation to eliminate two of the components from two of the equations (19) and thus to establish the equation of order four

$$
F(\xi, \tau) u = 0
\tag{22a}
$$

for all components of the electromagnetic vector.

Cauchy's problems for Maxwell's equations (7) and for (22) in general are closely related: We show that the Cauchy problem for Maxwell's equations, which consists in giving the vectors $\mathfrak{E}$ and $\mathfrak{H}$ for $t = 0$, can be represented in terms of solutions of (22) with Cauchy data in the special form

$$(22') \quad v(0, x) = 0, \qquad v_t(0, x) = 0, \qquad v_{tt}(0, x) = 0,$$
$$v_{ttt}(0, x) = g(x).$$

If we set $w^*(t, x) = \int_0^t (t - s)v(s, x)\, ds$, then

$$(23) \qquad D(\xi, \tau)w^* = 0,$$

and

$$(23') \quad \begin{cases} w^*(0, x) = \dfrac{\partial^i}{\partial t^i} w^*(0, x) = 0 & (i = 1, 2, 3, 4), \\ \dfrac{\partial^5}{\partial t^5} w^*(0, x) = g(x). \end{cases}$$

If this solution is denoted by

$$w^*(t, x) = U\{g\},$$

then the solution of the Cauchy problem

$$(24) \qquad D(\xi, \tau)w(t, x) = 0$$

$$(24') \quad w(0, x) = g_0(x), \qquad \frac{\partial^i}{\partial t^i} w(0, x) = g_i(x) \qquad (i = 1, \cdots, 5),$$

is clearly given by

$$\begin{aligned}
(25) \quad w(t, x) =\ & U\{g_5\} + \frac{\partial}{\partial t} U\{g_4\} + \frac{\partial^2}{\partial t^2} U\{g_3\} + U\{-\Psi(\xi)g_3\} \\
& + \frac{\partial^3}{\partial t^3} U\{g_2\} + \frac{\partial}{\partial t} U\{-\Psi(\xi)g_2\} \\
& + \frac{\partial^4}{\partial t^4} U\{g_1\} + \frac{\partial^2}{\partial t^2} U\{-\Psi(\xi)g_1\} + U\{\rho^2\Phi(\xi)g_1\} \\
& + \frac{\partial^5}{\partial t^5} U\{g_0\} + \frac{\partial^3}{\partial t^3} U\{-\Psi(\xi)g_0\} + \frac{\partial}{\partial t} U\{\rho^2\Phi(\xi)g_0\}.
\end{aligned}$$

Now, in order to solve the initial value problem for the system of

Maxwell's equations, we merely have to verify that the Cauchy data for the system immediately supply Cauchy data for (24). The system (7) gives the time derivatives of the components w in terms of the space derivatives, which in turn are given by the Cauchy data. The higher initial time derivatives are obtained by first differentiating the system with respect to t, etc. Thus the Cauchy problem for Maxwell's equations can be solved if we are able to solve the fourth-order equation (22) with Cauchy data of the special form (22'). We shall treat the latter problem in §14a.

6. Magnetohydrodynamics.[1] The motion of ionized gases or fluids subject to electromagnetic forces is a subject of increasing importance. It gives rise to a great variety of problems involving hyperbolic operators. For us at this point it will be of interest that even simple instances of such motion lead to characteristic surfaces of relatively complicated structure which, however, fall under the general theoretical aspects developed in this chapter. We confine ourselves to the simplest case, a *perfectly conducting fluid in the presence of a magnetic field*. Let u denote the vector of the flow velocity, B the magnetic vector, J the density of the vector of the electric current, and μ the magnetic permeability. We also use the notation ∇ for gradient and $\times$ for the vectorial product; the scalar product is indicated by · when necessary.

The equations of motion of such a fluid form a hyperbolic system which is both anisotropic and nonlinear. The simplifying assumption is made that all flow velocities are small compared to the velocity of light, so that relativistic effects, such as the displacement current in Maxwell's equations, may be neglected. Then we have

$$(13) \qquad \mu J = \operatorname{curl} B,$$

where μ denotes the magnetic permeability. The further assumption of infinite electrical conductivity permits us to express the electric field vector E:

$$(13') \qquad E = -u \times B.$$

The two remaining Maxwell equations, together with Euler's

[1] For more details about this field see F. de Hoffmann and E. Teller [1], K. O. Friedrichs and H. Kranzer [1], H. Grad [1], J. Bazer and O. Fleischman [1]; see also F. G. Friedlander [2], M. J. Lighthill [1] and H. Weitzner [1]. The notations in this article correspond to those used in the physical literature rather than those used in §3.

equations of motion for the fluid, yield the following system of partial differential equations for B, u, and the density ρ of the fluid as functions of position $x = (x, y, z)$ and time t:

$$(14) \quad \begin{cases} \operatorname{div} B = 0 \\ B_t - \operatorname{curl} (u \times B) = 0 \\ \rho u_t + \rho(u \cdot \nabla)u + \operatorname{grad} p - \mu^{-1}(\operatorname{curl} B) \times B = 0 \\ \rho_t + \operatorname{div} (\rho u) = 0. \end{cases}$$

The last term in the third equation represents the force $J \times B$ which the magnetic field exerts on a unit volume of fluid. As in the ordinary Maxwell equations the first equation has the character of an initial condition.

We shall consider these equations in the linearized form, omitting all terms of second order. Then it is easily seen that the system is equivalent to a system which is symmetric hyperbolic in the sense of §3, 8.

We consider first the case of an *incompressible fluid*, $\rho = $ const. The last equation of (14) then becomes

$$\operatorname{div} u = 0,$$

and the pressure p may be taken as a dependent variable. To obtain the characteristic condition, we consider an arbitrary manifold $\phi(x, t) = 0$ and set

$$(15) \qquad v = \phi_t + u \cdot \nabla \phi = \dot{\phi}.$$

($v = \dot{\phi}$ is the rate of change in time of the quantity ϕ observed from a point which moves with the fluid, since $(d/dt)\phi(t, x(t)) = \phi_t + u \cdot \nabla \phi$.) Now, if $\phi = 0$ is a characteristic manifold, v satisfies a determinant equation which, when expanded, is easily found to be

$$(16) \qquad \mu \rho v(\mu \rho v^2 - (B \nabla \phi)^2)^2 = 0.$$

Thus we have a single characteristic speed corresponding to $v = 0$ and two characteristic speeds of multiplicity 2 corresponding to each of the two relative speeds $\pm v = B \nabla \phi (\mu \rho)^{-1/2}$.

As in article 2, the characteristic $v = 0$ represents a streamline of the flow. The other factor of (16) represents forward and backward *Alfvèn waves*. With $\tau = \phi_t$, $\xi = \nabla \phi$, their normal surfaces are the planes given by $\sqrt{\mu \rho}(\tau + u \xi) \pm (B \xi) = 0$.

There is no loss of generality if we set $u = 0$, i.e., utilize a coordinate system which moves with the velocity of the fluid, since our differential equations are invariant under such translations. The reciprocal normal surface for a forward Alfvèn wave (sometimes considered instead of the genuine normal surface) is just a sphere of diameter

$$b = |B|(\mu\rho)^{-1/2}$$

passing through the origin (see Figure 50). Its pole is the ray surface, consisting of the single point $x = bB/|B|$ and projected from the origin by an isolated ray. Hence a forward Alfvèn wave propagates in the B-direction alone; its velocity b relative to the fluid is called the *Alfvèn speed*. Similarly, a backward Alfvèn wave propagates with relative speed b in the direction opposite to B.

For a compressible fluid (14) still applies, but now p is a given function $p(\rho)$. Since the *sound speed*

$$a = \sqrt{p'}$$

as well as the Alfvèn speed b enters, the situation is now more complicated. Defining symbols as in the case of incompressible flow, we find the characteristic condition to be

$$(17) \qquad v(v^2 - (b\nabla\phi)^2)[v^4 - (a^2 + b^2)v^2 + a^2(b\nabla\phi)^2] = 0,$$

with $a = \sqrt{p'}$, $b = |B|(\mu\rho)^{-1/2}$. The two first factors again represent the streamline characteristic $v = 0$ and the Alfvèn characteristics $v = \pm B\nabla\phi(\mu\rho)^{-1/2}$.

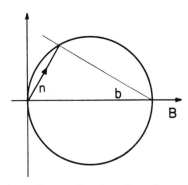

FIGURE 50. Reciprocal normal surface for a forward Alfvèn wave.

The loci corresponding to the factor of fourth degree in (17), i.e., the normal surfaces and ray surfaces, are more interesting and more complicated than the Alfvèn loci. To simplify the discussion without restricting generality we again assume that the velocity u vanishes, at least at the point under consideration (our differential equations are invariant under translation of the coordinate system at constant speed). As before, in the characteristic differential equation we replace $\nabla\phi$ by the vector ξ and ϕ_t by τ.

The normal surfaces consist in general of three sheets for which we may substitute their cross section in the ξ_1 ,ξ_2-plane, since there is rotational symmetry about the axis of the magnetic vector B. The inner sheet of the normal cone, enclosing the core, is convex. The outer sheets extend to infinity; they have the form of planes orthogonal to the axis (of B) dented at the axis (see Figure 51).

For the geometrical discussion one should distinguish three cases, according to the relation between sound speed a and Alfvèn speed b:

(a) $\qquad\qquad\qquad a^2 < b^2,$

(b) $\qquad\qquad\qquad a^2 = b^2,$

(c) $\qquad\qquad\qquad a^2 > b^2.$

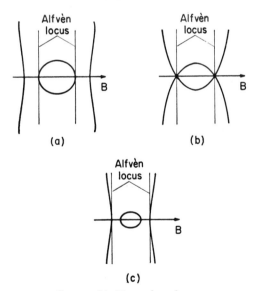

Figure 51. Normal surface.

The normal surface is shown in Figure 51.   In cases (a) and (c) the normal surface has two double points, namely the points at distance $1/b$ from the origin in the direction of B.   There is also a quadruple point at infinity.   It is noteworthy that two of the sheets which extend to infinity correspond to the fourth-degree factor of the characteristic equation.   Thus the normal cone does not consist of nested closed surfaces.

In case (b) there are two triple points, which correspond to a common point of intersection of the Alfvèn locus and two sheets of the fourth-degree locus.   Here we have an example in which the convex core of the normal cone is composed of fragments of sheets which intersect and extend to infinity.   The points of intersection are conical points of the normal surface.

The corresponding reciprocal normal surfaces are obtained by inverting the loci (Figure 51 (a), (b), (c)) with respect to the unit sphere.   Figure 52 (b), (c) show that the outer sheet of the reciprocal normal surface need not be convex, even though the core of the genuine normal surface is convex.

The ray surface is shown in Figure 53 (a), (b), (c).   It has a higher algebraic degree than the normal surface.   Case (a) shows that even though the normal cone and normal surface are regular, the ray cone may have singularities.   In this case the triangular-shaped cusped figures are obtained from the harmless-looking ovals in Figure 52 (a), (c).   The cusps correspond to the points of inflection and to the points at infinity of the outer sheets of the normal surface.   In case 53 (b), the outer shell of the ray cone is no longer convex and therefore it does not coincide with its convex hull.   The rays of the Alfvèn locus lie outside the rest of the ray cone, even though the Alfvèn locus of normal speeds, that is, the reciprocal

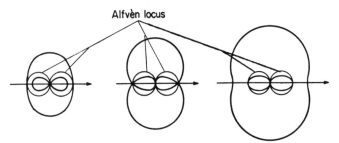

Alfvèn locus

FIGURE 52. Reciprocal normal surface.

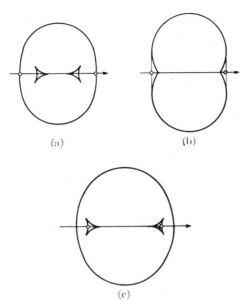

(a)   (b)

(c)

FIGURE 53. Ray surface

normal surface, always lies on or inside the other normal speed locus. This is easily explained if we realize that normal velocities and the velocities along rays are different in general.

Since the ray cone has cusps, we might expect that disturbances which originally have smooth boundaries eventually develop cusps. This is indeed true, as is indicated in Figure 54.

These examples indicate that the ray conoid may have a complicated structure, and may exhibit singularities even though the normal cone is regular. Correspondingly, other characteristic surfaces may have unexpected singularities. These singularities point to interesting questions, involving the relationship between algebraic geometry and the theory of partial differential equations.

FIGURE 54

In this section we have considered the geometric shape of characteristic elements, in particular wave fronts. Later, when we discuss initial value problems and wave propagation in detail, we shall recognize the significance of the preceding analysis for problems of wave propagation. (See, e.g., §§4, 7, 12, 14, 15.)

§4. Propagation of Discontinuities and Cauchy's Problem

1. Introduction. For smooth initial data the solution of Cauchy's problem will be constructed in §§8, 9 and 10 by the method of "energy integrals", a method not directly based on the concept of characteristics. In the present and in the next section, however, we shall realize that the structure of the solutions is nevertheless dominated by characteristic surfaces and rays. The point of departure will be the analysis of propagation of discontinuities in characteristic surfaces along rays (see also §§1 and 2). But this analysis will lead to a somewhat broader approach; as a result we shall obtain at least approximate solutions for wide classes of problems by integrating merely ordinary differential equations along rays.

Discontinuities of the initial data or of derivatives of these initial functions will mostly be restricted to $(n - 1)$-dimensional manifolds in the initial space; elsewhere these data are assumed smooth to any desired degree (i.e., to have continuous derivatives of any desired order).

As will be pointed out in detail, the analysis of propagation of discontinuities is based on the assumption that the characteristics under consideration are distinct, or, in case of multiple characteristics, that the multiplicity is the same for all points and all characteristic elements. This limitation excludes from the scope of the general theory some relevant problems of mathematical physics. While these cases are amenable to specific treatment, the imperfection of the general theory presents a challenge.[1] Numerous examples and comments in the following sections will illustrate the situation.

2. Discontinuities of First Derivatives for Systems of First Order. Transport Equation. We first discuss systems of linear differential equations[2] of first order

$$(1) \qquad L[u] = \sum_{i=0}^{n} A^i u_i + Bu = 0.$$

---

[1] Some progress in the theory for multiple characteristics has been made, e.g. by D. Ludwig [3]. See also R. Lewis [1].

[2] As to quasi-linear systems, see article 9.

Previously (see Chapter III and Ch. VI, §3) we saw that surfaces $C: \phi(x) = 0$ across which u is continuous, while the first or possibly higher derivatives suffer jumps, are characteristic. The jump relation[1] across C

$$(L[u]) = \sum_{i=0}^{n} A^i(u_i) = 0$$

implies

$$\sum_{i=0}^{n} A^i\phi_i(u_\phi) = A(u_\phi) = 0,$$

since $u_i = u_\phi \phi_i +$ tangential derivatives, and since the tangential derivatives are assumed continuous across C. Hence the matrix $A = \sum A^i\phi_i$ is singular, and the jump of the outward derivative u_ϕ across C

(2) $$g = (u_\phi) = \sigma r$$

is a right nullvector of the characteristic matrix A, as defined in §3, σ being a scalar factor.

Unless explicitly stated otherwise, we assume as in §3 that the characteristic matrix A has the rank $k - 1$ and therefore possesses within a scalar factor σ one and only one right nullvector r and one left nullvector l such that $lA = Ar = 0$.

Every solution ϕ of the partial differential equation

$$| A | = Q(\phi_0, \cdots, \phi_n : x) = 0$$

represents a family of characteristic surfaces

$$C_c : \phi(x) = c = \text{const.},$$

and the nullvectors r, l are defined for all these surfaces.

Now we state the theorem: The scalar σ which controls the jump

$$(u_\phi) = \sigma r$$

is propagated along bicharacteristic rays in C according to the ordinary differential equation (transport equation)

(3) $$\dot{\sigma} + P\sigma = 0$$

(the dot denotes differentiation with respect to a parameter along

[1] Here as elsewhere the symbol (f) denotes the jump discontinuity of f across C.

the ray) with

(4) $$P = lL[r].$$

To prove this fact briefly[1] we account for the discontinuity of u_ϕ across $C: \phi = 0$ by representing the solution, with $g = (u_\phi) = \sigma r$, in the form

$$u = \tfrac{1}{2} \, | \, \phi \, | \, g(x) + R(x),$$

where $g(x)$ and $R(x)$ possess continuous first derivatives for which also the interior derivatives in C are continuous. We set $h(\phi) = \tfrac{1}{2}$ for $\phi > 0$, and $h(\phi) = -\tfrac{1}{2}$ for $\phi < 0$. Then, because $Ag = 0$, we have $L[u] = h(\phi)Ag + | \, \phi \, | \, L[g] + L[R] = | \, \phi \, | \, L[g] + L[R] = 0$ on both sides of C. Multiplying by the nullvector l yields $| \, \phi \, | lL[g] + lL[R] = 0$. $lL[R]$ is a tangential operator since $l \sum A^i \phi_i = lA = 0$. Hence by assumption the first derivatives of $lL[R]$ are continuous across C. Therefore differentiating the last equation with respect to ϕ and taking the jump of the differentiated equation across C we obtain immediately the interior differential equation in the characteristic surface $C: lL[g] = l \sum A^i g_i + lBg = 0$. Moreover, substituting $g = \sigma r$ yields

$$l \sum A^i r \sigma_i + l \Big(\sum A^i r_i + Br \Big) \sigma = 0$$

or, because of the lemma in §3, 11

$$\dot{\sigma} + lL[r] \, \sigma = 0,$$

as stated.

Incidentally, we also could have derived the result in analogy with §2, 3 by first assuming ϕ in the special form $\phi = x_n$ and then making use of the invariance of characteristics and characteristic differentiation.

3. Discontinuities of Initial Values. Introduction of Ideal Functions. Progressing Waves. As pointed out already in Ch. V, §9, 1, an adequate mathematical representation of physical reality requires the admission of generalized solutions with more severe discontinuities, e.g., solutions u, which themselves suffer a jump $(u) \neq 0$ across C. However, we cannot ascribe physical relevance to all

---

[1] In article 4 we shall analyze discontinuities for systems (1) in **greater** detail and generality.

mathematically conceivable discontinuities.[1] Generalized solutions should be restricted by the requirement that they and their relevant derivatives represent idealizations by limiting processes from smooth solutions. Following the pattern indicated above, we shall admit as physically meaningful solutions those which can be expressed as ideal functions, also called generalized functions or *distributions*.

Such distributions, in particular Dirac's delta-function, have been used before in this book as symbolic abbreviations. Here and subsequently we shall define and use them more systematically. While a coherent general theory of distributions is given in the appendix to the present chapter, the following relevant points should be formulated here:

Distributions $S(\phi)$ or generalized functions depending on a variable ϕ may be defined in a finite region of the variable ϕ as symbolic derivatives of continuous functions $W(\phi)$ of a "phase variable" ϕ,

$$S(\phi) = D^\alpha W(\phi),$$

where α is a positive number and D means $d/d\phi$. We may then differentiate distributions as if they were ordinary functions. We may also form linear combinations where the coefficients are ordinary functions at least α times differentiable, and thus obtain other distributions. Moreover, we may substitute distributions in linear differential operators and treat these idealized functions as if they were ordinary functions.

We shall consider in addition to the distribution $S = S_0$ the distributions $S_\nu(\phi)$ such that

(5) $$S_\nu'(\phi) = DS_\nu = S_{\nu-1}.$$

If $i > \alpha$ then S_i may be defined as the corresponding $(i - \alpha)$-times iterated integral of W with zero as lower boundary, so that then $S_i(0) = 0$.

For $\phi \neq 0$ let us assume that $S(\phi)$ is a regular function; that means that $W(\phi)$ possesses ordinary derivatives for $\phi \neq 0$. Then obviously the functions $S_\nu(\phi)$ are increasingly less singular for increasing ν, are continuous for $\nu \geq \alpha$ and are arbitrarily smooth for suffi-

---

[1] For example, consider two different solutions u^1, u^2 of $L[u] = 0$, and identify a "discontinuous solution" u with u^1 on one side of an arbitrary surface C, with u^2 on the other side. Then u suffers a jump (u) across C, and neither C nor (u) is in any way distinguished.

ciently large ν. Moreover, for a bounded interval and $\nu > \alpha$ we have

$$| S_\nu(\phi) | < \frac{M}{(\nu - \alpha)!} | \phi |^{\nu-\alpha},$$

where M is a constant.

As examples we consider

$$W(\phi) = \tfrac{1}{2}(| \phi | + \phi), \qquad DW = \eta(\phi),$$

$$D^2 W = \delta(\phi), \qquad D^\alpha W = \delta^{(\alpha-2)}(\phi),$$

where $\eta(\phi)$ is the *Heaviside function*: $\eta(\phi) = 1$ for $\phi > 0$, $\eta(\phi) = 0$ for $\phi < 0$, or

$$W(\phi) = \tfrac{1}{2} | \phi |, \qquad DW = \eta(\phi) - \tfrac{1}{2} = h(\phi),$$

$$D^2 W = \delta(\phi), \qquad D^\alpha W = \delta^{(\alpha-2)}(\phi),$$

or

$$W(\phi) = \sqrt{| \phi |}, \qquad DW(\phi) = \tfrac{1}{2} | \phi |^{-1/2}, \qquad \text{etc.},$$

or

$$W(\phi) = \log | \phi |, \qquad D(\phi) = \frac{1}{\phi}, \qquad D^2(\phi) = \frac{1}{\phi^2}, \qquad \text{etc.}$$

At the singular points, here $\phi = 0$, the identification of the ordinary functions with the symbolic generalized functions denoted by the same symbol is meaningless.[1] Yet, as stated, such generalized functions, or combinations of them, may be inserted in linear differential equations and treated as if they were ordinary functions.[1]

To represent functions $u(t, x)$ with a singularity on the surface $C: \phi(x, t) = 0$ we observe that subtracting a singular term may leave a remainder with a milder discontinuity. This motivates the introduction of discontinuous (idealized) functions of the following form:

$$(6) \qquad u(x, t) = \sum_{\nu=0}^{N} S_\nu(\phi) g^\nu(x, t) + R(x, t),$$

where N can be chosen at our convenience, where the coefficients g^ν are as regular as we desire, and where the remainder R likewise is to be regular to any desired degree. If u is a vector then the coefficients g^ν and R also are vectors, while $S_0, S_1, \cdots$ are scalars.

[1] See the Appendix.

One should observe that such a representation is not unique, if we merely want to account for a singularity at $\phi = 0$. Except along this surface C: $\phi = 0$, we may modify the coefficients g^ν arbitrarily, assuming $S(\phi)$ regular except for $\phi = 0$. The modifications can then always be absorbed in the remainder R. If we do not need to pay attention to a refined resolution of the singularity of u, we might also combine the expansion (6) into one or two terms. For instance, if $W(\phi) = \log |\phi|$, hence $S(\phi) = S_0(\phi) = \text{const.}$ $\phi^{-\alpha}$, then for integral α the representations (6) can be contracted into the form

$$u = \frac{1}{\phi^\alpha} G(x, t) + \log |\phi| \, G^*(x, t)$$

with regular coefficients G, G^*.

Nevertheless, expansions of the form (6) will prove to be most suitable for the analysis.

It will often be convenient not to exhibit in (6) the number N and the remainder R and to write simply

$$(6') \qquad u(x, t) \sim \sum_{\nu=0}^{\infty} S_\nu(\phi) g^\nu(x, t), \qquad S_0 = S,$$

meaning that this formal expansion should be broken off after a suitable number N of terms and that a remainder R of prescribed degree of smoothness should be added.

If the series (6') in fact terminates after N terms with a remainder zero we call u (with a view to later discussions of the wave concept in §18) a *progressing wave of degree N*, or if it can be extended as a convergent series we call u a *complete progressing wave*; otherwise u is called an *approximate progressing wave*.

As mentioned above, distributions of the type described may be considered as (weak) limits for $\epsilon \to 0$ of functions $D^\alpha W^\epsilon(\phi)$, where $W^\epsilon(\phi)$ has derivatives of all relevant orders. Then u is considered as a limit of regular functions u^ϵ, which show a "break" across but not along C, and for which $L[u^\epsilon]$ tends to $L[u]$. Instead of operating with these functions u^ϵ and passing to the limit $\epsilon \to 0$ in the end, we avail ourselves of simple rules of manipulation with the δ-function and more generally with symbolic functions $D^\alpha W(\phi)$ (see Appendix).

The use of the progressing wave form (6') is further motivated by the need to account for detailed features of the discontinuity of u across C.

For example: A jump discontinuity of u may be simply represented by one term $\eta(\phi)g(x, t)$; to exhibit simultaneously a jump discontinuity of the normal derivative u_ϕ we use two terms such as $\eta(\phi)g(x, t) + |\phi| g^1(x, t)$, etc.

Some assumptions may be restated in advance: First, we always assume that the coefficients of the differential operator and the data are as smooth as necessary for the validation of statements made about the smoothness of the coefficients g^ν and the remainder R. Second, except for a singularity at $\phi = 0$ and possibly a number of other values of ϕ, the singularity $S(\phi)$ is likewise assumed to be a suitably smooth function.

To construct generalized solutions $u(x, t)$ of the progressing wave type (6) or (6') we simply insert these expansions in the differential equation, differentiating the generalized functions as if they were ordinary functions. Moreover, we stipulate that the resulting smooth factors of each of the distributions S_ν vanish separately. This will lead to a sequence of differential equations for the coefficients g^ν, and these equations can be reduced to quite simple ordinary differential equations.

If the number N is sufficiently large, then the remaining terms can be made as smooth as desired; this will allow us to complete the solution of Cauchy's problem by referring to the construction of §10.

In case the series (6') terminates or converges one does not have to account for remainder terms; the procedure then covers *arbitrary functions* $S(\phi) = S_0(\phi)$ whether they are singular or smooth. This observation (see also §18) will make it possible to achieve the complete construction of the solution for important classes of data without recourse to the existence theorems of §10.

4. Propagation of Discontinuities for Systems of First Order. To carry out the procedure, we note, indicating regular terms by dots, that

$$(7) \qquad u_i = S_{-1}(\phi)\phi_i\, g + \sum_{\nu=0}^{N-1} S_\nu(\phi)\{g_i^\nu + g^{\nu+1}\phi_i\} + \cdots$$

$(g^0 = g)$ and also

$$(8) \qquad \begin{aligned} u_{ij} &= S_{-2}\,\phi_i\,\phi_j\, g + S_{-1}\{\phi_{ij}\, g + \phi_i\, g_j + \phi_j\, g_i + g^1\phi_i\,\phi_j\} \\ &+ \sum_{\nu=0}^{N-2} S_\nu\{g_{ij}^\nu + g_i^{\nu+1}\phi_j + g_j^{\nu+1}\phi_i + g^{\nu+1}\phi_{ij} + g^{\nu+2}\phi_i\,\phi_j\} + \cdots, \end{aligned}$$

and so forth. As said before, the formal expansions, if broken off after a suitable number of terms, leave an arbitrarily smooth remainder.

Inserting (7) in the operator of first order (1) we have (with $A = A^i \phi_i$)

$$L[u] = S_{-1} A g^0 + \sum_{\nu=0}^{N-1} S_\nu (A g^{\nu+1} + L[g^\nu])$$

(9)

$$+ (S_N L[g^N] + L[R]) = 0,$$

where R is regular for sufficiently large N. We stipulate that all the factors of S_{-1}, S, S_1, $\cdots$ as well as $S_N L[g^N] + L[R]$ vanish not only for $\phi = 0$ but also on $C_c : \phi = c \neq 0$ in an adjacent x-space. Thus

(10) $Ag = 0,$

(10′) $L[g^\nu] + A g^{\nu+1} = 0,$ $(\nu = 0, 1, \cdots, N - 1),$

(10″) $S_N L[g^N] + L[R] = 0.$

Hence, as before, $|A| = Q(D\phi) = 0,$

$$g^0 = g = \sigma r.$$

Consequently $\phi = $ const. represents a family of characteristic surfaces C_c. Multiplying (10′) by the left nullvector l yields

(11) $lL[g^\nu] = 0$

and in particular, for $\nu = 0$,

(11′) $lL[g] = lL[\sigma r] = \sum_\nu l A^\nu r \sigma_\nu + lL[r] \sigma = 0$

or (see the lemma of §3, 11)

(12) $lL[\sigma r] = \dot{\sigma} + lL[r] \sigma = 0.$

This is the fundamental ordinary differential equation which determines $g^0 = g$, as stated before in article 2.

The factors g^1, g^2, $\cdots$ are now determined recursively; we consider $g^{\nu+1}$ as a solution of a system of linear equations (10′); by its origin they are compatible[1] even though their matrix A is singular. As a consequence we have:

(13) $g^{\nu+1} = \sigma^{\nu+1} r + h^{\nu+1}.$

Here $h^{\nu+1}$ is known (modulo r) if $L[g^\nu]$ is known, and $\sigma^{\nu+1}$ is a scalar factor.

---

[1] The compatibility is, moreover, an immediate consequence of the subsequent relations (12′).

Substituting (13) in (11) and writing $\nu + 1$ instead of ν, we again obtain an ordinary differential equation along a ray:

$$(12') \qquad\qquad \dot{\sigma} + lL[r]\sigma + k^{\nu} = 0 \qquad\qquad \sigma = \sigma^{\nu+1},$$

where k^{ν} is known if $L[g^{\nu}]$ is known.

These ordinary "*transport*" *differential equations* (12), (12') therefore determine recursively the functions g^{ν} on $C_c = \phi = \mathrm{const.} = c$ as soon as their initial values along the intersection of C_c with a transversal manifold, say $x_0 = 0$, are known. Then g^{ν} is determined in a $(n + 1)$-dimensional part of the x-space, filled by the characteristic surfaces $\phi = \mathrm{const.} = c$.

For the case of jump discontinuities of u: $S(\phi) = \eta(\phi)$, i.e., $(u) = g$, the relations (7), (8) allow us immediately to express the jumps of the derivatives (u_i), (u_{ij}), $\cdots$ on the discontinuity surface $C:\phi = 0$ in terms of the coefficients g^{ν} and their derivatives on C. Since the jumps of $\delta(\phi)$, $\delta'(\phi)$, $\cdots$ are zero we have

$$(u) = g$$
$$(14) \qquad (u_i) = g_i + \overset{1}{g}\phi_i$$
$$(u_{ij}) = g_{ij} + \overset{1}{g_i}\phi_j + \overset{1}{g_j}\phi_i + \overset{1}{g}\phi_{ij} + \overset{2}{g}\phi_i\phi_j$$

$$\cdots\cdots\cdots\cdots\cdots\cdots\cdots\cdots\cdots\cdots\cdots.$$

Conversely the jumps of the derivatives of u successively determine $g, \overset{1}{g}, \overset{2}{g}, \cdots$ along C_0 (it should be kept in mind that the continuation of $g, \overset{1}{g}, \cdots$ outside of C_0 is not determined by the jumps of u, u_i, etc.).

5. Characteristics with Constant Multiplicity. As stated in §3, for symmetric hyperbolic systems multiple characteristic elements do not necessarily cause serious difficulties. While most equations in mathematical physics are symmetric, it remains desirable to extend the preceding analysis to the important case of equations with multiple characteristics where the multiplicity is constant, that is, does not change with the direction of the normal, nor from point to point; the system may be symmetric or not. We assume that there are s-independent right nullvectors $r^1, r^2, \cdots, r^s$ and s-independent left nullvectors $l^1, l^2, \cdots, l^s$: $Ar^i = l^jA = 0, 1 \leq i, j \leq s$. Because of $A(u) = 0$ the jump (u) must be a linear combination of $r^1, r^2, \cdots, r^s$:

$$(15) \qquad\qquad (u) = \sigma_1 r^1 + \sigma_2 r^2 + \cdots + \sigma_s r^s.$$

In general we derive differential equations for the scalar factors σ_i : Substituting (15) into $lL[u] = 0$, we find for $l = l^j$,

$$(11'') \quad l^j \sum_{i=1}^{s} \sum_{\nu=0}^{n} A^\nu r^i \sigma^i_\nu + l^j \sum_{i=1}^{s} \sum_{\nu=0}^{n} (A^\nu r^i_\nu + B r^i) \sigma^i = 0 \ (j = 1, \cdots, s).$$

Again, these equations represent a system of s partial differential equations for σ_1, σ_2, $\cdots$, σ_s within C. Thus, the differentiations for $s > 1$ might not reduce to ordinary differentiation along bicharacteristic rays; discontinuities initially sharply localized might spread over the surface C. (Examples will be given below.) However, this cannot happen and as before the discontinuities spread along rays if the characteristic surfaces have the same multiplicity s throughout x-space. Indeed, referring to §3, 10, we define such s-fold multiplicity as follows: The algebraic equation $Q(\phi_0, \cdots, \phi_n) = 0$ defines, to an arbitrary set of values ϕ_1, $\phi_2 \cdots$, ϕ_n (in a certain n-dimensional domain), a multiple root $\phi_0 = f(\phi_1 \cdots, \phi_n)$ such that the matrix $A = A^\nu \phi_\nu$ possesses s independent right nullvectors r^i $(i = 1, 2, \cdots, s)$ and the same number of independent left nullvectors l^i.

To prove our statement we revert to the lemma of §3, 11. Accordingly, in the differential equations $(11'')$ all the quantities σ^i are differentiated in the same bicharacteristic direction

$$\dot{x}_\nu : \dot{x}_0 = -f_\nu , \qquad\qquad \nu > 0,$$

and thus the statement is proved. As is immediately seen, these bicharacteristic rays belong to the irreducible factor of Q which defines the sheet under consideration.

5a. Examples for Propagation of Discontinuities Along Manifolds of More Than One Dimension. Conical Refraction. If the assumption of article 5 is not satisfied, that is, if a characteristic surface C is of higher multiplicity but does not belong to a family of characteristic surfaces which have the same multiplicity for all directions of the normal in space at all points in space, then it may well happen that initial discontinuities spread from a point in C along two dimensional or higher dimensional manifolds in C.

This can be seen immediately by almost obvious examples. Let us consider a system of three equations and assume that the first equation does not contain differentiation with respect to the variable x_3 . Then the planes $x_3 =$ const. are characteristic surfaces with the multiplicity two (or possibly three) since there is at least one linear combination of the last two equations for which differentiation with respect to x_3 does not occur. Now, an initial discontinuity in $C:x_3 = 0$ is easily seen to spread in two dimensions along C. It is sufficient to consider a typical example. We write x, y, z for the independent variables and u, v, w for u^1, u^2, u^3, and consider the system

$$u_x = 0$$

$$v_y - u = 0$$

$$w_z - v = 0.$$

The characteristic plane $z = 0$ corresponds to the two left independent nullvectors $(1, 0, 0)$ and $(0, 1, 0)$; this plane C is covered not by one, but by two sets of bicharacteristic curves $x = $ const. and $y = $ const. The component v satisfies the differential equation $v_{xy} = 0$, and obviously discontinuities of v across the plane $z = 0$ spread as solutions of the same equation. A discontinuity initially localized at a point spreads between the two bicharacteristics $x = $ const. and $y = $ const. through this point.

In the preceding example the deviation from "normal" behavior may be considered as minor. However, there are cases significant for physics, in which the difference from normal behavior is more incisive. In these cases, the number of independent variables is more than three, and the characteristic curves through an initial point of the characteristic surface under consideration form a two-dimensional manifold, a cone. The most famous example is that of crystal optics (*conical refraction*). According to the analysis of the differential equations of crystal optics in §3a, the plane wave fronts $\alpha_1 x_1 + \alpha_2 x_2 + \alpha_3 x_3 - \eta t = 0$ belonging to these differential equations are characteristic surfaces if the normal $\alpha = \xi/|\xi|$ and the speed $\eta = -\tau/|\xi|$ are connected by the characteristic equation (9b) of §3a. This equation defines in general two speeds for a given normal direction and one characteristic ray for each of the two speeds. (The construction of the two rays for the given normals is contained in §3, 4.)

There is, however, an exception to this relation, namely when the normal α points in the direction of an optic axis of the crystal. Then the two speeds coincide, and instead of the two associated rays through the point we have a circular cone of rays. The phenomenon of "conical refraction", first discovered theoretically by Hamilton and later confirmed experimentally, means that an incoming ray after entering the crystal in a direction of the axis is split into the totality of the generators of that cone, and a discontinuity carried by the incoming ray is propagated primarily along the conical surface, but also (greatly attenuated) inside this cone. Analytically this fact is seen by applying the procedure of article 5 and suitably transforming the resulting differential equations; the strength of the discontinuity then is seen to satisfy the wave equation for two space dimensions and time.

A similar situation of conical refraction occurs in magnetohydrodynamics (see §3a, 6) if an incoming electromagnetic wave front from a vacuum impinges on a magnetized fluid in the direction of the magnetic field,[1] and the sound speed of the fluid and the Alfvén speed coincide.

6. Resolution of Initial Discontinuities and Solution of Cauchy's Problem. In this article we use the preceding theory to reduce Cauchy's problem with discontinuous initial values to one with smooth data. Existence and uniqueness of the solution of this latter problem will be proved in subsequent sections. This reduction rests on the fact that we can construct a finite wave w of the form (6) which absorbs the given initial discontinuities and for which $L[w]$ is as smooth as desired.

We assume $x_0 = t = 0$ as the space-like initial manifold $\mathcal{S}$. Then from an $(n - 1)$-dimensional manifold $\mathcal{S}_0$ defined by

$$\phi(0, \, x_1, \, \cdots, x_n) = \phi(x) = c = \text{const. in } \mathcal{S},$$

there originate k characteristic surfaces C^κ: $\phi^\kappa(t, \, x) = \text{const.} = c$ ($\kappa = 1, \, \cdots, \, k$) fanning out into the x, t-space. Whether or not some of them coincide, the hyperbolic character of $L[u]$ implies the existence of k linearly independent right nullvectors r^κ of the characteristic matrix.

The initial values with a discontinuity S across the $(n - 1)$-dimensional manifold $\mathcal{S}_0$: $\phi(x) = 0$ are assumed to have the form

$$u(0, \, x) = u_0(x) = \sum_{\nu=0}^{N} S_\nu(\phi(x)) g^\nu(x) + R(x).$$

The first step in solving Cauchy's problem is to resolve the initial values $u_0(x)$ into k components each belonging to one of the k

---

[1] For a detailed analysis see D. Ludwig [2].

characteristic surfaces $\phi^\kappa(t, x) = c$ of the fan through $\phi(x) = 0$. Correspondingly we decompose the solution u into k components

$$u(t, x) = \sum_{\kappa=1}^{k} U^\kappa(t, x).$$

In the notation of article 3 we require

$$U^\kappa \sim \sum_\nu S_\nu(\phi^\kappa(t, x))g^{\nu,\kappa}(t, x),$$

$$L[U^\kappa] = 0 \quad \text{and} \quad \sum_\kappa U^\kappa(0, x) = u_0(x).$$

Accordingly, since $\phi^\kappa(0, x) = \phi(x)$ for all κ we stipulate for each ν

(16) $$\sum_\kappa g^{\nu,\kappa}(0, x) = g^\nu(x),$$

considering $g^\nu(x)$ as known from the initial data. According to article 4, the functions $g^{\nu,\kappa}$ are expressed in terms of scalars $\sigma^{\nu,\kappa}$ and the known nullvectors r^κ by (13). This leads for each ν to a system of k linear equations for the initial values of $\sigma^{\nu,\kappa}$ on $\mathcal{I}$.

In particular for an initial jump discontinuity, i.e., $S_0(\phi) = S(\phi) = \eta(\phi)$, $\eta(\phi)$ again denoting the Heaviside function, $g^0(\phi) = g(\phi)$ for $\phi = 0$ measures the jump of the vector u across C.

By article 4 the jump of u across $C^\kappa:\phi^\kappa(t, x) = 0$ is given by an expression $(u^\kappa) = \sigma^\kappa(t, x)r^\kappa(t, x)$; hence we have in the initial plane at $\phi = 0$

(17) $$\sum (u^\kappa) = (u_0) = \sum_{\kappa=1}^{k} \sigma^\kappa(0, x)r^\kappa(0, x) = g(0, x).$$

This system of k linear equations is nonsingular and hence determines the scalars $\sigma^\kappa(0, x)$ uniquely, since by the assumption of hyperbolicity the vectors $r^\kappa(0, x)$ are linearly independent. Thus the initial discontinuity (u_0) is resolved into k components each attached to one of the k characteristics of the fan through the initial discontinuity.

While this resolution is compulsory for the initial manifold $\phi = 0$, we now stipulate it according to (17) for all characteristics C_c^κ of the fan through $\phi^\kappa(t, x) = c = \text{const.}$, in agreement with remarks in article 4. As a consequence of the ordinary transport differential equations (12') initial values $\sigma^\kappa(0, x)$ now determine σ^κ and thus factors $g^{0,\kappa}$ for all the characteristic surfaces C_c^κ even though for $c \neq 0$ the surfaces C_c^κ do not carry discontinuities.

Using (13) we can in a similar way determine by a number of obvious steps the resolution of the coefficients $g^\nu(x)$ and their continuations along the characteristic C_c^κ. For the scalar $\sigma^{\nu,\kappa}(0, x)$ the equations (16) lead immediately to a system of linear equations of the form

$$\sum_\kappa \sigma^{\nu,\kappa} r^\kappa = M^\kappa,$$

where M^κ is known if $g^{\mu,\kappa}$ is known for $\mu < \nu$. Thus, step by step, first the initial values of $\sigma^{\nu,\kappa}$ are determined and then by (12') these scalars and hence $g^{\nu,\kappa}$ are determined along C_c^κ. Thereby these functions are defined in the $(n + 1)$-dimensional neighborhoods of C_c^κ for suitable neighborhoods of $c = 0$.

The preceding decomposition of g^ν applies not merely to jump discontinuities but to arbitrary singularities $S(\phi)$. For $S(\phi) = \eta(\phi)$, however, we can go a step further by relating the coefficients g^ν to the jumps of the derivatives of u_0. As the reader will easily ascertain, this is done on the basis of the formulas (14) of article 4.

After these preparations the solution of Cauchy's problem is readily completed: Taking N sufficiently large, we form

$$U = \sum_{\kappa=1}^k \sum_{\nu=1}^N S_\nu(\phi^\kappa(x, t)) g^{\nu,\kappa}(x, t).$$

Then $L[U] = G(x, t)$ is as smooth as desired, and so is $w = u - U$. Hence, $L[u] = 0$ implies a differential equation

$$L[w] = -G(x, t)$$

with smooth initial condition and smooth right-hand side. The unique solution of problems of this type will be constructed in §10. Therefore $u = U + w$ is the unique solution of the Cauchy problem posed in the present article.

6a. Characteristic Surfaces as Wave Fronts. The following remarks about *waves* and *wave fronts* should be made at this point. In §3

characteristic surfaces were considered as potential carriers of discontinuities, or wave fronts, of solutions u (called "waves") of the
equation $L[u] = 0$. Now the preceding results show: *Any characteristic surface* $\phi(t, x) = 0$ *is a wave front* for a suitably constructed
wave u. One may therefore *define* characteristic surfaces as wave
fronts.

Moreover, Huyghens' construction of wave fronts as envelopes of
families of other wave fronts $\phi(t, x, \alpha)$, depending on parameters
$\alpha, \alpha_1, \alpha_2, \cdots$, is reflected in the following theorem: If the wave
$u(t, x, \alpha)$ is a solution of $L[u] = 0$ depending on the parameters α
and singular on a wave front $\phi(t, x, \alpha) = 0$ then the superposition
$u(t, x) = \int u(t, x, \alpha) \, d\alpha$ is a wave whose singularities are restricted
to the envelope of the fronts $\phi(t, x, \alpha) = 0$. For the proof of this
fact reference is made to the argument in §15, 3.[1]

7. Solution of Cauchy's Problem by Convergent Wave Expansions.
If the progressing wave expansion terminates after N terms or converges to a complete progressing wave, then the preceding construction solves Cauchy's problem even without recourse to the existence
proof in §10. As said before, in this case the expansion represents a
solution of $L[u] = 0$ for arbitrary $S(\phi)$ no matter whether S is
singular or not.

We consider the case of complete progressing waves. To con-
struct all the coefficients g^γ we presuppose that the coefficients A^ν,
B possess derivatives of all orders; precisely, we assume that they are
analytic (e.g., constants or polynomials). Then the following remarkable theorem holds:

Let initial values be given by a series

$$u(0, x) = \sum_{\nu=0}^{\infty} S_\nu(\phi(x)) g^\nu(x)$$

so that $\sum_{\nu=0}^{\infty} (1/\nu!)(\phi)^\nu g^\nu(x)$ is analytic in x, i.e., has a uniformly converging power series expansion in a certain neighborhood of $x = 0$,
$\phi = 0$. Then the infinite series

$$u(t, x) = \sum_{\kappa=1}^{k} \sum_{\nu=0}^{\infty} S_\nu(\phi^\kappa(t, x)) g^{\kappa,\nu}(t, x)$$

as defined by article 4 likewise converges for sufficiently small $| x |$

---

[1] See also D. Ludwig [1].

and t and solves Cauchy's problem. Here $S(\phi)$ is an arbitrary distribution.

One could attempt the proof by estimating the coefficients $g^{\kappa,\nu}$ and their derivatives on the basis of the construction of article 4. However, we merely refer to an elegant proof by D. Ludwig[1] who succeeded in reducing the statement to the Cauchy-Kowalewsky existence theorem.

8. Systems of Second and Higher Order. For solutions of the form (6) of a system of equations of order m a remarkably general result is obtained by the same procedure as for first order systems. It may be stated as follows: The phase function ϕ is characteristic, and the factors g^ν are given by components of the form

$$g^\nu = \sigma^\nu r + h^\nu,$$

where σ^ν is a scalar, h^ν is known if $g^{\nu-1}$, $g^{\nu-2}$, $\cdots$ is known, $h^0 = 0$, and r is the right nullvector of the characteristic matrix A.

On the characteristic surface $C_c : \phi = c = $ const. the scalar $\sigma = \sigma^\nu$ satisfies the following ordinary differential transport equation along the bicharacteristic rays:

$$(18) \qquad \dot{\sigma} + P\sigma + k^\nu = 0,$$

where k^ν is known if $g^{\nu-1}$, $\cdots$ are known, $k^0 = 0$, and where, independently of ν,

$$(19) \qquad P = \frac{1}{(m-1)!}\, lL[(\phi - c)^{m-1}r]$$

on C_c. l denotes the left nullvector of A.

The proof of this general theorem can easily be given by explicit calculation, e.g., for $m = 2$, that is, for a system of k equations of second order:

$$(20) \qquad L[u] \equiv \sum_{i,j=0}^{n} A^{ij}u_{ij} + \sum_{i=0}^{n} A^{i}u_{i} + Bu = 0,$$

where $A^{ij} = A^{ji}$, A^{i} and B are $k \times k$ matrices (see §3, 1). These matrices are assumed to possess sufficiently many derivatives with respect to x (in the quasi-linear case also with respect to u and u_i). $A = \sum A^{ij}\phi_i\,\phi_j$ is the characteristic matrix.

[1] See D. Ludwig [1].

Because of $A_{\phi_i} = 2\sum_{j=0}^{n} A^{ji}\phi_j$, inserting (7), (8) in (20) yields

$$L[u] = S_{-2}\, Ag^0 + S_{-1}[Ag^1 + A_{\phi_i}\, g_i^0 + (A^{ij}\phi_{ij} + A^i\phi_i)g^0]$$

$$+ \sum_{\nu=0}^{N-2} S_\nu[Ag^{\nu+2} + A_{\phi_i}\, g_i^{\nu+1} + (A^{ij}\phi_{ij} + A^i\phi_i)g^{\nu+1} + Lg^\nu]$$

$$+ \cdots.$$

Again we stipulate[1] for $i \leq N - 2$ that all the factors of S_i vanish not only on $C = C_0$ but for $C_c : \phi = c = $ const. in an $(n + 1)$-dimensional neighborhood of C_0. Therefore we have on these surfaces $Ag = 0$. Hence $|A| = 0$; thus the whole family C_c consists of characteristics, and $g = \sigma r$, as seen before in article 4.

The factors of S_{-1}, S_0, $\cdots$ yield on C_c

(21) $Ag^1 + A_{\phi_i}\, g_i^0 + (A^{ij}\phi_{ij} + A^i\phi_i)g^0 = 0,$

(21') $Ag^{2+\nu} + A_{\phi_i}\, g_i^{1+\nu} + (A^{ij}\phi_{ij} + A^i\phi_i)g^{1+\nu} + L[g^\nu] = 0,$

$$(\nu = 0, 1, \cdots)$$

· ·

Now multiplying with the left nullvector l of A we obtain, because of $lA = 0$, exactly as in article 4:

(22) $lA_{\phi_i}\, g_i^0 + l(A^{ij}\phi_{ij} + A^i\phi_i)g^0 = 0,$

(22') $lA_{\phi_i}\, g_i^\nu + l(A^{ij}\phi_{ij} + A^i\phi_i)g^\nu + lL[g^{\nu-1}] = 0,$

·   .

Again, these ordinary differential equations are interior in the characteristic surfaces $\phi = $ const., since there $\sum_{j=1}^{n} lA_{\phi_j}\, \phi_j = 2lA = 0$. According to the lemma of §3, 11 the characteristic ray vector $\dot{x}_i = Q_{\phi_i}$ is proportional to $lA_{\phi_i}\, r$. As a consequence, inserting $g = \sigma r$ in (22), we obtain for σ the ordinary differential equation along rays in $C_c : \dot{\sigma} + P\sigma = 0$, with $P = lL[r(\phi - c)]$.

The inhomogeneous ordinary differential equations governing the singularity functions g^ν for $\nu > 0$ also follow exactly as in article 4: (21') is a system of linear equations for $g^{\nu+2}$ with the singular matrix A. Hence, since (21') is a compatible system (see article 4), we have

---

[1] On $\phi = 0$ this is rather a consequence of $L[u] = 0$, if S_i is sufficiently singular.

$$(23) \qquad\qquad g^{\nu+2} = \sigma^{\nu+2}r + h^{\nu+2},$$

where $h^{\nu+2}$ is known if $g, g^1, \cdots, g^{\nu+1}$ and their derivatives are known, and where $\sigma^{\nu+2}$ is a scalar. Writing ν instead of $\nu + 2$ and substituting (23) in (22′) we obtain on $\phi - c = 0$ for σ^{ν} the inhomogeneous ordinary differential equation (transport equation)

$$\dot{\sigma}^{\nu} + P\sigma^{\nu} = K^{\nu},$$

with the same homogeneous part as before. Thus step by step all the factors g^{ν} in (6) are determined provided their initial values are known on an $(n - 1)$-dimensional initial manifold in C_c which intersects all the rays.

For systems of higher order the proof is essentially the same. The preceding calculation could be simplified by using the invariance properties of the characteristics as stated in §3. A detailed presentation will be omitted.

9. Supplementary Remarks. Weak Solutions. Shocks. As discussed in Ch. V, §9 for two independent variables, generalized solutions which admit singularities can be obtained by the concept of "weak solutions" if we replace the differential equation $L[u] = 0$ by an integral relation

$$\int uL^*[v]\, dx = 0.$$

Here L^* is the adjoint operator and v is an arbitrary smooth test function having "compact support", that is, vanishing identically outside a sufficiently large sphere. For n independent variables the argument remains exactly the same and need not be repeated here. In any case, it is closely related to the concept of distributions (see Appendix).

But a remark may be added regarding *quasi-linear equations* (1). For them the concept of characteristics and bicharacteristics remains of course the same as in the linear case. Also the results concerning discontinuities in the first derivatives remain valid.

However, the theory of the preceding section breaks down if discontinuities of the function u itself are to be considered. Just as in Ch. V, §9 we can introduce such discontinuities, called "shocks", in case the system of first order has the form of a set of *conservation laws*, that is, if it can be expressed in the form

$$\sum_{i=1}^{n} \frac{\partial}{\partial x_i} P^{ij}(x, u) = 0 \qquad j = 1, \cdots, k.$$

Then for a surface $C : \phi(x) = 0$ across which u suffers a jump, the interpretation of these laws in the weak sense leads, exactly as in Chapter V, to the *shock conditions*:

$$\sum (P^{ij}(u))\phi_i = 0 \qquad (j = 1, \cdots, k),$$

in which the jumps of u and the discontinuity surfaces $C : \phi = 0$ are interconnected and thus the *shock surface* C no longer is characteristic.

The most important example is given by the equations of hydrodynamics[1] already discussed in Chapter V for the case of two independent variables.

§5. Oscillatory Initial Values.[2] Asymptotic Expansion of the Solution. Transition to Geometrical Optics

1. Preliminary Remarks. Progressing Waves of Higher Order. The procedure of §4 applied to oscillatory initial values clarifies the phenomenon of wave propagation along rays in "geometrical optics". For fast oscillations it provides an asymptotic solution of Cauchy's problem making use merely of solutions of ordinary differential equations.

It seems natural to construct the solution by superposition of solutions of initial value problems with discontinuous initial functions, that is, with $\psi(x) = \delta(x - \xi)$, sharply localized at the point $x = \xi$ (see also §15). However, one might just as well resolve the initial values into plane waves by Fourier integrals. Accordingly, one would first seek oscillatory solutions, and then combine them into the complete solution of Cauchy's problem, a procedure analogous to that for constant coefficients (see Ch. III, §5).

In the light of the theory of §4, we start by considering "progressing waves" expressed formally by

(1) $$\sum_{j=1}^{\infty} T_j(\phi(x, t))g^j(x, t),$$

---

[1] For a detailed analysis see R. Courant and K. O. Friedrichs [1], pp. 116–172.

[2] See P. Lax [4].

where T_j does not have to be a singular function of the phase function $\phi(x, t)$, and where again $T'_j = T_{j-1}$. As in §4 we insert (1) formally into $L[u] = 0$ and stipulate that all the factors of T_{-1}, T_0, T_1, $\cdots$ vanish separately. Obviously the resulting formal relations for the factors g^i are the same ones obtained in §4 for singular $T_j = S_j$.

Using complex notation we choose, in particular

$$T_0(\phi(x, t)) = e^{i\xi\phi(x,t)}, \qquad T_j = \frac{1}{(i\xi)^j} e^{i\xi\phi(x,t)},$$

with a large parameter ξ.

2. Construction of Asymptotic Solutions. Irrespective of such motivating remarks we now aim at constructing asymptotic solutions of the hyperbolic system

$$(2) \qquad L[u] = u_t + \sum_{j=1}^{n} A^j u_{x_j} + Bu = u_t + N(u) = 0,$$

as a sum of k expressions of the form[1]

$$(3) \qquad U^\kappa(x,t;\xi) = e^{i\xi\phi^\kappa(x,t)} \left\{ u_0^\kappa + \frac{u_1^\kappa}{\xi} + \cdots + \frac{u_\nu^\kappa}{\xi^\nu} + \cdots \right\},$$

where the function vectors u_0, u_1, $\cdots$, u_ν do not depend on ξ, and where the expansion up to the term u_ν/ξ^ν is expected to leave a remainder of the order of magnitude $\xi^{-\nu-1}$.

We want to solve (2) with Cauchy data

$$u(0, x) = e^{i\xi\phi(x)}\psi(x).$$

Thus we set

$$\phi^\kappa(x, 0) = \phi(x);$$

initial data for U^κ will be assigned later.

We now omit in (3) the superscript κ and analyze the structure of a single expression (3). Substituting (3) into $L[u] = 0$, denoting the unit matrix by I, and introducing the characteristic matrix $A = I\phi_t + \sum_{j=1}^{n} A^j \phi_j$, we obtain formally from (3),

$$0 = e^{-i\xi\phi(x,t)} L[u] = i\xi A \sum_{\nu=1}^{\infty} \frac{u_\nu}{\xi^\nu} + \sum_{\nu=0}^{\infty} \frac{L[u_\nu]}{\xi^\nu}.$$

[1] The omission of the factor $(i)^j$ in the denominator is inessential.

As in §4, comparison of the coefficients of $\xi^{-\nu}$, for $\nu = -1, 0, 1 \cdots$, on both sides leads successively to the relations:

$$(5) \qquad\qquad A u_0 = 0,$$

$$(6) \qquad\qquad i A u_{\nu+1} + L[u_\nu] = 0 \qquad \text{for } \nu = 0, 1, \cdots .$$

A solution u of (5) implies again $Q = |A| = 0$. Consequently, $\phi = \text{const.} = c$ is a family of characteristic manifolds with the prescribed initial values $\phi(x, 0) = \phi(x)$. Assume that the rank of A is $k - 1$ and denote right and left nullvectors of A by r and l respectively: $A r = 0$, $l A = 0$. Then we have the relation

$$(7) \qquad\qquad u_0 = \sigma r,$$

with a scalar σ. Multiplying (6) for $\nu = 0$ with l we obtain for u_0, or, equivalently, for σ, the relation

$$(8) \qquad\qquad l L[u_0] = l L[\sigma r] = 0$$

which is identical with (11′) in §4. This relation determines σ by means of an ordinary differential equation along the rays belonging to the characteristic surface $\phi = \text{const.}$, if the initial values of σ for $t = 0$ are known.

Furthermore, if we assume u_0 known, then (6) for $\nu = 0$ is a system of linear equations with the singular matrix A for the components of the vector u. As in §4, it follows that

$$(9) \qquad\qquad u_1 = \sigma^1 r + h^1$$

with a scalar σ^1 and an expression for h^1 known in terms of $L[u_0]$. To determine σ^1 we multiply (6) for $\nu = 1$ by l and obtain the "transport equation"

$$(10) \qquad\qquad l L[u_1] = l L[r\sigma^1] + l L[h^1] = 0,$$

again an ordinary differential equation (this time inhomogeneous), for the scalar σ^1 along a ray belonging to the characteristic manifold $\phi = \text{const.}$

In the same manner we find

$$(11) \qquad\qquad u_j = \sigma^j r + h^j,$$

$$(12) \qquad l L[u_j] = 0 \quad \text{or} \quad l L[r\sigma^j] + l L[h^j] = 0,$$

where h^j is known if the functions $u_0, u_1, \cdots, u_{j-1}$ are known from the preceding ordinary differential equations for the scalars σ^j.

Writing

$$(13) \qquad {}^J U = \alpha e^{i\xi\phi(x,t)} \left\{ u_0 + \frac{u_1}{\xi} + \cdots + \frac{u_J}{\xi^J} \right\},$$

we have, with bounded P^J,

$$(14) \qquad L[{}^J U] = \alpha e^{i\xi\phi(x,t)} \sum_{\nu=-1}^{J} \left[iA u_{\nu+1} + L[u_\nu] \right] \frac{1}{\xi^\nu} = \frac{P^J}{\xi^J}$$

an expression of order ξ^{-J} in ξ, that is, an asymptotic expression for U.

Turning now to Cauchy's problem for (2) we remember that the hyperbolic character of (2) implies (see §3): From each initial manifold of the family $\phi(x) = c = $ const., k characteristic surfaces $\phi^\kappa(x, t) = c$ fan out, all corresponding to the same initial values $\phi^\kappa(x, 0) = \phi(x)$; all of these ϕ^κ satisfy the characteristic differential equation

$$Q = | \phi_t I + \sum_{j=1}^{n} A^j \phi_j | = 0.$$

Even if some of these surfaces should coincide, the k left nullvectors ${}^1 l,\ {}^2 l, \cdots,\ {}^k l$ and the k right nullvectors ${}^1 r,\ {}^2 r, \cdots,\ {}^k r$ are linearly independent systems.

For each of these characteristic surfaces ϕ^κ we consider asymptotic solutions U^κ as before. We shall accordingly have to consider in (7), (11) k scalar factors σ, which we distinguish as ${}^1\sigma,\ {}^2\sigma, \cdots,\ {}^k\sigma$, omitting the reference index j of u_j. Then we adjust initial values of the components $u_0^\kappa(x, 0)$, $u_j^\kappa(x, 0)$, $\cdots$ by stipulating

$$(15) \qquad \sum_{\kappa=1}^{k} u_0^\kappa(x, 0) = \psi(x)$$

$$(16) \qquad \sum_{\kappa=1}^{k} u_j^\kappa(x, 0) = 0 \qquad\qquad (j \geq 1),$$

with arbitrary sufficiently smooth $\psi(x)$.

In (15) and (16) we substitute from (7), (11) $u_0^\kappa = {}^\kappa\sigma^0 r^\kappa$ and $u_j^\kappa = {}^\kappa\sigma^j r^\kappa + {}^\kappa h^j$ for $j > 1$, with the scalars ${}^\kappa\sigma^j$, and then consider these relations for $t = 0$. The linear independence of the vectors r^κ assures that for each j the initial values of the k scalars ${}^\kappa\sigma^j$ are uniquely determined. Thus Cauchy's problem for (2) with the initial values $u(x, 0) = e^{i\phi(x)}\psi(x)$ is indeed solved by a sum of k sums of the form (3).

From the fact that according to (14) the components U^κ are asymptotically expressed by the finite power series for $^J U^\kappa$ in $1/\xi$, we finally conclude: $^J U = \sum_{\kappa=1}^{k} {}^J U^\kappa$ is indeed an asymptotic approximation to the function u with the initial values $e^{i\phi(x)}\psi(x)$. Indeed because of $L[u] = 0$, we have for $^J V = {}^J U - u$, by (14),

$$L[^J V] = \frac{P^J}{\xi^J}$$

while $^J V$ has the initial values zero. Therefore Duhamel's integral representation (see, e.g., §10, 1) for $^J V$ shows that $^J U$ is of order ξ^{-J} in ξ and so our expansion is asymptotic.

Moreover, similar estimates are valid for the asymptotic expressions of derivatives up to an order which can be chosen arbitrarily high if the data possess a sufficient number of continuous derivatives.

The preceding asymptotic solutions lead to an approximate—often rather precise—solution of the initial value problem. It is obtained by solving only ordinary differential equations; for determination of ϕ^k as well as the successive determination of $u_0, u_1, \cdots, u_j$ depends only on ordinary differential equations and algebraic processes.

3. Geometrical Optics. The asymptotic theory of geometrical optics indicated here goes back to Green, Liouville (1837), A. Sommerfeld and J. Runge.[1]

Now we are in a position to understand the relationship between *wave optics*, represented by the partial differential equation (2), and *geometrical optics*, describing propagation phenomena in terms of the geometry of rays. In optics, we deal with high frequency oscillations; then the exact solution of the partial differential equation is in the first approximation given by the first term of its asymptotic expansion. This term, as well as the further terms of the expansion, is determined by means of ordinary differential equations along rays. Naturally there is a close connection between geometrical optics and the expansions of §4 and the present section.

Usually in optics one is not concerned with initial value problems, but rather with mixed or boundary value problems, under the assumption that the data and solution are independent of time except for a factor $e^{i\omega t}$. Moreover, the coefficients of the differential equation are assumed independent of time. Consider the first order system

[1] For literature see J. B. Keller [1], J. B. Keller, R. M. Keller, and B. D. Seckler [1]; M. Kline [1], [2]; P. D. Lax [4]; and R. K. Luneburg [2], [1].

(17) $\quad L[u] \equiv u_t + M[u] \equiv u_t + \sum_{\nu=1}^{n} A^\nu(x) \frac{\partial u}{\partial x_\nu} + B(x)u = 0.$

We may separate the space and time variables by writing

$$u(x, t) = e^{i\omega t} v(x).$$

Thus $v(x)$ must satisfy the "reduced equation":

(18) $\qquad\qquad\qquad M[v] + i\omega v = 0.$

On the other hand, we know that (17) has solutions of the form

(19) $\qquad\qquad\qquad u = \sum_{j=0}^{\infty} T_j(\phi) g^j(x, t),$

where $T_j(\phi) = e^{i\omega\phi}/(i\omega)^j$. In view of the fact that the coefficients of the differential equation are independent of t, we consider solutions where the coefficients $g^j(x, t)$ are independent of t. If we set

$$\phi(x, t) = t + \psi(x),$$

then we have

$$u(x, t) = e^{i\omega t} \cdot e^{i\omega\psi} \sum_{j=0}^{\infty} \frac{g^j(x)}{(i\omega)^j};$$

hence

(20) $\qquad\qquad\qquad v(x) = e^{i\omega\psi} \sum_{j=0}^{\infty} \frac{g^j(x)}{(i\omega)^j}.$

Thus we are led by the techniques of this section to an asymptotic expansion of the reduced equation (18). The coefficients $g^j(x)$ must satisfy the relations

(21) $\qquad (I\phi_t + \sum_{\nu=1}^{n} A^\nu\phi_\nu)g^{j+1} + L(g^j) = 0 \qquad (j = -1, 0, \cdots),$

or equivalently,

$$(I + \sum_{\nu=1}^{n} A^\nu\psi_\nu)g^{j+1} + Mg^j = 0 \qquad (j = -1, 0, \cdots).$$

As seen before in §4 and the present section, equations (21) can be solved using merely ordinary differential equations along the rays. Hence the expansion (20) can be found by means of ordinary differential equations. These ordinary differential equations along rays are the equations of geometrical optics.

There is a similar correspondence between the theory of §4 and the inhomogeneous reduced equation (18). Consider the equation

$$(22) \qquad L[u] \equiv u_t + Mu = e^{i\omega t}f(x).$$

If $u(t, x) = e^{i\omega t}v(x)$, then $v(x)$ must satisfy

$$Mv + i\omega v = f(x).$$

We define $w(x, t)$ by the conditions

$$L[w] = 0 \qquad\qquad \text{for } t > 0,$$

$$w(x, 0) = f(x).$$

Then Duhamel's principle leads to the formal solution

$$(23) \qquad v(x, \omega) = \int_0^\infty e^{-i\omega s}w(x, s)\, ds.$$

The asymptotic expansion of v for large ω is determined from knowledge of the singularities of $w(x, t)$, which depend upon the singularities of $f(x)$; according to §4 these singularities propagate along rays. Thus, as before, the asymptotic expansion of $v(x)$, corresponding to geometrical optics, is determined by means of ordinary differential equations along rays.

§6. *Examples of Uniqueness Theorems and Domain of Dependence for Initial Value Problems*

The method of energy integrals was first used by S. Zaremba.[1] It was rediscovered and extended by A. Rubinowicz[2] and by K. O. Friedrichs and H. Lewy[3] and was used for the treatment of symmetric hyperbolic systems. Various later publications by K. O. Friedrichs[4] as well as J. Schauder[5] have clearly established the power and versatility of the method.

In this section we consider some typical examples of uniqueness and determinacy of the solution of Cauchy's problem, primarily for some equations of second order, so that the motivation of the general theory in §8 can be better appreciated (see also Ch. V, §4).

[1] See S. Zaremba [1].
[2] See A. Rubinowicz [1] and [2].
[3] See K. O. Friedrichs and H. Lewy [1].
[4] See, e.g., K. O. Friedrichs [2].
[5] See J. Schauder [4].

Existence and uniqueness of the solutions of Cauchy's problem for arbitrary initial data are topics closely related to each other (see Ch. III, §6); they will be discussed in §§8, 9, 10 with the aid of certain quadratic mean values of the solutions and their derivatives, called "energy integrals". The solution of the problem at a point P is uniquely determined if the Cauchy data are known merely in bounded "domains of dependence" associated with the point P. It is the object of §§6, 7 to clarify the significant facts of uniqueness and determinacy.

1. The Wave Equation. For the two-dimensional wave equation

$$(1) \qquad L[u] \equiv u_{tt} - u_{xx} - u_{yy} = 0,$$

the following uniqueness proof deviates somewhat from the corresponding argument in Chapter V. Let S be an arbitrary space-like initial surface $\phi(x, y, t) = 0$, with

$$\phi_t^2 - \phi_x^2 - \phi_y^2 > 0$$

or

$$t_\nu^2 - x_\nu^2 - y_\nu^2 > 0,$$

where x_ν, y_ν, t_ν denote the components of the unit normal to the surface, i.e.,

$$x_\nu = \frac{\phi_x}{\sqrt{\phi_x^2 + \phi_y^2 + \phi_t^2}}, \qquad y_\nu = \frac{\phi_y}{\sqrt{\phi_x^2 + \phi_y^2 + \phi_t^2}},$$

$$t_\nu = \frac{\phi_t}{\sqrt{\phi_x^2 + \phi_y^2 + \phi_t^2}}.$$

Suppose that a solution u of the differential equation vanishes on a subdomain B' of S together with its first derivatives. We assert: *u vanishes at all points P such that the characteristic cone through P intersects the initial surface inside B'.* See Figure 55. We let G de-

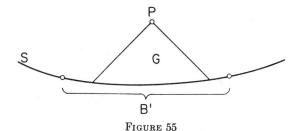

FIGURE 55

note the region bounded by B' and the characteristic cone with vertex at P. The characteristic cone is the cone in x,y,t-space whose elements are inclined at an angle of 45 degrees to the plane $t = 0$, i.e., are characteristic rays.

The proof is based on the identity

$$
(2) \quad
\begin{aligned}
&2u_t\, L[u] \\
&\equiv -2(u_t\, u_x)_x - 2(u_t\, u_y)_y + (u_x^2)_t + (u_y^2)_t + (u_t^2)_t = 0,
\end{aligned}
$$

integrated over the domain G. Since the expression (2) is a divergence, Gauss' integral theorem and the initial condition yield

$$
0 = \iint_M (u_x^2\, t_\nu + u_y^2\, t_\nu + u_t^2\, t_\nu - 2u_t\, u_x\, x_\nu - 2u_t\, u_y\, y_\nu)\, dS
$$

$$
= \iint_M \frac{1}{t_\nu} [(u_x\, t_\nu - u_t\, x_\nu)^2 + (u_y\, t_\nu - u_t\, y_\nu)^2]\, dS,
$$

where M is the part of the surface of the cone which belongs to the boundary of G, dS is the surface element of M, and the relation $t_\nu^2 - x_\nu^2 - y_\nu^2 = 0$ on M is taken into account. Hence the last integral over the cone vanishes; therefore, the integrand likewise vanishes. In other words, $u_x\, t_\nu - u_t\, x_\nu = 0$ and $u_y\, t_\nu - u_t\, y_\nu = 0$ everywhere on M; thus on M two linearly independent interior derivatives of u vanish. Hence the solution u is a constant on M, and hence identically zero because of the initial condition. From this the vanishing of u at the point P follows.

At the same time this argument exhibits the *domain of dependence* for the differential equation in the following sense: *The values of a solution u at the point P, with prescribed initial values on S, depend only on the initial values in that part of S which is cut out by the backward characteristic ray cone through P.*

The question of uniqueness and domain of dependence for three and more independent variables may be similarly answered by the method of Ch. V, §4 for the more general differential equation $u_{tt} - \Delta u + au_x + bu_y + cu_t + du = 0$ in which the coefficients a, b, c, d are arbitrary continuous functions of t and the space variables. (See the general theory in §8.)

We next prove the *uniqueness for the "characteristic initial value problem of the wave equation"*. Here the initial values are no longer given along a space-like initial manifold with $\phi_t^2 - \phi_x^2 - \phi_y^2 > 0$, but are prescribed instead along a special characteristic manifold namely a characteristic half-cone K:

$$(3) \qquad (t - t_0)^2 - (x - x_0)^2 - (y - y_0)^2 = 0 \qquad (t \geq t_0).$$

In agreement with earlier considerations, we cannot prescribe the function u and an outward derivative (i.e., all derivatives) arbitrarily but merely the values of the function u itself. We suppose these initial values given as the values attained on the surface of the cone by a function continuously differentiable in a neighborhood of the cone which includes the vertex. Then: *The values of u on the half-cone K defined by* (3) *determine u uniquely everywhere in the interior of the half-cone*, i.e., for

$$(t - t_0)^2 - (x - x_0)^2 - (y - y_0)^2 > 0, \qquad t > t_0.$$

See Figure 56. The proof follows from the formulas above: Suppose that the initial values u of a solution vanish on the characteristic cone, and integrate the expression (2) over a domain G which is bounded by this cone and by the characteristic cone through P. If M_1 and M_2 are the corresponding parts of the surfaces of the cones, we at once have, with the preceding notation,

$$\iint_{M_2} \frac{1}{t_\nu} [(u_x t_\nu - u_t x_\nu)^2 + (u_y t_\nu - u_t y_\nu)^2] \, dS = 0.$$

The integral over the lower cone vanishes, since in the integrand only

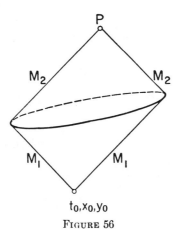

t_0, x_0, y_0

FIGURE 56

interior derivatives of u appear on this cone and they, by hypothesis, are all zero. Hence the independent interior derivatives $u_x \, t_\nu - u_t \, x_\nu$, $u_y \, t_\nu - u_t \, y_\nu$ likewise vanish on the cone corresponding to P. In other words, on the surface of this cone u is constant and therefore zero, since u vanishes on the intersection of the surfaces of the two cones.

2. The Differential Equation $u_{tt} - \Delta u + \dfrac{\lambda}{t} u_t = 0$ **(Darboux Equation).**[1] Another variant of the method concerns the differential equation

$$(4) \qquad L[u] \equiv u_{tt} + \frac{\lambda}{t} u_t - \Delta u = 0,$$

named after Darboux, which contains a singular term. Here λ is·an arbitrary non-negative continuously differentiable function of the variables x_i and t. The characteristic equation is again

$$(5) \qquad \phi_t^2 - \phi_{x_1}^2 - \phi_{x_2}^2 - \cdots - \phi_{x_n}^2 = 0$$

or

$$(5') \qquad \left(\frac{\partial t}{\partial \nu}\right)^2 - \sum_{i=1}^{n} \left(\frac{\partial x_i}{\partial \nu}\right)^2 = 0,$$

and the characteristic cones $\phi(x, t) = (t - \tau)^2 - \sum_{i=1}^{n} (x_i - \xi_i)^2 = 0$ are the same as in article 1. We show: *If a twice continuously differentiable solution u of equation* (4) *and the derivative u_t vanish on the base B in the plane $t = 0$ of a characteristic cone with vertex at $P(t > 0)$, then u vanishes at P and everywhere in the interior G of the cone.*

Proof: We have

$$0 = -2u_t \, L[u] \equiv 2 \sum_{i=1}^{n} (u_t \, u_{x_i})_{x_i} - \left(\sum_{i=1}^{n} u_{x_i}^2 + u_t^2\right)_t - \frac{2\lambda}{t} u_t^2 \,.$$

Therefore, if we integrate over the domain G with the volume element dv, take into account the initial condition on B and apply Gauss' theorem to the divergence on the right, we have

$$0 = \iiint_G \frac{2\lambda}{t} u_t^2 \, dv + \iint_M \left[-2u_t \sum_{i=1}^{n} u_{x_i} \frac{\partial x_i}{\partial \nu} + \left(u_t^2 + \sum_{i=1}^{n} u_{x_i}^2 \right) \frac{\partial t}{\partial \nu} \right] dS,$$

[1] See §13.

where M denotes the surface of the cone and dS the surface element on M. The integrand in the integral over M can be written in the form

$$\frac{1}{t_\nu} \sum_{i=1}^{n} \left(u_{x_i} \frac{\partial t}{\partial \nu} - u_t \frac{\partial x_i}{\partial \nu} \right)^2$$

in view of the characteristic relation (5′) on M. Since $\lambda > 0$, the relation $u_t = 0$ follows immediately everywhere in G. Thus u is identically zero in G, as asserted.

3. Maxwell's Equations in Vacuum. As an example of a system of differential equations with four independent variables, we again consider the system of Maxwell's equations[1] (cf. Ch. III, §2), setting the velocity of light $c = 1$:

(6) $$\mathfrak{E}_t - \operatorname{curl} \mathfrak{H} = 0, \qquad \mathfrak{H}_t + \operatorname{curl} \mathfrak{E} = 0.$$

For this system of differential equations we consider the initial value problem for the plane $t = 0$, assuming the initial values of the vectors $\mathfrak{E}$ and $\mathfrak{H}$ prescribed. Our object is to prove: *If the initial values of $\mathfrak{E}$ and $\mathfrak{H}$ vanish, then the vectors $\mathfrak{E}$ and $\mathfrak{H}$ vanish identically.* In fact, the following uniqueness proof (as well as a similar one for the differential equations of crystal optics) becomes much more transparent as a case of the general result of §8. Nevertheless, since historically the proof in the special case has stimulated the development of the general theory, it will be given here.

To any point P of four-dimensional x,y,z,t-space belongs a characteristic cone, whose intersection with the initial plane $t = 0$ is a three-dimensional sphere B. By means of a plane $t = h$ parallel to the initial plane, we cut off a truncated cone G_h of this four-dimensional cone G, bounded by B, and also a part M_h of the surface of the cone and the spherical portion D_h of the boundary on $t = h$ (cf. Figure 57). Maxwell's equations imply

$$0 = 2\mathfrak{E}(\mathfrak{E}_t - \operatorname{curl} \mathfrak{H}) + 2\mathfrak{H}(\mathfrak{H}_t + \operatorname{curl} \mathfrak{E})$$

$$= (\mathfrak{E}^2 + \mathfrak{H}^2)_t + 2 \operatorname{div} [\mathfrak{E} \times \mathfrak{H}]$$

---

[1] Maxwell's equations are supplemented by the relations

$$\operatorname{div} \mathfrak{E} = 0, \qquad \operatorname{div} \mathfrak{H} = 0.$$

On the basis of (6), it is easily shown that these relations are satisfied everywhere, if they are satisfied in the initial space $t = 0$. Thus they have the character of initial conditions.

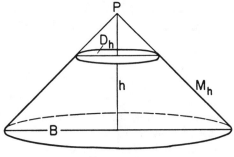

in view of the vector relation $\mathfrak{H}$ curl $\mathfrak{E} - \mathfrak{E}$ curl $\mathfrak{H} = \operatorname{div} [\mathfrak{E} \times \mathfrak{H}]$. We now integrate over G_h first with respect to x, y, z, holding t fixed, and then with respect to t between the limits 0 and h. Taking into account the initial condition $\mathfrak{E} = \mathfrak{H} = 0$ on $t = 0$, we obtain at once

$$\begin{aligned}
(7) \quad &\iiint_{M_h} (\mathfrak{E}^2 + \mathfrak{H}^2) t_\nu \, dS + 2 \iiint_{M_h} [\mathfrak{E} \times \mathfrak{H}] \underline{r}_\nu \, dS \\
&\qquad\qquad + \iiint_{D_h} (\mathfrak{E}^2 + \mathfrak{H}^2) \, dx \, dy \, dz = 0,
\end{aligned}$$

where $\underline{r}_\nu$ is the normal vector in three-dimensional x,y,z-space of the sphere with radius t and center at the projection of P, and where $t_\nu = \frac{1}{2}\sqrt{2}$ is the t-component of the normal of M_h.

By virtue of the characteristic condition $t_\nu^2 = \underline{r}_\nu^2$ on M_h we have

$$(\mathfrak{E}^2 + \mathfrak{H}^2) t_\nu^2 + 2 t_\nu \, \underline{r}_\nu [\mathfrak{E} \times \mathfrak{H}] = \mathfrak{E}^2 t_\nu^2 + 2 \mathfrak{E} [\mathfrak{H} \times \underline{r}_\nu] t_\nu + \mathfrak{H}^2 \underline{r}_\nu^2$$

on M_h. Since $[\mathfrak{H} \times \underline{r}_\nu]^2 = \mathfrak{H}^2 \underline{r}_\nu^2 - (\mathfrak{H} \cdot \underline{r}_\nu)^2$, the right-hand side is equal to

$$(\mathfrak{E} t_\nu + [\mathfrak{H} \times \underline{r}_\nu])^2 + (\mathfrak{H} \underline{r}_\nu)^2.$$

From (7) then it follows that

$$\begin{aligned}
0 = &\iiint_{D_h} (\mathfrak{E}^2 + \mathfrak{H}^2) \, dx \, dy \, dz \\
&+ \iiint_{M_h} \frac{1}{t_\nu} \{ (\mathfrak{E} t_\nu + [\mathfrak{H} \times \underline{r}_\nu])^2 + (\mathfrak{H} \cdot \underline{r}_\nu)^2 \} \, dS,
\end{aligned}$$

implying that $\mathfrak{E} = \mathfrak{H} = 0$ on D_h, and hence everywhere in G, which is what we wanted to prove.

For the differential equations of crystal optics (7) of §3a, uniqueness and domain of dependence could be obtained similarly; however we omit this discussion referring instead to the general theory of §8.

§7. Domains of Dependence for Hyperbolic Problems

1. Introduction. As indicated by the examples of §6, an important feature of hyperbolic initial value problems is: The solution $u(P)$ at a point P is uniquely determined by Cauchy data only in bounded domains $\Gamma = \Gamma_P$ belonging to P. Cauchy data outside of such domains have no influence on the value $u(P)$. Equivalently, $u(P) = 0$ if the Cauchy data in Γ_P vanish. This feature corresponds to the fact that hyperbolic problems are associated with wave propagation at finite speeds.

Obviously the existence of such *domains of dependence* Γ_P is expressed by uniqueness theorems which link $u(P)$ with the data in Γ_P. We shall derive and formulate those uniqueness theorems in §8. In the present section we confine ourselves to describing domains of dependence Γ_P geometrically.

In passing we observe that the concept of domain of dependence Γ_P has as a corollary the concept of *domain of influence* for an initial domain D. This is the set of all points P whose domains of dependence have points in common with D. Physically speaking, the phenomena outside this domain of influence I_D are not affected by the data in D, that is, the medium outside I_D remains "unaware" of the initial state in D.

A certain vagueness of the concepts should be pointed out in advance: The statement "Γ_P is a domain of dependence for $u(P)$" has a negative character; it merely asserts that data outside Γ *do not influence* the value $u(P)$. Any domain Γ^* which includes Γ falls into the same category. Therefore it would seem reasonable to define a *domain of dependence in the precise sense*, as the *smallest* point set Γ such that $u(P)$ is uniquely determined by arbitrary data in Γ. But this domain Γ is difficult to characterize in a general way. It may be a domain in x-space having gaps,[1] or in other cases it may consist only of the boundary of such domains (see later discussions of Huyghens' principle). Therefore we seek a compromise by de-

---

[1] The occurrence of such gaps is exemplified in the case of elastic waves (see §13a) and crystal optics. A penetrating general study of this phenomenon is due to Petrovskii [3].

fining a domain of dependence Γ as small as can be conveniently done by natural and general geometrical descriptions, without necessarily aiming at the optimal point set Γ. Such descriptions will be given in the following article.

2. Description of the Domain of Dependence. Again we shall single out the time $t = x_0$ and denote the space variables $x_1, \cdots, x_n$ by x, assuming that the x-space is space-like and thus that t is a genuine time variable. We shall consider closely related concepts: The domain of dependence Γ_P and the hull of the ray conoid or the outer spherical wave front centered at P, called Γ in §3. This conoid drawn from P towards decreasing values of t is the "backward conoid of dependence"; the conoid drawn towards increasing t is the "forward" one. If the equation is inhomogeneous, then the domain of dependence consists of all points of the $(n + 1)$-dimensional x, t-space inside and on Γ_P and between P and the initial manifold. If the equation is homogeneous then the domain of dependence consists merely of those points on and inside Γ_P which lie on the initial manifold I. We shall take the liberty of denoting the domain of dependence again by the symbol Γ_P, or, in the homogeneous case, γ_P.

Repeating and amplifying the definitions given in §3, we first consider operators $L[u]$ whose principal parts have constant coefficients. The hyperbolic character of $L[u]$ then assures that the core of the normal cone is convex (see §3, 7). Its dual cone Γ is also convex; at each of the points on Γ the normals belong to the boundary of the core of the normal cone. Hence the surface elements of Γ separate space-like surface elements from others. As said above, $\Gamma = \Gamma_P$ is defined as the conoid of dependence.

We may also define Γ_P as follows: Let P have coordinates $(\tau, \xi_1. \cdots, \xi_n)$. Consider for every α the characteristic plane

$$v(t - \tau) - \alpha \cdot (x - \xi) = 0,$$

for which the velocity v is maximal. Γ_P is the intersection of, that is, the set of points common to, all half-spaces $v(t - \tau) - \alpha \cdot (x - \xi) \leq 0$, where α varies over the unit sphere.

For nonconstant coefficients the definition is analogous: Through the linear manifold $(\alpha x) = 0$, or $(\alpha(x - \xi)) = 0$ in the space-like plane $t = \tau$, we draw towards $t < \tau$ the "plane-like" characteristic surfaces. One of them, whose "negative" normal directions point into the boundary of the core of the normal cone, separates space-like elements from others and corresponds to a maximal local velocity at P. Then Γ_P is the intersection of the corresponding half-spaces, where α varies over the unit sphere. Γ_P may again be considered as a hull of the backward ray conoid through P.

The definition of "time-like" in §3, 7 leads immediately to the following interpretation: *The interior of the conoid of dependence Γ_P is the set of points which can be connected with P by everywhere time-like curves.*

In the next section we shall restrict ourselves to *symmetric hyperbolic operators* of first order

$$(1) \qquad L[u] = u_t + \sum_{i=1}^{n} A^i u_i + Bu = u_t + Mu.$$

Then $t = $ const. is space-like: The surfaces R_τ consisting of the plane $t = \tau$ inside the conoid plus the "mantle" R_τ^* of the conoid Γ between $t = 0$ and $t = \tau$, $(0 \le t \le \tau)$, form a "space-like lens" L_τ. If $t = \tau$ is the time coordinate of P, then this lens coincides with the interior of the conoid for $0 \le t \le \tau$.

If $\phi(t, x) = 0$ represents the boundaries of the lens L_τ, then the characteristic form $A = I\phi_t + \sum_i^n A^i \phi_i$ is simply equal to the unit matrix I on $t = 0$ and $t = \tau$, and *the characteristic matrix A is non-negative on the mantle R^*.*[1]

Indeed, on a plane-like characteristic surface corresponding to a maximal velocity, we see immediately that the characteristic matrix is non-negative. Since every point on the boundary of Γ_P is a tangent point of such a characteristic surface, the characteristic matrix is non-negative on the boundary of Γ_P. Hence for a surface whose elements are just on the border between the set of space-like and non-space-like elements the matrix A is certainly not negative.

Naturally, the concepts of domain of dependence and domain of influence apply in the same way more generally to any point set Π, e.g., a domain in the plane $t = $ const. > 0. The domain of dependence C_Π then consists of the closure of the set of all points $t \ge 0$ which can be reached from Π by time-like curves.

---

[1] The part R^* is "weakly space-like"; we take the liberty of sometimes calling R^* space-like. The main point is that on R^* the inequality $A \ge 0$ still holds.

As emphasized above, the concept of domain of dependence Γ_P allows a leeway inasmuch as Γ_P may be replaced by a wider domain $\bar{\Gamma}_P$ which contains Γ_P. In particular, we may consider a "tetrahedral" domain $\bar{\Gamma}_P$ bounded by three plane-like characteristic surfaces with maximal velocity through P. Γ_P is then the domain common to all these tetrahedral domains $\bar{\Gamma}_P$.

Again: The justification of the concepts of this section is implicit in the subsequent proof of uniqueness and existence.[1]

§8. Energy Integrals and Uniqueness for Linear Symmetric Hyperbolic Systems of First Order

1. Energy Integrals and Uniqueness for the Cauchy Problem. In this section we restrict ourselves to linear essentially symmetric hyperbolic systems of first order as defined before (see e.g. §3, 8 and p. 651). They have the form

$$(1) \qquad L[u] = \sum_{i=0}^{n} A^i u_i + Bu = 0,$$

where all the matrices A^i are symmetric, but where B need not necessarily be symmetric.

According to §3, 7 a surface $S:\phi(x) = 0$ is space-like for L if on S the characteristic matrix

$$A = A^i \phi_i$$

is positive definite. We assume that the system (1) is hyperbolic, i.e., that it admits space-like surfaces. Without restricting the generality we again assume that the hyperplanes $x_0 = t = $ const. are space-like, i.e., that the matrix A^0 is positive definite.

The discussions of this section are based on Gauss' formula, which is the immediate consequence of the following "divergence" representation of $uL[u]$:

$$(2) \qquad 2uL[u] = (u, A^j u)_j + 2(u, \hat{B}u) = 0$$

---

[1] Special explicitly accessible cases show that it may not always be necessary to replace the ray conoid by its convex hull. For example, take $n = 2$, $k = 2$, $L_1 = D_0^2 - D_1^2 - 4D_2^2$, $L_2 = D_0^2 - 4D_1^2 - D_2^2$. The ray surface in the x,y-plane consists of two intersecting ellipses. The system $L_1[u_1] = 0$, $L_2[u_2] = 0$ obviously has a unique solution if initial data for u_1 and u_2 are given in one ellipse each; there is no need for considering the convex hull. But if two equations are coupled by terms of order below 2 containing both u_1 and u_2, then the ray surface remains unchanged, but the data must be given in the convex hull.

with

$$(3) \qquad \hat{B} = B - \tfrac{1}{2} \sum_{j=0}^{n} A_j^j \, .$$

By a simple change of variables we can always transform L into an operator for which $(u, \hat{B}u)$ is positive definite: We introduce instead of u the function

$$(4) \qquad v = e^{-\mu x_0} u,$$

with a positive constant μ. This implies $L[u] = e^{\mu x_0}\{L[v] + \mu A^0 v\} = 0$ and hence

$$(1a) \qquad \sum A^j v_j + B^* v = 0,$$

with

$$(5) \qquad B^* = \hat{B} + \mu A^0.$$

Since by assumption A^0 is positive definite, the quadratic form $(u, B^* u)$ is also positive definite in any given region if μ is chosen sufficiently large. Specifically we may stipulate

$$(6) \qquad (u, u) \leq 2(u, B^* u).$$

Let D be a lens-shaped domain as described in §7, i.e., a domain bounded by two surfaces S_0 and S both of which are space-like and are joined along their common boundary. Then the following *uniqueness theorem*[1] holds: *If a solution u vanishes on S_0 it also vanishes on each surface S which forms with S_0 a space-like lens.*

The proof follows almost immediately by integrating (2) over the lens D. Gauss' formula, thus obtained, is

$$0 = \int_S (u, Au) \, dS - \int_{S_0} (u, Au) \, dS + \iint_D (u, \hat{B}u) \, dx,$$

where dS denotes the surface element, $dx = dx_0 \, dx_1 \, dx_2 \cdots$ denotes the volume element, and A is the characteristic matrix on the boundary. Since both (u, Au) and $(u, \hat{B}u)$ are positive definite it follows that u vanishes in D.

We remark that in this proof we have made use of the space-like character of S but not that of S_0.

If B vanishes identically the operator L is called *conservative*. In

---

[1] The reader should compare this discussion with the discussion of Holmgren's theorem in Ch. III, App. 2. The proof of Holmgren's theorem is based on the existence of the solution of Cauchy's problem for the adjoint differential equation but does not make use of symmetry or hyperbolicity.

this case the "energy"

$$\int (u, Au)\, dS$$

contained on the two surfaces S_0 and S is the same, a fact which re-calls "conservation of energy" and motivates the name "energy integral".

2. Energy Integrals of First and Higher Order. We shall dis-tinguish the variable $x_0 = t$ as time and, as before, assume the hyper-planes $t = $ const. to be space-like. According to §3, new unknowns u can be introduced in terms of which the differential equation takes the simplified form

$$(1b) \qquad L[u] = u_t + \sum_1^n A^i u_{x^i} + Bu = 0,$$

the matrices A^i remaining symmetric.

We consider special lens-shaped domains, or, more precisely, one-parameter families of lens-shaped domains constructed as follows: Let P be a point in t,x-space, with t positive. Let Γ_P be the back-ward-drawn conoid of dependence of P (see §7). Let R_h denote the intersection of Γ_P with the hyperplane $t = h$. Then we consider the lens-shaped domain bounded by $S_0 = R_0$ and $S_1 = R_h + M_h$, where M_h is the portion of the boundary (mantle) of the conoid contained between the hyperplanes $t = 0$ and $t = h$. As we saw in §7, M_h is "weakly" space-like, i.e., (u, Au) is non-negative on M_h. Since R_0 and R_h are space-like, the uniqueness theorem stated before follows:

If u satisfies the equation $L[u] = 0$ in Γ_P and vanishes on R_0, then u vanishes throughout Γ_P.

This result implies the important fact:

The value of u at P is uniquely determined by the value of $L[u] = f$ in Γ_P and by the Cauchy data of u on R_0; it is therefore not influenced by the data outside of Γ_P.

This last fact indeed justifies the name "conoid of dependence" for Γ_P.

If we introduce the notation[1]

$$(7) \qquad \| u(h) \| = \left\{ \int_{R_h} u^2(x, h)\, dx \right\}^{1/2},$$

---

[1] The notation here is at variance with that used for the "maximum norm" in Ch. V, §§6, 7.

the proof of the uniqueness theorem in the previous article implies (under the assumption $(u, \hat{B}u) \geq 0$) the *energy inequality*

$$(8) \qquad \| u(h) \| \leq \| u(0) \|.$$

Analogous inequalities are satisfied by *energy integrals* $\| u(h) \|_r$ of order r, defined as follows:

$$(9) \qquad \| u(h) \|_r^2 = \int_{R(h)} \sum_{|p| \leq r} | D^p u |^2 \, dx,$$

where the summation on the right extends over all partial derivatives $D^p u$ of order $| p | \leq r$ with respect to the x variables (we shall denote the set of all these partial derivatives as the vector V).

The *energy inequality of order* r states:

$$(10) \qquad \| u(h) \|_r \leq c \| u(0) \|_r \, .$$

Here c is a constant which depends on bounds for the magnitude of the coefficients of L and the derivatives of these coefficients up to order r. (It is always assumed that according to (4) an exponential factor with sufficiently large μ is extracted.)

The proof is the same as for the ordinary energy inequality (8). One merely has to apply the same reasoning to the system of differential equations for the vector V obtained by differentiating $L[u] = 0$ with respect to all x-variables. Thus we form a system of equations having the form $L[V] + MV = 0$; here the same operator L applies to every single derivative $D^l u$, and MV denotes linear combinations of the components of V.

The differential equation (1b) and the equations obtained from it by differentiation with respect to t and x express all r-th order derivatives of u in terms of the x-derivatives of u of order $\leq r$. Combining this with the higher order energy inequality (10) we obtain

$$(11) \qquad \| u(h) \|_r^* \leq c' \| u(0) \|_r \, ,$$

$$(9') \qquad \| u(h) \|_r^{*2} = \int_{R_h} \sum_{|p| \leq r} | D^p u |^2 \, dx,$$

the summation being now extended over *all* partial derivatives of $u(x, t)$ of order $\leq r$, not merely the x-derivatives.

Finally: *For solutions of an inhomogeneous equation* $L[u] = f$ *the following energy inequality holds*:

$$(12) \qquad \| u(h) \|^2 \leq 2 \| u(0) \|^2 + \int_0^h \| f(t) \|^2 \, dt,$$

where

$$\| f(t) \|^2 = \int_{R(t)} f^2(x, t) \, dx .$$

To prove (12), we combine the identity (2) with the inequality

$$2uf \leq u^2 + f^2$$

and use inequality (6):

$$2(u, \hat{B}u) - u^2 \geq 0.$$

The following analogous result holds for the higher order energy integrals:

(12a) $$\| u(h) \|_r^2 \leq c \| u(0) \|_r^2 + c \int_0^h \| f(t) \|_r^2 \, dt.$$

As it will become apparent later, all these inequalities are of significance for the construction and analysis of the solution, not merely for the proof of uniqueness.

3. Energy Inequalities for Mixed Initial and Boundary Value Problems. The energy expressions of article 2 allow us immediately to extend uniqueness proofs to some important classes of problems, that is, problems for symmetric but not necessarily hyperbolic equations and for mixed boundary and initial conditions (see Ch. V, §6). In the present article we indicate briefly such an extension.[1] We restrict ourselves to mixed hyperbolic problems, and for that matter to *mixed initial and boundary value problems* in which the variable $t = x_0$ is singled out and the form (1b) of the differential equations assumed. These problems are to find solutions u of equation (1b), defined for all x in a given domain G of the x-space and for all positive values of t, with the initial values of $u(x, 0)$ prescribed in G; further, u must satisfy suitable conditions at the boundary S of G. (See also Ch. V, §6, 4.) We assume these boundary conditions to be linear homogeneous relations between the components of u on S. They may vary from point to point on S, and they may depend on the time t. Physically they may express the restrictions imposed on the system by confining it to the domain S (such as kinematic conditions); or they may express the interaction between the system and the mechanism whereby the system is constrained, for example re-

---

[1] Reference is made to a paper by K. O. Friedrichs [3] in which uniqueness and existence for symmetric systems, not necessarily hyperbolic, is treated. See also P. D. Lax and R. S. Phillips [1] and G. F. D. Duff [3].

flection, refraction, clamping, damping, cooling, evaporation, radiation, etc.

Our discussion will concern merely uniqueness; general existence proofs for mixed problems, while given for Cauchy's problem in §10, will remain outside the scope of this volume. (See, however, Ch. V, §6, 4, and Appendix for the case of two independent variables.)

We shall take the liberty of denoting by $S : \gamma(x_1, \cdots, x_n) = 0$ either the boundary of G or the vertical boundary of the cylinder Z perpendicular to $t = 0$ on the boundary of G. On S the characteristic matrix $A = \sum_{i=0}^k \xi_i A^i$ is simply

$$A = \sum_{i=1}^k \xi_i A^i \qquad (\xi_i = \gamma_{x_i})$$

since the t-component of the normal vector on Z vanishes.

The main object of the present article is to formulate suitable boundary conditions on Z (or S) for $t > 0$ which together with Cauchy data on G guarantee uniqueness of the solution and may be expected also to be sufficient for existence of the solution.

The relevant boundary conditions, just as in Ch. V, §6, 4 will be expressed as follows: At each point x, t of the cylindrical boundary S of Z, the vector $u(x, t)$ belongs to a linear space N of $r = k - p$ dimensions defined by a number p of linear homogeneous relations

$$(13) \qquad (u, m^j) = 0 \qquad (j = 1, \cdots, p)$$

with p independent vectors m^j. The space N is restricted by the stipulation: For vectors u belonging to N *the quadratic form*

$$Q(u, u) = uAu$$

is non-negative on Z.

The motivation for the boundary condition is given by Gauss' formula (following from (2))

$$\iiint_D (u, \hat{B}u) \, dt \, dx + \iint_S (u, Au) \, dS = 0,$$

where $A = \sum_{i=0}^n A^i \phi_i$ is the characteristic matrix on the surface of D. If $(u\hat{B}u)$ is a positive definite form, then this formula immediately assures uniqueness for the solution of (1) in D, provided that boundary conditions are prescribed which imply the non-negative character of the quadratic form $(uAu) = Q(u)$ at every point on S.

It is of interest that this observation holds whether or not (1) is hyperbolic.

In particular, if we choose for D the cylindrical domain over G between $t = 0$ and $t = \tau$, we obtain for a solution of (1b)

$$(14) \quad \int_G (u, u) \, dx \Big|_{t=0}^{t=\tau} + \int_0^\tau \int_S (u, Au) \, ds \, dt + \iint_D (u, \hat{B}u) \, dx \, dt = 0.$$

The boundary conditions for u then indeed imply uniqueness for a mixed problem exactly as for Cauchy's problem in article 1. This observation is the main point of this article.

However, we want to add a few remarks to illuminate the situation further and to provide an applicable criterion.

First: For given A, the space N may be chosen in a variety of ways. For example, in the simple case $k = 2$ and $(u, Au) = u_1^2 - u_2^2$, each linear space given by $u_2 - \sigma u_1 = 0$ with $|\sigma| < 1$ obviously satisfies the definiteness condition.

Second: We cannot expect a solution to exist if too many boundary conditions are imposed, i.e., if fewer boundary conditions would guarantee the uniqueness of the solution. Since the boundary space N is larger when we impose fewer conditions, we are led to introduce the following concept as a proper requirement for existence: *The boundary space N is maximal non-negative, i.e., it cannot be enlarged to a larger linear space over which the quadratic form (u, Au) is still everywhere non-negative.* Although we shall not give the existence proof we shall stipulate this *maximality of N*.

The most relevant question concerning N is: What is dim N, the dimension of N, i.e., how many linearly independent conditions (13) define a positive maximal space N?

Let us assume for simplicity that the cylinder Z is nowhere characteristic, i.e., that no eigenvalue of A vanishes.[1] Then we state: *The number of dimensions r of a maximal non-negative space N is equal to the number of positive eigenvalues of A.* The number r and the number $p = k - r$ of conditions $(u, m^j) = 0$ to be imposed on u do not depend on the specific choice of the maximal space N (for two independent variables cf. Ch. V, §6).

The criterion provided by this theorem is often directly applicable.

The proof is based on the fact that in a d-dimensional linear space of k-vectors (that is, vectors with k components) one always can find a nonvanishing vector u which satisfies s linear homogeneous conditions provided that $d + s < k$.

---

[1] It is easily seen that this convenient assumption is not essential.

Now consider a complete set of normed eigenvectors $u_1, \cdots, u_k$ of A with the eigenvalues $\lambda_1, \cdots, \lambda_k$: Assume the first eigenvalues $\lambda_1, \cdots, \lambda_r$ positive, the others negative. If

$$u = \alpha_i u_i$$

then

$$(u, Au) = \alpha_i^2 \lambda_i .$$

First we show: Dim $N \leq r$. Indeed if N had more than r dimensions then we could find in N a vector u so that the r orthogonality conditions $\alpha_1 = \cdots = \alpha_r$ are satisfied, hence $(u, Au) = \sum_{i=r+1}^{k} \alpha_i^2 \lambda_i$ < 0 in contradiction to the assumption that A is positive for u in N.

Next we show that a space N is not maximal if $d = \dim N < r$. Indeed, the conditions

$$(u, Av) = 0 \qquad \text{stipulated for any vector } v \text{ of } N$$

and

$$(u, u_i) = 0 \qquad \text{for } i = r + 1, \cdots, k$$

represent $d + k - r$ linear homogeneous equations for a vector u so that, $d + k - r$ being less than k, a nonvanishing solution u exists. (The first condition implies that u is not in N.) Then, for any vector of the form $w = \alpha u + v$ with v in N and α any positive constant we have

$$(w, Aw) = \alpha^2(u, Au) + (v, Av) > 0.$$

This would mean that all vectors w belong to N which contradicts the assumption of maximality of N since u is not in N.

Therefore $\dim N = r$ is the only alternative left, and the statement is proved.

Similarly, one could define maximal negative spaces N^* of dimension $p = k - r$ which are positive spaces for the form $-A$.

So far we assumed $|A| \neq 0$. If on the contrary $|A| = 0$, then some eigenfunctions u which satisfy the equation $Au = 0$ exist. Nothing essentially is changed: Only that N and N^* have in common an additional, say g-dimensional, null space of these eigenfunctions.

As a final remark it may be stated that for the adjoint backward problem for $L^*[u] = 0$ with $u(x, \tau)$ given in G and sought for $t < \tau$, the proper boundary condition on Z is: u must belong to a maximal nonpositive space N^*.

4. Energy Integrals for Single Second Order Equations. In this article we derive an energy inequality, and thereby uniqueness theorems, for a single second order hyperbolic equation. In §3, article 8 we have reduced second order hyperbolic equations to first order symmetric hyperbolic systems. Hence such energy inequalities follow from the results of the previous articles of this section. Nevertheless an independent derivation in the case of second order (which was historically first[1]) is worthwhile, because it provides a general simple basis for the examples of §6 and because it serves as model for the higher order case, which will be outlined in the next section.

As in §3, we consider a second order equation of the form

$$(16) \qquad L[u] = u_{tt} - \sum a^{ij} u_{ij} + \cdots = 0.$$

We assume that this equation is hyperbolic and that the hyperplanes $t = $ const. are space-like. According to §3 this is the case if and only if the quadratic form a^{ij} is positive definite. We express $u_t L[u]$ as a divergence[2]:

$$2u_t L[u] = (u_t^2)_t - 2\sum (a^{ij} u_i u_t)_j + \sum (a^{ij} u_i u_j)_t + Q,$$

Q denoting terms, at most quadratic, which do not contain second order derivatives. If we assume for convenience that the operator L contains no term of order zero, then Q is a quadratic form in the first order partial derivatives of u.

This identity is now integrated over a lens D bounded by two surfaces S_0 and S_1 ; if u is a solution of $L[u] = 0$, we find, by Gauss' formula

$$(17) \qquad \int_{S_1} q \, dS - \int_{S_0} q \, dS = \iint_D Q \, dx \, dt,$$

where

$$q = \tau u_t^2 - 2\sum a^{ij} u_i u_t \xi_j + \sum a^{ij} u_i u_j \tau.$$

Here τ and ξ_j denote the t- and x_j-components of the normals to S_0 and S_1 drawn in the positive t-direction.

Now take S_0 to be the domain $R(0)$ cut out of the initial hyperplane $t = 0$ by the backward ray conoid Γ through a point P, and let S_1 consist of $R(h)$, cut out of the hyperplane $t = h$ by the ray conoid, plus $M(h)$, the mantle of the conoid contained between the hyperplanes $t = 0$ and $t = h$. The integral relation (17) then can be rewritten as follows:

$$\int_{R(h)} q \, dS + \int_{M(h)} q \, dS - \int_{R(0)} q \, dS = \iint Q \, dx \, dt.$$

---

[1] See footnote §6, p. 642.
[2] Q in this article, and in §9, does not of course mean the characteristic form.

As stated in §3, the quantity q regarded as a quadratic form in u_t and u_x is non-negative on $R(h)$, $R(0)$ and $M(h)$. Then, denoting the integrals over $R(h)$ by $E(h)$, we obtain the energy inequality

$$E(h) \leq E(0) + \int\int Q \, dx \, dt,$$

where

$$E(h) = \int_{R(h)} q \, dS = \int \{u_t^2 + \sum a_{ij} u_i u_j\} \, dx.$$

The quadratic form Q is dominated by a sufficiently large multiple of q:

$$Q \leq Cq.$$

Substituting this into the energy inequality above we have

$$E(h) \leq E(0) + C \int_0^h E(t) \, dt.$$

An easy consequence of this integral inequality is

$$E(h) \leq E(0)e^{ch}$$

which is the desired final form of the energy inequality.

Similar inequalities for higher order energy integrals follow as above after differentiating (16).

§9. Energy Estimates for Equations of Higher Order

1. Introduction. The results of §8 concerning symmetric first order systems are sufficiently general for most wave propagation problems arising in physics. Still it remains desirable to develop a theory for other hyperbolic problems. For such problems a satisfactory theory has been developed[1] under the severe condition that all the characteristics remain distinct, so that the sheets of the normal cone can be separated by sheets of another cone.

We indicate briefly Leray's method[2] for a single equation of m-th order. The method applies without any modification to weakly coupled systems of hyperbolic equations having the same principal part. Since (see Ch. I, §2, 2) an arbitrary hyperbolic system with distinct characteristics can be reduced to a weakly coupled system,

---

[1] Such a general theory was originated by I. G. Petrovskii [5] and [4].

[2] J. Leray [2]. Leray's beautiful argument is here presented in a version due to L. Gårding [3].

Leray's method effectively covers such hyperbolic systems of equations.

The key to the theorems of uniqueness in §8 and existence (§10) are energy inequalities. We therefore confine ourselves to indicating how energy estimates can be obtained also without the assumption of symmetry.

2. Energy Identities and Inequalities for Solutions of Higher Order Hyperbolic Operators. Method of Leray and Gårding. The method used for the second order case in §8 was to multiply the equation $L[u] = 0$ by $\partial u/\partial t$, integrate this quadratic expression over a lens-shaped domain, and reduce it by integration by parts to a sum of a domain- and boundary-integrals of quadratic forms in the first derivatives of u. The final step was to show that the boundary integrals are positive definite. To generalize this procedure to higher order equations we must 1) choose an appropriate multiplier, 2) perform the reduction by integrating by parts, and 3) recognize that the boundary integrals are positive definite.

Let L be a hyperbolic operator of order m. The multiplier will be of the form Nu, where N is an operator of order $m - 1$. Before specifying the choice of N, we express the product $NuLu$ as a gradient plus a quadratic form in derivatives of order less than m. This formal identity is independent of the hyperbolic character of L. For its derivation we may confine ourselves to the terms of order $m - 1$ and m in N and L respectively since the products involving lower order terms can be absorbed in the form Q introduced below.

Let G be a domain and B its boundary with the components γ_m of the normal unit vector. Then there exists a pair of forms $q(u)$ and $Q(u)$ quadratic in the $(m - 1)$-st and lower order partial derivatives of u such that

$$(1) \qquad \iint_G NuLu \; dx = \int_B q(u) \; dS + \iint_G Q(u) \; dx,$$

where dS is the surface element on B, and dx the volume element of G. It is sufficient to prove (1) when N and L are monomials, $N = aD_1 \cdots D_{m-1}$, $L = bD_m \cdots D_{2m-1}$, where the symbols D_j, $j = 1, \cdots, 2m - 1$, denote partial differentiations with respect to the variables (not necessarily different variables for different j), and a and b are functions of x. Integrating by parts with respect to D_m, we obtain

$$\iint_G ab\{D_1 \cdots D_{m-1} u\}\{D_m \cdots D_{2m-1} u\}\, dx$$

$$= -\iint_G ab\{D_1 \cdots D_m u\}\{D_{m+1} \cdots D_{2m-1} u\}\, dx + O\dot{O},$$

where $O O$ denotes terms to be absorbed in Q or q since they are integrals of quadratic form as stipulated, specifically here

$$-\iint_G D_m\{ab\}\{D_1 \cdots D_{m-1} u\}\{D_{m+1} \cdots D_{2m-1} u\}\, dx$$

$$+ \int_B ab\{D_1 \cdots D_{m-1} u\}\{D_{m+1} \cdots D_{2m-1} u\}\gamma_m\, dS.$$

The first term can again be transformed by integrating successively by parts with respect to D_1, $\cdots$ and thus successively exchanging the differentiations D_i between the first bracket $\{\cdots\}$ and the second. After an odd number $2m - 1$ of these operations N and L are interchanged and an identity of the following type results:

$$\iint_G N[u]L[u]\, dV = -\iint_G L[u]N[u]\, dV + 2\int_B q(u)\, dS + 2\iint_G Q(u)\, dV,$$

where q and Q are quadratic forms in the $(m - 1)$-st order partial derivatives of u. This yields the desired identity (1). We remark that $Q \equiv 0$ if the coefficients of both L and N are constant and if L and N do not contain lower order terms.

The quadratic form $q(u)$, associated with the operators L and N and the surface B, is not uniquely determined when the order of L is greater than two and when there are more than two independent variables. But any two such forms differ by a surface divergence, i.e., the value of $\int_B q(u)\, dS$ is independent of the specific choice of q.

If L is hyperbolic, and G is a lens-shaped domain bounded by two space-like hypersurfaces S_1 and S_2 with a common boundary, we try to choose N so that the boundary integral $\int q\, dS$ in (1) is positive definite over S_1 and negative definite over S_2. This is accomplished if, following Leray, we choose N so that the sheets of *its characteristic normal cone separate* those of L, in the following sense: Let $L(\xi)$ and $N(\xi)$ denote the characteristic forms associated with

L and N. We call the direction ζ *space-like* for the operator L if for every direction θ not parallel to ζ the characteristic form $L(\lambda\zeta + \theta)$ vanishes for m distinct real values of λ. The characteristics of N are said to separate those of L if $N(\lambda\zeta + \theta)$ vanishes for $m - 1$ distinct real values of λ which separate[1] the roots of L.

For a given hyperbolic operator L, one can construct an operator N of one order lower whose characteristics separate those of L. We can, e.g., choose

$$N(\xi) = \frac{d}{d\lambda} L(\xi + \lambda\zeta)|_{\lambda=0} .$$

The separation property is an immediate consequence of the classical theorem that the roots of a polynomial are separated by the roots of its derivative.

We shall now formulate more precisely the assertion about the positive definiteness of the boundary integrals in (1). We assume the simplest situation: Let S_1 and S_2 be two hyperplanes $t = 0$ and $t = T$, and let the coefficients of L and N be constants. Also, we assume that the solution u under consideration vanishes for sufficiently large values of the space variables (i.e., the variables in the hyperplanes S). Then we assert:

If the operator L is hyperbolic and its characteristics are separated by those of N, and if the hyperplanes $t = $ const. are space-like, then the quadratic functional $\int q(u)\, dx$ associated with N, L and the hyperplanes $t = $ const. is *positive definite*, that is, for all smooth functions which vanish for large $|x|$ we have

$$(2) \qquad \int q(u)\, dx \geq \text{const.} \int \sum |D^\alpha u|^2\, dx,$$

where the sum on the right side is extended over all partial derivatives of u of order not exceeding $m - 1$.

In the corresponding situation for operators of order 2 we verified the definiteness of the associated quadratic functional $\int q\, dx$ by showing that the integrand $q(u)$ is a positive definite quadratic form of the first order partial derivatives of u. In the general case this is not always true, and hence it is impossible to carry out such a veri-

---

[1] It is not difficult to show that if the operator N has this separation property with respect to one space-like direction ζ, then it separates with respect to all space-like directions ζ.

fication.   Instead, we apply a criterion for definiteness due to Gård-ing.[1]

Denote by $U_\nu(\xi)$ the Fourier transform of $\partial^\nu u/\partial t^\nu$ with respect to the space variables.   Let $D^\alpha u$ be any partial derivative of u of order $m - 1$; write it as $D_1 \cdots D_{m-1-\nu} \partial^\nu u/\partial t^\nu$, where $D_1, \cdots, D_{m-1-\nu}$ again denote partial differentiation with respect to the space variables. (According to the well-known rule, the Fourier transform of $D^\alpha u$ is $i^{m-1-\nu}\xi_1 \cdots \xi_{m-1-\nu}U_\nu$).   Using Parseval's formula we can express the space integral $\int q\, dx$ as an integral over the ξ-space of an associated Hermitian form h:

$$\int q(u)\, dx = \int \sum h_{\nu\mu}(\xi)U_\nu \bar U_\mu\, d\xi,$$

where $h_{\nu\mu}$ is a polynomial in ξ of degree $2m - 2 - \nu - \mu$.

Similarly we have

$$\int \sum |D^\alpha u|^2\, dx = \int \sum |\xi|^{2m-2-2\nu} |U_\nu|^2\, d\xi,$$

where $|\xi|$ stands for $(\xi_1^2 + \cdots + \xi_m^2)^{1/2}$.   The desired inequality (2) is thus equivalent to

$$(3) \qquad \int \sum h_{\nu\mu} U_\nu \bar U_\mu\, d\xi \geq \text{const.} \int \sum |\xi|^{2m-2-2\nu} |U_\nu|^2\, d\xi.$$

Since the functions U_ν, $\nu = 0, \cdots, m - 1$, are independent of each other, it is easily seen that such an inequality is satisfied if and only if *the Hermitian form h is positive definite for all ξ.*

Thus, in order to prove (2) all that remains is to verify that under the stated conditions the Hermitian form h associated with q is positive definite.   For this we refer the reader to Gårding, loc. cit.

We summarize our results for equations with constant coefficients as a law of conservation of energy:

The positive definite quantity $\int q(u)\, dx$, which may be considered as an "energy integral", is independent of t for all solutions of the equation $L[u] = 0$.

This result can be extended to solutions of equations with variable coefficients without essential difficulties.   An analogous "energy integral" can be constructed which, although no longer independent of t, has a bounded rate of growth; i.e., its value at time $t = T$ is at

---

[1] See L. Gårding [3].

most e^{CT} times its value at $t = 0$. The size of the constant C depends on the coefficients of L and on the size of their first derivatives.[1]

3. Other Methods. Attention should be called to a rather different recent approach due to Calderón and Zygmund. Their investigations[2] on singular integral equations have provided a flexible tool for the problems discussed in the present section and for the study of other problems of linear partial differential equations. The key is the fact that by applying suitable integral operators one can symmetrize partial differential operators and thus provide other methods than that by Leray for the derivation of energy integrals.[3]

Finally we sketch another proof for energy inequalities due to G. Peyser[4] and based on the normal form of a linear hyperbolic operator in two variables x and t as discussed in Ch. V, §8, 2, 3.

We first consider a hyperbolic operator $L[u]$ of order m for a function $u(x, t)$ of two variables x, t in the formal form

$$(4) \qquad L[u] = D_1 D_2 \cdots D_m u + R,$$

where the differential operator R contains merely derivatives of order less than $m - 1$ and where all the m operators of characteristic differentiation $D_i = \partial/\partial t + \tau^i \partial/\partial x$ are distinct. (The notations differ slightly from those in Ch. V, §8.) We introduce the operators $V_j = (D_1 \cdots D_m)_j$ of order $m - 1$ where in the product on the right-hand side the factor D_j is omitted. By the results of Ch. V, §8 all the derivatives of order $m - 1$ and lower can be linearly expressed in terms of the quantities $V_j u$. We consider a solution u of $L[u] = 0$ in a slab $\sum : 0 \leq t \leq T$ and assume that u has in $\sum$ "compact support", that is, vanishes for values of $| x |$ above a fixed bound. Now we introduce the operator of order $m - 1$

$$(5) \qquad N[u] = \sum_{j=1}^{m} V_j[u]$$

which can be obtained from L by formally differentiating the principal part with respect to the symbol $\partial/\partial t$. Immediately we find the key identity

$$(6) \qquad N[u]\, L[u] = \frac{1}{2} \sum_j D_j (V_j u)^2 + Q[u],$$

---

[1] See Gårding [3]. We shall return to Gårding's inequality in Volume III.
[2] See A. P. Calderón and A. Zygmund [2], [1].
[3] See also more recent results by S. Mizohata [1] and [2].
[4] See G. Peyser [1].

where $Q[u]$ is a quadratic form in the derivatives of orders less than $m - 1$. Integrating over the slab, remembering that u is assumed to have compact support and considering t as the curve parameter on each of the k characteristics we have with the positive definite quadratic form $q(u) = (1/2)\sum (V_j u)^2$ the integral relation

$$\iint_\Sigma N[u]L[u]dx\, dt$$

$$= \int (q(u,\, T) - q(u,\, 0))\, dx + \iint Q(u(x,\, t))dx\, dt = E(T) - E(0).$$

Introducing the "energy integral"

$$(7) \qquad\qquad E(t) = \int q(x,t)\, dt$$

and estimating the lower order derivatives as indicated above one obtains almost immediately $\iint Q(x,t)dx\, dt \le \text{const.} \int_0^T E(t)\, dt$. Hence for our solutions u of $L[u] = 0$ an *energy inequality*

$$(8) \qquad\qquad E(T) \le E(0) + \text{const.} \int_0^T E(t)\, dt$$

is established with the positive quadratic functional $E(t)$ in the $(m - 1)$-st derivatives of u.[1]

The next step is the extension of this result to n variables $x_1,\, \cdots,\, x_n$. For this purpose we assume that L has constant coefficients. Then, briefly summarized, the procedure is first to decompose a solution u by Fourier's integral theorem into plane waves $v(t,\, y;\, \alpha)$ depending on a unit vector α in the x-space: $u(x,\, t) = \int_{|\alpha|^2 = 1} v(t,\, y;\, \alpha)$, $y = (\alpha,\, x)$, where the plane wave v satisfies as a function of x and t the equation $L[v] = 0$. The compactness of the support of u assures the possibility of this decomposition and that of all derivatives occurring in L. Then the previous energy relations applied to the function v of two variables y and t and afterwards integrated with respect to α yield directly the desired energy relation for the function $u(t,\, x)$ by Parseval's integral relations.

---

[1] It should be noted that, as indicated by the formal derivation of (1), a formula $\iint N[u]L[u]dx\, dt = \int (q(u(x,T)) - q(u(x,0)))\, dx + \iint Q(u(x,t))\, dx\, dt$ arises through repeated intergration of $N[u]L[u]$ by parts with respect to the original variables x and t; the function $q(u)$ thus obtained can differ from the present one only by a divergence expression so that the value of the functional $E(t)$ remains the same.

§10. *The Existence Theorem*

1. Introduction. In this section we shall use the energy inequalities derived in §§8, 9 to prove the existence of solutions of symmetric hyperbolic systems of equations

$$(1) \qquad L[u] = u_t + M[u] = 0$$

with arbitrarily prescribed smooth initial data.

We recall that the solution of Cauchy's problem for the inhomogeneous equation

$$(1') \qquad L[u] = f(x, t),$$

e.g., with the initial condition $u(x, 0) = 0$, is an immediate consequence of *Duhamel's principle.* If $U(x, t; \tau)$ is the solution of (1) for $t > \tau$ with $U(x, \tau; \tau) = f(x, \tau)$ then u is given by

$$u(x, t) = \int_0^t U(x, t; \tau) \, d\tau.$$

Thus we may restrict ourselves to the existence proof for the homogeneous equation (1)

For an adequately general formulation of the result it will be convenient to use the terminology of Hilbert space in the following manner: We consider in the x-space a domain R possibly depending on a parameter t and in R "smooth functions" u with continuous derivatives of order up to r. The domains R (or R_t) need not be chosen as plane sections of conoids of dependence Γ_P for a point P. They could, e.g., refer to domains of dependence for an arbitrary plane domain $t = $ const. in the strip $\sum$ (see §7). The functions u are not restricted to being solutions of the differential equation. Now we define the Hilbert space $H_r(t)$ or $H_r^*(t)$ as the completion[1] of our space of smooth functions u with the r-norm $\| u(t) \|_r$ or $\| u(t) \|_r^*$ as defined in §8, 2. Then the energy inequalities of §8 state: If the initial values are in a space $H_r(0)$ or $H_r^*(0)$ then the values u for suitably small positive t of a solution of (1) are in a corresponding Hilbert

---

[1] This means: If $u_1, u_2, \cdots, u_j, \cdots$ is a sequence of functions with continuous derivatives up to order r in R, with uniformly bounded norm $\| u_j \|_r$, and if, furthermore, the functions u_j converge in the r-norm, i.e., if $\| u_i - u_j \|_r \to 0$ for $i \to \infty, j \to \infty$, then we attribute to the sequence $\{u_j\}$ an ideal element u as a limit. In much the same way one introduces real numbers by completion of the system of rational numbers. The norm of the limit function is defined by

$$\| u \|_r = \lim_{j \to \infty} \| u_j \|_r .$$

space $H_r(t)$ or $H_r^*(t)$ and their norms are uniformly bounded as long as t is bounded. Then: The validity of the energy inequalities with suitable constants for arbitrary domains R could either be proved directly or by piecing together some domains R_t of the type discussed above.

In specific cases, the elements of the abstract spaces H_r have to be related to ordinary functions. Such a relation is formulated in Sobolev's lemma (see Ch. III, App. 1), which implies that for r large enough the elements of H_r are smooth functions. Specifically: If R is a smoothly bounded domain and if $r \geq n/2 + 1$, then

$$\underset{x \,\text{in}\, R}{\text{Max}} |u(x)| \leq \text{const.} \| u \|_r,$$

where the value of the constant depends only on R and r. Furthermore: For $r \geq n/2 + s + 1$, all elements of H_r are functions with bounded partial derivatives up to order s.

In addition to the energy inequality a constructive element is needed, which is supplied here[1] by the Cauchy-Kowalewski theorem, following an idea of Schauder.[2]

We shall give the proof in detail for *symmetric* hyperbolic systems of first order and indicate briefly the argument for nonsymmetric systems and equations of higher order. Furthermore it should be pointed out that the remarks in §4, 7, and later in §15, lead to a different construction of the solution.

2. The Existence Theorem. *The initial value problem for $L[u] = 0$, with $u(x, 0) = \psi(x)$, has a smooth solution u in $R(t)$, provided that the operator L is hyperbolic and its coefficients, as well as the initial function ψ, are sufficiently differentiable. If u belongs to H_r over $R(0)$, then u belongs to H_r in every section $R(t)$ of the conoid of dependence.*

It should be added: If ψ is merely continuous or not sufficiently differentiable to insure smoothness of u, then nevertheless a *generalized solution* u is defined by closure as an element $u = \lim u_n$ in a Hilbert space with the norm H_r.

The proof is divided into three steps: First we construct a solution in case the coefficients of L, as well as the initial data ψ, are analytic;

---

[1] For other constructions see K. O. Friedrichs [2] using finite difference equations, and P. D. Lax [6] using orthogonal projection in Hilbert space.

As to existence theorems for mixed initial and boundary value problems for second order equations see M. Krzyżański and J. Schauder [1] and O. Ladyzenskaya [1]. A general class of mixed boundary value problems for equations which are not necessarily hyperbolic is discussed by K. O. Friendrichs [3] see also P. D. Lax and R. S. Phillips [1].

[2] See J. Schauder [4].

next we pass to the case when the initial data belong to the Hilbert space H_r ; and finally we remove the restriction of analyticity of the coefficients of L.

1) Let ψ be any function in H_r . We approximate ψ in the r-norm by a sequence of analytic functions, say polynomials, ψ_l , ($l = 1$, 2, $\cdots$). The initial value problem for analytic $L[u]$ with analytic (polynomial) initial values ψ_l was solved in Ch. I, §7 by the Cauchy-Kowalewski method of power series. This analytic solution u_l , which by the uniqueness theorem is the only one, was constructed in a suitably narrow strip $\Sigma: 0 \leq t \leq T$. The width T of this strip is *independent* of the particular polynomials[1] ψ_l , although it does depend on the analytic structure of the coefficients of the equation.

2) Since the sequence ψ_l of initial values converges to ψ in the r-norm over $R(0)$, the energy estimates, applied to $u_l - u_m$, imply: All partial derivatives of order not greater than r of the corresponding solutions u_l converge in the r-norm over any section $R(t)$ in the strip Σ_T for $t \leq T$. Obviously the energy estimates remain valid and the convergence is true if R is extended to a finite part of the horizontal plane $t = $ const. in Σ.

It follows then from Sobolev's lemma that the sequence of functions u_n and all their derivatives of order less than $r - n/2$ converge uniformly in the strip Σ_T . If r is greater than $n/2 + 1$, the limit function u has continuous first derivatives and satisfies the differential equation (1). Furthermore, since H_r is complete, the solution $u(x, t)$ thus obtained belongs to H_r over $R(t)$ for $t \leq T$.

We can now repeat this process; given in the strip $0 \leq t \leq T$ a solution u whose values over $R(T)$ belong to H_r , we construct a solution u in the section[2] $T \leq t \leq 2T$ with prescribed values on $R(T)$, etc. Thus we can construct in the strip Σ_T a solution with prescribed initial values. Obviously, by choosing instead of P a point P' whose conoid of dependence $\Gamma_{P'}$ includes Γ_P we can extend u into a wider domain beyond Γ_P .

3) For coefficients A^i, B of (1) which are not analytic but merely sufficiently smooth—say they possess continuous derivatives up to the order $r + 1$—the proof can easily be generalized by using the energy estimates again. A brief indication suffices: We assume $r > n/2 + 1$, and we approximate in the strip Σ the coefficients

---

[1] See Ch. I, §7.

[2] The uniformity of the width of the strip in which analytic solutions exist follows from the *Cauchy-Kowalewski* theorem.

of L and their derivatives uniformly by analytic functions A_n^i, B_n, $(n = 1, 2, \cdots)$, and their derivatives. Thus we replace the operator L by an analytic approximating operator L_n. For this operator the initial value problem is solved in Σ by a function u_n with $L_n[u_n] = 0$, $u_n(x, 0) = \psi(x)$.

Assume at first that ψ belongs to H_{r+1}. Then according to the energy inequality, $u_n(t)$ also belongs to H_{r+1}, and $\| u_n(t) \|_{r+1} \leq$ const. $\| \psi \|_{r+1}$, where the value of the constant does not depend on n. The difference $u_n - u_m$ obviously satisfies the inhomogeneous equation

$$L_n(u_n - u_m) = (L_m - L_n)u_m = f_{nm}.$$

Since $u_n - u_m$ vanishes for $t = 0$, we have the energy inequality

$$\| u_n - u_m \|_r \leq \text{const.} \| f_{nm} \|_r.$$

Since the coefficients of L_n and all their partial derivatives up to order r approximate uniformly those of L, it follows that $\| f_{nm} \|_r \leq \epsilon_{nm} \| u_m \|_{r+1}$, where ϵ_{nm} tends to zero with increasing n and m. We have already shown that $\| u_m \|_{r+1}$ is uniformly bounded. Hence $\| f_{nm} \|_r$ tends to zero, and therefore, by the energy inequality, so does $\| u_n - u_m \|_r$.

According to Sobolev's lemma, the sequence of the functions u_n and their first partial derivatives also converges uniformly. The limit $u(t)$ of u_n belongs to H_r for every value of t. Furthermore, since the functions u_n satisfy the energy inequality

$$\| u_n(t) \|_r \leq \text{const.} \| \psi \|_r$$

with a constant independent of n, the same inequality holds for the limit function u.

So far we have assumed that ψ belongs to H_{r+1}. To construct a solution for any ψ in H_r, we approximate ψ by a sequence ψ_n in H_{r+1}. According to the preceding result these initial value problems have unique solutions u_n which belong to H_r, and the energy inequality is valid for $u_n - u_m$:

$$\| u_n - u_m \|_r \leq \text{const.} \| \psi_n - \psi_m \|_r.$$

This shows that the sequence u_n converges in the r-norm to a limit u, which is the desired solution.

3. Remarks on Persistence of Properties of Initial Values and on Corresponding Semigroups. Huyghens' Minor Principle. Imagine the initial values $\psi(x) = u(x, 0)$ given over the whole hyperplane $t = 0$.

According to the existence theorem of article 2, the solution of (1) is uniquely determined at all later times $t > 0$. Furthermore, if $\psi(x)$ belongs to H_r over the whole x-space, then the solution $u(x, t)$ is in $H_r(t)$ at all later times. This result can be expressed as follows:

The relation of solutions of $L[u] = 0$ at time $t > 0$ to their values at $t = 0$ is a mapping of the function space H_r into itself. Specifically the initial value problem is solved by a linear operator T which depends on t, and expresses u in the form

$$u(x, t) = T(t, t_0)\psi$$

if u has the initial values ψ for $t = t_0$.

Obviously, instead of proceeding in one step, e.g., from the initial values for $t = t_0 = 0$ to the values of u for t, one could first interpose an intermediate value t_1, find $u(x, t_1)$ and then solve the initial value problem with $u(x, t_1)$ instead of $u(x, t_0) = \psi(x)$. The solution then would be given by a linear operator

$$u(x, t) = T(t, t_1)u(x, t_1)$$

and thus by the composition of the two operators

$$u(x, t) = T(t, t_1)T(t_1, t_0)\psi.$$

Hence by the uniqueness theorem, the operators T must satisfy the composition rule

$$(2) \qquad T(t, t_0) = T(t, t_1)T(t_1, t_0) \qquad \text{for } t > t_1 > t_0.$$

If we assume that the coefficients of L do not depend on t, then the operator $T(t, t_0)$ depends only on the difference $t - t_0$, and (2) has the form

$$(2a) \qquad T'(t - t_0) = T(t - t_1)T(t_1 - t_0) \quad \text{for } t > t_1 > t_0.$$

This then is simply a group relation. Since it is stipulated here only for $t \geq t_1 \geq t_0$ one says: The operators T form a *semigroup*.[1]

Hadamard has called attention to the relation (2), or (2a), and named the underlying fact "Huyghens' minor principle". Whenever the operator T can be explicitly represented as an integral operator, the relation (2) leads to interesting identities for the kernels of these operators. This, incidentally, is true also of boundary value

---

[1] By considering the "backward" Cauchy problem, we see: The operator T has a unique inverse $T(t_0, t_1) = T(t_1, t_0)^{-1}$. Therefore the semigroup in fact is extended to a group proper.

problems for elliptic and parabolic equations solved in terms of Green's function.[1]

One calls a property P of initial functions *persistent* if: Whenever an initial function ψ has the property P, the value of the corresponding solution $u(x, t)$ at any other time also has the property P. Our previous results can be restated in this language: *The property of having finite r-norm is persistent.*[2]

As we shall see by examples in the next article, conditions of existence and continuity of derivatives of $\psi(x)$ are not persistent; some of these differentiability properties may be lost under the operation T. Since Huyghens' minor principle is obviously a physically reasonable postulate, it is an essential insight that the physically relevant persistency conditions are the existence of energy norms, rather than differentiability properties.[3]

4. Focussing. Example of Nonpersistence of Differentiability. Usually in mathematical physics the differential equations $L[u] = 0$ have completely regular, e.g., constant, coefficients, and the initial data are likewise regular, e.g., arbitrarily often differentiable, except possibly at an initial manifold $\phi(x_1, \cdots, x_{n-2}) = 0$, where u or derivatives of u suffer discontinuities as described in §4, 3. These discontinuities propagate along characteristics, according to §4. Yet the solution was constructed there only "in the small", i.e. in a

---

[1] In these cases the boundary value problems have solutions in a half-space only, say in $t \geq 0$, and the corresponding solution operators form only a semigroup.

[2] For nonsymmetric equations, this property need not be persistent. For example, consider the system

$$u_t - v_x = 0, \qquad v_t = 0,$$

with Cauchy data

$$u(0, x) = 0, \qquad v(0, x) = \begin{cases} \sqrt{x} & \text{for } x > 0 \\ 0 & \text{for } x \leq 0. \end{cases}$$

Then, for $x > 0$,

$$u(t, x) = \frac{t}{2\sqrt{x}}, \qquad v(t, x) = \sqrt{x}.$$

Obviously, taken over any finite interval $\int (u^2 + v^2) \, dx \,|_{t=0} < \infty$, but for $t > 0$, $\int (u^2 + v^2) \, dx$ does not exist. Another example is provided by the initial values

$$u(0, x) = 0, \qquad v(0, x) = \sqrt{|x|} \, e^{-x}.$$

Then

$$u(t, x) = t \left(\frac{1}{2} \frac{x}{|x| \sqrt{|x|}} - \sqrt{|x|} \right) e^{-x}, \qquad v(t, x) = \sqrt{|x|} \, e^{-x}.$$

[3] This fact was first recognized by K. O. Friedrichs and H. Lewy [1].

strip $0 \leq x_n \leq \lambda$ with sufficiently small λ, which can be extended as far as the rays issuing from S_0 generate smooth characteristic surfaces C. If, however, on C these rays have an envelope, a "caustic", which may be an edge of regression of C, the initial singularities may be enhanced on this caustic. While this process, called *focussing*, does not prevent the solution from being extended beyond the caustic,[1] it reduces locally the differentiability or continuity properties of the solutions.

The following example may be interpreted in this sense even though the caustic consists merely of one point. We consider the initial value problem for the wave equation, say in three dimensions,

$$u_{tt} - u_{xx} - u_{yy} - u_{zz} = 0,$$

with initial values $u(0, x, y, z) = 0$,

$$u_t(0, x, y, z) = \begin{cases} (1 - r^2)^{3/2} & \text{for } r^2 \leq 1 \\ 0 & \text{for } r^2 \geq 1. \end{cases}$$

Here $x^2 + y^2 + z^2 = r^2$. These initial values imply continuity of the initial derivatives up to the second order. The explicit solution of the initial value problem is given along the t-axis by[2]

$$u(t, 0, 0, 0,) = \begin{cases} t(1 - t^2)^{3/2} & \text{for } t \leq 1 \\ 0 & \text{for } t > 1. \end{cases}$$

We have $u_t = 0$, $u_{tt} = 0$ for $t > 1$, while for $t < 1$

$$u_t = (1 - t^2)^{3/2} - 3t^2(1 - t^2)^{1/2},$$

$$u_{tt} = -3(1 - t^2)^{1/2} - 6t(1 - t^2)^{1/2} + 3t^2(1 - t^2)^{-1/2}.$$

Obviously, u_t is continuous for $t = 1$ but u_{tt} is not.[3]

[1] See for details and proofs D. Ludwig [1].

[2] The formulas of Ch. III, §3, show that spherically symmetric solutions of the wave equation have the form $u(r, t) = (f(t + r) - f(t - r))/r$. We can see directly from this formula that $u(0, t)$ has one derivative less than f.

[3] In the text we considered a single differential equation of higher order. If we set $w = u_t + u_x$, $v^1 = -u_y$, $v^2 = -u_z$, then w, v^1 and v^2 satisfy $w_t - w_x + v_y^1 + v_z^2 = 0$, $v_t^1 + v_x^1 + w_y = 0$, $v_t^2 + v_x^2 + w_z = 0$ with the initial conditions

$$w(0, x, y) = \begin{cases} (1 - r^2)^{3/2} & (r^2 \leq 1) \\ 0 & (r^2 > 1) \end{cases}$$

$$v^1(0, x, y) = 0$$

$$v^2(0, x, y) = 0.$$

Here w is continuous for $t = 1$, but w_t is not. Thus continuity of first derivatives of the initial data does not guarantee continuity of first derivatives of the solution.

5. **Remarks about Quasi-Linear Systems.**[1] It should be pointed out briefly that the theory can be extended to nonlinear systems by an iteration scheme. If the system (1) is quasi-linear, i.e., if the coefficient matrices depend on the unknown functions u, then the method of iteration of Ch. V, §7 can be extended to the case of n variables. Thus we obtain the proof of uniqueness as well as existence of a solution u as the limit $u = \lim_{n \to \infty} u^n$, where u^{n+1} is the solution of the linear initial value problem

$$u_t^{n+1} + \sum_{\nu=1}^{m} A^{\nu}(x, u^n)u_{x_\nu}^{n+1} + B(x, u^n)u^{n+1} = 0.$$

Here the initial values for all functions u^n are the given values $\psi(x)$.

6. **Remarks about Problems of Higher Order or Nonsymmetric Systems.** As emphasized before, most hyperbolic problems of mathematical physics refer to symmetric systems for which the preceding theory of Cauchy's problem is satisfactory. Yet the problem for single differential equations of higher order or nonsymmetric systems of first order is of great mathematical interest. It is a remarkable fact that the preceding energy estimates allow the generalization of the proofs to such problems.

The energy estimates of §8, 4, immediately lead to the solution of Cauchy's problem for any single hyperbolic equation of second order. (Alternately this problem is also solved by reducing it to a symmetric hyperbolic system; this was done in §3, 8.)

To solve Cauchy's problem for a single hyperbolic equation of order m or for a system of such equations with the same principal part, one can use the general energy estimates of §9. The existence proof proceeds thereafter along the lines of the first articles of the present section.

The estimates of §9 and the conclusions drawn from them apply likewise to any hyperbolic system of first order with distinct characteristics, because any system of this type can, by direct elimination (see Ch. I §2, 2), be reduced to a system of k equations of higher order m, where each principal part of the operator, i.e., the highest order terms in each of the k equations, is the same and refers to a single one of the unknown functions $u^1, u^2, \cdots, u^k$. The coupling of the equations takes place only through lower order terms. As in

---

[1] For second order quasi-linear systems this procedure has been carried out by J. Schauder [4].

Further contributions are due to F. Frankl [1]. Cauchy's problem for nonlinear hyperbolic equations of arbitrary order has been treated by J. Leray [2].

Chapter V, such systems are amenable to almost the identical treatment as a single equation of order k.

Again it should be pointed out that the method as developed assumes hyperbolicity in the strict sense, i.e., the sheets of the characteristic normal cone are separated (and hence so are the sheets of the ray cone). The case of general equations with multiple characteristics is not covered by the preceding theories and presents an open challenge.[1]

PART II

Representation of Solutions

§11. *Introduction*

1. Outline. Notations. For linear hyperbolic differential equations, in particular for those with constant coefficients, the solution of Cauchy's problem can be represented by more or less explicit formulas.[2] (See also Ch. III, §5 and Chapter V.) Such representations as linear functionals of the data not only lead to many attractive formal relations,[3] but, what is perhaps more important, they allow a study of specific properties. They are based on the decomposition of solutions, and, for that matter, other arbitrary functions, into *plane waves* (see also Ch. III, §3) and sometimes into *spherical waves*. A plane wave was defined as a solution of the differential equation which depends only on the time t and one plane space coordinate. A spherical wave is a solution with spherical symmetry around a point in space.

Oscillatory plane waves used in the Fourier decomposition of a function are not necessarily the simplest tools. A somewhat more direct approach is sometimes possible. But always the use of plane waves fails to exhibit clearly the domains of dependence and the role of characteristics. This shortcoming, however, is compensated by the elegance of the explicit results.

In §12 and §13 we shall consider single equations of second order; in §13a we shall insert as an example the integration of the equations for elastic waves. In §14 we shall discuss problems with constant coefficients of any order, and in §15 we shall develop a representation

---

[1] See further remarks in §4 and §15 and D. Ludwig [3].

[2] Among other literature see in particular a comprehensive monograph by F. John [4].

[3] This subject has been covered very extensively in the literature. We restrict ourselves, somewhat subjectively, to aspects of interest in the general framework of this book.

for arbitrary hyperbolic problems with not necessarily constant coefficients. We shall rely on the existence and uniqueness theorems proved in the first part of this chapter and we shall assume that the data possess the continuity properties stipulated there (§10). It may, however, be stated that the explicit expressions of Part II could be used not merely as a representation but also for a constructive existence proof by direct verification.

In §§16, 17 we shall discuss "ultrahyperbolic" differential equations, and discuss problems not "well posed" for hyperbolic equations. A final section will summarize some features of the transmission of signals as governed by hyperbolic problems.

As before, we single out the time variable $t = x_0$ and combine the space variables x_1, x_2, $\cdots$, x_n into a vector x with the absolute value

$$| x | = \sqrt{x_1^2 + x_2^2 + \cdots + x_n^2}.$$

The unit sphere in n dimensions will be denoted by ω_n or ω, its surface element by $d\omega_n$ or $d\omega$, its surface area[1] denoted by

$$\omega_n = 2\sqrt{\pi^n}/\Gamma(n/2).$$

The sphere of radius r will be called Ω_r or Ω, its surface element is $d\Omega = r^{n-1} d\omega$. Volume elements are also $dx = dx_1 dx_2 \cdots dx_n = | x |^{n-1} d\omega_n d | x | = | r |^{n-1} d\omega_n dr$. Unit vectors are denoted by α, β, and the surface element of the unit sphere is sometimes simply written $d\alpha$, $d\beta$ instead of $d\omega$. We shall on occasion use $ad\ hoc$ notations if convenient.

2. Some Integral Formulas. Decomposition of Functions into Plane Waves. It is useful for the subsequent sections to assemble a few formulas of integral calculus in n dimensions, mostly concerning integrals over spheres.

We first consider integrals of a function $f(x) = f(x_1, \cdots, x_n)$ over the interior of spheres $| x | \le r$ with r fixed. With the notation $x_1^2 + \cdots + x_{n-1}^2 = \rho^2 = r^2 - p^2$, $x_n = p$, we write

$$
\begin{aligned}
(1) \quad K(r) &= \iint_{|x| \le r} f(x)\, dx \\
&= \int_{-r}^{+r} dp \int_{\rho^2 \le r^2 - p^2} f(x_1, \cdots, x_{n-1}, p)\, dx_1 \cdots dx_{n-1}.
\end{aligned}
$$

The surface integrals over the sphere Ω_r, i.e., $| x | = r$, are expressed by

---

[1] Cf. Ch. IV, §1, 1.

(2) $$\iint_{|x|=r} f \, d\Omega = \iint_{|x|=r} f r^{n-1} \, d\omega = \frac{d}{dr} K(r).$$

If f does not depend on $x_n = p$ we have

(3) $$\iint_{|x|=r} f \, d\Omega = \iint_{\rho \leq r} r f(x_1, \cdots, x_{n-1}) \frac{dx_1 \cdots dx_{n-1}}{\sqrt{r^2 - \rho^2}}.$$

Of even greater importance will be the case where f depends only on a single linear variable x_1 or more generally on the inner product of x and a unit vector β written in the notation

$$(x, \beta) = (x\beta) = p.$$

Because of the rotational symmetry of our integrals, we may choose $p = x_n$. Then (1) and (2) yield for $r = 1$ the important formula —we write $d\beta$ instead of $d\omega_n$

(4) $$\iint_{\beta^2=1} f(x\beta) \, d\beta = \omega_{n-1} \int_{-1}^{+1} f(p \mid x \mid)(1 - p^2)^{(n-3)/2} \, dp.$$

Indeed, we consider the spherical intersection of $\mid x \mid \leq r$ with the plane $x_n = p$, introduce polar coordinates and assume $f = f(p)$, with $p = (x\beta)$, for example $p = x_n$. Then we have in this intersection $dx_1 \cdots dx_{n-1} = \rho^{n-2} \, d\rho \, d\omega_{n-1}$. The integral of f over the cross section is $[1/(n - 1)](r^2 - p^2)^{(n-1)/2} \omega_{n-1} f(p)$ and then (4) follows immediately.

Attention should be given to the fact that (4) exhibits a different behavior for odd and even n inasmuch as the factor $(1 - p^2)^{(n-3)/2}$ in the integrand is rational in p for odd n, and irrational for even n.

We note some particular cases: $f = 1$ yields

$$\omega_n = \omega_{n-1} \int_{-1}^{+1} (1 - p^2)^{(n-3)/2} \, dp = \frac{\Gamma\left(\dfrac{n-1}{2}\right)\sqrt{\pi}}{\Gamma\left(\dfrac{n}{2}\right)} \, \omega_{n-1}.$$

$f = \log (\mid \beta x \mid)$ yields

$$\iint \log \mid x\beta \mid d\beta = \omega_{n-1} \int_{-1}^{+1} \log (p \mid x \mid)(1 - p^2)^{(n-3)/2} \, dp$$

or, easily evaluated,

(4a) $$\iint \log \mid x\beta \mid d\beta = 2\pi^{n/2} \frac{\omega_n}{\Gamma\left(\dfrac{n}{2}\right)} \log \mid x \mid + c$$

with a constant c. For $f = |x\beta|$ we obtain

$$\iint |x\beta| \, d\beta = 2\omega_{n-1} \int_0^1 |x| \, p(1 - p^2)^{(n-3)/2} \, dp$$

from which

(4b) $$\iint_{\beta^2=1} |x\beta| \, d\beta = \frac{2\omega_{n-1}}{n-1} |x|$$

follows. These formulas exhibit $|x|$ and $\log |x|$ as superpositions of plane waves.

The formulas (4a) and (4b) in conjunction with a basic formula of potential theory lead to the decomposition of an arbitrary function $f(x_1, \cdots, x_n)$ into plane waves, i.e., functions which depend only on a linear combination (αx) of the space variables $x_1, \cdots, x_n$ with a unit vector α.

This decomposition is expressed as follows:

(5) $$4(2\pi)^{n-1}(-1)^{(n-1)/2} f(z) = \Delta_z^{(n+1)/2} \iint f(x) \, |\, ((x-z)\alpha)\, |\, d\alpha \, dx$$

for odd n and

(5') $$(2\pi)^n (-1)^{(n-2)/2} f(z) = \Delta_z^{n/2} \iint f(x) \log |\, ((x-z)\alpha)\, |\, d\alpha \, dx \, ,$$

for even n, where Δ_z^ν denotes the ν-times iterated Laplacian operator with respect to the variables z, and where the integration is extended over the unit sphere $\alpha^2 = 1$ and over the whole x-space. The function $f(x)$ may be assumed to vanish for large $|x|$, so that no difficulties of convergence arise from $|x| \to \infty$.

These formulas (5) indeed decompose $f(z)$ into plane waves, depending only on $(x - z, \alpha)$.

For the proof we recall Poisson's theorem of potential theory (see Ch. IV, §2). The function

(6) $$w(z) = \frac{1}{(2-n)\omega_n} \int_{-\infty}^{\infty} \cdots \int f(x) \, |z - x|^{2-n} \, dx \qquad \text{for } n > 2,$$

(6') $$w(z) = \frac{1}{2\pi} \int_{-\infty}^{\infty} \cdots \int f(x) \log |z - x| \, dx \qquad \text{for } n = 2$$

solves Poisson's equation $\Delta_z w = f(z)$. Here Δ_z denotes the Laplace operator with respect to the independent variables z.

Now, a short elementary calculation shows

(7) $$\Delta^{(n-1)/2} |x| = (-1)^{(n-1)/2} \frac{(n-1)!}{2-n} |x|^{2-n} \qquad \text{for } n \text{ odd}$$

$$(7') \quad \Delta^{(n-2)/2} \log|x| = \frac{2^{n-2}(-1)^{(n-2)/2}}{2-n}\left(\left(\frac{n-2}{2}\right)!\right)^2 |x|^{2-n} \text{ for } n \text{ even.}$$

These formulas, together with (4a) and (4b), result immediately in the representations (5), (5′) if the constant factors are properly combined. We simply have to express $|x|^{2-n}$ in Poisson's formula by (7), (7′), after replacing $|x|$ by $|z-x|$, and then write the iterated Laplace operators in front of the integrals.

By using the language of idealized functions we can bring the relations proved above into a remarkably concise form, which we insert here for use later in §15:

First we recall that (6) and (6′) can be written as

$$(6^*) \qquad \delta(x) = \begin{cases} \Delta \dfrac{1}{(2-n)\omega_n} \dfrac{1}{|x|^{n-2}} & \text{for } n > 2 \\[2ex] \dfrac{1}{2\pi}\Delta \log x & \text{for } n = 2, \end{cases}$$

where $\delta(x) = \delta(x_1, \cdots, x_n)$ denotes the delta-function in n dimensions; hence (7) and (7′) imply

$$(8) \qquad \begin{aligned} \delta(x_1, \cdots, x_n) &= \delta(x) \\ &= \frac{(-1)^{(n-1)/2}}{2(2\pi)^{n-1}} \int_{\alpha^2=1} \delta^{(n-1)}(x, \alpha)\, d\alpha \quad \text{for } n \text{ odd,} \end{aligned}$$

$$(9) \qquad \begin{aligned} \delta(x_1, \cdots, x_n) &= \delta(x) \\ &= \frac{(-1)^{(n+2)/2}}{(2\pi)^n} \int_{\alpha^2=1} \log^{(n)}|x, \alpha|\, d\alpha \quad \text{for } n \text{ even.} \end{aligned}$$

Here $\delta(x\alpha)$ is the delta-function of the single variable (x, α), $\delta^{(n-1)}$ its $(n-1)$-st derivative, and

$$(10) \qquad \log^{(n)}|z| = \frac{d^n}{dz^n}\log|z|$$

is to be interpreted as a distribution.

The following elegant procedure due to F. John[1] and I. M. Gelfand and G. Shilov[2] combines the expression for odd and even n. For real values of z we consider the principal value $\log z = \log|z| + i\pi(1 - \eta(z))$, η denoting the Heaviside function. Then the succes-

---

[1] See F. John [4].
[2] See I. M. Gelfand and G. Shilov [1].

sive derivatives, considered as distributions with $z = 0$ as singular point, are written

$$\frac{d \log z}{dz} = \log^{(1)}(z) = \frac{1}{z} - \pi i \delta(z)$$

$$\frac{d^2 \log z}{dz^2} = \log^{(2)}(z) = -\frac{1}{z^2} - \pi i \delta'(z).$$

With this notation the representations (8) and (9) can be combined in the formula

$$(11) \qquad \delta(x_1, \cdots, x_n) = \delta(x) = \frac{-1}{(2\pi i)^n} \int_{|\alpha|=1} \log^{(n)}(x, \alpha) \, d\alpha,$$

which is valid[1] for arbitrary n. It is a highly interesting and useful decomposition of the delta-function into plane waves.

The proof consists simply of a suitable interpretation of the preceding results: Poisson's formula expresses $\delta(x)$ in terms of $\Delta \mid x \mid^{2-n}$. Hence $\delta(x)$ is obtained by repeatedly applying the Laplacian to the left-hand sides of (7), (7′), that means to the left-hand sides of (4a), (4b).   According to the rules for distributions this may be done under the integral sign.   At the same time we observe that for any function or generalized function f of one variable (αx), we have

$$\Delta f(\alpha x) = f''(\alpha x).$$

This follows by direct differentiation since $\alpha^2 = 1$.   Then (8), (9) result finally by properly accounting for the various constant coefficients.

§12. Equations of Second Order with Constant Coefficients

1. **Cauchy's Problem.**   According to Ch. III, §3, all hyperbolic differential equations of second order with constant coefficients are completely covered if we consider the specific differential equation

$$(1) \qquad u_{tt} - \Delta u - cu = 0$$

with the initial conditions

$$(1') \qquad u(x, 0) = 0, \qquad u_t(x, 0) = \psi(x),$$

where $\psi(x)$ denotes a function of $x_1, x_2, \cdots, x_n$ with continuous

---

[1] Because of symmetry of the region of integration, the imaginary part vanishes.

derivatives up to at least $(n + 1)/2$-th order if n is odd and at least $(n + 2)/2$-th order if n is even. (For the motivation of this assumption see §8 and the explicit expressions below.)

As said before, if u is the solution of this initial value problem, then $v = u_t$ is the solution of a corresponding initial value problem for which $v(x, 0) = \psi(x)$ and $v_t(x, 0) = 0$ are the initial conditions. Therefore, in order to solve the initial value problem for arbitrarily prescribed initial values of u and u_t, it is sufficient, by the principle of superposition, to find the solution of the initial value problem $(1')$.

In the present section we shall use Fourier's integral to find a formal construction of the solution of Cauchy's problem. The verification of the result obtained by formal manipulation with the Fourier integral will be omitted because in the subsequent sections the result will be derived in various ways by different methods.

We first concentrate on the wave equation $u_{tt} - \Delta u = 0$ with the initial conditions $u(x, 0) = 0$, $u_t(x, 0) = \psi(x)$.

The explicit solution, to be obtained in the present article, is

$$(2) \qquad u(x, t) = \frac{1}{(n-2)!} \frac{\partial^{n-2}}{\partial t^{n-2}} \int_0^t (t^2 - r^2)^{(n-3)/2} \, rQ(x, r) \, dr,$$

with

$$(3) \qquad Q(x, r) = \frac{1}{\omega_n} \iint_{\beta^2 = 1} \psi(x + \beta r) \, d\beta.$$

It has the same form for even and odd numbers n of space dimensions. But the following representations of the solution show a different behavior of the solutions for even and for odd n:

$$(4) \qquad u(x, t) = \frac{\sqrt{\pi}}{2\Gamma\left(\dfrac{n}{2}\right)} \left(\frac{\partial}{\partial t^2}\right)^{(n-3)/2} (t^{n-2} Q) \qquad \text{for } n \text{ odd}$$

$$(5) \qquad u(x, t) = \frac{t}{\Gamma\left(\dfrac{n}{2}\right)} \left(\frac{\partial}{\partial t^2}\right)^{(n-2)/2} (t^{n-3} H) \qquad \text{for } n \text{ even,}$$

with

$$(6) \qquad H(x, t) = \int_0^t \frac{rQ(x, r)}{\sqrt{t^2 - r^2}} \, dr,$$

$$\frac{\partial}{\partial t^2} = \frac{1}{2t} \frac{\partial}{\partial t}.$$

If we wish a unified expression for (4) and (5), valid for even and odd n alike, we have to make use of the concept of fractional differentiation. Then (4) covers both cases. (See §13, 2.)

2. Construction of the Solution for the Wave Equation. In line with Ch. III, §5, we tentatively write the desired solution in the form

$$(7) \qquad u = \int_{-\infty}^{\infty} \cdots \int A(a) e^{i(ax)} \sin \rho t \, da,$$

where a denotes the vector $a_1, \cdots, a_n$,

$$\rho = \sqrt{a_1^2 + a_2^2 + \cdots + a_n^2} = |a|,$$

and where (a, x) is the inner product of a and x, and $da = da_1 \cdots da_n$. From the initial condition for $t = 0$ we then find

$$\psi(x) = \int_{-\infty}^{\infty} \cdots \int \rho A(a) e^{i(ax)} \, da,$$

by differentiating under the integral sign. The interchange of operations here and later is justified by a direct verification of the result.[1] Fourier's inversion formula yields immediately

$$(8) \qquad \rho A = \frac{1}{(2\pi)^n} \int_{-\infty}^{\infty} \cdots \int \psi(\xi) e^{-i(a\xi)} \, d\xi.$$

Inserting this expression for A in equation (7) and changing the order of integration leads formally to

$$u = \frac{1}{(2\pi)^n} \int_{-\infty}^{\infty} \cdots \int \psi(\xi) \, d\xi \int_{-\infty}^{\infty} \cdots \int e^{+i(a(x-\xi))} \frac{\sin \rho t}{\rho} \, da.$$

However, for $n > 2$ the inner integral in this expression is divergent, since in polar coordinates we have $da = da_1 \, da_2 \cdots da_n = \rho^{n-1} \, d\omega_n \, d\rho$. We avoid this formal difficulty by the following device:[2] For odd $n \geq 3$ we consider the expression

$$(9) \qquad v(x, t) = \int_{-\infty}^{\infty} \cdots \int \frac{A(\alpha)}{\rho^{n-2}} e^{i(ax)} \cos \rho t \, da,$$

[1] We might also make use of the concept of distributions, used elsewhere in this chapter, but it is not needed for the present purpose.

[2] Essentially our previous use of δ-functions and other symbolic functions originated from related devices.

and for even $n \geq 2$

$$(9') \qquad w(x, t) = \int_{-\infty}^{\infty} \cdots \int \frac{A(\alpha)}{\rho^{n-2}} e^{i(\alpha x)} \sin \rho t \, da.$$

For odd $n \geq 3$ formal differentiation yields

$$(9a) \qquad u(x, t) = (-1)^{(n-1)/2} \frac{\partial^{n-2}}{\partial t^{n-2}} v(x, t)$$

and for even $n \geq 2$,

$$(9a') \qquad u(x, t) = (-1)^{(n-2)/2} \frac{\partial^{n-2}}{\partial t^{n-2}} w(x, t).$$

After inserting the expression (8) for $A(\alpha)$ in (9) and (9') we may invert the order of integration with respect to ξ and α. As the result we find:

$$(10) \qquad u = \frac{\partial^{n-2}}{\partial t^{n-2}} \int_{-\infty}^{\infty} \cdots \int \psi(x + \xi) K_n(r, t) \, d\xi$$

$$(d\xi = d\xi_1 \cdots d\xi_n)$$

where $r = \sqrt{\xi_1^2 + \xi_2^2 + \cdots + \xi_n^2}$, and

$$K_n = \begin{cases} \dfrac{\pi \omega_{n-1}}{(2\pi)^n} \dfrac{1}{r} \left(\dfrac{t^2}{r^2} - 1 \right)^{(n-3)/2} & \text{for } r < t \\ 0 & \text{for } r > t. \end{cases}$$

To prove this result, we calculate the inner integral for odd n only. (The procedure for even n is similar; moreover, the result for even n will be easily deduced from that for odd n.) First, in place of $a_1, a_2, \cdots, a_n$ we introduce the polar coordinates $\rho = \sqrt{a_1^2 + a_2^2 + \cdots + a_n^2}$ and the n-dimensional unit vector β with components $\beta_i = \rho^{-1} a_i$, so that $da = \rho^{n-1} \, d\rho \, d\beta$. Substitution of (8) in (9) and interchange of orders of integration yields for the inner integral in (9):

$$(11) \qquad S_n(r, t) = \frac{\omega_n}{(2\pi)^n} \int_0^{\infty} M(\rho r) \cos \rho t \, d\rho,$$

where $M(r)$ is the mean value

$$(12) \qquad M(r) = \frac{1}{\omega_n} \int_{\Omega_n} \cdots \int e^{i(\beta \xi)} \, d\omega_n.$$

Using equation (4) of §11, we obtain[1]

$$(13) \qquad M(r) = \frac{\omega_{n-1}}{\omega_n} \int_{-1}^{1} (1 - \lambda_1^2)^{(n-3)/2} \, e^{i\lambda_1 r} \, d\lambda_1 \, .$$

In (11) we now set $\rho r = s$ and obtain

$$S_n(r, t) = \frac{\omega_n}{r(2\pi)^n} \int_0^\infty M(s) \cos s \frac{t}{r} \, ds = \frac{\omega_n}{2r(2\pi)^n} \int_{-\infty}^\infty M(s) \, e^{ist/r} \, ds$$

or, by (13), after a simple transformation

$$S_n(r, t) = \frac{\omega_{n-1}}{r(2\pi)^n} \lim_{N \to \infty} \int_{-1}^{1} (1 - \lambda_1^2)^{(n-3)/2} \frac{\sin N \left(\lambda_1 + \dfrac{t}{r} \right)}{\lambda_1 + \dfrac{t}{r}} \, d\lambda_1.$$

By the well-known properties of the "Dirichlet integral" on the right-hand side (cf. Vol. I, p. 78), we have

$$(14) \qquad S_n(r, t) = \begin{cases} \dfrac{\pi \omega_{n-1}}{r(2\pi)^n} \left(1 - \dfrac{t^2}{r^2} \right)^{(n-3)/2} & \text{for } r > t \\[2ex] 0 & \text{for } r < t. \end{cases}$$

But v was given by the representation

$$v = \int_{-\infty}^\infty \cdots \int \psi(x + \xi) \, S_n(r, t) \, d\xi.$$

Introducing polar coordinates

$$r = \sqrt{\xi_1^2 + \xi_2^2 + \cdots + \xi_n^2} \, , \qquad \alpha_1, \alpha_2, \cdots, \alpha_n$$

and introducing for the mean value of the function ψ on the sphere of radius r and center x

$$(15) \qquad Q(x_1, x_2, \cdots, x_n \, ; r) = \frac{1}{\omega_n} \int_{\Omega_n} \cdots \int \psi(x + \alpha r) \, d\alpha,$$

we have because of (14)

---

[1] By Courant-Hilbert, Vol. I, p. 482, we thus incidentally obtain

$$M(r) = 2^{(n-2)/2} \, \Gamma \left(\frac{n}{2} \right) \frac{J_{(n-2)/2}(r)}{r^{(n-2)/2}} \, .$$

$$v = \omega_n \int_0^\infty Q(x, r) r^{n-1} S_n(r, t) \, dr$$

$$= \frac{\pi \omega_{n-1} \omega_n}{(2\pi)^n} \int_t^\infty (r^2 - t^2)^{(n-3)/2} r Q(x, r) \, dr.$$

For odd $n \geq 3$ we use the identity

$$(15') \qquad \frac{\partial^{n-2}}{\partial t^{n-2}} \int_0^\infty (r^2 - t^2)^{(n-3)/2} r Q(x, r) \, dr = 0.$$

Since the integrand is a polynomial of degree $n - 3$ in t, its derivative of order $n - 2$ with respect to t vanishes. Subtracting $(15')$ from $\partial^{n-2} v / \partial t^{n-2}$, we obtain for $u = (-1)^{(n-1)/2} \partial^{n-2} v / \partial t^{n-2}$ the representation

$$u(x, t) = C_n \frac{\partial^{n-2}}{\partial t^{n-2}} \int_0^t (t^2 - r^2)^{(n-3)/2} r Q(x, r) \, dr,$$

which is equivalent to (10). The constant C_n can be found either from the above formulas or simply by considering the special case $\psi = 1, u = t$, for which $Q = 1$. We find $C_n = 1/(n - 2)!$. Thus the solution for the initial value problem is given by the formula

$$(16) \qquad u(x, t) = \frac{1}{(n - 2)!} \frac{\partial^{n-2}}{\partial t^{n-2}} \int_0^t (t^2 - r^2)^{(n-3)/2} r Q(x, r) \, dr.$$

The same formula is valid for even n and can be obtained by starting with $(9')$. However, the calculation in this case is somewhat more tedious. We therefore prefer to extend (16) to the case of an even number of dimensions by the direct step described in the next article.

3. Method of Descent. The method of descent[1] is based on the observation that from the solution of our problem for n independent space variables one can, by specialization, obtain the solution for $n - 1$ or fewer space variables. In doing this one "descends" from the "higher" problem to the "lower".

In virtue of the uniqueness theorem, we obtain the solution for $n - 1$ space variables from the formula for n space variables, by making and exploiting the assumption that the initial function $\psi(x_1, x_2, \cdots, x_n)$ does not depend on the variable x_n. The solution u is then also independent of x_n and hence solves the initial value

[1] See J. Hadamard [2]. See also Ch. III, §4, 4.

problem for $n - 1$ space variables. Similarly, we can descend from n space variables to $n - 2$ space variables by introducing in our formulas the assumption that ψ depends only on x_1, x_2, $\cdots$, x_{n-2}, etc.

We now shall use the process of descent to derive (16) for an even number of dimensions if we assume it proved for an odd number of dimensions.

We start with a space of $n + 1$ variables, x_1, x_2, $\cdots$, x_{n+1}. In this space we consider for a function $\psi(x_1$, x_2, $\cdots$, $x_n)$ of only n variables the spherical mean over a sphere Ω_{n+1} of radius r

$$Q_{n+1}(x; r) = \frac{1}{\omega_{n+1}} \int \cdots \int_{\Omega_{n+1}} \psi(x + \alpha r)\, d\omega_{n+1},$$

where α is a unit vector. Since ψ does not depend on the $(n + 1)$-st space variable, this surface integral is, according to §11, given by an integral over an n-dimensional sphere. Now

$$Q_{n+1}(x, r) = \frac{\omega_n}{\omega_{n+1}\, r^{n-1}} \int_0^r \rho^{n-1}\, d\rho \int \frac{\psi(x + \rho\alpha)}{\omega_n \sqrt{r^2 - \rho^2}}\, d\omega_n.$$

Since $\omega_n = 2(\sqrt{\pi})^n / \Gamma(n/2)$ we have

$$(17) \qquad Q_{n+1}(x, r) = \frac{\Gamma\left(\dfrac{n + 1}{2}\right)}{\sqrt{\pi}\,\Gamma\left(\dfrac{n}{2}\right)} \frac{1}{r^{n-1}} \int_0^r \frac{\rho^{n-1} Q_n(x, \rho)}{\sqrt{r^2 - \rho^2}}\, d\rho,$$

where

$$Q_n(x, \rho) = \frac{1}{\omega_n} \int \cdots \int \psi(x + \alpha\rho)\, d\omega_n$$

denotes the corresponding mean value of ψ in the n-dimensional space.

Now the descent of one step leads to the desired result by the following short calculation. Replacing n in (16) by $n + 1$ and substituting the expression (17) for $Q_{n+1}(x, r)$, we have after changing the order of integration and differentiating,

$$u = C \frac{\partial^{n-2}}{\partial t^{n-2}} \left[t \int_0^t \rho^{n-1} Q_n(x, \rho)\, d\rho \int_\rho^t \frac{(t^2 - r^2)^{(n-4)/2}}{r^{n-2}\sqrt{r^2 - \rho^2}}\, dr \right],$$

where C is a constant. We introduce a new variable of integration z, by $1 - \rho^2/r^2 = (1 - \rho^2/t^2)z$, and obtain easily

$$(18) \quad \int_\rho^t \frac{(t^2 - r^2)^{(n-4)/2}}{r^{n-2}\sqrt{r^2 - \rho^2}}\, dr = \frac{(t^2 - \rho^2)^{(n-3)/2}}{2t\rho^{n-2}} \int_0^1 z^{-1/2}(1 - z)^{(n-4)/2}\, dz.$$

Hence

$$u = C_n \frac{\partial^{n-2}}{\partial t^{n-2}} \int_0^t (t^2 - \rho^2)^{(n-3)/2}\rho Q_n(x, \rho)\, d\rho.$$

The constant C_n is again found, by specialization, to be

$$C_n = 1/(n - 2)!.$$

Thus the solution retains its form when we descend to lower values of n. It is therefore sufficient to derive (16) only for odd n, since we can descend to even values from larger odd ones.

4. Further Discussion of the Solution. Huyghens' Principle. One can briefly put the solution (16) into different forms (4), (5) which exhibit the significant fact concerning Huyghens' principle.[1]

Denoting by $G(t)$ any λ-times continuously differentiable function of t, we consider the functions

$$(19) \qquad U_\lambda(t) = \frac{1}{(2\lambda + 1)!} \frac{\partial^{2\lambda+1}}{\partial t^{2\lambda+1}} \int_0^t (t^2 - r^2)^\lambda rG(r)\, dr$$
$$(\lambda = 0, 1, \cdots).$$

For (19) the recursion formula

$$(20) \qquad U_\lambda(t) = \frac{1}{2\lambda + 1}\, (tU'_{\lambda-1} + 2\lambda U_{\lambda-1})$$

is easily proved. $U_0 = tG(t)$ implies

$$(21) \qquad U_\lambda = t \sum_{\nu=0}^\lambda a_{\lambda,\nu}\, t^\nu G^{(\nu)}(t),$$

where the quantities $a_{\lambda,\nu}$ are numerical constants. In symbolic form, if $P_\lambda(t)$ denotes the polynomial

$$P_\lambda(t) = \sum_{\nu=0}^\lambda a_{\lambda,\nu}\, t^\nu,$$

[1] See Ch. III, §4, 6.

then we have

$$(22) \qquad U_\lambda(t) = tP_\lambda(tG),$$

where the powers of G are to be replaced by the corresponding derivatives. Thus the solution (16) of the initial value problem can now be written for odd $n = 2\lambda + 3$ in the form

$$(23) \qquad u = tP_{(n-3)/2}(tQ),$$

with $G(t) = Q(t) = Q(x, t)$ defined before by (15).

From this result we obtain a representation of the solution also for an even number n of space dimensions by descending from the odd dimension $n + 1$ to n; we have at once

$$(23a) \qquad u = tP_{(n-2)/2}(tG),$$

with $G = \dfrac{1}{\omega_{n+1}} \displaystyle\int \cdots \int \psi(x + \beta t)\, d\omega_{n+1}$, that is, by article 3,

$$(23') \qquad G(t) = \frac{2}{\sqrt{\pi}} \frac{\Gamma\left(\dfrac{n+1}{2}\right)}{\Gamma\left(\dfrac{n}{2}\right)} \frac{1}{t^{n-1}} \int_0^t \frac{r^{n-1}Q(x, r)}{\sqrt{t^2 - r^2}}\, dr.$$

The solution is therefore represented in terms of the derivatives up to order $(n - 2)/2$ with respect to t of this function $G(x, t)$.

In analogy with the discussion for odd dimension we also could easily give a slightly different representation in terms of the derivatives of the expression

$$(24) \qquad H(x, t) = \int_0^t \frac{rQ(x, r)}{\sqrt{t^2 - r^2}}\, dr,$$

which itself satisfies the wave equation for $n = 2$. We set

$$(25) \qquad U_{\lambda-1/2} = Z_\lambda = \frac{1}{(2\lambda)!} \frac{\partial^{2\lambda}}{\partial t^{2\lambda}} \int_0^t (t^2 - r^2)^{\lambda-1/2} rQ(r)\, dr$$

and obtain for Z the recursion formula

$$(26) \qquad \begin{cases} Z_\lambda = \dfrac{1}{2\lambda}\left[(2\lambda - 1)Z_{\lambda-1} + tZ'_{\lambda-1}\right], & (\lambda = 1, 2, \cdots), \\[2mm] Z_0 = U_{\lambda-1/2} = H. \end{cases}$$

From this follows a formula of the form

$$(27) \qquad Z_\lambda = \sum_{\nu=0}^{\lambda} b_{\lambda,\nu}\, t^\nu H^{(\nu)}(t),$$

with constants $b_{\lambda,\nu}$. In terms of the polynomials

$$\Pi_\lambda(t) = \sum_{\nu=0}^{\lambda} b_{\lambda,\nu}\, t^\nu,$$

we can write symbolically

$$(28) \qquad Z_\lambda = \Pi_\lambda(tH).$$

Hence the solution of the initial value problem for even dimensions is

$$(29) \qquad u = \Pi_{(n-2)/2}(tH),$$

with

$$H(x, t) = \int_0^t \frac{rQ(x, r)}{\sqrt{t^2 - r^2}}\, dr.$$

Formulas (23) and (29) show that u is continuous if the initial function ψ has continuous derivatives up to order $(n - 2)/2$ for even n or up to order $(n - 3)/2$ for odd n. To assure the differentiability of u two more times, as required by the differential equation (1), we assume that ψ has continuous derivatives up to order $(n + 1)/2$ if n is odd and up to order $(n + 2)/2$ if n is even.[1]

Some facts concerning Huyghens' principle follow from the preceding representations. Huyghens' principle asserts: *The value of the solution of Cauchy's problem for the wave equation depends only on the boundary of the domain of dependence in the plane $t = 0$; i.e., it depends only on the initial values of u and its derivatives on the boundary of the base $r = t$ of the characteristic cone, but not on the initial values in the interior of this base.* We recall that Huyghens' principle is valid for the wave equation in three space dimensions but not in two dimensions.[2] Our present formulas show this fact to be a special case of the following general rule: *Huyghens' principle is valid for the wave equation when the number n of space dimensions is odd, but not when n is even.*[3]

---

[1] Independently of our representation formula it was shown in §10 that such an assumption is indeed appropriate for the initial value problem.

[2] See Ch. III, §4 and Ch. VI, §2.

[3] This was apparently first clearly recognized by Vito Volterra [1] and O. Tedone [1].

Finally: The recursion formulas (20), (26) lead easily to the representations (4), (5). If we introduce instead of U or Z the functions $R_\lambda = (2/\sqrt{\pi})\Gamma(2\lambda + 3/2)U_\lambda$ or $S_\lambda = (1/t)\lambda!Z_\lambda$ respectively, we obtain a simplified recursion scheme:

$$R_\lambda = \lambda R_{\lambda-1} + t^2 \frac{d}{dt^2} R_{\lambda-1}, \qquad R_0 = tQ(x, t)$$

$$S_\lambda = \lambda S_{\lambda-1} + t^2 \frac{d}{dt^2} S_{\lambda-1}, \qquad S_0 = \frac{1}{t} H(x, t),$$

where $d/dt^2 = 1/2t \, d/dt$. The recursion problem is immediately solved by

$$R_\lambda = \left(\frac{d}{dt^2}\right)^\lambda (t^{2\lambda+1}Q)$$

and

$$S_\lambda = \left(\frac{d}{dt^2}\right)^\lambda (t^{2\lambda-1}H)$$

which implies the representations (4), (5).

5. The Nonhomogeneous Equation. Duhamel's Integral. For the integration of the nonhomogeneous equation

$$(30) \qquad\qquad u_{tt} - \Delta u = f(x, t)$$

with given $f(x, t)$, under the initial condition

$$(30') \qquad\qquad u(x, 0) = u_t(x, 0) = 0,$$

we refer again to the *method of variation of parameters* or "Duhamel's integral" discussed, e.g., in Ch. III, §4, 3. The function f is again supposed to have continuous derivatives up to order $(n + 1)/2$ or $(n + 2)/2$, respectively. Let the function $v(x, t; \tau)$, depending on the parameter τ, be a solution of the homogeneous differential equation

$$v_{tt} - \Delta v = 0$$

under initial conditions

$$v(x, 0; \tau) = 0, \qquad v_t(x, 0; \tau) = f(x, \tau).$$

Then

$$(31) \qquad\qquad u = \int_0^t v(x, t - \tau; \tau) \, d\tau.$$

From this "Duhamel integral" the solution of the initial value problem for the nonhomogeneous equation (30) with the initial conditions (30′) follows at once. With the mean value

$$Q(x, r; \tau) = \frac{1}{\omega_n} \int \cdots \int f(x + \beta r, \tau) \, d\omega_n$$

we have

$$v(x, t; \tau) = \frac{1}{(n-2)!} \frac{\partial^{n-2}}{\partial t^{n-2}} \int_0^t (t^2 - r^2)^{(n-3)/2} r Q(x, r; \tau) \, dr$$

and therefore

$$u(x, t) = \frac{1}{(n-2)!} \int_0^t d\tau \, \frac{\partial^{n-2}}{\partial t^{n-2}} \int_0^{t-\tau} [(t-\tau)^2 - r^2]^{(n-3)/2} r Q(x, r; \tau) \, dr$$

or, with an easily justified change of differentiation and integration,

$$(32) \qquad u(x, t) = \frac{1}{(n-2)!} \frac{\partial^{n-2}}{\partial t^{n-2}} \int_0^t d\tau \int_0^\tau (\tau^2 - r^2)^{(n-3)/2} r Q(x, r; t - \tau) \, dr$$

For $n = 2$ and $n = 3$ the expressions

$$(33) \qquad u = \frac{1}{2\pi} \int_0^t d\tau \iint_{\rho \le \tau} \frac{f(\xi, \eta; t - \tau)}{\sqrt{\tau^2 - \rho^2}} \, d\xi \, d\eta$$

$$(\rho^2 = (x - \xi)^2 + (y - \eta)^2),$$

and

$$(34) \qquad u = \frac{1}{4\pi} \iiint_{\rho \le t} \frac{1}{\rho} f(\xi, \eta, \zeta; t - \rho) \, d\xi \, d\eta \, d\zeta$$

$$(\rho^2 = (x - \xi)^2 + (y - \eta)^2 + (z - \zeta)^2),$$

follow, in agreement with the result in Chapter III.

6. Cauchy's Problem for the General Linear Equation of Second Order. On the basis of article 5, the initial value problem for the general second order hyperbolic differential equation can now be solved easily by the method of descent. It suffices to consider first the differential equation

$$(35) \qquad \Delta u + c^2 u = u_{tt}$$

with the initial conditions

$$(35') \qquad u(x, 0) = 0, \qquad u_t(x, 0) = \psi(x).$$

According to Ch. III, §3 the general case can be reduced to this one.

The explicit solution is again obtained by the method of descent. We artificially increase the number of "space" variables to $n + 1$ by setting $x_{n+1} = z$, and consider the initial value problem of the differential equation

$$(36) \qquad \Delta v = v_{tt}$$

for $v(x_1, x_2, \cdots, x_{n+1}, t)$ with the initial conditions

$$(36') \quad v(x, 0) = 0, \qquad v_t(x, 0) = \psi(x_1, \cdots, x_n)e^{cx_{n+1}} = \psi(x)e^{cz}.$$

Introducing

$$(37) \qquad v = e^{cz}u(x_1, x_2, \cdots, x_n)$$

then yields u as the solution of the initial value problem (35). Indeed, our previous representations (16) show that the solution v of the initial value problem (36) has the form $e^{cz}u(x_1, x_2, \cdots, x_n)$. But if we insert the function v in equation (36), we immediately obtain for u the original initial value problem (35), (35'). By the uniqueness theorems proved in §8, there can exist only a single solution u, and therefore $u = ve^{-cz}$.

We can now represent v, and hence also u, by the expressions of article 4. Thus for *even n*, and hence odd $n + 1$, we have

$$v = tP_{(n-2)/2}(tG^*)$$

with

$$G^*(t) = \frac{e^{cz}}{\omega_{n+1}} \int \cdots \int_{\Omega_{n+1}} \psi(x + \beta t)e^{ct\beta_{n+1}} \, d\omega_{n+1}.$$

Hence $u = tP_{(n-2)/2}(tG)$, with

$$G(t) = \frac{1}{\omega_{n+1}} \int \cdots \int_{\Omega_{n+1}} \psi(x + \beta t)e^{ct\beta_{n+1}} \, d\omega_{n+1}.$$

As before, $d\omega_{n+1}$ denotes the surface element on the $(n + 1)$-dimensional unit sphere. Since the function ψ does not depend on the last variable $z = x_{n+1}$, we obtain, in view of

$$t^{n-1} \, d\omega_{n+1} = \frac{1}{\sqrt{t^2 - \rho^2}} \, d\xi_1 \, d\xi_2 \cdots d\xi_n = \frac{1}{\sqrt{t^2 - \rho^2}} \, \rho^{n-1} \, d\omega_n \, d\rho,$$

the relation

$$G(x, t) = \frac{\omega_n}{\omega_{n+1}} \frac{1}{t^{n-1}} \int_{-t}^{t} \frac{\rho^{n-1}}{\sqrt{t^2 - \rho^2}} e^{c\sqrt{t^2 - \rho^2}} Q(x, \rho)\, d\rho$$

or

$$(38) \qquad G(x, t) = \frac{2\Gamma\left(\dfrac{n+1}{2}\right)}{\sqrt{\pi}\,\Gamma\left(\dfrac{n}{2}\right)} \frac{1}{t^{n-1}} \int_{0}^{t} \frac{\rho^{n-1}}{\sqrt{t^2 - \rho^2}}$$

$$\cosh c\sqrt{t^2 - \rho^2}\, Q(x, \rho)\, d\rho$$

with

$$Q(x, \rho) = \frac{1}{\omega_n} \int \cdots \int_{\Omega_n} \psi(x + \beta\rho)\, d\omega_n.$$

Similarly, for *odd* n we obtain the solution

$$(39) \qquad u = \Pi_{(n-1)/2}(tH),$$

with

$$(40) \qquad H(x, t) = \frac{n}{t^n} \int_{0}^{t} \rho^{n-1} J_0\left(ic\sqrt{t^2 - \rho^2}\right) Q(x, \rho)\, d\rho,$$

where J_0 is the Bessel function of order zero. Indeed, in this case

$$v = \Pi_{(n-1)/2}(tH^*),$$

with

$$H^*(x, t) = \frac{2e^{cz}}{\omega_{n+2}\, t^n} \int \cdots \int_{\rho^2 + \xi_{n+1}^2 \leq t^2} \frac{\psi(x + \xi)}{\sqrt{t^2 - \rho^2 - \xi_{n+1}^2}}$$

$$\cdot e^{c\xi_{n+1}}\, d\xi_1 \cdots d\xi_{n+1}.$$

In the inner integral, integration with respect to ξ_{n+1} from $-\sqrt{t^2 - \rho^2}$ to $+\sqrt{t^2 - \rho^2}$ yields the expression $\pi J_0(ic\sqrt{t^2 - \rho^2})$, so that

$$H(x, t) = H^*(x, t)e^{-cz} = \frac{2\pi}{\omega_{n+2}} \frac{1}{t^n}$$

$$\cdot \int \cdots \int_{\rho \leq t} \psi(x + \xi) J_0(ic\sqrt{t^2 - \rho^2})\, d\xi_1\, d\xi_2 \cdots d\xi_n,$$

in accordance with (40).

To apply the result to the *telegraph equation*

$$(41) \qquad \Delta u = u_{tt} + u_t$$

with the initial condition $u(x, 0) = 0$, $u_t(x, 0) = \psi(x)$, we set $u = ve^{-t/2}$ and obtain for v the differential equation

$$(42) \qquad v_{tt} = \Delta v + \frac{v}{4}$$

which is equation (35) for $c = \frac{1}{2}$.

For example, the solution of the telegraph equation for $n = 1, 2, 3$ is

$$u = e^{-t/2} \int_0^t J_0\left(\frac{i}{2}\sqrt{t^2 - \rho^2}\right) Q(x, \rho)\, d\rho,$$

$$Q(x, \rho) = \tfrac{1}{2}[\psi(x + \rho) + \psi(x - \rho)] \qquad \text{for } n = 1,$$

$$u = e^{-t/2} \int_0^t \rho\, \frac{\cosh \tfrac{1}{2}\sqrt{t^2 - \rho^2}}{\sqrt{t^2 - \rho^2}}\, Q(x, \rho)\, d\rho,$$

$$(43)$$

$$Q(x, \rho) = \frac{1}{2\pi} \int_{\Omega_2} \psi(x_1 + \rho\beta_1, x_2 + \rho\beta_2)\, d\omega_2 \qquad \text{for } n = 2,$$

$$u = e^{-t/2}\frac{1}{t}\frac{\partial}{\partial t} \int_0^t \rho^2 J_0\left(\frac{i}{2}\sqrt{t^2 - \rho^2}\right) Q(x, \rho)\, d\rho,$$

$$Q(x, \rho) = \frac{1}{4\pi} \iint_{\Omega_3} \psi(x_1 + \rho\beta_1, x_2 + \rho\beta_2, x_3 + \rho\beta_3)\, d\omega_3 \quad \text{for } n = 3.$$

7. The Radiation Problem.[1]   The result of article 5 leads to the solution of the *radiation problem* for the general wave equation in n space dimensions by means of a simple limiting process.[2]   We formulate this radiation problem as follows: *For $t > 0$ we seek a solution of the homogeneous wave equation $u_{tt} - \Delta u = 0$ which vanishes together with its derivative u_t for $t = 0$ except at the origin of the x-space, and which for $r = \sqrt{x_1^2 + x_2^2 + \cdots + x_n^2} = 0$, i.e., on the "time axis", is singular in such a way that*

$$(44) \qquad \lim_{\epsilon \to 0} \int \cdots \int \frac{\partial u}{\partial \nu}\, ds = -g(t).$$

---

[1] For a related exposition, see A. Weinstein [2].

[2] See §15 for a more general and systematic discussion.

Here the integration is at the time t over a sphere K_ϵ about the origin of the x-space. $\partial/\partial\nu$ denotes differentiation along the normal of the sphere, ds is the surface element, and the radius ϵ of the sphere is contracted to zero. The function $g(t)$ is the prescribed *intensity of radiation* as a function of the time.

We may formulate the radiation problem more concisely as that of the nonhomogeneous "differential equation" [1]

$$(44') \qquad u_{tt} - \Delta u = g(t)\delta(x, y, z)$$

with the homogeneous initial conditions $(30')$; here $\delta(x, y, z)$ is the three-dimensional delta function introduced before.

We are led to attempt construction of the solution by a limiting process from solutions of nonhomogeneous equations. Specifically, let $u = u_h$ be the solution of equation (30) with $f(x, t) = \psi(x)g(t)$, where

$$\psi = 0 \quad \text{for } r > h \qquad \text{and} \qquad \psi \geq 0 \quad \text{for } r < h$$

and $\int \cdots \int_{r \leq h} \psi \, dx_1 \, dx_2 \cdots dx_n = 1$. Here $r^2 = x_1^2 + x_2^2 + \cdots + x_n^2$.

The desired radiation function is constructed as the limit of the solution u_h as $h \to 0$. The result is

$$u_h = \frac{1}{(n-2)!} \frac{\partial^{n-2}}{\partial t^{n-2}} \int_0^t g(t - \tau) \, d\tau \int_0^\tau (\tau^2 - s^2)^{(n-3)/2} sQ(x, s) \, ds$$

with $Q(x, s) = \dfrac{1}{\omega_n} \displaystyle\int \cdots \int_{\Omega_n} \psi(x + \beta s) \, d\omega_n$. If we write the inner integral in the form

$$\int_0^\tau (\tau^2 - s^2)^{(n-3)/2} sQ(x, s) \, ds = \frac{1}{\omega_n} \int \cdots \int_{s \leq \tau} \frac{(\tau^2 - s^2)^{(n-3)/2}}{s^{n-2}} \psi(\xi) \, d\xi$$

$$(s^2 = (x_1 - \xi_1)^2 + (x_2 - \xi_2)^2 + \cdots + (x_n - \xi_n)^2),$$

and take the limit for $h \to 0$, we obtain

$$\lim_{h \to 0} \int_0^\tau (\tau^2 - s^2)^{(n-3)/2} sQ(x, s) \, ds = \begin{cases} 0 & \text{for } r \geq \tau, \\[2mm] \dfrac{(\tau^2 - r^2)^{(n-3)/2}}{\omega_n \, r^{n-2}} & \text{for } r < \tau. \end{cases}$$

---

[1] We shall later (see §§14 and 15) make systematic use of the δ-function, but here we present the slightly more conventional procedure.

Hence the desired solution of the radiation problem is

$$u = 0 \qquad\qquad \text{for } r \geq t,$$

and for $r < t$

$$u = \frac{1}{\omega_n(n-2)!} \frac{1}{r^{n-2}} \frac{\partial^{n-2}}{\partial t^{n-2}} \int_r^t g(t-\tau)(\tau^2 - r^2)^{(n-3)/2}\, d\tau$$

or

$$(45) \qquad u = \frac{1}{\omega_n(n-2)!} \frac{1}{r^{n-2}} \frac{\partial^{n-2}}{\partial t^{n-2}} \int_0^{t-r} g(\tau)[(t-\tau)^2 - r^2]^{(n-3)/2}\, d\tau.$$

For $n = 2$ and $n = 3$ (see Ch. III, §4) we again find

$$(46) \qquad u = \frac{1}{2\pi} \int_r^t \frac{g(t-\tau)}{\sqrt{\tau^2 - r^2}}\, d\tau$$

and

$$(47) \qquad u = \frac{1}{4\pi r} g(t-r)$$

respectively. In the case $n = 5$ we find

$$u = \frac{1}{8\pi^2} \frac{1}{r^3} [(g(t-r) + rg'(t-r)]$$

and in the case $n = 4$,

$$u = \frac{1}{4\pi^2} \frac{1}{r} \frac{\partial}{\partial t} \int_r^t \frac{g(t-\tau)}{\sqrt{\tau^2 - r^2}}\, d\tau.$$

By a procedure quite similar to that of article 4 the radiation solution can again be written[1] in a concise form

$$(48) \qquad u(x,t) = \frac{(-1)^{(n-3)/2}}{4\pi^{(n-1)/2}} \left(\frac{d}{dr^2}\right)^{(n-3)/2} \frac{g(t-r)}{r} \qquad \text{for } n \text{ odd}$$

$$(49) \qquad u(x,t) = \frac{(-1)^{(n-2)/2}}{2\pi^{n/2}} \left(\frac{d}{dr^2}\right)^{(n-2)/2} H(t-r) \qquad \text{for } n \text{ even}$$

with $H(t-r) = \int_r^t [g(t-\tau)/\sqrt{\tau^2 - r^2}]\, d\tau$. It should be stated

---

[1] For a detailed discussion see §18, 3. We also note that by the use of idealized functions and fractional derivatives the formulas (48) and (49) can be unified.

that results of this article would remain valid for *inward radiation* if we were to replace $t - r$ by $t + r$. It also should be noted that the radiation solutions (48), (49) can be expressed as sums of the form

$$(50) \qquad u = \frac{r^{2-n}}{4\pi^{(n-1)/2}} \sum_{\nu=1}^{(n-3)/2} a_{\nu,n}(2r)^{\nu} g^{\nu}(t - r) \qquad \text{for odd } n$$

and

$$(51) \qquad u = \frac{r^{2-n}}{2^{n-1}\pi^{n/2}} \sum_{\nu=1}^{(n-2)/2} b_{\nu,n}(2r)^{\nu} H^{\nu}(r, t) \qquad \text{for even } n$$

with constants $a_{\nu,n}$, $b_{\nu,n}$.

Expressions of the form (50) are called progressing waves of higher degree $(n - 3)/2$. (See §4, and later §18.)

Again, by inspection of our representations the remarkable fact is confirmed: *Huyghens' principle is valid for the radiation problem with odd space dimensions n.*[1] The effect, at a point x at the time t, of a disturbance radiating from the origin depends only on the character of this disturbance at a single previous instant of time, the time $t - r$. The disturbance, advancing from the origin with unit velocity, reaches the point x at just the time t. Impulses at the origin that are sharply defined in time, i.e., represented by functions $g(t)$ which are different from zero only in a short time interval, are thus perceived at any point at the distance r only during a correspondingly short time interval beginning r time units later.

Huyghens' principle does not, however, hold for radiation in an even number of space dimensions. This is immediately apparent from the form of (49). It is true that in this case a disturbance at the origin sharply defined in time will not be noticeable at a point of distance r from the origin until r time units later. However, after this moment the effect at this point remains noticeable. In other words, the solution generally remains different from zero for $t > r$. Thus in a space of even dimension sharp reception of transmitted signals which satisfy the wave equation would not be possible. Instead, reception of signals would always be blurred. This fact and further considerations (in §18) show that the three-dimensional space in which we exist is distinguished in that signals can be transmitted sharply without distortion.

---

[1] Of course the statements of Huyghens' principle for Cauchy's problem and for the radiation problem are equivalent.

§ 13. *Method of Spherical Means. The Wave Equation and the Darboux Equation*

The results of §12 will now be put on a firm basis by using a different method[1] for the solution of Cauchy's problem.

1. Darboux's Differential Equation for Mean Values. For a twice continuously differentiable function $\psi(x_1, x_2, \cdots, x_n) = \psi(x)$ of the n variables x_i, we consider the mean value

$$(1) \quad v(x_1, x_2, \cdots, x_n, r) = v(x, r) = \frac{1}{\omega_n} \int \cdots \int \psi(x + \beta r) \, d\omega$$

of this function over a sphere Ω_r of radius r about the center x. In this integral β denotes a unit vector with the components $\beta_1, \beta_2, \cdots, \beta_n$; furthermore $d\omega = d\omega_n = d\beta$ is the surface element of the unit sphere, and $d\Omega = r^{n-1} \, d\omega$ is the surface element of the sphere of radius r and center x. Then the mean value functions v satisfy the *Darboux differential equation*[2] (cf. §6, 2)

$$(2) \qquad\qquad v_{rr} + \frac{n-1}{r} v_r - \Delta v = 0$$

with the initial conditions

$$(2') \qquad\qquad v(x, 0) = \psi(x), \qquad v_r(x, 0) = 0.$$

Hence v can be uniquely continued for negative r as a continuously differentiable solution of the differential equation by the definition

$$v(-r) = v(r).$$

We express this fact by the statement: *v is a solution of the Darboux equation which is even with respect to the variable r.*

For the proof we form, using Gauss' integral theorem,

$$v_r(x, r) = \frac{1}{\omega_n} \int \cdots \int_{\Omega_n} \left(\sum_{i=1}^{n} \beta_i \frac{\partial \psi}{\partial x_i} \right) d\omega = \frac{1}{\omega_n r^{n-1}} \int \cdots \int \frac{\partial \psi}{\partial \nu} \, d\Omega_r$$

$$= \frac{1}{\omega_n r^{n-1}} \iint_{G_r} \cdots \int \Delta \psi \, dx,$$

---

[1] Fritz John [4].

[2] See A. Weinstein [2] for the extension of the theory of the Darboux equation to the case when $n - 1$ is replaced by an arbitrary parameter λ.

where G_r is the interior of the sphere Ω_r, where dx is the volume element $dx = dx_1 \, dx_2 \cdots dx_n$, and $\partial/\partial\nu = \sum_{i=1}^{n} \beta_i(\partial/\partial x_i)$ denotes differentiation in the direction of the outer normal on the sphere Ω_r. Further differentiation with respect to r yields

$$v_{rr} = -\frac{n-1}{\omega_n r^n} \int \cdots \int_{G_r} \Delta\psi \, dx + \frac{1}{\omega_n r^{n-1}} \int \cdots \int_{\Omega} \Delta\psi \, d\Omega$$

$$= -\frac{n-1}{r} v_r + \Delta v,$$

from which our assertion follows.

If we assume in particular that the function

$$\psi(x_1, x_2, \cdots, x_n) = \phi(x_1)$$

depends only on a single variable $x = x_1$ and that it is twice differentiable with respect to x, then according to §11, 2 the mean value can be written in the form

$$(3) \qquad v(x, r) = \frac{\omega_{n-1}}{\omega_n} \int_{-1}^{1} \phi(x + r\mu)(1 - \mu^2)^{(n-3)/2} \, d\mu$$

and satisfies the differential equation

$$(4) \qquad v_{rr} + \frac{n-1}{r} v_r - v_{xx} = 0.$$

From §6, 2 we infer: Formulas (1) and (3) solve the initial value problem for the Darboux equations (2) and (4), respectively, with the initial conditions (2').

2. Connection with the Wave Equation. There is a one-to-one relation between the solutions of Darboux equation and the wave equation: Differentiating formula (3) twice with respect to x, we have

$$v_{xx} = \frac{\omega_{n-1}}{\omega_n} \int_{-1}^{1} \phi''(x + r\mu)(1 - \mu^2)^{(n-3)/2} \, d\mu.$$

Hence the Darboux equation implies

$$\frac{n-1}{r} v_r + v_{rr} = \frac{\omega_{n-1}}{\omega_n} \int_{-1}^{1} \phi''(x + r\mu)(1 - \mu^2)^{(n-3)/2} \, d\mu.$$

In this formula the parameter x, which is added to $r\mu$, is inessential. Accordingly,

If the functions $v(r)$ and $\phi(r)$ are connected by the integral transformation

$$(5) \qquad v(r) = \int_{-1}^{1} \phi(r\mu)(1 - \mu^2)^{(n-3)/2} \, d\mu,$$

then

$$(6) \qquad v'' + \frac{n-1}{r} v' = \int_{-1}^{1} \phi''(r\mu)(1 - \mu^2)^{(n-3)/2} \, d\mu.$$

Thus to the operation ϕ'' on $\phi(r)$ there corresponds the operation $v'' + (n-1)v'/r$ on $v(r)$.

We may apply (6) if $\phi(s)$ depends not only on a single parameter x, but on n parameters $x_1, \cdots, x_n$; then we consider the functions

$$v(r, x_1, \cdots, x_n) = \int_{-1}^{1} \phi(x_1, \cdots, x_n, r\mu)(1 - \mu^2)^{(n-3)/2} \, d\mu.$$

Now we obtain the connection between the wave equation and Darboux's equation as follows. Let $u(x_1, \cdots, x_n, t)$ be a twice continuously differentiable solution of the wave equation with $u_t(x, 0) = 0$, that is, a solution which is "even in t". Then, replacing t by r we have the even solution

$$v(x, r) = v(x_1, \cdots, x_n, r)$$

$$= \frac{\omega_{n-1}}{\omega_n} \int_{-1}^{1} u(x_1, \cdots x_n, r\mu)(1 - \mu^2)^{(n-3)/2} \, d\mu,$$

with

$$v(x, 0) = u(x, 0) = \psi(x), \qquad v_r(x, 0) = 0.$$

If we employ the result obtained above, we have the differential equation

$$\Delta v = \frac{\omega_{n-1}}{\omega_n} \int_{-1}^{1} \Delta u(x, r\mu)(1 - \mu^2)^{(n-3)/2} \, d\mu$$

$$= \frac{\omega_{n-1}}{\omega_n} \int_{-1}^{1} u_{tt}(x, r\mu)(1 - \mu^2)^{(n-3)/2} \, du$$

$$= v_{rr} + \frac{n-1}{r} v_r,$$

for which the initial conditions $v(x, 0) = \psi$, $v_r(x, 0) = 0$ are satisfied in virtue of the definition of v as a spherical mean.

According to article 1, this initial value problem for the Darboux equation is solved by

$$v(x_1, \cdots, x_n, r) = \frac{1}{\omega_n} \int \cdots \int_{\Omega_n} \psi(x + r\beta)\, d\omega.$$

Thus the solution u of the wave equation, assumed even in $t = r$, must satisfy the relation

(7)
$$\frac{2\omega_{n-1}}{\omega_n} \int_0^1 u(x_1, \cdots, x_n, r\mu)(1 - \mu^2)^{(n-3)/2}\, d\mu$$
$$= \frac{1}{\omega_n} \int \cdots \int_{\Omega_n} \psi(x + r\beta)\, d\omega.$$

Conversely, from this relation we now obtain the solution u of the wave equation in a unique manner. The problem amounts to inverting the functional equation (5): The substitution $r = \sqrt{s}$, $r\mu = \sqrt{\sigma}$, in (5) yields

$$v(\sqrt{s})s^{(n-2)/2} = \int_0^s \frac{\phi(\sqrt{\sigma})}{\sqrt{\sigma}} (s - \sigma)^{(n-3)/2}\, d\sigma.$$

With the abbreviations $w(s) = v(\sqrt{s})s^{(n-2)/2}$, $\chi(\sigma) = \dfrac{\phi(\sqrt{\sigma})}{\sqrt{\sigma}}$, this becomes

(8)
$$w(s) = \int_0^s \chi(\sigma)(s - \sigma)^{(n-3)/2}\, d\sigma.$$

If n is odd, the solution is given uniquely by

(9)
$$\left(\frac{n-3}{2}\right)! \chi(s) = \left(\frac{d}{ds}\right)^{(n-1)/2} w(s).$$

Hence

(10)
$$\phi(r) = \frac{1}{\left(\dfrac{n-3}{2}\right)!} r \left(\frac{d}{dr^2}\right)^{(n-1)/2} (r^{n-2}v(r)).$$

If n is even, the inversion is performed by "fractional" differentiation of order $(n-1)/2$. We obtain

$$(11) \qquad \chi(s) = \frac{1}{\sqrt{\pi}\Gamma\left(\dfrac{n-1}{2}\right)} \left(\frac{d}{ds}\right)^{n/2} \int_0^s \frac{w(\sigma)}{\sqrt{s-\sigma}}\, d\sigma.$$

Hence

$$(12) \qquad \phi(r) = \frac{2r}{\sqrt{\pi}\Gamma\left(\dfrac{n-1}{2}\right)} \left(\frac{d}{dr^2}\right)^{n/2} \int_0^r \frac{\rho}{\sqrt{r^2-\rho^2}}\, (\rho^{n-2}v(\rho))\, d\rho.$$

Thus, setting $\phi(r) = u(x, r)\omega_{n-1}/\omega_n$, we obtain as the solution of the initial value problem for the wave equation $\Delta u - u_{tt} = 0$ for *even* n the formula

$$(13) \qquad u(x, t) = \frac{1}{\Gamma\left(\dfrac{n}{2}\right)} \left(\frac{d}{dt^2}\right)^{n/2} \int_0^t \frac{r}{\sqrt{t^2-r^2}}\, (r^{n-2}Q(x, r))\, dr$$

with

$$(14) \qquad Q(x, r) = \frac{1}{\omega_n} \int \cdots \int \psi(x + r\beta)\, d\omega,$$

and for *odd* n the formula

$$(15) \qquad u(x, t) = \frac{\sqrt{\pi}}{\Gamma\left(\dfrac{n}{2}\right)} t \left(\frac{\partial}{\partial t^2}\right)^{(n-1)/2} (t^{n-2}Q(x, t)).$$

We have thus found new explicit representations for the solution of the wave equation. They are easily identified with the expressions of §12 if we take into account the modified initial condition. Thus the various formulas of §12, 1, which all were reduced to equation (10) of that section, are now established. Incidentally, it would be sufficient to restrict ourselves to the proof of (15) for odd n only.

3. The Radiation Problem for the Wave Equation. From the results derived in article 2, we can also obtain an interesting derivation of those solutions of the wave equation which correspond to *radiation phenomena*. First we ask for solutions of the wave equation depending only on $r^2 = \sum_{i=1}^n x_i^2$ and the time t. These solutions $u(r, t)$ must be even in r and satisfy the differential equation (see Ch. III, §3)

$$(16) \qquad u_{tt} - \frac{n-1}{r} u_r - u_{rr} = 0$$

which is just the Darboux equation (2) with the role of the space variable x and the time variable t interchanged. According to article 1, its solution is given by

$$(17) \qquad u(r, t) = \int_{-1}^{1} \phi(t + r\mu)(1 - \mu^2)^{(n-3)/2} \, d\mu$$

for arbitrary ϕ. For odd n, expansion of the power under the integral by the binomial theorem yields with binomial coefficients c_ν

$$(18) \quad u(r, t) = \frac{1}{r^{n-2}} \sum_{\nu=0}^{(n-3)/2} c_\nu r^{n-3-2\nu}(-1)^\nu \int_{-r}^{r} \phi(t + \rho)\rho^{2\nu} \, d\rho.$$

If g is any function such that $(d^{n-2}/dx^{n-2})g(x) = \phi(x)$, then we may integrate by parts. We obtain

$$(19) \quad u(r, t) = \frac{1}{r^{n-2}} \sum_{\nu=0}^{(n-3)/2} A_\nu r^\nu ((-1)^\nu g^{(\nu)}(t + r) - g^{(\nu)}(t - r)),$$

where $g^{(\nu)}$ denotes the ν-th derivative of g and where the coefficients A_ν can be determined,[1] for instance, by insertion into (16). Moreover, the individual terms

$$\frac{1}{r^{n-1}} \sum (-1)^\nu A_\nu r^\nu g^{(\nu)}(t + r) \quad \text{and} \quad \frac{1}{r^{n-2}} \sum A_\nu r^\nu g^{(\nu)}(t - r)$$

themselves satisfy the wave equation, since the value of g at the point $t + r$ is independent of its value at the point $t - r$ except when $r = 0$.

4. Generalized Progressing Spherical Waves. The spherical progressing waves of the form $u = \phi(t - r)/r$ in three space dimensions possess a counterpart for any odd number of space dimensions.[2] Denoting the wave operator $(\partial^2/\partial t^2) - \Delta$ by $\square$ we state: For odd n the functions

$$u = \square^{(n-1)/2}\phi(t - r),$$

with arbitrary waveforms ϕ, are solutions of the wave equation

$$\square u = 0.$$

---

[1] We obtain (cf. §12, 7)

$$A_\nu = \binom{\dfrac{n-3}{2}}{\nu} \frac{2^\nu}{\nu!} \bigg/ \binom{n-3}{\nu}.$$

[2] This interesting observation is due to K. O. Friedrichs.

For $n = 1, 3, 5$, we find solutions $\phi(t - r), 2(\phi'(t - r)/r) = f(t - r)/r$ with $f = 2\phi'$, and $(8/r^3)\phi'(t - r) + (8/r^2)\phi''(t - r)$. The last of these is not simply a relatively undistorted progressing wave, but the superposition of two such waves.

The statement follows immediately from the identity[1]

$$\square^{(n+1)/2}\phi(t - r) \equiv 0,$$

which can be verified directly but may better be derived from a more general identity concerning the operator

$$(20) \qquad L_n[u(r, t)] = u_{rr} + \frac{n - 1}{r} u_r - u_{tt}.$$

For n odd, and for arbitrary g both the function $g(t + r)$ and the function $g(t - r)$ satisfy the $(n + 1)/2$-fold iterated Darboux equation

$$(21) \qquad L_n^{(n+1)/2}[u] = 0.$$

To prove this assertion, we show first that for arbitrary ϕ and for an integer $\nu \geq 0$, the relation

$$(22) \qquad L_n \int_{-1}^{1} \phi(t + r\mu)(1 - \mu^2)^\nu \, d\mu$$
$$= d_{n,\nu} \int_{-1}^{1} \phi''(t + r\mu)(1 - \mu^2)^{\nu+1} \, d\mu$$

holds,[2] where $(\nu + 1)d_{n,\nu} = (n - 3)/2 - \nu$. If the integral on the left is denoted by $w(r, x)$, then by article 1

$$L_{2\nu+3}[w] = 0,$$

and, since $L_n = L_{2\nu+3} + [n - (2\nu + 3)/r](d/dr)$,

$$L_n[w] = \frac{n - (2\nu + 3)}{r} w_r$$
$$= \frac{n - (2\nu + 3)}{r} \int_{-1}^{1} \phi'(t + r\mu)(1 - \mu^2)^\nu \mu \, d\mu.$$

We obtain the desired form for the right-hand side by integrating by parts.

[1] On various occasions A. Weinstein has pointed out generalizations of this identity. See for example his article [3].

[2] The result of article 1 is obtained as a special case for $\nu = (n - 3)/2$.

By repeated application of formula (22) we obtain

$$L_n^{(n-1)/2} \int_{-1}^{1} \phi(t + r\mu) \, du$$

$$= d_{n,0} \, d_{n,1} \cdots d_{n,(n-3)/2} \int_{-1}^{1} \phi^{n-1}(t + r\mu)(1 - \mu^2)^{(n-1)/2} \, d\mu.$$

Here the right-hand side vanishes since $d_{n,(n-3)/2} = 0$. Hence for any ϕ with continuous derivatives up to order $n - 1$

$$(23) \qquad L_n^{(n-1)/2} \int_{-1}^{1} \phi(t + r\mu) \, d\mu = 0.$$

For $\phi = g''$ we have

$$\int_{-1}^{1} \phi(t + r\mu) \, du = \frac{g'(t + r) - g'(t - r)}{r}$$

$$= \frac{1}{n - 1} L_n[g(t + r) + g(t - r)],$$

and thus, by (23), the statement (21) follows.

Our original assertion is therefore proved, since in case of rotational symmetry Darboux's operator is identical with $\square$.

§13a. *The Initial Value Problem for Elastic Waves Solved by Spherical Means*

The method of spherical mean values in three space dimensions (cf. §13) leads to an elegant solution of the fourth order initial value problem for waves in isotropic elastic media.[1]

The small elastic deformation of the point x of the infinite medium into the point ξ may be represented by

$$\xi_i = x_i + u^i(x_1, x_2, x_3, t)$$

or, in vector notation,

$$\xi = x + u(t).$$

The stress-strain relations (see Vol. I, Ch. IV) for the stress tensor t^{ij} are

$$t^{ij} = \lambda \delta^{ij} \theta + \mu(u_{x_j}^i + u_{x_i}^j),$$

[1] In this section we take the liberty of using *ad hoc* notations which facilitate comparison with the literature on elasticity.

where λ, μ are elastic constants, δ^{ij} is Kronecker's symbol, and

$$\theta = \operatorname{div} u = \sum_i u^i_{x_i}.$$

The equations of motion are

(1) $$\sum_j t^{ij}_{x_j} = \mu \, \Delta u^i + (\lambda + \mu)\theta_{x_i} = \rho u^i_{tt}$$

or

(2) $$\mu \, \Delta u + (\lambda + \mu) \operatorname{grad} \theta = \rho u_{tt},$$

where ρ is the density. Introducing the velocities c_1, c_2 defined by

$$c_1^2 = \frac{\lambda + 2\mu}{\rho}, \qquad c_2^2 = \frac{\mu}{\rho},$$

and introducing the second order operators

$$L_1 = \frac{\partial^2}{\partial t^2} - c_1^2 \, \Delta, \qquad L_2 = \frac{\partial^2}{\partial t^2} - c_2^2 \, \Delta,$$

we easily ascertain for θ

$$L_1[\theta] = 0.$$

For the vector u we obtain the fourth order differential equation

(3) $$L_1 L_2[u] = 0.$$

Thus for the stress tensor we have the equation

$$L_1 L_2[t^{ij}] = 0 \qquad\qquad (i, j = 1, 2, 3).$$

In passing we mention special solutions of (3) under the assumption

$$t^{ij} = 0 \qquad\qquad \text{for } i \neq j.$$

Then for

$$\sum_i t^{ii} = p = (3\lambda + 2\mu)\theta,$$

we find

$$L_1[\theta] = L_1[p] = 0.$$

These solutions represent compression waves with purely normal pressure p and no shear forces.

On the other hand, shear waves without compression are represented by solutions with

$$\theta = 0, \qquad L_2[u] = 0$$

or

$$L_2[t^{ik}] = 0.$$

Since L_1 and L_2 are both wave operators, the characteristic ray cone for (3) consists of two concentric circular cones. The initial value problem for (3) with given initial vectors

$$(4) \qquad u(x, 0) = F_0(x), \qquad u_i(x, 0) = F_1(x)$$

can, remarkably, be solved and analyzed by spherical means, as was the case for the wave equation in three dimensions:

First, we obtain from (2) and (4) initial values at $t = 0$ for u_{tt}, u_{ttt} simply by substituting $t = 0$ in

$$u_{tt} = c_2^2 \, \Delta u + (c_1^2 - c_2^2) \text{ grad div } u$$

and in the equation obtained from this by differentiation:

$$(5) \quad \begin{aligned} u_{tt}(x, 0) &= c_2^2 \, \Delta F_0 + (c_1^2 - c_2^2) \text{ grad div } F_0 = F_2(x), \\ u_{ttt}(x, 0) &= c_2^2 \, \Delta F_1 + (c_1^2 - c_2^2) \text{ grad div } F_1 = F_3(x). \end{aligned}$$

Now we introduce spherical means over spheres about the point x with radius r as follows:

$$(6) \qquad I(x, r, t) = \frac{1}{4\pi} \int_{|\xi|=1} u(x + r\xi, t) \, d\omega_\xi \,,$$

$$(7) \qquad \phi_i(x, r) = \frac{1}{4\pi} \int_{|\xi|=1} F_i(x + r\xi) \, d\omega_\xi \quad (i = 0, 1, 2, 3).$$

I and ϕ are considered as even functions of r. We have

$$I(x, 0, t) = u(x, t), \qquad \phi_i(x, 0) = F_i(x).$$

By our previous calculation (§13, 1) we have

$$\Delta(rI) = (rI)_{rr}\,, \qquad \Delta(r\phi_i) = (r\phi_i)_{rr}\,;$$

moreover for $t = 0$ we have the initial values

$$\frac{d^i}{dt^i} I = \phi_i \qquad\qquad (i = 0, 1, 2, 3).$$

By taking the mean value in (3), we thus immediately obtain

(3a)
$$\left(\frac{\partial^2}{\partial t^2} - c_1^2 \frac{\partial^2}{\partial r^2}\right)\left(\frac{\partial^2}{\partial t^2} - c_2^2 \frac{\partial^2}{\partial r^2}\right)(rI) = 0.$$

The integration of (3a) under the initial conditions (7) can now be effected explicitly. Indeed, for every fixed x the function $rI(r, t)$, which according to the previous theory is uniquely determined, can be represented in the form

$$rI(r, t) = G_1(r + c_1 t) + G_2(r - c_1 t) + G_3(r + c_2 t) + G_4(r - c_2 t).$$

The four spherical waves G are determined by the initial conditions for $t = 0$:

$$G_1 + G_2 + G_3 + G_4 = r\phi_0 ,$$

$$c_1 G_1' - c_1 G_2' + c_2 G_3' - c_2 G_4' = r\phi_1 ,$$

$$c_1^2 G_1'' + c_1^2 G_2'' + c_2^2 G_3'' + c_2^2 G_4'' = r\phi_2 ,$$

$$c_1^3 G_1''' - c_1^3 G_2''' + c_2^3 G_3''' - c_2^3 G_4''' = r\phi_3 .$$

We first consider the case where

$$F_0 = F_1 = F_2 = 0, \qquad F_3 = F,$$

so that

$$\phi_0 = \phi_1 = \phi_2 = 0, \qquad \phi_3 = \phi .$$

For G_1 we find the even function

$$G_1(r) = \frac{1}{4c_1(c_1^2 - c_2^2)} \int_0^r (r - s)^2 s\phi(s)\, ds,$$

and

$$G_2(r) = -G_1(r), \qquad G_3(r) = -\frac{c_1}{c_2} G_1(r), \qquad G_4(r) = \frac{c_1}{c_2} G_1(r).$$

Denoting the solution with initial data 0, 0, 0, F by $U(F)$, we thus have

$$U(F) = U(x, t) = I(x, 0, t) = 2G_1'(c_1 t) - \frac{2c_1}{c_2} G_1'(c_2 t).$$

Hence

$$U(F) = V(F) - W(F),$$

with

$$V(F) = \frac{1}{c_1(c_1^2 - c_2^2)} \int_0^{c_1 t} (c_1 t - s)s\phi(s)\,ds$$

$$W(F) = \frac{1}{c_2(c_1^2 - c_2^2)} \int_0^{c_2 t} (c_2 t - s)s\phi(s)\,ds.$$

Here V and W satisfy

$$\frac{\partial^2}{\partial t^2} V(F) = c_1^2 \,\Delta V(F) = V(c_1^2 \,\Delta F)$$

$$\frac{\partial^2}{\partial t^2} W(F) = c_2^2 \,\Delta W(F) = W(c_2^2 \,\Delta F).$$

This corresponds to decomposition of the wave into an irrotational or dilatational part (curl $U = 0$) and an equivoluminous part (div $U = 0$) propagating with velocities c_1 and c_2 respectively.

Moreover, we may also write

$$U(F) = \frac{1}{c_1(c_1^2 - c_2^2)} \left[\int_0^{c_2 t} \phi(s) \frac{c_1 - c_2}{c_2} s^2\,ds \right.$$

$$\left. + \int_{c_2 t}^{c_1 t} \phi(s)(c_1 st - s^2)\,ds \right].$$

Hence

$$4\pi c_1(c_1^2 - c_2^2)U(F) = \iiint\limits_{|\xi| < c_2 t} \frac{c_1 - c_2}{c_2} F(x + \xi)\,d\xi$$

$$+ \iiint\limits_{c_2 t < |\xi| < c_1 t} \frac{c_1 t - |\xi|}{|\xi|} F(x + \xi)\,d\xi.$$

The solution u corresponding to the general initial data F_0, F_1, F_2, F_3 is given by

$$U(F_3 - (c_1^2 + c_2^2)\,\Delta F_1) + \frac{d}{dt} U(F_2 - (c_1^2 + c_2^2)\,\Delta F_0)$$

$$+ \frac{d^2}{dt^2} U(F_1) + \frac{d^3}{dt^3} U(F_0).$$

Hence, by (5)

$$u = U(-c_1^2 \Delta F_1 + (c_1^2 - c_2^2) \operatorname{grad} \operatorname{div} F_1)$$

$$+ \frac{d}{dt} U(-c_1^2 \Delta F_0 + (c_1^2 - c_2^2) \operatorname{grad} \operatorname{div} F_0) + \frac{d^2}{dt^2} U(F_1)$$

$$+ \frac{d^3}{dt^3} U(F_0).$$

The following remarkable fact can be read off directly: There is no contribution to the solution from the interior of the sphere of radius $c_2\, t$, since the expression $-c_1^2 \Delta F_1 + (c_1^2 - c_2^2) \operatorname{grad} \operatorname{div} F_1$ is a divergence. Hence *the domain of dependence in the strict sense does not contain the interior of the inner sphere of radius $c_2\, t$ around x.* In fact, it is the spherical shell between this sphere and an outer sphere of radius $c_1\, t$. (Cf. §15, 4.)

§14. *Method of Plane Mean Values. Application to General Hyperbolic Equations with Constant Coefficients*

We turn in this section to a general method for treating any linear hyperbolic equation or system with constant coefficients. In accordance with the introductory remarks in §11, 1 the solution of Cauchy's problem will be achieved by representation of the solution as a superposition of plane waves,[1] which avoids difficulties arising from interchanging the order of operations. A more penetrating analysis in the next section will, moreover, remove the assumption of constant coefficients.[2]

1. General Method. Let $L[u]$ be any linear hyperbolic differential operator of order k in the function $u = u(x_1, \cdots, x_n, t) = u(x, t)$. We consider Cauchy's problem in the form

$$(1) \qquad L[u] = g(x_1, \cdots, x_n, t),$$

where g is at least $k/2$ times differentiable. Initial values are given for the function u and its first $(k - 1)$ time derivatives:

$$(2) \quad \frac{\partial^i u}{\partial t^i} = u_{t^i}(x_1, \cdots, x_n, 0) = h_{(i)}(x_1, \cdots, x_n)$$

$$\text{for } i = 0, 1, \cdots, k - 1,$$

---

[1] See R. Courant and A. Lax [1], p. 501.
[2] Although the discussion in §15 is more general than that given here, the present version of the method seems justified since it is more direct.

where the functions h_i are smooth in the x-space. Since $L[u]$ is hyperbolic, the domain of dependence of any fixed point $(x_1, \cdots, x_n, t)$ is bounded. Without loss of generality we may therefore assume that the functions g, $h_{(i)}$, and u vanish outside some large sphere in x-space.

One can reduce the problem for (1) to problems for hyperbolic equations in only two independent variables. Such a reduction is suggested by integrating (1) with respect to $n - 1$ of the x-variables, e.g., $x_2, \cdots, x_n$, then under our assumptions all the terms containing derivatives with respect to any of these variables are cancelled. What remains, therefore, is a differential equation for the function of two variables

$$U(x_1, t) = \int \cdots \int u \, dx_2 \cdots dx_n .$$

More generally: We select an arbitrary unit vector

$$\alpha = (\alpha_1, \cdots, \alpha_n)$$

and introduce instead of $x_1, \cdots, x_n$ a new orthogonal coordinate system $y_1, \cdots, y_n$, where

$$(3) \qquad y_1 = (\alpha x) = \sum_{\nu=1}^{n} \alpha_\nu x_\nu = p$$

and $y_2, \cdots, y_n$ are other linear combinations of the x_ν consistent with this choice of y_1. We then consider the integrals of the function u (or other functions) over the planes $p = $ const. with normals α. These integrals are denoted by

$$(4) \qquad I(p, t, \alpha) = I(p, t, \alpha, u) = \iint_{(x\alpha)=p} u(x, t) \, dS_\alpha ,$$

where dS_α is the surface element of the plane. One may also write

$$(4a) \qquad I(p, t, \alpha) = \frac{\partial}{\partial p} \iiint_{(x\alpha) \geq p} u(x, t) \, dx,$$

where the integral is extended over the half-space $(x\alpha) \geq 0$. Or using Dirac's δ-function we have

$$(4b) \qquad I(p, t, \alpha) = \iiint u(x, t) \, \delta((x\alpha) - p) \, dx,$$

where the integral is formally extended over the whole x-space. If the hyperplane contains a fixed point z we have $p = (z, \alpha)$ and

$$(4c) \quad I(p, t, \alpha) = \iint_{(x-z,\alpha)=0} u \, dS_\alpha = \iiint u(x, t) \, \delta((x - z) \, \alpha) \, dx.$$

Now, after (1) is integrated over $y_2, \cdots, y_n$, all space derivatives except those with respect to p drop out because of the vanishing of u and its derivatives at infinity. We thus obtain a differential equation for the new dependent function $I(p, t, \alpha)$ of only two variables, p, t (α being a parameter):

$$(5) \quad L^\alpha(I(p, t, \alpha)) = g(p, t, \alpha).$$

The right-hand side of (5) is given by

$$(6) \quad g(p, t, \alpha) = I(p, t, \alpha, g(x, t))$$

and the initial values for $t = 0$ are

$$I(p, \alpha) = I(p, 0, \alpha, h(x))$$

$$(7) \quad \cdots\cdots\cdots\cdots\cdots\cdots\cdots\cdots\cdots\cdots$$

$$\frac{\partial^{k-1}}{\partial t^{k-1}} I(p, t, \alpha) = I(p, 0, \alpha, h_{k-1}(x)).$$

On the basis of the definition of hyperbolicity given in Chapters III, V, VI, the reduced equation (5) is easily seen to be hyperbolic for every choice of α. For each α, therefore, (5) and (7) form a solvable Cauchy problem in two independent variables for the unknown function $I(p, t, \alpha)$. Hence these functions can be constructed by the methods of Chapter V. We assume that this has been done for each unit vector α, and furthermore that the resulting solutions $I(p, t, \alpha)$ depend continuously on α.[1]

This leaves us with the problem of recovering a function $u(x, t)$ if its plane integrals I^α are known.[2] We proceed here as follows: We form for $p = (z\alpha)$ plane integrals $I((z\alpha), t)$ for all planes $x\alpha = z\alpha = p$ through a fixed point O with the coordinates $z_1, \cdots, z_n$;

[1] The latter assumption is nontrivial; there exist cases in which it is not satisfied. It should, however, be emphasized that the assumption is trivially satisfied if we presuppose existence and continuity of the solution u and aim merely at a representation.

[2] See F. John [4], pp. 7–13.

then we integrate over the unit sphere $\alpha^2 = 1$. Thus we obtain a function

$$(8) \qquad V(z, t) = V(z_1, \cdots, z_n, t) = \iint_{\alpha^2=1} I^\alpha(z\alpha, t) \, d\alpha,$$

where the plane integral I^α refers to the plane $(x\alpha) = (z\alpha) = p$.

Obviously, therefore, V is expressible as a weighted space average of u, with the weight w depending only on the distance between x and z: $|x - z| = [(x_1 - z_1)^2 + \cdots + (x_n - z_n)^2]^{1/2} = r$,

$$(9) \qquad \begin{aligned} &V(z, t) \\ &= \iint_{-\infty}^{\infty} \cdots \int u(x_1, \cdots, x_n, t) \, w(|x - z|) \, dx_1 \cdots dx_n, \end{aligned}$$

where $w = w(r)$ is a positive function. Precisely,

$$(10) \qquad V(z, t) = \omega_{n-1} \int \cdots \int \frac{u(x, t)}{|x - z|} \, dx,$$

where the integral is extended over the whole space.

To prove (10) (see also §15, 2), we replace u in (4) by any function $f(r)$. Then we have

$$I^\alpha = \omega_{n-1} \int_0^\infty r^{n-2} f(r) \, dr,$$

where ω_{n-1} is the surface of the $(n - 1)$-dimensional unit sphere. Integrating with respect to α, we obtain, according to (6),

$$V = \omega_n \omega_{n-1} \int_0^\infty r^{n-2} f(r) \, dr.$$

On the other hand, we have

$$V = \omega_n \int_0^\infty f(r) r^{n-1} w(r) \, dr.$$

Since $f(r)$ is arbitrary, we infer that $w(r) = \omega_{n-1}(1/r)$.

Now we assume that the number n of space dimensions is odd. Then the solution u is easily found by applying an iterated Laplacian operator to both sides of (10). Observing that the Laplacian (with respect to z or x) of any power $|z - x|^k$ is simply a constant (depending on n and k) multiplied by $|z - x|^{k-2}$, we obtain

$$(11) \quad \Delta^{(n-3)/2} V(x, t) = b_n \int_{-\infty}^{\infty} \cdots \int \frac{u(z_1, \cdots, z_n, t)}{|z - x|^{n-2}} dz_1 \cdots dz_n,$$

where b_n depends only on n. Poisson's integral formula in Ch. IV, §2 shows that the Laplacian of the right-hand side is given by

$$(12) \quad \Delta_x \iint_{-\infty}^{\infty} \cdots \int \frac{u(z_1, \cdots, z_n, t)}{|z - x|^{n-2}} dz_1 \cdots dz_n = a_n u(z_1, \cdots, z_n, t),$$

where a_n is another constant depending on n, and where Δ_x means the Laplacian with respect to x. Substituting from (11) and combining constants, we find for odd n the following explicit solution u of our original Cauchy problem:

$$(13) \quad u(z, t) = C_n \Delta_x^{(n-1)/2} V(z, t).$$

The factor C_n is easily determined as

$$(13a) \quad C_n = \frac{1}{2(2\pi i)^{n-1}} = \frac{1}{2(2\pi)^{n-1}} (-1)^{(n-1)/2}.$$

The form of (13) suggests that it might also be valid for even n provided that fractional powers of the Laplacian operator were suitably defined. We shall not elaborate on this subject, since we always can descend from odd n to even n; moreover a more general discussion for even and odd dimensions of x will be given in §15.

Finally it should be emphasized that the method and results of this section apply likewise to hyperbolic systems of differential equations: If $L[u]$ is a vectorial differential operator with constant coefficients, operating on a vector u, and if g likewise is a vector, then the result remains literally the same. In particular $L[u] = g$ may be a system of equations of the first or second order. Of course, the initial data must correspond to the order of $L[u]$.

2. Application to the Solution of the Wave Equation. The method of article 1 will now be tested by constructing again the solution of the wave equation

$$(14) \quad L[u] \equiv \Delta u - u_{tt} = 0$$

under the initial conditions

$$(15) \quad u(x, t) = 0, \quad u_t(x, t) = \psi(x),$$

considered earlier in §12. As pointed out in article 1, there is no loss of generality in assuming that $\psi(x)$ vanishes identically outside some large fixed sphere S in x-space.

In all transformed coordinate systems described (see article 1) the function $I(p,\ t)\ =\ \int\int_{-\infty}^{\infty} \cdots \int u(p,\ y_2,\ \cdots,\ y_n,\ t)dy_2 \cdots dy_n$ satisfies the same differential equation

$$(16) \qquad L^{\alpha}[I_{\alpha}] \equiv I_{pp} - I_{tt} = 0$$

and the initial conditions

$$(17) \qquad I^{\alpha}(p, 0) = 0, \qquad I_t^{\alpha}(p, 0) = \chi^{\alpha}(p),$$

where

$$(18) \qquad \chi^{\alpha}(p) = \int_{-\infty}^{\infty} \cdots \int \psi(p, y_2, \cdots, y_n)\ dy_2 \cdots dy_n.$$

The solution of (16) is (see Ch. I, §7)

$$(19) \quad I^{\alpha}(p, t) = \frac{1}{2} \int_{p-t}^{p+t} \chi(\eta)\ d\eta = \frac{1}{2} \int \cdots \int_{(p-t \leq \eta \leq p+t)} \psi\ d\eta\ dy_2 \cdots dy_n.$$

Since ψ vanishes outside of S, this solution must depend continuously on α if ψ is piecewise continuous. Thus we may form the surface integral

$$(20) \qquad V(x_1, \cdots, x_n, t) = \iint_{|\alpha|=1} I^{\alpha}(\alpha \cdot x, t)\ d\alpha$$

and obtain the solution $u(x,\ t)$ by applying the iterated Laplacian:

$$(21) \qquad u(x, t) = C_n \cdot \Delta^{(n-1)/2} V(x, t),$$

where $C_n = 1/2(2\pi i)^{n-1}$.

We carry this out in detail only for odd n; the case of even n[1] may then be treated by Hadamard's method of descent. From (19), (20), and (21), we find

$$u(x, t) = C_n \frac{\partial^{n-1}}{\partial t^{n-1}} \iint_{|\alpha|=1} I^{\alpha}(\alpha x, t)\ d\alpha$$

---

[1] See also the treatment in §15, which is related to the procedure of the present section.

$$(22) \qquad = \frac{C_n}{2} \frac{\partial^{n-1}}{\partial t^{n-1}} \iint_{|\alpha|=1} d\alpha \iint_{|\alpha\xi|\leq t} \psi(x + \xi) \, d\xi$$

$$= \frac{C_n}{2} \frac{\partial^{n-1}}{\partial t^{n-1}} \iint_{-\infty}^{\infty} \cdots \int \psi(x + \xi) K(\xi, t) \, d\xi,$$

where $K(\xi, t)$ is defined by the surface integral

$$(23) \qquad K(\xi, t) = \int \cdots \int_{\substack{|\alpha|=1 \\ |\alpha\xi| \leq t}} d\alpha$$

and represents, geometrically, the surface area of a zone of height $2t/|\xi|$ on the n-dimensional unit sphere.

This surface integral is easily evaluated; it is

$$(24) \qquad K(\xi, t) = \chi\left(\frac{t}{|\xi|}\right)$$

with the function $\chi(s)$ given by

$$(25) \qquad \chi(s) = \begin{cases} 2\omega_{n-1} \int_0^1 (1 - \lambda^2)^{(n-3)/2} \, d\lambda = \text{const.} & \text{for } |s| > 1, \\ 2\omega_{n-1} \int_0^{|s|} (1 - \lambda^2)^{(n-3)/2} \, d\lambda & \text{for } |s| < 1. \end{cases}$$

Substituting (24) into (22), and writing the mean value of ψ on the surface of a sphere of radius r about the point x

$$(26) \qquad Q(x, r) = \frac{1}{\omega_n} \iint_{|\xi|=1} \psi(x + r\xi) \, d\xi,$$

we obtain the solution $u(x, t)$ in the form

$$u(x, t) = \omega_n \frac{C_n}{2} \frac{\partial^{n-1}}{\partial t^{n-1}} \int_0^{\infty} Q(x, r) \chi\left(\frac{t}{r}\right) r^{n-1} \, dr.$$

Performing one differentiation explicitly yields

$$u(x, t) = \omega_n \frac{C_n}{2} \frac{\partial^{n-2}}{\partial t^{n-2}} \int_0^{\infty} Q(x, r) \chi'\left(\frac{t}{r}\right) r^{n-2} \, dr,$$

or, in view of (25),

$$(27) \qquad u(x, t) = A_n \frac{\partial^{n-2}}{\partial t^{n-2}} \int_t^{\infty} Q(x, r)(r^2 - t^2)^{(n-3)/2} r \, dr,$$

where $A_n = \omega_{n-1} \omega_n C_n$ is again a constant depending only on n.

As in §12, 2 (27) can be rewritten as

$$(28) \quad u(x, t) = (-1)^{(n-1)/2} A_n \frac{\partial^{n-2}}{\partial t^{n-2}} \int_0^t Q(x, r)(t^2 - r^2)^{(n-3)/2} r \, dr.$$

Both of these expressions involve the initial function ψ only through the values of Q and of its first $(n-3)/2$ derivatives with respect to r for $r = t$, in keeping with Huyghens' principle.

As in §12, 2 the constant factor is determined by specializing $\psi = 1$, $Q = 1$, $u = t$. Thus we obtain for odd n the solution

$$(29) \quad u(x, t) = \frac{1}{(n-2)!} \frac{\partial^{n-2}}{\partial t^{n-2}} \int_0^t Q(x, r)(t^2 - r^2)^{(n-3)/2} r \, dr,$$

as stated in §12, 2 (16). It was proved in §12, 3 that the same formula is true for even values of n.

§14a. *Application to the Equations of Crystal Optics and Other Equations of Fourth Order*[1]

1. Solution of Cauchy's Problem. In line with §3a Cauchy's problem for the system of differential equations governing crystal optics[2] (see §3, 3) can easily be reduced to Cauchy's problem for the same single fourth-order differential equation which is satisfied by each component of the electric or the magnetic vector. This equation has the form

$$(1) \qquad \tau^2 P(\xi, \tau) w(x, t) = 0,$$

where P is a homogeneous polynomial of fourth degree in the symbols $\partial/\partial x_i = \xi_i$ and $\partial/\partial t = \tau$, such that $P(0, \tau) = \tau^4$. We have $\tau^2 P = Q$, where Q is the characteristic form of the original system; the solutions of (1) always generate solutions of the equation

$$(1a) \qquad P(\xi, \tau) u = 0$$

with $u = \tau^2 w$.

In this article we shall quite generally solve the Cauchy problem for (1a), a fourth order equation not necessarily that of crystal optics, with the initial conditions

$$(2) \qquad \tau^\kappa u(x, 0) = \begin{cases} 0 & \text{for } \kappa = 0, 1, 2 \\ f(x) & \text{for } \kappa = 3 \end{cases}$$

[1] See F. John [4] for a simplified presentation of the fundamental work of G. Herglotz.

[2] Again we use in this article some *ad hoc* notations.

for any hyperbolic equation of the form (1).[1] As it is easily seen, the Cauchy problem for $(1a)$ with arbitrary data as well as the Cauchy problem for the original system can be solved by suitable combinations of such solutions and their derivatives. (See §3a, 5.)

Considering a vector $\xi = (\xi_1, \xi_2, \xi_3)$ and a scalar λ, the characteristic equation corresponding to (1) is

$$(3) \qquad P(\xi, \lambda) = 0.$$

For every real ξ, equation (3) must have four real roots λ if (1) is to be hyperbolic. For simplicity, we confine our consideration to the case where these roots are distinct for all except possibly a finite number of unit vectors ξ, and where even for these special ξ no roots of multiplicity higher than two may occur.[2] Then we shall be able to solve Cauchy's problem explicitly. The special case

$$(4) \qquad P(\xi, \tau) = F(\xi, \tau) = \tau^4 - \Psi(\xi)\tau^2 + \rho^2\Phi(\xi)$$

corresponding to crystal optics satisfies these conditions (see §3a, article 3).

As before, we assume that $f(x)$ vanishes for x outside some sphere. Transforming to the y-variables, $y_1 = (\alpha x)$, we find that the plane integral

$$(5) \qquad I^\alpha(y, t) = \int_{-\infty}^{\infty} \int u(y, y_2, y_3, t)\, dy_2\, dy_3$$

satisfies the equation

$$(6) \qquad P(\alpha\eta, \tau)I^\alpha(y, t) = 0,$$

where

$$\eta = \partial/\partial y.$$

[1] The reader may find it useful to express the calculations by using the δ-function as in the procedure given later in §15. Also one may solve the problem by a straightforward application of the method of the Fourier integral. (See Ch. III, §5, and this chapter §12.)

[2] That this restriction is not mandatory was shown by F. John [4], Ch. II. However, it should be repeated that the case of multiple characteristic elements presents difficulties which have not yet been overcome in general, so that one must be contented with the analysis of more special classes of problems.

In addition, I^α satisfies the initial conditions

$$(7) \qquad \tau^\kappa I^\alpha(y, 0) = \begin{cases} 0 & \text{for } \kappa = 0, 1, 2 \\ f^\alpha(y) & \text{for } \kappa = 3, \end{cases}$$

where

$$(8) \qquad f^\alpha(y) = \iint\limits_{-\infty}^{\infty} f(y, y_2, y_3) \, dy_2 \, dy_3.$$

We denote the four solutions λ of the equation $P(\alpha, \lambda) = 0$ by $\lambda_j = \lambda_j(\alpha)$, where $j = 1, \cdots, 4$. Because of the homogeneity of P it is then easily seen that any linear combination of the form

$$(9) \qquad I^\alpha(y, t) = \sum_{j=1}^{4} w_j(y + \lambda_j t),$$

with arbitrary functions w_j, satisfies equation (6). To satisfy (7), we first consider those α for which the four roots λ_j are distinct. Newton's identities

$$(10) \qquad \sum_{j=1}^{4} \frac{(\lambda_j)^\kappa}{P_\lambda(\alpha, \lambda_j)} = \begin{cases} 0 & \text{for } \kappa = 0, 1, 2, \\ 1 & \text{for } \kappa = 3, \end{cases}$$

for any fourth degree polynomial in λ with the highest coefficient 1 and distinct roots, lead to the choice

$$(11) \qquad w_j(y) = \frac{g^\alpha(y)}{P_\lambda(\alpha, \lambda_j)},$$

where $g^\alpha(y)$ is any function such that

$$(12) \qquad \eta^3 g^\alpha(y) = f^\alpha(y).$$

Hence in the case of distinct roots λ_j the solution of (6), (7) is given by

$$(13) \qquad I^\alpha(y, t) = \sum_{j=1}^{4} \frac{g^\alpha(y + \lambda_j t)}{P_\lambda(\alpha, \lambda_j)}.$$

It would appear from (13) that this solution I^α depends on the particular choice of the function g^α. This is not so, however, since (12) allows g^α to vary only by addition of a quadratic function of y, and (10) insures that changing g^α by a quadratic function does not affect the value of I^α given by (13).

Furthermore, the definition (8) implies $f^\alpha(y) = 0$ uniformly in α for large $|y|$. Therefore there exist solutions $g^\alpha(y)$ of (12) which vanish for large positive y, and also other solutions which vanish for large negative y. Since I^α does not depend on the specific choice of g^α it follows that, for any fixed t and all α, $I^\alpha(y, t)$ is identically zero if $|y|$ is sufficiently large. By similar reasoning, one concludes that I^α is a continuous function of α as long as the values λ_j remain distinct.

We can now find I^α for those isolated values of α which correspond to double roots λ_j by making use of this continuity. Suppose, for example, that for a particular value of $\alpha = \alpha^*$ the roots λ_1 and λ_2 coincide. Then for α near α^*, there exists a root λ' of $P_\lambda(\alpha, \lambda) = 0$ between λ_1 and λ_2. We can therefore rewrite the first two terms of (13) in the form

$$\frac{g^\alpha(y + \lambda_1 t) - g^\alpha(y + \lambda' t)}{P_\lambda(\alpha, \lambda_1)} + \frac{g^\alpha(y + \lambda_2 t) - g^\alpha(y + \lambda' t)}{P_\lambda(\alpha, \lambda_2)}$$

$$+ g^\alpha(y + \lambda' t) \left[\frac{1}{P_\lambda(\alpha, \lambda_1)} + \frac{1}{P_\lambda(\alpha, \lambda_2)} \right].$$

If we do this for each α near α^* and pass to the limit $\alpha \to \alpha^*$, we obtain for the singular value of $\alpha = \alpha^*$ the following function I^α:

$$I^\alpha(y, t) = \frac{2\mathrm{th}^\alpha(y + \lambda_1 t)}{P_{\lambda\lambda}(\alpha, \lambda_1)} - \frac{2g^\alpha(y + \lambda_1 t) P_{\lambda\lambda\lambda}(\alpha, \lambda_1)}{3 P_{\lambda\lambda}(\alpha, \lambda_1)^2}$$

(14)
$$+ \sum_{j=3}^{4} \frac{g^\alpha(y + \lambda_j t)}{P_\lambda(\alpha, \lambda_j)},$$

where $h^\alpha(y) = \eta g^\alpha(y)$.

We must still verify that this function I^α satisfies the differential equation (6) and the initial conditions (7). Direct substitution shows that (6) is satisfied. I^α also satisfies (7) and is independent of the choice of g^α as follows from the limiting form

(15)
$$\frac{2\kappa(\lambda_1)^{\kappa-1}}{P_{\lambda\lambda}(\alpha, \lambda_1)} - \frac{2(\lambda_1)^\kappa P_{\lambda\lambda\lambda}(\alpha, \lambda_1)}{3 P_{\lambda\lambda}(\alpha, \lambda_1)^2} + \sum_{j=3}^{4} \frac{(\lambda_j)^\kappa}{P_\lambda(\alpha, \lambda_j)}$$

$$= \begin{cases} 0 & \text{for } \kappa = 0, 1, 2 \\ 1 & \text{for } \kappa = 3 \end{cases}$$

of Newton's identities in the case $\lambda_2 = \lambda_1$. Since, by construction,

(14) joins continuously onto (13), all conditions required for the validity of the method of §14, 1 are met.[1]

Applying the general method of §14, 1 to (10) and remembering that $n = 3$, we find

(16)
$$u(x, t) = -\frac{1}{8\pi^2} \Delta \iint\limits_{|\alpha|=1} I^\alpha(\alpha x, t) \, d\alpha$$

$$= -\frac{1}{8\pi^2} \iint\limits_{|\alpha|=1} \left(\sum_{j=1}^{4} \frac{\Delta g^\alpha(\alpha x + \lambda_j t)}{P_\lambda(\alpha, \lambda_j)} \right) d\alpha.$$

If, in conformity with (12), we choose

(17)
$$g^\alpha(y) = \frac{1}{2} \int_{-\infty}^{y} (y - q)^2 f^\alpha(q) \, dq,$$

then for $|\alpha| = 1$ and any p we have

(18)
$$\Delta g^\alpha(\alpha x + p) = \int_{-\infty}^{\alpha x + p} f^\alpha(q) \, dq,$$

or, in view of (8),

(19)
$$\Delta g^\alpha(\alpha x + p) = \iiint\limits_{\alpha z \leq p} f(x + z) \, dz_1 \, dz_2 \, dz_3 .$$

Substituting this expression into (16), we find

(20)
$$u(x, t) = \iiint\limits_{-\infty}^{\infty} f(x + z) K(z, t) \, dz_1 \, dz_2 \, dz_3 ,$$

with the kernel

(21)
$$K(z, t) = -\frac{1}{8\pi^2} \sum_{j=1}^{4} \iint\limits_{\substack{|\alpha|=1 \\ \alpha z \leq \lambda_j(\alpha) t}} \frac{d\alpha}{P_\lambda(\alpha, \lambda_j(\alpha))} .$$

In the case of crystal optics, and, more generally, whenever $P(\xi, 0) \neq 0$ for all real $\xi \neq 0$, the roots λ_1 and λ_2 are positive for all

---

[1] It is obvious how (14) must be modified if also the roots λ_3 and λ_4 coalesce. This behavior—the pairwise occurrence of double roots—is shown by the equations of crystal optics.

α, while λ_3 and λ_4 are negative. In any such case, the normal surface N corresponding to the differential equation (1), i.e., the surface

$$(22) \qquad P(\xi, 1) = 0,$$

is intersected by each ray from the origin in two (possibly coincident) points. Thus N consists of two sheets, N_1 and N_2. These sheets are given parametrically by $\xi = \alpha/\lambda_1(\alpha)$ and $\xi = \alpha/\lambda_2(\alpha)$, respectively, where the parameter α varies over the surface of the unit sphere. By our assumptions, the inner sheet N_1 and the outer sheet N_2 are distinct and can have only a finite number of points in common.

We proceed to investigate some properties of this kernel. First of all, it follows directly from (10) that

$$(23) \qquad K(z, 0) = 0.$$

Secondly, if we arrange the roots $\lambda_j(\alpha)$ in descending order:

$$\lambda_1(\alpha) \geq \lambda_2(\alpha) \geq \lambda_3(\alpha) \geq \lambda_4(\alpha)$$

and observe

$$(24) \qquad \lambda_3(\alpha) = -\lambda_2(-\alpha), \qquad \lambda_4(\alpha) = -\lambda_1(-\alpha),$$

we can easily reduce K to a form which involves only λ_1 and λ_2, viz:

$$(25) \quad K(z, t) = -\frac{1}{8\pi^2} \sum_{j=1}^{2} \left[\iint_{\substack{|\alpha|=1 \\ \alpha z \leq \lambda_j(\alpha)t}} \frac{d\alpha}{P_\lambda(\alpha, \lambda_j(\alpha))} - \iint_{\substack{|\alpha|=1 \\ \alpha z \geq \lambda_j(\alpha)t}} \frac{d\alpha}{P_\lambda(\alpha, \lambda_j(\alpha))} \right].$$

2. Further Discussion of the Solution. Domain of Dependence. Gaps.[1] We now transform the "fundamental solution" K, given by (25) as an integral over a portion of the unit sphere, into an integral over N. Let dN be the element of area on N, and denote by grad P the vector whose i-th component is $P_{\xi_i}(\xi, 1)$. It is evident by projection from the origin that

$$(26) \qquad dN = |\xi|^2 \frac{|\operatorname{grad} P|}{|\alpha \cdot \operatorname{grad} P|} d\alpha.$$

But by the homogeneity of P

$$(27) \qquad \sum_{i=1}^{3} \xi_i P_{\xi_i}(\xi, 1) + P_\lambda(\xi, 1) = 4P(\xi, 1) = 0,$$

[1] General conditions for existence of gaps in the domain of dependence are given in I. G. Petrovskii [3].

for ξ on N. Hence for ξ on N_j

$$| \alpha \text{ grad } P | = \left| \sum_{i=1}^{3} \alpha_i P_{\xi_i}(\xi, 1) \right| = | \xi |^{-1} \left| \sum_{i=1}^{3} \xi_i P_{\xi_i}(\xi, 1) \right|$$

$$(28) \qquad = | \xi |^{-1} | P_\lambda(\xi, 1) | = | \xi |^2 \left| P_\lambda \left(\alpha, \frac{1}{| \xi |} \right) \right|$$

$$= | \xi |^2 | P_\lambda(\alpha, \lambda_j(\alpha)) | = \epsilon | \xi |^2 P_\lambda(\alpha, \lambda_j(\alpha)),$$

where

$$\epsilon = \begin{cases} 1 & \xi \text{ on } N_1, \\ -1 & \xi \text{ on } N_2. \end{cases}$$

Therefore

$$(29) \qquad \frac{d\alpha}{P_\lambda(\alpha, \lambda_j(\alpha))} = \frac{\epsilon \, dN}{| \text{grad } P |}.$$

Moreover, for $j = 1, 2$ the quantity $(\lambda_j t - \alpha z)$ has the same sign as $(t - \xi z)$. Thus (25) becomes[1]

$$(30) \qquad K(z, t) = -\frac{1}{8\pi^2} \iint_N \frac{\epsilon \, sgn(t - \xi z) \, dN}{| \text{grad } P(\xi, 1) |}.$$

If $\xi z \leq t$ for all ξ on N, $K(z, t)$ is identically equal to the constant

$$(30a) \qquad -\frac{1}{8\pi^2} \iint_N \frac{\epsilon \, dN}{| \text{grad } P(\xi, 1) |}.$$

This occurs under the condition that the plane $z\xi = t$ does not intersect the outer sheet N_2, or under the equivalent condition that z/t lies inside the surface S_2. S_2 is the polar reciprocal with respect to the unit sphere of the convex hull of N_2.

Conversely, if the plane $z\xi = t$ intersects the inner sheet N_1, the kernel $K(z, t)$ is zero. To prove this, let ξ_0 be any vector on this plane and interior to N_1. In (30) put $\xi = \xi' + \xi_0$. The integral for K is then taken over the surface N' given by

$$(31) \qquad P(\xi' + \xi_0, 1) = 0.$$

This surface is obtained from N by translation, so that the gradient

---

[1] Under our assumptions, the denominator in the integrand vanishes only to first order at the singular points. Hence this integral converges.

and surface element at corresponding points are the same. Hence

$$(32) \qquad K(z, t) = -\frac{1}{8\pi^2} \iint_N \frac{\epsilon \, sgn(-\xi'z) \, dN'}{|\,\text{grad } P(\xi' + \xi_0, 1)\,|} \,.$$

But this last integral is just $\bar{K}(z, 0)$, where $\bar{K}$ is the kernel which would be obtained from (21) if the polynomial $P(\xi, \lambda)$ in the original differential equation (1) were replaced by the polynomial

$$(33) \qquad \bar{P}(\xi, \lambda) = P(\xi + \xi_0, \lambda).$$

Since ξ_0 is inside N_1, any line through ξ_0 intersects N in four points. Hence $\bar{P}$ has four real roots λ_j for each ξ. Therefore relation (23) is applicable to $\bar{K}$, and our contention is proved.

Since N is a fourth-order surface and any line intersecting N_1 must intersect N_2 at least twice, it follows that no line can intersect N_1 in three points. Hence N_1 is convex.[1] Therefore we can state the preceding result in the following form: $K(z, t) = 0$ if z/t lies outside the polar reciprocal S_1 of N_1 with respect to the unit sphere.

Returning to the original solution $u(x, t)$, we take the point (x, t) as vertex and construct two conical sheets C_1 and C_2 with axes in the backward t-direction. The cone C_j is defined by the property that its section at unit distance from its vertex is congruent to S_j. Clearly, C_2 is contained in the interior of C_1. From (20), we may now conclude that the domain of dependence of the point (x, t) is precisely the intersection of the interior of C_1 with the initial plane $t = 0$. Furthermore, the subset of this domain which is also interior to C_2 has the remarkable property that each of its points makes the same contribution (30a) to the solution.

A substantial simplification occurs if the normal surface N is symmetrical about the origin (as it is for the equations of crystal optics). In this case,

$$\iint_{\substack{N \\ \xi z \geq t}} \frac{\epsilon \, dN}{|\,\text{grad } P(\xi, 1)\,|} = \iint_{\substack{N \\ -\xi z \geq t}} \frac{\epsilon \, dN}{|\,\text{grad } P(\xi, 1)\,|} = \iint_{\substack{N \\ \xi z \leq -t}} \frac{\epsilon \, dN}{|\,\text{grad } P(\xi, 1)\,|} \,,$$

so that (30) yields

$$(34) \qquad K(z, t) = -\frac{1}{8\pi^2} \iint_{\substack{N \\ |\xi z| \leq t}} d^*N_\xi \,,$$

---

[1] This argument is a repetition from §3.

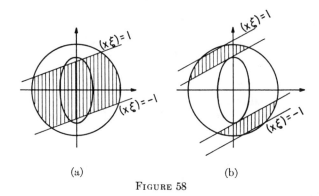

(a)   (b)

FIGURE 58

where $d^*N_\xi = \epsilon\, dN/\,|\,\text{grad}\, P(\xi, 1)\,|$ is called the element of "pseudo-area" of the normal surface N. Thus the "fundamental solution" K is geometrically interpreted as the "pseudo-area" of the portion of N lying between the two parallel planes $z\xi = \pm t.$[1] See Figure 58.

Finally we call attention to the fact that our fourth-order differential equation of crystal optics must be differentiated twice with respect to t to obtain the sixth-order equation which is the complete result of eliminating from the original system (7) of §3a all components but one of the electromagnetic vector. Assuming (7a) of §3a, it is not difficult to represent the solution of Cauchy's problem for this vector in terms of differential operators on solutions of the fourth-order equation (1); thereby it is seen:

The interior sheet of the ray cone, which corresponds to the exterior sheet of the normal cone, contributes to the solution of Cauchy's problem for our fourth-order equation a value constant in time for every point in space.

The vector solution of Cauchy's problem of crystal optics at any point x does not depend on the initial values inside the interior sheet of the cone of dependence; the interior is a gap.[2] This result is natural; it expresses the fact that the effect of a local disturbance at a point O in x-space arrives at a point x according to the faster speed of propagation and stops at a time corresponding to the slower speed of characteristic propagation; after the passing of the slower wave front no effect remains.

---

[1] This name was introduced by Herglotz, see F. John [4], p. 23 ff.

[2] See also §15, 4.

On the other hand the previous calculations allow without difficulty the conclusion: Every initial point in the convex hull of the backward ray cone, but not in the inner core, does contribute to the solution a value not constant in time. That means: The full zone between the convex hull Γ and the inner sheet of the ray cone actually constitutes the *precise domain of dependence* for the solutions and their time derivatives.

§15. The Solution of Cauchy's Problem as Linear Functional of the Data. Fundamental Solutions

1. Description. Notations. In the preceding sections and before in Ch. V, §5, the solution of Cauchy's problem—which is a linear functional of the data—was represented as an integral of a product; one factor is given by the data in the domain of dependence, while the other factor, the "kernel", depends only on the differential operator. The present section is concerned with a wide generalization of such an explicit "Riemann representation" for linear hyperbolic initial value problems in any number of independent variables and of any order, without assuming constant coefficients.[1] If the order of the differential equation is sufficiently high relative to the number of independent variables (as in crystal optics, or in magneto-hydrodynamics), then the kernel is an ordinary function. Otherwise, however, it must be expressed as a distribution, that is, obtained from continuous functions by differentiation processes.[2]

We aim at the construction and analysis of this Riemann kernel. As in Chapter V this representation throws light on the finer structure of the solution of Cauchy's problem; this is its main significance. Existence and uniqueness of the solutions of Cauchy's problem is assumed[3] in the present section on the basis of §§4, 8–10 where solutions with discontinuities along characteristic surfaces were constructed.

---

[1] Such a representation for a single equation of second order is the subject of Hadamard's famous theory (see article 5). Attention to the general case was first given by P. Lax. See R. Courant and P. D. Lax [2].

[2] The role of distributions for more than two independent variables was illustrated before by the solution $\delta(t - r)/r$ of the wave equation in three space dimensions (see §12) with a singularity along the ray cone.

It should be stated that the use of distributions, while convenient and elegant, can be avoided, as was done in §§12, 13, 14 and in R. Courant and P. D. Lax [2].

[3] See, however, the remark at the end of article 7.

To be specific, we consider again a hyperbolic system of first order equations for a function vector $u: u^1, \cdots, u^k$ in the form

$$(1) \qquad L[u] = u_t + A^i u_i + Bu = f(x, t) \qquad (i = 1, \cdots, n)$$

for $t \geq 0$ under the initial condition $u(0, x) = \psi(x)$ with sufficiently smooth matrix coefficients A^i, B and data ψ and f. (Systems of higher order will be treated similarly afterwards.) We may assume that the vectors ψ and f vanish identically for large values of their arguments. Then by the uniqueness theorem u also vanishes for fixed t and sufficiently large values of the space variables x.

Our starting point is the identity

$$(2) \qquad vL[u] - L^*[v]u = \sum_{i=1}^{n} \frac{\partial}{\partial x^i} (vA^i u) + \frac{\partial}{\partial t} (vu).$$

As before, L^* is the operator adjoint to L. We integrate (2) over the slab $0 \leq t \leq \tau$, and obtain

$$(3) \qquad \iint_{0 \leq t \leq \tau} (vL[u] - L^*[v]u) \, dx \, dt = \int_{t=\tau} vu \, dx - \int_{t=0} vu \, dx.$$

This holds provided that u or v or both vanish for large values of $|x|$. If we substitute for v a solution of $L^*[v] = 0$, and if $L[u] = f$, then (3) becomes

$$(3a) \qquad \int_{t=\tau} vu \, dx = \int_{t=0} vu \, dx + \iint_{0 \leq t \leq \tau} vf \, dx \, dt.$$

For heuristic purposes we shall apply (3) even if the integrands are distributions. Afterwards we shall show that, for the distributions considered in this section, this procedure is justified.[1] In particular, we consider v for the strip below τ as a matrix $R(x, t; \xi, \tau)$ satisfying

$$(4) \qquad\qquad L^*[R] = 0 \qquad\qquad \text{for } t < \tau,$$

which on $t = \tau$ has the "end values"

$$(4a) \qquad\qquad R(x, \tau; \xi, \tau) = \delta(x - \xi)I.$$

Here δ is the n-dimensional delta-function and I is the unit matrix.

---

[1] In passing it might be stated that (3a) could be used to give an abstract definition of generalized solutions.

With this choice of v, (3a) becomes

$$u(\xi, \tau) = \int R(x, 0; \xi, \tau)\psi(x)\, dx$$
(5)
$$+ \iint_{0 \le t \le \tau} R(x, t; \xi, \tau)f(t, x)\, dx\, dt,$$

where $\psi(x)$ are the initial values of $u(x, t)$.

For the homogeneous equation, that is, for $f = 0$, we have, with initial values $\psi(\tau_0, x) = \psi(x)$,

$$(5') \qquad u(\xi, \tau) = \int R(x, \tau_0; \xi, \tau)\psi(x)\, dx$$

from which (5) could immediately be retrieved by Duhamel's principle.

Similarly, we may introduce a matrix $S(x, t; \xi_1, \tau_1)$ instead of $u(x, t)$ which satisfies in the strip above $t = \tau_1$ the equation

$$(6) \qquad\qquad L[S] = 0 \qquad\qquad t > \tau_1$$

with the initial condition for $t = \tau_1$

$$(6a) \qquad S(x, \tau_1; \xi_1, \tau_1) = \delta(x - \xi_1)I.$$

Integration of (2) over the slab $\tau_1 \le t \le \tau$ then yields, formally,

$$\int R(x, \tau; \xi, \tau)S(x, \tau; \xi_1, \tau_1)\, dx$$

$$- \int R(x, \tau_1; \xi, \tau)S(x, \tau_1; \xi_1, \tau_1)\, dx = 0.$$

Using (4a) and (6a) we obtain the *law of symmetry*

$$(7) \qquad S(\xi, \tau; \xi_1, \tau_1) = R(\xi_1, \tau_1; \xi, \tau).$$

The representation formula (5) implies that R, considered as a function of ξ and τ, is a fundamental solution corresponding to the differential operator L:

$$(8) \qquad L_{\xi,\tau}[R(x, t; \xi, \tau)] = \delta(x - \xi, t - \tau).$$

Similarly,

$$(9) \qquad L^*_{\xi_1,\tau_1}[S(x, t; \xi_1, \tau_1)] = \delta(x - \xi_1, t - \tau_1).$$

Here we have on the right-hand side the $(n + 1)$-dimensional delta function. Thus $R(\xi_1, \tau_1 ; \xi, \tau)$ represents the effect at (ξ_1, τ_1) of backward radiation from a source located at (ξ, τ), and $S(\xi, \tau; \xi_1, \tau_1)$ represents the effect at (ξ, τ) of forward radiation from a source at (ξ_1, τ_1). The law of symmetry asserts that these effects are equal, if forward radiation is governed by the operator L, and backward radiation is governed by L^*.

In the next section we shall construct R and justify our procedure. In addition, we shall prove two important properties of R: First, $R(x, t; \xi, \tau)$ is regular except along the bicharacteristic rays emanating from (ξ, τ), and secondly, $R(x, t; \xi, \tau)$ vanishes identically outside the conoid of dependence of (ξ, τ). Hence in the expression (5) for the solution of $L[u] = f$ with $u(0, x) = \psi(x)$, the integration need not be extended outside the interior of the conoid of dependence Γ_P issuing backwards from $P(\xi, \tau)$.

For differential operators $L[u]$ of higher order m the situation is quite similar. Again, Riemann's radiation kernels R or S may be characterized by the simple initial conditions corresponding to (4a) or (6a) which involve merely the δ-function in the space variables.[1] For example for $m = 2$ we have

$$(10) \qquad \begin{aligned} L^*[R] &= 0 \quad \text{for } t < \tau; \\ R &= 0, \qquad R_t = \delta(x - \xi) \quad \text{for } t = \tau \end{aligned}$$

$$(10') \qquad \begin{aligned} L[S] &= 0 \quad \text{for } t > \tau; \\ S &= 0, \qquad S_t = \delta(x - \xi) \quad \text{for } t = \tau. \end{aligned}$$

Then (5) represents the solution corresponding to initial values

$$u(0, x) = 0, \qquad u_t(0, x) = \psi(x).$$

The solution for initial values if the Cauchy data u, u_t, $\cdots$ are given arbitrarily is of course likewise easily represented by applying Green's formula. For arbitrary m no further explanation is needed.

2. Construction of the Radiation Function by Decomposition of the Delta Function. We now replace the heuristic arguments of article 1 by a direct construction of the Riemann matrix R (or equivalently S). The basic idea is to reduce the point singularity of R to more

---

[1] Incidentally, the initial or the end condition singles out the radiation functions among all fundamental solutions, as follows from uniqueness.

easily manageable singularities spread over characteristic surfaces and to use the theory of Cauchy's problem as developed in §4 for such singularities. The following method is closely related to that of §14.

Let α be an arbitrary unit vector in the n-dimensional space, and let δ be the one-dimensional delta-function. We solve the backward Cauchy problem

$$L^*[V(x, t; \xi, \tau, \alpha)] = 0,$$

for $t < \tau$, with the end condition

$$V(x, \tau; \xi, \tau, \alpha) = \delta((x - \xi) \cdot \alpha)I.$$

If

$$L[u] = 0 \qquad\qquad (t > 0)$$

and $u(x, 0) = \psi(x)$, then (3a) implies that

$$I(\xi, \tau; \alpha; u) = \int u(x, \tau)\delta((x - \xi) \cdot \alpha)\, dx = \int V(x, 0; \xi, \tau, \alpha)\psi(x)\, dx$$

This formula expresses the plane integrals I, defined in §14, in terms of the initial values of u. We can retrieve u from its plane integrals according to §14, and thus by implication arrive at an expression for R.

A slightly different and more elegant procedure which covers equally even and odd values of n seems preferable. It is based on the decomposition of the n-dimensional delta function $\delta(x)$ into plane waves given in §11, 1 by formula (11). Accordingly, we may compose the function $R(x, t; \xi, \tau)$ (which is characterized by the differential equation $L^*_{x,t}R = 0$ and the end condition $R(x, \tau; \xi, \tau) = \delta(x - \xi)$) as a superposition of functions $U(x, t; \xi, \tau, \alpha)$ which satisfy the same differential equation $L^*_{x,t}U = 0$ but the simpler end condition

$$U(x, \tau; \xi, \tau) = \overset{\cdot}{\log}^{(n)}((x - \xi), \alpha)I$$

on the n-dimensional plane $t = \tau$, with an arbitrary unit vector α.

These functions $U(x, t; \xi, \tau, \alpha) = U(\alpha)$ are constructed by the method of §4. Considering the functions $U(\alpha)$ as known we therefore find immediately the desired representation for R

$$(11) \qquad R(x, t; \xi, \tau) = -\frac{1}{(2\pi i)^n} \int_{\alpha^2=1} U(x, t; \xi, \tau; \alpha)\, d\alpha.$$

Of course, a similar formula can be written for the adjoint Riemann matrix S: Construct by the method of §4 the function $V(x, t; \xi, \tau; \beta)$ with an arbitrary unit vector β which satisfies the initial condition $V(\beta) = V(x, t; \xi, \tau, \beta) = \log^{(n)}(x - \xi, \beta)I$ for $t = \tau_0$, and for $t > \tau_0$ solves the differential equation $L_{x,t}[V] = 0$.   Then

$$(11a) \qquad S(x, t; \xi, \tau) = -\frac{1}{(2\pi i)^n} \int_{\beta^2=1} V(\beta) \, d\beta.$$

(11) and (11a) express our general result.   Of course, it can easily be written in a less symbolic language by tracing back the steps that led to the representation formula (11) of §11, 1:   The boundary values $\log^{(n)}(\alpha(x - \xi))$ can, by use of the Laplace operator, be expressed in the form

$$\log^{(n)}((x - \xi)\cdot\alpha) = \Delta_\xi^{(n+2)/2} \log^{(-2)}((x - \xi)\cdot\alpha) \qquad (n \text{ even})$$

$$\log^{(n)}((x - \xi)\cdot\alpha) = \Delta_\xi^{(n+3)/2} \log^{(-3)}((x - \xi)\cdot\alpha) \qquad (n \text{ odd}),$$

where the functions $\log^{(-2)}$ and $\log^{(-3)}$ are continuous and differentiable.   Thus we may represent U and hence R by iterated Laplace operators with respect to ξ applied to integrals of continuously differentiable matrices.

The expression of ideal functions by differentiation of continuous ones can also be used to justify the reciprocity relation (7) as a consequence of (3) by substituting $v = V(\beta)$ and $u = U(\alpha)$ and replacing $t = 0$ by $t = \tau_1$.   Since $U(\alpha)$ and $V(\beta)$ are distributions with respect to two different variables α and β, there is indeed no objection to this substitution.   Then integration with respect to α and β results immediately in the symmetry law (7).

It should be observed that a high order for L implies high regularity for the initial values for U, hence also for the Riemann matrix itself. For this reason, for example, the Riemann matrix for the fourth-order differential equation of crystal optics is an ordinary function and does not have to be interpreted as a distribution.

We should finally remark that the uniqueness theory of §8, 9, or Ch. III, App. 3, implies: *The Riemann matrix vanishes identically outside the conoid of dependence issuing from the singular end point or initial point respectively.*   Indeed the uniqueness theorem assures us, for example, that the function S vanishes in all points of the x, t-space whose domain of dependence does exclude the singularity.

The preceding theory applies almost literally if L is a system of operators of higher order.   Consider, for example, the second-order system

$$L[u] \equiv u_{tt} + \sum_{\nu=1}^{n} \sum_{\mu=0}^{n} A^{\nu\mu} u_{\nu\mu} + \sum_{\mu=0}^{n} A^{\mu} u_{\mu} + Bu.$$

Then the radiation matrix $R(x, t; \xi, \tau)$ is defined by the conditions:

(12) $LR = 0$ if $t > \tau$

(13) $R(x, \tau; \xi, \tau) = 0$

(13′) $R_t(x, \tau; \xi, \tau) = \delta(x - \xi)I,$

and can be constructed in exactly the same way as for a system of equations of first order.

3. Regularity of the Radiation Matrix. From the symbolic form of the preceding results it is easy to deduce important specific properties[1] and concrete expressions of the solution. In this article we show that the matrix R—and thus S—is a distribution whose singularities are concentrated merely on the ray conoid through the vertex P and which elsewhere is a continuous function with continuous derivatives of all orders compatible with the regularity properties of the operator L.[2]

For brevity we consider a first-order system (1) and assume that the coefficients A and B have continuous derivatives of any desired order. Then we show: *The radiation functions R and S are regular, i.e., have continuous derivatives of any order desired at all points x, t which cannot be reached by a characteristic ray from the vertex P: ξ, τ.*

We have to deal with an integral of the form

(14) $R(x, t; \xi, \tau) = \displaystyle\int_{|\alpha|=1} U(x, t; \xi, \tau, \alpha) \, d\alpha.$

The singular part of U consists of a sum of terms of the form[3]

(15) $S_i(\phi(x, t; \xi, \tau, \alpha)) g_i(x, t; \xi, \tau, \alpha),$

[1] Attention should be given to the following fact. If isolated or multiple rays occur in the ray cone, some of the corresponding statements in the present article have to be modified. For multiple characteristics a general analysis of the singularities of the radiation matrix on the ray conoid as well as on its hull has as yet not been achieved; one must study individually specific classes of examples, as we did implicitly before. As to recent progress in the subject reference should be made to D. Ludwig [3] and J. Leray [3].

[2] A slight modification of the following argument shows that R and S are analytic except on the ray conoid if the coefficients are analytic. See a forthcoming paper by D. Ludwig.

[3] The letter S is not used here in the meaning of articles 1 and 2.

where $S = S_i$ is an ideal function of the variable ϕ with singularity at the origin, and $g = g_i$ is regular in all of its arguments. It is sufficient to examine the smoothness of a single term of the form

$$(16) \qquad K = \int_{|\alpha|=1} S(\phi(x, t; \xi, \tau, \alpha))g(x, t; \xi, \tau, \alpha) \, d\alpha.$$

Let us fix $(x, t; \xi, \tau)$ and consider the set Ω consisting of all points on the sphere $|\alpha| = 1$ which satisfy the equation

$$\phi(x, t; \xi, \tau, \alpha) = 0.$$

Ω is compact, and hence the following alternative holds: Either (a) the gradient of ϕ with respect to the coordinates on the sphere vanishes at some point x, t of Ω, or (b) the gradient of ϕ with respect to the coordinates on the sphere is bounded away from zero at every point of Ω.

If condition (a) is satisfied, then (x, t) is a point of the envelope of the characteristic surfaces through ξ, τ given by

$$\phi(x, t; \xi, \tau, \alpha) = 0.$$

The theory of first-order partial differential equations shows that the envelope must contain a ray which passes through (x, t). This ray intersects the plane $t = \tau$ in a point P at which

$$\phi(x, \tau; \xi, \tau, \alpha) = (x - \xi) \cdot \alpha.$$

The only point of the envelope of the planes $(x - \xi) \cdot \alpha = 0$ is the point $x = \xi$. Thus $P = (\xi, \tau)$, and (x, t) and (ξ, τ) lie on the same ray. We conclude that condition (a) is satisfied only if (x, t) and (ξ, τ) lie on the same ray.

If condition (b) is satisfied, then we can introduce ϕ as a local coordinate on the sphere at each point of Ω. Thus we can break up the domain of integration in (3) in such a way that, in each subdomain, either (i) ϕ is a local coordinate, or (ii) ϕ is bounded away from zero. In subdomains of type (ii) the integrand is smooth, and hence the integral taken over such subdomains is smooth. In subdomains of type (i), ϕ is a local coordinate, and we can integrate by parts with respect to ϕ as often as we like. By means of repeated integration by parts we arrive at expressions as smooth as we like. Hence, if condition (b) is satisfied, K is a smooth function of x, t.

Thus we have proved the following alternative: Either (a) (x, t)

and (ξ, τ) lie on the same ray, or (b) $R(x, t; \xi, \tau)$ is a smooth function[1] of x, t as well as ξ, τ.

3a. The Generalized Huyghens Principle. The result of article 3 illuminates a fact which is of great importance for the role of hyperbolic systems as models for transmission of signals (see also §18): The classical "principle of Huyghens", discussed at various places (see e.g., §12 and §18 of this chapter) asserts that sharp signals are transmitted as sharp signals, that is that the solution of a differential equation, representing the propagation of signals emitted at the time $t = 0$, depends merely on the data at the boundary of the conoid of dependence, not on the data inside. This principle is valid only under quite exceptional conditions, essentially only for the wave equation in 3, 5, 7, $\cdots$ space dimensions. However, the result of the preceding article 3 can be interpreted as a *generalized principle of Huyghens*[2] which states that in an approximate, and for that matter, usually satisfactory, sense any hyperbolic system does preserve initially sharp signals as sharp signals, though in general slightly blurred.

We first state: The solution $u(P)$ of the system (1) is smooth in the neighborhood of a point P if all the rays through P meet the initial manifold $(t = 0)$ in a closed region G^* in which the initial data are smooth.

More quantitatively: Assuming that the coefficients of $L[u]$ are sufficiently smooth and that the initial data and their derivatives up to an order r are bounded in a region G^* described above, then the solution u and its derivatives up to the order r remain bounded in a point P, as defined above; the bounds depend on a bound for t and on the bounds of the initial values.

We may also describe the Huyghens principle in a generalized sense[3] as follows: The singularities of the solution u at the point P depend only on the singularities of the initial data and only in so far as these data are presented on the ray conoid through P.

We can express the principle more sharply if we make specific assumptions about the initial data: For example, if these data are

---

[1] R has derivatives of every order if the coefficients of $L[u]$ have derivatives of every order.

[2] Pointed out by P. D. Lax. See R. Courant and P. D. Lax [2] and P. D. Lax [4].

[3] As said before, in case of multiple characteristic elements modifications of the statements are necessary.

smooth except for jump discontinuities along a smooth curve C. Then $u(P)$ is smooth except at points P such that the ray conoid through P is tangent to C. If the conoid has a contact of higher order with C, then "focussing" takes place, and the solution will in general have fewer derivatives than the initial data.[1] Qualitatively the generalized Huyghens principle asserts that the solution is particularly sensitive to the transmission of initial discontinuities (or other irregularities such as violent changes in values) along characteristic rays.[2]

4. **Example. Special Linear Systems with Constant Coefficients. Theorem on Gaps.** We consider a system (1) with constant coefficients where $B = 0$ and $f = 0$. Then the solution $U(x, t, \xi, \tau; \alpha)$ is given by §4, 6 simply as a superposition of k plane waves reducing merely to the first term of the progressing wave expansion:

$$U(\alpha) = \sum_{\kappa=1}^{k} \sigma^\kappa r^\kappa \log^{(n)} (\lambda^\kappa t + \alpha x)$$

(assuming $\xi = 0$), where we write $\lambda^\kappa = \lambda^\kappa(\alpha)$ for the eigenvalues (normal velocities) of the matrix $\lambda - \sum A^\nu \alpha_\nu$ (see §3), where r^κ are the right eigenvectors of this characteristic matrix and where the scalars σ^κ are determined according to §4, 6. If n is odd, then $U(\alpha)$ is given within a constant factor by

$$\sum_\kappa \sigma^\kappa r^\kappa \delta^{(n-1)} (\lambda^\kappa t + \alpha x).$$

Riemann's fundamental solution is obtained by integration with respect to α. Now: Assuming that none of the speeds λ^κ is zero and assuming that x/t is in the inner core of the ray cone, it is evident that the fundamental solution vanishes in this core, since for each α in this core the functions $\delta^{(n-1)}(\lambda^\kappa t + \alpha x)$ vanish.

Thus we obtain the remarkable general theorem: *For a system (1) with constant coefficients with $B = 0$ and nowhere vanishing propagation speed in a space of odd dimensions, the inner core of the ray cone represents a gap, that is, does not belong to the domain of dependence in the strict sense of the vertex.*

[1] See D. Ludwig [1].

[2] The dependence of u on the initial values can easily be analyzed if one expresses the kernel R by derivatives of smooth functions with respect to a variable $\phi = 0$ (for which ϕ represents a mantle of the characteristic conoid): Then iterated integration by parts removes the singularities of the kernel and introduces corresponding derivatives of the initial values.

This theorem remains true even if some of the speeds λ^κ vanish, provided that the corresponding factors σ^κ then also vanish. This is true as can be easily ascertained in the case of crystal optics (see §14, 2), provided that the divergence conditions are satisfied.

5. Example. The Wave Equation. The general theory of article 2 contains as special cases such solutions of Cauchy's problem as we have found explicitly in previous sections. Thus we can easily identify the representations of §§14, 14a with the present one, and we confine ourselves here to a few simple examples. First we consider the problem of the wave equation.

Of course the formulas (48) and (49) of §12 yield immediately an explicit expression of R or S—the functions are identical because of the self-adjoint character of the wave operator—if we replace $g(\lambda)$ in these formulas by the Dirac function $\delta(\lambda)$.

However, we want to derive these or equivalent expressions from the general representation (11) in article 2.

Let $S^{(n)}$ be the Riemann function for the wave equation with $n + 1$ independent variables: $x^1, \cdots, x^n, t$. The distribution $S^{(n)}(x, t; \xi, \tau)$ satisfies the following conditions:

$$(17) \qquad S_{tt}^{(n)} - \sum_{\nu=1}^{n} S_{\nu\nu}^{(n)} = 0 \qquad\qquad (t > \tau),$$

$$(18) \qquad S^{(n)}(x, \tau; \xi, \tau) = 0,$$

$$(19) \qquad S_t^{(n)}(x, \tau; \xi, \tau) = \delta(x - \xi).$$

Since the wave operator has constant coefficients, there is no loss of generality if we set $\tau = 0$ and $\xi = 0$. $S^{(n)}$ may be expressed in the form

$$(20) \qquad S^{(n)}(x, t) = \frac{-1}{(2\pi i)^n} \int_{|\alpha|=1} V^{(n)}(x, t; \alpha)\, d\alpha,$$

where $V^{(n)}$ satisfies the conditions:

$$(21) \qquad V_{tt}^{(n)} - \sum_{\nu=1}^{n} V_{\nu\nu}^{(n)} = 0 \qquad\qquad (t > 0),$$

$$(22) \qquad V^{(n)}(x, 0; \alpha) = 0,$$

$$(23) \qquad V_t^{(n)}(x, 0; \alpha) = \log^{(n)}(x \cdot \alpha).$$

Here $\log^{(n)}$ is the n-th derivative of the log function, in the sense of distributions.

Obviously,

$$(24) \quad V^n(x, t; \alpha) = \tfrac{1}{2}[\log^{(n-1)}(t + x \cdot \alpha) - \log^{(n-1)}(-t + x \cdot \alpha)].$$

Hence

$$(25) \quad S^{(n)}(x, t) = \frac{-1}{2(2\pi i)^n} \int_{|\alpha|=1} [\log^{(n-1)}(t + x \cdot \alpha)$$
$$- \log^{(n-1)}(-t + x \cdot \alpha)] \, d\alpha.$$

Formula (4) of §11 implies that if $n \geq 2$,

$$(26) \quad S^{(n)}(x, t) = \frac{-\omega_{n-1}}{2(2\pi i)^n} \int_{-1}^{1} [\log^{(n-1)}(t + rp)$$
$$- \log^{(n-1)}(-t + rp)](1 - p^2)^{(n-3)/2} \, dp,$$

where $\omega_{n-1} = 2\pi^{(n-1)/2}/\Gamma((n-1)/2)$ and $r = |x|$. Changing notations slightly, we consider $S^{(n)}$ as a function of r and t.

Differentiating (26), we have

$$(27) \quad \frac{\partial S^{(n)}}{\partial r} = \frac{-\omega_{n-1}}{2(2\pi i)^n} \int_{-1}^{1} [\log^{(n)}(t + rp)$$
$$- \log^{(n)}(-t + rp)]p(1 - p^2)^{(n-3)/2} \, dp.$$

Now we can integrate by parts; since there is no contribution from the ends of the interval of integration, we have

$$(28) \quad \frac{\partial S^{(n)}}{\partial r} = \frac{-r\omega_{n-1}}{(n - 1) \cdot 2 \cdot (2\pi i)^n} \int_{-1}^{1} [\log^{(n+1)}(t + rp)$$
$$- \log^{(n+1)}(-t + rp)](1 - p^2)^{(n-1)/2} \, dp.$$

An easy calculation shows that[1]

$$(29) \qquad\qquad S^{(n+2)} = -\frac{1}{\pi} \frac{\partial S^{(n)}}{\partial (r^2)}.$$

Thus $S^{(n)}$ can be determined recursively once $S^{(2)}$ and $S^{(3)}$ are known. From (25),

$$(30) \quad S^{(2)} = \frac{-1}{2(2\pi i)^2} \frac{\partial}{\partial t} \int_{|\omega|=1} [\log(x \cdot \omega + t) + \log(x \cdot \omega - t)] \, d\omega.$$

[1] See also §12.

Introducing θ as an angular coordinate on the unit circle, we find

$$(31) \quad S^{(2)} = \frac{\partial}{\partial t} \frac{1}{(2\pi)^2} \int_0^\pi [\log |r \cos \theta + t| + \log |r \cos \theta - t|] \, d\theta.$$

By a well-known integral formula one finds

$$(32) \qquad S^{(2)} = \frac{1}{2\pi} \frac{1}{\sqrt{t^2 - r^2}} \qquad (t^2 > r^2).$$

We may express $S^{(2)}$ in terms of a fractional integral of the δ-function. It follows immediately from the definition that

$$(33) \qquad S^{(2)} = \frac{1}{2\sqrt{\pi}} \delta^{(-1/2)}(t^2 - r^2).$$

The evaluation of $S^{(3)}$ is given almost immediately by equation (26). We have

$$(34) \quad S^{(3)} = \frac{-2\pi}{2(2\pi i)^3} \int_{-1}^1 [\log^{(2)}(rp + t) - \log^{(2)}(rp - t)] \, dp.$$

Now, by definition, on the real axis

$$(35) \qquad Im \log z = \pi(1 - \eta(z)),$$

where η is the Heaviside function. Hence

$$(36) \qquad Im \log_{-2} z = -\pi \delta'(z),$$

and

$$(37) \quad S^{(3)} = \frac{1}{2(2\pi)^2} \int_{-1}^1 - \pi[\delta'(rp + t) - \delta'(rp - t)] \, dp,$$

or

$$(38) \qquad S^{(3)} = \frac{1}{4\pi r} (\delta(t - r) - \delta(t + r)).$$

It follows from the definition of the δ-function that

$$(39) \qquad \delta(t^2 - r^2) = \frac{1}{2r} [\delta(t - r) - \delta(t + r)].$$

Hence

$$(40) \qquad S^{(3)} = \frac{1}{2\pi} \delta(t^2 - r^2).$$

Using (29), (33) and (40), we have

$$(41) \qquad S^{(n)} = \frac{1}{2\pi^{(n-1)/2}} \, \delta^{((n-3)/2)}(t^2 - r^2).$$

The formula (41) represents the solution for all n.[1] In particular, if n is greater than or equal to 3 and odd, $S^{(n)}$ vanishes except on the cone $t^2 - r^2 = 0$. This is the strict form of Huyghens' principle.

6. Example. Hadamard's Theory for Single Equations of Second Order. The modern theory of Cauchy's problem was greatly inspired by Hadamard's pioneering work on hyperbolic equations of second order.[2] Although the use of distributions and other features of the present chapter allow a broader and simplified approach, basic ideas of Hadamard's method underlie much of the discussion of this section and will now be reviewed briefly, even though the theory of the preceding articles 2, 3 applied to the case of a single second order equation covers the subject of Hadamard's theory.

Hadamard's achievement was in the first place the construction of the fundamental solution; this construction proceeds directly without the benefit of the simplification which is attained in article 2 by first decomposing the n-dimensional δ-function into plane waves and then integrating over a unit sphere. Secondly Hadamard, not having the technique of ideal functions available, used "finite parts" of divergent integrals to evaluate expressions such as $\int_a^b [A(x)/(b - x)^{3/2}] \, dx$ in which the integrands are not interpreted as ideal but as regular functions. Here we shall not elaborate on the concept of "finite parts of an integral" (see, however, Appendix), since it is indeed obviated by the use of ideal functions. Instead we give a modified account of Hadamard's construction of the fundamental solution.

Guided by the results for the wave equation (see article 4), we consider a general analytic equation of second order

$$(42) \qquad L[u] = \sum_{i,k=0}^{n} a_{ik} u_{ik} + \sum_{i=0}^{n} b_i u_i + cu = 0.$$

We look for solutions with a singularity along the ray conoid

$$\Gamma(x, \xi) = 0.$$

---

[1] This formula is derived and discussed in Gelfand and Shilov [1], pp. 288–290. Naturally it is equivalent to the results in §12, in particular to the expressions (14) and (5) in §12.

[2] See J. Hadamard [2].

$\Gamma(x, \xi)$ denotes the square of the distance between x and ξ measured in the Riemannian metric defined by the equation. Γ satisfies the equation

$$(43) \qquad \sum_{i,k=0}^{n} a_{ik}\, \Gamma_i\, \Gamma_k = 4\Gamma,$$

and hence satisfies the characteristic equation only on the ray conoid.[1]

In analogy to the procedure of §4 we seek a solution of $L[u] = 0$ in the form

$$(44) \qquad u(x, \xi) \approx \sum_{\nu=0}^{\infty} S_\nu(\Gamma) g^\nu(x, \xi),$$

where

$$(45) \qquad \frac{d}{d\Gamma} S_\nu(\Gamma) = S_{\nu-1}(\Gamma) \qquad\qquad \text{for all } \nu;$$

the distribution $S_0(\Gamma)$ and the functions $g^\nu(x, \xi)$ will be specified presently. Several differences between the method of Hadamard and the method of §4 should be pointed out. First, since Γ does not satisfy the characteristic equation identically, the coefficients $g^\nu(x, \xi)$ will depend on the choice of S_0. Secondly, the surface $\Gamma = 0$ has a singularity at the vertex of the ray conoid ($x = \xi$). Such singularities are excluded in the considerations of §4.

Here, the singularity at the vertex leads after some calculation to the condition

$$(46) \qquad \Gamma S_{-2}(\Gamma) + \left(\frac{n+1}{2}\right) S_{-1}(\Gamma) = 0$$

at $\Gamma = 0$; there is no loss of generality if we require that (46) be satisfied for all values of Γ. Then the coefficients $g^\nu(x, \xi)$ can be determined by a procedure quite similar to that in §4 as solutions of ordinary differential equations along the geodesics issuing from the vertex $x = \xi$. The condition that g^ν be regular at the vertex $x = \xi$ determines u within a multiplicative constant.

Having determined all of the coefficients in the expansion (44), we may combine the terms. By using merely equations (45) and (46), we find

$$(47) \qquad \frac{d^\nu}{(d\Gamma)^\nu} (\Gamma S_{\nu-2}(\Gamma)) = \Gamma S_{-2} + \nu S_{-1} = \left(\nu - \frac{n+1}{2}\right) S_{-1}.$$

[1] See §1 and Ch. II, §9.

Hence, choosing constants of integration properly,

$$(48) \qquad \Gamma S_{\nu-2} = \left(\nu - \frac{n+1}{2} \right) S_{\nu-1} \qquad (\nu = 1, \cdots).$$

Setting $\mu = (n-1)/2$, it follows that

$$(49) \qquad S_\nu(\Gamma) = \frac{\Gamma^\nu}{(1-\mu) \cdots (\nu-\mu)} S_0(\Gamma),$$

provided that the denominator does not vanish. However, if n is odd, the denominator will vanish for some $\nu = \mu$ and we introduce a suitable idealized function

$$(50) \qquad T(\Gamma) = S_\mu(\Gamma)$$

to be specified below. Then, for $\nu > \mu$, again choosing constants of integration properly, we have

$$(51) \qquad S_\nu(\Gamma) = \frac{\Gamma^{\nu-\mu}}{(\nu-\mu)!} T(\Gamma).$$

Now we can use formulas (49) and (51) to combine terms in our expansion. If n is even, then

$$(52) \qquad u(x, \xi) = S_0(\Gamma) U(x, \xi),$$

where

$$(53) \qquad U(x, \xi) = \sum_{\nu=0}^{\infty} \frac{\Gamma^\nu}{(1-\mu) \cdots (\nu-\mu)} g^\nu(x, \xi).$$

If n is odd, then

$$(54) \qquad u(x, \xi) = S_0(\Gamma) U(x, \xi) + T(\Gamma) V(x, \xi),$$

where

$$(55) \qquad U(x, \xi) = \sum_{\nu=0}^{\mu-1} \frac{\Gamma^\nu}{(1-\mu) \cdots (\nu-\mu)} g^\nu(x, \xi),$$

$$(56) \qquad V(x, \xi) = \sum_{\nu=\mu}^{\infty} \frac{\Gamma^{\nu-\mu}}{(\nu-\mu)!} g^\nu(x, \xi).$$

Only now we specify the functions $S_0(\Gamma)$ and $T(\Gamma)$. Hadamard concluded from (46) that, within a constant multiple,

$$(57a) \qquad S_0(\Gamma) = \Gamma^{(1-n)/2}.$$

However, in accordance with the example of the wave equation in article 4, and intending to satisfy the condition (10) from article 2, we prefer to start with the singularity function

(57b) $$S_0(\Gamma) = \delta^{((n-3)/2)}(\Gamma).$$

Since $\delta(\Gamma)$ satisfies the equation

(58) $$\Gamma\delta(\Gamma) = 0,$$

and $\delta^{(-1/2)}(\Gamma) = \dfrac{1}{\sqrt{\pi}}\dfrac{1}{\sqrt{\Gamma}}$ satisfies the equation

(59) $$\Gamma\frac{d\delta^{(-1/2)}}{d\Gamma}(\Gamma) + \tfrac{1}{2}\delta^{(-1/2)}(\Gamma) = 0,$$

it follows by differentiation that not only (57a) but also our choice (57b) of $S_0(\Gamma)$ satisfies (46). (This ambiguity in determining S_0 is due to the fact that equation (46) is singular at $\Gamma = 0$.)

How shall we choose between (57a) and (57b)? Let us first examine the situation when n is even. In this case, (a) and (b) are identical, except that in (b), S_0 is to be interpreted as a distribution. The only difference appears when integrals of these functions are evaluated over domains which include the ray conoid $\Gamma = 0$. In case (a) such integrals in general diverge. This difficulty caused an essential obstacle,[1] and led to Hadamard's invention of "finite parts" of such divergent integrals. These "finite parts" are, however, simply the expressions obtained by integrating the ideal function (b). In fact, Hadamard's invention of the finite part may be regarded as an important motivation for the modern theory of distributions.

The situation is similar for odd n. In this case, Hadamard had to introduce the "logarithmic part" of integrals involving his solution. These expressions are the same as we would obtain by use of the delta-function.

Therefore we may write our solution for n even or odd alike in the form

(60) $$u(x, \xi) = \delta^{((n-3)/2)}(\Gamma)U(x, \xi) + \eta(\Gamma)V(x, \xi),$$

where $\eta(\Gamma)$ is the Heaviside function, and V vanishes if n is even. It follows from the properties of the δ-function and its derivatives that $u(x, \xi)$ vanishes except on the conoid $\Gamma = 0$ if and only if n is odd and $V(x, \xi) \equiv 0$. This is the strict form of *Huyghens' principle*.

[1] An interesting account is given in Hadamard [3].

It is not difficult to compare (60) with Hadamard's expression, in which log Γ appears, and not $\eta(\Gamma)$. To verify that the fundamental solution has indeed the form (60) can be done by using the definitions of article 2 and may be left to the reader.

7. Further Examples. Two Independent Variables. Remarks.
Attention also should be given to the fact that the generalization of Riemann's representation of Chapter V to single differential equations of higher order or systems of k equations of first order in two independent variables is an almost trivial special case of the general representation in this section.

As to degenerate equations, one typical example of the operator $u_{x_1 x_2 x_3}$ is worth mentioning: Here

$$\eta(x_1 - \xi_1)\eta(x_2 - \xi_2)\eta(x_3 - \xi_3)$$

with the Heaviside η function, is the Riemann function as seen from its characterization in article 2.

Generally speaking it appears that on multiple characteristic elements, isolated rays and caustics the Riemann matrix is more highly singular than elsewhere on the ray surface or its hull. A thorough study of these singularities would throw more light on the generalized Huyghens principle.[1]

Finally a previous remark concerning existence and uniqueness should be repeated: While existence proofs in §10 were based on energy integrals, the theorems about convergence of the progressing wave expansion together with the construction of the present section produces in the case of analytic operators L the Riemann matrix independently of energy integrals, no matter whether or not the operator is symmetric. The Riemann matrix then immediately yields the solution of Cauchy's problem and moreover exhibits the differentiability properties of the solution. Holmgren's theorem of Ch. III, App. 2 secures the uniqueness of the solution. Thus an approach to Cauchy's problem different from that of §§8–10 is open.

§16. *Ultrahyperbolic Differential Equations and General Differential Equations of Second Order with Constant Coefficients*

1. The General Mean Value Theorem of Asgeirsson.[2] The method of spherical means leads to a simple yet powerful mean value theorem,

---

[1] For steps in this direction see, e.g., D. Ludwig [3].
[2] Cf. L. Asgeirsson [2] and [3].

due to L. Asgeirsson, for arbitrary linear differential equations of second order with constant coefficients. According to Ch. III, §3, 1 homogeneous differential equations of this type, if not parabolically degenerate, can always be brought into the form

$$u_{x_1 x_1} + \cdots + u_{x_n x_n} = u_{y_1 y_1} + \cdots + u_{y_m y_m} - cu,$$

by making a suitable linear transformation of the coordinates and, if necessary, by cancelling an exponential factor. We can also eliminate the coefficient c formally (in case it is positive) by introducing an artificial new variable x_{n+1} and setting $u = ve^{\sqrt{c} x_{n+1}}$. The differential equation takes on the form

$$u_{x_1 x_1} + \cdots + u_{x_{n+1} x_{n+1}} = u_{y_1 y_1} + \cdots + u_{y_m y_m},$$

where we write again u instead of v. In the special case where u depends on none of the variables y we obtain the potential equation. If u depends on only one of the variables y, we obtain the wave equation.

Moreover, by assuming that the function u is independent of certain of the variables x or y, we can, without loss of generality, write the differential equation in the form

(1) $$\Delta_x u = \Delta_y u,$$

i.e.,

(1') $$\sum_{i=1}^{m} u_{x_i x_i} = \sum_{i=1}^{m} u_{y_i y_i},$$

with an equal number m of x- and y-variables, some of them artificial. We call differential equations of this type *ultrahyperbolic*.

We now assume that throughout the region of the x,y-space under consideration the function u is a twice continuously differentiable solution of equation (1). With such a solution u we then form the following integrals over the unit sphere $\beta_1^2 + \cdots + \beta_m^2 = 1$ with the surface area ω_m and the surface element $d\omega_m = d\beta$:

(2) $$\mu(x, y, r) = \frac{1}{\omega_m} \int \cdots \int u(x + \beta r; y) \, d\beta$$

and

(3) $$\nu(x, y, r) = \frac{1}{\omega_m} \int \cdots \int u(x; y + \beta r) \, d\beta.$$

Thus $\nu(x, y, r)$ is the mean value of the function u over the surface of a sphere of radius r about the y-point in y-space, for a fixed value of x, and similarly $\mu(x, y, r)$ is the mean value in x-space for a fixed y. We suppose ν and μ to be continued as even functions for negative r.

We also form the mean value

$$w(x, y; r, s)$$

(4)
$$= \frac{1}{\omega_m^2} \int \cdots \int_{\alpha^2=1} \int \cdots \int_{\beta^2=1} u(x + \beta r; y + \alpha s) \, d\beta \, d\alpha,$$

that is, the average over a sphere of radius r in x-space and a sphere of radius s in y-space. Clearly we have

$$\mu(x, y, r) = w(x, y; r, 0),$$
$$\nu(x, y, r) = w(x, y; 0, r).$$

Asgeirsson's mean value theorem is then expressed simply by the equation

(5) $$\mu(x, y, r) = \nu(x, y, r),$$

or, in words:

The average for fixed x over a sphere of radius r in y-space is the same as the average for fixed y over a sphere of radius r in x-space; that is, $\mu(x, y; r) = \nu(x, y; r)$. More generally,

(6) $$w(x, y; r, s) = w(x, y; s, r);$$

i.e., the double mean value is symmetric in the radii r and s.

We prove the special theorem first. By §13, both mean values ν and μ satisfy Darboux's equations

(7)
$$\Delta_x \mu - \frac{m-1}{r} \mu_r - \mu_{rr} = 0,$$

$$\Delta_y \nu - \frac{m-1}{r} \nu_r - \nu_{rr} = 0,$$

where y occurs as the parameter in the first equation, and x as the parameter in the second. From (1) we have $\Delta_x \mu = \Delta_y \mu$ and $\Delta_y \nu = \Delta_x \nu$. Hence

$$\Delta_x \nu - \frac{m-1}{r} \nu_r - \nu_{rr} = 0.$$

In addition, we have by definition

$$\mu(x, y, 0) = u(x, y), \qquad \mu_r(x, y, 0) = 0,$$

$$\nu(x, y, 0) = u(x, y), \qquad \nu_r(x, y, 0) = 0.$$

Thus μ and ν, as functions of x and r with parameters y, are solutions of the same initial value problem for Darboux's equation. Hence they are identical by the uniqueness theorem of §6, 2.

Similarly, the general relation (6) is obtained as follows: For the double mean value we have $\Delta_x w = \Delta_y w$, and also Darboux's equations

$$\Delta_x w = \frac{m-1}{r} w_r + w_{rr},$$

$$\Delta_y w = \frac{m-1}{s} w_s + w_{ss},$$

so that

(8) $$\frac{m-1}{r} w_r + w_{rr} = \frac{m-1}{s} w_s + w_{ss}.$$

For $s = 0$ we consider $w(x, y; r, 0)$ as known; we also know $w_s(x, y; r, 0) = 0$. These initial conditions determine $w(x, y; r, s)$ uniquely, as we see by the argument of §6, 2.

If $w(x, y; r, s) = u(r, s)$ and $w(x, y; s, r) = v(r, s)$, then the functions u and v satisfy the same differential equation (8) with the initial conditions $u(r, 0) = w(r, 0)$; $u_s(r, 0) = 0$ and $v(r, 0) = w(0, r)$; $v_s(r, 0) = 0$. In consequence of the special mean value theorem $w(0, r) = w(r, 0)$. The uniqueness theorem then implies that the solutions must be identical:

$$v(r, s) = w(r, s) = w(x, y; r, s).$$

Since the mean value equation (5.) holds for the surface of every sphere of radius r, we at once obtain a corresponding *mean value theorem for the interior of spheres* by integrating with respect to r:

(5') $$\int \cdots \int_{\rho \leq r} u(x + \xi; y) \, d\xi = \int \cdots \int_{\rho \leq r} u(x; y + \xi) \, d\xi;$$

the integration on both sides is extended over the interior of the sphere $\rho^2 = \xi_1^2 + \cdots + \xi_m^2 \leq r^2$.

This leads to a corresponding theorem for $n > m$, i.e., for the differential equation

$$u_{x_1 x_1} + \cdots + u_{x_n x_n} = u_{y_1 y_1} + \cdots + u_{y_m y_m} .$$

Suppose that in $(5')$ the solution u is independent of the variables $y_{m+1}, \cdots, y_n$ with $m < n$. Then we can more generally state

$$\int \cdots \int_{\rho \leq r} u(x_1 + \xi_1, \cdots, x_n + \xi_n ; y_1, \cdots, y_m) \, d\xi_1 \cdots d\xi_n$$

$(5'')$
$$= \frac{\omega_{n-m}}{n-m} \int \cdots \int_{\rho_1 \leq r} (r^2 - \rho_1^2)^{(n-m)/2} \, u(x; y + \xi) \, d\xi,$$

with $\rho_1^2 = \xi_1^2 + \cdots + \xi_m^2$ and $\rho^2 = \xi_1^2 + \cdots + \xi_n^2$.

2. Another Proof of the Mean Value Theorem. A second proof of the general mean value theorem under stronger differentiability assumptions is obtained by setting

$$(9) \quad w(r, s) = \int_{-1}^{1} \int_{-1}^{1} v(\alpha r, \beta s)(1 - \alpha^2)^{(m-3)/2}(1 - \beta^2)^{(m-3)/2} \, d\alpha \, d\beta$$

for some function $v(a, b)$ and by assuming w sufficiently differentiable so that the even function v belonging to w can be uniquely determined according to §13. We have, from §13, 2

$$\int_{-1}^{1} \int_{-1}^{1} (v_{aa}(\alpha r, \beta s) - v_{bb}(\alpha r, \beta s))(1 - \alpha^2)^{(m-3)/2} (1 - \beta^2)^{(m-3)/2} \, d\alpha \, d\beta$$

$$= \frac{m-1}{r} w_r + w_{rr} - \frac{m-1}{s} w_s - w_{ss} ;$$

and, using (8),

$$v_{aa} = v_{bb} ;$$

hence

$$v(a, b) = g(a + b) + h(a - b).$$

If we substitute this for $v(a, b)$ in (9), we obtain an expression which is symmetric in r and s, and hence the desired equation

$$w(r, s) = w(s, r)$$

follows.

3. Application to the Wave Equation. We solve once again the initial value problem for the wave equation $u_{tt} - \Delta u = 0$ with the initial conditions $u(x, 0) = \psi(x)$, $u_t(x, 0) = 0$. Accordingly we interpret the wave equation as a special case of the ultrahyperbolic differential equation (1) with $y_1 = t$, under the additional condition that the solution u be independent of $y_2, \cdots, y_n$. We now apply the mean value theorem for an arbitrary point x of x-space and the origin $y = 0$ of y-space. This yields

$$(10) \quad \frac{1}{\omega_m} \int \cdots \int_{\beta^2=1} u(x + \beta t; 0) \, d\beta = \frac{1}{\omega_m} \int \cdots \int_{\alpha^2=1} u(x; \alpha_1 t) \, d\alpha,$$

where on the right-hand side u depends only on the component α_1 of the unit vector α. The expression on the left is thus the mean value $Q(x, t)$ of the initial function ψ, which is known from the initial conditions. Since the integrand in the mean value on the right depends only on the single quantity $\alpha_1 t$ in addition to the parameters x, according to §11 this mean value may be written as

$$\frac{2\omega_{m-1}}{\omega_m \, t^{m-2}} \int_0^t u(x_1, \cdots, x_m, \rho)(t^2 - \rho^2)^{(m-3)/2} \, d\rho.$$

Thus the mean value theorem yields the integral relation

$$(11) \qquad Q(x, t) = \frac{2\omega_{m-1}}{\omega_m \, t^{m-2}} \int_0^t u(x, \rho)(t^2 - \rho^2)^{(m-3)/2} \, d\rho,$$

which is identical with that considered in §13 and was solved there by means of fractional or ordinary differentiation.

Our procedure amounts essentially to treating space coordinates and the time parameter in a symmetric manner by introducing additional artificial "time parameters" of which the physical phenomena are independent.

4. Solutions of the Characteristic Initial Value Problem for the Wave Equation. A further application of Asgeirsson's theorem is the following method of solving the characteristic initial value problem for the wave equation

$$(12) \qquad\qquad u_{tt} - u_{xx} - u_{yy} - u_{zz} = 0$$

in three dimensions, which has been formulated in §6, 1. We suppose that the values u are prescribed on the cone

$$K = x^2 + y^2 + z^2 - t^2 = 0,$$

so that

$$u(x, y, z, \sqrt{x^2 + y^2 + z^2}) = \psi(x, y, z)$$

is known. We try to find a solution of equation (12) (regular for $K \leq 0$) which assumes the prescribed values on $K = 0$.

We first construct the solution of this problem on the axis $x = y = z = 0$ of the cone. To do this we apply the mean value theorem, and find

$$2\pi t \int_0^{2t} u(0, 0, 0, r)\, dr = t^2 \iint_\Omega \psi(\alpha t, \beta t, \gamma t)\, d\omega$$

or

$$4\pi \int_0^t u(0, 0, 0, r)\, dr = t \iint_\Omega \psi\left(\frac{\alpha}{2} t, \frac{\beta}{2} t, \frac{\gamma}{2} t\right) d\omega,$$

where the integral on the right is taken over the surface of the unit sphere in α,β,γ-space. By differentiation we obtain

$$u(0, 0, 0, t) = \frac{1}{4\pi} \iint_\Omega \psi\left(\frac{\alpha}{2} t, \frac{\beta}{2} t, \frac{\gamma}{2} t\right) d\omega$$

$$+ \frac{t}{8\pi} \iint_\Omega (\alpha\psi_x + \beta\psi_y + \gamma\psi_z)\, d\omega,$$

where the arguments $\frac{\alpha}{2} t$, etc., are to be inserted in ψ_x, ψ_y, ψ_z in the second integral as well. If for a point P not on the t-axis K is still negative, the solution $u(P)$ is again given immediately. We have only to transform P into a point of the time axis by a *Lorentz transformation*, i.e., a transformation which leaves the characteristic cone unchanged. If, for example, P has the coordinates $x = x_0$, $y = 0$, $z = 0$, $t = t_0$ with $x_0 < t_0$, this transformation is furnished by the substitution

$$(14) \quad \begin{cases} x = \dfrac{t_0}{\sqrt{t_0^2 - x_0^2}} x' + \dfrac{x_0}{\sqrt{t_0^2 - x_0^2}} t', \\[2ex] y = y', \\[1ex] z = z', \\[1ex] t = \dfrac{x_0}{\sqrt{t_0^2 - x_0^2}} x' + \dfrac{t_0}{\sqrt{t_0^2 - x_0^2}} t', \end{cases}$$

and the image P' of P has coordinates

$$x' = y' = z' = 0,$$

$$t' = \sqrt{t_0^2 - x_0^2}\,.$$

Since the differential equation is invariant under the Lorentz transformation (14), formula (13) remains applicable to the function

$$v(x', y', z', t') = u(x, y, z, t)$$

with the boundary values

$$\chi(x', y', z') = \psi(x, y, z),$$

and accordingly we have

$$u(x_0, 0, 0, t_0) = v\left(0, 0, 0, \sqrt{t_0^2 - x_0^2}\right)$$

$$= \frac{1}{4\pi} \iint_\Omega \chi \left(\frac{\alpha}{2} \sqrt{t_0^2 - x_0^2}, \frac{\beta}{2} \sqrt{t_0^2 - x_0^2}, \frac{\gamma}{2} \sqrt{t_0^2 - x_0^2}\right) d\omega$$

$$+ \frac{t'}{8\pi} \iint_\Omega (\alpha\chi_{x'} + \beta\chi_{y'} + \gamma\chi_{z'})\, d\omega.$$

If χ is then expressed in terms of ψ, we easily find

$$u(x_0, 0, 0, t_0) = \frac{1}{4\pi} \iint_\Omega \psi \left(\frac{1}{2}(x_0 + \alpha t_0), \frac{\beta}{2}\sqrt{t_0^2 - x_0^2}, \frac{\gamma}{2}\sqrt{t_0^2 - x_0^2}\right) d\omega$$

$$+ \frac{1}{8\pi} \iint_\Omega (x_0 + \alpha t_0)\, \psi_x\, d\omega + \frac{\sqrt{t_0^2 - x_0^2}}{8\pi} \iint_\Omega (\beta\psi_y + \gamma\psi_z)\, d\omega.$$

Here the arguments in ψ_x, ψ_y, ψ_z are the same as in ψ, namely:

$$\frac{1}{2}(x_0 + \alpha t_0), \qquad \frac{\beta}{2}\sqrt{t_0^2 - x_0^2}, \qquad \frac{\gamma}{2}\sqrt{t_0^2 - x_0^2}.$$

Therefore, at every point P for which $t > 0$ and $Q < 0$, u is uniquely determined by ψ; we simply imagine the point P transformed by a rotation of the coordinate system into the plane $y = z = 0$. Thus: *The value of u at a point P depends only on the initial values v along the ellipse cut out of a plane by the characteristic cone. This ellipse is identical with the intersection of the initial cone with the characteristic cone through P.*

The reader may try to verify that the function so constructed actually solves the problem under suitable assumptions at infinity. We also remark that the characteristic initial value problem for ultrahyperbolic differential equations can be discussed in a similar way.   Notice that this method only furnishes the values of u *inside* the characteristic cone on which the data are given.   If a solution exists outside as well, its values at any point depend on the data on the whole characteristic cone.[1]

5. Other Applications.   The Mean Value Theorem for Confocal Ellipsoids.   Other familiar mean value theorems are special cases of Asgeirsson's theorem.   For example, we obtain the mean value theorem of potential theory by interpreting a potential function $u(x_1, \cdots, x_m)$ as a particular solution of the differential equation (1) which does not depend on any of the variables y.   Application of the mean value relation (5) for an arbitrary point x and $y = 0$ immediately yields the mean value theorem of potential theory.   This theorem also follows from the more general relation (5″) for $m = 0$.

A less trivial mean value theorem of potential theory is obtained as follows: Let $u(x_1, \cdots, x_m)$ be a solution of the potential equation $\Delta u = 0$.   In place of the m coordinates x_i we artificially introduce $2m$ new coordinates ξ_i and η_i through a system of equations

$$x_i = \xi_i \cosh \alpha_i + \eta_i \sinh \alpha_i \qquad (i = 1, \cdots, m)$$

in which the quantities $\alpha_1, \cdots, \alpha_m$ can be arbitrary. $u(x)$ is thus transformed into a function $\hat{u}(\xi, \eta)$ of the variables

$$\xi_1, \cdots, \xi_m; \qquad \eta_1, \cdots, \eta_m,$$

and the differential equation $\Delta u = 0$ becomes the ultrahyperbolic differential equation

$$\Delta_\xi \hat{u} = \Delta_\eta \hat{u}.$$

We now apply Asgeirsson's mean value theorem, in the form (5′), for the point $\xi_i = \eta_i = 0$ and for the sphere $K_1 : \xi_1^2 + \cdots + \xi_m^2 \leq r^2$ in ξ-space and the corresponding sphere $K_2 : \eta_1^2 + \cdots + \eta_m^2 \leq r^2$ in η-space.   To these spheres correspond the two confocal ellipsoids

$$S_1 : \sum_{i=1}^{m} \frac{x_i^2}{\cosh^2 \alpha_i} \leq r^2$$

and

$$S_2 : \sum_{i=1}^{m} \frac{x_i^2}{\sinh^2 \alpha_i} \leq r^2$$

[1] See F. John [4], pp. 114–120.

in x-space. The mean values over the spheres are transformed into the mean values of the function $u(x)$ over the interior of the corresponding two ellipsoids. Since, with our formulas, any pair of confocal ellipsoids can be represented by a suitable choice of the quantities α_i and r, we immediately obtain the following theorem:

The mean value of a regular potential function over the interior of an ellipsoid is the same for each member of an entire family of confocal ellipsoids.

We also remark that the last application can be approached from a general point of view: There exists a group of linear transformations transforming the ultrahyperbolic differential equation (1) into itself. These are the linear transformations ("ultra-Lorentz" transformations) which transform the characteristic form

$$\sum_{i=1}^{m} (x_i^2 - y_i^2)$$

into itself, up to a factor, and which therefore leave the characteristic cone of the differential equation invariant. Naturally, this group (which merits further study[1]) has as a subgroup not only the similarity transformations but also Lorentz' transformations in correspondingly lower space dimensions.

Performing substitutions of the "ultra-Lorentz" group and then applying Asgeirsson's theorem is a source of further mean value theorems for solutions of special cases of the ultrahyperbolic equation.[2]

---

[1] For $m = 3$, this group is equivalent to the group of transformations of the totality of lines of three-space into themselves; see e.g., Felix Klein [1].

The equivalence of these two groups is seen more concretely in an application, due to F. John, of the Asgeirsson mean value theorem to the ultrahyperbolic equation

$$u_{x_1 y_2} = u_{x_2 y_1}$$

with $m = 2$. (See F. John [5].) We interpret x_1, x_2, y_1, y_2 as line coordinates of the lines of a three-dimensional ξ,ζ-space in the sense of line geometry. Then the general solution, which is defined in the entire space and satisfies certain regularity assumptions at infinity, is given by the integral of an arbitrary function of ξ, ζ over the lines of the space. With respect to the two families of lines of an arbitrary hyperboloid of one sheet in ξ,ζ-space, Asgeirsson's mean value theorem can be formulated as follows: The integral of any solution u over the lines of one of the families is equal to the integral over the lines of the other family.

[2] See L. Asgeirsson [2].

§17. *Initial Value Problems for Non-Space-Like Initial Manifolds*

The mean value theorem of §16 illuminates the situation regarding initial value problems for ultrahyperbolic differential equations and for hyperbolic differential equations with non-space-like initial manifolds. In particular, we shall see why initial value problems of this kind are not meaningful or "well posed" in the sense of Ch. III, §6.

1. Functions Determined by Mean Values over Spheres with Centers in a Plane. The integral of a function $f(x, t) = f(x_1, \cdots, x_n, t)$ over the sphere of radius r and center $(x, 0)$ in x, t-space is given by

$$(1) \qquad g(x, r) = \int_{\xi^2 + \tau^2 = r^2} f(x + \xi, \tau) \, dS = Q[f].$$

Obviously $Q[f]$ only depends on the even part $f(x, t) + f(x, -t)$ of f. We wish to determine $f(x, t) + f(x, -t)$ for prescribed $g(x, r)$. For this purpose we form the integral of f over the solid sphere of radius r and center $(x, 0)$:

$$(2) \qquad G(x, r) = \int_0^r g(x, \rho) \, d\rho = \int_{\xi^2 + \tau^2 < r^2} f(x + \xi, \tau) \, d\xi \, d\tau.$$

Differentiating G with respect to one of the x variables, x_i, we find

$$(3) \qquad G_{x_i} = \int_{\xi^2 + \tau^2 < r^2} f_{\xi_i}(x + \xi, \tau) \, d\xi \, d\tau$$

$$= \frac{1}{r} \int_{\xi^2 + \tau^2 = r^2} f(x + \xi, \tau) \xi_i \, dS.$$

Hence

$$Q[f(x, t) x_i] = \int_{\xi^2 + x^2 = r^2} f(x + \xi, \tau)(x_i + \xi_i) \, dS$$

$$= x_i g(x, r) + r G_{x_i}(x, r) = x_i g(x, r) + r \frac{\partial}{\partial x_i} \int_0^r g(x, \rho) \, d\rho$$

$$= D_i g,$$

where D_i is a linear operator acting on functions $g(x, r)$. Applying the same argument to the function $x_i f$ in place of f and repeating this process indefinitely we see that for any polynomial $P(x_1, \cdots, x_n)$

the integral of the function Pf over the sphere of radius r and center $(x, 0)$ is given by

$$Q[Pf] = P[D_1, \cdots, D_n)g$$

and hence is known when g is known. We have also

(4) $$Q[Pf] = \int_{(\eta-x)^2 < r^2} P(\eta)(f(\eta, \tau) + f(\eta, -\tau)) \frac{r \, d\eta}{\tau},$$

where

$$\tau = \sqrt{r^2 - (\eta - x)^2}.$$

In virtue of the completeness of the polynomials P in the sphere the function

(5) $$\frac{f(\eta, \tau) + f(\eta, -\tau)}{\tau} \qquad (\tau = \sqrt{r^2 - (\eta - x)^2})$$

and hence f itself is determined uniquely by the known expressions $P(D_1, \cdots, D_n)g$.[1]

We now observe the following important fact: To calculate the operator $D_i g$ for any system of values $x_1^0, \cdots, x_n^0, r^0$ it is sufficient to know the mean value $g(x, r)$ of $f(x, t)$ for

$$0 \le r \le r^0,$$

(6)

$$\sum_{i=1}^{n} (x_i - x_i^0)^2 \le \epsilon^2,$$

where ϵ may be arbitrarily small. The same is true for the calculation of all polynomials $P(D_1, \cdots, D_n)g$.

From this follows: The values of g in the region characterized by (6) uniquely determine the even part of the function f in the entire solid sphere

$$\sum_{i=1}^{n} (x_i - x_i^0)^2 + t^2 \le r_0^2.$$

This, in turn, uniquely determines the integral $g(x, t, r)$ over any

---

[1] The fact that the function (5) is singular for $\tau = 0$ does not affect this conclusion. We only have to smooth out the function (5) in a neighborhood of the sphere $|\eta - x| = r$ and choose for the polynomials P a sequence approximating uniformly this smoothed function.

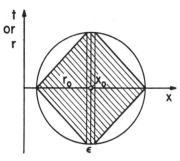

FIGURE 59

sphere with center at $t = 0$ in the interior of the sphere in which f is known, provided that

$$(7) \qquad r + \sqrt{\sum_{i=1}^{n} (x_i - x_i^0)^2} \leq r_0 .$$

Thus we have shown:

g is uniquely determined in the entire double cone (7) *by its values in the cylindrical region* (6) *of arbitrarily small thickness* ϵ. See Figure 59.

2. Applications to the Initial Value Problem. We consider the ultrahyperbolic equation

$$(8) \qquad \sum_{i=1}^{l} u_{y_i y_i} = \sum_{i=1}^{n} u_{x_i x_i} + u_{tt} ,$$

in which we single out $x_{n+1} = t$ and suppose $n \geq 2$, but do not necessarily assume $l = n + 1$. We try to determine a solution by prescribing its values on the plane $t = 0$. Thus for $t = 0$, let

$$u(x, y, 0) = \psi(x, y) \qquad \text{and} \qquad u_t(x, y, 0) = \phi(x, y)$$

be given. We consider the initial values in a domain of the x,y-space, where y is assumed to lie in some domain G of the y-space, while x varies inside the small sphere

$$(9) \qquad \sum_{i=1}^{n} (x_i - x_i^0)^2 \leq \epsilon^2$$

in x-space. The domain in which the initial values are prescribed is thus the "product domain" of a small sphere in the x-space and an arbitrary domain G in y-space. Consider the solution u as a function of x, t with y as parameter. Then our prescribed values ψ determine the integrals of u over the surfaces of those spheres in x,t-space whose centers x_i, t lie in

$$t = 0, \qquad \sum_{i=1}^{n} (x_i - x_i^0)^2 \leq \epsilon^2,$$

and whose radii are not greater than the radius r^0 of the largest sphere about the point y in y-space which still lies entirely in G.

This follows immediately from the mean value theorem of §16 for $n > l$. If $n < l$, this mean value theorem at first yields only the integrals

$$\iint_{V_r} u(x, t) \left(r^2 - t^2 - \sum_{i=1}^{n} x_i^2 \right)^{(l-n)/2} dx \, dt$$

over the interior V_r of every sphere in x,t-space with radius $r \leq r^0$ and center $x_1, \cdots, x_n$, $t = 0$; here $\sum_{i=1}^{n} (x_i - x_i^0)^2 \leq \epsilon^2$. If we denote the integral of u over the surface of such a sphere of radius r by $I(r)$, the integral above can be written as

$$\int_0^r I(\rho)(r^2 - \rho^2)^{(l-n)/2} \, d\rho.$$

But if this expression is known for $r < r^0$, then $I(r)$ is also uniquely determined for $r \leq r^0$. This follows from our earlier discussion (cf. §16) by solving an Abelian integral equation. Our assertion is thus proved also for $l > n$.

By article 1 the even function $u(x, y, t) + u(x, y, -t)$ is uniquely determined in the entire sphere

$$\sum_{i=1}^{n} (x_i - x_i^0)^2 + t^2 \leq (r^0)^2$$

by the given values of ψ. Similarly the even function

$$u_t(x, y, t) + u_t(x, y, -t)$$

is determined by ϕ. It follows immediately that $u(x, y, t)$ is determined uniquely. In particular, the initial values $u(x, y, 0)$ are determined for $t = 0$ in the sphere

$$(10) \qquad \sum_{i=1}^{n} (x_i - x_i^0)^2 \leq (r^0)^2$$

of the n-dimensional initial space R_n, and we therefore obtain the remarkable result:

If the initial values of a solution u of the ultrahyperbolic equation (8) *are known for y in G and for x in an arbitrarily small sphere*

$$\sum (x_i - x_i^0)^2 \leq \epsilon^2$$

(cf. article 1), *then the initial values are uniquely determined everywhere in the larger sphere*

$$\sum_{i=1}^{n} (x_i - x_i^0)^2 \leq (r^0)^2,$$

where r^0 is defined as above.

A consequence of this result is: *One cannot arbitrarily prescribe initial values $u(x, y, 0)$.*

For example, if, with given a, one prescribes initial values $u(y_1, y_2, x, 0)$ to solve the equation

(11) $$u_{y_1 y_1} + u_{y_2 y_2} - u_{xx} - u_{tt} = 0$$

in a thin cylindrical disk

$$t = 0, \qquad (y_1 - y_1^0)^2 + (y_2 - y_2^0)^2 \leq a^2, \qquad |x - x^0| \leq \epsilon,$$

then $u(y_1, y_2, x, 0)$ is *a priori* uniquely determined in the double cone

$$t = 0, \qquad \sqrt{(y_1 - y_1^0)^2 + (y_2 - y_2^0)^2} + |x - x^0| \leq a.$$

See Figure 60.

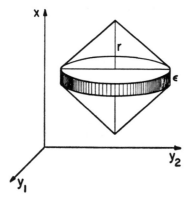

FIGURE 60

Similarly, consider the wave equation

(12) $$u_{yy} - u_{x_1x_1} - y_{x_2x_2} - u_{tt} = 0$$

in which, however, the roles of the space variable y and the time variable t are interchanged. If the function $u(y, x_1, x_2, t)$ is prescribed in the thin cylinder

$$t = 0, \qquad (x_1 - x_1^0)^2 + (x_2 - x_2^0)^2 \leq \epsilon^2, \qquad |y - y^0| \leq a$$

parallel to the y-axis, the initial value $u(y, x_1, x_2, 0)$ is at once uniquely determined in the double cone

$$\sqrt{(x_1 - x_1^0)^2 + (x_2 - x_2^0)^2} + |y - y^0| \leq a.$$

See Figure 61.

Thus we see: *On a non-space-like plane, it is not possible to prescribe arbitrarily the initial values for the solution of the wave equation.*

If, in the case of the general equation (8), the initial value $u(y_1, y_2, \cdots, y_l; x_1, \cdots, x_n, 0)$ is prescribed for

$$\sum_{i=1}^{l} (y_i - y_i^0)^2 \leq a^2, \qquad \sum_{i=1}^{n} (x_i - x_i^0)^2 \leq \epsilon^2,$$

then it is *a priori* known for the region

$$\sqrt{\sum_{i=1}^{l} (y_i - y_i^0)^2} + \sqrt{\sum_{i=1}^{n} (x_i - x_i^0)^2} \leq a,$$

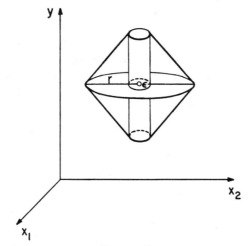

FIGURE 61

and the solution $u(y, x, t)$ is uniquely determined for

$$\sqrt{\sum_{i=1}^{l} (y_i - y_i^0)^2} + \sqrt{\sum_{i=1}^{n} (x_i - x_i^0)^2 + t^2} \leq a.$$

For the potential equation $(l = 0)$

$$(13) \qquad u_{x_1 x_1} + \cdots + u_{x_n x_n} + u_{tt} = 0,$$

this means: *If a solution u is an even function of t, then the values $u(x, 0)$ in an arbitrarily small sphere*

$$\sum_{i=1}^{n} x_i^2 \leq \epsilon^2$$

uniquely determine for arbitrary a the values of the solution in the domain

$$\sum_{i=1}^{n} x_i^2 + t^2 \leq a^2.$$

In particular, for $t = 0$, the values $u(x, 0)$ uniquely determine the initial values of u. The statement for the initial values is again true without restriction to even solutions.

This result concerning the potential equation might have been expected from the analytic character of its solutions, which was already known to us. In the case of hyperbolic and ultrahyperbolic differential equations, however, the relations obtained above between the values of a solution on the initial plane are not so obvious. In fact, these initial functions may very well not even be analytic. Thus, in investigating the values of the solutions of hyperbolic and ultrahyperbolic differential equations along non-space-like planes, we are dealing with the remarkable phenomenon of functions which are not necessarily analytic, yet whose values in an arbitrarily small region determine the function in a substantially bigger domain.[1]

§18. Remarks About Progressing Waves, Transmission of Signals and Huyghens' Principle

1. Distortion-Free Progressing Waves. While the term "wave" was used in this book quite generally for any solution of a hyperbolic

---

[1] Cf. F. John [3] where, by a different method, even more extensive results are obtained for general linear equations with analytic coefficients.

problem,[1] there are certain specific classes of waves of particular interest, for example "standing waves", represented as products of a function of time and a function of the space variables. In the present section we want to comment further on the importance of another such class, the progressing waves, discussed in Chapter III for differential equations with constant coefficients and more generally in §4 of the present chapter. This concept is a key for the theory of transmission of signals, indeed a central subject in the theory of hyperbolic differential equations. For brevity we shall consider a single equation $L[u] = 0$.

In keeping with Ch. III, §3 we define a *family of undistorted progressing waves* as a family of solutions of $L[u] = 0$ depending on an *arbitrary function* $S(\phi)$ and having the form

$$(1) \qquad u = S(\phi(x, t)),$$

where S is called the *wave form* and $\phi(x, t)$ is a fixed *phase function* of the space variables x and the time $t = x_0$. Such a phase function might be

$$\phi(x, t) = \chi(x) - t.$$

The solution u represents the undistorted motion of the wave form S through space.

Using the arbitrariness of $S(\phi)$, we conclude that ϕ must satisfy

$$L[\phi] = 0,$$

and the characteristic equation

$$Q(D\phi) = 0.$$

The first equation is obtained by the special substitution $S(\phi) = \phi$; the second follows if we choose $S = \delta(\phi - c)$ with an arbitrary constant c (see §4). Thus we may state: The phase function ϕ is a characteristic function, i.e., the phase surfaces $\phi = $ const. are characteristic wave fronts.

In spite of this overdeterminacy of ϕ, some differential equations $L[u] = 0$ exist which do admit families of undistorted progressing waves. This is the case, e.g., for linear differential equations $L[u] = 0$ with constant coefficients containing only the highest order terms, in particular for the wave equation (see Ch. III, §3). However, in

---

[1] To avoid confusion we have consistently reserved the name "wave front" for surfaces of discontinuity which satisfy not the original differential equation but the associated characteristic equation of first order.

general the conditions for ϕ are not compatible. It is therefore appropriate to introduce the less restrictive concept of *"relatively undistorted"* *progressing wave families* having the form

$$(2) \qquad\qquad u = g(x, t)S(\phi),$$

where again $S(\phi)$ is arbitrary and where not only the phase function $\phi(x, t)$, but also the distortion factor g, is specific. Such waves still can serve as suitable carriers of signals inasmuch as the factor g simply represents an attenuation. Spherical waves in three space dimensions, e.g., $\dfrac{S(t - r)}{r}$ or $\dfrac{S(t + r)}{r}$, are typical examples of such relatively undistorted wave families. Concentric spheres in space define the moving characteristic phase surfaces.

Again the conditions restrict (2) to a characteristic function; they imply an overdetermined system of differential equations for the distortion factor g. We recognize this simply, e.g., by substituting (2) in the differential equation and by realizing that the arbitrariness of S implies the vanishing of all the coefficients of S, S', S'', $\cdots$.

Hence, an equation $L[u] = 0$ has the desirable property of possessing relatively distortion-free families of solutions only in exceptional cases. Of course, if a differential equation $L[u] = 0$ does have this property, an entire class of equivalent differential equations possesses the same property. Two differential equations $L[u] = 0$ and $L^*[u^*] = 0$, for two functions $u(x)$ and $u^*(x)$, are called equivalent if they can be transformed into each other by a transformation of the form $x_i^* = \alpha_i(x_0, x_1, \cdots, x_n)$, $u^* = f(x)u$.

The question of determining all operators L which allow such families of solutions has hardly been touched.[1]

---

[1] A special fact, easily proved, is:

In the case of two variables $x_1 = x$, $t = x_0 = y$, *the only differential equations of second order which admit relatively distortion-free progressing wave families in both space directions are* $u_{xy} = 0$ *and equations equivalent to it.*

Certainly, the differential equation is equivalent to an equation of the form $2u_{xy} + Bu_x + Cu = 0$, where B and C are functions of x and y, where $x + y$ and $x - y$ represent the time and space coordinates, respectively, and where $x = $ const., $y = $ const. are the characteristics. The existence of the wave family $u = g(x, y)S(y)$ requires that $g_x = 0$ holds as well as $2g_{xy} + Bg_x + Cg = 0$ and hence $C = 0$. If, in addition, a wave family $u = h(x, y)S(x)$ advancing in another direction is to exist, then $2h_y + Bh = 0$ must be satisfied together with $2h_{xy} + Bh_x = 0$ so that $B_x = 0$ follows. But the equation

$$2u_{xy} + B(y)u_x = 0$$

is equivalent to the equation $u_{xy} = 0$.

There is a related problem for which a solution is known: Consider the wave equation with four independent variables. What are the possible wave fronts of relatively undistorted progressing waves? The answer is that all such wave fronts are cyclides of Dupin,[1] which include planes and spheres as special cases.

In general, in order to mitigate or to eliminate the overdeterminacy of the distortion factor g, one must introduce more such factors, as indeed we did in §4, defining progressing waves of higher degree, or even complete progressing waves. According to §§4 and 5 such waves provide an important step towards the construction of solutions, although they do represent a distortion of the initial shape of the signal.

2. Spherical Waves. The problem of transmission of signals is further clarified by the concept of "spherical waves" which generalize the spherical solutions of the three-dimensional wave equation. We confine ourselves to the case of linear differential equations $L[u] = 0$ of second order and consider a time-like[2] line Λ given in the form $x_i = \xi(\lambda)$ with a parameter λ. (The time variable is not emphasized here.) With the point $\xi(\lambda)$ as vertex we consider the characteristic conoid or spherical wave front $\Gamma(x; \xi) = 0$.

For given x, we may determine λ as a function of x from the equation $\Gamma(x, \xi(\lambda)) = 0$; we write $\lambda = \phi(x)$. The characteristic conoid with vertex $\xi(\phi(x))$ is given by the equation $\phi(x) = \text{const.}$ *A family of relatively undistorted spherical waves issuing from* Λ *may* then be defined as a solution u of the second order differential equation in the form

$$u(x) = g(x)S(\phi(x))$$

with specific g and arbitrary S.

Little is known about the scope of this concept, which obviously relates spherical waves to the problem of transmitting with perfect fidelity signals in all directions. All we can do here is to formulate a conjecture which will be given some support in article 3: *Families of spherical waves for arbitrary time-like lines* Λ *exist only in the case of two and four variables, and then only if the differential equation is equivalent to the wave equation.*

A proof of this conjecture would show that the four-dimensional

[1] See F. G. Friedlander [1] and M. Riesz [1].
[2] See §3, 7.

physical space-time world of classical physics enjoys an essential distinction.

Here we merely emphasize that for the wave equation using the t-axis as the time-like line Λ and with $r^2 = x^2 + y^2 + z^2$, we have such waves with $\phi = t - r$ and $g = 1/r$. For other straight time-like lines, spherical waves are obtained by Lorentz transformations.

In the case of an even number of independent variables $n + 1 = 2\nu + 4$ ($\nu = 1, 2, \cdots$), solutions exist in the form of families of progressing waves[1] of higher order. The explicit solutions given in §12, 4 or §15, 4 no longer enjoy the property of freedom from distortion, but still represent a progressing phenomenon.

As to equations of higher order, it is worth noting, as an example, that for all even values of $n + 1$ the $(n + 1)/2$-th iterated wave equation of order $n + 1$

$$L^{(n+1)/2}[u] = \left(\frac{\partial^2}{\partial t^2} - \Delta\right)^{(n+1)/2} u = 0,$$

possesses undistorted families of spherical waves

$$u = S(t - r), \qquad u = S(t + r),$$

although the wave equation itself

$$L[u] \equiv \left(\frac{\partial^2}{\partial t^2} - \Delta\right) u = 0$$

does not. This fact is simply another interpretation of the theorem proved in §13, 4. It indicates that higher order equations admit various possibilities not existing for second order.

Finally it should be recalled that *individual progressing spherical waves* of higher degree with specific S, not necessarily families with arbitrary S, occur and are of importance (§4). In particular, the fundamental solution of §15, e.g., Hadamard's expression for the fundamental solution of single second order equations (§15, 6) is represented by such waves:

$$R = S(\Gamma)g(x, t) + S_1(\Gamma)g^1(x, t) + \cdots,$$

where $S(\Gamma)$ is a specific distribution.

3. Radiation and Huyghens' Principle. Huyghens' principle dis-

---

[1] The notation is slightly different from that above.

cussed on various occasions in this volume, stipulates that the solution at a point ξ, τ does not depend on the totality of initial data within the conoid of dependence (see §7) but only on data on the characteristic rays through ξ, τ. (Again we emphasize $x_0 = t$ and $\xi_0 = \tau$.) The principle is tantamount to the statement that the radiation matrix of §15 vanishes identically except on the rays through ξ, τ. Equivalently we may state: A sharp signal issued at the time τ and the location ξ is transmitted as a sharp signal along the rays and remains unnoticeable outside the ray conoid. The principle does not, however, state that signals are transmitted without distortion.

For single differential equations of second order with constant coefficients we have seen: Only for the wave equation in 3, 5, 7, $\cdots$ space dimensions, and for equivalent equations, is Huyghens' principle valid. For differential equations of second order with variable coefficients *Hadamard's conjecture*[1] states that the same theorem holds even if the coefficients are not constant. Examples to the contrary show that this conjecture cannot be completely true in this form,[2] although it is highly plausible that somehow it is essentially correct.[3]

Altogether, the question of Huyghens' principle for second order equations should be considered in the light of the much more comprehensive problem of the exact domains of dependence and influence for any hyperbolic problem (see §7), a problem which is still completely open.

Concerning the transmission of signals which not only remain sharp but are undistorted, the conjecture in article 2 stated that this phenomenon is possible only in three space dimensions. For an isotropic homogeneous medium, i.e., for constant coefficients (and second order equations), the proof of this conjecture is contained in the preceding discussions. Thus our actual physical world, in which acoustic or electromagnetic signals are the basis of communi-

---

[1] This famous conjecture in fact was not categorically asserted by Hadamard.

[2] An example to the contrary for seven space dimensions was recently given by K. L. Stellmacher [1].

[3] Hadamard has identified the condition of validity of Huyghens' principle with the vanishing of the logarithmical term in his expression of the fundamental solution for odd number n of space dimensions. In our version Huyghens' principle means that the series (44) in §15 does not contain terms with the Heaviside function and its integrals.

cation, seems to be singled out among other mathematically conceivable models by intrinsic simplicity and harmony.

Yet, at least in an approximate sense, any hyperbolic system implies preservation of sharpness of signals in the sense of the generalized Huyghens' principle (see §15, 3). This generalized principle is therefore significant for the mathematical understanding of transmission of signals. This is true all the more since the validity of Huyghens' principle is at best a highly unstable property of a differential operator; this property is destroyed by infinitesimal variations of the coefficients of the operator. It would seem, therefore, that the generalized Huyghens' principle should be considered as the proper expression of physical reality.

Appendix to Chapter VI

Ideal Functions or Distributions

§1. *Underlying Definitions and Concepts*

1. Introduction. In this appendix we shall discuss the concept of distributions or "ideal functions".[1] The specific use of these ideal functions in the preceding chapters will be justified within a more general framework.

It should be understood that "function" may mean a function-vector with k components. The functions involved may be complex-valued, but the independent variable x is always a real vector with n components.

Much of the substance of the theory has long played a role in physical literature and elsewhere.[2] But only since the publication

---

[1] The name "distributions" indicates that ideal functions, such as Dirac's delta-function and its derivatives, may be interpreted by mass distributions, dipole distributions, etc., concentrated in points, or along lines or on surfaces, etc. However, the term "ideal functions" seems much more indicative of the true role of this concept as it is used in connection with differential equations and in mathematical analysis generally. This role is indeed that of functions, almost as the role of real numbers is that of ordinary numbers.

[2] For example, attention might be called to a paper by S. L. Sobolev [1], which long preceded the present flurry of literature.

of Laurent Schwartz's comprehensive book on distributions,[1] has the topic been treated systematically in a multitude of monographs,[2] some of which go far in the direction of refinement.[3] The present appendix concentrates on the elementary core of the theory as far as it is relevant for our study of linear differential equations. We omit a detailed discussion of the much-treated applications to the theory of Fourier transformations. (See, however, §4, 4.)

2. Ideal Elements. "Distributions" are most appropriately introduced as ideal elements in function spaces. It is one of the very basic procedures of mathematics to extend a given set or "space" S of mathematical objects by additional new "ideal elements" not defined as entities in the original set S, and not defined descriptively but defined merely by relationships such that in the extended set $\bar{S}$ the original rules for basic operations are preserved. The purpose of the extension always is to remove restrictions prevailing in the original set S.

Thus, in projective geometry ideal "points at infinity" are defined by sets of parallel lines. In other cases, the ideal elements are introduced by completion of the original set S by "strong" limit processes with a suitable *norm*: For example, the real numbers are defined by[4] convergent sequences of rational numbers r_n for which the norm $|r_n - r_m|$ converges to zero if n and m tend to infinity. Lebesgue-integrable and also square-integrable functions similarly can be defined by sequences of continuous functions $f_n(x)$ for which in the respective x-domains the integrals $\int |f_n - f_m| \, dx$ and $\int |f_n - f_m|^2 \, dx$ converge to zero.—Functions in Hilbert spaces are ideal elements represented by sequences of suitably smooth functions f_n for which underlying positive quadratic forms $Q(f_n - f_m)$ converge to zero.

In these examples, the extended spaces $\bar{S}$ are complete, that is,

---

[1] See L. Schwartz [1].

[2] See, e.g., I. M. Gelfand and G. E. Shilov [1].

Also, a recent short book by Lighthill should be mentioned especially for its emphasis on Fourier analysis. Lighthill's book partly follows a publication by G. Temple. See M. J. Lighthill [2] and G. Temple [1], and further literature quoted in these publications.

[3] See, e.g., a series of papers by L. Ehrenpreis [1].

[4] Often the desire for descriptive definitions of ideal elements has led to logical twists such as the assertion: "A real number *is* a Dedekind cut in the set of rational numbers." It seems that little is gained by attempts to avoid the need for defining ideal objects by relationships instead of substantive descriptions.

they cannot be enlarged by using convergence under the same norm. In contrast, the following definitions of ideal functions will not emphasize completion by convergence under a suitable norm.[1]

Ideal functions are introduced in order to widen the scope of elementary linear calculus by removing the straitjacket of differentiability conditions. Operating with distributions as distinct objects (as Laurent Schwartz first did with clarity[2]), instead of using *ad hoc* devices, was a most fortunate turn of thought; furthermore, by considering these objects as "functions" one attains a definite simplification of certain otherwise more cumbersome procedures.[3]

For the purposes of this book it would suffice to introduce (as in Ch. VI, §4) ideal functions by linear differential operators on continuous functions and by a simple rule of operation. However, it is useful to present two other definitions and show the equivalence of all three. Before doing so in §§2, 3, we recall and supplement some notations from Chapter VI.

3. Notations and Definitions. For a pair of vectors y, z we define $y < z$ if one of the components of y is smaller and none larger than the corresponding component of z.

As in Ch. VI, §3, we denote by r a vector with n integral nonnegative components $r_1, \cdots, r_n$, by $|r|$ the sum $r_1 + \cdots + r_n$; we sometimes write $(-1)^r$ instead of $(-1)^{|r|}$. Also, we sometimes write $r + 1$ for a vector with the components $r_i + 1$, etc.

Furthermore $r \to \infty$ means that all components of r tend to infinity. Also, as in §3, we define $r! = r_1! \; r_2! \cdots r_n!$. For any vector ξ in the n-space, ξ^r was defined as the product $\xi_1^{r_1} \cdots \xi_n^{r_n}$.

By $\mathcal{I}$, $\mathcal{I}'$, $\mathcal{I}^*$, etc., we denote rectangular domains[4] in the x-space,

---

[1] They are "weak definitions". It should, however, be mentioned that for the ideal functions the principle of "strong" extension by convergence under a suitable norm is also applicable. (See a remark later in §4, 4.) The connection or equivalence between weak and strong extensions has been emphasized by Friedrichs [4].

[2] See Laurent Schwartz [1].

[3] See e.g. I. M. Gelfand and G. E. Shiloff [1].—Formal simplifications thus attainable must not create the illusion that the core of intrinsic difficulties can thereby be mastered rather than merely isolated and clarified. Often the genuine difficulty is shifted to the final task of ascertaining in what sense a result obtained in terms of ideal functions is indeed expressible by ordinary functions.

[4] It is convenient but not essential that $\mathcal{I}$ be rectangular.

say $-\alpha < x_i < A$, or also the whole x-space. Usually we denote by $\bar{g}$, $\bar{g}'$, etc., corresponding closed domains.

The inner product (g, h) of two functions g and h is, of course, defined as the integral of gh over the *base domain* g which may be the whole space.

By $D^r = D_1^{r_1} \cdots D_n^{r_n}$ we denote differentiation, where D_i means $\partial/\partial x_i$; by D_z we denote the corresponding differentiation if we want to emphasize z as the independent vector.

Sometimes it is useful to define operators of integration by symbols D_i^{-1}, D^{-s}, etc., without always specifying initial points for the respective integrals.

By C^r (or C^∞) we denote a space of functions ψ for which $D^r\psi$ (or $D^\rho\psi$ for all ρ) is continuous or at least piecewise continuous.

Finally we recall the definitions of *maximum norm* or *maximum r-norm* $\|\phi\|$ or $\|\phi\|_r$ of a function ϕ in g, i.e., the least upper bound in g for $|\phi|$ or any of the derivatives $|D^{r'}\phi|$ of ϕ with $r' \leq r$.

4. Iterated Integration. With a parameter point z in a rectangle g, say $0 < x_i < 1$, we set in the sector $x_i < z$, called Σ, of the x-space

$$(1) \qquad q_r(x; z) = q(x; z) = \frac{1}{r!} (z - x)^r,$$

$$q_r(x; z) = 0$$

outside this sector. Then, for a continuous function $h(x)$ which vanishes in Σ for large values of $|x|$ set

$$(2) \qquad G(z) = \int \cdots \int q_r(x; z) h(x) \, dx;$$

we have according to elementary calculus

$$(3) \qquad D^{r+1} G(z) = h(z).$$

5. Linear Functionals and Operators—Bilinear Form. We recall the general concept of a linear functional $\Lambda[\phi]$ which depends on a function $\phi(x)$ defined in a base domain g and otherwise restricted to membership in a linear "space" $\mathfrak{D}$ of *test functions* ϕ; for example $\mathfrak{D}$ may be the set of functions each continuous in a subdomain g' of g and zero outside g'. The basic propery of a linear functional is expressed by the identity $\Lambda[c_1\phi_1 + c_2\phi_2] = c_1\Lambda[\phi_1] + c_2\Lambda[\phi_2]$ for two test functions ϕ_1 and ϕ_2 with two arbitrary constants c_1 and c_2. A consequence under assumptions of continuity (see article 6) is the identity

$$(4) \qquad \Lambda[\psi(x)] = \int \Lambda[\phi(x; \xi)] \, d\xi$$

if

$$(4') \qquad \psi(x) = \int \phi(x; \xi) \, d\xi,$$

and if the test function $\phi(x; \xi)$ depends continuously on the parameters ξ in a ξ-domain of integration.

If the functional $\Lambda[\phi(x); y]$ depends on a parameter y as well as on the test function $\phi(x)$, then Λ represents a *linear operator* or *linear transformation*

$$\Lambda[\phi(x); y] = \omega(y)$$

(sometimes briefly denoted by $\Lambda[\phi] = \omega$) of $\phi(x)$ into $\omega(y)$. Usually we consider cases in which the variable y ranges over the same domain $\mathcal{J}$ as the variable x.

Whenever the inner product over $\mathcal{J}$ of the function $\omega(y) = \Lambda[\phi]$ with a test function $\psi(y)$ can be formed, one calls

$$(\omega, \psi) = B[\phi; \psi] = (\Lambda[\phi], \psi) = \int_{\mathcal{J}} \Lambda[\phi(x); y]\psi(y) \, dy$$

the *bilinear form* or the *bilinear functional* attached to the operator Λ. Obviously for fixed ϕ it is a linear functional of ψ and for fixed ψ it is a linear functional of ϕ. If there is a linear operator $\Lambda^*[\psi]$ such that B can be expressed in the form

$$(5) \qquad B[\phi; \psi] = (\Lambda[\phi], \psi) = (\phi, \Lambda^*[\psi]),$$

then Λ^* is called the *adjoint operator* to Λ.[1]

Derivatives, and more generally differential operators, are linear operators of a special character in so far as they are sharply *localized*; that is, their value does not depend on the values of the argument function $\phi(x)$ in a whole x-domain but only on the values in the infinitesimal neighborhood of a point $x = y$. (In our context such operators should perhaps be written more circumstantially in the form

[1] See Ch. III, App. 1, § 2 for the concept of adjoint operator.

$\Lambda[\phi(y) ; x]$, where the role of x and y is interchanged. However, we take the liberty of using the ordinary notation and simply write $L[\phi(x)]$ for differential operators, as we have been doing throughout this book.)

It may be noted that, in contrast, the operator of fractional differentiation is not localized, as the formulas for fractional differentiation exhibit immediately.

6. Continuity of Functionals. Support of Test Functions. Once and for all we shall assume that the test functions $\phi(x)$ are continuous and furthermore that each is identically zero outside a finite *domain $\mathcal{J}^*$ of support*, or that they have *bounded support*. We sometimes say: ϕ is in $\mathcal{J}^*$.

If, moreover, all derivatives $D^{r'}\phi$ for $r' \leq r$ are continuous (or if ϕ is "of the class $C^{r'}$") then we say that the functions ϕ form the class $\mathfrak{D}_r$; if *all* derivatives of ϕ are assumed continuous (or if ϕ is in C^{∞}), then they form the narrower class $\mathfrak{D}_{\infty}$ or briefly $\mathfrak{D}$. Obviously $\mathfrak{D}_{r'}$ includes $\mathfrak{D}_r$ if $r' < r$. Unless otherwise stated, we shall usually assume that ϕ is in C^{∞}, or $\mathfrak{D}$.

We consider a base domain $\mathcal{J}$ for x. A functional $\Lambda[\phi]$ (and similarly a linear operator) is called continuous if, uniformly for test functions ϕ with support in any closed subdomain $\bar{\mathcal{J}}^*$ of $\mathcal{J}$ and a common bound: $\| \phi \| < m$,

$$\Lambda[\phi_\epsilon] \to \Lambda[\phi];$$

here $\phi_\epsilon(x)$ denotes a sequence of such test functions which converge for $\epsilon \to 0$ uniformly in x to the test function $\phi(x)$. Because of the linearity, the following stipulation is equivalent: For all continuous test functions with support in a closed subregion $\bar{\mathcal{J}}^*$ of $\mathcal{J}$ there exists a fixed positive number λ such that $| \Lambda[\phi]| \leq \lambda \| \phi \|$.

A less restrictive concept of continuity is defined as follows: The functional $\Lambda[\phi]$ is called *r-continuous* in $\bar{\mathcal{J}}^*$ if with a fixed positive λ we have

$$| \Lambda[\phi]| < \lambda m$$

for all test functions ϕ in $\bar{\mathcal{J}}^*$ of $\mathfrak{D}_r$ for which the *maximum r-norm* of ϕ, $\| \phi \|_r$, is bounded by a bound m.

Obviously r'-continuity for $\Lambda[\phi]$ is a less restrictive stipulation than r-continuity if $r' > r$. Continuity as defined above merely demanding boundedness of $\| \phi \|$ is a sharper condition for $\Lambda[\phi]$

than r-continuity for any $r > 0$. We might call this "ordinary" continuity *zero-continuity*.

Sometimes it is necessary to consider in an open base domain $\mathcal{g}$ a functional $\Lambda[\phi]$ which in any closed subdomain $\bar{\mathcal{g}}^*$ is r-continuous for some index r (possibly depending on $\mathcal{g}^*$). We shall call such a functional *mildly continuous* in $\mathcal{g}$. This concept suffices to cover all relevant cases.

For the benefit of formal generality it is sometimes considered desirable to stipulate the apparently less restrictive condition which refers in the whole domain $\mathcal{g}$ to the most restricted class $\mathfrak{D}$ or $\mathfrak{D}_\infty$ of test functions: Let m_ν be an arbitrary sequence of positive numbers, and $\bar{\mathcal{g}}^*$ any closed subdomain of $\mathcal{g}$. Then we stipulate: There exists a bound λ (depending on $\mathcal{g}^*$ and the sequence m_ν) such that $|\Lambda[\phi]| \leq \lambda$ for any test function ϕ of $\mathfrak{D}_\infty$ with support in $\bar{\mathcal{g}}^*$ and restricted by $\|D^\nu\phi\| \leq m_\nu$.

It should be noted that in this definition the boundedness with respect to ν of the infinitely many bounds m_ν is not demanded. Therefore this definition cannot be stated in terms of a maximum norm such as in the case of r-continuity.[1]

We shall sometimes call this least demanding type of continuity ∞-*continuity* or ω-*continuity* of $\Lambda[\phi]$.

As we shall see in the following article, this concept is not more general than that of mild continuity.

7. Lemma About r-Continuity. The following almost trivial lemma should be inserted[2] here for a finer comparison of r-continuity and mild continuity with ω-continuity: If $\Lambda[\phi]$ is ω-continuous in the open domain $\mathcal{g}$ then for each closed subdomain $\bar{\mathcal{g}}^*$ there exists a finite index r such that Λ in $\bar{\mathcal{g}}^*$ is even r-continuous for all test functions with support in $\bar{\mathcal{g}}^*$. The index r, of course, may depend on $\mathcal{g}^*$ and may have to be increased if we extend $\mathcal{g}^*$ (as examples will show). In other words, ω-continuity and mild continuity of $\Lambda[\phi]$ in $\mathcal{g}$ are the same.

The proof of the lemma is indirect and amounts to a simple evaluation of the meaning of the definition: Assume that $\Lambda[\phi]$ is not r-continuous in the closed domain $\bar{\mathcal{g}}^*$ no matter how large the index

[1] The space $\mathfrak{D}_\infty$ is not metrized by a maximum norm.
[2] It is strictly a statement of existence and not at all essential in applications to mathematical physics.

r is chosen. Then there would exist for every r an admissible test function ϕ_r such that $|\Lambda[\phi_r]| > |r|$, while $\|\phi_r\|_r$ becomes arbitrarily small, e.g., $\|\phi_r\|_r \leq 1/(|r| + 1)$. (As before, $\|\phi\|_r$ denotes the maximum norm of order r for the closed domain $\bar{\jmath}^*$.) Now the sequence ϕ_r obviously converges uniformly to zero in $\bar{\jmath}^*$ for $r \to \infty$, and so does every derivative $D^{r'}\phi_r$ for fixed r'. By assumption of ω-continuity in $\mathfrak{D}$ with $m_r = \max_{\text{all } r'} \|\phi_{r'}\|_r$ the values of $\Lambda[\phi_r]$ must therefore be bounded, which contradicts $|\Lambda[\phi_r]| > |r|$. This completes the indirect proof.

8. Some Auxiliary Functions. For reference in §2 we construct some specific functions of C^∞ with bounded support: We first consider one independent variable x and set for positive α

$$(6) \qquad p(x;\alpha) = \begin{cases} e^{z(x;\alpha)} & \text{for } -\alpha \leq x \leq \alpha, \\ 0 & \text{for } x < -\alpha, \\ 1 & \text{for } x > \alpha, \end{cases}$$

with

$$(6a) \qquad z(x;\alpha) = -e^{x/(x^2-\alpha^2)}.$$

This function p is zero for $x < -\alpha$, identically equal to one for $x > \alpha$, and is in C^∞. The product

$$(6b) \qquad P(x;\alpha) = p(x_1;\alpha)p(x_2;\alpha) \cdots p(x_n;\alpha)$$

for n variables x is in C^∞; it is zero for $x \leq -\alpha$ (that means $x_i \leq -\alpha$) and identically one in the section $x \geq \alpha$.

Similarly the function of one variable x

$$(7) \qquad \delta_\epsilon(x) = \frac{d}{dx} p(x;\epsilon)$$

is in C^∞ and vanishes outside the interval $-\epsilon < x < \epsilon$. For each value of ϵ we have

$$(8) \qquad \int_{-\infty}^{\infty} \delta_\epsilon(x)\, dx = 1.$$

Setting $\epsilon = 1/n$ we take the liberty of writing δ_n instead of δ_ϵ and have for any continuous function $\phi(x)$ in $\jmath$

$$(9) \qquad \phi(x) = \lim \int_{\jmath} \delta_n(x-\xi)\phi(\xi)\, d\xi = \lim\, (\phi, \delta_n(x-\xi))$$

as is well known. The limit always means limit for $n \to \infty$ or $\epsilon \to 0$.

9. Examples. Some examples illustrate our concepts of continuity: Let x be a single variable, $\mathcal{g}$ the open interval $0 < x < 1$; then $\phi(0)$ is a continuous functional of $\phi(x)$ for $\phi(x)$ in $\mathfrak{D}_0$. The quantity $D^r(\phi(0)) = \phi^r(0)$ is an r-continuous functional of $\phi(x)$ in $\mathcal{g}$.

As the reader may confirm, the symbol

$$(10) \qquad \sum_{\nu=0}^{\infty} \frac{\phi^\nu(0)}{\nu!}$$

does not, however, represent a functional of ϕ continuous in the sense defined in article 6.

In contrast, the series

$$(11) \qquad \sum_{\nu=1}^{\infty} \phi^\nu\left(1 - \frac{1}{\nu}\right) = \sum_\nu D^\nu \phi\left(1 - \frac{1}{\nu}\right)$$

converges if the support of ϕ is in a closed subinterval $\bar{\mathcal{g}}^*$, e.g., in the interval $0 \le x \le 1 - 1/r$; there it is r-continuous. The index r increases as $\bar{\mathcal{g}}^*$ is extended towards $\mathcal{g}$, and for the whole open x-domain $\mathcal{g}$ the expression (11) is obviously not r-continuous; yet it is ω-continuous with respect to the comprehensive space $\mathfrak{D}_\infty$.

Differently, the expression $\sum_{\nu=1}^{\infty} \phi'\left(1 - \frac{1}{\nu}\right)$ represents a 1-continuous functional for the domain $\mathcal{g}: 0 < x < 1$ with respect to the space $\mathfrak{D}_1$ of functions whose support is in $\mathcal{g}$. Similarly $\sum_{-\infty}^{\infty} \phi(\nu)$ is continuous for the whole x-axis as base domain $\mathcal{g}$.

§2. Ideal Functions

1. Introduction. Turning to the essential points of the theory we give three variants of a definition[1] of ideal functions and then show their equivalence. These extensions of the concept of ordinary functions are based on the principle of *weak definition* and of *weak convergence*: Instead of characterizing a continuous function f in $\mathcal{g}$ by the store of its values we could characterize it just as well by the *totality of the inner products* (f, ϕ) with all test functions ϕ of a class $\mathfrak{D}$ whose support is in $\mathcal{g}$.[2] For continuous functions f this "weak" definition is equivalent to the ordinary definition if $\mathfrak{D} = \mathfrak{D}_0$ consists,

---

[1] As said before, these definitions are not descriptive.

[2] We recall: "A test function is in $\mathcal{g}$" means that its support is in $\mathcal{g}$; "a functional $\Lambda[\phi]$ in $\mathcal{g}$" means: The test functions ϕ should have their support in $\mathcal{g}$.

e.g., of continuous functions. However, it is this weak definition which opens the way to generalization: We disregard the original definition of a "function" as represented by the store of its values at the location x, and instead consider values of the inner products (f, ϕ) with ordinary test functions ϕ as the exhaustive characterization of the function symbol f: In order to introduce a "function" f as a symbolic factor in the inner product the main step is to give a proper definition of the inner product (f, ϕ), a definition, that is, which does not presuppose knowing f otherwise. The three definitions differ slightly in the way in which these inner products are defined. As said before, the main objective is to establish unrestricted differentiability in the extended set $\bar{S}$ of ideal functions and, incidentally, to represent linear functionals generally as inner products.

The three definitions are based a) on linear differential operators L; b) on weak limits of continuous functions; c) on the general concept of linear functionals. In each case we need to give the definitions only for a finite closed subdomain $\bar{g}^*$ of g and can then easily extend them to domains $\hat{g}$ containing $\bar{g}^*$. For the sake of brevity we assume our domains g, g^*, etc., as rectangular.[1]

As we shall see in article 6, the *ideal functions form a linear space*, that is, sums and suitable linear combinations of ideal functions are again ideal functions.

2. Definition by Linear Differential Operators. For any pair of adjoint linear differential operators L and L^* of order r with continuously differentiable coefficients in a closed subdomain $\bar{g}^*$ of g and any continuous—or merely piecewise continuous—function W in $\bar{g}^*$, we denote by the symbol

$$f = L[W]$$

an ideal function in $\bar{g}^*$ and give this symbol a meaning by the following *definition* of the inner product:

(1) $(f, \phi) = (L[W], \phi) = (W, L^*[\phi])$,

where the test functions ϕ are in C_r and have their support in $\bar{g}^*$.

Inner products (f, g) always refer to an underlying base domain g. Also, one may define: A functional has its support in g if it vanishes for all test functions which vanish in g.

[1] Since other closed domains can always be covered by squares, there is no difficulty in admitting more general domains. But for our purposes we might just as well avoid the slight inherent complications.

Two such linear operators L and M of orders r and s respectively define, with ordinary functions W and V, the same ideal function f in $\bar{g}^*$ if for all ϕ in $\bar{g}^*$, $(W, L^*[\phi]) = (V, M^*[\phi])$.

If, moreover, in a domain $g^\wedge$ containing $\bar{g}^*$ the operators M and M^* as well as the function V are defined, then by the symbol $M[V]$ the ideal function f is extended[1] into $g^\wedge$. If such an ideal function f is defined for any closed subdomain of an open base domain g, then the ideal function f is defined for the whole base domain g.

This *definition* a)—which paraphrases the discussions of Ch. VI, §4—is motivated by Green's identity

$$(W, L^*[\phi]) = (\phi, L[W])$$

which is valid for ordinary functions W in C^r because the support of ϕ is in $\bar{g}^*$. Again: Only in the context of such inner products independently defined do the ideal functions f occur and have a meaning.

Of particular significance are the adjoint operators

$$L[W] = D^r W, \qquad L^*[\phi] = (-1)^{|r|} D^r \phi.$$

Of course, in the case of a single variable x the successive derivatives $D^r W = f$ of a continuous function W are ideal functions characterized according to definition a) by the inner products (1), D^r denoting differentiation with respect to x.

For any number n of x-variables and any index s, the derivative $D^s f = D^s L[W]$ is defined by a linear differential operator $N = D^s L$. Assuming that N and N^* still have continuous coefficients we see: *Ideal functions may be differentiated without restriction.* The assumption about the coefficients is in fact irrelevant, for, as we shall see in article 5, every ideal function f in $\bar{g}^*$ can be represented in the special form $f = D^r W$ with a suitably large index r and a continuous "generating function" W.

We add a significant observation: Obviously the *inner product* (f, ϕ) *is an r-continuous linear functional of the test function* ϕ *in* $\bar{g}^*$.

Examples of §1 may be confronted with the present definitions:

For one variable x we denote by $.g(x)$ a function equal to $g(x)$ for $x > 0$ and equal to zero for $x < 0$; then we have for the operator $L = D^2$ and the function $W = .x^2/2$

$$D^3 W = \delta(x),$$

---

[1] For the larger domain an index s larger than r may be needed.

IDEAL FUNCTIONS OR DISTRIBUTIONS

where δ denotes the delta-function. Or, in obvious notation, with $W(x) = (\frac{1}{2}) \sum_{\nu=1}^{\infty} \cdot (x - x_\nu)^2$,

$$D^3 W = \sum_{\nu=1}^{\infty} \delta(x - x_\nu)$$

provided that $x_\nu \to \infty$ for $\nu \to \infty$. Here the domain $\mathcal{I}$ is the infinite x-axis. If the values x_ν^2 converge to a limit, say 1, then we choose for $\mathcal{I}$ the open interval $-1 < x < 1$.

For

$$W(x) = \sum_{\nu=1}^{\infty} \cdot (x - \nu)^\nu \frac{1}{\nu!}$$

we cannot form an ideal function by a single operator D^r for fixed r in the whole infinite domain $\mathcal{I}$, say $-\infty < x < \infty$, but we have for the interval $\mathcal{I}^*: -\frac{1}{2} - r < x < \frac{1}{2} + r$ a corresponding expression

$$D^{r+2} W = \delta^r (x - 1) + \delta^{r-1}(x - 2) + \cdots + \delta(x - r).$$

Of course, all these ideal functions remain the same in a fixed sub-interval $\bar{\mathcal{I}}^*$, even if r increases indefinitely.

Other examples illustrating the above concepts can easily be provided by the reader.

3. Definition by Weak Limits. We consider a sequence of functions f_n which are continuous in the base domain $\mathcal{I}$. For a closed sub-domain $\bar{\mathcal{I}}^*$ and test functions ϕ in $\bar{\mathcal{I}}^*$ we assume that the inner products

$$((f_n - f_l), \phi)$$

converge to zero as $n \to \infty$ and $l \to \infty$ uniformly for all functions ϕ which for some fixed index r are bounded by the same maximum r-norm m: $\|\phi\|_r \leq m$. Then we call the sequence f_n *weakly r-convergent* in $\mathcal{I}^*$. We now attach to such a sequence an *ideal limit* for $n \to \infty$: $f = \lim f_n$ and characterize this ideal function f in $\mathcal{I}^*$ by *defining* the inner product

$$(2) \qquad (f, \phi) = \lim (f_n, \phi).$$

Frequently this *definition b)* is valid with the same index r for the whole base domain $\mathcal{I}$. For the sake of generality, however, one must allow that the index r may be replaced by a larger index and thus a weaker demand would be made for convergence of f_n if we enlarge the domain $\mathcal{I}^*$. Thus we define an ideal function f in the whole base domain $\mathcal{I}$ as *the mildly weak limit* of functions f_n continuous in $\mathcal{I}$

by stipulating that for each closed subdomain $\bar{\mathcal{J}}^*$ there exist an index r for which the sequence f_n is weakly r-convergent in $\bar{\mathcal{J}}^*$.

It should be emphasized again that weak r-convergence of f in $\mathcal{J}^*$ implies r'-convergence for any index r' larger than r since the demand for r-convergence is more restrictive than the demand for r'-convergence with $r' > r$.

The most restrictive definition of weak convergence therefore is zero-convergence, that is, weak convergence in which uniform boundedness of $\|\phi\|$ only is stipulated. In contrast, weak convergence for the base domain $\mathcal{J}$ as defined by the mild stipulation above covers a much wider set of sequences f_n.[1]

The stipulation for weak convergence in $\mathcal{J}$ which restricts the sequence f_n least may be called weak ω-convergence; it demands weak convergence of f_n uniformly for all test functions ϕ of C^∞ with bounded support in each subdomain $\bar{\mathcal{J}}^*$, provided that $\|\phi\|_\nu \le m_\nu$, where m_ν are arbitrarily chosen positive bounds. However, as above, there is equivalence between mild weak convergence and weak ω-convergence (see §1, 7), so that the latter concept need not be used at all.

Of importance is the fact: The weak limit (f, ϕ) in $\bar{\mathcal{J}}^*$ is an r-continuous linear functional of ϕ. For the whole base domain $\mathcal{J}$ the limit $(f, \phi) = \lim (f_n, \phi)$ is a continuous linear functional of ϕ in the sense explained above.

We point out that for $n = 1$ the delta-function $\delta(x)$ is the weak limit ($r = 0$) of the functions $\delta_n(x)$ defined in §1, 8 and of course, of every sequence f_n of functions which tend to zero outside $x^2 < 1/n$, which are non-negative, and for which $\displaystyle\int_{-1/n}^{1/n} f_n \, dx \to 1$.

4. Definition by Linear Functionals. The third variant c) of the definition arises if we reverse conclusions and assumptions in the definitions a) and b) and define: To each linear functional $\Lambda[\phi]$ which is r-continuous or merely mildly continuous in the base domain $\mathcal{J}$ we attach an ideal function f simply by defining

$$(3) \qquad\qquad (f, \phi) = \Lambda[\phi]$$

as the inner product (f, ϕ). This means, for admissible test functions ϕ in a closed subdomain $\bar{\mathcal{J}}^*$ of $\mathcal{J}$ in which $\Lambda[\phi]$ is r-continuous, the relation (3) constitutes the weak definition of f.

---

[1] Incidentally, one could also introduce weak r-convergence for negative indices r; but for our specific purposes here we would not gain much thereby.

Incidentally, the lemma of §1, 7 shows that we might just as well have started by supposing $\Lambda[\phi]$ to be ω-continuous in $\mathcal{g}$.

At any rate, we avoid the existentialist connotation of the lemma and proceed from the assumption that for each closed subdomain $\bar{\mathcal{g}}^*$ of $\mathcal{g}$ the functional $\Lambda[\phi]$ is r-continuous with a suitably chosen index r.

We recall: If $\Lambda[\phi]$ is r-continuous in $\bar{\mathcal{g}}^*$ we may substitute for ϕ test functions ψ in $\mathcal{g}^*$ which need have continuous derivatives $D^{r'}\psi$ only for $r' \leq r$. Such functions ψ with their derivatives up to $D^r\psi$ can be uniformly approximated by means of ϕ_n in C^∞; therefore by the stipulated r-continuity, the value of $\Lambda[\psi]$ is defined as the corresponding limit value of the numbers $\Lambda[\phi_n]$ for $n \to \infty$.

The somewhat abstract character of the definition c) will be immediately made tangible by the following considerations.

5. Equivalence. Representation of Functionals. All three above definitions are equivalent.

First, as stated, a) and b) define functionals (f, ϕ), which enjoy the property c). Furthermore, from definition c) the property of definition b) is easily deduced. This is seen if we write, e.g., $\Lambda[\phi] = \lim \int \phi(\xi)\Lambda[\delta_n(x - \xi)]\,d\xi$, in accordance with (8) from §1, 9. Since the function $\delta_n(x - \xi)$ is in C^∞, we have then for $n \to \infty$

$$(4) \qquad \Lambda[\phi] = \lim (f_n, \phi)$$

with

$$(5) \qquad f_n(x) = \Lambda[\delta_n(x - \xi)],$$

in accordance with definition b).

Hence we only need establish the equivalence of definitions c) and a).

To this end, we assume the functional to be $(r - 2)$-continuous[1] in a closed subdomain $\bar{\mathcal{g}}^{**}$ of $\mathcal{g}$. For simplicity,[2] we assume the domain $\bar{\mathcal{g}}^{**}$ as the square $-\alpha \leq x_i \leq 1 + \alpha$ with a small positive value of α and the domain $\bar{\mathcal{g}}^*$ as the square $0 \leq x_i \leq 1$, while the base domain $\mathcal{g}$ should include the larger square $-2\alpha < x_i < 1 + 2\alpha$.

Then the equivalence of c) and a) follows from the main repre-

---

[1] This, of course, implies $r \geq 2$ (i.e. $r_i \geq 2$ for $i = 1, \ldots, n$) since the index of continuity of $\Lambda[\phi]$ is assumed non-negative.

[2] As said before, the special rectangular shapes of $\mathcal{g}$, $\mathcal{g}^*$, and $\mathcal{g}^{**}$ are not essential, merely convenient.

sentation theorem: *If the functional $\Lambda[\phi]$ is $(r - 2)$-continuous with $r - 2 \geq 0$ in a domain $\bar{\mathcal{I}}^{**}$ which includes $\bar{\mathcal{I}}^{*}$ then we can construct for $\bar{\mathcal{I}}^{*}$ a continuous function W such that for all test functions ϕ with support in $\bar{\mathcal{I}}^{*}$*

$$(6) \qquad \pm \Lambda[\phi] = (W, D^r\phi),$$

where $\pm$ stands for $(-1)^{|r|}$, and where

$$(6a) \qquad W = \Lambda[\psi]$$

will be explicitly defined by the function ψ below.

This is the property of definition a) with a special simple *normal form D^r of the linear differential operator L.*[1]

To prove (6) and to construct W we may substitute in $\Lambda[\phi]$ test functions which need not be in C^∞ but merely are $(r - 2)$-continuous in $\mathcal{I}^{**}$. (See §2, 3.) Specifically, using the functions $p(x; \alpha)$ and $q_r(x; z)$ defined in §1, 4 and §1, 8, we define such a test function ψ with the parameter point z in $\mathcal{I}^{*}$:

$$(6b) \qquad \psi(x, z) = p(x; \alpha)q_{r-1}(x; z).$$

For a given index n of approximation, and the definition (5) of $f_n(x)$, we have in

$$(f_n, \psi) = W_n(z)$$

a continuous function of the parameter z in $\bar{\mathcal{I}}^{*}$; furthermore, because of (5), the limit function

$$(7) \quad W(z) = \lim W_n(z) = \lim (f_n, \psi) = \lim (pf_n, q) = \Lambda[\psi]$$

is continuous in z. Moreover, because of the basic property of the function q_r as stated in §1, 4, formula (3), the functions W_n have continuous derivatives of orders up to the index r such that $D^r(W_n) = pf_n$; and since $p = 1$ in $\bar{\mathcal{I}}^{*}$ we have

$$(8) \qquad D^r W_n = f_n(z) \qquad \text{in } \bar{\mathcal{I}}^{*}.$$

Finally we turn back to the functional $\Lambda[\phi] = \lim (f_n, \phi)$ for any test function ϕ of C^∞ having support only in $\bar{\mathcal{I}}^{*}$. Using integration

---

[1] The incidental conclusion is: For any such differential operator L there exists in $\mathcal{I}^{*}$ a linear operator T (which is essentially the integral operator $D^{-r}L)^{-1}$) such that $LT = D^r$ for some r. This fact, of course, could be easily verified explicitly as an exercise in elementary calculus.

by parts we obtain[1]

$$\Lambda[\phi] = \lim (D^r W_n, \phi) = \pm \lim (W_n, D^r \phi).$$

In the last expression we may pass to the limit under the integral sign and obtain

$$\Lambda[\phi] = \pm (W, D^r \phi),$$

with $W = \Lambda[\psi]$, as the theorem stated.

It should again be emphasized that this relation with the "generating function" W constructed above holds also if W is replaced by $W + V$, where V is a function in C_r, or even a suitable ideal function, for which $D^r V$ vanishes identically.

Also we emphasize: *If in the preceding definitions of ideal functions*

$$f_n(z) = \Lambda(\delta_n(x - z)) \to f(z),$$

where f is a continuous function, then for the generating function W the derivative $D^r W$ exists in the ordinary sense, and $D^r W = f$.

Likewise, if in the case of definitions a) or b) $L[W] = f$ exists in a domain $\mathfrak{z}'$ as a regular continuous differential operator, or if f_n converges uniformly to a continuous function f, then in $\mathfrak{z}'$ the ideal function is identified with the continuous function f. (See article 8.)

6. Some Conclusions. From our equivalent definitions the following facts become evident:

The sum of two ideal functions f and g is again an ideal function. If f and g are $(r - 2)$-continuous then so is the sum $(r - 2 \geq 0)$.

The product of an ideal function f and an ordinary function g of C^∞ is again an ideal function. If in a domain $\mathfrak{z}^*$ merely $(r - 2)$-continuity is assumed for f then the product is again an $(r - 2)$-continuous ideal function if g is merely in C_{r-1}.

Localization and Decomposition. An ideal function $f(x)$, while formally not defined in isolated points, can be restricted to an arbitrarily small closed domain $\bar{\mathfrak{z}}^*$ by restricting the test functions ϕ to those whose support is in $\bar{\mathfrak{z}}^*$. Moreover, as we shall discuss in article 8 any ideal function f can be decomposed into a sum of two or more ideal functions each of which is zero except in a closed domain, so that these domains cover the base domain $\mathfrak{z}$.

7. Example. The Delta-Function. The preceding general concepts are once more illustrated by Dirac's delta-function. For one

---

[1] Here and elsewhere the sign $\pm$ means $(-1)^{|r|}$.

variable x it is defined by $\delta(x) = D^2(.x)$, where again the dot in front of a function means that for negative values of the independent variable the function is defined as identically zero. The associated linear functional $\Lambda[\phi] = (\delta(x), \phi) = \phi(0)$ is obviously zero-continuous.

Incidentally, this example shows that the span from $r - 2$ to r in the representation theorem of article 5 is proper for the general representation of $(r - 2)$-continuous functionals by derivatives $D^r W$ of a continuous function W.

The n-dimensional delta-function $\delta(x)$ may be defined by

$$\delta(x) = D^2 W \qquad \text{for} \qquad W = (.x_1)(.x_2) \cdots (.x_n)$$

or also simply as the product

$$\delta(x) = \delta(x_1)\delta(x_2) \cdots \delta(x_n).$$

Again it corresponds to the functional $\Lambda[\phi] = \phi(0)$.

The delta-functions and their derivatives (see below) are the simplest ideal functions which have their sole support concentrated in one point O and are identified (see article 8) with the regular function identically zero outside of this point O.

Of course, the derivatives $D^s \delta(x)$ are defined by the corresponding derivatives $D^{s+2} W$ of the generating function W or also as limits of derivatives of the functions f_n which define δ as a weak limit.

In a natural transfer of notation from W or f_n we call the *delta-function* an *even function*

$$\delta(-x) = \delta(x)$$

and the derivatives alternatingly odd or even.

We add a few formulas representing the one-dimensional delta-function as a weak limit:

$$(9) \qquad \delta(x) = \lim_{\epsilon \to 0} \frac{1}{\sqrt{\pi}\epsilon} e^{-x^2/\epsilon^2}$$

$$(10) \qquad \delta(x) = \frac{1}{\pi} \lim \frac{\sin nx}{x}.$$

These formulas express facts well known from the discussions of the heat equation and the Fourier series respectively. The second formula can also be written

$$2\pi\delta(x) = \lim \int_{-n}^{+n} e^{i\xi x} d\xi$$

or briefly (see §4, 3)

$$(11) \qquad 2\pi\delta(x) = \int_{-\infty}^{\infty} e^{ix\xi}\, d\xi$$

as the "Fourier transform of the function identically 1". (Incidentally, formula (9) is an example for zero-convergence and (10) an example of one-convergence.) Another interesting representation follows directly from Poisson's integral formula for Green's function of potential theory for the upper half plane; obviously $\delta(x)$ simply symbolizes the boundary value at $y = 0$ of Green's function:

$$(12) \qquad \pi\delta(x) = \lim_{\nu \to 0} \frac{y}{x^2 + y^2}.$$

8. Identification of Ideal with Ordinary Functions. Maximum generality in the theory of ideal functions is not always the appropriate framework for attacking individual problems. In most instances the restriction to narrower, but more tangible, classes of ideal functions is advantageous, specifically to such ideal functions which in some subdomains of $\mathcal{g}$ can be identified with ordinary functions: If in a closed subdomain $\bar{\mathcal{g}}^*$ of $\mathcal{g}$ the generating function[1] W possesses continuous derivatives up to the order r, then in $\bar{\mathcal{g}}^*$ the ideal function D^rW should and always will be identified with the ordinary continuous function $f(x)$. Equivalently this identification is made if in $\bar{\mathcal{g}}^*$ the generating sequence[1] converges uniformly, not merely weakly, to a continuous function $f(x)$ (or, with regard to definition c) if $f_n = \Lambda[\delta_n(x - \xi)]$ converges in $\mathcal{g}^*$ to a continuous function f).

We insert an incidental remark: If $\mathcal{g}$ is covered by a finite number of domains $\mathcal{g}_1, \mathcal{g}_2, \cdots$, then we can always decompose any admissible test function ϕ into a sum $\phi = \phi_1 + \phi_2 + \cdots$ of admissible test functions such that ϕ_ν has its support only in $\mathcal{g}_\nu$; therefore any ideal function f in $\mathcal{g}$ is the sum of ideal functions $f_1 + f_2 + \cdots = f$ such that f_ν is identically zero outside of $\mathcal{g}_\nu$.

In particular, in most individual cases one is concerned with ideal functions $f = D^rW$ which are ordinary functions everywhere except at *singularities localized* in a subdomain G^* of a larger domain G such that outside of G^* the generating function W possesses continuous derivatives D^rW, or equivalently such that the generating sequence f_n

---

[1] It is useful to observe that we may always replace in our definitions the stipulation of continuity of W or f_n by the weaker condition of piecewise continuity.

converges uniformly, not just weakly, to a continuous function f. Frequently the point set G^* consists of isolated points, lines, etc. The function D^rW at these singular points may have singularities in the ordinary sense, but if we consider f as an ideal function these singularities enter only in operations according to the rules established for ideal functions (i.e., according to the definition of the inner product); outside the set G^* the ideal character of D^rW need not be emphasized. In this sense we have operated in §4 of Chapter VI with the singularity function $S(\phi)$ whose singularities are concentrated on a manifold[1] $\phi = 0$ and which appear as ideal functions of the single variable ϕ.

We list again a few functions W of one variable x and their derivatives which have isolated singularities at $x = 0$:

$$W_{\cdot} = x \log |x| - x, \quad DW = \log |x|, \quad D^2W = \frac{1}{x}, \quad D^3W = -\frac{1}{x^2},$$

$$W = .(x \log x - x), \quad DW = . \log x, \quad D^2W = .\frac{1}{x}, \quad D^3W = -.\frac{1}{x^2},$$

$$W = .x^\alpha, \qquad DW = .\alpha x^{\alpha-1}, \quad D^2W = .\alpha(\alpha - 1)x^{\alpha-2}, \cdots$$

$$(0 < \alpha < 1).$$

At the singular points the interpretation of these expressions as ideal functions differs essentially from their interpretation as ordinary functions. No infinite values are attributed at $x = 0$ to ideal functions; on the contrary, the weak definitions smooth the effect of the singularity. (See article 9.)

The consistency of these definitions is immediately evident from the discussion in article 5.

As a consequence we prove easily: *An ideal function f which has its support only in one point, say the origin, is a linear combination of the n-dimensional delta-function and its derivatives up to a certain order.*

Indeed, as is easily seen, the construction in article 5 produces for the generating function W of f a polynomial of degree less than r inside the "positive" space segment $x_i > 0$ for $i = 1, \cdots, n$, while W is identically zero outside this segment of the x-space. Indeed, outside the small square Q_ϵ: $|x_i| < \epsilon$ the functions f_n in that construction vanish; hence $W_n(z) = 0$ for z outside the segment $z_i \geq -\epsilon$ since for each such point z in the defining inner product $W_n(z) = (f_n, \psi)$

---

[1] No confusion with the notation ϕ need occur.

one of the two factors vanishes. Furthermore, for $z_i \geq 0$ the factor $\psi(x; z)$ is a polynomial of degree less than r in x. Because of $W_n(z) \to W(z)$ the statement about $W(z)$ follows. Therefore W is the sum of monomials $(.x_1^{r_1'}) \cdots (.x_n^{r_n'})$ with $r_i' \leq r_i - 1$ or $r' \leq r - 1$, where some exponents r_i' may be zero. Obviously the contribution to $D^r W$ obtained by applying the operator D^r to this monomial results in a product of derivatives $D_i^{r_i}(.x_i)^{r_i'} = D_i^{s_i} D_i D^{r_i'}(.x_i)^{r_i'} = D^{s_i}\delta(x_i)$. This product, except for a constant factor, can be written as $D^s\delta(x)$, where $s = r - r' - 1$ is a non-negative index, that is, a system of non-negative integers. Adding all the corresponding results for the different monomials yields the result as stated.

We give another variant of the proof of the last theorem, proceeding without reference to the construction of article 5. The assumption is that the functional (f, ϕ) vanishes for all test functions ϕ which vanish in a neighborhood of the origin and have continuous derivatives $D^{r'}\phi$ for $r' < r$. Then we prove that (f, ϕ) for arbitrary ϕ depends only on the values of ϕ and the derivatives $D^{r'}\phi$ at the origin, or, what is the same, that (f, ϕ) vanishes if ϕ and these derivatives of ϕ vanish at the origin. Now, with the function $P(x; \alpha)$ of §1, 7 and with $\alpha = 1/n$, we form $Q = 1 - P$ and $\phi_n = Q\phi$. For $n \to \infty$ the functions ϕ_n and their derivatives of order $<r$ tend uniformly to ϕ and the corresponding derivatives of ϕ. Hence, because (f, ϕ) is an $(r - 1)$-continuous linear functional of ϕ and $(f, \phi_n) = 0$, we have $(f, \phi) = \lim_{\alpha \to 0} (f, \phi_n) = 0$, as stated. Since (f, ϕ) only depends on a finite number of the derivatives of ϕ at the origin, it must be a linear combination of these derivatives. But this fact expresses our theorem.

9. Definite Integrals. Finite Parts. We now turn to the "definite integral of ideal functions". The representation $f = D^r W$ immediately gives meaning to the concept of *primitive functions* or *indefinite integrals* $D^{-s}f$ of an ideal function $f = D^r W$. They could simply be defined as functions $D^r D^{-s} W$ for $s \leq r$. From article 8 we see in which way a primitive function remains undetermined.

To discuss the *definite integral* over a domain G of an ideal function $f = D^r W$ we restrict ourselves in the present article to functions f which are regular continuous functions except in a closed subdomain $\bar{G}^*$ of G and which are continuous up to the (smooth) boundary Γ of G.

The result will be: *For such ideal functions the relations between the primitive function and definite integral, that is, the fundamental integral theorems of calculus, persist.*

We first consider functions f of a single variable x and take G as the interval $-1 < x < 1$; we assume that $D^r W = f$ is continuous in a neighborhood of the end points $x = 1$ and $x = -1$. What value Z do we have to attach to the symbol $\int_{-1}^{+1} f(x)dx$? To find an answer, we consider a test function ϕ which is identically equal to 1 in G, hence has all derivatives zero on the boundary of G and otherwise is extended arbitrarily into the region $G^\wedge = \mathcal{G} - G$ complementary to G such that its support is bounded. Now, integrating by parts, we have

$$\pm(f, \phi) = \pm^{\cdot}\!\int (\phi D^r W)\, dx = \int_{G+G^\wedge} W \cdot D^r\phi\, dx = \int_{G^\wedge} W D^r\phi\, dx$$

$$= \pm \int_{G^\wedge} \phi D^r W\, dx + D^{r-1}W \bigg|_{x=-1}^{x=+1},$$

where again ± 1 stands for $(-1)^r$. On the other hand naturally we have to define $Z = (f, \phi) - \int_{G^\wedge} f\phi\, dx$. Thus we obtain the result: The integral over an interval $G: x^2 < 1$ of an ideal function $f = D^r W$ which is regular in the neighborhood of the end points is properly defined by

$$(13) \qquad Z = \int_{-1}^{+1} f(x)\, dx = D^{r-1}W \bigg|_{-1}^{+1}.$$

This indeed is exactly the formula which would link the definite integral with the indefinite integral in the case of an everywhere ordinary function f. Incidentally, for ideal functions $f = W'$, i.e., $r = 1$, the result remains valid even if W is not differentiable at the boundary.

This value (13) for Z has been introduced as the *finite part* of the integral of f and used as a basic tool by Hadamard in his ingenious investigations of Cauchy's problem. (See Ch. VI, §15.)

Of course, a similar formula, slightly more cumbersome, arises if we consider the integrand in the form $f = g D^r W$, where $g(x)$ is a regular r-continuous factor. While this form of f is reducible to that above, we can just as well proceed directly according to the preceding pattern, successively integrating by parts. The result is

$$\int_{-1}^{+1} gD^r W \, dx$$

$$= (gD^{r-1}W - DgD^{r-2}W + \cdots \mp D^{r-2}gDW) \Big|_{-1}^{+1}$$

$$\pm \int_{-1}^{+1} (W \cdot D^{r-1}g) \, dx.$$

Thus the integral of the ideal function is reduced to boundary terms and to the integral of the continuous function $W \cdot D^{r-1}g$.

As an example we evaluate the integral $f = \displaystyle\int_{-1}^{+1} (1/x^2) dx$ with $1/x^2$ interpreted not as an ordinary function but as the ideal function $-D^2 \log |x|$. We find immediately since $DW = -1/x$

$$\int_{-1}^{+1} \frac{1}{x^2} \, dx = -2.$$

Another example is the integral of the ideal function defined by $f = -4D^2(.x^{\frac{1}{2}})$. Except for the singularity at $x = 0$ this is identified with $x^{-\frac{3}{2}}$ for $x > 0$ and with 0 for $x < 0$. Its integral over $-1 \le x \le 1$ can therefore be evaluated as

$$-4 \int_{-1}^{+1} D^2(.x^{\frac{1}{2}}) \, dx = \int_0^1 x^{-\frac{3}{2}} \, dx = -2,$$

where the result means the "finite part" of the singular integral.

Also for n variables similar results are obtained corresponding to fundamental integral theorems of calculus. This is, for example, true of *Gauss' integral formula* if the integrand is composed of ideal functions which are merely first derivatives of continuous functions; in particular if the integrand is the divergence of a continuous vector field, then Gauss' integral formula remains valid as in classical calculus.

If, however, the integrand $f = D^r W$ is generated by higher differentiation, then, as said before, we restrict ourselves to the case where the ideal function $D^r W$ or at least one of the n functions $D_i^{-1} D^r W$ is a continuous function in the neighborhood of the smooth boundary Γ of the domain of integration G with the outward normal vector ξ and the surface element dS.

We use the same notation G, G^*, $G^\wedge$ as above and choose a test function ϕ which is identically one in G and may be arbitrarily continued into the complementary domain $G^\wedge = \mathcal{G} - G$ subject to the appropriate conditions of smoothness and the condition of bounded support. Since $f = D^r W$ is defined by inner products with test functions, and since all derivatives of our particular test function vanish in G and on Γ, we are justified in defining the integral of f over G on the basis of the following self-explanatory relations:

$$(f, \phi) = \iiint_{G+G_\wedge} \phi D^r W \, dx = \iiint_G f \, dx + \iiint_{G_\wedge} \phi D^r W \, dx$$

$$= \pm \iiint_{G_\wedge} W D^r \phi \, dx = \mp \iiint_{G_\wedge} \phi D^r W$$

$$\pm \iint_\Gamma (D_i^{-1} D^r W) \xi_i \, dS.$$

Thus we have to set

$$\iiint_G f \, dx = \iint_\Gamma (D_i^{-1} D^r W) \xi_i \, dS,$$

emphasizing the index i. Symmetry is obtained if we introduce an ideal vector A with the components $A_i = D_i^{-1} D^r W$. Then the result is a *Gauss formula*

$$\iiint_G \operatorname{div} A \, dx = \iint_\Gamma A \xi \, dS.$$

In a similar way the other integral theorems of calculus can easily be generalized.

§3. Calculus with Ideal Functions

Introducing ideal functions may appear as a sweeping extension of ordinary calculus. Yet, in the realm of ideal functions not all operations of classical calculus can be carried out. Thus the advantage of securing unrestricted differentiability is partly offset by the loss of freedom in multiplying functions or in forming composite functions. It is not even completely true that an ideal function of several variables becomes an ideal function of fewer variables if some of the others are kept constant in a domain of definition. Without examining systematically the balance between gain and loss in scope of calculus we shall in the present and the next section emphasize a few pertinent key points.

1. Linear Processes. If in a closed domain $\bar{\mathfrak{z}}^*$ we set $f = D^r W$ with a piecewise continuous generating function W, we have easily ascertained: Linear combinations of such r-continuous ideal functions with ordinary suitably smooth factors are always again r-continuous ideal functions.

If the generating function W possesses continuous derivatives with respect to parameters α then the ideal function f may be differentiated with respect to these parameters. For example, for the function $f = \delta(\alpha x)$, where α is a vector, we have $f_{\alpha_i} = x_i \delta'(\alpha x)$.

Likewise, an integral of an ideal function may be differentiated with respect to parameters under the integral sign if the above assumptions for W are made. (Compare the examples in Ch. 6, §15.)

If the piecewise continuous generating functions W_n tend to a piecewise continuous function W then we have in the sense of weak r-convergence $D^r W_n = f_n \to f = D^r W$.

Correspondingly we may differentiate infinite series term by term or interchange limiting processes such as differentiation with respect to different parameters.

As to integration, we refer to the discussions in §2, 8.

2. Change of Independent Variables. We recall the fact that one may localize ideal functions to rectangular, or generally, piecewise smoothly bounded domains $\bar{\mathfrak{z}}^*$, simply by modifying the original piecewise continuous generating function W, i.e., by continuing W as identically zero outside $\bar{\mathfrak{z}}^*$. Therefore, to introduce new variables y instead of x by $x = g(y)$ we may confine ourselves to suitably small domains $\bar{\mathfrak{z}}^*$. We assume that the variables x as functions of the variables y possess in $\bar{\mathfrak{z}}^*$ continuous derivatives of orders up to say r, and that the Jacobian $\partial(y)/\partial(x)$ is bounded away from zero, such that the same continuity property holds for the inverse transformation. In $\bar{\mathfrak{z}}^*$ thereby $f = \lim f_n$ is transformed into an ideal r-continuous function of the variables y. This follows almost directly from the second definition of f by an r-convergent sequence of continuous functions f_n, applied to test functions ϕ with support in $\bar{\mathfrak{z}}^*$. Indeed, the derivatives with respect to y of order $r' \leq r$ of test functions ϕ are bounded by bounds for the derivatives of ϕ with respect to x up to the order r'. Thus, according to §2, an ideal function $\lim f_n$ of y is defined by $\lim (f_n, \phi)$.

As an incidental consequence we note: The ideal functions S of one variable introduced in Ch. VI, §4 are likewise ideal functions in the n variables x.

Further explanation is not needed to show that for differentiation of composite ideal functions formed by admissible coordinate transformations the rules of calculus remain valid. However, it must be realized that our statements about transformations to new coordinates and composite functions depend on sufficient smoothness of the transformation functions.

Emphatically, one must be aware of the fact that the concept of an ideal function of ideal functions is meaningless. Furthermore, while an ideal function of a smooth ordinary function can be formed, the concept of an ordinary function of ideal functions does not make sense. Even such simple operations as squaring an ideal function or multiplying two ideal functions of the same variables x are inadmissible. Here lies a principal limitation of a "calculus with ideal functions".

3. Examples. Transformations of the Delta-Function. The most important concrete examples concern the delta-function. As is easily seen, we have with constant a, b and a single variable x

$$(1) \qquad \delta(ax - b) = \frac{1}{a}\delta\left(x - \frac{b}{a}\right) \qquad (a \neq 0).$$

Generally, for a function $p(x)$ with $p'(0) \neq 0$ and $p(0) = 0$ we have in a sufficiently small interval $\bar{\jmath}^*$ about the origin

$$\delta(p(x)) = p'(0)^{-1}\delta(x).$$

Assume that the zeros of a smooth function $p(x) = 0$ are $a_1, a_2, \cdots, a_n, \cdots$ and that their number m is finite or that a_m tends to infinity and that for all m we have $p'(a_m) \neq 0$. Then one ascertains easily

$$(2) \qquad \delta(px) = \sum_{\nu} p'(a_\nu)\delta(x - a_\nu).$$

In particular one has (see also e.g. Ch. VI, §15, 5)

$$(3) \qquad \delta(x^2 - a^2) = \frac{1}{2a}\left(\delta(x - a) - \delta(x + a)\right).$$

As an exercise the reader may prove and interpret the formula

$$(4) \qquad \delta(\sin x) = \sum_{\nu=-\infty}^{+\infty} (-1)^\nu \delta(x - \nu).$$

As an incidental observation we state:

The delta-function is homogeneous of degree -1 in the variable x (likewise $\delta(x)$ in n dimensions is homogeneous of degree $-n$). Accordingly we have, for one variable the relations

$$(5) \qquad\qquad x\delta'(x) + \delta(x) = 0.$$

As to the *transformation to polar coordinates* of the n-dimensional delta-function, the general formulas of article 2 are not applicable in the neighborhood of the origin. Nevertheless the often cited formula (ω_n denoting the surface area of the n-dimensional unit sphere)

$$(6) \qquad\qquad \delta(x) = \frac{1}{\omega_n}\frac{\delta(|x|)}{|x|^{n-1}}$$

is well motivated as the expression of the fact that $\phi(0)$ represents the functional $\iint \delta(x)\phi(x)dx$. Yet, one should realize that strictly speaking, to express the n-dimensional ideal function $\delta(x)$ by (6) requires some slight adaptation of our general definitions, such as the stipulation $\int_0^\infty \phi(x)\delta(x)dx = \tfrac{1}{2}\phi(0)$.

4. Multiplication and Convolution of Ideal Functions. While the product of two ideal functions $f(x)$ and $g(x)$ as a general concept is meaningless, products $f(x)g(y)$ of ideal functions of two *different sets* x *and* y *of independent variables* are by our definitions ideal functions of the $2n$ variables x and y. Indeed, if e.g., $f(x) = D_x^r W(x)$ and $g(y) = D_y^s V(y)$, then in obvious notation

$$(7) \qquad\qquad f(x)g(y) = D_x^r D_y^s VW.$$

For ideal functions $f(x) = D^r W$ and $g(x) = D^s V$ of the same variables x, the process of *convolution*[1] of the two functions always is meaningful and results in a new ideal function $F(x)$ provided that one of the two factors, say $g(x)$, has bounded support. The convolution is defined by

$$(8) \qquad F(x) = f*g = g*f = \iint f(x-\xi)g(\xi)\,d\xi$$

$$= \iint f(\xi)g(x-\xi)\,d\xi,$$

[1] In German, "Faltung".

where we assume the base domain g of integration as the whole x-space.

(The importance of the concept of convolution is illustrated by its occurrence in the representation of solutions of differential equations by fundamental solutions and also by the fact that *every function is the convolution of itself with the delta-function.*)

Our statement is expressed by the following definition: Writing $f = D^r W$ and $g = D^s V$ with continuous W and V and bounded support for V, then we set

$$F(x) = D^r D^s (W * V).$$

An important property of the convolution operator is that it commutes with differentiation.

It may be stated that significant applications of convolutions are based on the obvious fact: Any linear operator $L[u(x)]$ can be represented as the convolution $(L[u(\xi)], \delta(x - \xi))$; therefore $L[u(x)] = 0$ is equivalent to the formal integral equation for u

$$(9) \qquad (u(\xi), L_\xi^* \delta(x - \xi)) = 0.$$

If we now approximate the ideal delta-function and its derivatives by suitable smooth regular functions, we arrive at approximations of the functional equation $L[u] = 0$ by genuine integral equations. (Reference may be made to the forthcoming Volume III of this work.)

§4. *Additional Remarks. Modifications of the Theory*

1. Introduction. As indicated above, there remains leeway for generalizing the concept of function. Such modifications have a useful role in mathematical physics; also from a purely theoretical viewpoint they are an attractive topic.[1] In this section we shall briefly discuss some of these modifications. Except for the example in article 5, they concern the behavior of ideal functions for the whole base domain g—as influenced by boundary conditions or rather conditions at infinity for the test functions.

It will be convenient to consider complex-valued functions f, g, ϕ, ψ and to define for them the inner product by the self-explanatory formula

$$(1) \qquad (f, g) = \int_g f(x)\overline{g(x)} \, dx = \overline{(g, f)}.$$

---

[1] Compare, e.g., J. Berkowitz and P. D. Lax [1].

2. Different Spaces of Test Functions.   The Space $\mathfrak{S}$.   Fourier Transforms.   While the space of test functions with bounded support seems entirely satisfactory for the purposes of this book, it is sometimes advantageous to use slightly wider spaces (thereby slightly narrowing the "dual" space of ideal functions), without essentially modifying the definitions and procedures of §§2, 3.   In particular, with the whole x-space as base domain $\mathfrak{g}$, we consider the space $\mathfrak{S}$ of test functions ϕ in C^{∞} which with all derivatives tend to zero at infinity more sharply than any power of $|x|$, that is, for which

$$|x|^{N}\,|\phi| \to 0, \qquad |x|^{N}\,|D^{r}\phi| \to 0$$

for all indices r no matter how large the positive number N is chosen.[1]

With respect to the space $\mathfrak{S}$ of test functions ϕ, the inner product (W, ϕ) is defined as an ordinary integral over the whole x-space $\mathfrak{g}$ of the product $W\bar{\phi}$ for all functions W which are continuous in $\mathfrak{g}$ and for which there is a positive index M such that $|W|\,|x|^{-N} \to 0$ for $|x| \to \infty$ and $N > M$, in other words, functions W which do not increase at infinity more strongly than a polynomial.   In agreement with the definitions of §2 we may then introduce for any integral index r ideal functions

$$(2) \qquad\qquad f = D^{r}W$$

and the inner products f, ϕ defined as

$$(3) \qquad\qquad (f, \phi) = (-1)^{|r|} \int WD^{r}\bar{\phi}\, dx.$$

Obviously every ideal function thus defined over the space $\mathfrak{S}$ of test functions is also an ideal function as defined in §2 over the space $\mathfrak{D}$ of test functions with bounded support.   However, the converse is not always true.[2]

The space $\mathfrak{S}$ of test functions is useful in the theory of *Fourier transformations* of functions $g(x)$ into functions

$$(4) \qquad\qquad \hat{g}(x) = \frac{1}{(\sqrt{2\pi})^{n}} \int_{\mathfrak{g}} \bar{g}(\xi)e^{ix\xi}\, d\xi = f(x).$$

---

[1] For example, the space of all functions in C^{∞} which are of order not exceeding e^{-x^2} is contained in this space $\mathfrak{S}$.

[2] Let us consider, e.g., the ordinary function $f = e^{x^4}$.   It can be interpreted as an ideal function over the space $\mathfrak{D}$ since its inner product with any test function ϕ of bounded support is immediately given as an ordinary integral. Yet with respect to the space $\mathfrak{S}$ this definition collapses, since e.g. the function $\phi = e^{-x^2}$ is an admissible test function in $\mathfrak{S}$, and the inner product of W and ϕ does not exist.

If g is a function of $\mathfrak{S}$ then, as is easily seen, the transform $\hat{g}$ is again in $\mathfrak{S}$, and Fourier reciprocity

$$(5) \qquad\qquad g(x) = \hat{f}(x), \qquad f(x) = \hat{g}(x)$$

is perfect: The space $\mathfrak{S}$ is mapped onto itself in a one-to-one manner by the Fourier transformation.

We note the important *Parseval formula*

$$(6) \qquad\qquad (f, g) = (\hat{f}, \hat{g})$$

or

$$(6a) \qquad\qquad \int_s f\bar{g}\, dx = \int_s \hat{f}\hat{g}(x)\, dx$$

which is easily proved for functions f and g of the class $\mathfrak{S}$ since the integrals with respect to x converge rapidly.

We are now in a position to give a satisfactory definition of the Fourier transform $\hat{f}$ of a function f not necessarily in $\mathfrak{S}$ but defined by $f = D^r W$ (equation (2)); that is, we define the Fourier transform of those ideal functions f "over $\mathfrak{S}$" arising by differentiation from continuous functions W which increase at infinity at most as a polynomial.[1]

First we realize again that if ϕ belongs to $\mathfrak{S}$ then so does $\psi = \hat{\phi}$ and vice versa. Then we use the Parseval formula (6) not as a theorem but as a weak definition of $\hat{f}$: We set

$$(8) \qquad\qquad (\hat{f}, \psi) = (\hat{f}, \hat{\phi}) = (f, \phi),$$

realizing that the right-hand side is *a priori* known according to (3) to be $\pm(W, D^r\phi)$. Therefore the transform $\hat{f}$ is justifiably defined by the inner products (8), where ψ and ϕ range over the space $\mathfrak{S}$.

This definition of $\hat{f}$ can be used as the point of departure for a deeper probe into the theory of Fourier transformations.

Almost immediately one realizes the important fact: The Fourier transform of $D^s f = f$ is $(-ix)^s \hat{f}$ for any index s; more generally, for any linear differential operator $L[f] = \sum_\rho a_\mu D^\rho f$ with constant coefficients a_ρ, the Fourier transform is

$$\hat{L}[f] = P(-ix)^s \hat{f},$$

where P is the polynomial $\sum_\rho a_\rho(-ix)^\rho$ associated with the differential operator L. (See Ch. III, App. 1, § 2.)

---

[1] For this wide extension of the scope of Fourier transformations compare S. Bochner [1] and L. Schwartz [1].

We desist from further discussions and merely state that the remarks made in §2 about $\delta(x)$ and the function identically one as being Fourier transforms of each other can be easily fitted into the present framework.

3. Periodic Functions. Frequently it is useful to restrict oneself to periodic generating functions W and periodic test functions ϕ with the same period, say 2π, in all variables x_i and with the base domain $\mathcal{J}$: $0 \leq x_i \leq 2\pi$. Then, using complex notation, we take as the space $\mathfrak{P}$ of test functions the space spanned by the trigonometric functions $e^{-i\nu x} = \phi_\nu$ where ν is a vector with integral components. We also call ideal functions defined by $f = D^r W$ periodic (cf. the definition in §2). Moreover, slightly generalizing the definition in §2 we admit as generating functions W square integrable functions.

Through these modifications we gain freedom from the complications inherent in boundary conditions for infinite domains. For differential equations considered in a finite domain one can often extend the coefficients and solutions periodically beyond the domain under consideration without restricting generality.[1] Thus one obtains the advantage of rather explicit procedures.

As usual we define the Fourier coefficients a_ν of W by

$$(9) \qquad a_\nu = (W, \phi_\nu) = \int W e^{-i\nu x} \, dx;$$

here and afterwards we shall assume (without in fact restricting generality) that the coefficient a_0 of the constant term in the Fourier expansions vanishes. Then we have *Parseval's theorem* in the form

$$(9a) \qquad \|W\|^2 = (W, W) = \int |W|^2 \, dx = \pi^{2n} \sum_{\nu=-\infty}^{\infty} |a_\nu|^2.$$

For the ideal functions $f = D^r W$ we define the inner products according to §2 by $(f, \phi) = \pm (W, D^r\phi)$. This allows the definition of the Fourier coefficients c_ν of the ideal function $f = D^r W$ as

$$(9b) \qquad c_\nu = (f, \phi_\nu) = (D^r W, \phi_\nu)' = \pm (i\pi)^r (W, \phi_\nu) = \pm (i\pi)^r a_\nu .$$

Analogously we may consider the integrals $D^{-r}W$ and their Fourier coefficients

$$(10) \qquad d_\nu = (+i\pi\nu)^{-r} a_\nu = (D^{-r} W, \phi_\nu),$$

always assuming the coefficients of the constant term to be zero.

---

[1] Such an artifice was successfully introduced by P. Lax in [6], and has been subsequently used by others.

Parseval's formula (9a) suggests that we define not only for W but also for $D^r W$ and for $D^{-r} W$ an r-norm and a $-r$-norm respectively simply by the ordinary norm of W as follows:

$$(11) \qquad \|D^r W\|^2_{-r} = \|D^{-r} W\|^2_r = \|W\|^2 = \pi^{2n} \sum |a_\nu|^2.$$

At any rate, the Fourier coefficients of an ideal function f, just like those of an ordinary function, provide a tangible "*concrete*" or "*explicit*" *representation* of f: An ideal periodic function f is represented by a sequence of numbers c_ν for all indices ν (assuming $c_0 = 0$) such that there is a fixed exponent index r for which

$$(12) \qquad \sum_\nu |c_\nu|^2 |x^{-2r}| = \sum |a_\nu|^2$$

is convergent; then the numbers $(i\pi\nu)^{-r} = c_\nu = a_\nu$ represent the Fourier coefficients of a square integrable function W.

Of course, all operations with periodic ideal functions could easily be based on this representation and carried out explicitly.

4. Ideal Functions and Hilbert Spaces. Negative Norms. Strong Definitions. In §2 and above in article 2 we have introduced ideal functions f by means of "weak definitions", that is, by defining "inner products" with functions ϕ of a space $\mathfrak{D}$ and considering the ideal functions f as elements in the "dual space" just as *in projective geometry planes and points correspond dually to each other* with respect to the inner product of their coordinates. It is of considerable interest[1] that one may also define ideal functions by *strong definitions* through *completion* of dense sets of smooth functions *by convergence in a quadratic Hilbert norm*. Such a procedure, which underlies basic operations in the classical direct calculus of variations, has been successfully applied by P. Lax[2] among others. Only a brief indication should be given here:

In the case of *periodic functions* as discussed in article 3 the situation is explicit and very simple. First we consider in the Hilbert space of periodic functions W with the norm $\|W\|$ in the dense subspace of functions W with derivatives of orders up to r; then we form the continuous functions $D^r W = f$ and close this set with respect to the Hilbert norm $\|W\|$. To this closed Hilbert space we thereby attach

---

[1] Reference might be made, e.g., to applications to the construction of solutions of boundary value problems by variational methods (see Volume III).

[2] See, e.g., P. Lax [6].

the limits f by closure as ideal elements and call $\|W\| = \|f\|_{-r}$ the *negative* $-r$-*norm of* f. Of course, for periodic functions W these norms and their relationships are explicitly given in article 2.

At any rate, the negative norms allow a *strong definition*, that is definition by closure with respect to a quadratic norm, of ideal elements. It is not difficult to see that these strongly defined ideal functions are essentially equivalent to the ideal elements obtained before by "weak definitions".

The assumption of periodicity is not at all essential for these considerations: For functions W and f in the whole x-space we might consider test functions in $\mathfrak{D}$ and let W range over the Hilbert space obtained by completing with the norm $\|W\|$ the space of functions W in C^∞ with bounded support. Then for each index r the completion of W leads to a Hilbert space of ideal functions $D^r W = f$ with the negative norm $\|f\|_{-r} = \|W\|$. By letting r range over all indices r and forming the union of all ideal elements thus defined we arrive at a definition of ideal functions essentially (though not completely) isomorphic to that given in §2.

A more explicit analogy with the definition by Fourier coefficients in the case of periodicity is naturally provided by Parsevel's theorem (6) for Fourier integrals.

5. Remark on Other Classes of Ideal Functions. As an illustration of the degree of arbitrariness permissible for introducing ideal elements, the following brief remark may be added: One could define a useful class of ideal functions as the boundary values, say on the unit circle, of regular harmonic functions inside (or, generally, of a regular solution of an elliptic differential equation). These boundary values may not exist in the ordinary sense but are represented as ideal elements by the regular analytic *harmonic* functions inside. In §2 we have observed that the delta-function could be defined in such a way by the boundary values of Green's function. Yet, as the reader will easily see by examples, boundary values of regular harmonic functions u need not be ideal functions at all in the sense of §2 or the previous articles in the present section.[1] On the other hand this new type of

[1] For example, written in polar coordinates ρ and θ in the unit circle $\rho < 1$, the series

$$u = \sum_{\nu=1}^{\infty} a_\nu \rho^\nu \cos \nu\theta$$

ideal functions enjoys properties which might make them a useful tool in analysis. If instead of harmonic functions in the unit circle we consider regular analytic functions of the complex variable $z = x + iy$ we gain even further advantages, e.g., the possibility of defining in a natural manner products and functions of ideal functions.

Notwithstanding the merits of the theory developed in this appendix, the above remark should call attention to the need for further study of other less well explored possibilities of generalizing the concept of function by introducing suitable ideal elements. The value of all such concepts should be measured not by their formal generality but by their usefulness in the broader context of analysis and mathematical physics.

---

with

$$a_\nu = e^{\sqrt{\nu}}$$

is harmonic for $\rho < 1$; the norm $\pi^2 \sum |a_\nu|^2 \nu^{-2r}$ converges for no integer r (compare article 3) while the series for u converges uniformly for $0 \leq \rho < 1 - \epsilon$ with any positive ϵ, and is harmonic.

Bibliography

(Page references in parentheses are to this volume.)

Agmon, S., A. Douglis, and L. Nirenberg
 [1] Estimates near the boundary for solutions of elliptic partial differential equations satisfying general boundary conditions, I. Communs. Pure and Appl. Math., Vol. 12 (1959), pp. 623-727. (pages 331, 345, 350, 369, 401)

Ahlfors, L., and L. Bers
 [1] Riemann's mapping theorem for variable metrics. Ann. of Math., Vol. 72 (1960), pp. 385-404. (page 350)

Aleksandrov, A. D.
 [1] Dirichlet's problem for the equation
$$\text{Det } \| z_{ij} \| = \phi(z_1, \cdots, z_n, z, x_1, \cdots, x_n),$$
I. Vestnik Leningrad. Univ., Ser. Mat. Mekh. i Astron., Vol. 13, No. 1 (1958), pp. 5-24. (In Russian; English summary.) (page 367)

Aronszajn, N.
 [1] A unique continuation theorem for solutions of elliptic partial differential equations or inequalities of second order. J. de Math., T.36 (1957), pp. 235-249. (page 383)

Asgeirsson, L.
 [1] Some hints on Huygens' principle and Hadamard's conjecture. Communs. Pure and Appl. Math., Vol. 9, No. 3 (1956), pp. 307-326. (page 576)
 [2] Über eine Mittelwertseigenschaft von Lösungen homogener linearer partieller Differentialgleichungen 2. Ordnung mit konstanten Koeffizienten. Math. Ann., Vol. 113 (1936), p. 321 ff. (pages 744, 753)
 [3] Über Mittelwertgleichungen, die mehreren partiellen Differentialgleichungen 2. Ordnung zugeordnet sind. Studies and Essays, Interscience, New York, 1948, pp. 7-20. (page 744)

Bader, R., see Germain, P., and R. Bader
Banach, S.
 [1] Théorie des Opérations Linéaires. Warsaw, 1932. Reprinted by Chelsea Publishing Co., New York, 1933. (page 333)

Bazer, J., and O. Fleischman
 [1] Propagation of weak hydromagnetic discontinuities. Rep. No. MH-10, Institute of Mathematical Sciences, New York University, 1959. (page 612)

Beckenbach, E. F., and L. K. Jackson
 [1] Subfunctions of several variables. Pacific J. Math., Vol. 3 (1953), pp. 291-313. (page 342)

Bergman, S.
[1] Linear operators in the theory of partial differential equations. Trans. Am. Math. Soc., Vol. 53 (1943), pp. 130–155. (page 401)

Berkowitz, J., and P. D. Lax
[1] Functions of a real variable. To be published by Wiley, New York. (page 792)

Bernstein, S.
[1] Sur la nature analytique des solutions de certaines équations aux dérivées partielles du second ordre. Math. Ann., Vol. 59 (1904), pp. 20–76. (page 347)

Bers, L., see also Ahlfors, L., and L. Bers
[1] An outline of the theory of pseudoanalytic functions. Bull. Am. Math. Soc., Vol. 62 (1956), pp. 291–331. (page 375)
[2] Lectures on elliptic equations. Summer Seminar in Applied Mathematics, University of Colorado, Boulder, Colorado, 1957. Lectures in Applied Mathematics, Vol. IV, to be published by Interscience, New York. (page 240)
[3] Mathematical aspects of subsonic and transonic gas dynamics. Wiley, New York, 1958. (pages 367, 390)
[4] Non-linear elliptic equations without non-linear entire solutions. J. Ratl. Mech. Anal., Vol. 3 (1954), pp. 767–787. (page 400)
[5] Results and conjectures in the mathematical theory of subsonic and transonic gas flows. Communs. Pure and Appl. Math., Vol. 7 (1954), p. 79. (page 162)

Bers, L., and A. Gelbart
[1] On a class of functions defined by partial differential equations. Trans. Am. Math. Soc., Vol. 56 (1944), pp. 67–93. (page 389)

Bers, L., and L. Nirenberg
[1] On a representation theorem for linear elliptic systems with discontinuous coefficients and its applications. Convegno Internazionale sulle Equazioni lineari alle derivate parziali. Edizioni Cremonese, Rome, 1955, pp. 111–140. (pages 367, 394)
[2] On linear and nonlinear elliptic boundary value problems in the plane. Convegno Internazionale sulle Equazioni derivate parziali. Edizioni Cremonese, Rome, 1955, pp. 141–167. (pages 269, 367, 395)

Birkhoff, G. D., and O. D. Kellogg
[1] Invariant points in function space. Trans. Am. Math. Soc., Vol. 23 (1922), pp. 96–115. (page 357)

Bochner, S.
[1] Vorlesungen über Fouriersche Integrale. Akademische Verlagsges., Leipsig, 1932. (page 794)

Boyarskii, B. V.
[1] Generalized solutions of systems of differential equations of first order and elliptic type with discontinuous coefficients. Mat. Sbornik, Vol. 43 (85) (1957), pp. 451–503. (In Russian.) (pages 394, 399)

Brelot, M.
[1] Lectures on potential theory. Tata Institute of Fundamental Research, Bombay, 1960. (page 305)

Buchwald, V. T.
[1] Elastic waves in anisotropic media. Proc. Roy. Soc. London, Ser. A, Vol. 253 (1959), pp. 563–580. (page 599)

Burgatti, P.
[1] Sull' estensione del metodo d'integrazione di Riemann all' equazioni lineari d'ordine n con due variabili independenti. Rend. reale accad. lincei, Ser. 5ª, Vol. 15, 2° (1906), pp. 602–609. (page 454)

Calderón, A. P.
[1] Uniqueness in the Cauchy problem for partial differential equations. Am. J. Math., Vol. 80 (1958), pp. 16–36. (pages 238, 383)

Calderón, A. P., and A. Zygmund
[1] Singular integral operators and differential equations. Am. J Math., Vol. 79 (1957), pp. 901–921. (page 666)
[2] Singular integral operators and differential equations. Am. J. Math., Vol. 80 (1958), pp. 16–36. (page 666)

Carathéodory, C.
[1] Variationsrechnung und partielle Differentialgleichungen erster Ordnung. B. G. Teubner, Leipzig and Berlin, 1935. (page 62)

Carleman, T.
[1] Sur les systèmes linéaires aux dérivées partielles du premier ordre à deux variables. Compt. rend., Vol. 197 (1933), pp. 471–474. (page 383)
[2] Sur un problème d'unicité pour les systèmes d'équations aux dérivées partielles à deux variables independentes. Ark. Mat., Astr. Fys., Vol. 26B, No. 17 (1939), pp. 1–9. (page 238)

Cauchy, A. L.
[1] Mémoire sur l'intégration des équations linéaires aux différentielles partielles et à coefficients constants. Oevres Complètes, Ser. 2, T. 1, 1823. (page 217)

Cinquini-Cibrario, M.
[1] Un teorema di esistenza e di unicita per un sistema di equazioni alle derivate parziali. Ann. di mat. (4), Vol. 24 (1945), pp. 157–175. (page 464)

Cohen, P.
[1] The non-uniqueness of the Cauchy problem. Technical Rep. No. 93, Applied Mathematics and Statistics Laboratory, Stanford University, 1960. (page 383)

Copson, E. T.
[1] On the Riemann-Green function. J. Ratl. Mech. Anal., Vol. 1 (1958), pp. 324–348. (page 461)

Cordes, H. O.
[1] Über die eindeutige Bestimmtheit der Lösungen elliptischer Differentialgleichungen durch Anfangsvorgaben. Nachr. Akad. Wiss. Göttingen, Math.-Phys. Kl., Jahre 1956, pp. 239–258. (page 383)

[2] Über die erste Randwertaufgabe bei quasilinearen Differentialgleichungen zweiter Ordnung in mehr als zwei Variablen. Math. Ann., Vol. 131
(1956), pp. 273–312. (page 367)

Courant, R.
[1] Differential and integral calculus, Vol. II. Interscience, New York,
1936. (page 111)
[2] Dirichlet's principle, conformal mapping and minimal surfaces. Interscience, New York, 1950. (pages 167, 160, 225)
[3] Cauchy's problem for hyperbolic quasi-linear systems of first order
partial differential equations in two independent variables. Communs.
Pure and Appl. Math., Vol. 14, No. 3 (1961), pp. 257–265. (page 477)

Courant, R., and K. O. Friedrichs
[1] Supersonic flow and shock waves. Interscience, New York, 1948. (pages
38, 429, 432, 490, 636)

Courant, R., K. O. Friedrichs, and H. Lewy
[1] Über die partiellen Differenzengleichungen der Physik. Math. Ann.,
Vol. 100 (1928–1929), pp. 32–74. (page 231)

Courant, R., and A. Lax
[1] Remarks on Cauchy's problem for hyperbolic partial differential equations with constant coefficients in several independent variables. Communs. Pure and Appl. Math., Vol. 8 (1955), pp. 497–502. (page 711)

Courant, R., and P. D. Lax
[1] On nonlinear partial differential equations with two independent variables. Communs. Pure and Appl. Math., Vol. 2 (1949), pp. 255–273.
(page 464)
[2] The propagation of discontinuities in wave motion. Proc. Natl. Acad.
Sci. U.S., Vol. 42, No. 11 (1956), pp. 872–876. (pages 727, 735)

Doetsch, G.
[1] Handbuch der Laplace Transformation. 3 Vols. Birkhäuser, Basel,
1950. (page 535)
[2] Theorie und Anwendung der Laplace Transformation. Springer, Berlin,
1937. (page 535)

Douglis, A., see also Agmon, S., A. Douglis, and L. Nirenberg
[1] Some existence theorems for hyperbolic systems of partial differential
equations in two independent variables. Communs. Pure and Appl.
Math., Vol. 5, No. 2 (1952), pp. 119–154. (pages 464, 478, 494)
· [2] A criterion for the validity of Huygens' principle. Communs. Pure and
Appl. Math., Vol. 9, No. 3 (1956), pp. 391–402. (page 576)

Duff, G. F. D.
[1] Mixed problems for linear systems of first order equations. Can. J.
Math., Vol. 10 (1958), pp. 127–160. (page 56)
[2] The Cauchy problem for elastic waves in an anisotropic medium. Philos.
Trans. Roy. Soc. London, Ser. A, Vol. 252 (1960), pp. 249–273. (page
599)
[3] Mixed problems·for linear systems of first order equations. Can. J.
Math., Vol. 10 (1958), pp. 127–160. (page 656)

Dunford, N., and J. T. Schwartz
[1] Linear operators, Part I: General theory. Interscience, New York, 1958.
(pages 333, 406)

Ehrenpreis, L.
[1] Solutions of some problems of division. I: Am. J. Math., Vol. 76 (1954),
pp. 883–903; II: Vol. 77 (1955), pp. 286–292; III: Vol. 78 (1956), pp. 685–
715. (pages 154, 767)

Finn, R.
[1] Isolated singularities of nonlinear partial differential equations. Trans.
Am. Math. Soc., Vol. 75 (1953), pp. 385–404. (page 400)
Finn, R., and D. Gilbarg
[1] Three dimensional subsonic flows and asymptotic estimates for elliptic
partial differential equations. Acta Math., Vol. 98 (1957), pp. 265–296.
(p. 367)
Fleischman, O., see Bazer, J., and O. Fleischman
Frankl, F.
[1] On the initial value problem for linear and nonlinear hyperbolic partial
differential equations of the second order. Mat. Sbornik (N.S.), Vol. 2
(44) (1937), pp. 793–814. (In Russian.) (page 675)
Friedlander, F. G.
[1] Simple progressive solutions of the wave equation. Proc. Cambridge
Phil. Soc., Vol. 43 (1946), pp. 360–373. (page 763)
[2] Sound pulses in a conducting medium. Proc. Cambridge Phil. Soc.,
Vol. 55, Pt. IV (1959), pp. 341–367. (page 612)
Friedman, A.
[1] On the regularity of the solutions of nonlinear elliptic and parabolic
systems of partial differential equations. J. Math. and Mech., Vol. 7
(1958), pp. 43–60. (page 347)
Friedrichs, K. O., see also Courant, R., and K. O. Friedrichs; Courant, R.,
K. O. Friedrichs, and H. Lewy
[1] Nonlinear hyperbolic differential equations for functions of two independent variables. Am. J. Math., Vol. 70 (1948), pp. 555–588. (page 464)
[2] Symmetric hyperbolic linear differential equations. Communs. Pure
and Appl. Math., Vol. 7 (1954), pp. 345–392. (pages 642, 669)
[3] Symmetric positive linear differential equations. Communs. Pure and
Appl. Math., Vol. 11 (1958), pp. 333–418. (pages 656, 669)
[4] The identity of weak and strong extensions of differential operators.
Trans. Am. Math. Soc., Vol. 55, No. 1 (1944), pp. 132–151. (page 768)
Friedrichs, K. O., and H. Kranzer
[1] Notes on magneto-hydronamics VIII. Nonlinear wave motion. Rep.
No. NYO-6486, AEC Computing & Applied Mathematics Center, Institute of Mathematical Sciences, New York University, 1958. (page 612)
Friedrichs, K. O., and H. Lewy
[1] Über die Eindeutigkeit und das Abhängigkeitsgebiet der Lösungen beim

Anfangswertproblem linearer hyperbolischer Differentialgleichungen. Math. Ann., Vol. 98 (1928), pp. 192–204. (pages 642, 673)

Garabedian, P.
[1] Partial differential equations. To be published by Wiley, New York. (page 506)
[2] Partial differential equations with more than two independent variables in the complex domain. J. Math. and Mech., Vol. 9 (1960), pp. 241–272. (page 506)

Garabedian, P., H. Lewy, and M. Schiffer
[1] Axially symmetric cavitational flow. Ann. of Math., Ser. 2, Vol. 56 (1952), pp. 560–602. (page 348)

Garabedian, P., and H. M. Lieberstein
[1] On the numerical calculation of detached bow shock waves in hypersonic flow. J. Aeronaut. Sci., Vol. 25 (1958), pp. 109–118. (page 499)

Gårding, L.
[1] Dirichlet problem for linear elliptic partial differential equations. Math. Scand., Vol. 1 (1953), pp. 55–72. (page 240)
[2] Linear hyperbolic partial differential equations with constant coefficients. Acta Math., Vol. 85 (1951), pp. 1–62. (pages 187, 216, 479, 483, 592, 599)
[3] Solution directe du problème de Cauchy pour les équations hyperboliques. Comptes rendus du colloque pour les équations aux dérivées partielles, Colloque International, CNRS, Nancy (1956), pp. 71–90. (pages 661, 665. 666)
[4] The solution of Cauchy's problem for two totally hyperbolic linear differential equations by means of Riesz integrals. Ann. of Math., Vol. 48 (1947), pp. 785–826. (page 599)

Gelbart, A., see Bers, L., and A. Gelbart

Gelfand, I. M.
[1] Some questions in the theory of quasilinear equations. Uspehi Mat. Nauk., Vol. 14, No. 2(86) (1959), pp. 87–158. (In Russian.) (page 490)

Gelfand, I. M., and G. E. Shilov
[1] Obobshchennye funkstii i deistviia nad nimi. Gos. Izd. Fiz.-mat. Lit., Moskva (1958) (in Russian). Verallagemeinerte Funktionen. Deutscher Verlag, Berlin, 1960. (pages 680, 740, 767, 768)

Germain, P.
[1] Remarks on the theory of partial differential equations of mixed type and applications to the study of transonic flow. Communs. Pure and Appl. Math., Vol. 7 (1954), pp. 117–144. (page 162)

Gilbarg, D., see Finn, R., and D. Gilbarg

Germain, P., and R. Bader
[1] Unicité des écoulements avec chocs dans la mécanique de Burgers. Office Nationale des Études et des Recherches Aeronautiques, Paris, 1953, pp. 1–13. (page 151)

Giorgi, E. de
[1] Sulla differenziabilità e l'analiticità delle estremali degli integrali multipli

regolari. Mem. accad. sci. Torino. Cl. sci. fis. mat. nat., Ser. 3, Vol. 3, Pt. 1 (1957), pp. 25–43. (page 347)

Grad, H.

[1] Propagation of magnetohydrodynamic waves without radial attenuation. Rep. No. NYO-2537, Magneto-Fluid Dynamics Division, Institute of Mathematical Sciences, New York University, 1959. (page 612)

Haar, A.

[1] Über Eindeutigkeit und Analytizität der Lösungen partieller Differenzialgleichungen. Atti del congr. intern. dei mat., Bologna, Vol. 3 (1928), pp. 5–10. (page 146)

Hadamard, J.

[1] Equations aux dérivées partielles. Les conditions définies en général. Le cas hyperbolique. Enseignement math., Vol. 35 (1936), pp. 5–42. (page 221)

[2] Le problème de Cauchy et les équations aux dérivées partielles linéaires hyperboliques. Hermann et Cie, Paris, 1932. (pages 39, 56, 221, 491, 495, 505, 686, 740)

[3] Psychology of invention in the mathematical field. Dover, New York, 1954. (page 743)

Hamel, G.

[1] Eine Basis aller Zahlen und die unstetigen Lösungen der Funktionalgleichung $f(x + y) = f(x) + f(y)$. Math. Ann., Vol. 60 (1905), pp. 459–462. (page 280)

Hartman, P., and A. Wintner

[1] On hyperbolic differential equations. Am. J. Math., Vol. 74 (1952), pp. 834–864. (pages 464, 494)

Heinz, E.

[1] On certain nonlinear elliptic differential equations and univalent mappings. J. d'Analyse Math., Vol. 5 (1956–1957), pp. 197–272. (page 367)

[2] Über gewisse elliptische Systeme von Differentialgleichungen zweiter Ordnung mit Anwendung auf die Monge-Ampèresche Gleichung. Math. Ann., Vol. 131 (1956), pp. 411–428. (pages 348, 367)

Hellwig, G.

[1] Partielle Differenzialgleichungen. Teubner, Stuttgart, 1960. (page 154)

Hilbert, D.

[1] Grundzüge einer allgemeinen Theorie der linearen Integralgleichungen. Göttingen Nachrichten, 1910; Leipzig, 1924. (page 362)

de Hoffmann, F., and E. Teller

[1] Magneto-hydrodynamic shocks. Phys. Rev., Vol. 80 (1950), p. 692. (page 612)

Holmgren, E.

[1] Über Systeme von linearen partiellen Differentialgleichungen. Ofversigt af kongl. Vetenskapsakad. Förh., Vol. 58 (1901), pp. 91–103. (pages 54, 237)

[2] Sur les systèmes linéaires aux dérivées partielles du premier ordre à charactéristiques réelles et distinctes. Ark. Mat., Astr. Fys., Vol. 6, No. 2 (1909), pp. 1–10. (page 454)

Hopf, E.

[1] A remark on linear elliptic differential equations of second order. Proc. Am. Math. Soc., Vol. 3 (1952), pp. 791–793. (page 328)

[2] Elementare Bemerkungen über die Lösungen partieller Differentialgleichungen zweiter Ordnung vom elliptischen Typus. Sitber. preuss. Akad. Wiss. Berlin, Vol. 19 (1927), pp. 147–152. (page 326)

[3] The partial differential equation $u_t + uu_x = \mu u_{xx}$. Communs. Pure and Appl. Math., Vol. 3 (1950), pp. 201–230. (page 152)

[4] Über den funktionalen, insbesondere den analytischen Charakter der Lösungen elliptischer Differentialgleichungen zweiter Ordnung. Math. Z., Vol. 34, 2 (1931), pp. 194–233. (pages 345, 347)

Hörmander, L.

[1] Differential operators of principal type. Math. Ann., Vol. 140 (1960), pp. 124–146. (page 54)

[2] On the interior regularity of the solutions of partial differential equations. Communs. Pure and Appl. Math., Vol. 11 (1958), pp. 197–218. (page 348)

[3] On the regularity of the solutions of boundary problems. Acta Math., Vol. 99 (1958), pp. 225–264. (page 348)

[4] On the theory of general partial differential operators. Acta Math., Vol. 94 (1955), pp. 161–248. (pages 154, 348)

[5] On the uniqueness of the Cauchy problem. (I) Math. Scand., Vol. 6 (1958), pp. 213–225; (II) Math. Scand., Vol. 7 (1958), pp. 177–190. (page 238)

Jackson, L. K., see Beckenbach, E. F., and L. K. Jackson

John, F.

[1] General properties of solutions of linear elliptic partial differential equations. Proceedings of Symposium on Spectral Theory and Differential Problems, Oklahoma Agricultural and Mechanical College, Stillwater, Oklahoma (1951), pp. 113–175. (pages 362, 367)

[2] Numerical solution of the equation of heat conduction for preceding times. Ann. di mat. Ser. 4, T. 40 (1955), pp. 129–142. (page 231)

[3] On linear partial differential equations with analytic coefficients—Unique continuation of data. Communs. Pure and Appl. Math., Vol. 2 (1949), pp. 209–253. (pages 239, 760)

[4] Plane waves and spherical means applied to partial differential equations. Tract 2. Interscience, New York, 1955. (pages 216, 240, 591, 676, 680, 699, 713, 718, 719, 726, 752)

[5] The ultrahyperbolic differential equation with four independent variables. Duke Math. J., Vol. 4 (1938), pp. 300–322. (page 753)

Kalashnikov, A. S.

[1] Construction of generalized solutions of quasi-linear equations of first

order without convexity conditions as limits of solutions of parabolic
equations with a small parameter. Doklady Akad. Nauk, S.S.S.R.,
Tom 127, No. 1 (1959), pp. 27–30. (page 153)

Kasahara, K., see Yamaguti, M., and K. Kasahara

Keller, J. B.

[1] "A geometrical theory of diffraction." Calculus of variations and its
applications, ed., L. M. Graves. Proceedings of Symposia on Applied
Mathematics, Vol. 8, pp. 27–52. American Mathematical Society, Providence, 1958. (page 640)

Keller, J. B., R. M. Lewis, and B. D. Seckler

[1] Asymptotic solution of some diffraction problems. Communs. Pure and
Appl. Math., Vol. 9 (1956), pp. 207–265. (page 640)

Kellogg, O. D., see also Birkhoff, G. D., and O. D. Kellogg

[1] Foundation of potential theory. Springer, Berlin, 1929. (page 240)

[2] On the derivatives of harmonic functions on the boundary. Trans. Am.
Math. Soc., Vol. 33 (1931), pp. 486–510. (page 336)

Kiselev, A. A., and O. A. Ladyzhenskaya

[1] On the existence and uniqueness of the weak solutions of the Navier-
Stokes equations. Akad. Nauk, S.S.S.R., Izvest. Ser. Mat. 21 (1957),
pp. 655–680. (page 367)

Klein, F.

[1] Vorlesungen über Höhere Geometrie. 3rd ed. Springer, Berlin, 1926.
(page 753)

Kline, M.

[1] An asymptotic solution of Maxwell's equation. Communs. Pure and
Appl. Math., Vol. 4 (1951), pp. 225–263. (page 640)

[2] Asymptotic solution of linear hyperbolic partial differential equations.
J. Ratl. Mech. Anal., Vol. 3 (1954), pp. 315–342. (page 640)

Koppenfels, W. von

[1] Der Faltungssatz und seine Anwendung bei der Integration linearer
Differentialgleichungen mit konstanten koeffizienten. Math. Ann., Vol.
105 (1931), pp. 694–706. (page 530)

Korn, A.

[1] Über Minimalflachen, deren Radkurven wenig von evenen Kurven ab-
weichen, Anhang II. Abhandl. Königl. preuss. Akad. Wiss., Berlin,
Jahre 1909, 37 pp. (page 401)

Kranzer, H., see Friedrichs, K. O., and H. Kranzer

Krzywoblocki, M. Z.

[1] Bergman's linear integral operator method in the theory of compressible
fluid flow. Österrlich. Ing.-Arch., Vol. 6 (1952), pp. 330–360. (page
401)

Krzyżański, M., and J. Schauder

[1] Quasilineare Differentialgleichungen zweiter Ordnung vom hyperbolischen Typus. Gemischte Randwertaufgaben. Studia Math., T. 6
(1936), pp. 162–189. (page 669)

Kunzi, H. P.

[1] Quasikonform Abbildungen. Springer, Berlin, 1960. (page 392)

Ladyžhenskaya, O. A., see also Kiselev, A. A., and O. A. Ladyžhenskaya;
Visik, M. I., and O. A. Ladyžhenskaya
[1] The mixed problem for a hyperbolic equation. Gos. Izd. Tekh.-Teor. Lit.,
Moscow, 1953. (page 669)

Lax, A., see also Courant, R., and A. Lax
[1] On Cauchy's problem for partial differential equations with multiple
characteristics. Communs. Pure and Appl. Math., Vol. 9, No. 2 (1956),
pp. 135–169. (pages 483, 592, 599)

Lax, P. D., see also Berkowitz, J., and P. D. Lax; Courant, R., and P. D. Lax
[1] A stability theorem for solutions of abstract differential equations, and
its application to the study of the local behavior of solutions of elliptic
equations. Communs. Pure and Appl. Math., Vol. 9, No. 4 (1956), pp.
747–766. (page 383)
[2] Hyperbolic systems of conservation laws II. Communs. Pure and Appl.
Math., Vol. 10 (1957), pp. 537–566. (pages 153, 490)
[3] Weak solutions of nonlinear hyperbolic equations and their numerical
computation. Communs. Pure and Appl. Math., Vol. 7 (1954), pp. 159–
193. (page 152)
[4] Asymptotic solutions of oscillatory initial value problems. Duke Math.
J., Vol. 24 (1957), pp. 627–646. (pages 636, 640, 735)
[5] Nonlinear hyperbolic equations. Communs. Pure and Appl. Math.,
Vol. 6, No. 2 (1953), pp. 231–258. (page 464)
[6] On Cauchy's problem for hyperbolic equations and the differentiability
of solutions of elliptic equations. Communs. Pure and Appl. Math.,
Vol. 8 (1955), pp. 615–633. (pages 669, 795, 796)

Lax, P. D., and R. S. Phillips
[1] Local boundary conditions for dissipative symmetric linear differential
operators. Communs. Pure and Appl. Math., Vol. 13, No. 3 (1960), pp.
427–456. (pages 656, 669)

Leray, J.
[1] Uniformisation de la solution du problème linéaire analytique de Cauchy
près de la variété qui porte les données de Cauchy. Bull. Soc. Math.
France, Vol. 85 (1957), pp. 389–429. (pages 56, 65)
[2] Hyperbolic differential equations. Lecture Notes, Institute for Advanced Study, Princeton, 1951–1952. (pages 661, 675)
[3] Lectures on hyperbolic equations. Institute for Advanced Study,
Princeton, 1952. (page 733)

Leray, J., and J. Schauder
[1] Topologie et équations fonctionelles. Ann. école norm. supérieur 51
(1934), pp. 45–78. (page 357)

Levi, E. E.
[1] Caratteristiche multiple e problema di Cauchy. Ann. di mat., Ser. 3ª,
Vol. 16 (1909), pp. 161–201. (page 464)
[2] Sulla problema di Cauchy per le equizioni a caratteristiche reali e distinte.
Rend. reale accad. lincei, Ser. 5ª, Vol. 18, 1º (1908), pp. 331–339. (page
464)

[3] Sulla problema di Cauchy per le equazioni Lineari in due variabili a caratteristiche reali. I and II. Rend. Ist. Lombardo, Ser. 2, Vol. 41 (1908), pp. 408–428 and 691–712. (page 483)

[4] I problemi dei valori al contorno per le equazioni lineari totalmente ellittiche alle derivati parziali. Mem. soc. It. delle sc., Ser. 3, Vol. 16 (1909), pp. 3–113. (page 362)

Lewis, R., see also Keller, J. B., R. M. Lewis, and B. D. Seckler

[1] Discontinuous initial value problems for symmetric hyperbolic linear differential equations. J. Math. and Mech., Vol. 7 (1958), pp. 571–592. (page 618)

Lewy, H., see also Friedrichs, K. O., and H. Lewy; Garabedian, P., H. Lewy, and M. Schiffer

[1] An example of a smooth linear partial differential equation without solution. Ann. of Math., (2) 66 (1957), pp. 155–158. (page 54)

[2] On the reflection laws of second order differential equations in two independent variables. Bull. Am. Math. Soc., Vol. 65, No. 2 (1959), pp. 37–58. (pages 348, 502, 505, 507)

[3] A priori limitations for solutions of Monge-Ampère equations I. Trans. Am. Math. Soc., Vol. 37 (1935), pp. 417–434. (page 502)

[4] A priori limitations for solutions of Monge-Ampère equations II. Trans. Am. Math. Soc., Vol. 41 (1937), pp. 365–374. (page 502)

[5] Eindeutigkiet der Lösung des Anfangsproblems einer elliptischen Differentialgleichung zweiter Ordnung in zwei veränderlichen. Math. Ann., Vol. 104 (1931), pp. 325–339. (page 502)

[6] Neuer Beweis des Analytischen Charakters der Lösungen elliptischer Differentialgleichungen. Math. Ann., Vol. 101 (1929), pp. 609–619. (pages 499, 501, 505)

[7] Über das Anfangswert Problem einer hyperbolischen nichtlinearen Differentialgleichung zweiter Ordnung mit zwei unabhängigen Veränderlichen. Math. Ann., Vol. 97 (1927), pp. 179–191. (pages 491, 495)

Lichtenstein, L.

[1] Neuere Entwicklung der Theorie partieller Differentialgleichungen zweiter Ordnung vom Elliptischen Typus. Encyklopädie der mathematischen Wissenschaften, Bd. II-3, Haft 8, Leipzig, 1924, pp. 1277–1334. (page 240)

Lieberstein, H. M., see Garabedian, P., and H. M. Lieberstein

Lighthill, M. J.

[1] Studies on magnetohydrodynamic waves and other anisotropic wave motions. Phil. Trans. Roy. Soc. (London) Ser. A (Math.-Phys. Sci.), Vol. 252 (1960), pp. 397–430. (page 612)

[2] An introduction to Fourier analysis and generalized functions. Cambridge University Press, New York, 1958. (page 767)

Lippman, B. A., and J. Schwinger

[1] Variational principles for scattering processes I. Phys. Rev., Vol. 79 (1950), p. 409. (page 319)

Ludwig, D.

[1] Exact and asymptotic solutions of the Cauchy problem. Communs.

Pure and Appl. Math., Vol. 13, No. 3 (1960), pp. 473–508. (pages 56, 632, 633, 674, 736)

[2] Conical refraction in crystal optics and hydromagnetics. NYO Rep. No. 9084, AEC Computing and Applied Mathematics Center, Institute of Mathematical Sciences, New York University, 1960. (page 629)

[3] The singularities of the Riemann function. NYO Rep. No. 9351, AEC Computing and Applied Mathematics Center, Institute of Mathematical Sciences, New York University, 1960. (pages 618, 676, 733, 744)

Luneburg, R. K.

[1] Mathematical theory of optics. Brown University Press, Providence, 1944. (page 640)

[2] Propagation of electromagnetic waves. New York University, New York, 1948 (mimeographed). (page 640)

Magenes, E., and G. Stampacchia

[1] I problemi al contorno per le equazioni differenziali di tipo ellittico. Ann. scuola norm. sup. Pisa, 1958–1959. (page 240)

Magnus, W., and F. Oberhettinger

[1] Formulas and theorems for the special functions of mathematical physics. Chelsea Publishing Co., New York, 1949. (page 461)

Malgrange, B.

[1] Existence et approximation des solutions des équations aux dérivées partielles et des équations de convolution. Ann. inst. Fourier, T. 6 (1955–1956), pp. 271–355. (pages 154, 383)

[2] Sur une classe d'opérateurs différentials hypoelliptiques. Bull. Soc. Math. France, Vol. 85, No. 3 (1957), pp. 283–306. (page 348)

Mikusiński, J.

[1] Operational calculus. Pergamon Press, New York, 1959. (page 508)

Miranda, C.

[1] Equazioni alle derivate parziali di tipo ellittico. Springer, Berlin, 1955. (pages 240, 331)

Mizohata, S.

[1] Le problème de Cauchy pour les systèmes hyperboliques et paraboliques. Mem. Coll. Sci. Univ. Kyoto Ser. A, Vol. 32 (1959), pp. 181–212. (page 666)

[2] Systèmes hyperboliques. J. Math. Soc. Japan, Vol. 11 (1959), pp. 205–233. (page 666)

Morrey, C. B.

[1] Multiple integral problems in the calculus of variations and related topics. University of California Publications in Mathematics (N. S.), Vol. 1 (1943), pp. 1–130. (page 347)

[2] On the analyticity of the solutions of analytic non-linear elliptic systems of partial differential equations, I and II. Am. J. Math., Vol. 80, No. 1 (1958), pp. 198–237. (page 347)

[3] On the solutions of quasi-linear elliptic partial differential equations. Trans. Am. Math. Soc., Vol. 43 (1938), pp. 126–166. (pages 347; 367, 394)

[4] Second order elliptic systems of differential equations. Annals of Mathematics Studies, Princeton, No. 33 (1954), pp. 101–159. (pages 346, 367)

Moser, J.

[1] A new proof of de Giorgi's theorem concerning the regularity problem for elliptic differential equations. Communs. Pure and Appl. Math., Vol. 13 (1960), pp. 457–468. (page 347)

[2] On Harnack's theorem of elliptic differential equations. Communs. Pure and Appl. Math., Vol. 14, No. 3 (1961), pp. 577–591. (page 269)

Müller, C.

[1] On the behavior of the solutions of the differential equation $\Delta u = F(x, u)$ in the neighborhood of a point. Communs. Pure and Appl. Math., Vol. 7 (1954), pp. 505–515. (page 238)

Nash, J.

[1] Continuity of solutions of parabolic and elliptic equations. Am. J. Math., Vol. 80 (1958), pp. 931–954. (page 347)

Newman, M. H. A.

[1] Topology of plane sets of points. Cambridge University Press, New York, 1951. (page 312)

Nirenberg, L., see also Bers, L., and L. Nirenberg

[1] On a generalization of quasi-conformal mappings and its application to elliptic partial differential equations. Annals of Mathematics Studies, Princeton, No. 33 (1954), pp. 95–100. (pages 346, 367)

[2] On elliptic partial differential equations. Ann. scuola norm. sup. Pisa, Ser. 3, Vol. 13, fasc. 2 (1959), pp. 115–162. (pages 234, 240, 347)

[3] On nonlinear elliptic partial differential equations and Hölder continuity. Communs. Pure and Appl. Math., Vol. 6, No. 1 (1953), pp. 103–156. (pages 346, 367)

[4] Remarks on strongly elliptic partial differential equations. Communs. Pure and Appl. Math., Vol. 8 (1955), pp. 649–675. (page 348)

Nitsche, J.

[1] Elementary proof of Bernstein's theorem on minimal surfaces. Ann. of Math., Vol. 66 (1957), pp. 543–544. (page 400)

[2] Zu einem Satze von L. Bers über die Lösungen der Minimalflächengleichung. Arch. Math., Vol. 9 (1958), pp. 427–429. (page 400)

[3] Über Unstetigkeiten in den Ableitungen von Lösungen quasilinearer hyperbolischer Differentialgleichungssyteme. J. Ratl. Mech., Vol. II (1953), pp. 291–297. (page 427)

Oberhettinger, F., see Magnus, W., and F. Oberhettinger

Oleinik, O. A.

[1] On Cauchy's problem for nonlinear equations in the class of discontinuous functions. Doklady Akad. Nauk S.S.S.R., Vol. 95 (1954), p. 451. (page 152)

[2] On the discontinuous solutions of nonlinear differential equations. Doklady Akad. Nauk S.S.S.R., Vol. 109 (1956), p. 1098. (page 151)

[3] On the uniqueness of a generalized solution of the Cauchy problem for a non-linear system of equations encountered in mechanics. Uspehi Mat. Nauk, Vol. 12, No. 6 (78) (1957), pp. 167–176. (In Russian.) (page 151)

[4] Discontinuous solutions of nonlinear differential equations. Uspehi Mat. Nauk., Vol. 12 (75) (1957), pp. 3–73. (In Russian.) (page 490)

Osserman, R.

[1] Proof of a conjecture of Nirenberg. Communs. Pure and Appl. Math., Vol. 12 (1959), pp. 229–232. (page 400)

Perron, O.

[1] Eine neue Behandlung der Ranwertaufgabe für $\Delta u = 0$. Math Z., Vol. 18 (1923), pp. 42–54. (page 306)

Petrovskii, I. G.

[1] Lectures on partial differential equations. 2nd ed. Gos. Izd. Tekh.-Teor. Lit., Moscow, 1953. English translation, Interscience, New York, 1954. (page 240)

[2] On some problems of the theory of partial differential equations. Uspehi Mat. Nauk, (N.S.), Vol. 1, No. 3–4 (1946), pp. 44–70; Am. Math. Soc. translation 12. (page 154)

[3] On the diffusion of waves and the lacunas for hyperbolic equations. Rec. Math., Mat. Sbornik, Vol. 17 (59) (1945), pp. 289–370. (pages 649, 723)

[4] Über das Cauchysche Problem für ein System linearer partieller Differentialgleichungen im Gebiete der nichtanalytischen Funktionen. Bull. Univ. Moscou, Sér. Int., Sect. A, Vol, 1, No. 7 (1938), pp. 1–74. (page 661)

[5] Über das Cauchysche Problem für Systeme von partiellen Differentialgleichungen. Mat. Sbornik, 2 (44) (1937), pp. 815–866. (page 661)

Phillips, R. S., see Lax, P. D., and R. S. Phillips

Peyser, G.

[1] Energy integrals for the mixed problem in hyperbolic partial differential equations of higher order. J. Math. and Mech., No. 6 (1957), pp. 641–653. (page 666)

Pizetti, P.

[1] Sulla media die valori che una funzioni dei punti dello spazio assume alla superficie di una sfera. Rend. reale accad. lincei, Ser. 5ª, Vol. 18 (1909), pp. 309–316. (page 287)

Pliś, A.

[1] A smooth linear elliptic differential equation without any solution in a sphere. Communs. Pure and Appl. Math., Vol. 14, No. 3 (1961), pp. 599–617. (page 383)

[2] Characteristics of nonlinear partial differential equations. Bull. Acad. Polon. Sci. Cl III.2 (1954), pp. 419–422. (page 147)

[3] Non-uniqueness in Cauchy's problem for differential equations of elliptic type. J. Math. and Mech., Vol. 9 (1960), pp. 557–562. (page 383)

Privaloff, J.

[1] Sur les fonctions conjugées. Bull. Soc. Math. France, Vol. 44 (1916), pp. 100–103. (page 401)

Protter, N. H.

[1] The periodicity problem for pseudoanalytic functions. Ann. of Math. (2), Vol. 64 (1956), pp. 154–174. (page 388)

Pucci, C.

[1] Proprietà di massimo e minimo delle soluzioni di equazioni a derivate parziali del secondo ordine di tipo ellittico e parabolico. Rend. accad. naz. lincei (Classe di scienze fisiche, mat. e naturali), Ser. 8, Vol. 23, fasc. 6, December, 1957; Vol. 24, fasc. 1, January, 1958. (page 328)

[2] Studio col metodo delle differenze di un problema di Cauchy relativo ad equazioni a derivate parziali del secondo ordine di tipo parabolico. Ann. scuola norm. sup. Pisa, Ser. 3, Vol. 7 (1953), pp. 205–215. (page 231)

Rellich, F.

[1] Zur ersten Randwertaufgabe bei Monge-Ampèreschen Differentialgleichungen vom elliptischen Typus; differentialgeometrische Anwendungen. Math. Ann., Vol. 107 (1933), pp. 505–513. (page 324)

[2] Verallgemeinerung der Riemannschen Integrationsmethode auf Differentialgleichungen n-ter Ordnung in zwei Veränderlichen. Math. Ann., Vol. 103 (1930), pp. 249–278. (page 454)

Riemann, G. F. B.

[1] Über die Fortpflanzung ebener Luftwellen von endlicher Schwingungsweite, Abhandl. Königl. Ges. Wiss. Göttingen, Vol. 8 (1860).(page 449)

Riesz, M.

[1] A special characteristic surface. Rep. No. 25, Dept. of Mathematics, University of Maryland, 1957. (page 763)

[2] A geometric solution of the wave equation in space-time of even dimension. Communs. Pure and Appl. Math., Vol. 13 (1960), pp. 329–351. (page 576)

Rubinowicz, A.

[1] Eindeutigkeit der Lösungen der Maxwellschen Gleichungen. Physik. Z. Vol. 27 (1926), pp. 707–710. (page 642)

[2] Herstellung von Lösungen gemischter Randwertprobleme bei hyperbolischen Differentialgleichungen zweiter Ordnung durch Zusammenstückelung aux Lösungen einfacherer gemischter Randwertaufgaben. Monatsch. Math. Phys., Vol. 30 (1920), pp. 65–79. (page 642)

Sauer, R.

[1] Anfangsvertprobleme bei partiellen Differenzialgleichungen. 2nd ed. Springer, Berlin, 1958. (page 407)

Schapiro, Z.

[1] Sur l'existence des représentations quasi-conformes. Doklady Akad. Nauk S.S.S.R., Vol. 30, No. 8 (1941), pp. 690–692. (page 399)

Schauder, J., see also Krzyżański, M., and J. Schauder; Leray, J., and J. Schauder

[1] Numerische Abschätzungen in elliptischen linearen Differentialgleichungen. Studia Math., Vol. 5 (1937), pp. 34–42. (page 331)

[2] Über lineare elliptische Differentialgleichungen zweiter Ordnung. Math. Z., Vol. 38, No. 2 (1934), pp. 257–282. (page 331)

[3] Cauchysches Problem für partielle Differentialgleichungen erster Ordnung. Anwendung einiger sich auf die Absolutbeträge der Lösungen beziehenden Abschätzungen. Comment. Math. Helv., Vol. 9 (1936–1937), pp. 263–283. (page 464)

[4] Das Anfangswertproblem einer quasilinearen hyperbolischen Differentialgleichung zweiter Ordnung. Fund. Math., Vol. 24 (1935), pp. 213–246. (pages 642, 669, 675)

Schiffer, M., see Garabedian, P., H. Lewy, and M. Schiffer

Schmidt, E.

[1] Bemerkung zur Potentialtheorie, Mathematische Abhandlungen Hermann Amandus Schwarz zu seinem fünfzigjährigen Doktorjubiläum am 6. August 1914, gewidmet von Freunden und Schülern. Julius Springer, Berlin, 1914, pp. 365–383. (page 258)

Schwartz, L.

[1] Théorie des distributions, Vols. 1 and 2. Hermann et Cie, Paris, 1950–1951. (pages 767, 768, 794)

Schwinger, J., see also Lippman, B. A., and J. Schwinger

[1] On the charge independence of nuclear forces. Phys. Rev., Vol. 78 (1950), p. 135. (page 319)

[2] The theory of quantized fields V. Phys. Rev., Vol. 93 (1954), p. 615. (page 319)

Seckler, B. D., see Keller, J. B., R. M. Lewis, and B. D. Seckler

Serrin, J.

[1] On the Harnack inequality for linear elliptic equations. J. d'Analyse Math., Vol. 4 (1956), pp. 292–308. (page 269)

Shilov, G. E., see Gelfand, I. M., and G. E. Shilov

Simonoff, N.

[1] Über die erste Randwertaufgabe der nichtlinearen elliptischen Gleichung. Bull. Math. Univ. Moscou, Sér. Int., Vol. 2, No. 1 (1939), pp. 3–18. (page 342)

Sobolev, S. L.

[1] Méthode nouvelle à résoudre le problème de Cauchy pour les équations linéarires hyperboliques normales. Mat. Sbornik (N.S.), Vol. 1 (1936), pp. 39–71. (page 766)

Stampacchia, G., see Magenes, E., and G. Stampacchia

Stellmacher, Karl L.

[1] Ein Beispiel einer Huygensschen Differentialgleichung. Nachr. Akad. Wiss. Göttingen, Math.-Phys. Kl. No. 10 (1953), pp. 133–138. (pages 576, 765)

Tautz, G.

[1] Zur Theorie der elliptischen Differentialgleichungen II. Math. Ann., Vol. 118 (1943), pp. 733–770. (page 342)

Tedone, O.

[1] Sull'integrazione dell'equazione $\partial^2\phi/\partial t^2 - \sum_{i}^{m} \partial^2\phi/\partial x_i^2 = 0$. Ann. di mat., Ser. 3, Vol. 1 (1889), p. 1. (page 690)

Teller, E., see de Hoffmann, F., and E. Teller

Temple, G.
[1] Generalized functions. Proc. Roy. Soc. London Ser. A., Vol. 228 (1955), pp. 175–190. (page 767)

Tricomi, F.
[1] Sulle equazioni lineari alle derivate parziali di seconde ordine, di tipo misto. Rend. reale accad. lincei, Ser. 5, Vol. 14 (1923), pp. 134–247. (pages 162, 390)

Ungar, P.
[1] Single higher order equations and first order systems. Thesis, New York University, 1958. (page 484)

Vekua, I.
[1] Generalized analytic functions. Gos. Izd. Fiz.-Mat. Lit., Moskva (1959). (In Russian; an English translation is in preparation by the American Mathematical Society.) (page 375)
[2] New methods for solving elliptic equations. OGIZ, Moscow, 1948. (In Russian.) (page 401)
[3] Systems of partial differential equations of first order of elliptic type, and boundary value problems with applications to the theory of shells. Mat. Sbornik, Vol. 31, No. 73 (1952), pp. 217–314. (page 375)

Visik, M. I., and O. A. Ladyžhenskaya
[1] Boundary value problems for partial differential equations and certain classes for operator equations. Uspehi. Mat. Nauk (N.S.), Vol. 11, No. 6 (72) (1956), pp. 41–97; American Mathematics Society Translation Ser. 2, Vol. 10 (1958), pp. 223–281. (page 240)

Volterra, Vito
[1] Sur les vibrations des corps elastiques isotropes. Acta Math., Vol. 18 (1894), pp. 161–232. (page 690)

Weinstein, A.
[1] Conformal representation and hydrodynamics. Proceedings of the First Canadian Mathematical Congress, 1945, University of Toronto Press, Toronto (1946), pp. 355–364. (page 226)
[2] The generalized radiation problem and Euler-Poisson-Darboux equation. Instituto Brasileiro de Educação, Cienciae Cultura, 1955, pp. 126–146. (pages 695, 699)
[3] The method of axial symmetry in partial differential equations. Convegno Internazionale sulle Equazioni Lineari alle Derivate Parziali, Trieste (1954) pp. 86–96. Edizioni Cremonese, Roma, 1955. (page 705)

Weitzner, H.
[1] On the Green's function for two-dimensional magnetohydrodynamic waves. Bull. Am. Phys. Soc., Ser. 2, Vol. 5 (1960), p. 321. (page 612)

Weyl, H.
[1] Ausbreitung elektromagnetischer Wellen über einem ebenen Leiter. Ann. Physik Ser. 4, Vol. 60 (1919), pp. 481–500. (page 196)

[2] Die Idee der Riemannschen Fläche. 3rd ed. B. G. Teubner, Stuttgart, 1954. (page 160)

Wiener, N.
[1] Certain notions in potential theory. J. Math. and Phys., Vol. 3 (1924), pp. 24–51. (page 306)
[2] The Dirichlet problem. J. Math. and Phys., Vol. 3 (1924), pp. 127–146. (page 305)

Wintner, A., see Hartman, P., ano A. Wintner

Yamaguti, M., and K. Kasahara
[1] Sur le système hyperbolique à coefficients constants. Proc. Japan Acad., Vol. 35 (1959), pp. 547–550. (page 596)

Zaremba, S.
[1] Sopra un teorema d'unicita relativo alla equazione delle onde sferiche. Rend. reale accad. lincei, Ser. 5ᵃ, Vol. 24 (1915), pp. 904–908. (page 642)

Zygmund, A., see Calderón, A. P., and A. Zygmund

Symposia and Colloquia

[1] Colloqua International No. 71. La Théorie des équations aux dérivées partielles, Nancy, April 9–15, 1956. (page 240)
[2] Convegno Internazionale sulle Equazioni derivate parziali. Trieste, 1954. Edizioni Cremonese, 1955. (page 240)
[3] Transactions of the Symposium on partial differential equations. Berkeley, California, June 1955. Reprinted in Communs. Pure and Appl. Math., Vol. 9, No. 3, 1956. (page 240)

Supplementary Titles

Beltrami, E.—Sul principio di Huygens. Rend. Ist. Lombardo Ser. 2, Vol. 22 (1889), pp. 428–438.

Bieberbach, L.—Theorie der Differentialgleichungen. Springer, Berlin, 1930 (Grundledren der Math. Wiss., Bd. 6).

Bureau, F.—Quelques questions de Géométrie suggérées par la théorie des équations aux dérivées partielles totalement hyperboliques. Colloque de Géométrie Algébrique, Liège, 1949.

Carson, J. R.—Electrical circuit theory and the operational calculus. 1st ed. McGraw-Hill, New York, 1926.

Forsythe, G. E., and P. C. Rosenbloom—Numerical analysis and partial differential equations. Wiley, New York, 1958.

Friedrichs, K. O.—Asymptotic phenomena in mathematical physics. Bull. Am. Math. Soc., Vol. 61 (1955), pp. 485–504.
—On the differentiability of the solutions of linear elliptic differential equations. Communs. Pure and Appl. Math., Vol. 6, No. 3 (1953), pp. 299–325.
—On differential operators in Hilbert spaces. Am. J. Math., Vol. 61, No. 2 (1939), pp. 523–544.

Gårding, L.—Hyperbolic equations. Lecture Notes, University of Chicago, 1957.

Goursat, É.—Cours d'analyse mathématique, Vols. 1–3 (especially Vols. 2 and 3). 5th ed. Gauthier-Villars, Paris, 1956.

—Leçons sur l'intégration des équations aux derivées partielles du premier ordre. 2nd ed. Hermann, Paris, 1921.

Hadamard, J.—Leçons sur la propagation des ondes et les équations de l'hydrodynamique. Hermann, Paris, 1903.

—Problèmes à apparence difficile. Mat. Sbornik (N.S.), Vol. 17 (1945), pp. 3–7.

Halpern, S.—Lacunae of non-hyperbolic equations. Soviet Math., Vol. 1 (1960), pp. 680–683.

—Sur les conditions pour que le problème de Cauchy pour un système compatible d'équations linéaires aux dérivées partielles soit correctement posé. Mat. Sbornik (N.S.), Vol. 7 (1940), pp. 111–141. (In Russian; French summary.)

Holmgren, E.—Sur les systémes linéaires aux dérivées partielles du premier ordre à deux variables indépendantes à caractéristiques réelles et distinctes. Ark. Math., Astr. Fys., Vol. 5, No. 1, 13 pp.

—Sur l'extension de la méthode d'integration de Riemann. Ark. Math., Astr. Fys., Vol. 1 (1903), pp. 317–326; Vol. 5, No. 16, 12 pp.

—Über die Existenz der Grundlösung bei einer linearen Differentialgleichung der zweiten Ordnung. Ark. Math., Astr. Fys., Vol. 1 (1903), pp. 209–224.

—Über Randwertaufgaben bei einer linearen Differentialgleichung der zweiten Ordnung. Ark. Math., Astr. Fys., Vol. 1 (1903), pp. 401–417.

Jaffe, G.—Unstetige und mehrdeutige Lösungen der hydrodynamischen Gleichungen. Z. angew. Math. Mech., Vol. 1 (1921), pp. 398–410.

Jeffreys, H.—Operational methods in mathematical physics. Cambridge Tracts No. 23, Cambridge University Press, New York, 1927.

Kamke, E.—Differentialgleichungen reeller Funktionen. Akademische Verlagsgesellschaft, Leipzig, 1930.

Kasahara, K., and M. Yamaguti—Strongly hyperbolic systems of linear partial differential equations with constant coefficients. Mem. Coll. Sci., Kyoto, Ser. A (33), Mathematics No. 1 (1960).

Leray, J.—La solution unitaire d'un opérateur différentiel linéare. Bull. Soc. Math. France, Vol. 86 (1958), pp. 75–96.

—Le calcul differentiel et intégral sur une variété analytique complexe. Bull. Soc. Math. France, Vol. 87 (1959), pp. 81–180.

Levi-Civita, T.—Caratteristiche dei sistemi differenziali e propagazione ondosa. Zanichelli, Bologna, 1931.

Mizohata, S.—Analyticité des solutions élémentaires du système hyperbolique à coefficients constants. Mem. Coll. Sci. Kyoto, Ser. A (32) (1959), pp. 213–234.

Petrovskii, I. G.—Vorlesungen über die Theorie der gewöhnlichen Differentialgleichungen. B. G. Teubner, Leipzig, 1954.

Picard, É.—Traité d'analyse, Vols. 1–3. 3rd ed. Gauthier-Villars, Paris, 1922–1928.

Poincaré, H.—Théorèmes généraux sur le potential Newtonien, Leçon 1 in: Figures d'équilibre d'une masse fluide. Leçons professées à la Sorbonne en 1900, Naud, Paris, 1902.

Rellich, F.—Über die Reduktion gewisser ausgearteter Systeme von partiellen Differentialgleichungen. Math. Ann., Vol. 109 (1934), pp. 714–745.

Riesz, M.—L'intégrale de Riemann-Liouville et le problème de Cauchy. Acta Math., Vol. 81 (1949), pp. 1–223.

Stellmacher, K.—Zum Anfangswertproblem der Gravitationsgleichungen. Math. Ann., Vol. 115 (1938), pp. 136–152.

Thomas, T. Y., and E. W. Titt—Systems of Partial Differential Equations and Their Characteristic Surfaces. Ann. of Math., Vol. 34 (1933), pp. 1–80.

Weinstein, A.—Les conditions aux limites introduites par l'hydrodynamique. L'enseignement mathématique (1936), pp. 107–125.

—On the wave equation and the equation of Euler-Poisson. The Fifth Symposium in Applied Mathematics, McGraw-Hill, New York, 1954, pp. 137–147.

Yamaguti, M.—Le problème de Cauchy et les opérateurs d'intégral singulière. Mem. Coll. Sci., Kyoto, Ser. A (32)(1959), pp. 121–151.

INDEX

(See the bibliography for an index of names.)